The Availability of Spatial and Environmental Data in the European Union

Energy and Environmental Law & Policy Series
Supranational and Comparative Aspects

VOLUME 10

The aim of the Editor and the Editorial Board of this series is to publish works of excellent quality that focus on the study of energy and environmental law and policy.

Through this series the Editor and Editorial Board hope:

- to contribute to the improvement of the quality of energy/environmental law and policy in general and environmental quality and energy efficiency in particular;
- to increase the access to environmental and energy information for students, academics, non-governmental organizations, government institutions, and business;
- to facilitate cooperation between academic and non-academic communities in the field of energy and environmental law and policy throughout the world.

KLUWER LAW INTERNATIONAL

The Availability of Spatial and Environmental Data in the European Union: At the Crossroads between Public and Economic Interests

by Katleen Janssen

Wolters Kluwer
Law & Business

AUSTIN BOSTON CHICAGO NEW YORK THE NETHERLANDS

Published by:
Kluwer Law International
PO Box 316
2400 AH Alphen aan den Rijn
The Netherlands
Website: www.kluwerlaw.com

Sold and distributed in North, Central and South America by:
Aspen Publishers, Inc.
7201 McKinney Circle
Frederick, MD 21704
United States of America
Email: customer.service@aspenpublishers.com

Sold and distributed in all other countries by:
Turpin Distribution Services Ltd.
Stratton Business Park
Pegasus Drive, Biggleswade
Bedfordshire SG18 8TQ
United Kingdom
Email: kluwerlaw@turpin-distribution.com

Printed on acid-free paper.

ISBN 978-90-411-3287-1

Printed in Great Britain.

Author Information

Katleen Janssen obtained her law degree at the Katholieke Universiteit Leuven in 2001 and her postgraduate degree in corporate law at the Katholieke Universiteit Brussel in 2004. She joined the Interdisciplinary Centre for Law and ICT of the Katholieke Universiteit Leuven in 2001.

Dr Janssen specializes in access to and use of public sector information and legal issues related to geographic data and spatial data infrastructures. She has worked on several national and European projects on these matters and publishes regularly in national and international journals.

In 2009, Dr Janssen also worked as a policy advisor on geographic data for the Ministry of the Flemish Government. After obtaining her PhD in December 2009, she returned to the Katholieke Universiteit Leuven as a postdoctoral researcher.

Summary of Contents

Table of Contents

Preface

This book is the result of the author's doctoral research, completed at the Katholieke Universiteit Leuven in 2009. The promoter of this research was Prof. Dr Jos Dumortier, and the members of the 'jury' included Prof. Dr Marc Boes, Prof. Dr Joep Crompvoets, Prof. Mr Jan Kabel, Prof. Dr Jules Stuyk, and Prof. Dr Peggy Valcke.

The research was supported by the Spatialist project (Spatial Data Infrastructures and Public Sector Innovation), funded by the Agentschap voor Innovatie door Wetenschap en Technologie Vlaanderen (IWT-Vlaanderen), and by the Instituut voor Breedband Technologie.

The information in this book is current as of February 2010.

Acknowledgement

Many people have contributed to this work, and the fact that I managed to finish it is just as much their merit as it is mine. Therefore, my warm gratitude goes out to the following people:

My promoter, Prof. Dr Jos Dumortier, for giving me the opportunity to pursue my dream to become an academic and for giving me the trust and freedom to develop as a researcher and as a person.

The members of my jury, for taking the time and effort to read this work and for challenging me to think beyond traditional boundaries.

Dr Joep Crompvoets, for his continuous encouragement, both as a colleague and a friend, and his invaluable scientific and personal integrity.

My colleagues at the Interdisciplinary Centre for Law and ICT (ICRI), for the wonderful times that we had together and for their great support. Eva, for being the perfect roommate and kindred spirit in this journey. Peggy, for being such a strong woman and being the example that we can all look up to. David, for making me laugh and for putting things into perspective. Eleni, for being right about the good times starting in 2007. Hannelore, for all the things we've shared over the years. Nicole, for taking care of us with her heart and soul. Aleksandra and Matthias, for helping me out with the projects.

Danny Vandenbroucke, for being my mentor in the spatial data infrastructure community; it has been an honour to work with him. The researchers and promoters of the Spatialist project, for giving me the time and space to focus on my PhD research. Hans Dufourmont, Clare Hadley, Ian Jackson, and Beatrice Eiselt, for the enriching discussions, their faith in my capabilities and, most of all, the fun at our drafting team meetings.

Fiona and Ingrid, for the much needed relaxing evenings. Gabriella and Kirsten, for the reality checks and the constant encouragement in the background. Cindy, for being my best friend and the best support system anyone could wish for.

Johan and Auke, my wonderful brother and gorgeous niece, for understanding that I wasn't around more often. My mother and father, for always being supportive and concerned without being overbearing – for simply being the best parents anyone could wish for.

Finally, Raf, for your infinite care and never-ending belief in me. You mean everything, my love.

– Katleen

List of Abbreviations and Short References

Aarhus Convention:	Convention on Access to Information, Public Participation in Decision-making and Access to Justice in Environmental Matters
Aarhus Implementation Guide:	United Nations Economic Commission for Europe, *The Aarhus Convention: An implementation guide. Part 2*
Access Directive:	Directive 2003/4/EC of the European Parliament and of the Council of 28 January 2003 on public access to environmental information and repealing Council Directive 90/313/EEC
APPSI:	Advisory Panel for Public Sector Information
CFI:	Court of First Instance
Directive 90/313/EEC:	Council Directive 90/313/EEC of 7 June 1990 on the freedom of access to information on the environment
ECHR:	Convention for the Protection of Human Rights and Fundamental Freedoms
ECJ:	European Court of Justice
eEurope Communication:	Communication to the Council, the European Parliament, the Economic and Social Committee and the Committee of the Regions. eEurope 2002: creating a EU framework for the exploitation of public sector information
EU:	European Union
GMES:	Global Monitoring for Environment and Security

GEO:	Group on Earth Observation
GEOSS:	Global Earth Observation System of Systems
Green Paper on Public Sector Information:	Public sector information: a key resource for Europe. Green Paper on public sector information in the information society
ICCPR:	International Covenant on Civil and Political Rights
INSPIRE Directive:	Directive 2007/2/EC of the European Parliament and of the Council of 14 March 2007 establishing an Infrastructure for Spatial Information in the European Community (INSPIRE)
IPR:	intellectual property rights
KNMI:	Royal Dutch Meteorological Office (Koninklijk Nederlands Meteorologisch Instituut)
OPSI:	Office of Public Sector Information
OFT:	Office of Fair Trading
PSI Directive:	Directive 2003/98/EC of the European Parliament and of the Council of 17 November 2003 on the re-use of public sector information
SDI:	Spatial data infrastructure
SEIS:	Shared Environmental Information System
Synergy Guidelines:	Guidelines for improving the synergy between the public and the private sectors in the information market
Transparency Directive:	Commission Directive 2006/111/EC of 16 November 2006 on the transparency of financial relations between Member States and public undertakings as well as on financial transparency within certain undertakings
UN:	United Nations
UNECE:	United Nations Economic Commission for Europe
UNESCO:	United Nations Educational, Scientific and Cultural Organization

Chapter 1
Introduction

I. SETTING

A. Spatial Data

Use of spatial data – Spatial data are part of our daily lives. Many of us use a GPS or an online route planner to find our way to meetings. We check the weather forecast to decide whether to take an umbrella when leaving home. On the weekends or on holiday, we use cycling or hiking maps or tourist guides. We check Google Earth or Google Street View for photographs of our house or streets, and so on. All of these applications are based on spatial data, that is, data that refer to specific locations or geographic areas. Such spatial data are also needed for professional reasons by, for example, architects creating plans or models, engineers building bridges or drilling pipelines, telecommunication companies putting new cables in the ground, farmers organizing the fertilization of their land, and so on. All of these people need accurate spatial data to avoid making wrong decisions or damaging other people or property.

Geographic information systems – Traditionally, spatial data took the form of maps. The development of information technology has enabled the digitization of these maps and the creation of spatial databases in which data are stored in a logical manner. Geographic information systems (GIS) make it possible to combine different spatial data sets and visualize them on a computer screen. For instance, a user can combine data on cadastral parcels with soil data to check the ground stability for a new house or with pollution data to determine whether it is safe to live at a certain place.

Spatial data and environmental data – While spatial data are used for many different purposes, the origin of their value lies in environmental decision making, particularly on the international level and in the European Union (EU). The development of environmental policies initiated the need for accurate data on the condition of a particular place or area in order to make appropriate decisions. Such data are inherently spatial, but they also relate to the environment. Therefore, a considerable part of spatial data can also be seen as environmental data and vice versa. The particular relationship between environmental data and spatial data in the EU legal framework will be explained in Chapter 2.[1]

B. Public Sector Spatial Data

Use of spatial data for policy purposes – The main producers and users of spatial data are the government and the public sector. They need spatial data to prepare, implement, and evaluate their policies with regard to land use and planning, environmental policy, transport, agriculture, public utilities, national defence, emergency services, and so on. Examples of applications include the selection of sites for industrial development, the evaluation of sites for waste disposal, the inventory and management of roads, the provision of weather forecasts, the mapping of territories of animal and plant species, and the navigation of ships and airplanes.[2]

Spatial data infrastructure – Hence, public bodies are important creators of spatial data and users of GIS. However, as many government policies involve public bodies of several hierarchical levels or countries, the exchange of these data between public bodies is of vital importance. The need for the sharing of spatial data has led to the movement from GIS to spatial data infrastructures (SDIs), particularly for the purpose of environmental policy. The objective of an SDI is to facilitate and coordinate the exchange and sharing of spatial data between all stakeholders in the spatial data community.[3] Its primary purpose is to support decision making. An SDI includes not only the technology for exchanging data (software, standards, web services, and so on), but also a number of non-technical components, such as legislation and policies, expertise, funding, coordination, human resources, and partnerships.[4]

1. Cf. *infra*, Ch. 2, s. III.
2. Harlan Onsrud et al., 'The Future of the Spatial Information Infrastructure', in *A Research Agenda for Geographic Information Science*, ed. Robert B. McMaster & E. Lynn Usery (Boca Raton: CRC Press, 2004), 225.
3. Joep Crompvoets et al., 'Assessing the Worldwide Developments of National Spatial Data Clearinghouses', *International Journal of Geographical Information Science* 18 (2004): 665.
4. Many definitions of SDI exist, but one of the most often cited definitions of a (national) SDI was part of former US President Clinton's Executive Order 12906 establishing the National SDI in the United States: 'the technology, policies, standards, and human resources necessary to acquire, process, store, distribute, and improve utilization of geospatial data'. This definition highlights not only that technology is needed to develop an SDI, but also that organizational and human aspects also play an indispensable role. People, organizational capacity building, and coordination are critical components of an SDI that are often overlooked. (William J. Clinton, 'Executive Order 12906 – Co-ordinating Geographic Data Acquisition and Access: The National Spatial Data Infrastructure', 1994, <http://govinfo.library.unt.edu/npr/library/direct/orders/20fa.html>, 22 Feb. 2010; Lukasz Grus, Joep Crompvoets & Arnold K. Bregt, 'Multi-view SDI Assessment Framework', *International Journal of Spatial Data Infrastructures Research* 2 (2007): 36–37,

C. Value of Public Sector Spatial Data

Value for the public sector – From the examples given above, one can see that spatial data have considerable value for governmental and public bodies. They are indispensable for governments to prepare, implement, and evaluate their policies and to provide efficient services to citizens with regard to the environment, but also in other policy domains. Spatial data help to ensure sustainable development, the understanding of climate change, and the conservation of biodiversity. Other examples include the understanding of factors affecting human health and well-being, management of energy resources, and improving the availability of water sources.

Value for citizens – However, spatial data held by the public sector are also of great value outside of the public sector. First, citizens can be direct beneficiaries of the availability of public sector spatial data. For instance, a citizen intending to buy a piece of property can find out whether the ground has been polluted, whether there is a risk of flooding, or whether the property is close to public transportation to schools. He or she can check whether the local authority has followed the appropriate land-use plans in allowing a factory to be built near his or her house, or made the right decision on the expropriation of his or her property for the expansion of an airport or industrial terrain in the vicinity, and so on.

Value for companies – Companies as well as citizens can benefit from the availability of public sector spatial data. The economic value for companies is two-fold. On the one hand, spatial data may be needed for decision making. For instance, a company may want to use spatial data for its business strategies, such as determining locations for new company branches by analyzing spatial data in combination with consumer statistics, environmental regulations, public transportation arrangements, and so forth. On the other hand, public sector spatial data can also be used to create information products and services for the market, such as navigation systems, weather services, environmental reports, real-estate services, marketing, tourism services, and so on.[5]

<http://ijsdir.jrc.ec.europa.eu/index.php/ijsdir/article/view/27/21>, 22 Feb. 2010; Max Craglia, 'Introduction to the International Journal of Spatial Data Infrastructures Research', *International Journal of Spatial Data Infrastructures Research* 1 (2006): 1, <http://ijsdir.jrc.ec.europa.eu/index.php/ijsdir/article/view/14/8>, 22 Feb. 2010.

5. A number of attempts at estimating the economic value of public sector spatial data have been made. For instance, the 2000 PIRA study on the commercial exploitation of public sector information estimated the economic value of public sector information at euro (EUR) 68.5 billion, with the sector of spatial data representing EUR 36 billion out of that amount (PIRA International, 'Commercial Exploitation of Europe's Public Sector Information. Final Report', 2000, 9, 15, <http://ec.europa.eu/information_society/policy/psi/docs/pdfs/pira_study/commercial_final_report.pdf>, 22 Feb. 2010. The geo-information market in Germany alone was valued at EUR 1.5 billion in 2007, and the market in Europe for weather and climatological services was estimated around EUR 530 million in 2006 (Micus Management Consulting Gmbh, Press Release, 'New Markets for Geo-Information in Times of Google Earth: MICUS-Study Names Market Information, Market Volume and Success Factors', 2008, <www.micus.de/pdf/Press_Release-GIW-study.pdf>,

D. The Role of the Public Sector in Making Spatial Data Available

Making data available to citizens – The value of public sector spatial data entails that there is great interest in obtaining them by other public bodies, by citizens, and by companies. In this regard, the public body holding the data is confronted with different needs and interests. First, there is a growing trend towards openness and transparency of the functioning of public bodies, which includes access to public sector data. Public bodies increasingly face obligations to provide information to the public, either upon request or of their own motion. One of the main examples of this trend is the adoption of the Aarhus Convention on Access to Information, Public Participation in Decision-Making and Access to Justice in Environmental Matters (hereafter referred to as the 'Aarhus Convention').[6] This Convention forms the international recognition of the idea that citizens need public sector information in order to participate in decision making with regard to the environment.[7] While the Convention only addresses environmental information, it is embedded in growing national and European traditions of access and freedom of information legislation, which applies to all data held by the public sector.

Making data available to other public bodies – A second trend that public bodies are confronted with is the growing requirement of efficiency and cooperation within the public sector, including that of the collection and exchange of spatial data. Data should be collected only once in the appropriate place and be used by other public bodies when possible. This should save the public bodies money, time, and effort and also improve cooperation within and between countries. For instance, the easy exchange of spatial and environmental data between public bodies enables a swift reaction to emergencies and calamities, as, for example, the exchange of population data, data on water levels and road makes it easier to organize the evacuation of a village under a threat of flooding. Many global, European, and national initiatives are currently being undertaken to improve this exchange of spatial data, such as the development of the Infrastructure for Spatial Information in Europe (INSPIRE) and the United Nations' SDI.

Making data available to companies – Third, the economic value of spatial data held by the public sector makes them an interesting resource for the

22 Feb. 2010; Richard E.W. Pettifer, 'Towards a Stronger European Market in Applied Meteorology', *Meteorological Applications* 15 no. 2 (2008): 305–312).

6. United Nations Economic Commission for Europe, 'Convention on Access to Information, Public Participation in Decision-making and Access to Justice in Environmental Matters', 25 Jun. 1998, <www.unece.org/env/pp/documents/cep43e.pdf>, 22 Feb. 2010.
7. As environmental information almost always refers to the condition of the environment in a certain area or location and hence can be considered as spatial data, this Convention is of great importance for spatial data.

information industry and private companies. A number of characteristics of these data make them particularly valuable. They enjoy the assumption of reliability and sustainability, and they often have a strong reputation for high quality.[8] In addition, there are significant sunk costs in the collection of data, and spatial data in particular are very expensive to collect.[9] Therefore, the private sector will obtain the data from another source rather than create them by itself. With such public sector spatial data, the private sector can create information services and products and sell them against a profit on the market. In doing so, it has an interest in keeping its costs low and obtaining the data resources from the public bodies at a small price.

USING DATA TO GAIN REVENUES – The fourth interest that plays a role in the availability of public sector spatial data is that of the public bodies themselves. Some of the public bodies have realized that they hold a resource with considerable economic value, and they would like to use the data to gain some extra revenue in addition to their budgets received from central government. Some public bodies are even required to obtain such revenues because they are no longer (completely) funded by the government. They can do this in two ways. On the one hand, a public body can charge or impose conditions for the spatial data that it provides to users, including companies, citizens, and public bodies. In this way, it may try to recover the costs of collecting and processing the data or to provide financing for other projects. On the other hand, it may create its own information services or products based on its data and sell these added-value products to users, for example, hiking maps, environmental reports, dedicated weather services for leisure boating, navigation services, and so on. In such cases, it might be in competition with the private sector, which might offer the same or comparable services on the market. This could create tension between the public and private sectors, as a public agency that sells commercial products and services will not be eager to pass on its data resources to competing providers. Since public agencies often have a monopoly, or at least a dominant position, distortion of competition becomes a genuine risk. Temptation will be great for a public agency to sell at high prices to the competition or to eliminate this competition by charging very low prices to the consumers for end products.[10]

LEGAL INITIATIVES – Against this background, a number of legal initiatives have been taken in the EU to ensure the availability of public sector spatial and environmental data. These legal instruments are the main focus of this book.

8. Office of Fair Trading, 'The Commercial Use of Public Information (CUPI)', 2006, 64, <www.oft.gov.uk/shared_oft/reports/consumer_protection/oft861.pdf>, 22 Feb. 2010.
9. *Ibid.*
10. Katleen Janssen & Jos Dumortier, 'The Legal Framework for a European Spatial Data Infrastructure: Uncrossing the Wires', in *Research and Theory in Advancing Spatial Data Infrastructure Concepts*, ed. Harlan Onsrud (Redlands California: ESRI Press, 2007), 236.

E. The Legal Framework for the Availability of Public Sector Spatial Data

Three directives – EU institutions have taken a number of initiatives to set up a framework for the availability of public sector spatial data in the Member States. Three directives are in the centre:

- Directive 2007/2/EC of the European Parliament and of the Council of 14 March 2007 establishing an Infrastructure for Spatial Information in the European Community (hereafter the 'INSPIRE Directive')[11];
- Directive 2003/4/EC of the European Parliament and of the Council of 28 January 2003 on public access to environmental information and repealing Council Directive 90/313/EEC (hereafter the 'Access Directive')[12]; and
- Directive 2003/98/EC of the European Parliament and of the Council of 17 November 2003 on the re-use of public sector information (hereafter the 'PSI Directive')[13]

INSPIRE directive – First, the INSPIRE Directive provides the legal basis for setting up an European Community (EC) SDI for environmental policy and obliges the Member States and public bodies to share data with other public bodies for the performance of their public tasks relating to the environment. Hence, under this directive, public bodies have to make their data available, under certain conditions, to other public bodies for the development of their environmental policies on pollution, climate change, sustainability, and so on. In some cases, the availability of the data may be limited or charges and licenses can be imposed to ensure the interests of the public bodies that supply the data.

Access directive – Second, the Access Directive imposes an obligation on public bodies to make their environmental information available to the public, either upon the public's request or on their own initiative. This directive ensures that citizens can obtain environmental information (including many spatial data sets) in order to participate in the decision-making process and to check up on the activities of the government that they have elected. Limitations to the right of access are included to ensure the protection of a number of legitimate interests, and the public bodies' sustainability can be protected by the levy of charges for the information.

PSI directive – Third, the PSI Directive addresses the economic value of public sector data, including spatial data held by public bodies. It intends to harmonize the conditions for re-use of public sector data by the private sector and lays

11. Directive 2007/2/EC of the European Parliament and of the Council of 14 Mar. 2007 establishing an Infrastructure for Spatial Information in the European Community (INSPIRE), *OJ L* 108, 25 Apr. 2007, 1.
12. Directive 2003/4/EC of the European Parliament and of the Council of 28 Jan. 2003 on public access to environmental information and repealing Council Directive 90/313/EEC, *OJ L* 41, 14 Feb. 2003, 26.
13. Directive 2003/98/EC of the European Parliament and of the Council of 17 Nov. 2003 on the re-use of public sector information, *OJ L* 245, 31 Dec. 2003, 90.

down a set of minimum rules for public bodies to make their data available for any further use for commercial and non-commercial purposes. On the one hand, it facilitates the use of data by the private sector by imposing conditions on the format of delivery of the data, the charges for the data, and so on. On the other hand, it tries to ensure a level playing field by imposing a number of conditions on public bodies when these create their own information products or services based on their data.

OUTLINE – The relationship between these three directives is the main focus of this book. As is indicated in the next section, the three directives should provide a clear framework for the availability of public sector spatial data, yet this book questions whether it actually does.

II. THE PROBLEMATIC CHARACTER OF THE EU LEGAL FRAMEWORK FOR THE AVAILABILITY OF PUBLIC SECTOR SPATIAL DATA

DISTINCTIONS BETWEEN LEGAL INSTRUMENTS – The EU legal framework for the availability of public sector spatial data is formed by the Access Directive, the INSPIRE Directive, and the PSI Directive. As these three directives are all applicable to spatial data, it is important to know which rules need to be applied in a particular situation. The distinctions between these rules are based on the purpose of the use for which the public sector spatial data are made available. Three types of use can be distinguished. Access involves obtaining public sector spatial data for the purpose of exercising one's democratic rights or obligations. Next, re-use is defined as any use of public sector spatial data for commercial or non-commercial purposes outside of the public task, requiring the processing of the data beyond access needs. Finally, sharing is any use of public sector spatial data for the purpose of performing public tasks. These types of use determine which rules have to be taken into account by the public bodies making the data available and by the users requesting the data.

IMPORTANCE OF CLEAR DISTINCTIONS – As the rules for access, re-use, and sharing are different, understanding the field of application of each type of use and its respective rules in the Access Directive, the INSPIRE Directive, and the PSI Directive is vital. Without such understanding, users will be discouraged from requesting public sector spatial data from public bodies. Users need to know how they can obtain the data, what they can use the data for, how much money they can be charged for the data, and to whom they need to address their requests and complaints. The lack of clarity in the legal framework leads to uncertainty and reluctance of users to request spatial data and to reticence of public bodies to make spatial data available.

PROBLEMATIC NATURE OF THE DISTINCTIONS – This book questions whether the distinctions between access, re-use, and sharing are sufficiently clear for the EU legal framework for the availability of public sector spatial data to properly ensure this availability. I believe that this is not the case. The hypothesis can be formulated as follows: *The distinctions between access, re-use, and sharing on which the EU legal framework for the availability of public sector spatial data is built are problematic and hinders the availability of these data.* I believe that this problematic nature is due to the criteria that are used to define the distinctions between the different rules that are applicable to the availability of spatial data. There are two levels to this problem: first, it is not possible to have a framework that completely fits without overlaps because it is impossible to make a clear difference between the types of use in practice. Second, there is also a conceptual problem: it will never be possible to properly determine the distinctions between the types of use, as one of the main concepts that underpins the framework is the *public task*. I believe that, as this concept cannot be defined on a European level, it causes the legal framework for the availability of public sector spatial data to be ineffective in stimulating the availability of public sector spatial data for all types of use.

III. STRUCTURE AND OUTLINE OF THE FOLLOWING CHAPTERS

A. Structure

Analysis of legal sources – This book starts with an analysis of the concepts that are the main elements that determine the field of application of the EU legal framework for the availability of public sector spatial data. Next, this book uses the traditional legal sources of legislation and policy documents, case law, and legal literature. These are mainly focused on the EU as a whole, and documents that apply to national situations are only considered insofar as they provide useful information at the EU level.

Analysis of other sources – The importance of the availability of public sector spatial data for the development of society and the economy also requires the examination of literature from the domain of economics and the fields of social and political science. When necessary, such literature has also been included in order to give a more complete overview of the particular issue that is discussed.

Structure – The set-up of the chapters is based on two axes. On the one hand, this book is built around the three concepts mentioned above: access, re-use, and sharing. On the other hand, the problematic character of the legal framework is examined on two levels: the making available of public sector spatial data by public bodies and the provision of information services based on these data by the public bodies. All legislation, policy documents, and literature are examined in light of these two axes, entailing that some of these documents will be addressed multiple times, but each time with a different concept or level in mind.

B. Outline

Chapter 2 – This chapter begins with a delineation of the scope of this book, which determines what we understand by the EU legal framework for the availability of public sector spatial data. The scope of these public sector spatial data is examined from the point of view of the data themselves and the organizations that hold these data. Next, we examine the distinctions between the instruments that make up the EU legal framework and the criteria on which these distinctions are based.

CHAPTER 3 – Chapter 3 gives an overview of the rules of the EU legal framework for the availability of public sector spatial data to provide an idea of the impact of the lack of effective distinctions between the different rules that are applicable. This overview is based on the text of the directives and on the evolution towards these texts in earlier initiatives and preparatory documents.

CHAPTER 4 – The overview from the previous chapter enables us to elaborate on the importance of the distinction between the different instruments in the legal framework and to introduce the problems in making the distinctions between them. This problematic character of the legal framework is set forth in Chapter 4, which submits that the distinction between access, re-use, and sharing is important for the EU legal framework to work effectively but that it cannot be maintained.

CHAPTERS 5 AND 6 – These two chapters address the making available of public sector spatial data by public bodies for access, re-use, and sharing. They further examine the distinctions between the three types of use that are distinguished in the EU legal framework and whether or not these distinctions can actually be enforced. First, we look at the legal obligations that are applicable to all the Member States of the EU and that may give indications about the obligations of the Member States to provide data for access, sharing, or re-use purposes. For this, the legal and policy instruments with regard to the availability of public sector spatial data that are applicable to all Member States of the EU are examined. From there on, we try to find a way for a public sector body to determine when it is providing data for re-use, access, or sharing purposes. Can the differences between access, re-use, and sharing be seen from the type of data that is requested? Can they be shown from the goals for which data are provided to citizens? Or from the level of processing that the data have undergone before they are made available? I will argue that this is not the case.

CHAPTERS 7 AND 8 – While the previous two chapters examined the availability of public sector spatial data, Chapters 7 and 8 focus on the provision of spatial information services based on these data by public bodies. Such services complicate the distinctions between the instruments of the EU legal framework for the availability of public sector spatial data even further, because the services might be provided as part of the public task or outside of it. This determines the applicability of the legal framework. Therefore, the public task to provide information services needs to be examined further. First, we check the obligations in the legal and policy instruments with regard to the availability of public sector spatial data to provide information services. Next, we try to determine whether there are criteria to determine whether or not a particular information service is a public task and whether inspiration can be found in the domain of competition law, which has had to deal with comparable issues.

CHAPTER 9 – The final chapter contains conclusions and an evaluation of whether or not the hypothesis that was made in the beginning of this book is valid. Next, it will try to put these findings in a broader context outside of the particular field of public sector spatial data.

Chapter 2
Scope and Definitions

I. INTRODUCTION

Defining the scope in terms of the legislation – Determining the scope of this book has three aspects. Firstly, we will address the area of law that will be the book's focus. Public sector spatial data are, just like any other object, structure, or person, subject to many different laws and regulations, which all have their own field of application and their own focus. It would be impossible to include all of these in this work. Therefore, the first section of this chapter includes the first limitation of the scope, that is, the EU legislation on the availability of spatial data held by public bodies.

Definition of the terms – Secondly, it is important that the terms and concepts are clearly defined so that there is no risk of misunderstanding or misinterpreting the findings of this book. The sections on the type of data and the type of organizations that are covered each contain two types of clarifications. On the one hand, we examine a number of terms and concepts that are often found when discussing the legal framework for the availability of public sector spatial or environmental data in the EU. Some of these terms have well-outlined definitions or meanings, while others are still the subject of debate among scholars and practitioners. They are included to give some background to the discussion and to give the reader an overview of the different terms and concepts that he or she might come across in the debate on the availability of public sector spatial data. On the other hand, we examine the definitions that are used throughout the book. These definitions are based on the existing legal framework, as this framework is the main object of this work, and they are adapted to working definitions for the sake of making the arguments more clear.

Categories of availability – Lastly, after identifying the common denominator in the field of application of the EU legislation, the distinctions between the three main pieces of EU legislation stimulating the availability of public sector spatial data will be set out. These distinctions form the basis for the rest of the chapters.

II. SCOPE FROM THE POINT OF VIEW OF THE LEGISLATION

TWO TYPES OF LEGISLATION AND POLICIES – The EU legal framework for public sector spatial data contains two categories of legislation. These two categories are not always in agreement and sometimes contradict each other.

PROMOTION OF AVAILABILITY – On the one hand, certain data policies make spatial data, by themselves or as a subset of a wider category of data, such as public sector data or environmental data, available to interested parties. These interested parties can be public bodies, citizens, or undertakings. In the EU, the most important rules are laid down in the PSI Directive, the Access Directive, and the INSPIRE Directive. These three legal regimes each have their own purpose and target group, but they all aim to increase the availability of public sector spatial data.[1]

LIMITATION OF AVAILABILITY – On the other hand, public sector spatial data are also regulated by legislation and policies that can impede their availability, often for a legitimate reason. These mainly stem from concerns about, among others, privacy, intellectual property rights, security, and liability. The laws, regulations, and practices that protect these interests bring a decrease in the spatial data that are available to the outside world.

PERSONAL DATA – Firstly, considering the wide definition of *personal data* that is given in the 1995 Directive on the protection of personal data[2] and its transposing legislation in the Member States, a considerable amount of public sector spatial data can also be classified as personal data. Hence, the processing of these data, in particular their dissemination, is subject to the limitations of national legislation on the processing of personal data. The privacy issues that arise when dealing with spatial data have been discussed extensively in the literature.[3]

1. Katleen Janssen & Jos Dumortier, 'The Legal framework for a European Spatial Data Infrastructure: Uncrossing the Wires', in *Research and Theory in Advancing Spatial Data Infrastructure Concepts*, ed. Harlan Onsrud (Redlands California: ESRI Press, 2007), 237.
2. Directive 95/46/EC of the European Parliament and of the Council of 24 Oct. 1995 on the protection of individuals with regard to the processing of personal data and on the free movement of such data, *OJ L* 281, 23 Nov. 1995, 31.
3. Harlan J. Onsrud, 'Geographic Information Legal Issues', 2004, <www.spatial.maine.edu/~onsrud/pubs/GILegalIssues.html>, 22 Feb. 2010; Harlan J. Onsrud, Jeff P. Johnson & Xavier Lopez, 'Protecting Personal Privacy in Using Geographic Information Systems', *Photogrammetric Engineering and Remote Sensing* (1994): 1083–1095, <www.spatial.maine.edu/%7eonsrud/pubs/privacy16.pdf>, 22 Feb. 2010; Bastiaan Van Loenen & Jaap A. Zevenbergen, 'The Impact of the European Privacy Regime on Location Technology Development', *Journal of Location Based Services* (2007): 165–178; Bastiaan Van Loenen & Jitske De Jong, 'SDIs and Privacy: Conflicting

INTELLECTUAL PROPERTY RIGHTS – Secondly, whether public sector spatial data are subject to intellectual property rights (IPR) has also been the subject of many discussions. Generally speaking, spatial data are protected by copyright if they fulfil the condition of originality. Hence, the protection is not self-evident, as spatial data are representations of facts, and facts are not protected by copyright.[4] However, in the cases where copyright is claimed, the need for permission of the copyright holder can cause restrictions on the availability of spatial data.[5] In addition, spatial databases are often protected by database rights.[6]

SECURITY – Thirdly, public sector spatial data can hold information on critical points for national security, for example, on the location of utilities or telecommunications infrastructures. The concerns about the threats of terrorism that have emerged since 9/11 entail that public bodies are more careful about which data they

---

Interests of the Spatially Enabled Society', in *Towards a Spatially Enabled Society*, ed. Abbas Rajabifard (Melbourne: University of Melbourne, 2007), 271–284; Bastiaan Van Loenen et al., 'Privacy Versus National Security: The Impact of Privacy Law on the Use of Location Technology for National Security Purposes', in *The European Information Society: Leading the Way with Geo-Information Lecture Notes in Geoinformation and Cartography*, ed. Sara I. Fabrikant & Monica Wachowicz (Berlin: Springer, 2007), 135–152; Jaap A. Zevenbergen, 'European Privacy Law and Its Effect on Location Information', in *Location Privacy Workshop: Individual Autonomy as a Driver of Designweb*, ed. Harlan Onsrud & Silvia Nittel (Orono, Maine: University of Maine, 2004), 1–11; Mark Burdon, 'Commercializing Public Sector Information: Privacy and Security Concerns', *IEEE Technology and Society Magazine* 28, no. 1 (2009): 34–40.

4. Harlan J. Onsrud & Xavier Lopez, 'Intellectual Property Rights in Disseminating Digital Geographic Data, Products, and Services: Conflicts and Commonalities among European Union and United States Approaches', 1998, <www.spatial.maine.edu/~onsrud/harlan/Course/CPOnsrudLopez.html>, 22 Feb. 2010.
5. For more information on IPR on spatial data, see Katleen Janssen & Jos Dumortier, 'The Protection of Maps and Spatial Databases in Europe and the United States by Copyright and the Sui Generis Right', *John Marshall Journal of Computer and Information Law* 24 (2006): 195–225; George Cho, *Geographic Information Science: Mastering the Legal Issues* (Chicester: John Wiley & Sons Ltd, 2005), 474; George Cho, *Geographic Information Systems and the Law, Mapping the Legal Frontiers* (New York: John Wiley & Sons Ltd, 2001), 358; Dennis S. Karjala, 'Copyright in Electronic Maps', *Jurimetrics Journal* 35 (1995): 395–415, <http://homepages.law.asu.edu/~dkarjala/Articles/Jurimetrics1995.html>, 22 Feb. 2010; Michelle R. Silverstein, 'The Copyrightability of Factual Compilations: An Interpretation of Feist Through Cases of Maps and Numbers', *Annual Survey of American Law* (1996): 147–218; Charlotte Waelde & Mags McGinley, 'Designing a Licensing Strategy for Sharing and Re-use of Geospatial Data in the Academic Sector', 2007, <http://edina.ac.uk/projects/grade/gradeDigitalRightsIssues.pdf>, 22 Feb. 2010; Harlan J. Onsrud & Xavier Lopez, 'Intellectual Property Rights in Disseminating Digital Geographic Data, Products, and Services: Conflicts and Commonalities among European Union and United States Approaches', 1998, <www.spatial.maine.edu/~onsrud/harlan/Course/CPOnsrudLopez.html>, 22 Feb. 2010; Ravi Bedrijvenplatform, *Auteursrecht op geo-informatie* (Amersfoort: Ravi, 1994), 41 [in Dutch]; Ravi Bedrijvenplatform, *Bescherming en verstrekking van geo-informatie* (Amersfoort: Ravi, 1995), 66 [in Dutch].
6. Katleen Janssen & Jos Dumortier, 'The Protection of Maps and Spatial Databases in Europe and the United States by Copyright and the Sui Generis Right', *John Marshall Journal of Computer and Information Law* 24 (2006): 195–225.

make available to the public, as these data might fall into the wrong hands and cause danger to society.[7]

LIABILITY – Finally, public bodies have created or acquired their spatial data for specific purposes. These data may not always be usable for other goals, and this might lead to errors or accidents, and subsequently to the occurrence of damage. Depending on the applicable legislation and possible liability waivers, the public body from which the data originate might be held liable for this damage. These concerns about liability, as well as the impossibility of checking what happens with spatial data after these are provided to other parties, may cause public bodies to limit the availability of their data.[8]

A COHERENT LEGAL FRAMEWORK – Reconciling all these competing policies in a coherent framework is an enormous challenge. Currently, even the policies promoting the availability of public sector spatial data seem to cause more impediments to this availability than they solve. In addition, one must also take into account that the conflicting legal policies are underpinned by questions concerning the relationship between the public and the private sector in the provision of spatial data and the role of the public sector in the information society and on the information market. As public bodies are often the sole or at least the most important supplier of certain spatial data, this might translate into a dominant position or a monopoly in competition law terms. Hence, European and national competition law may also be an undercurrent in the EU legal framework for public sector spatial data. Finally, procedural rules for the cooperation between several public bodies or between public bodies and the private sector must also be taken into account, such as rules on coordinating bodies or structures, procedures for cooperation agreements, public–private partnerships, and so on.

Figure 1 is a simplified representation of the EU legal framework for public sector spatial data.

7. John C. Baker et al., *Mapping the Risks: Assessing the Homeland Security Implications of Publicly Available Geospatial Information* (Santa Monica: RAND Corporation, 2004), 236; Daniel Z. Sui (ed.), *Geospatial Technologies and Homeland Security: Research Frontiers and Future Challenges* (Dordrecht: Springer Netherlands, 2008), 360; Patricia Salkin, 'GIS In an Age of Homeland Security: Accessing Public Information to Ensure a Sustainable Environment', *William and Mary Environmental Law and Policy Review* 33 (2005): 55–94.
8. Earl F. Epstein, 'Liability Insurance and the Use of Geographical Information', *International Journal of Geographical Information Science* 12, no. 3 (1998): 203–214; Earl F. Epstein, 'Legal Aspects of GIS', in *Geographical Information Systems. Principles and applications*, ed. David Maguire, Michael Goodchild, & David Rhind (Essex: Longman Scientific and Technical, 1993), 490; Harlan J. Onsrud, 'Liability in the Use of Geographic Information Systems and Geographic Data Sets', in *Geographic Information Systems: Principles, Techniques, Management, and* Applications, ed. David Maguire et al. (New York, Wiley, 1999), 643–652, <www.spatial.maine.edu/~onsrud/pubs/liabilityabstract40.htm>, 22 Feb. 2010; Jennifer Phillips, 'Information Liability: the Possible Chilling Effect of Tort Claims against Producers of Geographic Information Systems Data', *Florida State University Law Review* 26 (1999): 743–781; George Cho, *Geographic Information Science: Mastering the Legal Issues* (Chicester: John Wiley & Sons Ltd, 2005), 474.

Figure 1. EU Legal Framework for Public Sector Spatial Data (adapted from Janssen and Dumortier, 2007)

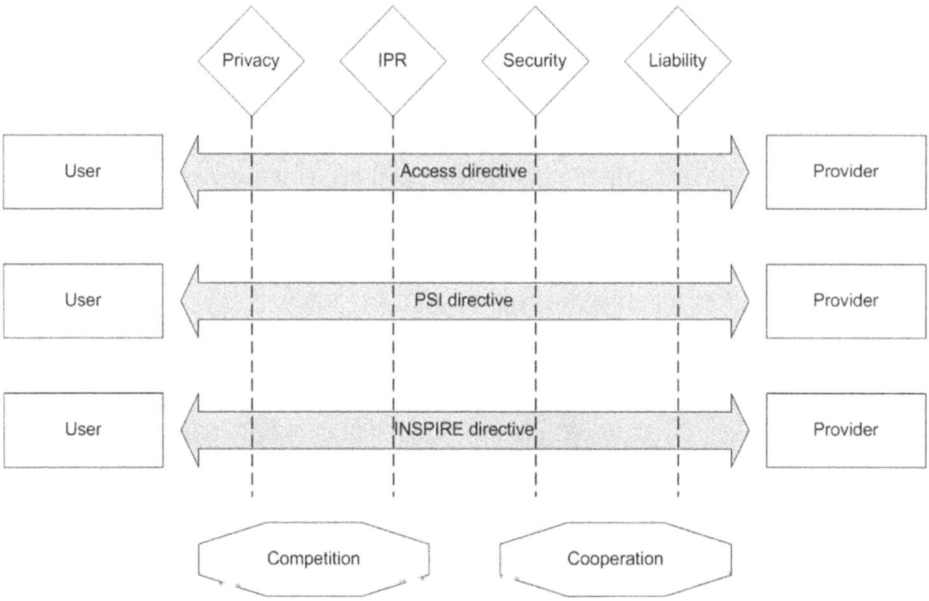

FOCUS ON LEGISLATION PROMOTING AVAILABILITY – The starting point and the main focus of this book are the legal policies that promote the availability of public sector spatial data, that is, the Access Directive, the PSI Directive and the INSPIRE Directive. The Access Directive is applicable to *environmental information held by public authorities*, the PSI Directive refers to *public sector documents held by public sector bodies*, and the INSPIRE Directive addresses *spatial data held by public authorities*. As will be discussed later on, these three categories have considerable overlap, at the cross-cut of which lie *public sector environmental spatial data*. I will argue that for this overlapping set of data, the legal framework regarding their availability is not clear, which causes problems in its application. This incoherence should be addressed before attempting to determine the interaction with the policies limiting the availability of public sector spatial data in further research.

FOCUS ON EUROPEAN UNION LEVEL – I have chosen to focus on the EU and only address the legislation and policies that are applicable to the EU as a whole.[9] Over the past ten years, the attention of the spatial data stakeholders has gone to the

9. Since the Treaty of Lisbon, the European Community has been replaced by the European Union. The Treaty establishing the European Community has become the Treaty on the Functioning of the European Union. However, we will still refer to the European Community when the official texts regarding the particular subject require us to do so.

development of a European SDI that facilitates the availability of public sector spatial data for cross-border issues in which the EU feels it might have a role to play, such as environmental planning, environmental safety, transport, security issues, interregional cooperation, industrial development, and so on. As the EU has adopted a number of directives that can be applicable to the availability of public sector spatial data, this domain has reached a high level of harmonization. However, as will be discussed later on, the directives leave a number of questions that may be problematic for this harmonization to have its full effect. Therefore, it is interesting to focus on the European level and detect those problems in order to see if this harmonization can be successful.

III. SCOPE FROM THE POINT OF VIEW OF THE TYPE OF DATA

OUTLINE – The focus of this work is spatial data that fall under the three directives that have been identified above, that is, spatial data that fall under one of the data themes mentioned in the INSPIRE Directive, which can also be qualified as environmental information and public sector documents. In order to identify this common denominator, the different terms of the directives are examined. However, this section will first discuss the term *spatial data* to provide a general idea of its meaning outside of the particular meaning that it has in the INSPIRE Directive.

A. SPATIAL DATA

1. Data

a. Data or Information

DISTINCTION BETWEEN DATA AND INFORMATION – In some literature and policy documents, the term spatial *data* is used, while others refer to spatial *information*. This leads us to the question of whether there is a distinction between data and information.

EVERYDAY LANGUAGE – The terms *data* and *information* are often used interchangeably. However, the difference between the terms should be highlighted.[10] First, we consider whether this difference comes to light in everyday language. The Merriam-Webster OnLine Dictionary defines *data* as:

1. factual information (as measurements or statistics) used as a basis for reasoning, discussion, or calculation;

10. P. Bernt Hugenholtz, *Auteursrecht op informatie: auteursrechtelijke bescherming van feitelijke gegevens en gegevensverzamelingen in Nederland, de Verenigde Staten en West-Duitsland: een rechtsvergelijkend onderzoek* (Deventer: Kluwer, 1989), 10 [in Dutch].

2. information output by a sensing device or organ that includes both useful and irrelevant or redundant information and must be processed to be meaningful;
3. information in numerical form that can be digitally transmitted or processed.[11]

From this we could assume that there is no significant difference between data and information. This is even more obvious from the definition that the Merriam-Webster Dictionary provides for *information:*

1. the communication or reception of knowledge or intelligence;
2. a: (1): knowledge obtained from investigation, study, or instruction (2): intelligence, news (3): facts, data; b: the attribute inherent in and communicated by one of two or more alternative sequences or arrangements of something (as nucleotides in DNA or binary digits in a computer program) that produce specific effects; c: (1): a signal or character (as in a communication system or computer) representing data (2): something (as a message, experimental data, or a picture) which justifies change in a construct (as a plan or theory) that represents physical or mental experience or another construct; d: a quantitative measure of the content of information; *specifically*: a numerical quantity that measures the uncertainty in the outcome of an experiment to be performed.[12]

INTERNATIONAL ORGANIZATION FOR STANDARDIZATION – Merriam-Webster apparently considers information to be synonymous with facts and data. Perhaps the descriptions of the International Organization for Standardization (ISO) can provide more insight. According to ISO standard 2382/1, *data* are 'a representation of facts, concepts or instructions in a formalized manner suitable for communication, interpretation or processing by human beings or by automatic means',[13] while *information* is 'the meaning that participants in the communication process give to the data that are passed on, according to the conventions that are applied to those data'.[14]

11. Merriam-Webster Online Dictionary, 'Data', <www.m-w.com/dictionary/data>, 22 Feb. 2010.
12. Merriam-Webster Online Dictionary, 'Information', <www.m-w.com/dictionary/information>, 22 Feb. 2010. Merriam-Webster Online also included '3: the act of informing against a person; and 4: a formal accusation of a crime made by a prosecuting officer as distinguished from an indictment presented by a grand jury', but as these definitions refer to a specific context unrelated to the topic of this work, they are disregarded in the discussion.
13. International Standards Organisation, 'Standard 2382-1. Information technology – Vocabulary – Part 1: Fundamental terms', <www.iso.org/iso/en/CatalogueDetailPage.CatalogueDetail?CSNUMBER=7229>, 22 Feb. 2010.
14. Egbert J. Dommering, 'Informatie, communicatie en technologie', in *Informatierecht. Fundamentele rechten voor de informatiesamenleving*, ed. Egbert J. Dommering et al. (Amsterdam: Otto Cramwinckel Uitgever, 2000), 459 [in Dutch]. See also Jean-Michel Bruguière, *Les données publiques et le droit* (Paris: Editions Litec, 2002), 2 [in French]; Caroline Uyttendaele, *Openbare informatie. Het juridisch statuut in een convergerende media-omgeving* (Antwerp: Maklu, 2002), 21 [in Dutch].

From data to information – This indicates that the term *data* is a neutral or objective notion, relating to observations or representations of facts, concepts, or instructions that are suitable for interpretation or communication.[15] Data are the resource from which information can emanate. Hence, data are potential information. They become information from the moment they are understood by a certain person or are placed in a particular context.[16] Van Loenen argues that in a spatial context, data are:

> the bits and bytes without meaning or context. These bits and bytes are transferred to information through the data processing system using a data model. When the data model is applied to the data results in a specific context, they become meaningful. The data model enables interpretation of the facts as well as providing meaning, which is then considered real information. Data can be considered information when someone recognizes it as such. Thus, the value of information can vary among people.[17]

Hence, information is *data with a meaning.*

Information as reduction in uncertainty – In line with this approach, economic theory views information as data with a meaning *to a particular recipient* of those data. In economics, information is often defined as a reduction in uncertainty.[18] This approach views information as something – anything – that alters the

15. Commissie Computercriminaliteit, *Informatietechniek en Strafrecht*'s Gravenhage: Commissie Computercriminaliteit, 1987), 15 [in Dutch].; P. Bernt Hugenholtz, *Auteursrecht op informatie: auteursrechtelijke bescherming van feitelijke gegevens en gegevensverzamelingen in Nederland, de Verenigde Staten en West-Duitsland: een rechtsvergelijkend onderzoek* (Deventer: Kluwer, 1989),10 [in Dutch]; Earl F. Epstein, 'Legal Aspects of GIS', in *Geographical Information Systems. Principles and applications*, ed. David Maguire, Michael Goodchild & David Rhind (Essex: Longman Scientific and Technical, 1993), 490; George Cho, *Geographic Information Systems and the Law: Mapping the Legal Frontiers* (Chicester: John Wiley & Sons Ltd, 2001), 28.
16. P. Bernt Hugenholtz, *Auteursrecht op informatie: auteursrechtelijke bescherming van feitelijke gegevens en gegevensverzamelingen in Nederland, de Verenigde Staten en West-Duitsland: een rechtsvergelijkend onderzoek* (Deventer: Kluwer, 1989), 10 [in Dutch]; Earl F. Epstein, 'Legal Aspects of GIS', in *Geographical Information Systems. Principles and applications*, ed. David Maguire, Michael Goodchild & David Rhind (Essex: Longman Scientific and Technical, 1993), 490; Roger Clarke, 'Freedom of Information? The Internet as Harbinger of the New Dark Ages', *First Monday* 4, no. 11 (1999), <http://firstmonday.org/htbin/cgiwrap/bin/ojs/index.php/fm/issue/view/112>, 22 Feb. 2010; Serge Guthwirth, 'Het juridisch statuut van informatie', in *Technologie en recht*, ed. Paul De Vroede (Antwerp: Kluwer Rechtswetenschappen, 1987), 200 [in Dutch].
17. Bastiaan Van Loenen, *Developing Geographic Information Infrastructures: The Role of Information Policies* (Delft: DUP Science, 2006), 12.
18. Kenneth J. Arrow, 'The Economics of Information', in *The Computer Age: A Twenty Year Review*, ed. Michael Dertouzos & Joe Moses (Cambridge: MIT Press, 1979), 306. See also John Cahir, 'Understanding Information Laws: A Sociological Approach', *Journal of Information, Law and Technology* (2002), <www2.warwick.ac.uk/fac/soc/law/elj/jilt/2002_3/cahir/>, 22 Feb. 2010; William Mock, 'On the Centrality of Information Law: A Rational Choice Discussion of Information Law and Transparency', *John Marshall Journal of Computer & Information Law* 17 (1999): 1072.

awareness of the recipient and makes that recipient perceive the world as a different place from the one it was before. This entails that there must be some interaction between the potential information and a recipient in order for the potential to become actual information.[19] According to Hugenholtz, this also indicates that there is a connection between information and improbability or unpredictability. The more unlikely the information is, the greater its informational value. Information implies a certain degree of novelty or surprise. When a receiver knows the content of the message beforehand, he or she does not receive information.[20] This entails that information is a subjective or relative notion.[21] A message can contain information for one person but not for another.

PREFERENCE FOR *SPATIAL DATA* – Hence, if we think of data as something independent from the user and merely a representation of facts, which only becomes information once it is understood by a certain person or in a certain context, it seems more appropriate for this work to refer to spatial *data* rather than spatial *information*. Spatial data can contain different information for different users, depending on their interpretation of the data. Therefore, in this book, I attempt to consistently use the term *data* with the meaning of a representation of facts that is independent from the user. However, in some specific contexts, referring to *information* may be necessary because it is used as part of a term that is defined by law and therefore should be used in that context, for example, environmental information.[22] In addition, I refer to *information* when spatial data are indeed put into a context and are as such offered to the user.[23]

b. *Information or Document*

REFERENCE TO A CARRIER – Some existing legislation, policy documents, and literature refer to a third term: *document*. While *information* refers to content, or to data

---

19. William Mock, 'On the Centrality of Information Law: A Rational Choice Discussion of Information Law and Transparency', *John Marshall Journal of Computer & Information Law* 17 (1999): 1072.
20. P. Bernt Hugenholtz, *Auteursrecht op informatie: auteursrechtelijke bescherming van feitelijke gegevens en gegevensverzamelingen in Nederland, de Verenigde Staten en West-Duitsland: een rechtsvergelijkend onderzoek* (Deventer: Kluwer, 1989), 6 [in Dutch].
21. John Cahir, 'Understanding Information Laws: A Sociological Approach', *Journal of Information, Law and Technology* (2002), <www2.warwick.ac.uk/fac/soc/law/elj/jilt/2002_3/cahir/>, 22 Feb. 2010.
22. We should keep in mind that data and information are only two links in the chain from data to wisdom, which also includes knowledge and intelligence. According to Cho, knowledge and intelligence derive from information, while wisdom entails the correct use of knowledge (George Cho, *Geographic Information Systems and the Law: Mapping the Legal Frontiers* (Chicester: John Wiley & Sons Ltd, 2001), 28), while Epstein believes that knowledge is the contribution of the human mind to information, and wisdom is the knowledge possessed by particular, recognized individuals in a society (Earl F. Epstein, 'Legal Aspects of GIS', in *Geographical Information Systems: Principles and Applications*, ed. David Maguire, Michael Goodchild & David Rhind (Essex: Longman Scientific and Technical, 1993), 490).
23. Cf. *infra*, Chs 7 and 8.

put into a context, *document* usually refers to information having been fixed on a carrier.[24] Traditionally, this carrier was paper, but many current legislative measures refer to any type of carrier, whether it is analogue or digital, for example, reports, letters, CDs, CD-ROMs, e-mails, maps, databases, audiotape, videotapes, DVDs, and so on.

REFERENCE TO CONTENT – However, *document* sometimes also refers to content, which would make it synonymous with *information.* For instance, in the PSI Directive, *document* is defined as '(a) any content whatever its medium (written on paper or stored in electronic form or as a sound, visual or audiovisual recording); (b) any part of such content'.[25]

2. Spatial

SPATIAL OR GEOGRAPHIC – Not only are the terms *data* and *information* often used to describe the same thing, but also, so are *spatial* and *geographic.* This seems to indicate that the difference between *spatial* and *geographic* – if there even is one – is negligible. To add to the confusion, sometimes, mostly in Anglo-Saxon literature and practice, a third term is used: *geospatial.*[26]

GSDI COOKBOOK – The Global Spatial Data Infrastructure Association (GSDI) Cookbook does not seem to show any preference between these terms, as it provides descriptions for geographic information, geospatial data, and spatial data. The first is information 'concerning phenomena implicitly or explicitly associated with a location relative to the Earth',[27] while geospatial data are defined virtually the same: 'data that identifies the geographic location and characteristics of natural or constructed features and boundaries on the Earth'.[28] Thirdly, spatial data are defined as 'data concerned with the size, area or position of any location, event or phenomenon'.[29]

RELEVANCE OF THE DIFFERENCE – This might indicate that all three terms are commonly used and that the possible slight differences in meaning do not form a problem for policy makers, practitioners, and the academic community to interact. However, according to some authors, the difference does have some relevance.

24. Frankie Schram, *Begrenzingen aan de openbaarheid van bestuur: een analyse en vergelijking van de internrechtelijke openbaarheidsregelingen in het Belgische recht. Doctoral Thesis K.U.Leuven* (Leuven: K.U.Leuven, 2002), 442 [in Dutch].
25. Article 2.3 of the PSI Directive.
26. Richard Groot, 'Spatial Data Infrastructure (SDI) for Sustainable Land Management', *ITC Journal* no. 3/4 (1997): 288. See e.g., Richard Groot & John Douglas McLaughlin, *Geospatial Data Infrastructure: Concepts, Cases, and Good Practice* (New York: Oxford University Press), 320; Michael J. De Smith, Michael F. Goodchild & Paul A. Longley, *Geospatial Analysis* (Leicester: Troubadour Publishing, 2007), 516.
27. GSDI Association, 'The SDI Cookbook vol. 2.0, 2004', 162, <www.gsdi.org/docs2004/Cookbook/cookbookV2.0.pdf>, 22 Feb. 2010.
28. *Ibid.*, 163. The definition was taken from William J. Clinton, 'Executive Order 12906 – Co-ordinating Geographic Data Acquisition and Access: The National Spatial Data Infrastructure', 1994, <http://govinfo.library.unt.edu/npr/library/direct/orders/20fa.html>, 22 Feb. 2010.
29. *Ibid.*, 167.

Van Loenen, for instance, believes that spatial data may include any space, not just space on Earth's surface, while geographic data can be considered as a subset of spatial data for Earth's surface and near surface.[30]

GROWING CONSENSUS – However, a growing consensus seems to be building, also on the political level, towards referring to *spatial* data instead of *geographic* data. For instance, the INSPIRE initiative[31] of the European Commission refers to *spatial* data sets, while the earlier attempt at creating an initiative in this field begun in 1995 was entitled GI (geographic information) 2000. The INSPIRE Directive defines *spatial data* as 'any data with a direct or indirect reference to a specific location or geographical area'.[32]

PREFERENCE FOR *SPATIAL DATA* – In the following chapters, I intend to use the term *spatial* as consistently as possible, mainly for two reasons. On the one hand, the difference between the two terms *spatial* and *geograpic* seems negligible, as they are often used interchangeably or are even combined in the term *geospatial*. Therefore, I will follow the global trend of referring to SDIs, and that inherently includes *spatial* data. On the other hand, if the difference would be of importance, it seems artificial to distinguish between data that are related to a location on Earth's surface and data that are not, as this would exclude both meteorological data (from the atmosphere) and some geological data (from under Earth's surface). As both of these types of data share a very important trait with *geographic* data, namely that they are related to a certain location, and as they are just as important to obtain the objectives of environmental, agricultural, and socio-economic policymaking, it would be artificial to exclude them from the field of application of this work.

3. Spatial Data

WORKING DEFINITION – In summary, the definition of the INSPIRE Directive will be used as a working definition for any future reference to spatial data: 'any data with a direct or indirect reference to a specific location or geographical area'.

B. SPATIAL DATA UNDER THE INSPIRE DIRECTIVE

OUTLINE – Now that it is clear what is meant by *spatial data*, the field of application of this work is narrowed down to the spatial data that fall under the INSPIRE

30. Bastiaan Van Loenen, *Developing Geographic Information Infrastructures: The Role of Information Policies* (Delft: DUP Science, 2006), 11–12.
31. Commission of the European Communities, *Proposal for a directive of the European Parliament and of the Council establishing an infrastructure for spatial information in the Community (INSPIRE)*, COM (2004) 516 final, 23 Jul. 2004; Commission of the European Communities, 'GI2000: Towards a European Policy Framework for Geographic Information. A Communication from the Commission to the European Parliament, to the Economic and Social Committee and to the Committee of the Regions. Draft', 1998, <www.ec-gis.org/docs/F20680/MICROSOFT%20WORD%20-%20GI2000V10_7_98.PDF>, 22 Feb. 2010.
32. Article 3.2 of the INSPIRE Directive.

Directive and that can also be classified as environmental information under the Access Directive and public sector document under the PSI Directive. The next sections examine these definitions.

ARTICLE 4 OF THE INSPIRE DIRECTIVE – The directive covers spatial data sets 'which fulfil the following conditions:

(a) they relate to an area where a Member State has and/or exercises jurisdictional rights;
(b) they are in electronic format;
(c) they are held by or on behalf of any of the following:
  (i) a public authority, having been produced or received by a public authority, or being managed or updated by that authority and falling within the scope of its public tasks;
  (ii) a third party to whom the network has been made available in accordance with Article 12;
(d) they relate to one or more of the themes listed in Annex I, II or III'.[33]

INSPIRE DATA THEMES – These Annexes contain thirty-four themes of spatial data, divided into groups in Annexes I, II, and III of the directive, which should all be considered to be relevant for the environmental policy of the EU, as this is the legal basis for the directive. Originally, about sixty different data sets were listed to be of importance for INSPIRE, but during the impact assessment and the consultation, it became clear that most stakeholders found this too ambitious. Therefore, the list was reduced to what eventually ended up in these thirty-four themes.[34]

SCOPE OF THE DATA THEMES – The difference between the Annexes lies in the timeframe within which the requirements of the INSPIRE Directive have to be implemented in the Member States for the respective categories of spatial data. It is not the aim of this book to determine the scope of the data sets that are included under the INSPIRE Directive. For more information, I refer to the work of the thematic working groups of the INSPIRE drafting team on data specifications.[35] However, it is interesting to see the wide scope of these data themes. Therefore, the list is reproduced here:

33. Article 4 of the INSPIRE Directive.

34. Commission of the European Communities, *Commission staff working document. Proposal for a Directive of the European Parliament and of the Council establishing an infrastructure for spatial information in the Community (INSPIRE) – Extended Impact Assessment*, SEC (2004) 980, 23 Jul. 2004, <http://ec.europa.eu/governance/impact/docs/ia_2004/sec_2004_0980_en.pdf>, 22 Feb. 2010; Commission of the European Communities, 'Consultation Paper on a Forthcoming EU Legal Initiative on Spatial Information for Community Policy-making and Implementation', 2003, <www.ec-gis.org/docs/F6017/INSPIRE-INTERNETCONSULTATIONPHASEII.PDF>, 22 Feb. 2010.

35. For an overview of activities and documents of the INSPIRE drafting team on data specifications, see <http://inspire.jrc.ec.europa.eu/implementingRulesDocs_ds.cfm>, 22 Feb. 2010.

Annex I of the INSPIRE Directive contains the following themes of spatial data[36]:

(1) Coordinate reference systems: systems for uniquely referencing spatial information in space as a set of coordinates (x, y, z) and/or latitude and longitude and height, based on a geodetic horizontal and vertical datum.
(2) Geographical grid systems: harmonized multi-resolution grid with a common point of origin and standardized location and size of grid cells.
(3) Geographical names: names of areas, regions, localities, cities, suburbs, towns or settlements, or any geographical or topographical feature of public or historical interest.
(4) Administrative units: units of administration, dividing areas where Member States have and/or exercise jurisdictional rights, for local, regional and national governance, separated by administrative boundaries.
(5) Addresses: location of properties based on address identifiers, usually by road name, house number, postal code.
(6) Cadastral parcels: areas defined by cadastral registers or equivalent.
(7) Transport networks: road, rail, air and water transport networks and related infrastructure. Includes links between different networks. Also includes the trans-European transport network as defined in Decision No 1692/96/EC of the European Parliament and of the Council of 23 July 1996 on Community Guidelines for the development of the trans-European transport network and future revisions of that Decision.
(8) Hydrography: hydrographic elements, including marine areas and all other water bodies and items related to them, including river basins and sub-basins. Where appropriate, according to the definitions set out in Directive 2000/60/EC of the European Parliament and of the Council of 23 October 2000 establishing a framework for Community action in the field of water policy and in the form of networks.
(9) Protected sites: area designated or managed within a framework of international, Community and Member States' legislation to achieve specific conservation objectives.

Annex II of the INSPIRE Directive contains the following themes of spatial data[37]:

(1) Elevation: digital elevation models for land, ice and ocean surface. Includes terrestrial elevation, bathymetry and shoreline.
(2) Land cover: physical and biological cover of the earth's surface including artificial surfaces, agricultural areas, forests, (semi-)natural areas, wetlands, water bodies.
(3) Orthoimagery: geo-referenced image data of the Earth's surface, from either satellite or airborne sensors.

36. Annex I of the INSPIRE Directive.
37. Annex II of the INSPIRE Directive.

(4) Geology: geology characterized according to composition and structure. Includes bedrock, aquifers and geomorphology.

Annex III of the INSPIRE Directive contains the following themes of spatial data[38]:

(1) Statistical units: units for dissemination or use of statistical information.
(2) Buildings: geographical location of buildings.
(3) Soil: soils and subsoil characterized according to depth, texture, structure and content of particles and organic material, stoniness, erosion, where appropriate mean slope and anticipated water storage capacity.
(4) Land use: territory characterized according to its current and future planned functional dimension or socio-economic purpose (e.g., residential, industrial, commercial, agricultural, forestry, recreational).
(5) Human health and safety: geographical distribution of dominance of pathologies (allergies, cancers, respiratory diseases, etc.), information indicating the effect on health (biomarkers, decline of fertility, epidemics) or well-being of humans (fatigue, stress, etc.) linked directly (air pollution, chemicals, depletion of the ozone layer, noise, etc.) or indirectly (food, genetically modified organisms, etc.) to the quality of the environment.
(6) Utility and governmental services: includes utility facilities such as sewage, waste management, energy supply and water supply, administrative and social governmental services such as public administrations, civil protection sites, schools and hospitals.
(7) Environmental monitoring facilities: location and operation of environmental monitoring facilities includes observation and measurement of emissions, of the state of environmental media and of other ecosystem parameters (biodiversity, ecological conditions of vegetation, etc.) by or on behalf of public authorities.
(8) Production and industrial facilities: industrial production sites, including installations covered by Council Directive 96/61/EC of 24 September 1996 concerning integrated pollution prevention and control and water abstraction facilities, mining, storage sites.
(9) Agricultural and aquaculture facilities: farming equipment and production facilities (including irrigation systems, greenhouses and stables).
(10) Population distribution – demography: geographical distribution of people, including population characteristics and activity levels, aggregated by grid, region, administrative unit or other analytical unit.
(11) Area management/restriction/regulation zones and reporting units: areas managed, regulated or used for reporting at international, European, national, regional and local levels. Includes dumping sites, restricted areas around drinking water sources, nitrate-vulnerable zones, regulated

38. Annex III of the INSPIRE Directive.

fairways at sea or large inland waters, areas for the dumping of waste, noise restriction zones, prospecting and mining permit areas, river basin districts, relevant reporting units and coastal zone management areas.

(12) Natural risk zones: vulnerable areas characterized according to natural hazards (all atmospheric, hydrologic, seismic, volcanic and wildfire phenomena that, because of their location, severity, and frequency, have the potential to seriously affect society), for example, floods, landslides and subsidence, avalanches, forest fires, earthquakes, volcanic eruptions.

(13) Atmospheric conditions: physical conditions in the atmosphere. Includes spatial data based on measurements, on models or on a combination thereof and includes measurement locations.

(14) Meteorological geographical features: weather conditions and their measurements; precipitation, temperature, evapotranspiration, wind speed and direction.

(15) Oceanographic geographical features: physical conditions of oceans (currents, salinity, wave heights, etc.).

(16) Sea regions: physical conditions of seas and saline water bodies divided into regions and sub-regions with common characteristics.

(17) Bio-geographical regions: areas of relatively homogeneous ecological conditions with common characteristics.

(18) Habitats and biotopes: geographical areas characterized by specific ecological conditions, processes, structure, and (life support) functions that physically support the organisms that live there. Includes terrestrial and aquatic areas distinguished by geographical, abiotic and biotic features, whether entirely natural or semi-natural.

(19) Species distribution: geographical distribution of occurrence of animal and plant species aggregated by grid, region, administrative unit or other analytical unit.

(20) Energy resources: energy resources including hydrocarbons, hydropower, bio-energy, solar, wind, etc., where relevant including depth/height information on the extent of the resource.

(21) Mineral resources: mineral resources including metal ores, industrial minerals, etc., where relevant including depth/height information on the extent of the resource.

PREPARATORY PROCESS: LIST OF DATA THEMES – The list of data themes has undergone a considerable number of changes since the INSPIRE process was started in 2002.[39] In the preparatory phase, the position papers of the INSPIRE Working Group on Reference Data and Metadata set out a first list of reference data themes

39. For a short overview of the history of the preparation and implementation of the INSPIRE data specifications, see INSPIRE Drafting Team on Data Specifications, *Deliverable D2.3: Definition of Annex Themes and Scope*, 2008, <http://inspire.jrc.ec.europa.eu/implementingRulesDocs_ds.cfm>, 22 Feb. 2010.

that the members of the Group believed were indispensable for setting up INSPIRE and that needed to be shared in order to ensure better environmental policies.[40] Simultaneously, the Working Group on Environmental and Thematic Data dealt with the need for other kinds of thematic data. The Working Groups' position papers highlighted the common spatial data components needed in different environmental policy areas, examining both the high-level needs connected to policy evaluation and the needs for the implementation of policies, including the management needs at local and regional levels.[41] Reference data that were needed included administrative boundaries, elevation and bathymetry,[42] hydrography, transport networks, geographical names, geodetic reference data,[43] and ortho-images. Other thematic data that needed to be made available included soil, river catchments, ecological regions, land cover, location of utilities, location of facilities, protected areas, land use plans/area regulation, and demographic data. Other data themes with environmental importance that also needed to be included in the INSPIRE infrastructure were monitoring sites, groundwater bodies, management and reporting units, satellite multipurpose data, natural risk vulnerability zones, technological risk vulnerability zones, local contaminated sites, economic data, and environment sector regulation. In total, sixty data sets were included under these data themes.

PREPARATORY PROCESS – The 2004 INSPIRE Scoping Task Force was assigned the task of reducing these sixty data sets to a more limited number. Based on an analysis of common reference and thematic data requirements for the development, monitoring, and review of environmental strategies and policies, the Task Force found that: 'It becomes clear that some thematic data is less frequently needed. It is therefore proposed to reduce the scope of INSPIRE at this stage by focussing only on those thematic data that are most often needed. INSPIRE can then if necessary be extended at a later stage to also cover the individual needs of individual policies'.[44] The work of the Task Force resulted in the data themes of the Annexes to the INSPIRE Directive that were mentioned above.

---

40. The main functional requirements of geographical reference data were described as: providing an unambiguous location for a user's information; enabling the merging of data from various sources; and providing a context to allow others to better understand the information that is being presented (INSPIRE Reference Data and Metadata Working Group, *Reference Data and Metadata Position Paper*, 2002, <http://inspire.jrc.ec.europa.eu/reports/position_papers/inspire_rdm_pp_v4_3_en.pdf>, 22 Feb. 2010).
41. INSPIRE Environmental Thematic Coordination Group, *Environmental Thematic User Needs Position Paper*, 2002, < http://inspire.jrc.ec.europa.eu/reports/position_papers/inspire_etc_pp_v2_3_en.pdf>, 22 Feb. 2010.
42. Bathymetry includes data about the underwater depth of lakes and oceans.
43. Geodesy is the science of measuring the size and shape of Earth and the locations of points on its surface, e.g., under the form of coordinates. Geodetic reference systems provide a reference for navigation and mapping.
44. Task Force Scoping, *INSPIRE Scoping Paper*, 2004, <http://inspire.jrc.ec.europa.eu/reports/inspire_scoping24mar04.pdf>, 22 Feb. 2010.

C. Environmental Data

Outline – After examining the specific scope that has been given to environmental information in the Access Directive, the definition of the Aarhus Convention is examined, as this Convention forms the basis for the Access Directive and can currently be seen as the most important legal instrument ensuring public access to environmental information. However, first the concept of environmental data in everyday language is examined briefly.

1. Environmental

Environment – An obvious definition of *environmental data* would be 'any data relating to the environment'. This would then require a definition of *environment*. However, the word *environment* has many different meanings. According to the Merriam-Webster OnLine Dictionary, *environment* can be defined as:

1. the circumstances, objects, or conditions by which one is surrounded;
2. (a) The complex of physical, chemical, and biotic factors (as climate, soil, and living things) that act upon an organism or an ecological community and ultimately determine its form and survival; b) the aggregate of social and cultural conditions that influence the life of an individual or community;
3. the position or characteristic position of a linguistic element in a sequence;
4. a computer interface from which various tasks can be performed <a programming environment>.[45]

No legal definition – The meaning that is intended for *environmental data* in this book relates first and foremost to the second definition. However, policy documents and legislation tend to not define *environment*. For instance, the Aarhus Convention does not contain a definition of *environment*. In its definition of *environmental information*, a number of elements are listed that are considered to be part of the environment, but at the same time it is also made clear that these elements are not exhaustive. Next, the Treaty on the Functioning of the European Union (the former EC Treaty), which determines the competences of the European institutions regarding the environment, does not indicate what is to be understood by the term *environment*. Moreover, it uses the term not only as it relates to the environment as it is meant in 'environmental information', but also in such combinations as 'working environment' and 'encouraging an environment favourable to cooperation between undertakings'.[46] Also, the glossary created by the Directorate-General on Environment of the European Commission

---

45. Merriam-Webster Online Dictionary, <www.webster.com/dictionary/environment>, 22 Feb. 2010.
46. For example, Art. 114 (former Art. 95), Art. 153 (former Art. 137), Art. 173 (former Art. 157) of the Treaty on the Functioning of the European Union.

contains over twenty terms relating to the environment, but none regarding the environment itself.[47]

FOCUS ON ENVIRONMENTAL DATA – This book does not intend to provide a definition of *environment*. What is of interest is what constitutes environmental *data*. Therefore, the next sections look at the definitions of *environmental data* from the Access Directive, one of the three pillars of the legal framework on the availability of public sector spatial data, and the Aarhus Convention, on which the Access Directive is based. These texts all refer to environmental *information*; therefore, this terminology is followed in this book in relation to these pieces of legislation. However, for any other uses in the next chapters, the term environmental *data* is maintained.

2. Environmental Information in the Access Directive

THE ACCESS DIRECTIVE – While the EU institutions have claimed that they are not competent to regulate matters of access to information held by the government or public bodies because this would go beyond the subsidiarity principle, an exception is made for environmental information. The EU competence on environmental policy and the participation of the EU as a party to the Aarhus Convention was used to create a Directive on environmental information under Article 175 of the EC Treaty (now Article 192 of the Treaty on the Functioning of the European Union).

ARTICLE 2.1 OF THE ACCESS DIRECTIVE – Article 2.1 of the Access Directive states that:

> For the purposes of this directive, environmental information shall mean any information in written, visual, aural, electronic or any other material form on:
>
> a) the state of the elements of the environment, such as air and atmosphere, water, soil, land, landscape and natural sites including wetlands, coastal and marine areas, biological diversity and its components, including genetically modified organisms, and the interaction among these elements;
> b) factors, such as substances, energy, noise, radiation or waste, including radioactive waste, emissions, discharges and other releases into the environment, affecting or likely to affect the elements of the environment referred to in (a);
> c) measures (including administrative measures), such as policies, legislation, plans, programmes, environmental agreements, and activities affecting or likely to affect the elements and factors referred to in (a) and (b) as well as measures or activities designed to protect those elements;
> d) reports on the implementation of environmental legislation;
> e) cost-benefit and other economic analyses and assumptions used within the framework of the measures and activities referred to in (c); and

47. Commission of the European Communities, 'Alphabetical Index of Terms Referring to the Environment', <http://ec.europa.eu/environment/abc.htm#E>, 22 Feb. 2010.

f) the state of human health and safety, including the contamination of the food chain, where relevant, conditions of human life, cultural sites and built structures inasmuch as they are or may be affected by the state of the elements of the environment referred to in (a) or, through those elements, by any of the matters referred to in (b) and (c).

CONFORMITY WITH AARHUS CONVENTION – The Access Directive ensures the transposition of the Aarhus Convention into EU law, so obviously, it had to look towards the Convention for determining the scope of the term *environmental information*. In the discussions between the Council and the European Parliament, continuous efforts were made to keep the definition as close as possible to the one in the Aarhus Convention so as to enable the proper ratification of the treaty by the EU.[48] Hence, the definition in the Directive is almost identical to the description of *environmental information* in the Aarhus Convention, except for a small number of extra clarifications.

EUROPEAN COURT OF JUSTICE INTERPRETATION OF ENVIRONMENTAL INFORMATION IN DIRECTIVE 90/313 – The European Court of Justice (ECJ) has not passed judgment on the scope of the definition of *environmental information* under the Access Directive yet, but it has had the chance to examine Directive 90/313/EEC[49] and its definition of *environmental information*. According to the ECJ, the scope of the definition and hence of the directive 'must be considered to have been intended to be wide. It thus covers all information which relates either to the state of the environment or to activities or measures which could affect it, or to activities or measures intended to protect the environment, without the list in that provision including any indication such as to restrict its scope'.[50] However, the ECJ had stated earlier that Directive 90/313/EEC was 'not intended, however, to give a general and unlimited right of access to all information held by public authorities which has a connection, however minimal, with one of the environmental factors mentioned' in the definition of *environmental information*. The information had to fall within the three categories mentioned.[51]

---

48. See e.g., Commission of the European Communities, *Amended proposal for a directive of the European Parliament and of the Council on public access to environmental information*, COM (2001) 303 final, 6 Jun. 2001; Council of the European Union, *Common Position no. 24/2002 adopted on 28 January 2002 with a view to adopting Directive 2002/ . . . /EC of the European Parliament and of the Council of . . . on public access to environmental information and repealing Council Directive 90/313/EEC, OJ C* 113 E, 14 May 2002, 1. See also Jean-Jacques Paradissis, *The Right to Access Environmental Information* (Saarbrucken: VDM Verlag, 2009), 82–83.
49. Council Directive 90/313/EEC of 7 Jun. 1990 on the freedom of access to information on the environment, *OJ L* 158, 23 Jun. 1990, 56.
50. ECJ, Case C-233/00, *Commission v. French Republic*, 26 Jun. 2003, para. 44, *European Court Reports* 2003, I-6625; ECJ, Case C-321/96, *Wilhelm Mecklenburg v. Kreis Pinneberg – der Landrat*, 17 Jun. 1998, *European Court Reports* 1998, I-3809.
51. ECJ, Case C-316/01, *Eva Glawischnig v. Bundesminister für soziale Sicherheit und Generationen*, 12 Jun. 2003, para. 25, *European Court reports* 2003, I-05995. See also Jean-Jacques Paradissis, *The Right to Access Environmental Information* (Saarbrucken: VDM Verlag, 2009), 61, 87–89.

VALIDITY OF ECJ INTERPRETATION SINCE THE ACCESS DIRECTIVE – Since the ECJ judgment, Article 255 was introduced in the EC Treaty by the Treaty of Amsterdam, which states in its first paragraph that: '[a]ny citizen of the Union, and any natural or legal person residing or having its registered office in a Member State, shall have a right of access to European Parliament, Council and Commission documents'. This Article was part of a process of openness embedded in the EU Treaty: openness became the rule and secrecy the exception.[52] In addition, the Aarhus Convention has promoted access by including a definition of *environmental information* that should be interpreted broadly.[53] Therefore, one can assume that the ECJ, were it confronted with a question of interpretation of the term *environmental information* under the Access Directive, would still advocate a wide interpretation. As Article 255 is even extended by the Lisbon Treaty to include documents of all the institutions and bodies of the EU in Article 15 of the Treaty on the Functioning of the European Union, this extensive interpretation will be even more evident. In the course of 2009, a number of references for preliminary rulings were made by national courts on the scope of the definition of *environmental information*, so this scope will become clearer in the next few years.

3. Environmental Information in the Aarhus Convention

VALUE OF DEFINITION – As the Access Directive is based on the Aarhus Convention and forms the transposition of the Convention for the EU as a whole, it is interesting to go back to the definition of the Convention itself for some background information. Moreover, this definition is important because the Member States also have to implement the Aarhus Convention directly, as they are also Parties to the Convention, separately from the EU.

ARTICLE 2.3 OF THE AARHUS CONVENTION – According to Article 2.3 of the Aarhus Convention:

> Environmental information means any information in written, visual, aural, electronic or any other material form on:
>
> a) The state of elements of the environment, such as air and atmosphere, water, soil, land, landscape[54] and natural sites,[55] biological diversity[56] and

52. Frankie Schram, 'Openbaarheid van Europese bestuursdocumenten', *Nieuw Juridisch Weekblad* no. 32 (2003): 582–592 [in Dutch].
53. Cf. *infra*, s. III.C.3.
54. These terms can be considered in their everyday meaning (Jean-Jacques Paradissis, *The Right to Access Environmental Information* (Saarbrucken: VDM Verlag, 2009), 80).
55. Natural sites may refer to any objects of nature that are of specific value, including not only officially designated protected areas, but also, for example, a tree or park that is of localized significance, having special natural, historic or cultural value (United Nations Economic Commission for Europe, 'The Aarhus Convention. An implementation guide. Part 2', <www.unece.org/env/pp/implementation%20guide/english/part2.pdf>, 22 Feb. 2010, hereafter *Aarhus Implementation Guide*).
56. Biological diversity is defined as 'the variability among living organisms from all sources including, inter alia, terrestrial, marine and other aquatic ecosystems and the ecological

its components, including genetically modified organisms,[57] and the interaction among these elements;

b) Factors, such as substances, energy, noise and radiation, and activities or measures, including administrative measures, environmental agreements, policies, legislation, plans and programmes,[58] affecting or likely to affect the elements of the environment within the scope of subparagraph (a) above, and cost-benefit and other economic analyses and assumptions used in environmental decision-making[59];

c) The state of human health and safety, conditions of human life,[60] cultural sites[61] and built structures,[62] inasmuch as they are or may be affected by the state of the elements of the environment or, through these elements, by the factors, activities or measures referred to in subparagraph (b) above.

DEFINITION OF 'WHAT ENVIRONMENTAL INFORMATION IS ABOUT' – The Implementation Guide to the Aarhus Convention gives an extensive overview of the elements of the definition that need closer attention. What constitutes environmental information is explained 'indirectly in terms of what environmental information can be about'.[63] A description of *environment* itself is not given. The illustrative lists that are given for each category of information are not exhaustive, and the definition is only a minimum requirement that can be

---

complexes of which they are part; this includes diversity within species, between species and of ecosystems'. Biodiversity includes, but is not limited to, ecosystem diversity, species diversity and genetic diversity. In addition, tangible entities identifiable as a specific ecosystem (a dynamic complex of plant, animal and micro-organism communities and their non-living environment interacting as a functional unit), are considered components of biodiversity (United Nations Economic Commission for Europe, *Aarhus Implementation Guide*, 28).

57. Council Directive 90/220/EEC of 23 Apr. 1990 on the deliberate release into the environment of genetically modified organisms (*OJ L* 117, 8 May 1990, 15) provides the following definition of genetically modified organisms: 'an organism in which the genetic material has been altered in a way that does not occur naturally by mating and/or natural recombination'.

58. Substances, energy, noise and radiation may be considered as physical or natural agents, while administrative measures, environmental agreements, policies, legislation, plans and programmes imply human action (United Nations Economic Commission for Europe, *Aarhus Implementation Guide*, 29).

59. As the results of the economic analyses may have a great impact on whether or not a particular project will go ahead, it is important to be able to examine the thinking that went into it. See also Jean-Jacques Paradissis, *The Right to Access Environmental Information* (Saarbrucken: VDM Verlag, 2009), 77–78.

60. Conditions of life in a general sense may include quality of air and water, housing and workplace conditions, relative wealth, and various social conditions (United Nations Economic Commission for Europe, *Aarhus Implementation Guide*, 30).

61. Cultural sites cover specific places or objects of cultural value (United Nations Economic Commission for Europe, *Aarhus Implementation Guide*, 30).

62. Built structures refer to man-made constructions. They are not limited to large buildings and objects such as dams, bridges, highways, etc. but also cover small constructions, and even landscaping or other transformation of the natural environment (United Nations Economic Commission for Europe, *Aarhus Implementation Guide*, 30).

63. United Nations Economic Commission for Europe, *Aarhus Implementation Guide*, 27.

extended by the Parties by their choice. Some of the terms in the definition have common-sense definitions and do not need technical definitions. For others, international agreements or other legislation might be relevant in order to determine their scope.[64]

Broad definition – The definition of the Aarhus Convention is broad and goes beyond information on the elements of the environment and their interactions to also include information on human and non-human factors and activities or measures that affect or are likely to affect the elements of the environment. Furthermore, the definition under (b) also includes economic analyses and assumptions used in environmental decision making. The effect of the factor, activity, or measure does not have to be immediately evident. It is enough if there is some probability that an effect on the environment might happen in the future.[65]

D. Public Sector Data

Outline – Next to spatial data and environmental information, the third type of data that needs to be discussed in this section is the *public sector documents* addressed in the PSI Directive. Before examining that definition, a brief mention is made of terms that are commonly used to indicate this type of data and of terms that are related but should be seen separately from public sector data.

1. Public Sector Information – Government Information – Public Information – Public Domain Information

Public sector information and government information – The terms *public sector information* and *government information* can both be found in literature referring to information created or held by organizations that are part of the government or the public sector, mostly contrasted with data that are created by the private sector. In many cases, however, a definition of this type of information is not given, most likely because of the uncertainty about the scope of *government* and *public sector*. One interesting attempt at a definition was made by the Dutch *Commissie Grondrechten in het Digitale Tijdperk* (Commission on Fundamental Rights in the Digital Age): all kinds of information that is physically held by the government, including not only information on administrative matters and information that has originated from the government, but also information that has been created by other organizations or persons that is specifically

64. For example, with respect to air and atmosphere, the definition can be compared to the definition of *ambient air* in Council Directive 96/62/EC of 27 Sep. 1996 on ambient air quality assessment and management (*OJ L* 296, 21 Nov. 1996, 55). *Ambient air* is described as 'outdoor air in the troposphere, excluding work places'. As the definition of the Aarhus Convention is broader, it is understood to mean indoor and workplace air as well.
65. United Nations Economic Commission for Europe, *Aarhus Implementation Guide*, 29.

designated for the government authority involved and that resides under its responsibility.[66]

PUBLIC INFORMATION – The scope of *public sector information* or *government information* overlaps with the domain of *public information*, but it is not the same. Uyttendaele defines public information as any message that is meant by the originator for an audience that is principally undetermined.[67] On the one hand, this scope is larger than that of public sector information, as the public sector is only one of the possible creators of information. On the other hand, not all public sector information is public information, as some of it will not be directed towards an undefined audience, but only to a known, limited number of people or organizations, for example internal communications and confidential preparatory documents.

PUBLIC DOMAIN INFORMATION – Another term that is often found together with *public sector information* or *government information* is *public domain information*. According to the United Nations Educational, Scientific, and Cultural Organization's (UNESCO's) Policy Guidelines for the Development and Promotion of Governmental Public Domain Information, 'On the one hand, "public domain information" can be defined as what is left outside the scope of any form of statutory protection including intellectual property rights, the protection of national security or public order, privacy laws and obligations of confidentiality'. On the other hand, it also refers to 'information of an intrinsically public nature; that is, certain types of information that are produced by public authorities ("government" in the broad sense) in the course of their duties, and that are seen as a public good'.[68] Hence, depending on the definition, *public domain information* can mean the same as *public sector information* or not.

2. Public Sector Documents under the PSI Directive

DOCUMENTS HELD BY PUBLIC SECTOR BODIES – While the PSI Directive is entitled Directive 2003/98/EC on the re-use of public sector information, in the actual text of the directive, the terms *public sector* and *information* are not used; rather, *public sector bodies* and *documents* are favoured. Hence, the directive is applicable to documents held by public sector bodies.[69] Considering the broad definition of

66. Commissie Grondrechten in het Digitale Tijdperk, 'Rapport Grondrechten in het Digitale Tijdperk', 2000, 191, <www.ivir.nl/dossier/grondrechten/bronnen/rapport_gdt_5-00.pdf>, 22 Feb. 2010 [in Dutch].
67. Caroline Uyttendaele, *Openbare informatie. Het juridisch statuut in een convergerende mediaomgeving* (Antwerp: Maklu, 2002), 4 [in Dutch]. See also Jean-Michel Bruguière, *Les données publiques et le droit* (Paris: Editions Litec, 2002), 4 [in French].
68. UNESCO, 'Policy Guidelines for the Development and Promotion of Governmental Public Domain Information' (Paris: UNESCO, 2004), 2–3, <http://unesdoc.unesco.org/images/0013/001373/137363eo.pdf>, 22 Feb. 2010. See also Egbert J. Dommering, P. Bernt Hugenholtz & Jan C. Kabel, 'De overheid en het publiek domein van informatie voor wetenschappelijk onderzoek', in *De publieke dimensie van kennis*, ed. Huub Dijstelbloem & Cornelius J.M. Schuyt (The Hague: WRR, 2002), 252–258 [in Dutch].
69. Article 1.1 of the PSI Directive.

document and the emphasis on the content rather than on the carrier, the difference between this term and *information* is negligible.[70] Under the PSI Directive, a public sector document is 'any content, whatever its medium, or part of that content, held by the State, regional or local authorities, bodies governed by public law and associations formed by one or several such authorities or one or several such bodies governed by public law'.[71]

PSI V. PUBLIC SECTOR DATA – In the parts of this book where the PSI Directive is specifically addressed, or policy documents that refer to the term public sector information, the commonly used PSI acronym will be maintained. However, in other areas, I intend to refer to public sector *data* for the reasons mentioned earlier.

IV. SCOPE FROM THE POINT OF VIEW OF THE ORGANIZATIONS HOLDING THE DATA

FOCUS ON PUBLIC SECTOR – This work limits itself to the legislation dealing with spatial data held by the public sector only, disregarding private sector data suppliers. Generally, the public sector is considered one of the most important creators or suppliers of spatial data.[72] In addition, while the private sector is bound by general rules of contract, tort law, and so on, it is not under any general obligation to make its spatial data available to others under terms and conditions imposed by law, as the public sector is. As the rights of the public to obtain spatial data are the focus of this book, spatial data supplied by the private sector remain outside the scope of this book.

OUTLINE – The question that follows naturally is what we should understand by the *public sector*. The answer to that is anything but straightforward. When using the term *public sector*, one instinctively thinks of it as opposite to the private sector and also associates it with other terms such as the *state, government, public service, public interest, public bodies, public authorities*, and so on. All these terms have their own meanings and implications, and it would be beyond the reach of this book to determine the exact scope of these terms and the possible differences between them. Moreover, as this book is focused on the common denominator between the PSI Directive, the Access Directive, and the INSPIRE Directive, this section concentrates on the terminology that these directives use to indicate the public character of the institutions involved in making spatial data available. The PSI Directive is applicable to *public sector bodies*, while the Access Directive and the INSPIRE Directive are aimed at the *public authorities* of the Member States. However, it may also be interesting to the reader to have an overview of some of the terms that are often used in the debate on the scope of the public sector.

70. Cf. *supra*, s. III.A.2.
71. Article 2.3 of the PSI Directive.
72. UNESCO, 'Policy Guidelines for the Development and Promotion of Governmental Public Domain Information' (Paris: UNESCO, 2004), 1, 3, <http://unesdoc.unesco.org/images/0013/001373/137363eo.pdf>, 22 Feb. 2010.

Therefore, the following section begins with a brief overview of a number of terms that are often encountered in literature regarding the public sector and its data policies.

A. PUBLIC SECTOR AND RELATED TERMS

1. Public Sector

THREE APPROACHES – According to the European Commission's 1998 Green Paper on Public Sector Information in the Information Society (hereafter referred to as the Green Paper on Public Sector Information), the definition of *public sector* varies from one European Member State to another. Three possible approaches emerge[73]:

- The functional approach, in which the public sector includes those bodies with state authority or public service tasks. This definition in its turn holds a number of elements that by themselves have given rise to considerable discussion, namely *state* and *public service*.[74]
- The legalist/institutional approach, whereby only bodies that are explicitly listed in the relevant law(s) have a public sector character. This definition would be too restrictive, as it would make the availability of spatial data dependent on the existence of legislation that explicitly defines its creator as a part of the public sector.
- The financial approach, whereby the public sector includes all bodies mainly financed by public funds (i.e., not operating under the normal rules of the market).[75]

ORGANIC OR FUNCTIONAL DEFINITION – Matthijs defines the public sector as 'the ensemble of government institutions, services and facilities', meaning all the organizations that are established by the Constitution or in the implementation thereof by the government, namely by the legislative powers, and that are supervised by government institutions.[76] Usually these organizations are financed exclusively or for the most part by public funding – taxes and social security levies. However, other means of financing are possible, including retributions paid for the provision of services such as public transport or the investment of private capital in some of these institutions. Maes and Jochmans not only consider the public sector from the

73. Commission of the European Communities, *Public sector information: a key resource for Europe. Green Paper on public sector information in the information society*, COM (1998) 585 final, 11, 20 Jan. 1999 (hereafter referred to as Green Paper on Public Sector Information).
74. Cf. *infra*, ss IV.A.2 and IV.A.4.
75. The Green Paper on Public Sector Information and the answers to the consultation round it launched will also be discussed in Chs 3, 5, and 7.
76. Herman Matthijs, *Bestuurswetenschappen. De overheid, haar structuren en haar werking* (Brussels: Politeia, 2003), 162 [in Dutch].

organic point of view, but they also look at the activities of the public sector. They point out the following characteristics[77]:

- creation 'under the responsibility of the representatives that are chosen by the people';
- funding by the community;
- the use of these community funds in order to serve the general interest;
- specific rules for the organization and activities of the services in the public sector.

ECONOMIC APPROACH – The concept of the public sector has also been investigated extensively by economic scholars in relation to the public budget and the reasons why items of expenditure should appear in this budget. Examples of these theories can be found in the work of Lane, who gives an interesting overview of the economic demarcation of the public sector.[78]

PUBLIC V. PRIVATE SECTOR – Another way of defining the public sector is by contrasting it to the private sector.[79] This often entails a functional view of both sectors. While the private sector is interested in making a profit and is directed towards its own interest, the public sector is responsible for providing services to society. It acts in the general interest and provides public services. The private sector is mostly directed towards economic criteria, such as profit or competition, while the public sector's priorities are legal certainty and equality, and its criteria are formulated in terms of well-being and prosperity.[80] As a consequence, the public sector's tasks are inherently different from those of the private sector. However, this distinction is not absolute. On the one hand, the demands on the public sector are leading to the use of more economic criteria to evaluate its performance. Moreover, budget constraints encourage public bodies to perform profitable activities. On the other hand, private parties are also included in performing public activities, for example, via outsourcing or public–private partnerships.

2. State and Government

IMPOSSIBLE TO DEFINE THE STATE? – In the debate on the availability of public sector spatial data or on the role of the public sector, often, the words *state* or *government* appear, as in the well-used term *access to government information*. Establishing an all-encompassing sustainable definition of *state* and *government* is an impossible feat within the scope of this book. Even as a research topic in itself, it would

77. Rudolf Maes & Kathleen Jochmans (eds), *Inleiding tot de bestuurskunde* (Brussels: Studiecentrum Open Hoger Onderwijs, 1996), 13 [in Dutch].
78. For more information on the budgetary approach to the public sector, see Jan-Erik Lane, *The Public Sector. Concepts, Models and Approaches* (London: SAGE Publications, 1995), 327.
79. Herman Matthijs, *Bestuurswetenschappen. De overheid, haar structuren en haar werking* (Brussels: Politeia, 2003), 94 [in Dutch].
80. *Ibid.*, 94–95.

provide any scholar with a great challenge. This was already the case in the middle of the nineteenth century, when Bastiat, a French political economist and liberal theorist, stated:

> I wish someone would offer a prize for a good, simple and intelligent definition of the word "government" . . . The Government! What is it? Where is it? What does it do? What ought it to do? All we know is, that it is a mysterious personage; and, assuredly, it is the most solicited, the most tormented, the most overwhelmed, the most admired, the most accused, the most invoked, and the most provoked of any personage in the world.[81]

Demonstrating the liberalist ideology of his time, Bastiat defined *government* as 'the great fiction through which everybody endeavours to live at the expense of everybody else'.[82]

State as community – According to Van Impe, a *state* can be described as the most general and complete community of all people inhabiting a certain geographic area (an area of land, including the soil, air, and sea territory).[83] It is a historically grown community, accepting the same government and therefore subject to the same legal system.[84] It is also an organization of power that has a system of enforcement measures at its disposal (army, police, and so on) and the exclusive right to use it. A state is an inclusive, encompassing community: the members of the community of the state belong to it completely, even if government does not regulate certain community or individual activities, as it can still decide at any moment to regulate these activities anyway. Van Impe claims that a state is, at the same time:

- An organization of power: the state consists of a group of services that have a monopoly over the essential tasks.
- A forced community: the population is involved and guided by the authorities. It has to obey the orders.
- A legal community: the government decides which rules are binding for the population. By their general character these rules influence the entire community and its behavioural patterns.

81. Frédéric Bastiat, 'L'Etat', 1884, <http://bastiat.org/fr/l_Etat.html>, 22 Feb. 2010 [in French]. The above text is a translation of 'Je voudrais qu'on fondât, un prix, non de cinq cents francs, mais d'un million, avec couronnes, croix et rubans, en faveur de celui qui donnerait une bonne, simple et intelligible définition de ce mot : l'Etat. Quel immense service ne rendrait-il pas à la société ! L'État! Qu'est-ce? où est-il? que fait-il? que devrait-il faire? Tout ce que nous en savons, c'est que c'est un personnage mystérieux, et assurément le plus sollicité, le plus tourmenté, le plus affairé, le plus conseillé, le plus accusé, le plus invoqué et le plus provoqué qu'il y ait au monde'.
82. Frédéric Bastiat, 'L'Etat', 1884, <http://bastiat.org/fr/l_Etat.html>, 22 Feb. 2010 [in French]. The above text is the translation of *'L'État*, c'est la grande fiction à travers laquelle *tout le monde* s'efforce de vivre aux dépens de *tout le monde'*.
83. Herman Van Impe, *De Staat, wat is dat eigenlijk? Staat, democratie en recht* (Antwerp: Kluwer Rechtswetenschappen, 1990), 6 [in Dutch].
84. *Ibid.*, 12.

- A sustainable community: every person is only a member of a community of which he follows the way of life that already existed before he was born and that will continue to exist after he dies. Society can change, but crosses periods of time.
- A civilization model: the population is subject to the same legal system, the same psychological policies (ideological and cultural), has the same experiences and finds itself in the same situations.[85]

WIDE OR STRICT MEANING OF STATE – Van Impe also notes that the word *state* is, however, not just used in this wide meaning, but also in a more limited sense as 'the state government as an organization of power', and even more limited, as 'the central state government'.[86] Delfosse has a comparable approach. He considers the state an abstract entity: it only exists by the multiplicity of the public institutions that it encloses, that is, Parliament (legislature), government, courts, administrations, and public services.[87] In a strict sense, he qualifies the state as the central state, involving the conglomerate of official institutions that exercise their competences on a national level. In a larger sense, it also involves decentralized public institutions, of which the competences are geographically limited, and supranational institutions to which the state has delegated part of its powers or its tasks.

STATE V. GOVERNMENT – If one considers the state only in its meaning of institutions exercising authority over the people, then there seems to be little difference between *state* and *government*. This is also shown in the distinction that Matthijs makes between both. He believes that *state* refers to the central state, while government contains all levels, namely the state, the local authorities (regional or local), and other institutions, such as public services and social security. *Government* refers to the need of society for a number of institutions that can exercise authority or that can oblige or forbid the members of society to perform certain activities.[88] Hence, in his vision, state is a part of government, while in the view of Van Impe, government is a part of the state.

PREFERENCE FOR PUBLIC SECTOR – As there is no clear distinction between government and state and as it is beyond the scope of this book to determine whether there should be a distinction and, if so, what it should be, this section only aims to give the reader some insight into the debate and to some of the terms that he or she may come across in the literature on this subject. In what follows, I will refrain from using these terms. In cases where they are part of a commonly used concept or phrasing, it should be assumed that they have the same scope as *public sector*.

85. *Ibid.*, 13.
86. *Ibid.*, 13.
87. François Delfosse, *L'état, pour quoi, pour qui?* (Louvain-la-Neuve: Centre de Recherches Sociologiques – UCL, 1986), 9 [in French].
88. Herman Matthijs, *Bestuurswetenschappen. De overheid, haar structuren en haar werking* (Brussels: Politeia, 2003), 163 [in Dutch].

3. Administration

FUNCTIONAL OR ORGANIC DEFINITION – Another word that one encounters when demarcating the public sphere from the private sphere in a political sense is *administration*. Like *public service, administration* can be defined from a functional or an organic point of view. In its widest sense, the functional interpretation of *administration* includes all activities undertaken by the government in its entirety to ensure that the public interest is being looked after.[89] To define *administration* in the organic sense, Schram sees two possibilities: either one can make a list of institutions and claim that all their activities are 'administration in the functional sense', or one can start from the functional concept of administration and define an *organic administration* as the institutions that provide administration in the functional sense.[90] In any case, the term *administration*, like *public service, state, government, public sector*, and so on, is problematic to define, and it has been subject to many different interpretations. The use of this term will be avoided from here on.

4. Public Service

FUNCTIONAL OR ORGANIC DEFINITION – A final term that is related to all the previous ones is *public service*. Based on the literature,[91] the content of the term is comparable to that of *administration, government*, and even *public sector*. It can also be seen from an organic or a functional point of view.[92] According to Matthijs, an organic public service is established and abolished by government and exists under government supervision. He lists the following main characteristics of an organic public service: the establishment is due to the will of the government; the statutes are determined by legislation; the members of the board are for the most part appointed by the government; the government can determine the activities of the service; and the public service has certain extraordinary competences, for example, expropriations.[93] De Terwangne has a simpler description: an administration, whether it is national, regional, or local, is an organic public service in charge of providing for public interests and collective needs.[94] Regarding the functional public service,

89. Frankie Schram, *Begrenzingen aan de openbaarheid van bestuur: een analyse en vergelijking van de internrechtelijke openbaarheidsregelingen in het Belgische recht. Doctoral Thesis K.U.Leuven* (Leuven: K.U.Leuven, 2002), 28 [in Dutch].
90. *Ibid.*
91. For an extensive list of literature, see Cécile De Terwangne, *Société de l'information et mission publique d'information, Doctoral Thesis FUNDP* (Namur: Faculté Universitaire de Notre Dame de Paix, 2001), 543 [in French].
92. *Ibid.*
93. Herman Matthijs, *Bestuurswetenschappen. De overheid, haar structuren en haar werking* (Brussels: Politeia, 2003), 163 [in Dutch].
94. Cécile De Terwangne, *Société de l'information et mission publique d'information, Doctoral Thesis FUNDP* (Namur: Faculté Universitaire de Notre Dame de Paix, 2001), 543, referring to André Buttgenbach, *Manuel de droit administrative, première partie*, 3rd edn (Brussels: Larcier, 1966), no. 59 [in French].

Matthijs states that a functional public service is provided by government services according to the principles of public interest. A private undertaking can also provide public services through means of a concession. Typical examples are electricity and garbage-disposal services. Such a concession has to fulfil two conditions: there has to be a one-sided permission of the government to perform a certain activity or task, and the performance of the public service has to respect the rules and the supervision of the government.[95]

PREFERENCE FOR PUBLIC SECTOR AND PUBLIC TASK – In what follows, I will not use the term *public service*. As was mentioned above, the organizations involved are referred to as the *public sector*. Next, public service must also not be confused with the question of the public task, in the sense of the issue of which activities should be performed by the public sector and which activities should be left to the market. This discussion will be referred to as the *public task* debate.

B. PUBLIC AUTHORITIES UNDER THE ACCESS DIRECTIVE AND THE INSPIRE DIRECTIVE

OUTLINE – As the focus of this book relates to the common denominator between the INSPIRE Directive, the Access Directive and the PSI Directive, the question is also what specific scope is given to the public sector under these directives. The Access and INSPIRE Directives refer to public authorities, while the PSI Directive addresses public sector bodies.

IDENTICAL DEFINITION – The Access Directive and the INSPIRE Directive are both applicable to data held by *public authorities*. The INSPIRE Directive has taken over the definition from the Access Directive, which in its turn implemented the Aarhus Convention and hence introduced the definitions of the Convention in EU law.

1. The Definition of Public Authority in the Access Directive

ARTICLE 2.2 OF THE ACCESS DIRECTIVE – Under Article 2.2 of the Access Directive:

> Public authority shall mean:
>
> a) government or other public administration, including public advisory bodies, at national, regional or local level;
> b) any natural or legal person performing public administrative functions under national law, including specific duties, activities or services in relation to the environment; and
> c) any natural or legal person having public responsibilities or functions, or providing public services, relating to the environment under the control of a body or person falling within (a) or (b).

95. Herman Matthijs, *Bestuurswetenschappen. De overheid, haar structuren en haar werking* (Brussels: Politeia, 2003), 163 [in Dutch].

> Member States may provide that this definition shall not include bodies or institutions when acting in a judicial or legislative capacity. If their constitutional provisions at the date of adoption of this Directive make no provision for a review procedure within the meaning of Article 6,[96] Member States may exclude those bodies or institutions from that definition.

BROAD INTERPRETATION – In the case of government and the other public authorities referred to in Article 2.2.(a), the European Commission wanted to move away from the approach in Directive 90/313/EEC, which only applied to public authorities with responsibilities and possessing information relating to the environment. Recital 11 of the Access Directive explains that:

> To take account of the principle in Article 6 of the [Aarhus] Treaty, that environmental protection requirements should be integrated into the definition and implementation of Community policies and activities, the definition of public authorities should be expanded so as to encompass government or other public administration at national, regional or local level whether or not they have specific responsibilities for the environment. The definition should likewise be expanded to include other persons or bodies performing public administrative functions in relation to the environment under national law, as well as other persons or bodies acting under their control and having public responsibilities or functions in relation to the environment.

Hence, the definition of *public authority* should be interpreted widely. As reason for this, the Commission stated that experience showed that the precise meaning of the expression 'with responsibilities relating to the environment' gave rise to frequent disputes. According to the Commission, narrow interpretations of this notion had led to the exclusion of certain bodies from the scope of the directive 'on the basis that they did not have responsibilities for the environment but rather for other matters such as transport or energy. Consequently, it has been argued that the information held by them is not covered by the Directive'.[97]

PREPARATORY PROCESS: SERVICES OF GENERAL ECONOMIC INTEREST? – The original proposal of the European Commission included legal persons entrusted with 'the operation of services of general economic interest which affect or are likely to affect the state of elements of the environment', because the Commission felt that 'services of general interest traditionally performed by public authorities or public utilities are increasingly being carried out by bodies which no longer form part of the public sector; those bodies hold information to which the public should continue to have a right of access'.[98] Differences between the Member States – for

96. Article 6 of the Access Directive concerns access to review procedures for acts or omissions of the public authorities.
97. Commission of the European Communities, *Proposal for a Directive of the European Parliament and of the Council on public access to environmental information*, COM (2000) 402 final, 29 Jun. 2000.
98. *Ibid.* Services of general economic interest are discussed in Ch. VIII, s. 4.4.

example, in one State the service is carried out by a public authority and in another by a private body – would be undesirable. The Environment Working Party of the Council felt, however, that 'general economic interest' was too broad and too general 'since it could cover all economic activities. In addition, it would be inappropriate to oblige private companies to act under the rules of public administrations'.[99] The Parliament was equally reluctant to include this, but it gave no specific reason for its amendment for deletion.[100]

PREPARATORY PROCESS: ADVISORY BODIES? – A second element of discussion regarding the definition of public authority in the Access Directive was whether advisory bodies should be included or not. According to the Council, this would extend the scope of the directive too much and would give rise to the serious problem of designating those bodies,[101] yet the Parliament included these bodies in both its first and second reading amendments.[102] Eventually, a compromise was reached: only 'public advisory bodies' were included in the final version of the directive.

2. The Definition of Public Authority in the INSPIRE Directive

ARTICLE 3.9 OF THE INSPIRE DIRECTIVE – Article 3.9 of the INSPIRE Directive contains a definition of *public authority* that is identical to the definition in the Access Directive, except for the last paragraph:

99. Environment Working Party, *Preparation of the Council Environment on 18–19 December 2000. Proposal for a directive on public access to environmental information*, doc.nr. 13709/00, 24 Nov. 2000, <http://register.consilium.europa.eu/pdf/en/00/st13/st13709.en00.pdf>, 22 Feb. 2010.

100. European Parliament, *Legislative resolution on the proposal for a Directive of the European Parliament and of the Council on public access to environmental information, OJ C* 343, 5 Dec. 2001, 177; Committee on the Environment, Public Health and Consumer Policy, *Report on the proposal for a directive of the European Parliament and of the Council on public access to environmental information*, doc. nr. A5-0074/2001, 28 Feb. 2001, <www.europarl.europa.eu/sides/getDoc.do?pubRef=-//EP//NONSGML+REPORT+A5-2001-0074+0+DOC+PDF+V0//EN&language=EN>, 22 Feb. 2010.

101. Council of the European Union, *Common position adopted with a view to the adoption of a Directive of the European Parliament and of the Council on public access to environmental information and repealing Council Directive 90/313/EEC. Statement of the Council's Reasons*, doc. nr. 11878/01, 28 Jan. 2002, 8, <http://register.consilium.europa.eu/pdf/en/01/st11/st11878-re01ad01.en01.pdf>, 22 Feb. 2010.

102. European Parliament, *Legislative resolution on the proposal for a Directive of the European Parliament and of the Council on public access to environmental information, OJ C* 343, 5 Dec. 2001, 177; European Parliament, *Position adopted at second reading on 30 May 2002 with a view to the adoption of European Parliament and Council Directive 2002/ . . . /EC on public access to environmental information and repealing Council Directive 90/313/EEC, OJ C* 187E, 7 Aug. 2003; Committee on the Environment, Public Health and Consumer Policy, *Report on the proposal for a directive of the European Parliament and of the Council on public access to environmental information*, doc. nr. A5-0074/2001, 28 Feb. 2001, <www.europarl.europa.eu/sides/getDoc.do?pubRef=-//EP//NONSGML+REPORT+A5-2001-0074+0+DOC+

Public authority means:

a) any government or other public administration, including public advisory bodies, at national, regional or local level;
b) any natural or legal person performing public administrative functions under national law, including specific duties, activities or services in relation to the environment; and
c) any natural or legal person having public responsibilities or functions, or providing public services relating to the environment under the control of a body or person falling within (a) or (b);

Member States may provide that when bodies or institutions are acting in a judicial or legislative capacity, they are not to be regarded as a public authority for the purposes of this Directive.

LINK WITH ACCESS DIRECTIVE – The INSPIRE Directive has the objective to promote the exchange of spatial data between public authorities for the performance of their public tasks relating to the environment, and the scope of INSPIRE data and environmental information overlaps to a great extent.[103] In addition, both directives contain provisions on access of citizens to data held by public authorities. Hence, these directives have such a close relationship that the European Commission chose to maintain the same definition of *public authority* for both directives in order to ensure the highest possible coherence.

PUBLIC BODIES HOLDING SPATIAL DATA? – Most of the traditional public bodies holding spatial data fall under one of the categories of public authorities defined in the INSPIRE and the Access Directives. For instance, national mapping agencies are in most cases agencies that are part of the government on the national level, and, in the cases where they are truly treated as bodies separate from the central government, they are still performing administrative functions under national law. If there is no legal basis for their activities, they can still be considered public authorities because they are providing a public service under the control of a public authority. However, in this last situation, the task must be related to the environment. Considering the very broad scope of tasks related to the environment, this still includes mapping agencies. The same reasoning can be followed for many public bodies holding spatial data. Ministries and regional and local authorities fall under the first category, so there is no question about their status. Other agencies that hold great quantities of spatial data are environmental agencies, meteorological offices, geological surveys, statistical offices, land registries, cadastral offices, and so on. Many of these bodies have a legal basis, so they are covered under the

PDF+V0//EN&language=EN>, 22 Feb. 2010; Committee on the Environment, Public Health and Consumer Policy, *Recommendation for second reading on the Council Common Position for adopting a European Parliament and Council directive on public access to environmental information and repealing Council Directive 90/313/EEC*, doc. nr. A5-0136/2002, 24 Apr. 2002, <www.europarl.europa.eu/sides/getDoc.do?pubRef=-//EP//TEXT+REPORT+A5-2002-0136+0+NOT+XML+V0//EN>, 22 Feb. 2010.

103. Cf. *supra*, ss III.B. and III.C.2.

second category of public authorities. Even if they do not have a legal basis, they still have public responsibilities and have to report to their supervising authority. However, these responsibility may not always be related to the environment. For those tasks, the entities involved may not be public authorities.

3. The Definition of Public Authority in the Aarhus Convention

ARTICLE 2.2. OF THE AARHUS CONVENTION – As was mentioned above, the Access Directive ensures the transposition of the Aarhus Convention into EU law. Under Article 2.2 of the Aarhus Convention:

> Public authority means:
>
> a) Government at national, regional and other level;
> b) Natural or legal persons performing public administrative functions under national law, including specific duties, activities or services in relation to the environment;
> c) Any other natural or legal persons having public responsibilities or functions, or providing public services, in relation to the environment, under the control of a body or person falling within subparagraphs (a) or (b) above;
> d) The institutions of any regional economic integration organization referred to in Article 17 which is a Party to this Convention.
>
> This definition does not include bodies or institutions acting in a judicial or legislative capacity.

FUNCTIONAL DEFINITION – The Implementation Guide of the Convention makes it clear that the definition of *public authority* is important in defining the Convention's scope: 'While clearly not meant to apply to legislative or judicial activities, it is nevertheless intended to apply to a whole range of executive or governmental activities, including activities that are linked to legislative processes. The definition is broken into three parts in order to provide as broad a coverage as possible'.[104] This should be seen as a functional definition to ensure the possible later inclusion of new public authorities on a national, local, or supranational level.[105]

THREE CATEGORIES OF PUBLIC AUTHORITIES – The first category of public authorities listed above, 'government at national, regional, and other level', includes agencies, institutions, departments, bodies, and so on of political power at all geographical and administrative levels. 'In a typical situation, national ministries and agencies and their local offices, State, regional or provincial ministries and agencies and their regional and local offices, as well as local or municipal government offices, such as those found in cities, towns or villages, would be

104. United Nations Economic Commission for Europe, *Aarhus Implementation Guide*, 24.
105. Vera Rodenhoff, 'The Aarhus Convention and its Implications for the "Institutions" of the European Community', *RECIEL* 11, no. 3 (2002): 350.

covered'.[106] Under category (b), public authorities also include natural or legal persons performing 'public administrative functions', that is, functions that are normally performed by governmental authorities as determined according to national law. What are considered to be such functions may vary from country to country. However, in any case, there needs to be a legal basis for the performance of these functions.[107] The third type of public authorities also deserves some more attention: 'persons having public responsibilities or functions, or providing services, in relation to the environment, under the control of the other categories of public authorities'. In contrast with category (b), these public authorities do not derive their authority from national legislation, but rather indirectly through control by those bodies defined in (a) and (b). Yet the scope of this category is limited: while the first two categories do not contain any limitation on the field of activities, in the third category, only persons having public responsibilities or functions or providing services 'in relation to the environment' can be public authorities. Natural or legal persons that are publicly owned are covered, but possibly so are privately owned entities in cases where service providers can oblige residents to pay fees or engage in particular activities, for example, waste collection.[108]

Excluded bodies – Bodies or institutions in a legislative or judicial capacity are not included in the definition of public authorities because of the fundamentally different character of decision making in both cases.[109] Hence, the exception is applicable not only to parliaments, courts, and local councils, but also to executive branch authorities when they perform legislative or judicial functions.

C. Public Sector Bodies under the PSI Directive

1. The Definition of the PSI Directive

Article 2 of the PSI Directive – Article 2 of the PSI Directive defines public sector bodies as 'the State, regional or local authorities, bodies governed by public law and associations formed by one or several such authorities or one or several bodies governed by public law'.[110] In turn, a body governed by public law is described as 'any body

(a) established for the specific purpose of meeting needs in the general interest, not having an industrial or commercial character; and
(b) having legal personality; and

106. United Nations Economic Commission for Europe, *Aarhus Implementation Guide*, 24.
107. *Ibid.*
108. *Ibid.*, 25.
109. *Ibid.* 27–28. When exercising legislative capacity, elected representatives are more directly accountable to the public through the election process, while tribunals have to apply the law impartially and professionally without regard to public opinion.
110. Article 2.1 of the PSI Directive.

(c) financed, for the most part by the State, or regional or local authorities, or other bodies governed by public law; or subject to management supervision by those bodies; or having an administrative, managerial or supervisory board, more than half of whose members are appointed by the State, regional or local authorities or by other bodies governed by public law'.[111]

DEFINITIONS ORIGINATING FROM PUBLIC PROCUREMENT – These definitions were taken from the Public Procurement Directives 92/50/EEC, 93/36/EEC, 93/37/EEC and 98/4/EC.[112] They do not cover public undertakings.[113] Since the PSI Directive was proclaimed at the end of 2003, the European framework for public procurement has been replaced by two new directives coordinating the procurement procedures, Directive 2004/17/EC and Directive 2004/18/EC.[114] The definitions of *public sector body* and *body governed by public law* did not change, so the case law and jurisprudence dating from before these new directives are still valid.

JUSTIFICATION OF CHOICE – The European Commission did not explain why it chose to use public sector bodies, and not, for instance, public authorities. One might assume that this choice is related to the economic slant of the directive and the policy of the Commission to distinguish re-use of PSI from any discussions on access of citizens to public sector data. The question was also raised in the Working Party on Telecommunications of the Council, so the Danish Presidency suggested to the Working Party that it also address the relationship between public procurement rules and the PSI Directive in the recitals in the following statement: 'Whilst public procurement rules deal with the way public sector bodies can act in the market when buying goods or services, this Directive deals with the way public sector bodies can act in the information services market by selling or otherwise providing information resources'.[115] However, this was not maintained.

111. Article 2.2. of the PSI Directive.
112. Council Directive 92/50/EEC of 18 Jun. 1992 relating to the coordination of procedures for the award of public service contracts, *OJ L* 209, 24 Jul. 1992, 1; Council Directive 93/36/EEC of 14 Jun. 1993 coordinating procedures for the award of public supply contracts, *OJ L* 199, 9 Aug. 1993, 1; Council Directive 93/37/EEC of 14 Jun. 1993 concerning the coordination of procedures for the award of public works contracts, *OJ L* 199, 9 Aug. 1993, 54; Directive 98/4/EC of the European Parliament and of the Council of 16 Feb. 1998 amending Directive 93/38/EEC coordinating the procurement procedures of entities operating in the water, energy, transport and telecommunications sectors, *OJ L* 101, 1 Apr. 1998, 1.
113. Recital 10 of the PSI Directive.
114. Directive 2004/17/EC of the European Parliament and of the Council of 31 Mar. 2004 coordinating the procurement procedures of entities operating in the water, energy, transport and postal services sectors, *OJ L* 134, 30 Apr. 2004, 1; Directive 2004/18/EC of the European Parliament and of the Council of 31 Mar. 2004 on the coordination of procedures for the award of public works contracts, public supply contracts and public service contracts, *OJ L* 134, 30 Apr. 2004, 114.
115. Presidency of the Council of the European Union, *Note to the Working Party on Telecommunications on the Proposal for a Directive of the European Parliament and of the Council on the re-use and commercial exploitation of public sector documents*, 7 Oct. 2002, doc. nr. 12761/02, 7 Oct. 2002, 3, <http://register.consilium.europa.eu/pdf/en/02/st12/st12761.en02.pdf>, 22 Feb. 2010.

Criticism from European Economic and Social Committee – The European Economic and Social Committee found that the European Commission lessened the impact and effectiveness of the directive by limiting the notion of public sector bodies to the meaning in the Public Procurement Directives. According to the Committee, the public sector 'should include any body, regardless of its legal form, which has been commissioned, by decision of a public authority and under its control, to carry out a service of public benefit' and 'should include not only publicly-owned enterprises whose legal form is private but also private enterprises acting by virtue of administrative delegation or authorisation, in managing that aspect of public services which excludes any commercial function, and the legislative and judicial authorities'.[116]

Final choice for public sector body – However, 'public sector body' remained in the final version of the directive, entailing that in order to determine the scope of this term, and of 'body governed by public law', we can find inspiration in the interpretation of these terms under the regime of the Public Procurement Directives by the ECJ, as the coherence of the legislation of the EU requires that the same terms are interpreted in the same manner. Instead of 'public sector body', the Public Procurement Directives refer to 'contracting authorities'. They are, however, defined in the same manner. Hence, the same three groups of institutions fall under the definition of *contracting authorities.* The scope of this term does not depend on the definition of public or administrative law bodies in the Member States, but it is an autonomous concept of European law.[117]

2. Contracting Authorities under the Public Procurement Directives

Categories of contracting authorities – According to the definition in the Public Procurement Directives, there are three groups of organizations that fall under the definition of *contracting authorities*: the State, regional, and local authorities; bodies governed by public law; and associations formed by one or several authorities or one or several bodies governed by public law.

a. State, Regional, and Local Authorities

Functional approach – First, there is the category of the more traditional state authorities, that is, the State and the regional and local authorities. They are considered to be contracting authorities, regardless of the nature of their activities.[118]

116. European Economic and Social Committee, *Opinion on the Proposal for a Directive of the European Parliament and of the Council on the re-use and commercial exploitation of public sector documents, OJ C* 85, 8 Apr. 2003, 25, referring to European Economic and Social Committee, *Opinion on Public sector information: a key resource for Europe – Green Paper on public sector information in the information society, OJ C* 169, 16 Jun. 1999, 30.
117. Sue Arrowsmith, *The Law of Public and Utilities Procurement* (London: Sweet & Maxwell, 2005), 250.
118. *Ibid.*, 253.

In one of the landmark cases on public procurement, *Beentjes*, the ECJ held that the term *State* should be interpreted in functional terms.[119] The case concerned a local land consolidation committee, which was considered to fall within the definition of a contracting authority, even though it was not an integral part of the State in formal terms.[120] This functional interpretation entailed that the characteristics and the function of the commissioning authority must be taken into account in determining whether an entity is to be regarded as a contracting authority within the meaning of the EU directives,[121] for example, whether it is regulated by law and whether it depends on the authorities for the appointment of its members. Later on, the ECJ also applied this functional approach in *Connemara* and *Commission v. Ireland.*[122]

All bodies with legislative, executive, and judicial power – In a 1998 case concerning the Flemish Parliament, referring to *Beentjes*, the ECJ added that the State necessarily encompasses all the bodies that exercise legislative, executive, and judicial power, as it 'would be inconsistent to hold that a legislative body does not fall within the definition of the State for the purposes of the Community directives on public works contracts, when a body which is not an integral part of the State in formal terms has been held to fall within that definition for the purposes of the application of one of those directives'.[123] Arrowsmith warns to not overestimate the importance of this statement, as in *Commission v. Belgium*, the Court was not dealing with the margins of the concept of the state, but with one of its core entities, that is, the Flemish Parliament.[124]

Bodies without legal personality – The definition in *Beentjes* should be considered more significant. In this case, the state was deemed to cover, at least, a body 'whose composition and functions are laid down by legislation and which depends on the authorities for the appointment of its members, the observance of the obligations arising out of its measures and the financing of the public works

119. ECJ, Case C-31/87, *Gebroeders Beentjes v. The Netherlands*, 20 Sep. 1988, paras 11–13, *European Court reports* 1988, 4635. See also ECJ, Case C-373/00, *Adolf Truley GmbH v. Bestattung Wien GmbH*, 27 Feb. 2003, para. 37, *European Court reports* 2003, I-1931

120. Ann-Lawrence Durviaux, *Logique de marché et marché public en droit communautaire. Analyse critique d'un système* (Brussels: Larcier, 2006), 135 [in French]; Christopher Bovis, *EC Public Procurement: Case Law and Regulation* (London: Oxford University Press, 2006), 364; Johanna C.E. Ackermans-Wijn, 'De overheidsonderneming en Europese aanbesteding', in *Overheid en onderneming. Tussen publiek en privaat ondernemen*, ed. Sebastianus C.J.J. Kortmann et al. (Deventer: W.E.J. Tjeenk Willink, 1998), 273 [in Dutch]; Elies Steiger, 'Een overheidsbegrip in het Europese recht?', *Nederlands Tijdschrift voor Bestuursrecht* no. 7 (2005): 225 [in Dutch].

121. Elisabetta R. Manunza, 'Privatised Services and the Concept of "Bodies Governed by Public Law" in EC Directives on Public Procurement', *European Law Review* 28 (2003): 275.

122. ECJ, Case C-306/97, *Connemara Machine Turf Co. Ltd v. Coillte Teoranta*, 17 Dec. 1998, *European Court reports* 1998, I-8761; ECJ, Case C-353/96, *Commission v. Ireland*, 17 Dec. 1998, *European Court reports* 1998, I-8565.

123. ECJ, Case C-323/96, *Commission v. Kingdom of Belgium*, 17 Sep. 1998, para. 27–28, *European Court reports* 1998, I-5063.

124. Sue Arrowsmith, *The Law of Public and Utilities Procurement* (London: Sweet & Maxwell, 2005), 254.

contracts which it is its task to award'.[125] This definition would also include bodies without a separate legal personality. This is probably because, otherwise, such bodies could not be qualified as contracting authorities, as legal personality is needed for a body to fall under the definition of a body governed by public law.[126]

PRIVATE LAW ENTITIES – The use of the functional approach by the ECJ also has as a consequence that an entity's private law status does not preclude it from being classified as a contracting authority.[127] The method in which the entity concerned has been set up is irrelevant in determining the applicability of public procurement law.[128]

b. Bodies Governed by Public Law

IMPORTANCE OF DEFINITION – While the category of the State, regional, and local authorities does not seem to cause many problems, the interpretation of a 'body governed by public law' has led to more questions. As many of the public bodies that produce and supply spatial data may fall under this category of public sector bodies, it is important to have insight into the meaning given to the term by the ECJ.

THREE CUMULATIVE CRITERIA – As was mentioned above, a body governed by public law is defined in the Public Procurement Directives as a body that has a legal personality, is established for the specific purpose of meeting needs in the general interest, not having an industrial or commercial character, and that either (a) is financed for the most part by the State, regional, or local authority or by some other body governed by public law; (b) is subject to management supervision by the State, regional or local authorities or by bodies governed by public law; or (c) has an administrative, managerial, or supervisory board more than half of whose members are appointed by the State, regional, or local authorities or by bodies governed by public law.[129] These three possibilities of finance, managerial supervision, and appointment have been described by the ECJ as embodying a

125. ECJ, Case C-31/87, *Gebroeders Beentjes v. The Netherlands*, 20 Sep. 1988, para. 11, *European Court reports* 1988, 4635.
126. Sue Arrowsmith, *The Law of Public and Utilities Procurement* (London: Sweet & Maxwell, 2005), 255; Elisabetta R. Manunza, 'Privatised Services and the Concept of "Bodies Governed by Public Law" in EC Directives on Public Procurement', *European Law Review* 28 (2003): 276.
127. Christopher H. Bovis, 'Developing Public Procurement Regulation: Jurisprudence and Its Influence on Law Making', *Common Market Law Review* 43, no. 2 (2006): 471; Elies Steiger, 'Een overheidsbegrip in het Europese recht?', *Nederlands Tijdschrift voor Bestuursrecht* no. 7 (2005): 225 [in Dutch].
128. ECJ, Case C-44/96, *Mannesmann Anlagenbau Austria AG and Others v. Strohal Rotationsdruck GmbH*, 15 Jan. 1998, para. 6, *European Court reports* 1998, I-73; ECJ, Case C-360/96, *Gemeente Arnhem and Gemeente Rheden v. BFI Holding BV*, 10 Nov. 1998, para. 61–62, *European Court reports* 1998, I-6821; ECJ, Case C-237/99, *Commission v. French Republic*, 1 Feb. 2001, paras 50, 60, *European Court reports* 2001, I-939; ECJ, Case C-214/00, *Commission v. the Kingdom of Spain*, 15 May 2003, paras 54–55, 60, *European Court reports* 2003, I-4667.
129. Article 2.1 of Directive 2004/17/EC; Art. 1.9 of Directive 2004/18/EC.

relationship of 'close dependency on a contracting authority'.[130] The three conditions – meeting needs in the general interest not having an industrial or commercial character, having legal personality, and a close dependency – have to be fulfilled cumulatively.

BROAD AND AUTONOMOUS INTERPRETATION – The Public Procurement Directives include a list of bodies in each Member State that could be considered as bodies governed by public law. This list is intended to be as exhaustive as possible, and it can be updated if necessary.[131] Such an update was done by the European Commission in December 2008.[132] However, the list is not sanctifying: if a body that is not in the list complies with the three conditions, it will be considered as a body governed by public law.[133] Thus, this list can be a starting point to determine which bodies fall under the application of the PSI Directive, but one will still have to look at the general conditions of the definition and its interpretation by the ECJ to determine whether an organization is a body governed by public law or not. In any case, according to the ECJ, the concept of bodies governed by public law must be interpreted broadly.[134] In addition, like the concept of a contracting authority, it is a concept of EU law that must be given an autonomous and uniform interpretation throughout the EU.[135]

OUTLINE – Below, the criteria of meeting needs in the general interest not having an industrial or commercial character, and the dependency test are elaborated.

i. Meeting Needs in the General Interest Not Having an Industrial or Commercial Character

NO DEFINITION – The first criterion by which to determine whether an organization is a body governed by public law is that it is established for the specific purpose of

130. Sue Arrowsmith, *The Law of Public and Utilities Procurement* (London: Sweet & Maxwell, 2005), 251, referring to ECJ, Case C-44/96, *Mannesmann Anlagenbau Austria AG and Others v. Strohal Rotationsdruck GmbH*, 15 Jan. 1998, para. 20, *European Court reports* 1998, I-73; ECJ, Case C-380/98, *The Queen v. HM Treasury Ex p. the University of Cambridge*, 3 Oct. 2000, para. 20, *European Court reports* 2000, I-8035.
131. Article 8 of Directive 2004/17/EC, Art. 1.9 of Directive 2004/18/EC.
132. Commission Decision of 9 Dec. 2008 amending the Annexes to Directives 2004/17/EC and 2004/18/EC of the European Parliament and of the Council on public procurement procedures, as regards their lists of contracting entities and contracting authorities, *OJ L* 349, 24 Dec. 2008, 1. See also Rhodri Williams, 'European Commission Decision 2008/963 Updating the Lists of Contracting Authorities under the Procurement Directives', *Public Procurement Law Review* (2009): 93–94.
133. Sue Arrowsmith, *The Law of Public and Utilities Procurement* (London: Sweet & Maxwell, 2005), 251; Christopher Bovis, *EC Public Procurement: Case Law and Regulation* (London: Oxford University Press, 2006), 385.
134. ECJ, Case C-373/00, *Adolf Truley GmbH v. Bestattung Wien GmbH*, 27 Feb. 2003, *European Court reports* 2003, I-1931.
135. ECJ, Case C-44/96, *Mannesmann Anlagenbau Austria AG and Others v. Strohal Rotationsdruck GmbH*, 15 Jan. 1998, para. 20–21, *European Court reports* 1998, I-73; ECJ, Case C-470/99, *Universale-Bau et al.*, 12 Dec. 2002, paras 51–53, *European Court reports* 2002, I-11617; ECJ, C-214/00, *Commission v. the Kingdom of Spain*, paras 52–53, 15 May 2003, paras 54–55, 60,

meeting needs in the general interest not having an industrial or commercial character. The ECJ has not defined this first criterion but has only formulated a number of indications that can aid in determining when there are 'needs in the general interest not having an industrial or commercial character'.[136] This criterion entails two issues: on the one hand, one needs to define whether the body has the purpose of meeting needs in the general interest, and on the other hand, whether these needs in the general interest have a non-commercial and non-industrial character.

GENERAL INTEREST – The notion of general interest is often used but defined less often, and it is often said that it is more of a political term than a true legal one. It is not an easy term to define and is usually done so by opposing it to other interests or illustrating it by examples.[137] In any case, the ECJ has once again emphasized that needs in the general interest should be interpreted in light of Community law and not according to the national laws in each of the Member States.[138]

RELEVANT LEGAL AND FACTUAL CIRCUMSTANCES – The ECJ has held that needs in the general interest not having an industrial or commercial character within the meaning of the Public Procurement Directives 'are generally needs which are satisfied otherwise than by the availability of goods and services in the marketplace and which, for reasons associated with the general interest, the State chooses to provide itself or over which it wishes to retain a decisive influence'.[139] According to Bovis, case law makes it clear that, in determining whether or not there exists a need in the general interest not having an industrial or commercial character, account must be taken of relevant legal and factual circumstances, such as those prevailing when the body concerned was formed and the conditions in which it carries on its activity. These conditions may include lack of competition on the market, the fact that the body's primary aim is not the making of profits, the fact that it does not bear the risks associated with the activity, and any public financing of the activity in question.[140]

ONLY ENTITIES PROVIDING SERVICES TO THE PUBLIC? – Entities providing services directly to the public often meet needs in the general interest, for example, the collection of household waste (for health and environmental reasons, the need

---

European Court reports 2003, I-4667; ECJ, Case C-283/00, *Commission v. Kingdom of Spain*, 16 Oct. 2003, para. 69, *European Court Reports* 2003, I-11697. See also Mustafa T. Karayigit, 'Under the Triangle Rules of Competition, State Aid and Public Procurement: Public Undertakings Entrusted with the Operation of Services of General Economic Interest', *European Competition Law Review* 30, no. 11 (2009): 553.

136. Elisabetta R. Manunza, 'Privatised Services and the Concept of "Bodies Governed by Public Law" in EC Directives on Public Procurement', *European Law Review* 28 (2003): 279.
137. Ann-Lawrence Durviaux, *Logique de marché et marché public en droit communautaire. Analyse critique d'un système* (Brussels: Larcier, 2006), 137–138 [in French].
138. ECJ, Case C-373/00, *Adolf Truley GmbH v. Bestattung Wien GmbH*, 27 Feb. 2003, para. 35, *European Court reports* 2003, I-1931.
139. ECJ, Case C-373/00, *Adolf Truley GmbH v. Bestattung Wien GmbH*, 27 Feb. 2003, para. 50, *European Court reports* 2003, I-1931.
140. Christopher Bovis, *EC Public Procurement: Case Law and Regulation* (London: Oxford University Press, 2006), 386.

might not be adequately fulfilled by private firms),[141] the activities of a funeral undertaker (acting in the interest of the state by handling death certificates and for health reasons),[142] running a university,[143] providing recreational and leisure facilities,[144] and organizing exhibitions and fairs to promote commerce.[145] However, the fact that the activities provide a direct benefit only to a limited number of undertakings or even only to one undertaking does not necessarily preclude their meeting needs in the general interest.[146] In addition, not only providing services directly to the public, but also providing support for the activities of other entities delivering public services or engaged in other government activities may be an activity that meets needs in the general interest.[147]

NO MARKET ACTIVITIES? – In *Agora*, the ECJ stated that if an activity is performed under significant competition in the marketplace, this may be indicative of the fact that there is no need in the general interest. However, competition in the market is not an absolute determining factor.[148] Moreover, whether or not an entity also participates in commercial activities, in addition to providing public services, is not relevant to it being considered a body governed by public law for all its procurements. The proportion between the activities that meet needs in the general interest not having an industrial or commercial character and the commercial activities is also of no relevance.[149] In short, the condition that the entity must have been established for the 'specific' purpose of meeting needs in the general interest not having an industrial or commercial character, does not mean that it should be entrusted exclusively with meeting such needs, nor that this task should be its dominant activity.[150]

---

141. ECJ, Case C-360/96, *Gemeente Arnhem and Gemeente Rheden v. BFI Holding BV*, 10 Nov. 1998, paras 61–62, *European Court reports* 1998, I-6821.
142. ECJ, Case C-373/00, *Adolf Truley GmbH v. Bestattung Wien GmbH*, 27 Feb. 2003, *European Court reports* 2003, I-1931.
143. ECJ, Case C-380/98, *The Queen v. HM Treasury Ex p. the University of Cambridge*, 3 Oct. 2000, *European Court reports* 2000, I-8035.
144. ECJ, Case C-306/97, *Connemara Machine Turf Co. Ltd v. Coillte Teoranta*, 17 Dec. 1998, *European Court reports* 1998, I-8761.
145. ECJ, Case 223/99, *Agorà Srl and Excelsior Snc di Pedrotti Bruna & C. v. Ente Autonomo Fiera Internazionale di Milano and Ciftat Soc. coop. arl.*, 10 May 2001, *European Court reports* 2001, I-3605.
146. *Ibid.*, para. 55.
147. ECJ, Case C-283/00, *Commission v. Kingdom of Spain*, 16 Oct. 2003, para. 69, *European Court Reports* 2003, I-11697.
148. Christopher Bovis, *EC Public Procurement: Case Law and Regulation* (London: Oxford University Press, 2006), 370–371; Ann-Lawrence Durviaux, *Logique de marché et marché public en droit communautaire. Analyse critique d'un système* (Brussels: Larcier, 2006), 145 [in French].
149. Philippe Flamme, Maurice-Andre Flamme & Claude Dardenne, *Les marchés publics européens et belges* (Brussels: Larcier, 2005), 79 [in French]; Sue Arrowsmith, *The Law of Public and Utilities Procurement* (London: Sweet & Maxwell, 2005), 273.
150. Elisabetta R. Manunza, 'Privatised Services and the Concept of "Bodies Governed by Public Law" in EC Directives on Public Procurement', *European Law Review* 28 (2003): 280; Ann-Lawrence Durviaux, *Logique de marché et marché public en droit communautaire. Analyse critique d'un système* (Brussels: Larcier, 2006), 141–142 [in French].

ESTABLISHED FOR MEETING NEEDS IN THE GENERAL INTEREST? – In *Universale Bau*, the ECJ further limited the impact of the requirement '*established* for the specific purpose of meeting needs in the general interest' (emphasis added). The ECJ judged an entity that was originally created to perform a commercial activity and not established to perform an activity meeting needs in the general interest, but that was later on entrusted with such activities, to be a body governed by public law, even though its bylaws had not been adapted to incorporate its new task.[151] The ECJ decided – according to Durviaux, completely in contradiction to the text of the directive[152] – that the moment on which the tasks meeting needs in the general interest are entrusted to the organization does not matter, and that it is sufficient that it has taken responsibility for such needs and has actually satisfied these needs since that time.[153] In light of the functional interpretation of a body governed by public law, one should take into account the activity of the organization at the moment its qualification needs to be determined, and not its intended activity at the moment of its creation.[154]

NOT HAVING AN INDUSTRIAL OR COMMERCIAL CHARACTER – As was mentioned above, the second aspect of the first requirement for a body governed by public law is that the needs in the general interest that are being met do not have a commercial or industrial character. The ECJ stated – unfortunately without explaining the reason why – that needs in the general interest not having an industrial or commercial character are a subcategory of needs in the general interest.[155] Arrowsmith states that while the actual wording of the criterion appears

151. ECJ, Case C-470/99, *Universale Bau et al.*, 12 Dec. 2002, paras 57–58, *European Court reports* 2002, I-11617.
152. Ann-Lawrence Durviaux, *Logique de marché et marché public en droit communautaire. Analyse critique d'un système* (Brussels: Larcier, 2006), 150 [in French].
153. ECJ, Case C-470/99, *Universale-Bau et al.*, 12 Dec. 2002, paras 57–58, *European Court reports* 2002, I-11617; Christopher Bovis, *EC Public Procurement: Case Law and Regulation* (London: Oxford University Press, 2006), 391; Mustafa T. Karayigit, 'Under the Triangle Rules of Competition, State Aid and Public Procurement: Public Undertakings Entrusted with the Operation of Services of General Economic Interest', *European Competition Law Review* 30, no. 11 (2009): 554.
154. Philippe Flamme, Maurice-Andre Flamme & Claude Dardenne, *Les marchés publics européens et belges* (Brussels: Larcier, 2005), 82 [in French]; Etienne Fatôme & Laurent Richier, 'Un triomphe de l'interprétation fonctionnelle ou comment devenir un organisme de droit public', *Actualités de la Commande et des Contrats Publics* 2003 : 63 [in French]; Sue Arrowsmith, *The Law of Public and Utilities Procurement* (London: Sweet & Maxwell, 2005), 274; Christopher H. Bovis, 'Developing Public Procurement Regulation: Jurisprudence and Its Influence on Law Making', *Common Market Law Review* 43, no. 2 (2006): 475.
155. ECJ, Case C-360/96, *Gemeente Arnhem and Gemeente Rheden v. BFI Holding BV*, 10 Nov. 1998, para. 36, *European Court reports* 1998, I-6821; ECJ, Case 223/99, *Agorà Srl and Excelsior Snc di Pedrotti Bruna & C. v. Ente Autonomo Fiera Internazionale di Milano and Ciftat Soc. coop. arl.*, 10 May 2001, para. 32, *European Court reports* 2001, I-3605; Christopher H. Bovis, 'Developing Public Procurement Regulation: Jurisprudence and Its Influence on Law Making', *Common Market Law Review* 43, no. 2 (2006): 473–474; Philippe Flamme, Maurice-Andre Flamme & Claude Dardenne, *Les marchés publics européens et belges* (Brussels: Larcier, 2005), 79 [in French].

to require the courts to identify whether the needs have a commercial or industrial character, in applying the test, the ECJ has mainly focused not on the nature of the needs, but on whether the particular entity carries out the activity on a commercial basis.[156] She criticizes this approach because the fact that the test looks at the particular market in which the entity operates and at the circumstances of the particular entity means that entities carrying out the same activity in different Member States may be treated differently under the directives.[157]

PRINCIPLES OF INTERPRETATION – The ECJ has not given a definition of needs in the general interest *not having an industrial or commercial character*, but it has provided the national courts with a number of principles of interpretation.[158] Firstly, competition with other undertakings is considered relevant.[159] However, market competition is not sufficient for an activity to be commercial.[160] Yet, if an entity operates in normal market conditions, aims to make a profit, and bears the risks of its activity, the ECJ considers it unlikely that the needs it aims to meet are not of an industrial or commercial nature.[161] However, aiming to make a profit is not always necessary[162]: in *Agora*, an entity managed according to criteria of performance, efficiency, and cost-effectiveness was not considered a body governed by public law, even though it was not a profit-making organization.[163] Regarding the risk of the activity, in *Commission v. Spain*, the ECJ stated that an entity could be considered to not bear the financial risk of its activities, because it was likely that the State would protect the financial viability of the entity by stepping in to save the entity from compulsory liquidation.[164] One might wonder whether such an assumption of the Court should

156. Sue Arrowsmith, *The Law of Public and Utilities Procurement* (London: Sweet & Maxwell, 2005), 269.
157. *Ibid.*
158. Philippe Flamme, Maurice-Andre Flamme & Claude Dardenne, *Les marchés publics européens et belges* (Brussels: Larcier, 2005), 80 [in French]; Elisabetta R. Manunza, 'Privatised Services and the Concept of "Bodies Governed by Public Law" in EC Directives on Public Procurement', *European Law Review* 28 (2003): 279.
159. ECJ, Case C-360/96, *Gemeente Arnhem and Gemeente Rheden v. BFI Holding BV*, 10 Nov. 1998, para. 48–49, *European Court reports* 1998, I-6821; ECJ, Case 223/99, *Agorà Srl and Excelsior Snc di Pedrotti Bruna & C. v. Ente Autonomo Fiera Internazionale di Milano and Ciftat Soc. coop. arl.*, 10 May 2001, para. 38, *European Court reports* 2001, I-3605; Philippe Flamme, Maurice-Andre Flamme & Claude Dardenne, *Les marchés publics européens et belges* (Brussels: Larcier, 2005), 80 [in French].
160. Christopher H. Bovis, 'Developing Public Procurement Regulation: Jurisprudence and Its Influence on Law Making', *Common Market Law Review* 43, no. 2 (2006): 475; Ann-Lawrence Durviaux, *Logique de marché et marché public en droit communautaire. Analyse critique d'un système* (Brussels: Larcier, 2006), 153 [in French].
161. ECJ, Case C-360/96, *Gemeente Arnhem and Gemeente Rheden v. BFI Holding BV*, 10 Nov. 1998, para. 51, *European Court reports* 1998, I-6821.
162. Sue Arrowsmith, *The Law of Public and Utilities Procurement* (London: Sweet & Maxwell, 2005), 270.
163. ECJ, Case 223/99, *Agorà Srl and Excelsior Snc di Pedrotti Bruna & C. v. Ente Autonomo Fiera Internazionale di Milano and Ciftat Soc. coop. arl.*, 10 May 2001, paras 40, 43, *European Court reports* 2001, I-3605.
164. ECJ, Case C-283/00, *Commission v. Kingdom of Spain*, 16 Oct. 2003, para. 91, *European Court Reports* 2003, I-11697.

be a sufficient basis to determine whether or not an entity is meeting needs in the general interest not having an industrial or commercial character. Lastly, the support of activities by other forms of public finance may also indicate the non-commercial character of an activity, for example, funding specific activities or projects carried out by the entity.[165]

ii. The Close Dependency Test

THREE NON-CUMULATIVE CRITERIA – The second requirement for classification as a body governed by public law is often described as the dependency test. To qualify as a body governed by public law, a body has to be financed for the most part by the State, a regional or local authority, or some other body governed by public law; be subject to management supervision by the State, regional, or local authorities or by bodies governed by public law; or have an administrative, managerial, or supervisory board more than half of whose members are appointed by the State, regional, or local authorities or bodies governed by public law. These three criteria are often described as the 'close dependency' test. They are not cumulative, so it is sufficient if there is financial dependency, management supervision, or appointment.[166]

FINANCIAL CRITERION – In *University of Cambridge*, the ECJ laid down a number of criteria that need to be taken into account when determining whether an entity is 'financed for the most part' by the State, a regional or local authority, or some other body governed by public law. As a starting point, the Court made it clear that:

> Whilst the way in which a particular body is financed may reveal whether it is closely dependent on another contracting authority, it is clear that that criterion is not an absolute one. Not all payments made by a contracting authority have the effect of creating or reinforcing a specific relationship of subordination or dependency. Only payments which go to finance or support the activities of the body concerned without any specific consideration therefore may be described as public financing.[167]

In the case of universities, this would entail that support in the form of awards or grants for research work or student scholarships may qualify as public financing, while payments for contract research and the organization of seminars and conferences most likely will not.[168] In addition, 'for the most part' has to be interpreted quantitatively and means more than half.[169] This percentage should be calculated by including all income, including that from commercial activity, in the timeframe of the entity's budgetary year.

165. Sue Arrowsmith, *The Law of Public and Utilities Procurement* (London: Sweet & Maxwell, 2005), 270.
166. Ann-Lawrence Durviaux, *Logique de marché et marché public en droit communautaire. Analyse critique d'un système* (Brussels: Larcier, 2006), 156 [in French]; Christopher Bovis, *EC Public Procurement: Case Law and Regulation* (London: Oxford University Press, 2006), 366.
167. ECJ, Case C-380/98, *The Queen v. HM Treasury Ex p. the University of Cambridge*, 3 Oct. 2000, para. 21, *European Court reports* 2000, I-8035.
168. *Ibid.*, paras 22–24.
169. *Ibid.*, para. 33.

MANAGEMENT SUPERVISION CRITERION – While the corporate governance criterion is also quantitative (appointment of more than half of the members of the administrative, managerial, or supervisory board), this is not the case for the exercise of management supervision.[170] The ECJ considered in *Commission v. France*, concerning entities responsible for providing social housing, that it is necessary to consider whether the controls to which entities are subject render them 'dependent on the public authorities in such a way that the latter are able to influence their decisions in relation to public contracts'.[171] The supervision exercised must give rise to dependence equivalent to that existing when the criteria of financing or appointment are met. Ownership by a contracting authority is sufficient for an entity to be classified as a body governed by public law, but management supervision can also be established with a lesser degree of ownership, depending on, for instance, the extent of the public ownership or the influence of other shareholders.[172]

RELEVANT FACTORS FOR MANAGEMENT SUPERVISION – Arrowsmith states that, in general, the power to actually intervene in the management decisions of an entity can probably constitute management supervision,[173] but the ECJ also established in *Commission v. France* that such a power is not a requirement, at least when there are 'detailed' rules of management that the entity must follow. In that case, supervision over the compliance of the entity with such rules may be sufficient.[174] The ECJ highlighted a number of factors that may be relevant in determining the presence of management supervision. However, the precise significance of each factor, and whether it is sufficient by itself to establish management supervision, is not clear.[175] These factors included that the activities of the entity and the objectives of its activities were closely circumscribed; that there was legal provision for the entity to be supervised by the ministers responsible for finance and for construction and housing, without any limitations of this supervision being specified; that the responsible minister had the power to wind up the entity or to suspend the management and appoint a liquidator and administrator; that the minister could intervene when the entity's activity fell below a 'minimum level of dynamism' or limit any excessive activity; that an inter-ministerial task force had the power to inspect the entity, to draw up proposals for action, and also to ensure that it was implementing measures adopted by government ministers.[176] In *Adolf Truley*, the ECJ stated that

170. Christopher Bovis, *EC Public Procurement: Case Law and Regulation* (London: Oxford University Press, 2006), 366; Philippe Flamme, Maurice-Andre Flamme & Claude Dardenne, *Les marchés publics européens et belges* (Brussels: Larcier, 2005), 77 [in French].

171. ECJ, Case C-237/99, *Commission v. French Republic*, 1 Feb. 2001, para. 52, *European Court Reports* 2001, I-939.

172. Sue Arrowsmith, *The Law of Public and Utilities Procurement* (London: Sweet & Maxwell, 2005), 263.

173. *Ibid.*, 259.

174. ECJ, Case C-237/99, *Commission v. French Republic*, 1 Feb. 2001, para. 52, *European Court Reports* 2001, I-939.

175. Sue Arrowsmith, *The Law of Public and Utilities Procurement* (London: Sweet & Maxwell, 2005), 259.

176. ECJ, Case C-237/99, *Commission v. French Republic*, 1 Feb. 2001, para. 52, *European Court Reports* 2001, I-939.

the supervision criterion was met when other contracting authorities supervise the annual accounts of the entity involved and its conduct from the point of view of 'proper accounting, regularity, economy, efficiency and expediency' and are able to inspect its business premises/facilities and to report the results to the owner of the entity (in this case another company, which, in turn, was wholly owned by the city of Vienna).[177] In contrast, mere review afterwards of the entity's actions and the annual accounts would not be sufficient, as the contracting authority would not be able to influence the entity's decisions.[178]

c. *Associations Formed by One or Several Authorities or One or Several Bodies Governed by Public Law*

No overlap with a body governed by public law – In *BFI Holding*, the ECJ made clear that there is no overlap between the definition of a body governed by public law and the provision on associations, and that the latter only fulfils a residual function.[179] As a consequence, it is first necessary to consider whether an entity is a body governed by public law, and only if it is not might it fall under the 'association' provision.[180] The provision is mainly relevant for entities without legal personality, as those with legal personality will most likely be regarded as bodies governed by public law. It is also applicable to 'mixed' associations made up of one or more authorities or bodies governed by public law and one or more other bodies.

d. *Relevance for the PSI Directive*

Usefulness for the PSI Directive – One can assume that by using the terminology of the Public Procurement Directives, the European institutions had the intention of interpreting the notion of *public sector bodies* in the same manner as *contracting authorities*. As has been noted above, this notion is interpreted quite widely by the ECJ, which may have as a possible consequence that many agencies that at first sight might not qualify as public sector bodies, actually might fulfil the necessary conditions and therefore fall under the obligations of the PSI Directive and the corresponding national legislation. The lists of public sector bodies in the Member States that have been included in the Annexes of the Public Procurement Directives can also be used as an aid to determine the character of an agency and its activities under the PSI Directive. While this list is very extensive, it should be re-emphasized that it is not exhaustive.

---

177. ECJ, Case C-373/00, *Adolf Truley GmbH v. Bestattung Wien GmbH*, 27 Feb. 2003, para. 74, *European Court reports* 2003, I-1931.
178. *Ibid.*, para. 70.
179. ECJ, Case C-360/96, *Gemeente Arnhem and Gemeente Rheden v. BFI Holding BV*, 10 Nov. 1998, para. 27, *European Court reports* 1998, I-6821.
180. Sue Arrowsmith, *The Law of Public and Utilities Procurement* (London: Sweet & Maxwell, 2005), 275.

PUBLIC BODIES HOLDING SPATIAL DATA? – Considering the wide definition given to public sector bodies under the PSI Directive and contracting authorities in the Public Procurement Directives, many public bodies holding spatial data will fall under this definition. In addition, the lists in the Annexes to the Public Procurement Directives already contain many public bodies holding spatial data, so the Member States have already indicated that these are considered public sector bodies. Bodies on the list include ministry departments, national mapping agencies, cadastral offices, statistical offices, meteorological offices, land registries, environmental agencies, and so on.[181] Sometimes these are considered bodies governed by public law; sometimes they are listed as part of central government.

D. COMMON DENOMINATOR BETWEEN PUBLIC SECTOR BODIES AND PUBLIC AUTHORITIES

PUBLIC AUTHORITIES V. PUBLIC SECTOR BODIES? – In order to define the scope of this book, it must be determined which agencies have to make data available. Is there a big difference between the scope of these directives, that is, is a *public sector body* also a *public authority* and vice versa? In any case, both terms have to be interpreted broadly. For public sector bodies, this is laid down by the jurisprudence of the ECJ, while the Implementation Guide to the Aarhus Convention proclaims the extensive interpretation of public authorities.[182]

COMPARISON OF 'TRADITIONAL' ORGANIZATIONS – As far as traditional organizations of the public sector are concerned, both definitions apply to the 'State, regional or local authorities' and to the 'government or other public administration, at national, regional or local level'. However, public advisory bodies that are part of the public authorities will only qualify as public sector bodies when they fulfil the requirements of a body governed by public law.

COMPARISON OF OTHER PUBLIC BODIES – As far as other public agencies are concerned, the extent of the overlap between the definitions is less clear. First, the definition of *public authority* is broader, as there is no requirement for any legal personality. Hence, a natural person can qualify as a public authority, while a public sector body must be regarded as part of the State, regional, or local authorities, as only legal persons can be bodies governed by public law. A second element that makes the definition of *public authority* broader is that under (b) of the definition in Article 2.2 of the Access Directive, it concerns any natural or legal person performing public administrative functions under national law, including specific duties, activities, or services, without any demand for financing, appointment, or supervision by the State, regional, or local authorities; and the persons mentioned under (c) – that is, having public responsibilities or functions, or providing public services – do not have to be under the control of the state, regional, or local authorities, but can also be under the control of a public authority under (b). Third,

181. See Annex III of Directive 2004/18/EC.

182. United Nations Economic Commission for Europe, *Aarhus Implementation Guide*, 32.

except for State, regional, and local authorities, a body governed by public law has to perform activities that meet needs in the general interest not having an industrial or commercial character, while public authorities have to perform administrative functions, have public responsibilities or functions, or provide public services. Assuming that public services will always have the purpose of meeting needs in the general interest, the scope of public authorities is once again broader, as there is no need for the public services or functions to have a non-commercial and non-industrial character. However, the activities that meet needs in the general interest that are of a commercial or industrial interest will most likely be few and far between. On the other hand, the definition of *public authority* is much more limited, because for the authorities under (c), it is required that they perform public administrative functions or have public responsibilities or functions *relating to the environment*. Obviously, this is not necessary for a public sector body, as it is aimed at public procurement for any type of contract, not just for contracts relating to the environment.

SUMMARY – In short, the most important differences seem to be in the stricter dependency test for bodies governed by public law on the one hand, and, of course, the limitation to activities relating to the environment for public authorities. However, considering the broad interpretation of bodies governed by public law by the ECJ and the broad scope of the term *environment* under the Access Directive, the field of application *ratione personae* for the three directives of the existing legal framework on the availability of public sector spatial data will still be quite extensive.

COMMON DENOMINATOR: PUBLIC BODIES – As this book focuses on the intersection between the INSPIRE Directive, the Access Directive, and the PSI Directive, it addresses the entities that are both public authorities and public sector bodies. The criteria mentioned above imply that many, if not all, public sector entities holding spatial data will be public authorities and public sector bodies. I will refer to such bodies as *public bodies*, only using *public sector bodies* or *public authorities* when specifically addressing the separate categories.

V. COMMON DENOMINATOR BETWEEN SPATIAL DATA, PUBLIC SECTOR DATA, AND ENVIRONMENTAL DATA

PUBLIC SECTOR SPATIAL ENVIRONMENTAL DATA: THEORETICAL REPRESENTATION – In the previous sections, we examined the descriptions of spatial data, public sector data, and environmental data, and their respective definitions in the existing legal framework. As these categories overlap, so will the conditions for their availability, which creates confusion for both the data suppliers and the users. This common denominator between the three categories of data – and hence the three legal regimes of the legislative framework – can be described as public sector spatial environmental data. One might initially expect this common denominator to be represented as shown in **Figure 2.**

Figure 2. Common Denominator between Public Sector Data, Spatial Data, and Environmental Data

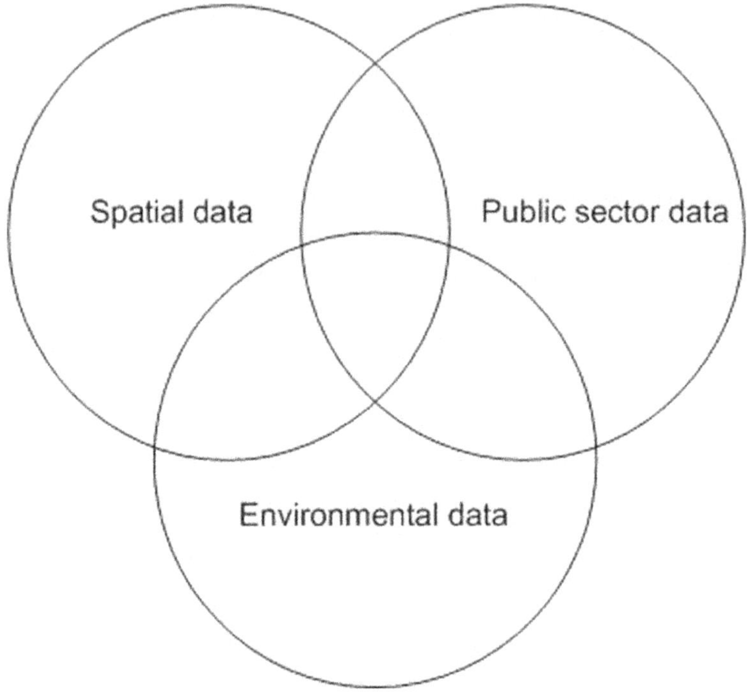

PUBLIC SECTOR SPATIAL ENVIRONMENTAL DATA: A MORE REALISTIC REPRESENTATION – However, one must first keep in mind that this subset of data still contains a very significant amount of data, both from a qualitative and quantitative point of view.[183] Second, as this book's findings are limited to only spatial data that are created or held by the public sector and examines the relationship between the PSI Directive, the Access Directive, and the INSPIRE Directive, the representation will have a very different look. This look will most likely correspond more with reality, as the environmental information of the Access Directive and the list of spatial data sets in the INSPIRE Directive will overlap to a great extent. Even though not all the categories of spatial data listed in the INSPIRE Directive will also qualify as environmental information and vice versa, the fact that the INSPIRE data sets are considered necessary for the public authorities to perform their public tasks relating to the environment indicates that most of them will be considered

183. A minimal amount of data that falls within the field of application of the INSPIRE Directive, might not be public sector data, as Art. 12 of the directive allows spatial data sets and services of third parties (i.e., any legal or natural person other than public authorities) to be linked to the network upon request. These third party datasets fall beyond the scope of this research and will not be addressed further.

environmental information under the broad definition of the Access Directive.[184] Therefore, the scope could be represented in the manner shown in **Figure 3.**

Figure 3. Common Denominator between the PSI Directive, the Access Directive, and the INSPIRE Directive

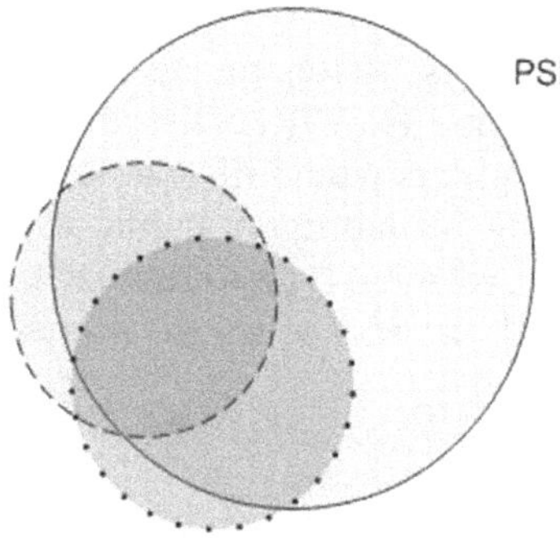

PREFERENCE FOR PUBLIC SECTOR SPATIAL DATA – Sticking to the proper terminology in indicating the field of application *ratione materiae* of this book would require the use of the term 'public sector environmental spatial data listed in the INSPIRE Directive and held by public authorities that are also public sector bodies'. However, for the ease of both the reader and the writer, the term public sector spatial data will be used in the remainder of this book. It is implied that this term refers to the common denominator that is set out in this section. Only when the different categories need to be addressed separately will the respective terms be used.

VI. SCOPE FROM THE POINT OF VIEW OF THE PURPOSE OF USE

CRITERION FOR DISTINCTION – The INSPIRE Directive, the Access Directive, and the PSI Directive all promote the availability of public sector spatial data. The three directives have a common field of application that includes a considerable amount of data. Yet, there has to be a criterion to determine the distinction between the three directives and when the respective directives apply. The criterion that underpins this legal framework is the purpose of the use for which the data are made available.

184. Recital 7 of the INSPIRE Directive takes a more modest approach: 'There is a degree of overlap between the spatial information covered by this Directive and the information covered by Directive 2003/4/EC of the European Parliament and of the Council of 28 January 2003 on public access to environmental information'.

TYPES OF USE – Public sector spatial data are made available by the public authorities to be used for different purposes by different parties. One can distinguish between three types of use, which are referred to in the respective directives as *access, re-use*, and *sharing*. It is submitted that access is regulated by the Access Directive and Chapter IV of the INSPIRE Directive on network services; re-use is addressed by the PSI Directive; and sharing is dealt with in Chapter V of the INSPIRE Directive.

OUTLINE – Hence, any discussion on access should not involve the PSI Directive, while debates on the INSPIRE Directive have nothing to do with re-use. Ideally, the three legal pillars would fit seamlessly into each other, creating a clear legal framework for the availability of public sector spatial data. However, it is argued that confusion between the types of use exists. In the next sections, each purpose is briefly explained, in order to set the scene for the next chapters.

A. ACCESS

ACCESS DIRECTIVE AND INSPIRE DIRECTIVE – Access to public sector spatial data is regulated by the Access Directive and the provisions of the INSPIRE Directive dealing with public access to network services. In essence, access ensures that citizens can participate fully in the democratic process.

A NATIONAL MATTER – A person's right to access to public sector spatial data is usually guaranteed by national freedom of information legislation or access legislation, or by sector-specific legislation referring to the conditions for the availability of a particular category of data. All Member States of the EU have national access legislation, except for Luxemburg and the Spanish national level.[185] The European institutions have made it clear that they do not intend to take any measures with the aim of harmonizing the various pieces of access legislation. They consider this to be a national matter that lies beyond the competence of the EU.[186]

EXCEPTION FOR ENVIRONMENTAL DATA – However, the European institutions have deviated from this point of view for one category of data: environmental data. This led to the Access Directive, which had the aim of guaranteeing the right of access to environmental information and to ensure that environmental information is progressively being made available and disseminated to the public.[187] The principles of the Access Directive will be addressed in Chapter 3.

185. Malta's Freedom of Information Legislation dates from July 2009 and enters into force in July 2010.

186. Commission of the European Communities, *Communication of 23 October 2001 to the Council, the European Parliament, the Economic and Social Committee and the Committee of the Regions. eEurope 2002: creating a EU framework for the exploitation of public sector information*, COM (2001) 607 final, 9–10 (hereafter referred to as *eEurope 2002 Communication*).

187. Article 1 of the Access Directive.

FREEDOM OF INFORMATION – Often, the terms *freedom of information* and *access* are used interchangeably, especially in common law countries. However, this can lead to confusion. Many national laws and regulations that provide access of citizens to documents held by public bodies carry the title 'Freedom of information Law' or a variant thereof. However, in the context of human rights, freedom of information has a different content and is usually considered to be part of freedom of expression. For instance, in Article 10 of the European Convention on Human Rights (ECHR),[188] the right of a person to freedom of expression includes the right to 'receive and impart information and ideas without interference by public authority and regardless of frontiers'. On the one hand, this right is less far-reaching than the rights conferred on the citizen under access legislation, as, in principle, it does not contain an obligation for public bodies to provide data to the citizen. On the other hand, its scope is much broader, as the right to receive and impart information is a human right, and it involves not only data originating from public bodies, but also data originating from private parties. As national freedom of information legislation generally does not refer to the freedom of information that is meant in Article 10 of the ECHR, but rather to access of citizens to data created or held by the public sector, the term *freedom of information* may give the wrong impression.[189]

TERMINOLOGY – In Dutch, a distinction is sometimes made between *openbaarheid*, which can be translated as *publicity*, and *toegang*, which corresponds to *access*. In French, the words *publicité* and *accès* apply. *Access* not only implies the access of the public to data held by public bodies, but it may also imply access of the data subject to his or her own personal data under legislation for the protection of personal data, or access of parties in a judicial procedure to documents regarding their case, for instance. *Publicity*, on the other hand, implies the availability of the data to anyone, without the need for any personal interest or involvement. As this distinction is not made in English, however, I will continue to use the term *access*.

ACTIVE AND PASSIVE RIGHT OF ACCESS – Public access itself is often described as having two different aspects: on the one hand, access means that a person can obtain access to a document *upon request*, without having to state a reason or interest. On the other hand, access to data held by a public body may also include obligations for that public body to disseminate data to the public. In continental Europe, this is often referred to as the difference between the *active* and *passive* right of access to data held by public bodies. In the United Kingdom, the terms *responsive openness* and *proactive openness* are sometimes used,[190] while the

188. Council of Europe, *Convention for the Protection of Human Rights and Fundamental Freedoms of 4 November 1950*, <http://conventions.coe.int/Treaty/en/Treaties/Html/005.htm>, 22 Feb. 2010.

189. Whether the availability of public sector data for access purposes falls under freedom of information under Art. 10 of the ECHR will be discussed in Ch. 5, s. I.B.

190. However, *openness* implies more than just access to data. It involves the mentality of the public sector and its servants to be prepared to listen to the citizens and to provide them with information. (See Monika Bargmann, Gerald Pfeifer & Boris Piwinger, 'A Citizen's

Victorian Economic Development and Infrastructure Committee refers to *push* and *pull* models.[191] However, as both access upon request and dissemination by public bodies of their own motion are usually included in national access legislation, I will use *access* to describe both, except when a distinction is necessary.

OBJECTIVE – Both the active and passive forms of access have the same objective. They provide natural or legal persons with data in order to exercise their democratic rights or obligations. Access is linked to the need for legitimacy of the government. This legitimacy is in turn based on three fundamental principles: democracy, the rule of law, and the efficiency and effectiveness of the operation of government.[192] A citizen can only exercise his or her democratic rights properly if there is some level of access to data held by the public bodies. The right of access can be used to check up on or control the activities of the government. This control can be indirect, through elections. In this sense, Daalder connects access to the citizen's right to vote: in order to make a well-considered choice for one candidate or another, he or she has to have at his or her disposal all necessary data, including the data that are held by public bodies.[193] Control by the citizen can also take more direct forms, by obtaining data about the public bodies' activities from their source. Access is important not only in the relationship between the citizen and government, but also among citizens. To properly exercise their rights and obligations towards other members of society, they need data from public bodies about what they are or are not entitled to do.

WORKING DEFINITION – For the further development of this theory, access is considered to be 'obtaining public sector spatial data for the purpose of exercising one's democratic rights or obligations'.

B. RE-USE

PSI DIRECTIVE – Re-use of public sector spatial data is regulated in the EU by the PSI Directive and its transposing national legislation. As a basic principle, re-use entails any use of data beyond the mere democratic need to be aware of them, and that is outside of the public task.

ECONOMIC VALUE OF PUBLIC SECTOR DATA – As was already indicated in Chapter 1, public sector spatial data do not only have a democratic value, but they also have an economic value. This economic value has two aspects. On the one hand, it ensures

---

Perspective on Public Sector Information', in *Public Sector Information in the Digital Age, Between Markets, Public Management and Citizens' Rights*, ed. Georg Aichholzer & Herbert Burkert (Cheltenham: Edward Elgar Publishing Limited, 2004), 255–262).

191. Economic Development and Infrastructure Committee, 'Inquiry into Improving Access to Victorian Public Sector Information and Data', 2009, 17–19, <www.parliament.vic.gov.au/edic/inquiries/access_to_PSI/final_report.html>, 22 Feb. 2010.

192. Ute Rosenthal et al., *Openbaar bestuur. Beleid, organisatie en politiek* (Alphen a/d Rijn: Samson H.H. Tjeenk Willink, 1996), 29–31 [in Dutch].

193. Eric J. Daalder, *Toegang tot overheidsinformatie. Het grensvlak tussen openbaarheid en vertrouwelijkheid* (The Hague: Boom Juridische Uitgevers, 2005), 12 [in Dutch].

that participants in society and on the market can make their decisions in a rational and informed manner, which can only benefit the market.[194] For instance, undertakings can use census data or environmental data combined with spatial data to determine their business strategies. No new data are created on the basis of the data obtained from public bodies, but the data have undergone some form of processing for purposes beyond the democratic 'checking up' on the government.[195] On the other hand, public sector spatial data can be used as a resource by the information industry to create added-value information products and services. From an efficiency point of view, an undertaking will prefer using data that have already been collected and checked by somebody else, in this case one or several public bodies, rather than spending time and money on duplicating work that has already been done. In this manner, the undertaking will be able to lower its production costs, provided that it can obtain the data at a cost that is lower than the production costs that it would incur. Both the above aspects constitute re-use.[196]

COMMERCIAL AND NON-COMMERCIAL PURPOSE OUTSIDE OF THE PUBLIC TASK – The previous paragraph might give the reader the idea that re-use only involves commercial activities. However, this is not the case. The definition of re-use in the PSI Directive is much broader: 'the use by persons or legal entities of documents held by public sector bodies, for commercial or non-commercial purposes other than the initial purpose within the public task for which the documents were produced. Exchange of documents between public sector bodies purely in pursuit of their public tasks does not constitute re-use'.[197] Hence, re-use can also involve non-commercial purposes or activities, for instance, non-governmental organizations using data to determine their position on an issue of interest to them and to create a publication that they disseminate to create awareness or to raise money for their cause. Re-use can also include private citizens who use maps on their websites or blogs, or students who use environmental pollution statistics from environmental agencies for writing papers or their theses. Hence, re-use is any use for commercial or non-commercial purposes outside of the public task. It can be stated that such use relates to a more economic purpose, as the user intends to gain a benefit from using the data, which is different from the democratic purposes of checking up on the government and public participation.

ACCESS V. RE-USE – It is submitted that the difference between access and re-use of public sector spatial data can be seen as follows. In general, re-use entails a level of processing of data that goes beyond mere access to the document in order to

---

194. Katleen Janssen & Jan J.C. Kabel, 'Commercialisering van overheidsinformatie door de overheid: rechtspraak en wetgeving in België en Nederland. De honden blaffen, maar de karavaan trekt voort', *Computerrecht* no. 3 (2005): 117 [in Dutch].

195. Katleen Janssen & Jos Dumortier, 'The Legal Framework for a European Spatial Data Infrastructure: Uncrossing the Wires', in *Research and Theory in Advancing Spatial Data Infrastructure Concepts*, ed. Harlan Onsrud (Redlands California: ESRI Press, 2007), 234.

196. Katleen Janssen & Jan J.C. Kabel, 'Commercialisering van overheidsinformatie door de overheid: rechtspraak en wetgeving in België en Nederland. De honden blaffen, maar de karavaan trekt voort', *Computerrecht* no. 3 (2005): 117 [in Dutch].

197. Article 2.4 of the PSI Directive.

learn its content.[198] For example, researchers need to perform some treatment on the data that they obtain from public authorities in order to get results, in many cases by combining those data with other data or test results. Merely accessing the data or reading the information will not be enough.

WORKING DEFINITION – *Re-use* can be defined as 'any use of public sector spatial data for commercial or non-commercial purposes outside of the public task', where it is assumed that this commercial or non-commercial use has the *economic purpose* of gaining a benefit and requires further processing beyond merely learning the content. The provisions on re-use imposed by the PSI Directive are discussed extensively in Chapter 3.

C. SHARING

INSPIRE Directive – The sharing of public sector spatial data is addressed by Chapter V of the INSPIRE Directive. Essentially, sharing involves the availability of public sector spatial data for the purpose of performing a public task.

PUBLIC TASK – Public sector spatial data are mostly created with the purpose of supporting the development and the implementation of public policy. Public bodies' first objective when they collect data is to use them for their own needs, that is, for the performance of their public tasks. Data that are collected in the framework of one public task or one government policy may also be useful outside of this original purpose for the performance of other public tasks. For instance, data collected for the purpose of town planning can also be useful to other public bodies for evaluating applications for subsidies, for environmental purposes, and so on.

INCREASE OF SHARING – In the past, public bodies were often not aware which data were already available from public bodies at other levels of competence or in other locations, or even from different departments of their own organizations. Currently, public bodies are increasingly realizing that such duplication of data costs a lot of effort and money that can be used for other public initiatives, and the call for sharing data is growing. This is especially the case for public sector spatial data, as they are often the backbone or the ground layer upon which other data are shown, putting these other data into a spatial perspective. Hence, public bodies are increasingly cooperating in the exchange of their data for public policy purposes. We refer to this delivery and obtaining of data for the purpose of performing a public task as sharing. Generally, sharing involves the exchange of data between public bodies. However, if private companies are entrusted with the provision of a public service, some of their activities using public sector spatial data may also fall under the concept of sharing.

198. Katleen Janssen & Jos Dumortier, 'The Legal Framework for a European Spatial Data Infrastructure: Uncrossing the Wires', in *Research and Theory in Advancing Spatial Data Infrastructure Concepts*, ed. Harlan Onsrud (Redlands California: ESRI Press, 2007), 234.

Working definition – *Sharing* is defined in this book as 'any use of public sector spatial data for the purpose of performing public task activities'. It will be discussed extensively in Chapter 3.

D. Interim Summary

Model – The description of access, re-use, and sharing shows that the differences between these three types of use of public sector spatial data are based on the purpose of the use. This can be represented in the manner shown in **Figure 4.**

Figure 4. Structure of the Legal Framework for the Availability of Public Sector Spatial Data Based on Purpose of Use

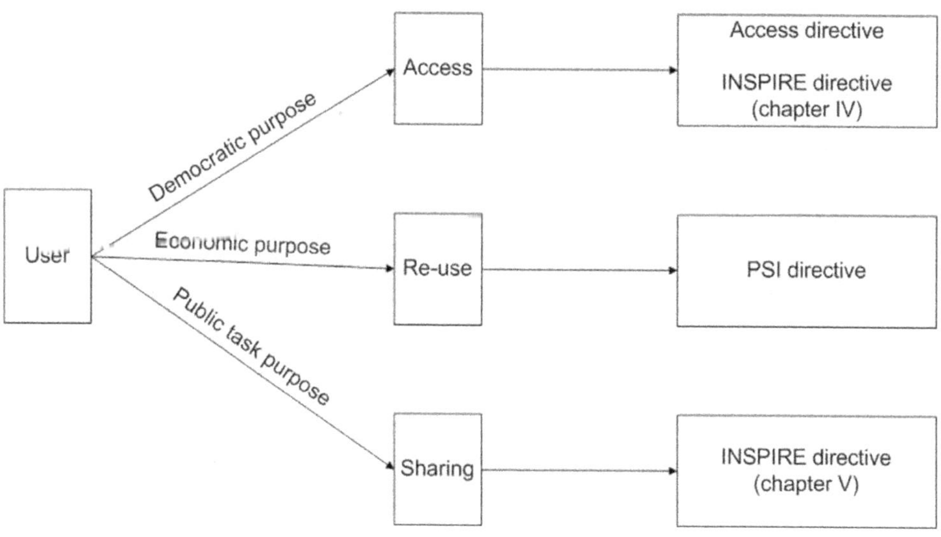

VII. A FIRST INDICATION OF THE PROBLEMATIC CHARACTER OF THE EU LEGAL FRAMEWORK FOR THE AVAILABILITY OF PUBLIC SECTOR SPATIAL DATA

Problematic character – The legal framework for the availability of public sector spatial data is problematic. As has been noted above, the distinction is based on the purpose of the use of the spatial data. However, the distinction between the three directives are problematic, mainly for two reasons. First, the distinctions are practically unenforceable. Second, the criterion of the public task as a distinction is not defined and is too contingent to be used.

The distinctions in practice – First, the description of the types of use has already indicated that it may not always be easy to distinguish between the three types of use. As the public task is contingent, what may be considered a task of a public body today may no longer be one tomorrow. Next, the examples above have also shown that the distinction between re-use and access is also difficult to make. How can a citizen exercising his or her democratic right be distinguished from a private person using data for his or her economic purpose of creating an economic benefit? Are some activities not a combination of both? Next, even if the distinction is clear, it may not always be enforceable, due to the volatile nature of data and information. It is very difficult to control where data go and for what purposes they are used.

Public task criterion – Second, from the three types of use that have been described in this section, it can be seen that the concept of the public task plays an important role. It determines the difference between data sharing and re-use: if an entity is using the data that it has obtained for public task purposes, the provisions on data sharing of the INSPIRE Directive are applicable. If the data are used for purposes outside of the public task, the PSI Directive and its rules on re-use apply. The rules for access also play a role. If a public body has to provide access to citizens under the Access Directive or the INSPIRE Directive, it is performing a public task. It may have to process these data in order for citizens to understand the information that they contain, or it may combine them with data from another public body. For this second data set, the relationship between the supplying body and the body using the data is regulated by the INSPIRE Directive. However, it may also be possible that these data are processed by a public body for purposes outside of the public task, as described above. In that case, the public bodies will actually be subject to the PSI Directive. In the next chapters, I will argue that the public task is not a suitable criterion for distinguishing access, re-use, and sharing.

Next chapter – The next chapter provides an in-depth discussion of the provisions of the Access Directive, the INSPIRE Directive, and the PSI Directive, in order to give a complete overview of the legal framework and to indicate the possible differences between these Directives.

Chapter 3

Existing Legal Framework for the Availability of Public Sector Spatial Data

I. INTRODUCTION

OUTLINE – This chapter provides an in-depth study of the three directives that were identified in the previous chapter as the data policies stimulating the availability of public sector spatial data in the EU: the Access Directive, the PSI Directive, and the INSPIRE Directive.

DEVELOPMENT PROCESS – For each directive, the relevant provisions for the availability of public sector spatial data are examined. Considerable attention is also given to the development process of the rules, both from the point of view of the documents that preceded the design of the Directives and the process of negotiation that led to the final text of the Articles. This may help the reader to grasp the underlying intentions of the legislator for some of the provisions and can indicate how the provisions should be interpreted.

DIVISION BY PURPOSE – Rather than just listing the directives, this chapter will follow the distinctions that were made in the previous chapter between access, re-use, and sharing in order to determine and group the different rules that are applicable to the three types of use. For the most part, this division follows the three directives: re-use is dealt with in the PSI Directive, and sharing is addressed in the INSPIRE Directive. However, for access, this is not the case. Access to public sector spatial data is dealt with by the Access Directive and the chapter of the INSPIRE Directive on network services. Therefore, these will be discussed together in the section on access.

II. ACCESS

A. Introduction

Outline – This section addresses the legislative actions of the EU directed towards access to public sector spatial data held by the Member States. This includes the INSPIRE Directive and the Access Directive, which both deal with a specific category of public sector data, respectively, spatial data and environmental information.

No initiatives on public sector data – The European legislator has not taken any initiatives directed towards the Member States regulating access to public sector data in general, and it has no intention of doing so. The European Commission stated this explicitly in a 2001 Communication on creating an EU framework for the exploitation of public sector information: 'With the exception of the environmental area, arrangements for access are first and foremost a national, regional and local responsibility. Therefore no harmonisation on the specific aspects of access to public sector information is envisaged'.[1] Even though the European Commission has addressed access to public sector data in a number of policy documents,[2] this has always been in relation to the question of re-use. Therefore, these policy documents will be discussed in the section on re-use.[3]

Exception for environmental data – Access to environmental data was considered closely related to environmental matters, for which the EU does have competence under Article 192 of the Treaty on the Functioning of the European Union (formerly Article 175 of the EC Treaty), so the Commission felt that action could be taken there.[4] As the INSPIRE Directive also falls under environmental competence and its scope forms a considerable overlap with the Access Directive, it could be justified that provisions on public access were included.

B. The Access Directive

Outline – Before addressing the provisions of the Access Directive, a brief overview is given of the previous Directive on public access to environmental information, Directive 90/313/EEC,[5] which was replaced by the Access Directive.

1. Commission of the European Communities, *eEurope 2002 Communication*, 11.
2. For example, Commission of the European Communities, *Green Paper on Public Sector Information*, 11–13; Commission of the European Communities, *eEurope 2002 Communication*, 9–10.
3. Cf. *infra*, s. III.
4. Katleen Janssen & Jos Dumortier, 'The Legal Framework for a European Spatial Data Infrastructure: Uncrossing the Wires', in *Research and Theory in Advancing Spatial Data Infrastructure Concepts*, ed. Harlan Onsrud (Redlands California: ESRI Press, 2007), 236.
5. For more information on the context and the history of Directive 90/313/EEC, see Ludwig Kramer, 'La directive 90/313/EEC sur l'accès à l'information en matière de l'environnement: genèse et perspectives d'application', in *The Right to Environmental Information*, ed. Marc

This directive is no longer in force, but it can provide some more background on the developments in the EU towards a greater availability of environmental data.

1. History: Directive 90/313/EEC

TIMELINE – Directive 90/313/EEC was adopted on 7 June 1990 and had to be implemented by the Member States of the European Economic Community by 31 December 1992. It represented a radical break with the tradition of official secrecy in most Member States and set the EU on a new course towards openness.[6]

OBJECTIVE – The objective of the directive was to ensure freedom of access to and dissemination of information on the environment held by public authorities and to set out the basic terms and conditions by which such information should be available.[7]

a. Access upon Request

DEFINITION: ENVIRONMENTAL INFORMATION – *Information relating to the environment* was defined as 'any available information in written, visual, aural or data-base form on the state of water, air, soil, fauna, flora, land and natural sites, and on activities (including those which give rise to nuisances such as noise) or measures adversely affecting, or likely to affect these, and on activities or measures designed to protect these, including administrative measures and environmental management programmes'.[8] According to Kramer, this definition was very wide and covered practically all information on the environment. He believed that finding aspects of the environmental domain that would fall outside of this definition would be very difficult.[9] As was mentioned before, in a judgment of 1999, the ECJ also considered the concept of environmental information to be 'a broad one, embracing both information and activities relating to the state of those aspects'.[10]

Pallemaerts (Brussels: Story-Scientia, 1991), 17–39 [in French]; Ralph E. Hallo, 'Directive 90/313/EEC on the Freedom of Access to Information on the Environment: Its Implementation and Implications', in *Access to Environmental Information in Europe: The Implementation and Implications of directive 90/313/EEC*, ed. Ralph E. Hallo (London: Kluwer Law International, 1996), 1–26.

6. Ralph E. Hallo, 'Directive 90/313/EEC on the Freedom of Access to Information on the Environment: Its Implementation and Implications', in *Access to Environmental Information in Europe: The Implementation and Implications of directive 90/313/EEC*, ed. Ralph E. Hallo (London: Kluwer Law International, 1996), 1.
7. Article 1 of Directive 90/313/EEC.
8. Article 2(a) of Directive 90/313/EEC.
9. Ludwig Kramer, 'La directive 90/313/EEC sur l'accès à l'information en matière de l'environnement: genèse et perspectives d'application', in *The Right to Environmental Information*, ed. Marc Pallemaerts (Brussels: Story-Scientia, 1991), 30 [in French].
10. ECJ, Case C-321/96, *Mecklenburg v. Kreis Pinneberg der Landrat*, 17 Jun. 1998, para. 19, *European Court Reports* 1998, I-3809. It was held that a statement of views given by a countryside protection authority in development consent proceedings would, if capable of influencing the outcome of those proceedings, constitute environmental information. See also Philip Coppel, 'Environmental Information: The New Regime', *Journal of Planning & Environment*

Definition: public authorities – Environmental information had to be held by *public authorities*: 'any public administration at national, regional or local level with responsibilities, and possessing information, relating to the environment with the exception of bodies acting in a judicial or legislative capacity'.[11] Hence, public authorities at all different levels were included, but only if they had responsibilities and held information relating to the environment. This left many possibilities for the public authorities to escape from the obligations of Directive 90/313/EEC by claiming that their responsibility was not related to the environment, but rather to, for example, transportation or housing.[12] However, Article 6 of the directive enlarged the field of application again by asking the Member States to ensure that information relating to the environment held by bodies with public responsibilities for the environment and under the control of public authorities was made available on the same terms and conditions. This was an attempt to counter the growing trend toward privatization of public functions.[13]

General principle – Any natural or legal person could make a request for information, so the right of access was not limited to citizens of the Member States or persons residing in the Member States. Nor did the applicant have to prove an interest or reason why he or she wanted to obtain the information. The Member States had to define the practical arrangements under which the information was effectively made available.[14]

Exceptions to the right of access – The Member States could include ten exceptions in their national legislation. Three of them could be considered as formal exceptions: 'A request of information may be refused where it would involve the supply of unfinished documents or data or internal communications, or where the request is manifestly unreasonable or formulated in a too general

---

Law (2005): footnote 27; William A. Wilcox, Jr., 'Access to Environmental Information in the United States and the United Kingdom', *Loyola of Los Angeles International and Comparative Law Review* 23 (2001): 137–138; Bernard Vanheusden, 'Vrije toegang tot milieu-informatie: een stand van zaken en de implicaties van het verdrag van Aarhus voor Europa en voor België', *Milieu- en Energierecht* (1999): 268 [in Dutch].

11. Article 2(b) of Directive 90/313/EEC.
12. Ralph E. Hallo, 'Directive 90/313/EEC on the Freedom of Access to Information on the Environment: Its Implementation and Implications', in *Access to Environmental Information in Europe: The Implementation and implications of directive 90/313/EEC*, ed. Ralph E. Hallo (London: Kluwer Law International, 1996), 10; Ralph E. Hallo, 'The Right of Access to Environmental Information: Directive 90/313/EEC and Other measures', in *Développements récents du droit européen de l'environnement: droits des citoyens et des associations dans le droit européen de l'environnement*, ed. Michel Paques & Patricia Van Pelt (Antwerp: Story-Scientia, 1998), 16–17; Ralph E. Hallo, *Public Access to Environmental Information: European Environment Agency's Experts Corner Number 1997/1* (Luxembourg: Office for Official Publications of the European Communities, 1997), 8–9.
13. Ralph E. Hallo, 'Directive 90/313/EEC on the Freedom of Access to Information on the Environment: Its Implementation and Implications', in *Access to Environmental Information in Europe: The Implementation and Implications of Directive 90/313/EEC*, ed. Ralph E. Hallo (London: Kluwer Law International, 1996), 10.
14. Article 3.1 of Directive 90/313/EEC.

manner'.[15] In addition, the Member States 'may provide for a request for information relating to the environment to be refused where it affects:

- the confidentiality of the proceedings of public authorities, international relations and national defence;
- public security;
- matters which are, or have been, sub judice, or under enquiry (including disciplinary enquiries), or which are the subject of preliminary investigation proceedings;
- commercial and industrial confidentiality, including intellectual property,
- the confidentiality of personal data and/or files;
- material supplied by a third party without that party being under a legal obligation to do so;
- material, the disclosure of which would make it more likely that the environment to which such material related would be damaged'.[16]

These interests had to be affected – not necessarily adversely, according to the text of the provision – but they did not have to be weighed against the public interest in disclosure of the information. The directive also did not stress that the exceptions had to be interpreted restrictively.

TIME LIMITS – The public authorities had to respond to a person requesting information as soon as possible and at the latest within two months.[17] Directive 90/313/EEC did not include a time limit for the delivery of the information if the request was accepted, nor any provisions on the form in which the access could be given or in which the information should be supplied. If the request was denied, the reasons for refusal had to be given.

CHARGES – A charge could be made for the supply of information, not exceeding a reasonable cost.[18] According to Vanheusden, this could include the cost of copying the information, but also the cost of the time and effort needed to look for the information.[19] The European Commission had suggested including that consultation on-site should be free of charge and the supply of copies should be at actual cost of reproduction,[20] but this was not maintained by the Council. In 1999, the ECJ made it very clear that the general principle of access should not be frustrated by the imposition of high costs: 'The term reasonable for the purposes of Article 5 of the Directive must be understood as meaning that it does not authorize Member States to pass on to those seeking information the entire amount

15. Article 3.3 of Directive 90/313/EEC.
16. Article 3.2. of Directive 90/313/EEC.
17. Article 3.4 of Directive 90/313/EEC.
18. Article 5 of Directive 90/313/EEC.
19. Bernard Vanheusden, 'Vrije toegang tot milieu-informatie: een stand van zaken en de implicaties van het verdrag van Aarhus voor Europa en voor België', *Milieu- en Energierecht* (1999): 2269–2270 [in Dutch].
20. Commission of the European Economic Community, *Proposal for a Council Directive on the Freedom of Access to Information on the environment*, COM (88) 484 final, 28 Nov. 1988.

of the costs, in particular indirect ones, actually incurred for the State budget in conducting an information search'.[21]

b. Dissemination by Public Authorities of their Own Motion

LIMITED OBLIGATIONS – Directive 90/313/EEC had very limited ambitions regarding the active dissemination of information relating to the environment by public authorities. In Article 7, the directive only required the Member States to 'take the necessary steps to provide general information to the public on the state of the environment by such means as the periodic publication of descriptive reports'. According to Hallo, it was clear that 'the Directive is primarily concerned with the passive right to information and does not pretend to address in any meaningful way the various aspects of active information'.[22] Directive 90/313/EEC was severely criticized for this lack of active commitment of the public authorities. Heyvaert stated that, in order to be able to effectively use their right of access, most people had to be made aware that they even had such a right, what information was available, and where they could find it. Moreover, the information that would be most useful to the public would require an active commitment of the public authorities for such activities as summaries and overviews or layperson's versions of technical and scientific reports.[23]

c. Follow-Up

EUROPEAN COMMISSION REPORT – Article 8 of Directive 90/313/EEC ordered reports on the experiences of the Member States by the end of 1996, which the European Commission had to compile into a general report. With some delay, due to the lateness of the national reports, the Commission presented its report on 29 June 2000.[24] This report showed the shortcomings of the directive in several areas: the limited interpretation of the definitions, the wide interpretations of the reasons for refusal, the time limits, the charges made for the supply of information, the

---

21. ECJ, Case C-217/97, *Commission of the European Communities v. Federal Republic of Germany*, 9 Sep. 1999, para. 48, *European Court Reports* 1999, I-5087. See also Maria Gavouneli, 'Access to Environmental Information: Delimitation of a Right', *Tulane Environmental Law Journal* 13 (2000): 316.
22. Ralph E. Hallo, 'Directive 90/313/EEC on the Freedom of Access to Information on the Environment: Its Implementation and Implications', in *Access to Environmental Information in Europe: The Implementation and implications of directive 90/313/EEC*, ed. Ralph E. Hallo (London: Kluwer Law International, 1996), 18.
23. Veerle Heyvaert, 'Wie wordt er wijzer van? Enkele bedenkingen bij het recht op toegang tot milieu-informatie', in *Ten Years of Access to Environmental Information in International, European and Belgian Law: Stock-Taking and Perspectives*, ed. Christine Larssen (Brussels: Bruylant, 2003), 59 [in Dutch].
24. Commission of the European Communities, *Report from the Commission to the Council and the European Parliament on the experience gained in the application of Council Directive 90/313/EEC of 7 June 1990 on freedom of access to information on the environment*, COM (2000) 400 final, 29 Jun. 2000.

review procedures, and the active supply of information. The Commission considered it desirable to replace Directive 90/313/EEC with a new directive, aiming to both correct the shortcomings of Directive 90/313/EEC and to align Community legislation with the Aarhus Convention.[25]

2. The Access Directive

a. Introduction

TIMELINE – Together with the report on Directive 90/313/EEC, the Commission presented a proposal for a new directive on public access to environmental information.[26] Discussions between the European Commission, the European Council, and the European Parliament took over two years, and the final directive was approved on 28 January 2003. The Access Directive had to be transposed by 14 February 2005.[27]

OBJECTIVE OF THE REVISION – The aim of the revision of Directive 90/313/EEC was three-fold: to correct the shortcomings of this directive; to pave the way towards ratification of the Aarhus Convention through the alignment of the proposal to the relevant provisions of the Convention; and to adapt the directive to developments in information technologies in order to 'reflect the changes in the way information is created, collected, stored and transmitted'.[28]

b. Objectives

TWOFOLD OBJECTIVE – The two objectives of the Access Directive follow the traditional distinction between the *passive* right of access upon request and the *active* dissemination of information by the public authorities of their own motion:

(a) 'to guarantee the right of access to environmental information held by or for public authorities and to set out the basic terms and conditions of, and practical arrangements for, its exercise; and

25. *Ibid.*, 10–12. See also Jean-Jacques Paradissis, *The Right to Access Environmental Information* (Saarbrucken: VDM Verlag, 2009), 54–55.
26. Commission of the European Communities, *Proposal for a Directive of the European Parliament and of the Council on public access to environmental information*, COM (2000) 402 final, 29 Jun. 2000.
27. A number of Member States did not make this deadline: Austria, France, Hungary, Ireland, Portugal, Belgium, Greece, Spain and Luxemburg. Austria and Ireland were convicted for non-transposition by the ECJ in 2007. The cases against Spain, Greece and Germany were removed from the Court's registries after these countries belatedly transposed the directive.
28. Commission of the European Communities, *Proposal for a Directive of the European Parliament and of the Council on public access to environmental information*, COM (2000) 402 final, 2, 29 Jun. 2000; Maguelonne Déjaent-Pons, 'Le Conseil de l'Europe et l'Accès à l' Information Environnementale: la Convention Européenne des Droits de l'Homme', in *Ten Years of Access to Environmental Information in International, European and Belgian Law: Stock-Taking and Perspectives*, ed. Christine Larssen (Brussels: Bruylant, 2003), 79 [in French].

(b) to ensure that, as a matter of course, environmental information is progressively made available and disseminated to the public in order to achieve the widest possible systematic availability and dissemination to the public of environmental information. To this end the use, in particular, of computer telecommunication and/or electronic technology, where available, shall be promoted'.[29]

PREPARATORY PROCESS: ACTIVE DISSEMINATION – The Council suggested making the original objective on active dissemination more proactive by changing the phrase 'ensuring that environmental information is made available and disseminated'[30] to 'promoting the widest possible systematic availability and dissemination to the public of environmental information',[31] but the European Commission and the European Parliament considered this to be a weaker obligation.[32] The final compromise cited above was reached during the conciliation procedure. To *promote* the availability of information requires an active effort of the Member State or the public authorities, but it does not guarantee anything about the result of this promotion, that is, whether information is actually available. *Ensuring* that information is available, however, does imply a certain result, a guarantee that information will be made available. From this point of view, the Commission and the Parliament were right to believe that their suggested wording was stronger.

PREPARATORY PROCESS: THIRD OBJECTIVE – The Committee on the Environment, Public Health and Consumer Policy of the European Parliament wanted to include a third objective: 'to establish the standard in relation to access to environmental information which must also apply to the institutions of the European Union either through the Regulation implementing Article 255 of the EC Treaty [the current Article 15 of the Treaty on the Functioning of the European Union] or through a

29. Article 1 of the Access Directive.
30. Commission of the European Communities, *Proposal for a Directive of the European Parliament and of the Council on public access to environmental information*, COM (2000) 402 final, 29 Jun. 2000.
31. Environment Working Party, *Preparation of the Council Environment on 18–19 December 2000. Proposal for a directive on public access to environmental information.* doc. nr 13709/00, 24 Nov. 2000, <http://register.consilium.europa.eu/pdf/en/00/st13/st13709.en00.pdf>, 22 Feb. 2010; Council of the European Union, *Common position adopted with a view to the adoption of a Directive of the European Parliament and of the Council on public access to environmental information and repealing Council Directive 90/313/EEC. Statement of the Council's Reasons*, doc. nr. 11878/01, 28 Jan. 2002, 15, <http://register.consilium.europa.eu/pdf/en/01/st11/st11878-re01ad01.en01.pdf>, 22 Feb. 2010.
32. Commission of the European Communities, *Communication to the European Parliament pursuant to the second subparagraph of Art. 251(2) of the EC Treaty concerning the common position of the Council on the adoption of a Directive of the European Parliament and of the Council on public access to environmental information*, SEC (2002) 103 final, 31 Jan. 2002; European Parliament, *Position adopted at second reading on 30 May 2002 with a view to the adoption of European Parliament and Council Directive 2002/ . . . /EC on public access to environmental information and repealing Council Directive 90/313/EEC, OJ C* 187E, 7 Aug. 2003.

specific Regulation'.[33] The Committee felt that the standard of access established in the Access Directive should equally apply to the EU institutions and that it was not acceptable that the institutions would apply a lower standard of access to environmental information than they required of the Member States. The plenary session of the European Parliament followed this argument and included this third objective into its amendments at the first reading.[34] However, the Commission and the Council did not accept this amendment because it fell 'clearly outside the scope of the directive which applies only to Member States',[35] and it was not taken up again in the second reading by the Parliament.

MINIMUM REQUIREMENTS – The Access Directive makes it clear that its disclosure obligations are not intended to dissuade Member States from providing more extensive disclosure regimes. In the words of Coppel, the directive imposes 'the minimum acceptable level of disclosure of environmental information: it does not set the bounds of permissible disclosure'.[36]

c. *Definitions*

OUTLINE – As the definitions of *environmental information* and *public authorities* were already discussed in Chapter 2,[37] this section will only address the definitions of information *held by or for* a public authority, the *applicant* and the *public*, in order to determine the full scope of the Access Directive.

HELD BY OR FOR – The directive provides a right of access to environmental information *held by or for* a public authority. This includes 'environmental information in its possession which has been produced or received' by the authority, or which is 'physically held by a natural or legal person on behalf of a public authority'.[38] Thus, the public authority cannot hide behind the fact that the information is not physically in its possession but is held elsewhere, as long as it is still held on its behalf.

33. Committee on the Environment, Public Health and Consumer Policy, *Report on the proposal for a directive of the European Parliament and of the Council on public access to environmental information.* doc. nr. A5-0074/2001, 28 Feb. 2001, <www.europarl.europa.eu/sidesSearch/sipadeMapUrl.do?PROG=REPORT&SORT_ORDER=D&REF_A=A5-2001-0074&L=EN#>, 22 Feb. 2010.
34. European Parliament, *Legislative resolution on the proposal for a Directive of the European Parliament and of the Council on public access to environmental information*, *OJ C* 343, 5 Dec. 2001, 177.
35. Commission of the European Communities, *Amended proposal for a directive of the European Parliament and of the Council on public access to environmental information*, COM (2001) 303 final, 6 Jun. 2001; Council of the European Union, *Common position adopted with a view to the adoption of a Directive of the European Parliament and of the Council on public access to environmental information and repealing Council Directive 90/313/EEC. Statement of the Council's Reasons*, doc. nr. 11878/01, 28 Jan. 2002, 11, <http://register.consilium.europa.eu/pdf/en/01/st11/st11878-re01ad01.en01.pdf>, 22 Feb. 2010.
36. Philip Coppel, 'Environmental Information: the New Regime', *Journal of Planning & Environment Law* (2005): 15–16.
37. Cf. *supra*, Ch. 2, s. III.C.2 and s. IV.B.1.
38. Article 2.3 of the Access Directive.

APPLICANT – The *applicant* can be any natural or legal person requesting environmental information. The Access Directive does not limit the right of access to environmental information to citizens of the Member States, natural persons residing in the Member States, or legal persons having their registered office in one of the Member States. Any person with any nationality or residing in any country can file a request to obtain environmental information from the public authorities in the EU Member States. The directive does not include, however, any reference to groups of natural or legal persons. This entails that environmental interest groups or any other groups without legal personality are not able to file a request for information. The members of these groups will have to do this in their own names.

PREPARATORY PROCESS: PUBLIC AUTHORITIES AS APPLICANTS – The Committee of Regions suggested adding the words 'including public authorities themselves' to the definition:

> It needs to be noted that in their leadership role some public authorities will themselves require environmental information to be made available from other organizations. This information will then, in due course, be used to provide information for the public. For example, it is important that local and regional government is able to gather information from other public authorities to enable the support of national government to meet objectives relating to climate protection.[39]

It is true that public authorities might also need access to environmental information themselves, but the definition of *applicant* should be sufficiently broad to cover this. In addition, the Committee of Regions was not just considering the need for public authorities to exercise their right of access, but it was also addressing the need for sharing between public authorities, which was not the purpose of the Access Directive.

PUBLIC – The Access Directive also contains a definition of the *public*: 'one or more natural or legal persons, and, in accordance with national legislation or practice, their associations, organizations or groups'.[40] This definition has been taken literally from the Aarhus Convention.[41] However, in the Aarhus Convention, the public is the beneficiary of both the active dissemination of information, and the right of access upon request, while in the Access Directive, the passive right of access is only directed towards applicants as defined above. This would entail that under the Aarhus Convention, interest groups without legal personality are able to file a request for information, at least if this is in accordance with national legislation, while under the Access Directive, this is not possible, as these interest groups are not applicants. This difference between both could cause problems for the Member States.

---

39. Committee of the Regions, *Opinion on the 'Proposal for a Directive of the European Parliament and the Council on Public Access to Environmental Information'*, *OJ C* 148, 18 May 2001, 9.
40. Article 2.6 of the Access Directive.
41. Article 2.4 of the Aarhus Convention.

d. Access to Environmental Information upon Request

OUTLINE – The first set of obligations on the Member States and the public authorities follow from the provisions on access to environmental information upon request of the Access Directive. These provisions contain the conditions under which public authorities have to make their information available to the applicants. These conditions involve time limits, possibilities to refuse access, charges, formats, and so on.

i. General Principle

PRINCIPLE – Article 3 of the Access Directive holds the general principle that 'public authorities are required, in accordance with the provisions of this Directive, to make available environmental information held by or for them to any applicant at his request and without his having to state an interest'.

NO STATEMENT OF INTEREST – The Commission suggested changing the wording of Directive 90/313/EEC – 'without his having to *prove* an interest' – because this had given rise to some difficulties. If this wording were taken literally, the applicant would not have to provide evidence for his or her interest, but that would not mean that he or she could not be required to state one. Therefore, and in order to be in alignment with the Aarhus Convention, the wording 'state an interest' was included.[42]

ii Time Limits

ONE MONTH – The requested information has to be made available to the applicant, having regard to any timeframe that he or she has specified, 'as soon as possible or, at the latest, within one month after the receipt by the public authority'.[43] This is in line with the time delay prescribed by the Aarhus Convention. At the same time, it corrects one of its shortcomings: the Convention did not indicate at what point the one-month time limit was to start. The Access Directive states clearly that the month starts at the receipt of the request by the public authority.[44]

DATA HAVE TO BE MADE AVAILABLE WITHIN THE TIME LIMIT – In comparison to Directive 90/313/EEC, two changes were made. The public authorities are not just required to *respond* to a request for access within the time limit, they also have to make the information available to the applicant. This avoids situations where a public authority responds within the time limit and promises to deliver the information but never does.[45] The second change moves up the deadline from two months after receiving the request to one month.

---

42. Commission of the European Communities, *Proposal for a Directive of the European Parliament and of the Council on public access to environmental information*, COM (2000) 402 final, 11, 29 Jun. 2000. See Jean-Jacques Paradissis, *The Right to Access Environmental Information* (Saarbrucken: VDM Verlag, 2009), 127–128.
43. Article 3.2 of the Access Directive.
44. *Ibid.*
45. Kurt Deketelaere & Frankie Schram, *Milieurechten van het publiek in de participatiedemocratie* (Bruges: Vanden Broele, 2002), 70–71 [in Dutch].

Shorter time limits – In some cases, the applicant may choose to state that he or she is requesting information for a specific purpose and specify a timeframe within which he or she needs the information. In such cases, public authorities should make reasonable efforts to enable the applicant to fulfil that purpose.[46] This is different from the Aarhus Convention and shows that the interest of a citizen can still play a role, even though it does not have to be stated. The specific interest does not matter for the decision whether or not access can be granted, but it does matter for determining the delay within which the information has to be made available.[47] The Commission actually wanted to include a separate paragraph regarding requests of information for specific purposes, mainly with public participation purposes in mind.[48] The Council did not agree: it considered that including this paragraph would create a 'two-speed system', whereby the applicants stating that the information was requested for a certain purpose were likely to receive more favourable treatment than those not having stated an interest. However, in order to take some account of the underlying idea behind the provision, the Council included the phrase 'having regard to any timescale specified by the applicant' at the beginning of the paragraph.[49]

Longer time limits – The time limit can also be extended to two months after the receipt of the request, if the volume or the complexity of the information is such that the one-month period cannot be complied with. The public authority has to inform the applicant of the extension and the reasons for it as soon as possible, and in any case before the end of the one-month period.

Preparatory process: shorter time limits – In its first reading, the European Parliament proposed an amendment reducing the time limit from one month to two weeks; and in its second reading, it introduced a time limit of fifteen working days.[50] The Committee on the Environment, Public Health and Consumer Policy

---

46. Commission of the European Communities, *Proposal for a Directive of the European Parliament and of the Council on public access to environmental information*, COM (2000) 402 final, 12, 29 Jun. 2000. See also Kurt Deketelaere & Frankie Schram, *Milieurechten van het publiek in de participatiedemocratie* (Bruges: Vanden Broele, 2002), 77 [in Dutch].
47. Kurt Deketelaere & Frankie Schram, *Milieurechten van het publiek in de participatiedemocratie* (Bruges: Vanden Broele, 2002), 77 [in Dutch].
48. Commission of the European Communities, *Proposal for a Directive of the European Parliament and of the Council on public access to environmental information*, COM (2000) 402 final, 12, 29 Jun. 2000.
49. Environment Working Party, *Proposal for a Directive on Public Access to Environmental Information. Outcome of Proceedings*, 18 Jan. 2001, doc. nr. 5451/01, <http://register.consilium.europa.eu/pdf/en/01/st05/st05451.en01.pdf>, 22 Feb. 2010; Commission of the European Communities, *Communication to the European Parliament pursuant to the second subparagraph of Art. 251(2) of the EC Treaty concerning the common position of the Council on the adoption of a Directive of the European Parliament and of the Council on public access to environmental information*, SEC (2002) 103 final, 31 Jan. 2002.
50. European Parliament, *Legislative resolution on the proposal for a Directive of the European Parliament and of the Council on public access to environmental information, OJ C* 343, 5 Dec. 2001, 177; European Parliament, *Position adopted at second reading on 30 May 2002 with a view to the adoption of European Parliament and Council Directive 2002/ . . . /EC on public*

had suggested the first change because '[t]he shorter time period is consistent with good practice; several Member States already require access to information in two weeks or less. A shorter time period is also consistent with the advances in information and communication technology which allow the retrieval and transmission of information to be much faster',[51] and it requested the second change because this time limit was consistent with Regulation 1049/2001 on public access to European Parliament, Council, and Commission documents.[52] The Committee of the Regions had also urged a more prompt service to the citizens,[53] while the Economic and Social Committee agreed in general that it was desirable to reduce response time but warned that this should not be 'at the expense of completeness and adequacy of information'. In addition, it felt that the possibility of extending the time limit to two months should not just be limited to voluminous and complex requests, but to 'all cases where such a situation may occur – duly justified and substantiated', such as when the legitimate interests whose confidentiality is safeguarded are difficult to assess or when different authorities are involved, such as legal or military authorities or international institutions.[54] Both attempts were unsuccessful, and the Commission and the Council stuck to the time limits that were originally planned and that were consistent with the Aarhus Convention.

iii. Requests Formulated in Too General a Manner

REQUEST FOR SPECIFICATION – If an applicant formulates his or her request in too general a manner, the public authority has to ask the applicant to specify the request as soon as possible and at least within the one-month time limit. The public authority has to assist the applicant in specifying his or her request, for example by providing information on the use of the public registers of information.[55] However, where they deem it appropriate, the public authorities can refuse a request. It is not clear from the text of the provision whether this refusal is only possible after the

---

access to environmental information and repealing Council Directive 90/313/EEC, *OJ C* 187E, 7 Aug. 2003.

51. Committee on the Environment, Public Health and Consumer Policy, *Report on the proposal for a directive of the European Parliament and of the Council on public access to environmental information*. doc. nr. A5-0074/2001, 28 Feb. 2001, <www.europarl.europa.eu/sides/getDoc.do?pubRef=-//EP//NONSGML+REPORT+A5-2001-0074+0+DOC+PDF+V0//EN&language=EN>, 22 Feb. 2010.
52. Committee on the Environment, Public Health and Consumer Policy, *Recommendation for second reading on the Council Common Position for adopting a European Parliament and Council directive on public access to environmental information and repealing Council Directive 90/313/EEC*, doc. nr. A5-0136/2002, 24 Apr. 2002, <www.europarl.europa.eu/sides/getDoc.do?pubRef=-//EP//TEXT+REPORT+A5-2002-0136+0+NOT+XML+V0//EN>, 22 Feb. 2010.
53. Committee of the Regions, *Opinion on the 'Proposal for a Directive of the European Parliament and the Council on Public Access to Environmental Information'*, *OJ C* 148, 18 May 2001, 9.
54. European Economic and Social Committee, *Opinion on the 'Proposal for a Directive of the European Parliament and of the Council on public access to environmental information'*, *OJ C* 116, 20 Apr. 2001, 43.
55. Article 3.3 of the Access Directive.

public authority has asked the applicant to specify his or her request and the applicant has failed to do so, or whether the public authority has the discretion to decide that there is no point in asking the applicant to rephrase his or her request and refuse it without asking for specification.

PREPARATORY PROCESS: REQUEST FOR SPECIFICATION – This provision was not included in the original proposal of the European Commission; it simply held that there was a possibility of refusal because a request was formulated in too general a manner, without requiring the public authorities to help the applicant specify his or her request first.[56] The European Parliament included a provision stating that '[w]here the request for information is formulated in too general a manner the public authority may seek clarification, but it must do so as soon as reasonably practicable after receipt of the request'.[57] The Council reformulated this to the final wording of Article 3.3, which was derived from Regulation 1049/2001.[58]

iv. Form or Format of the Information

FORMAT REQUESTED BY THE APPLICANT – If the applicant makes a request to obtain the environmental information in a specific form or format (including the form of copies), the public authority must deliver the information in that form, unless '(a) it is already available in another form or format . . . which is easily accessible by applicants; or (b) it is reasonable for the public authority to make it available in another form or format, in which case reasons shall be given for making it available in that form or format'.[59]

PREPARATORY PROCESS: READILY REPRODUCIBLE FORMAT – The European Parliament suggested making the process easier for the applicant by requiring public authorities to provide the environmental information in the form or format requested 'if it is readily reproducible in that form or format'.[60] This would put a considerable burden on the public authorities, as they could not refer the applicant to existing forms or formats, even though they might be easily available, but they would have to create the requested form or format if it was *readily reproducible*. This was also a reason for the European Commission not to accept this amendment: 'Public authorities should be granted a certain degree of flexibility to decide in

---

56. Commission of the European Communities, *Proposal for a Directive of the European Parliament and of the Council on public access to environmental information*, COM (2000) 402 final, 29 Jun. 2000.
57. European Parliament, *Legislative resolution on the proposal for a Directive of the European Parliament and of the Council on public access to environmental information, OJ C* 343, 5 Dec. 2001, 177.
58. Council of the European Union, *Common position adopted with a view to the adoption of a Directive of the European Parliament and of the Council on public access to environmental information and repealing Council Directive 90/313/EEC. Statement of the Council's Reasons*, doc. nr. 11878/01, 28 Jan. 2002, <http://register.consilium.europa.eu/pdf/en/01/st11/st11878-re01ad01.en01.pdf>, 22 Feb. 2010.
59. Article 2.4 of the Access Directive.
60. European Parliament, *Legislative resolution on the proposal for a Directive of the European Parliament and of the Council on public access to environmental information, OJ C* 343, 5 Dec. 2001, 177.

which form or format the information should be supplied. It has to be acknowledged that compliance with certain requests could be too burdensome for public authorities'.[61] The Council added to this that 'the applicants should be encouraged first to address the active information systems referred to in article 7'.[62]

PREPARATORY PROCESS: EASILY ACCESSIBLE – Another discussion between the Commission and the Council regarding this provision was whether the information should be already publicly available in a form or format that is easily accessible, or that is easily accessible *to the applicant*, as the Commission originally suggested. According to the Commission, the fact that information is easily accessible in general does not mean that it is easily accessible for the applicant who filed the request.[63] This could entail that the applicant may not be able to obtain the information, because even though it is claimed to be easily accessible, the applicant in fact cannot access it; for example, the information can be downloaded from a public authority's website, but the applicant does not have a computer and cannot go to a library to get access to the Internet. The European Parliament agreed with this reasoning and introduced an amendment to reinstate the Commission's proposal in its second reading.[64] The Council accepted the amendment during the conciliation procedure.

REFUSAL – If a public authority decides to invoke one of the reasons mentioned above not to provide the information in the requested form or format, it should give the applicant the reasons for this refusal within the general time limit of one month.

TECHNOLOGY – While the means and possibilities of the applicant should be taken into account, Article 3.4 also aims to make sure that the public authorities use modern technology as much as possible to make their environmental information available: 'Public authorities shall make all reasonable efforts to maintain environmental information held by or for them in forms or formats that are readily reproducible and accessible by computer telecommunications or by other electronic means'.

61. Commission of the European Communities, *Amended proposal for a directive of the European Parliament and of the Council on public access to environmental information*, COM (2001) 303 final, 6 Jun. 2001.

62. Council of the European Union, *Common position adopted with a view to the adoption of a Directive of the European Parliament and of the Council on public access to environmental information and repealing Council Directive 90/313/EEC. Statement of the Council's Reasons*, doc. nr. 11878/01, 28 Jan. 2002, 13, <http://register.consilium.europa.eu/pdf/en/01/st11/st11878-re01ad01.en01.pdf>, 22 Feb. 2010.

63. Commission of the European Communities, *Communication to the European Parliament pursuant to the second subparagraph of Art. 251(2) of the EC Treaty concerning the common position of the Council on the adoption of a Directive of the European Parliament and of the Council on public access to environmental information*, SEC (2002) 103 final, 7–8, 31 Jan. 2002.

64. European Parliament, *Position adopted at second reading on 30 May 2002 with a view to the adoption of European Parliament and Council Directive 2002/ . . . /EC on public access to environmental information and repealing Council Directive 90/313/EEC*, *OJ C* 187E, 7 Aug. 2003.

v. Facilitating Measures

ARTICLE 3.5 ACCESS DIRECTIVE – The European institutions recognized that merely giving citizens a right of access is not enough to make them use it to their fullest benefit. Measures to facilitate the exercise of the right of access are also needed. Article 3.5 of the Access Directive provides for a number of such measures. It requires the Member States to ensure that:

(a) officials are required to support the public in seeking access to information;
(b) lists of public authorities are publicly accessible; and
(c) the practical arrangements are defined for ensuring that the right of access to environmental information can be effectively exercised, such as
   - the designation of information officers;
   - the establishment and maintenance of facilities for the examination of the information required;
   - registers or lists of the environmental information held by public authorities or information points, with clear indications of where such information can be found.

In addition, the Member States have to ensure that public authorities adequately inform the public of the rights they enjoy as a result of the Access Directive and, to an appropriate extent, provide information, guidance, and advice to this end.[65]

ACTIVE V. PASSIVE – This Article implements the requirements of Article 5.2(b) of the Aarhus Convention. The Convention considered such practical arrangements as part of the active dissemination of information, while the Access Directive relates them to passive access upon request. This could show that it may not always be easy to determine where access upon request stops and active dissemination of information starts. On the one hand, the Member States are providing information of their own motion by making available lists of public authorities and of information. On the other hand, the availability of these lists could also be seen as inherent to promoting access upon request, as they provide the practical starting point for the applicant.

INDICATIVE LIST – The list of practical arrangements is only indicative, leaving it up to the Member States, in accordance with the subsidiarity principle, to determine which arrangements they want to put in place.[66] Originally, the European Commission and Council included the lists of public authorities in the indicative list of possible practical arrangements rather than as a separate category,[67] while the European Parliament wanted to make the entire Article 3.5 compulsory,

65. Article 3.5 of the Access Directive.
66. Commission of the European Communities, *Proposal for a Directive of the European Parliament and of the Council on public access to environmental information*, COM (2000) 402 final, 13, 29 Jun. 2000.
67. *Ibid.*

including the practical arrangements mentioned under (c).[68] They reached this compromise, inspired by the wording of the Aarhus Convention.

vi. Exceptions to the Right of Access

Two types of exceptions – Article 4 of the Access Directive contains the reasons why a public authority can refuse a request for environmental information. As in the Aarhus Convention, there are two types of exceptions: the first group could be considered to be 'formal exceptions', while the second group is based on the content of the information. None of the exceptions are mandatory, so the Member States can choose which exceptions they decide to implement in their national legislation.

Formal exceptions – First, Member States may provide for a request for information to be refused if:

(a) The information requested is not held by or for the public authority to which the request is addressed. In such a case, where that public authority is aware that the information is held by or for another public authority, it shall, as soon as possible, transfer the request to that other authority and inform the applicant accordingly or inform the applicant of the public authority to which it believes it is possible to apply for the information requested. The Access Directive follows the example of the Aarhus Convention and leaves the choice up to the public authorities whether to pass on the request or to inform the applicant that he or she needs to file his or her request with another public authority. The Commission's original proposal only held the first option: the public authority had to transfer the request and inform the applicant accordingly.[69] The Environment Working Party of the Council added the second option, upon request of the United Kingdom delegation, in order to be consistent with the Aarhus Convention.[70]

(b) The request is manifestly unreasonable. According to the Commission, manifestly unreasonable requests would 'include those, variously described in national legal systems as vexatious or amounting to an *abus de droit*. Moreover, compliance with certain requests could involve the public authority in disproportionate cost or effort or would obstruct or significantly interfere with the normal course of its activities. Authorities

68. European Parliament, *Legislative resolution on the proposal for a Directive of the European Parliament and of the Council on public access to environmental information*, *OJ C* 343, 5 Dec. 2001, 177.

69. Commission of the European Communities, *Proposal for a Directive of the European Parliament and of the Council on public access to environmental information*, COM (2000) 402 final, 13, 29 Jun. 2000.

70. Environment Working Party, *Proposal for a Directive on Public Access to Environmental Information. Outcome of Proceedings*, doc. nr. 5451/01, <http://register.consilium.europa.eu/pdf/en/01/st05/st05451.en01.pdf>>, 22 Feb. 2010.

should be able to refuse access in such cases in order to ensure their proper functioning'.[71]

(c) The request is formulated in too general a manner, taking into account Article 3(3). As was mentioned earlier, if the public authority considers that an applicant's request is formulated in too general a manner, the public authority first has to ask the applicant to specify his or her request and assist him or her in doing so. However, if it deems it appropriate, the public authority can refuse the request.

(d) The request concerns material in the course of completion or unfinished documents or data. In this case, the public authority has to state the name of the authority preparing the material and the estimated time needed for completion. The Commission emphasized that public authorities should have the necessary space to think in private, and, following the Aarhus Convention, it provided an exception for material in the course of completion.[72] The Presidency of the Council added *unfinished documents or data* to the exception, with a view to reintegrate these terms from Directive 90/313/EEC.[73] However, one could wonder what the difference is between material in the course of completion and unfinished documents, so the added value of restating this is not clear.

(e) The request concerns internal communications, taking into account the public interest served by disclosure. The Commission's 'space to think in private' argument also applies to internal communications. The European Parliament proposed deleting this exception in both its readings, but the Commission and the Council stood their ground and the exception was kept.[74,75]

Optional character of the exceptions – Even if the Member States decide to include all of these exceptions, it is important to remember that the public authorities can refuse a request based on one of the reasons above, but they do not have to. Thus, even if a request would be manifestly unreasonable or would refer to internal

71. Commission of the European Communities, *Proposal for a Directive of the European Parliament and of the Council on public access to environmental information*, COM (2000) 402 final, 13, 29 Jun. 2000. See Jean-Jacques Paradissis, *The Right to Access Environmental Information* (Saarbrucken: VDM Verlag, 2009), 149–152.

72. *Ibid.*

73. Environment Working Party, *Preparation of the Council Environment on 18–19 December 2000. Proposal for a Directive on Public Access to Environmental Information,* doc.nr 13709/00, 24 Nov. 2000, <http://register.consilium.europa.eu/pdf/en/00/st13/st13709.en00.pdf>, 22 Feb. 2010.

74. Article 4.1 of the Access Directive.

75. European Parliament, *Legislative resolution on the proposal for a Directive of the European Parliament and of the Council on public access to environmental information, OJ C* 343, 5 Dec. 2001, 177; European Parliament, *Position adopted at second reading on 30 May 2002 with a view to the adoption of European Parliament and Council Directive 2002/ . . . /EC on public access to environmental information and repealing Council Directive 90/313/EEC, OJ C* 187E, 7 Aug. 2003, 118.

communications, it is at the public authority's discretion as to whether or not it wants to invoke the exception.

EXCEPTIONS BASED ON THE CONTENT – Article 4.2 holds a second group of eight exceptions, taken almost literally from the Aarhus Convention. The Member States may 'provide for a request for environmental information to be refused if disclosure of the information would adversely affect':

(a) The confidentiality of the proceedings of public authorities, where such confidentiality is provided for by law. The Council did not want to restrict this exception to confidentiality that was *provided for by law*,[76] but the Commission agreed with the Parliament to include this, in alignment with the Aarhus Convention. While the Aarhus Convention addressed confidentiality provided for by *national* law,[77] the Access Directive refers to law in general, thereby incorporating both national and Community law.
(b) International relations, public security or national defence. The European Parliament wanted to restrict the confidentiality of international relations to cases of *vital interest*,[78] but the Commission could not accept this, because it deviated from the Aarhus Convention.[79] However, deviating from the Aarhus Convention by restricting the scope of the exceptions is possible, so the concerns of the Commission and the Council were not entirely valid.
(c) The course of justice, the ability of any person to receive a fair trial, or the ability of a public authority to conduct an enquiry of a criminal or disciplinary nature.[80]
(d) The confidentiality of commercial or industrial information where such confidentiality is provided for by national or Community law to protect a legitimate economic interest, including the public interest in maintaining statistical confidentiality and tax secrecy. The protection of statistical

76. Council of the European Union, *Common position adopted with a view to the adoption of a Directive of the European Parliament and of the Council on public access to environmental information and repealing Council Directive 90/313/EEC. Statement of the Council's Reasons*, doc. nr. 11878/01, 28 Jan. 2002, 9, <http://register.consilium.europa.eu/pdf/en/01/st11/st11878-re01ad01.en01.pdf>, 22 Feb. 2010. See Jean-Jacques Paradissis, *The Right to Access Environmental Information* (Saarbrucken: VDM Verlag, 2009), 158.

77. Article 4.4.(a) of the Aarhus Convention.

78. European Parliament, *Legislative resolution on the proposal for a Directive of the European Parliament and of the Council on public access to environmental information, OJ C* 343, 5 Dec. 2001, 177; European Parliament, *Position adopted at second reading on 30 May 2002 with a view to the adoption of European Parliament and Council Directive 2002/ . . . /EC on public access to environmental information and repealing Council Directive 90/313/EEC, OJ C* 187E, 118.

79. Commission of the European Communities, *Opinion pursuant to Art. 251(2), third subparagraph, point (c) of the EC Treaty, on the European Parliament's amendments to the Council's common position regarding the proposal for a Directive of the European Parliament and of the Council on public access to environmental information*, COM (2002) 498 final, 6, 5 Sep. 2002.

80. See Jean-Jacques Paradissis, *The Right to Access Environmental Information* (Saarbrucken: VDM Verlag, 2009), 166–170.

confidentiality and tax secrecy was not included in the Aarhus Convention, but the Council included this to clarify that this can also be part of commercial or industrial confidentiality.[81]

(e) Intellectual property rights. The European Parliament suggested removing this exception, but the Commission and the Council did not accept this. In general, the Commission could not 'accept the parts of the amendments which seek to depart from the exhaustive list of exceptions provided for in the Aarhus Convention', so this was also the case for the exception on IPR. The Commission made it sound as if the list of exceptions of the Aarhus Convention had to be completely implemented by the signatories to the Convention. However, while the list of exceptions of the Aarhus Convention is limitative, entailing that no other exceptions can be included, it is not mandatory. The Parties to the Convention can choose not to implement some of the exceptions.[82]

(f) The confidentiality of personal data and/or files relating to a natural person where that person has not consented to the disclosure of the information to the public, where such confidentiality is provided for by national or Community law. The Commission's proposal referred explicitly to Directive 95/46/EC on the processing of personal data,[83] for the sake of clarity and legal certainty, but it was overruled by the Council and the

81. Council of the European Union, *Common position adopted with a view to the adoption of a Directive of the European Parliament and of the Council on public access to environmental information and repealing Council Directive 90/313/EEC. Statement of the Council's Reasons*, doc. nr. 11878/01, 28 Jan. 2002, 16, <http://register.consilium.europa.eu/pdf/en/01/st11/st11878-re01ad01.en01.pdf>, 22 Feb. 2010.

82. One could argue that not implementing the full list of exceptions of the Aarhus Convention in the Access Directive would complicate the implementation by the Member States. In directly implementing the Aarhus Convention, the Member States can choose to include or exclude an exception. However, if the Access Directive would take one or more exceptions from the list, the Member States could not implement it, and they would be breaching Community law if they did so. Another interesting point concerns the position of the European Commission during the negotiations on the INSPIRE Directive. As will be discussed later on, the Commission changed its view on including IPR as a reason for refusal to grant access to spatial data under the INSPIRE Directive. In the negotiations on that directive, the Commission actually pushed for deletion of that exception, together with the European Parliament (cf. *infra*, s. II.C. 3.b)

83. Commission of the European Communities, *Proposal for a Directive of the European Parliament and of the Council on public access to environmental information*, COM (2000) 402 final, 14, 29 Jun. 2000. The Commission clarified that: 'Within this framework, if the public authority to which a request has been addressed considers that the public interest served by the disclosure outweighs the confidentiality of personal data, it must inform the competent national data protection supervisory authority, established by article 28 of Directive 95/46/EC, of the request received and of its intention of disclosing personal data to third parties. This referral to the national data protection authorities would of course only take place in cases where the public authority involved decided that there are grounds for granting access to environmental information including personal data. In other words, no referral would be necessary in cases where the public authority decided that the exception clearly applied, or where personal data can be separated out, and access to the remainder of the information can be given'.

European Parliament, which preferred to maintain the wording of the Aarhus Convention.

(g) The interests or protection of any person who supplied the information requested on a voluntary basis without being under, or capable of being put under, a legal obligation to do so, unless that person has consented to the release of the information concerned;

(h) The protection of the environment to which such information relates, such as the location of rare species.

ADVERSE EFFECT ON THE PROTECTED INTEREST – This second group of reasons for refusal can only be invoked if disclosure of the information would *adversely affect* the interests that are protected. Moreover, they have to be interpreted in a restrictive way, taking into account for each particular case the public interest served by disclosure. In every case, this public interest in disclosure has to be weighed against the interest served by the refusal. This is in line with the Aarhus Convention. In the case of the confidentiality of personal data, an additional safeguard is added: even if the public interest is considered of higher importance and environmental information containing personal data is made available, the requirements of Directive 95/46 on the processing of personal data have to be complied with. In any case, access cannot be refused because of the protection of an interest mentioned in Article 4.2. of the Access Directive if the disclosure of the information is required under another directive.[84]

PREPARATORY PROCESS: EMISSIONS – Information on emissions was a major point of discussion between the Commission, the Council, and the Parliament. Under the Aarhus Convention, information on emissions was considered to be a special category under the exception of confidentiality of commercial and industrial information. Information on emissions that is relevant for the protection of the environment has to be disclosed, even if it would qualify as confidential commercial or industrial information. The European Commission included this in its proposal,[85] but the European Parliament felt that this specific mention of emissions should apply to all exceptions, 'since the adverse effect of emissions is not dependent on the nature of the activities which give rise to them, whether it be a commercial or

84. ECJ, Case C-552/07, *Commune de Sausheim v. Pierre Azelvandre*, 17 Feb. 2009, *OJ C* 90/5, 18 Apr. 2009. See also Jean-Jacques Paradissis, *The Right to Access Environmental Information* (Saarbrucken: VDM Verlag, 2009), 63–64.

85. The Commission referred to information on emissions, discharges or other releases into the environment which are subject to provisions of Community legislation, because the definition of emissions under Community law was more narrow than the one from the Aarhus Convention. In the terminology of the Aarhus Convention, emission means 'all releases', while the Community meaning only concerns emissions into the air (Environment Working Party, *Preparation of the Council Environment on 18–19 December 2000. Proposal for a Directive on Public Access to environmental information*, doc. nr 13709/00, 24 Nov. 2000, <http://register.consilium.europa.eu/pdf/en/00/st13/st13709.en00.pdf>, 22 Feb. 2010. However, the suggestion of the Council to keep in line with the wording of the Aarhus Convention was accepted during the conciliation procedure.

other activity'.[86] The Council opposed this, but in the final stage of the conciliation procedure, a compromise was reached: the exceptions mentioned in subparagraphs (a), (d), (f), (g), and (h) cannot be invoked if the request relates to information on emissions into the environment. As Paradissis states it, in these cases, the balance between the different interests has already been struck by the legislator.[87]

PREPARATORY PROCESS: LIST OF CRITERIA – Whatever exceptions the Member States decide to implement in their national legislation, they may draw up a publicly accessible list of criteria on the basis of which public authorities can decide how to handle requests, that is, how they apply these exceptions.[88] The Parliament had introduced this provision as mandatory: if the Member States decided to implement exceptions, they were obliged to draw up a list of criteria and to submit it to the Commission for approval. This should help the public authorities to make a decision and to prevent arbitrary decisions.[89] The Commission did not agree with this, stating that, 'It is up to the courts and not to the Commission to give an interpretation of the exceptions provided for in the proposal',[90] and that it would be unduly burdensome for the public authorities. In addition, it went beyond the Aarhus Convention.[91] The Council had the same opinion: the requirement to make such lists 'would create new bureaucracy and . . . it is up to each administration, under the control of the courts, to interpret the exceptions in accordance with the national laws that will implement the Directive'.[92] The final compromise kept the provision but removed its mandatory character. The Member States can decide

86. Committee on the Environment, Public Health and Consumer Policy, *Report on the proposal for a directive of the European Parliament and of the Council on public access to environmental information*, doc. nr. A5-0074/2001, 28 Feb. 2001, <www.europarl.europa.eu/sides/getDoc.do?pubRef=-//EP//NONSGML+REPORT+A5-2001-0074+0+DOC+PDF+V0//EN&language=EN>, 22 Feb. 2010.
87. Jean-Jacques Paradissis, *The Right to Access Environmental Information* (Saarbrucken: VDM Verlag, 2009), 187.
88. Article 4.3 of the Access Directive.
89. Committee on the Environment, Public Health and Consumer Policy, *Report on the proposal for a directive of the European Parliament and of the Council on public access to environmental information*, doc. nr. A5-0074/2001, 28 Feb. 2001, 21, <www.europarl.europa.eu/sidesSearch/sipadeMapUrl.do?PROG=REPORT&SORT_ORDER=D&REF_A=A5-2001-0074&L=EN#>, 22 Feb. 2010.
90. Commission of the European Communities, *Amended proposal for a directive of the European Parliament and of the Council on public access to environmental information*, COM (2001) 303 final, 11, 6 Jun. 2001.
91. Commission of the European Communities, *Opinion pursuant to Art. 251(2), third subparagraph, point (c) of the EC Treaty, on the European Parliament's amendments to the Council's common position regarding the proposal for a Directive of the European Parliament and of the Council on public access to environmental information*, COM (2002) 498 final, 5 Sep. 2002.
92. Council of the European Union, *Common position adopted with a view to the adoption of a Directive of the European Parliament and of the Council on public access to environmental information and repealing Council Directive 90/313/EEC. Statement of the Council's Reasons*, doc. nr. 11878/01, 28 Jan. 2002, 13, <http://register.consilium.europa.eu/pdf/en/01/st11/st11878-re01ad01.en01.pdf>, 22 Feb. 2010.

freely if they want to create a list of criteria for the application of the exceptions that they have implemented.

SEPARATION OF INFORMATION – Article 4 contains two more provisions to bring the Access Directive in line with the Aarhus Convention. Article 4.4 states that, when it is possible to separate the environmental information falling under one of the exceptions from the rest of the information requested, public authorities have to make the rest of the information available, for example, by deleting or marking parts of a document.

NOTIFICATION OF REFUSAL – Article 4.5 follows the Aarhus Convention in requiring that a refusal to make information available has to be notified to the applicant in writing or electronically if the request was in writing or if the applicant requests a written reply. Obviously, this notification has to be done within the time limit. It has to state the reasons for the refusal and include information on the review procedure foreseen in the directive.

vii. Charges

ARTICLE 5 ACCESS DIRECTIVE – Article 5 of the Access Directive contains the rules under which access to information upon request can be charged for:

1. Access to any public registers or lists established and maintained as mentioned in Article 3(5) and examination *in situ* of the information requested shall be free of charge.
2. Public authorities may make a charge for supplying any environmental information but such charge shall not exceed a reasonable amount.
3. Where charges are made, public authorities shall publish and make available to applicants a schedule of such charges as well as information on the circumstances in which a charge may be levied or waived.

SUMMARY – Access to the registers[93] has to be free of charge. The same applies to examination of the information on site. If an applicant wants to obtain a copy of the information that he or she requested, the public authority can make a charge, but this charge cannot exceed a reasonable amount. In case such charges are made, they have to be accompanied by some extra information: a schedule of charges has to be made available, entailing that the public authority should in principle decide on its charges before making information available and in a general manner.

PREPARATORY PROCESS: ADVANCE PAYMENT – The Commission proposed including that the supply of information could not be made subject to the advance payment of a charge.[94] The Council did not agree because this was not mentioned in

93. These registers could be said to form metadata for the documents involved, as they include information on the public bodies holding the environmental information and on the location where it can be found.
94. Commission of the European Communities, *Proposal for a Directive of the European Parliament and of the Council on public access to environmental information*, COM (2000) 402 final, 29 Jun. 2000.

the Aarhus Convention[95] and suggested making it into an option, 'as applicants often forget or refuse to take the information after the effort of searching has been made'.[96] In the conciliation phase, any reference to advance payment disappeared, except in the recitals. According to Deketelaere and Schram, the fact that the Access Directive does not mention the possibility of asking for an advance payment, contrary to the Aarhus Convention, strengthens the interpretation that this is no longer possible.[97]

PREPARATORY PROCESS: REASONABLE CHARGES – Next to elaborate discussions on a logical order of Article 5, the main point of discussion was the reasonable character of the charges for the supply of information and what charges could be included. The European Parliament deemed it necessary that more detailed provisions be included and proposed an amendment that any charge 'should be reasonable, shall not exceed the actual cost of reproducing the material requested, and shall not include the costs of staff time spent on searches'. After the refusal of the Commission to include this provision and the sceptical reaction of the Council,[98] the Parliament dropped part of the amendment during the second reading, keeping only the statement that the charges should 'not exceed the actual cost of reproducing the material requested'.[99] However, the Commission still considered this to be beyond the Aarhus Convention and unduly burdensome for public authorities. After several attempts to find a compromise in the conciliation phase, the final version of Article 5 only demands that any charge does not exceed a reasonable amount, and the other concerns of the Parliament were moved to Recital 18:

> Public authorities should be able to make a charge for supplying environmental information but such a charge should be reasonable. This implies that, as a general rule, charges may not exceed actual costs of producing the

95. Environment Working Party, *Preparation of the Council Environment on 18–19 December 2000. Proposal for a directive on public access to environmental information*, doc. nr 13709/00, 24 Nov. 2000, <http://register.consilium.europa.eu/pdf/en/00/st13/st13709.en00.pdf>, 22 Feb. 2010.
96. Council of the European Union, *Common position adopted with a view to the adoption of a Directive of the European Parliament and of the Council on public access to environmental information and repealing Council Directive 90/313/EEC. Statement of the Council's Reasons*, doc. nr. 11878/01, 28 Jan. 2002, 17, <http://register.consilium.europa.eu/pdf/en/01/st11/st11878-re01ad01.en01.pdf> 22 Feb. 2010.
97. Kurt Deketelaere & Frankie Schram, *Milieurechten van het publiek in de participatiedemocratie* (Bruges: Vanden Broele, 2002), 132 [in Dutch].
98. Council of the European Union, *Common position adopted with a view to the adoption of a Directive of the European Parliament and of the Council on public access to environmental information and repealing Council Directive 90/313/EEC. Statement of the Council's Reasons*, doc. nr. 11878/01, 28 Jan. 2002, 14, <http://register.consilium.europa.eu/pdf/en/01/st11/st11878-re01ad01.en01.pdf>, 22 Feb. 2010.
99. European Parliament, *Position adopted at second reading on 30 May 2002 with a view to the adoption of European Parliament and Council Directive 2002/ . . . /EC on public access to environmental information and repealing Council Directive 90/313/EEC*, *OJ C* 187E, 7 Aug. 2003, 118.

material in question. Instances where advance payment will be required should be limited. In particular cases, where public authorities make available environmental information on a commercial basis, and where this is necessary in order to guarantee the continuation of collecting and publishing such information, a market-based charge is considered to be reasonable; an advance payment may be required.

COMMERCIAL BASIS – Hence, a *reasonable amount* can even include charges at a market level, if the public authorities operate on a commercial basis. However, what we should understand by *commercial* is not clear. Does it entail that the public authority intends to make a profit, or that it charges more than the actual cost? The underlying argument for this principle was that some public authorities have to provide for at least a part of their own funding, and they do this by *selling* their information to the public. If their activities are unsustainable without this commercial activity, they can charge market prices.

PREPARATORY PROCESS: EXCEPTIONS – The European Parliament also proposed that charges could not be made for requests for information for educational purposes, and that the public authorities could charge for reproducing information but not for searching for or compiling information.[100] The Commission refused to include the exception for educational purposes, as 'the Commission proposal already enables Member States to define . . . the cases when charges could be waived. This is after all a proposal for a framework directive. In accordance with the subsidiarity principle, Member States should be granted a certain degree of flexibility to transpose the directive into national law'.[101] Also, the Council was not enthusiastic: it could not accept 'the idea that information for educational purpose shall be free of charge: searches may [be] very time-consuming and costly, [and] freedom of charge may give rise to frivolous requests for information and the notion of education is very vague'.[102] During the second reading, the Parliament dropped this part of the amendment.

e. Access to Justice

LIMITED OBJECTIVE – Even though Article 6 of the Access Directive bears the general title *Access to Justice*, it does not intend to be the implementation of the third pillar of the Aarhus Convention, as access to justice is not a competence of the EU.

100. European Parliament, *Legislative resolution on the proposal for a Directive of the European Parliament and of the Council on public access to environmental information*, *OJ C* 343, 5 Dec. 2001, 177.

101. Commission of the European Communities, *Amended proposal for a directive of the European Parliament and of the Council on public access to environmental information*, COM (2001) 303 final, 11–12, 6 Jun. 2001.

102. Council of the European Union, *Common position adopted with a view to the adoption of a Directive of the European Parliament and of the Council on public access to environmental information and repealing Council Directive 90/313/EEC. Statement of the Council's Reasons*, doc. nr. 11878/01, 28 Jan. 2002, 14, <http://register.consilium.europa.eu/pdf/en/01/st11/st11878-re01ad01.en01.pdf>, 22 Feb. 2010.

It only addresses the means of redress concerning access to environmental information.

MEANS OF REDRESS – Any applicant who feels that his or her request for environmental information has been ignored, wrongfully refused (whether in full or in part), inadequately answered, or otherwise not dealt with in accordance with the Access Directive should have access to a procedure in which the acts or the omissions of the public authority concerned can be reconsidered by that or another public authority or reviewed administratively by an independent and impartial body established by law. Such procedures have to be expeditious and either free of charge or inexpensive. In addition, the applicant also should have access to a review procedure before a court of law or another impartial body established by law.

TWO POSSIBLE MEANS – Hence, two options for seeking redress should be open to the applicant. On the one hand, he or she should be able to demand that his or her request be reconsidered by a public authority, which may be, but is not necessarily the same public authority as the one that originally handled the request. On the other hand, the applicant should also have access to a procedure before a court of law or another body established by law that can review the acts or omissions of the public authority.[103]

FAILURE TO REPLY – Discussion arose between the European Parliament, the Commission, and the Council about the consequences of the failure of a public authority to reply to a request within the time limit of one month. Should this be seen as a negative decision against which the applicant can seek review by the court or the designated administrative body? The European Parliament saw it differently. In its first reading, it introduced an amendment stating that failure to provide the requested information within the relevant time limit or a reasoned refusal should be considered a positive response indicating that the information could be supplied. The Commission and the Council felt that there was no need to include this, as 'the proposal already foresees applicants may make use of the redress mechanisms provided for in cases of failure to reply within the deadline'.[104] Indeed, Article 6 includes the situation of a request being ignored or not dealt with, so any legal fiction of considering a failure to reply as either a positive or a negative decision in order to have a decision that can give rise to means of redress is unnecessary.

---

103. Commission of the European Communities, *Proposal for a Directive of the European Parliament and of the Council on public access to environmental information*, COM (2000) 402 final, 15, 29 Jun. 2000. See also Jean-Jacques Paradissis, *The Right to Access Environmental Information* (Saarbrucken: VDM Verlag, 2009), 192–197.

104. Commission of the European Communities, *Amended proposal for a directive of the European Parliament and of the Council on public access to environmental information*, COM (2001) 303 final, 11, 6 Jun. 2001; Council of the European Union, *Common position adopted with a view to the adoption of a Directive of the European Parliament and of the Council on public access to environmental information and repealing Council Directive 90/313/EEC. Statement of the Council's Reasons*, doc. nr. 11878/01, 28 Jan. 2002, 13, <http://register.consilium.europa.eu/pdf/en/01/st11/st11878-re01ad01.en01.pdf>, 22 Feb. 2010.

f. Dissemination of Environmental Information

GENERAL OBLIGATION – Contrary to Directive 90/313/EEC, and as a follow-up of the Aarhus Convention, the Access Directive contains an elaborate article on the active dissemination of environmental information by the Member States and the public authorities.[105] The first subparagraph of Article 7 contains a general obligation for the Member States to organize the dissemination of environmental information, while the next subparagraphs mostly refer to specific types of information or specific situations. Of course, for any dissemination, the reasons for refusal of a request for information can also apply.[106] If the information cannot be made available to an applicant, it certainly cannot be disseminated to the general public. As is the case for access upon request, the interests that are protected will have to be weighed against the public interest in disclosure.

ORGANIZATION OF ENVIRONMENTAL INFORMATION – Article 7.1 requires the Member States to 'take the necessary measures to ensure that the public authorities organize the environmental information which is relevant to their functions and which is held by or for them, with a view to its active and systematic dissemination to the public, in particular by means of computer telecommunications and/or electronic technology, where available'. While the Aarhus Convention required the public authorities to *possess* information relevant to their functions, the Access Directive makes them *organize* this information. Whereas *possessing* implies that the public authorities have to make sure that they have information, that is, they have to collect it, *organizing* seems to be limited to public authorities' preparing the information that they already have for dissemination. Public authorities only have to disseminate environmental information that is relevant to their functions, not environmental information in general. For any other information that they hold, citizens can still file a request for access.

ELECTRONIC INFORMATION – Public authorities are encouraged to disseminate information as much as possible by means of information technology, but the directive does not oblige public authorities to convert information into an electronic format: 'The information made available by means of computer telecommunication and/or electronic technology need not include information collected before the entry into force of this Directive unless it is already available in electronic form'.[107]

ELECTRONIC DATABASES – However, the Member States do need to ensure that environmental information progressively becomes available in electronic databases that are easily accessible through public telecommunication networks. The Commission clarifies that:

> If public authorities make increased use of new means of communication, like the Internet, and make environmental information accessible through those means, there should be a decrease in requests directly addressed to the

105. For a critical view on Art. 7 of the Access Directive, see Jean-Jacques Paradissis, *The Right to Access Environmental Information* (Saarbrucken: VDM Verlag, 2009), 107–108.
106. Article 7.5 of the Access Directive.
107. Article 7.1 of the Access Directive.

authorities because the public is able to have immediate access to the information they are seeking. On the other hand and as already highlighted, it should also be borne in mind that such a proactive approach contributes to raising public awareness in environmental matters. Public authorities may in some cases be confronted with increasingly complex follow-up requests for environmental information which may give rise to practical difficulties in handling them.[108]

INFORMATION TO BE INCLUDED IN THE DATABASES – Article 7.2 contains a list of information that has to be made available and disseminated at a minimum:

(a) texts of international treaties, conventions or agreements, and of Community, national, regional or local legislation, on the environment or relating to it;
(b) policies, plans and programmes relating to the environment;
(c) progress reports on the implementation of the items referred to in (a) and (b) when prepared or held in electronic form by public authorities;
(d) the reports on the state of the environment referred to in paragraph 3;
(e) data or summaries of data derived from the monitoring of activities affecting, or likely to affect, the environment;
(f) authorisations with a significant impact on the environment and environmental agreements or a reference to the place where such information can be requested or found in the framework of Article 3;
(g) environmental impact studies and risk assessments concerning the environmental elements referred to in Article 2(1)(a) or a reference to the place where the information can be requested or found in the framework of Article 3.

NON-EXHAUSTIVE LIST – This list is not exhaustive. The Member States can add other types of environmental information that they want to disseminate.

PREPARATORY PROCESS: INFORMATION TO BE INCLUDED – The last two categories of environmental information were added by the European Parliament because

> [e]nvironmental permits and agreements (formerly voluntary agreements), as well as impact and risk assessments should also be made available in electronic form. Ideally, draft permits and applications for permits should also be available. In particular, estimates of risks connected to anomalies in various functions may contain information much more valuable to citizens than the daily data on emissions which do not exceed the limits. Such risks could be found in sectors such as the transport of hazardous goods by road and rail.[109]

---

108. Commission of the European Communities, *Proposal for a Directive of the European Parliament and of the Council on public access to environmental information*, COM (2000) 402 final, 16, 29 Jun. 2000.
109. Committee on the Environment, Public Health and Consumer Policy, *Report on the proposal for a directive of the European Parliament and of the Council on public access to environmental information,* doc. nr. A5-0074/2001, 28 Feb. 2001, 28, <www.europarl.europa.eu/sidesSearch/sipadeMapUrl.do?PROG=REPORT&SORT_ORDER=D&REF_A=A5-2001-0074&L=EN#>, 22 Feb. 2010.

According to the Council, adding these two categories would require the public authorities to put 'on the Internet data that are so detailed and voluminous that they should rather be subject to passive information'.[110] In the conciliation negotiations, the Parliament managed to maintain the new categories but had to compromise on their scope, for example by only including authorizations with a *significant* impact, instead of any authorizations under Article 7.2(f).

REPORTING OBLIGATION – In addition to the list of environmental information that needs to be continuously available and updated, Article 7.3 also includes an obligation for periodic reporting: 'Member States shall take the necessary measures to ensure that national, and, where appropriate, regional or local reports on the state of the environment are published at regular intervals not exceeding four years; such reports shall include information on the quality of, and pressures on, the environment'.

THREATS TO HUMAN HEALTH OR ENVIRONMENT – Article 7.4 addresses the dissemination of information in the event of an imminent threat to human health or the environment, whether caused by human activities or due to natural causes. In this case, 'all information held by or for public authorities which could enable the public likely to be affected to take measures to prevent or mitigate harm arising from the threat' has to be 'disseminated, immediately and without delay'.

g. *Quality of Environmental Information: Article 8 of the Access Directive*

PREPARATORY PROCESS: QUALITY REQUIREMENTS – As far as it is within their power, the Member States have to ensure that any information that is compiled by them or on their behalf is up to date, accurate, and comparable. This provision on the quality of environmental information was introduced by the European Parliament during the first reading. The Committee on the Environment, Public Health and Consumer Policy argued that

> the right to information is worthless if the information received is out-dated, unclear or based on incorrect data or beliefs which do not stand up to scientific inspection, and is for any such reason misleading. The words 'so far as is within their power' are essential here, because this requirement cannot in practice be made mandatory for some types of environmental information (whether for historical reasons or because developments in research or measurement techniques)'.[111]

110. Council of the European Union, *Common position adopted with a view to the adoption of a Directive of the European Parliament and of the Council on public access to environmental information and repealing Council Directive 90/313/EEC. Statement of the Council's Reasons*, doc. nr. 11878/01, 28 Jan. 2002, 14, <http://register.consilium.europa.eu/pdf/en/01/st11/st11878-re01ad01.en01.pdf>, 22 Feb. 2010.

111. Committee on the Environment, Public Health and Consumer Policy, *Report on the proposal for a directive of the European Parliament and of the Council on public access to*

The Commission and the Council found this unduly burdensome on the public authorities,[112] but the Parliament maintained its argument and the new article, reworded to the liking of all parties, was included in the final compromise.

PREPARATORY PROCESS: REQUESTS FOR INFORMATION AND QUALITY – The Parliament had also added that, in case of requests for information on factors affecting or likely to affect the elements of the environment,[113] when the applicant requests it, the public authorities have to report 'on the place where information, if available, can be found on the measurement procedures, including methods of analysis, sampling, and pre-treatment of samples, used in compiling the information, or referring to a standardized procedure used'. According to the Committee on the Environment, Public Health and Consumer Policy:

> Another factor in the quality of information concerns the method by which the data are collected and analysed, including sampling and pre-treatment of samples. In order to make the information compatible and sufficiently accurate, the measuring procedure should be disclosed together with the actual emission information. Emission data may even mislead the public where such data are not compatible and comparable with recommended limit values or data collected from other countries, regions, industries or other actors.[114]

3. Review of the Access Directive

MEMBER STATE REPORTS – Article 9 of the Access Directive foresaw a review procedure: the Member States had to make a report on the experience gained in the application of the directive no later than 14 February 2009, and they had to communicate this report to the European Commission by 14 August 2009. The available national reports showed that, in general, the implementation of the Access Directive has had a positive influence on the availability of environmental

environmental information, doc. nr. A5-0074/2001, 28 Feb. 2001, 29, <www.europarl.europa.eu/sidesSearch/sipadeMapUrl.do?PROG=REPORT&SORT_ORDER=D&REF_A=A5-2001-0074&L=EN#>, 22 Feb. 2010.

112. Commission of the European Communities, *Amended proposal for a directive of the European Parliament and of the Council on public access to environmental information*, COM (2001) 303 final, 12, 6 Jun. 2001; Council of the European Union, *Common position adopted with a view to the adoption of a Directive of the European Parliament and of the Council on public access to environmental information and repealing Council Directive 90/313/EEC. Statement of the Council's Reasons*, doc. nr. 11878/01, 28 Jan. 2002, 14, <http://register.consilium.europa.eu/pdf/en/01/st11/st11878-re01ad01.en01.pdf>, 22 Feb. 2010.

113. Such as substances, energy, noise, radiation or waste, including radioactive waste, emissions, discharges and other releases into the environment.

114. Committee on the Environment, Public Health and Consumer Policy, *Report on the proposal for a directive of the European Parliament and of the Council on public access to environmental information*, doc. nr. A5-0074/2001, 28 Feb. 2001, 29, <www.europarl.europa.eu/sidesSearch/sipadeMapUrl.do?PROG=REPORT&SORT_ORDER=D&REF_A=A5-2001-0074&L=EN#>, 22 Feb. 2010.

information and public participation, but that it also has brought a considerable administrative burden for the public authorities.[115]

COMMISSION REPORT – Based on these reports from the Member States, and taking into account technological developments, the Commission has to report on the directive to the European Parliament and the Council and to propose revisions if needed. At the time of this writing in February 2010, the Commission had not started drafting its report yet.

C. INSPIRE DIRECTIVE

1. Introduction

FOCUS ON DATA SHARING – As was indicated in Chapter 3 and in the introduction to this chapter, the main purpose of the INSPIRE Directive is to provide a legal basis for the development of an EU Spatial Data Infrastructure, which should facilitate the availability of spatial data for the development of environmental policy and policies that may have an impact on the environment. This entails that the main focus of the INSPIRE Directive is on the availability of spatial data for the performance of public tasks relating to the environment, and hence on sharing. Therefore, the history leading to the INSPIRE Directive and its background are discussed in the section on sharing.

OBLIGATIONS ON ACCESS – However, some of the obligations imposed on the Member States also relate to access to spatial data. The European Commission's proposal even included this in its objectives: it intended to 'make interoperable spatial information readily available in support of both national and Community policy and to enable the public to access to this information'.[116] During the legislative procedure in the Council and the European Parliament, the reference

---

115. See e.g., Department for Environment, Food and Rural Affairs, 'Public Access to Environmental Information. Experience Gained in the Application of Directive 2003/4/EC', 2009, <www.defra.gov.uk/corporate/policy/opengov/eir/pdf/commission-report-uk.pdf>, 22 Feb. 2010; Bundesministerium für Umwelt, Naturschutz und Reaktorsicherheit, *Bericht der Bundesrepublik Deutschland über die bei der Anwendung der Richtlinie 2003/4/EG über den Zugang der Offentlichkeit zu Umweltinformationen gewonnenen Erfahrungen*, s.d., <www.bmu.de/umweltinformation/downloads/doc/45109.php>, 22 Feb. 2010; Ministry of the Environment, Energy and Climate Change, *National Report of Greece on Directive 2003/4/EC on Public Access to environmental information*, 2009, <www.minenv.gr/4/41/00/National%20Report%20on%20Directive%202003.pdf>, 22 Feb. 2010; Department of Environment, Heritage and Local Government, *Review of Implementation of EU Directive 2003/4/EC on Public Access to Information on the Environment. Report by Ireland*, 2009, <www.environ.ie/en/AboutUs/AccesstoInformationontheEnvironment/PublicationsDocuments/FileDownLoad,22174,en.doc>, 22 Feb. 2010.

116. Commission of the European Communities, *Proposal for a directive of the European Parliament and of the Council establishing an infrastructure for spatial information in the Community (INSPIRE)*, COM (2004), 516 final, 2, 23 Jul. 2004.

to the public was deleted from the INSPIRE objective, but the text maintained the obligations for the Member States with regard to access. These obligations relate to the network services that should be provided by the Member States. These services are the means through which the public can obtain access to public sector spatial data. As this chapter is organized according to the purpose of the availability of spatial data, and the obligations on the establishment of the network services provide the public with a right of access to public sector spatial data, these obligations are discussed in this section.

OUTLINE – In this section, the obligation for the Member States to establish network services is discussed, as it is important to know what types of services can be used by the public to obtain access to spatial data. Next, the limits on access are addressed, followed by the possibilities for charging and licensing conditions for access. First, however, the relationship between the INSPIRE Directive and the Access Directive is examined.

2. The INSPIRE Directive and the Access Directive

ARTICLE 2 INSPIRE DIRECTIVE – The examination of the definitions of *environmental information* under the Access Directive and the spatial data sets that fall under the INSPIRE Directive has shown that these two categories have a very large common denominator. The INSPIRE Directive also recognizes this: in Article 2, it states that it is without prejudice to the Access Directive. While theoretically it would be possible for the INSPIRE Directive to deviate from the earlier Access Directive, the Aarhus Convention still has to be taken into account. The EU is bound by this Convention and has to ensure that it is implemented within its jurisdiction. As the Access Directive was the means to implement the Convention, it cannot be deviated from (unless this deviation remains within the margins left by the Aarhus Convention). This was ensured by Article 2 of the Access Directive.

CONFLICTS BETWEEN THE TWO DIRECTIVES – Hence, on the occasions where the Access Directive and the INSPIRE Directive are both applicable, Article 2 applies. This relates precisely to the subset of spatial data and environmental information that is the focus of this book. This means that if the Access Directive and the INSPIRE Directive were to have irreconcilable provisions, the former would prevail. However, as long as the provisions of the INSPIRE Directive can be considered as within the margins left by the Access Directive, they can still apply.

3. Access to Network Services

OUTLINE – As was mentioned above, the availability of public sector spatial data for access is dealt with in the INSPIRE Directive by the provisions on network services. A short overview is given of what these services contain, followed by a description of the provisions determining the obligations of the public authorities with regard to these services.

a. Establishment of Network Services

TYPES OF SERVICES – Under Article 11 of the INSPIRE Directive, the Member States have to establish and operate a network of services for the spatial data sets and services that are part of the data themes listed in the annexes to the directive.[117] These services are considered necessary for sharing spatial data between and among the various levels of public authorities in the EU.[118] However, they also have to be available to the public.[119] For this, interoperability of the services is necessary, so that the services can interact without repetitive manual intervention. This is to be ensured by Commission regulations implementing the directive as regards the technical specifications of the services.[120] The following services have to be established:

(a) Discovery services making it possible to search for spatial data sets and services on the basis of the content of the corresponding metadata and to display the content of the metadata.[121] The main purpose of the discovery services is to find spatial data and enable their use. They should 'support the discovery, evaluation and use of spatial data and services through their metadata properties. Metadata is the information and documentation which makes these resources understandable and sharable for users over time';[122]

(b) View services making it possible, as a minimum, to display, navigate, zoom in/out, pan, or overlay viewable spatial data sets and to display legend information and any relevant content of metadata. A view service is 'a web service to provide a visual representation of geographic and thematic information by creating an image of these data using portrayal

117. These services have to be available for the spatial data sets and services for which metadata are available. These metadata have to be ready for the data themes in Annex I and II no later than two years after the adoption of the regulation implementing the Directive as regards to metadata, that is, 3 Dec. 2010. For the data themes of Annex III, the metadata have to be available five years after the adoption, that is, 3 Dec. 2013 (Arts 5 and 6 INSPIRE Directive). See also Commission Regulation 2008/1205/EC of 3 Dec. 2008 implementing Directive 2007/2/EC of the European Parliament and of the Council as regards metadata, *OJ L* 326, 4 Dec. 2008, 12.

118. Recital 17 of the INSPIRE Directive.

119. Article 11 1 of the INSPIRE Directive.

120. For an overview of activities and documents of the INSPIRE drafting team on data specifications, see <http://inspire.jrc.ec.europa.eu/implementingRulesDocs_ns.cfm>, 22 Feb. 2010.

121. The metadata that at least have to be provided concern the conformity with the rules on data specifications; the conditions applying to access to, and use of spatial data sets and services and, where applicable, corresponding fees; the quality and validity of spatial data sets; the public authorities responsible for the establishment, management, maintenance and distribution of spatial data sets and services; and limitations on public access and the reasons for such limitations (Art. 5 of the INSPIRE Directive).

122. INSPIRE Drafting Team on Network Services, *INSPIRE Network Services Architecture v. 3.0*, 2008, 9, <http://inspire.jrc.ec.europa.eu/reports/ImplementingRules/network/D3_5_INSPIRE_NS_Architecture_v3-0.pdf>, 22 Feb. 2010.

rules', making it possible to perform the actions on the spatial data sets that are mentioned.

(c) Download services, enabling copies of spatial data sets, or parts of such sets, to be downloaded and, where practicable, accessed directly. These services are 'the mechanisms by which users can have access to the full information content captured and transformed by [Member States] into their geographic data sets according to the INSPIRE defined themes. The download services provide the capability to access the information in order to fulfil a series of possible tasks including visualizing information in a variety of ways, integration with other information, and to allow deep analysis as a basis for knowledge and decision making'.[123] A download service can either download a pre-defined data set or pre-defined part of a data set, or give direct access to a data set, allowing the user to make his or her own selections.[124]

(d) Transformation services, enabling spatial data sets to be transformed with a view to achieving interoperability. The function of transformation services is to help other services in achieving compliance with the specifications imposed by the INSPIRE Directive and its implementing rules. Hence, they are not necessary if the other services are set up according to the INSPIRE rules.[125]

(e) Services allowing spatial data services to be invoked. These services also support the other services by allowing the combination of multiple services into a workflow and by enabling a user to run view or download services without needing the specific software on his or her own computer.[126]

ACCESS TO THE SERVICE – The services have to be easy to use, available to the public, and accessible via the Internet or any other appropriate means of telecommunications. Hence, under Article 11 of the INSPIRE Directive, the public is guaranteed access to the services, but not directly to the spatial data. However, as these services have to be made available for all the spatial data sets mentioned in the annexes to the directive, access to the services automatically entails access to the data.

ACTIVE V. PASSIVE – The network services are a good example of how the use of modern technology can make it difficult to distinguish between the *passive* access upon request and the *active* dissemination by public bodies of their data. On the one hand, the public bodies have to take action to make the services available to the public, whether or not requests are made for access to the data. On the other hand,

---

123. INSPIRE Drafting Team on Network Services, *D3.9 Draft Implementing Rules for Download Services v.2.0*, 2009, 5, <http://inspire.jrc.ec.europa.eu/documents/Network_Services/INSPIRE%20Draft%20Implementing%20Rules%20Download%20Services%20(version%203.0).pdf>, 22 Feb. 2010.

124. *Ibid.*, 8.

125. INSPIRE Drafting Team on Network Services, *INSPIRE Network Services Architecture v. 3.0*, 2008, 10, <http://inspire.jrc.ec.europa.eu/implementingRulesDocs_ns.cfm>, 22 Feb. 2010.

126. *Ibid.*

the network services still have to be accessed and queried by citizens to obtain the data, so there still has to be a request.

b. Limitation of Access to Network Services

ARTICLE 13.1 OF THE INSPIRE DIRECTIVE – Article 13.1 of the INSPIRE Directive contains a list of reasons why the Member States may limit public access to spatial data sets and services through the network services. This article was the subject of many discussions between the European Commission, the European Parliament, and the Council of Ministers. The end result is a distinction between discovery services and the other services, with fewer possibilities of restricting access to the former.

DISCOVERY SERVICES – Public access to the discovery services can be limited by the Member States 'where such access would adversely affect international relations, public security or national defence'. It is understandable that such an exception can only be invoked for very serious reasons involving international or national security. If one of these reasons is invoked, not only are the spatial data not available to the public, but also is the public not even entitled to know that the data exist. Limiting access to these services for any additional reasons would conflict with the very purpose of access legislation: enabling the participation of citizens in decision making and keeping citizens informed of the activities of their government. There is no need to refuse access to the discovery services when there is adequate protection of the interests concerned by the limitation of access to the other services.[127]

PREPARATORY PROCESS: DISCOVERY SERVICES – Originally, the European Commission had even proposed never limiting access to these services.[128] It felt that the public should always have the right to know that the data sets exist, even though it might not be able to see the data or to obtain a copy of them. Limiting access to the discovery services should not be possible, 'since this would mean that the public would not even be able to learn of the existence of the data'.[129] The Council of Ministers did not see a need to distinguish between the services and included in the political agreement that the entire list of exceptions should also

127. Katleen Janssen, 'INSPIRE and the PSI Directive: Public Task Versus Commercial Activities?', paper presented at the 11th EC-GIS Workshop. 'ESDI – Setting the Framework' (Alghero, 2005), <www.ec-gis.org/Workshops/11ec-gis/papers/303janssen.pdf>, 22 Feb. 2010.

128. Commission of the European Communities, *Proposal for a directive of the European Parliament and of the Council establishing an infrastructure for spatial information in the Community (INSPIRE)*, COM (2004), 516 final, 23 Jul. 2004.

129. Commission of the European Communities, *Communication to the European Parliament pursuant to the second subparagraph of Art. 251(2) of the EC Treaty concerning the common position of the Council on the adoption of a Directive of the European Parliament and of the Council establishing an infrastructure for spatial information in the Community (INSPIRE)*, COM (2006) 51 final, 10 Feb. 2006.

be applicable to the discovery services.[130] The European Parliament re-instated the distinction in order not to allow the limitation of public access to the discovery services.[131] In the final conciliation process, the institutions found a middle ground by allowing limitations to public access to the discovery services, but only for the three reasons mentioned earlier.

OTHER SERVICES – Access to the other network services can be limited for more reasons than is the case for discovery services. Article 13 of the INSPIRE Directive has included all possible limitations that were also listed in the Access Directive. For more information on these limitations, we can therefore refer to the section on the exceptions to access upon request in the Access Directive.[132] For the ease of the reader, the list itself is included once more below. According to Article 13, the Member States may limit public access to network services other than the discovery services, 'where such access would adversely affect any of the following':

(a) the confidentiality of the proceedings of public authorities, where such confidentiality is provided for by law;
(b) international relations, public security or national defence;
(c) the course of justice, the ability of any person to receive a fair trial or the ability of a public authority to conduct an enquiry of a criminal or disciplinary nature;
(d) the confidentiality of commercial or industrial information, where such confidentiality is provided for by national or Community law to protect a legitimate economic interest, including the public interest in maintaining statistical confidentiality and tax secrecy;
(e) IPR;
(f) the confidentiality of personal data and/or files relating to a natural person where that person has not consented to the disclosure of the information to the public, where such confidentiality is provided for by national or Community law;
(g) the interests or protection of any person who supplied the information requested on a voluntary basis without being under, or capable of being put under, a legal obligation to do so, unless that person has consented to the release of the information concerned;
(h) the protection of the environment to which such information relates, such as the location of rare species.[133]

---

130. Council of the European Union, *Common Position adopted by the Council on 23 January 2006 with a view to the adoption of a Directive of the European Parliament and of the Council establishing an Infrastructure for Spatial Information in the European Community (INSPIRE)*, *OJ C* 126E, 30 May 2006, 16.

131. European Parliament, *Position adopted at second reading on 13 June 2006 with a view to the adoption of Directive 2006/ . . . /EC of the European Parliament and of the Council establishing an Infrastructure for Spatial Information in the European Community (INSPIRE)*, <www.europarl.europa.eu/sides/getDoc.do?pubRef=-//EP//NONSGML+TA+P6-TA-2006-0252+0+DOC +PDF+V0//EN>, 22 Feb. 2010.

132. Cf. *supra*, s. II.B.2.d.

133. Article 13.2 of the INSPIRE Directive.

RESTRICTIVE INTERPRETATION – As is determined for environmental information in the Access Directive, the limitations of access must be interpreted in a restrictive way, taking into account for each case the public interest served by providing access. In every particular case, the public interest served by disclosure must be weighed against the interest served by limiting or conditioning the access. The prohibition of the Access Directive to refuse access to information on emissions into the environment is also included for reasons of confidentiality of the proceedings of public authorities, confidentiality of commercial or industrial information, confidentiality of personal data, the interest of persons who voluntarily supplied the data, and the protection of the environment.

PREPARATORY PROCESS: TWO EXCEPTIONS NOT INCLUDED? – The list of limitations in Article 13 brings the INSPIRE Directive completely in line with the Access Directive. However, this alignment was not as self-evident as one might assume. In its original proposal, the European Commission wanted the INSPIRE Directive to be without prejudice to the Access Directive 'except with respect to certain provisions that address the grounds for limiting access to the spatial data covered by this Directive, avoiding the possibility of unduly limitations of access to the spatial data covered by this Directive'.[134] The Commission's proposal for the INSPIRE Directive had left two exceptions off the list provided for in the Access Directive: IPR and the interests or protection of persons who had provided spatial data voluntarily. The Commission did not feel that access to the network services should be limited to protect these interests. However, the Environment Working Party of the Council did not see a 'reason to differentiate the treatment of access to this type of information from the general rules on public access to environmental information'. It found that all the exceptions of the Access Directive should be copied.[135]

PREPARATORY PROCESS: INTELLECTUAL PROPERTY RIGHTS – The Commission made no problem of re-introducing the exception with regard to the interests of voluntary suppliers of information. However, it felt that excluding the exception for IPR from the list did not 'in any way compromise the right of public authorities to assert intellectual property rights. It only requires that those rights are exercised in accordance with existing good practice by ensuring the provision of services as described'.[136] In addition, the directive 'rather provides clear, well-defined and

---

134. Commission of the European Communities, *Proposal for a directive of the European Parliament and of the Council establishing an infrastructure for spatial information in the Community (INSPIRE)*, COM (2004), 516 final, Recital 6, 23 Jul. 2004.

135. General Secretariat of the Council of the European Union, *Note to the Delegations. Proposal for a directive of the European Parliament and of the Council establishing an infrastructure for spatial information in the Community (INSPIRE)*, doc. nr. 14454/04, 13 Dec. 2004, <http://register.consilium.europa.eu/pdf/en/04/st14/st14454.en04.pdf>, 22 Feb. 2010; General Secretariat of the Council of the European Union, *Note to the Delegations. Proposal for a directive of the European Parliament and of the Council establishing an infrastructure for spatial information in the Community (INSPIRE)*, doc. nr. 5648/05, 1 Feb. 2005, <http://register.consilium.europa.eu/pdf/en/05/st05/st05648.en05.pdf>, 22 Feb. 2010.

136. General Secretariat of the Council of the European Union, *Note to the Delegations. Proposal for a directive of the European Parliament and of the Council establishing an infrastructure*

justified restrictions to the way those rights are exercised. Including this as a general ground for exemption is therefore superfluous and risks weakening' the provisions on public access.[137] This argument could not convince the Member States' delegations, and IPR were included in the Common Position to ensure that '[t]he extended list of grounds for limiting access set out in Article 13 is identical to that in Article 4(2) [of the Access Directive]'.[138] In its second reading, the plenary session of the European Parliament deleted IPR from Article 13 once more, based on the recommendation of the Committee on the Environment, Public Health and Food Safety, 'in order not to allow public authorities to use their [IPR], created with tax money, to limit public access to certain spatial data'.[139] The Commission agreed to include this in the amendment of its proposal,[140] but the Council managed to keep IPR in the text as part of the final conciliation agreement.

ARGUMENT FOR INCLUDING INTELLECTUAL PROPERTY RIGHTS – While the argument of consistency between the Access Directive and the INSPIRE Directive is a valid one, it should be noted that the approach of the European Commission – with no prejudice to the Access Directive except for some particular provisions – would be possible with regard to the list of limitations. This list is derived from the Aarhus

for spatial information in the Community (INSPIRE), doc. nr. 7488/05, 22 Mar. 2005, <http://register.consilium.europa.eu/pdf/en/05/st07/st07488.en05.pdf>, 22 Feb. 2010. It is interesting to note that during the negotiations for the Access Directive, the European Commission actually insisted on keeping the exception for intellectual property rights, while the European Parliament wanted to remove it (cf. *supra*, s. II.B.2.d.).

137. General Secretariat of the Council of the European Union, *Note to the Delegations. Proposal for a directive of the European Parliament and of the Council establishing an infrastructure for spatial information in the Community (INSPIRE)*, doc. nr. 7488/05, 22 Mar. 2005, <http://register.consilium.europa.eu/pdf/en/05/st07/st07488.en05.pdf>, 22 Feb. 2010.

138. Council of the European Union, *Common Position adopted by the Council with a view to the adoption of a Directive of the European Parliament and of the Council establishing an Infrastructure for Spatial Information in the European Community (INSPIRE). Statement of the Council's reasons*, doc. nr. 12064/2/05, 23 Jan. 2006, <http://register.consilium.europa.eu/pdf/en/05/st12/st12064-re02ad01.en05.pdf>, 22 Feb. 2010.

139. European Parliament, *Position adopted at second reading on 13 June 2006 with a view to the adoption of Directive 2006/ . . . /EC of the European Parliament and of the Council establishing an Infrastructure for Spatial Information in the European Community (INSPIRE)*, <www.europarl.europa.eu/sides/getDoc.do?pubRef=-//EP//NONSGML+TA+P6-TA-2006-0252+0+DOC +PDF+V0//EN>, 22 Feb. 2010; Committee on the Environment, Public Health and Food Safety, *Recommendation for second reading on the Council common position for adopting a directive of the European Parliament and of the Council establishing an Infrastructure for Spatial Information in the European Community (INSPIRE)*, doc. nr. A6-0081/2006, 2 May 2006, <www.europarl.europa.eu/RegWeb/application/registre/advancedSearch.faces>, 22 Feb. 2010.

140. Commission of the European Communities, *Opinion pursuant to Art. 251(2) third subparagraph, point (c) of the EC Treaty, on the European Parliament's amendments to the Council Common Position regarding the proposal for a Directive of the European Parliament and of the Council establishing an infrastructure for spatial information in the Community (INSPIRE)*, COM (2006) 484 final, 13 Sep. 2006.

Convention, but this Convention explicitly states that not all the exceptions have to be included. The Parties to the Convention can choose not to include a limitation of access, for example IPR or commercial interests. However, considering the large overlap with environmental information, it would cause confusion if access to environmental information that would also be spatial data under the INSPIRE Directive could not be refused because of IPR, while access to other environmental information could. From this point of view, consistency between the two directives is advisable.

ARGUMENT AGAINST INCLUDING INTELLECTUAL PROPERTY RIGHTS – However, there may be other points regarding why the possibility of limitation of access because of IPR reasons should be removed from the INSPIRE Directive. Proponents of the limitation of access for IPR reasons may state that there is sufficient protection of citizens' access because every access restriction has to be interpreted restrictively and has to be weighed against the interest served by providing access. If this reasoning is followed, the interests in protecting IPR have to be weighed against the interests of the public in obtaining the information. Two main interests of the public authorities in protecting their IPR by refusing access come to mind. First, a public authority may want to obtain financial compensation for making its data available. For some public authorities, these charges are essential for creating and maintaining their data. However, while this necessity may justify the charges that are made, it cannot be used as an argument to deny access to citizens. Moreover, from a practical point of view, denying access would be counterproductive: no income can be gained if no data are made available. A second reason to limit access to spatial data may be the concern that the data will be used for other, mostly commercial, reasons than those for which they were provided, without any control or supervision from the public authority. Once again, one could argue that limiting access to data for this reason is too far-reaching, as there are technical possibilities that would prevent such commercial abuse. As will be mentioned later on, public authorities can provide their spatial data via the view services in a form that would prevent their commercial use, so there is no need to limit the access.

c. Charges for Network Services

DISCOVERY SERVICES – As is the case for the limitations on access, the charges are also not the same for all types of network services. First, Article 14.1 of the INSPIRE Directive states that the discovery services have to be made available free of charge. Recital 19 adds that '[e]xperience in the Member States has shown that it is important, for the successful implementation of an infrastructure for spatial information, that a minimum number of services be made available to the public free of charge'.

VIEW SERVICES – Next, the view services also have to be free of charge. However, in some cases, the public authorities can impose charges on the public for access to these services, that is, 'where such charges secure the maintenance of spatial data sets and corresponding data services, especially in cases involving very large volumes of frequently updated data'.[141] This should ease the minds of the public authorities that create spatial data and that have to provide for their own funding under national law. These public authorities would not be able to collect, produce, or update spatial data if they could not recover any costs by charging the public for viewing their data. Hence, this is how the phrase *secure the maintenance* should be interpreted, that is, without the fees that are charged to the public, a public authority would not be able to do its work. Interpreting the wording in a more relaxed way, one might also argue that charges will always *secure* the maintenance since any money that comes in will make it possible for the public authority to perform its task in a better manner. However, this does not mean that the charges are *necessary* for the sustainability of the service. Such interpretation would render the provision meaningless. In addition, in line with the general principles of EU law, exceptions have to be interpreted restrictively, so the charges have to be necessary for the maintenance.[142] The question still remains what is meant by 'very large volumes of frequently updated data'. The main category of data that comes to mind is meteorological data, but for other types of data sets, it will be up to the courts to determine this.

OTHER SERVICES – Download services, transformation services, and invoke services can be charged for by public authorities without any restrictions.

PREPARATORY PROCESS: VIEW SERVICES FREE OF CHARGE? – The discussion on the possibility of levying charges between the Commission, the Council, and the European Parliament focused on the charges for the view services. The Commission never contested the right of the public authorities to charge the public for the use of the download services, the transformation services, or the services allowing spatial data services to be invoked. However, it did feel that the view services should be available free of charge, without exceptions for sustainability or any other reasons.[143] The Council had serious concerns about this: it felt that the difference between the view services and the download services could be small, so there were considerable possibilities for abuse of the services for commercial purposes, undermining the business models of some of the public authorities that make spatial data available.[144] Therefore, the Council suggested including the possibility of charging and licensing for the view services. The Commission

141. Article 14.2 of the INSPIRE Directive.

142. However, as the number of public authorities that complain on their sustainability is growing increasingly large, and includes many providers of basic data that will be needed to make most of the services comprehensible and interesting for the public, this provision may have more consequences than originally expected.

143. Commission of the European Communities, *Proposal for a directive of the European Parliament and of the Council establishing an infrastructure for spatial information in the Community (INSPIRE)*, COM (2004), 516 final, 23 Jun. 2004.

144. General Secretariat of the Council of the European Union, *Note to the Delegations. Proposal for a directive of the European Parliament and of the Council establishing an infrastructure for spatial information in the Community (INSPIRE)*, doc. nr. 14454/04, 13 Dec. 2004, <http://register.consilium.europa.eu/pdf/en/04/st14/st14454.en04.pdf>, 22 Feb. 2010.

responded that 'in the Internet age view services are commonly available free of charge, even from private commercial operators, therefore it is an essential component of any spatial data infrastructure'.[145]

PREPARATORY PROCESS: CHARGING FOR VIEW SERVICES? – The Committee on the Environment, Public Health and Food Safety of the European Parliament shared this concern about potential circumvention of the charges for commercial use, but it chose a more moderate measure to prevent this. It suggested an amendment stating that, '[i]n order to protect [IPR] held by public authorities in respect of spatial data, the data made available through the view services ... may be in a form preventing their re-use for commercial purposes, and a click-license may be included to restrict the use that can be made of the data'.[146] The Working Party on the Environment of the Council seemed inclined to accept this approach for a while, but some Member States still argued that data providers should be able to maintain charges for certain view services, for example for meteorological data.[147] The Common Position of the Council reflected the influence of these Member States: the view services should be 'as a rule, available to the public free of charge. However, in cases where charges and/or licenses are an essential precondition to maintain the spatial data sets and services to fulfil requirements of already existing international spatial data infrastructure in a sustainable way, Member States may apply charges and/or licenses either to the person providing the service to the public, or, where the service provider chooses, to the public itself'.[148] The Commission found that this would 'unduly limit public access, make implementation of the directive more difficult to achieve, and reduce the

---

145. General Secretariat of the Council of the European Union, *Note to the Delegations. Proposal for a directive of the European Parliament and of the Council establishing an infrastructure for spatial information in the Community (INSPIRE)*, doc. nr. 5648/05, 1 Feb. 2005, <http://register.consilium.europa.eu/pdf/en/05/st05/st05648.en05.pdf>, 22 Feb. 2010.

146. Committee on the Environment, Public Health and Food Safety, *Report on the proposal for a directive of the European Parliament and of the Council establishing an infrastructure of spatial information in the Community (INSPIRE)*, doc. nr. A6-0108/2005, 26 Apr. 2005, <www.europarl.europa.eu/sides/getDoc.do?pubRef=-//EP//NONSGML+REPORT+A6-2005-0108+0+DOC+PDF+V0//EN&language=EN>, 22 Feb. 2010.

147. General Secretariat of the Council of the European Union, *Note to the Working Party on the Environment. Proposal for a directive of the European Parliament and of the Council establishing an infrastructure for spatial information in the Community (INSPIRE)*, doc. nr. 7985/05, 3 May 2005, <http://register.consilium.europa.eu/pdf/en/05/st07/st07985.en05.pdf>, 22 Feb. 2010; General Secretariat of the Council of the European Union, *Note to the Working Party on the Environment. Proposal for a directive of the European Parliament and of the Council establishing an infrastructure for spatial information in the Community (INSPIRE)*, doc. nr. 8891/05, 13 May 2005, <http://register.consilium.europa.eu/pdf/en/05/st08/st08891-co01.en05.pdf>, 22 Feb. 2010.

148. Council of the European Union, *Common Position adopted by the Council on 23 January 2006 with a view to the adoption of a Directive of the European Parliament and of the Council establishing an Infrastructure for Spatial Information in the European Community (INSPIRE). Statement of the Council's reasons*, doc. nr. 12064/2/05, 23 Jan. 2006, <http://register.consilium.europa.eu/pdf/en/05/st12/st12064-re02ad01.en05.pdf>, 22 Feb. 2010.

availability of data and consequently the added value of the infrastructure'.[149] Later, in its response to the Parliament's second reading, which had re-introduced the free availability of viewing services, the Commission took a less stringent position: 'There are nevertheless some types of data, such as meteorological data, where it may not be cost-effective or reasonable for data providers to make the data available free of charge because they are too voluminous and/or are of interest for only a very limited period of time. The precise wording of this paragraph will need to reflect such cases'.[150]

PREVENTING COMMERCIAL RE-USE – Next to the possibility of charging, another concern of the public authorities regarding the viewing services was appeased by Article 14.3 of the INSPIRE Directive: data made available through the viewing services may be in a form preventing their re-use for commercial purposes, for example by watermarking, copy restrictions, and so on. The public authorities were concerned that the private information industry would build its own commercial services based on the view services, and in this way circumvent the PSI Directive and the corresponding national legislation. However, when re-use would be commercial has not been determined, and it will be difficult for public authorities to determine a form for the viewing services that prevents commercial re-use but not other re-use. In practice, this most likely means that any re-use, not just commercial, will be prevented, which goes beyond the purpose of Article 14.3 of the INSPIRE Directive. This does not necessarily cause problems for the possible re-user, as he or she would have to use the PSI Directive for his or her requests for re-use anyway.

E-COMMERCE SERVICES – Where the public authorities impose charges for access, the Member States have to 'ensure that e-commerce services are available. Such services may be covered by disclaimers, click-licenses or, where necessary, licenses'. The idea behind this provision is that if the public authorities want to impose charges, they have to enable citizens to obtain the spatial data and make the payment in an easy, harmonized, and automated way. The example that the European institutions had in mind with this provision was the British click-use license.[151]

GEO-PORTAL – Next, to also make it easier for citizens to locate the network services, the European Commission has to establish and operate an INSPIRE

---

149. Commission of the European Communities, *Communication to the European Parliament pursuant to the second subparagraph of Art. 251(2) of the EC Treaty concerning the common position of the Council on the adoption of a Directive of the European Parliament and of the Council establishing an infrastructure for spatial information in the Community (INSPIRE)*, COM (2006) 51 final, 10 Feb. 2006.

150. Commission of the European Communities, *Opinion pursuant to Art. 251(2) third subparagraph, point (c) of the EC Treaty, on the European Parliament's amendments to the Council Common Position regarding the proposal for a Directive of the European Parliament and of the Council establishing an infrastructure for spatial information in the Community (INSPIRE)*, COM (2006) 484 final, 13 Sep. 2006.

151. For more information on the click-use license, see <www.opsi.gov.uk/click-use/index>, 22 Feb. 2010.

geo-portal at the level of the EU, and the Member States have to provide access to their network services through the INSPIRE geo-portal. They can also decide to provide access through their own access points.[152]

III. RE-USE

DEFINITION – The second purpose for which public sector spatial data can be made available is re-use. This includes any use for commercial or non-commercial purposes outside of the public task and entails a processing of the data beyond the mere access for democratic purposes. Re-use is addressed by the PSI Directive.

OUTLINE – This section first takes a look at the long history of the discussion on re-use at the level of the EU. The fact that the EU has given a lot more attention to re-use than to access is related to its economic slant. Re-use policy can be framed in the EU ideal of a common market, while access is inherently linked to the democratic structures of the Member States, in which the EU should not interfere. After the road towards the PSI Directive is discussed, the provisions of the directive will be thoroughly examined, in order to fully understand the obligations of the Member States in making data available for re-use.

A. HISTORY

OUTLINE – The first roots of the PSI Directive can be traced back to the end of the 1980s. During the 1990s, many discussions were held on the appropriate approach to re-use on the level of the EU, which finally culminated in the 2003 PSI Directive. This process towards the PSI Directive is discussed in this section. This may give the reader insight into the origin of and reasons for the current provisions of the PSI Directive.

1. Guidelines for Improving the Synergy between the Public and Private Sectors in the Information Market[153]

IMPACT PROGRAMME – During the second half of the 1980s, the first signs of interest of the European Economic Community in public sector data started to show. In 1988, the Council of Ministers implemented the Information Market Policy Actions Programme (IMPACT) to help establish a common internal information market and improve competitiveness. One of the programme's objectives

152. Article 15 of the INSPIRE Directive.

153. Commission of the European Communities, *Guidelines for Improving the Synergy between the Public and the PrivateSsectors in the Information Market*, 1989, <http://ec.europa.eu/information_society/policy/psi/docs/pdfs/brochure/1989_public_sector_guidelines_en.pdf>, 22 Feb. 2010 (hereafter referred to as Synergy Guidelines).

was to help eliminate legal, technical, and administrative barriers to public sector data.[154]

INTRODUCTION OF THE SYNERGY GUIDELINES – The 1989 Guidelines for improving the synergy between the public and private sectors in the information market (hereafter referred to as Synergy Guidelines) were issued by the Directorate General for Telecommunications, Information Industries and Innovation as a result of a series of consultations and discussions with representatives of the public and private sectors of the European information market in all Member States and endorsed by delegations of all the Member States. The Synergy Guidelines were the result of a five-year thinking process on the interaction between the public and the private sector in the information market.[155]

TWO ROLES FOR THE PUBLIC BODIES – The Synergy Guidelines emphasised that more openness of public sector data was necessary to encourage synergy between the public and private sectors in establishing a Community information market and to reduce possible distortions of competition in the information market for public sector data.[156] They made a distinction between two roles of the public bodies: on the one hand they are producers of basic data, and on the other hand they can also be providers of electronic information services.

PUBLIC BODIES AS PRODUCERS OF DATA – With regard to public bodies' first role as producers of data, the Synergy Guidelines suggested to allow the basic information that they collected in the performance of their governmental functions to be used by the private sector and exploited by the information industry through electronic information services. Member States should publicize the availability of their data

154. Cheryl Buchwald, 'Canada in Context: an Overview of Information Policies in Four Industrialized Countries', 1995, <www3.fis.utoronto.ca/research/iprp/publications/wp/wp2.html>, 22 Feb. 2010.

155. For a detailed overview, see John Martin, 'Historical Background of EC Initiatives Relating to the Public/Private Sector Relationship in the Information Market', in *Report on the Workshop on Synergy between the Public and Private Sectors in the Information Market. Publaw 2 workshop*, ed. Commission of the European Communities (Luxembourg: DG XIII Legal Advisory Board, 1993), s.p.

156. Stephen Saxby, 'The Development of UK Government Policy Towards the Commercialization of Official Information', *International Journal of Law and Information Technology* 4, no. 3 (1996): 225; Harlan J. Onsrud & Xavier Lopez, 'Intellectual Property Rights in Disseminating Digital Geographic Data, Products, and Services: Conflicts and Commonalities among European Union and United States Approaches', 1998, <www.spatial.maine.edu/~onsrud/harlan/Course/CPOnsrudLopez.html>, 22 Feb. 2010; Katleen Janssen & Jos Dumortier, 'Towards a European Framework for the Re-use of Public Sector Information: A Long and Winding Road', *International Journal of Law and Information Technology* 11, no. 2 (2003): 186; Katleen Janssen, 'De exploitatie van overheidsinformatie: eindelijk een ontwerprichtlijn', *Mediarecht* no. 19 (2002): 108–109 [in Dutch]; Cheryl Buchwald, 'Canada in Context: an Overview of Information Policies in Four Industrialized Countries', 1995, <www3.fis.utoronto.ca/research/iprp/publications/wp/wp2.html>, 22 Feb. 2010; Charles D. Raab, 'Privacy Issues as Limits to Access', in *Public Sector Information in the Digital Age: Between Markets, Public Management and Citizens' Rights*, ed. Georg Aichholzer & Herbert Burkert (Cheltenham: Edward Elgar Publishing Limited, 2004), 31.

and compile and publicize guidelines defining the conditions of release, use, and exploitation of public sector data. Negotiation procedures and pricing principles should be harmonized across the public bodies as much as possible. The Synergy Guidelines acknowledged that pricing policies may vary depending on the nature of the information, but indicated that a price should be established that reflects the costs of preparing and passing the information to the private sector, but not necessarily including the full cost of collecting and handling the information in the course of routine administration. Exclusive agreements should not be granted if they lead to distortion of competition. If, however, an exclusive right would be deemed necessary for reasons such as the penetration of a new market or provision of a service in the public interest, it should be subject to regular review.

Public bodies as providers of information services – In addition to supplying basic data, public bodies can themselves also create and provide electronic information services based on public sector data. The Synergy Guidelines stated that the public bodies should, first of all, adopt policies and procedures that encourage investment by the private sector in the development of information services based on public sector data. If public bodies decide to provide information services directly themselves, they should avoid any practice that leads to distortion of competition. Before establishing a new electronic service or continuing an existing one, public bodies should first consider whether an existing private sector service can be used or adapted to meet their requirements. Existing electronic services should be reviewed regularly to see if their provision would be most appropriate by the public or the private sector. This emphasis on the needs of the private sector was the source of considerable criticism. The balance was said to be leaning very much towards the private information industry, with the role of the public bodies reduced to a mere supplier of data, except for particular circumstances.[157]

Relationship with access – In any case, the public bodies should, as much as possible, exempt from copyright texts of a legislative, administrative, or legal nature and official translations of such texts.[158] Also, while the Synergy Guidelines mainly addressed the information market and commercial re-use, they did emphasise that access should not be restricted. When public sector data was made available for private sector re-use, citizens' rights of access had to be preserved. Yet, some authors still claimed that not enough attention was paid to the relationship

157. Katleen Janssen & Jos Dumortier, 'Towards a European Framework for the Re-use of Public Sector Information: A Long and Winding Road', *International Journal of Law and Information Technology* 11, no. 2 (2003): 187; Jon Bing, 'Managing copyright in a digital environment', <www.portlandpress.com/pp/books/online/tiepac/session2/ch3.htm>, 22 Feb. 2010; Johan Pas & Bruno De Vuyst, 'Re-establishing the Balance Between the Public and the Private Sector: Regulating Public Sector Information Commercialization in Europe', *Journal of Information, Law and Technology* no. 2 (2004): 8, <www2.warwick.ac.uk/fac/soc/law/elj/jilt/2004_2/pasanddevuyst>, 22 Feb. 2010; Richard Groot, 'Spatial Data Infrastructure (SDI) for Sustainable Land Management', *ITC Journal* no. 3/4 (1997): 289.

158. Commission of the European Communities, *Synergy Guidelines*, 12.

between the issues of access to public sector data and the exploitation of the data.[159]

Impact of Synergy Guidelines – Since the Synergy Guidelines had no binding effect on the Member States and no sanctions were attached to them, they were not given sufficient attention by the Member States.[160] Moreover, the overall level of awareness of the guidelines among policy makers and the private sector was far from optimal. Yet, the European Commission considered the issue important enough to charge a research consortium in 1990 with studying the issue further, including the reasons why the Synergy Guidelines were unsuccessful.[161]

2. The Publaw Reports

Overview – Three sets of Publaw reports were executed between 1990 and 1995. The first addressed general access to information legislation in the European Community and the Member States,[162] while the second assessed the impact of the Synergy Guidelines on the information market in the Member States and explored the policies in the United States and Canada in the field of commercial exploitation

159. Johan Pas & Bruno De Vuyst, 'Re-establishing the Balance Between the Public and the Private Sector: Regulating Public Sector Information Commercialization in Europe', *Journal of Information, Law and Technology* no. 2 (2004): 8, <www2.warwick.ac.uk/fac/soc/law/elj/jilt/2004_2/pasanddevuyst>, 22 Feb. 2010.

160. Herbert Burkert, 'Public Sector Information: Towards a More Comprehensive Approach in Information Law?', *Journal of Law and Information Science* (1992): 50; Herbert Burkert, 'The Mechanics of Public Sector Information', in *Public Sector Information in the Digital Age, Between Markets, Public Management and Citizens' Rights*, ed. Georg Aichholzer & Herbert Burkert (Cheltenham: Edward Elgar Publishing Limited, 2004), 12; Marc De Vries, 'Implementatie van de EU-richtlijn hergebruik overheidsinformatie in de Wob. Beter ten halve gekeerd dan ten hele gedwaald', *Nederlands Juristenblad* 79, no. 39 (2004): 2039 [in Dutch]; Katleen Janssen & Jos Dumortier, 'Towards a European Framework for the Re-use of Public Sector Information: A Long and Winding Road', *International Journal of Law and Information Technology* 11, no. 2 (2003): 188; Katleen Janssen, Jo Steyaert & Roland Van Gompel, *Transparantie van overheidsinformatie in Vlaanderen. Beschouwingen en aanbevelingen* (Bruges: Die Keure, 2003), 129–131 [in Dutch]; Katleen Janssen, 'De exploitatie van overheidsinformatie: eindelijk een ontwerprichtlijn', *Mediarecht* no. 19 (2002): 109 [in Dutch]; Harlan J. Onsrud & Xavier Lopez, 'Intellectual Property Rights in Disseminating Digital Geographic Data, Products, and Services: Conflicts and Commonalities among European Union and United States Approaches', 1998, <www.spatial.maine.edu/~onsrud/harlan/Course/CPOnsrudLopez.html>, 22 Feb. 2010; Peter Seipel, 'The Technology of Insight: Computers and Informed Citizens', *Chicago-Kent Law Review* 69, no. 2 (1993): 448; David R. Worlock, 'Opportunities and Constraints of Public Data Commercialisation: A Private Company Viewpoint', paper given at the European Commission Conference: Access to Public Information. A Key to Commercial Growth and Electronic Democracy (Stockholm, 1996): 1.

161. Katleen Janssen, 'De exploitatie van overheidsinformatie: eindelijk een ontwerprichtlijn', *Mediarecht* no. 19 (2002): 109–110 [in Dutch].

162. Commission of the European Communities, *Publaw. Subject report: General Access to Information Legislation* (Luxembourg: CEC Legal Advisory Board, 1991), s.p.

of public sector data.[163] The third study followed this up by providing an update of law and practices in the area and assessing the need for further action.

PUBLAW 1: ACCESS – The first report did not look at access legislation from a democratic or government-citizen relationship perspective but examined access in the context of the information market. One of the main findings of the report was that there is tension between the individual's right to access and a market-oriented use of PSI.[164] However, this was not deemed due to the legislation itself, but to specific dissemination strategies.

PUBLAW 2: IMPACT OF THE SYNERGY GUIDELINES – As was mentioned above, the Publaw 2 study found that the impact of the Synergy Guidelines was minimal. A number of factors were believed to have contributed to this lack of awareness.[165] First, the Synergy Guidelines had not been effectively distributed. In some instances, the Synergy Guidelines had not been disseminated to the private sector, or only to high-level government officials rather than to public bodies directly responsible for the provision of information services. In addition, national policy guidelines either superseded the Synergy Guidelines or had assumed a higher profile on the national level. In some cases, the Synergy Guidelines had been distributed under another name.[166] Another key finding was that the stage of development of the information market and its supporting economic, technical, and human skills infrastructure had a high influence on the impact and applicability of the Synergy Guidelines on the national information markets.[167] When the market either was well advanced or extremely underdeveloped, the contents of the Synergy Guidelines were inappropriate. In other countries where there were more emerging markets, the Synergy Guidelines were useful and were used to influence decision making. For instance, the most advanced country was the United Kingdom, where there was a well-developed information market and there had been deliberate attempts for a number of years to develop national policies. Therefore, the Synergy Guidelines were of little use. The usefulness of the Synergy Guidelines in France was also debatable, since the discussion was already on a quite sophisticated level. In Spain, Greece, and Italy, the market was not developed enough because of the underdeveloped technical infrastructure and the fact that the

163. Commission of the European Communities, *Publaw 2. A Report to the Commission of the European Communities on an Evaluation of the Implementation of the Commission's Guidelines for Improving the Synergy between the Public and Private Sectors of the Information market* (Luxembourg: CEC Legal Advisory Board, 1993), 108.

164. Commission of the European Communities, *Publaw. Subject Report: General Access to Information Legislation* (Luxembourg: CEC Legal Advisory Board, 1991), s.p.

165. Commission of the European Communities, *Publaw 2. A report to the Commission of the European Communities on an evaluation of the implementation of the Commission's Guidelines for improving the synergy between the public and private sectors of the information market* (Luxembourg: CEC Legal Advisory Board, 1993), 6–8.

166. Ian Rowlands, 'Toward Public–Private Synergy in the European Information Services Market', *Journal of Government Information* 22, no. 3 (1995): 229–230.

167. *Ibid.*

information was still mostly maintained in paper form. A more basic information infrastructure was still needed.[168]

PUBLAW 2: WORKSHOP – The Publaw 2 report was accompanied by a workshop that brought together information service providers and users, publishers, representatives of the Member States and public bodies involved in the commercialization of public sector data and participants of several European Commission programmes and initiatives.[169] While the stakeholders all seemed to agree that the Synergy Guidelines provided a valuable framework for public-private synergy in the information market, the issues were found to be too complex to be directed by mere guidelines, and other initiatives were deemed to be necessary. However, opinions on the type of legal actions were divided. The public bodies denounced a legislative initiative of the Commission, while the private industry was mostly in favour of a directive, believing this instrument to be the only safeguard from the European information industry being trampled by global competition.[170]

PUBLAW 3: LEGAL INSTRUMENTS – As there was still no consensus, the European Commission ordered Publaw 3. The study, after examining the legal framework for re-use of public sector data in the Member States, advised the adoption of a legal instrument, preferably a directive, to provide a framework for the issues concerning re-use.[171] However, this advice still was not followed, as the same issue – adopt a directive or not – was raised once again in the Green Paper on Public Sector Information.[172]

3. The Stockholm Conference and the 1997 Bonn Ministerial Conference

TWO OBJECTIVES – In June 1996, the European Commission organized a conference in Stockholm entitled Access to Public Information: a Key to Commercial Growth and Electronic Democracy. The Commission stressed the importance of public sector data, both as an element in the democratic system and as a major vehicle for economic growth, providing a solution for one of Europe's biggest

168. Commission of the European Communities, *Publaw 2. A Report to the Commission of the European Communities on an Evaluation of the Implementation of the Commission's Guidelines for Improving the Synergy between the Public and Private Sectors of the Information Market* (Luxembourg: CEC Legal Advisory Board, 1993), 6.
169. *Ibid.*, 1.
170. Katleen Janssen & Jos Dumortier, 'Towards a European Framework for the Re-use of Public Sector Information: A Long and Winding Road', *International Journal of Law and Information Technology* 11, no. 2 (2003): 188–189; Ian Rowlands, 'Toward Public-private Synergy in the European Information Services Market', *Journal of Government Information* 22, no. 3 (1995): 235.
171. Commission of the European Communities, *Publaw 3. Final report* (Luxembourg: CEC Legal Advisory Board, 1995), 99.
172. Katleen Janssen & Jos Dumortier, 'Towards a European Framework for the Re-use of Public Sector Information: A Long and Winding Road', *International Journal of Law and Information Technology* 11, no. 2 (2003): 189.

problems: unemployment.[173] The objectives of the Commission were twofold: on the one hand, improving transparency for the benefit of citizens, and on the other hand, improving the synergy between the public and private sectors to help create a competitive European information industry.

CALL FOR A GREEN PAPER – During the conference, the need for a Green Paper on public sector information was expressed. In this way, a complex set of issues could be presented to a broad field of interested parties and the Commission could get a feel for the opinions on the matter from all the different players on the information market. Such a Green Paper was planned for the end of 1996, but it was only presented to the public in 1998.

MINISTERIAL CONFERENCE IN BONN – In the meantime, the issue of the availability of public sector data had been a topic at the 1997 Ministerial Conference in Bonn.[174] At this conference, the ministers of the Member States of the European Community, the European Free Trade Association (EFTA) countries, and the candidate Member States paid attention to the matters of access to and re-use of public sector data. They emphasised the importance of public sector data for the public bodies in the exercise of their tasks; for citizens to exercise their democratic rights and use public services; for the private sector, which needs information to determine its investment strategies; and for the information industry, which uses public sector data as resources to develop value-added products and services. The public and private sector both had a role to play in improving access to public sector data, and appropriate use should be made of the skills of the private sector in distributing data, as this would often prove to be cost-effective.

NEED FOR COMMON INITIATIVE – The differences between the Member States regarding access to public sector data and the practice of its re-use were considered an impediment for access of European citizens and industry, and for the possibilities of the information industry to create pan-European information products and services. Common initiatives were deemed necessary in order to benefit from the size and the economies of scale of the common market.[175]

4. The Green Paper on Public Sector Information

OBJECTIVE – In 1998, almost two years later than planned, the European Commission finally issued the Green Paper on Public Sector Information. The delay was due to discord between the Member States and also between the

173. Maeve McDonagh, 'Access to Public Information: A Key to Commercial Growth and Electronic Democracy – Stockholm Conference report', *Journal of Information, Law and Technology* no. 3 (1996): 2, <www2.warwick.ac.uk/fac/soc/law/elj/jilt/1996_3/mcdonagh/>, 22 Feb. 2010.

174. Katleen Janssen, Jo Steyaert & Roland Van Gompel, *Transparantie van overheidsinformatie in Vlaanderen. Beschouwingen en aanbevelingen* (Bruges: Die Keure, 2003), 132 [in Dutch].

175. Katleen Janssen, Jo Steyaert & Roland Van Gompel, *Transparantie van overheidsinformatie in Vlaanderen. Beschouwingen en aanbevelingen* (Bruges: Die Keure, 2003), 132–133 [in Dutch].

Directorates-General of the Commission.[176] The objective of the Green Paper on Public Sector Information was to 'undertake a broad public consultation involving all the actors concerned with a view to examining the main issues at stake and also to triggering a political discussion at European level'.[177]

NEED FOR A CONSISTENT POLICY ON PUBLIC SECTOR DATA – The Green Paper repeated the importance of public sector data, both for citizens and businesses, to improve the position of the European information industry on the world market and to provide new employment opportunities. To achieve this, a consistent EU-wide policy on re-use of public sector data was indispensable. However, the development of such a policy might raise a number of obstacles. These were tentatively summed up in the Green Paper on Public Sector Information.

CONSULTATION – The Green Paper launched a consultation, asking all stakeholders in the debate on the availability of public sector data for re-use to provide a response to a number of questions, addressing several issues that the Commission considered to be vital for developing a clear framework for public sector data. These questions concerned, for instance, the definition of public sector data, the impact of pricing policies, unfair competition, personal data, liability, barriers to re-use, and the actions that should be undertaken at the level of the EU.

RESPONSES TO THE CONSULTATION – The Green Paper was received with high interest among the Member State governments and public bodies, as well as citizens' organizations and industry members.[178] By the end of August 1999, a total of 185 responses to the consultation were sent to the Commission. The responses showed a general consensus that action was necessary at the level of the EU to overcome the barriers for re-use. However, the Green Paper was also criticized for putting too much emphasis on the re-use of public sector data, disregarding the importance of access for the exercise of fundamental and civil rights.[179]

PUBLIC HEARING – In May 1999 a public hearing was held in Brussels, where the relevance of the topic was recognized and again a consensus was reached that actions were needed to improve the access to and the exploitation of public sector data. As was the case in the Publaw 2 workshop,[180] private industry mostly seemed to be in favour of strong legislative action, while the Member States and public bodies were more reserved and advocated the introduction of common principles.[181]

---

176. Marc De Vries, 'Implementatie van de EU-richtlijn hergebruik overheidsinformatie in de Wob. Beter ten halve gekeerd dan ten hele gedwaald', *Nederlands Juristenblad* 79, no. 39 (2004): 2039 [in Dutch].
177. Commission of the European Communities, *Green Paper on Public Sector Information*, 1. See also Anne Fitzgerald, *Open Access Policies, Practices and Licensing: A Review of the Literature in Australia and Selected Jurisdictions* (Brisbane: Queensland University of Technology, 2009), 171–173.
178. Katleen Janssen & Jos Dumortier, 'Towards a European Framework for the Re-use of Public Sector Information: A Long and Winding Road', *International Journal of Law and Information Technology* 11, no. 2 (2003): 193.
179. Commission of the European Communities, *eEurope 2002 Communication. Annex*, 4.
180. Cf. *supra*, s. III.A.2.
181. Katleen Janssen, 'De exploitatie van overheidsinformatie: eindelijk een ontwerprichtlijn', no. 19 (2002): 113 [in Dutch]; Johan Pas & Bruno De Vuyst, 'Re-establishing the Balance

FOLLOW-UP – Even though the Green Paper on Public Sector Information was well received, the stakeholders had to wait another three years for the Commission's next action.[182] According to De Vries, the takeover of the position of Commissioner for the Information Society by the German Bangemann from the Finnish Liikanen caused the re-use issue to be back on the agenda after these years.[183] Yet, the next action was still not a legislative initiative, but a Commission Communication in the framework of the eEurope 2002 initiative.[184]

5. The eEurope 2002 Communication

FOCUS ON RE-USE – After a contemplative silence of over two years and the appointment of a new Information Society Commissioner, the European Commission decided that the time was right for the next step in the development of the framework for re-use of public sector data and issued a Communication on the subject.

Between the Public and the Private Sector: Regulating Public Sector Information Commercialization in Europe', *Journal of Information, Law and Technology* no. 2 (2004): 9, <www2.warwick.ac.uk/fac/soc/law/elj/jilt/2004_2/pasanddevuyst>, 22 Feb. 2010; Yvo Volman, 'Exploitation of Public Sector Information in the Context of the *eEurope* Action Plan', in *Public Sector Information in the Digital Age, Between Markets, Public Management and Citizens' Rights*, ed. Georg Aichholzer & Herbert Burkert (Cheltenham: Edward Elgar Publishing Limited, 2004), 101.

182. Nathalie Poupaert & Katleen Janssen, 'La directive du 17 Novembre 2003 concernant la réutilisation des informations du secteur public', *Revue du droit des technologies de l'information* no. 19 (2004): 31 [in French].
183. Marc De Vries, 'Implementatie van de EU-richtlijn hergebruik overheidsinformatie in de Wob. Beter ten halve gekeerd dan ten hele gedwaald', Nederlands Juristenblad 79, no. 39 (2004): 2039 [in Dutch].
184. The European Commission launched the eEurope initiative in 1999 to ensure that the EU would benefit fully from the changes brought on by the information society. Its key objectives were: 'bringing every citizen, home and school, every business and administration, into the digital age and online; creating a digitally literate Europe, supported by an entrepreneurial culture ready to finance and develop new ideas; and ensuring the whole process is socially inclusive, builds consumer trust and strengthens social cohesion'. The initiative covered ten action points, one of which was titled 'Government online'. The aim of this action was to make public information more easily accessible by extending and simplifying Internet access. This would also stimulate the development of new private sector services based on the new data sources that became available (Commission of the European Communities, *eEurope: An information society for all. Communication on a Commission Initiative for the Special European Council of Lisbon, 23 and 24 March 2000*, COM (99) 687 final, 8 Dec. 1999). The eEurope 2002 action plan revised the key target areas in the light of the Lisbon European Council of June 2000, but kept the essence of the objective (Council of the European Union and Commission of the European Communities, *eEurope 2002: An Information Society For All. Action plan prepared for the Feira European Council 19–20 June 2000*, <http://ec.europa.eu/information_society/eeurope/2002/action_plan/pdf/actionplan_en.pdf>, 22 Feb. 2010. This action plan was followed in 2002 by the eEurope 2005 action plan (Commission of the European Communities, *Communication to the Council, the European Parliament, the Economic and the Social Committee and the Committee of the Regions. eEurope 2005: An information society for all. An Action Plan to be presented in view of the Sevilla European Council, 21–22 June 2002*, COM (2002) 263 final, 28 May 2002).

In repetition, the Commission stressed the value of public sector data as an asset for the civic and democratic life of European citizens, businesses, and public bodies, but it emphasized that the Communication on PSI in particular addressed 'the issue of the exploitation of public sector information. No new proposals [were] made in relation to access to information'.[185]

Barriers for re-use – The Communication on PSI identified several types of barriers to a harmonized European information market. Many of them resulted from administrative rules or practices: differences in response time, the refusal to transmit the information in digital format (or in a specific pre-existing format), the need for the information industry to prove that it is directly concerned by the information, exclusive deals between public bodies and private companies, the requirement to consult data on the spot without obtaining a copy, and pricing policies.

Basic principle – The Communication's leading principle was a general right of re-use: whenever public sector data were generally accessible, commercial re-use should be possible. Fair trading and the prohibition of exclusive arrangements, as well as adequate and transparent pricing policies were also considered primary objectives.

Consultation – A new consultation round with the stakeholders was held, based on a working document distributed by the Directorate General Information Society in the beginning of 2002 that outlined a possible legal instrument for the re-use of public sector data.[186] One of the main objectives of the consultation was to examine which legal instrument would be the most appropriate to tackle these barriers.[187] Any such legal instrument should, according to the Communication, build upon the existing access regimes without proposing any changes to them. This legislative measure should be embedded in a coherent strategy, comprising the stimulation of concrete projects that exemplify the cross-border exploitation of PSI and the exchange of best practices between the different Member States.[188]

Responses to the consultation – A large number of responses from widely varying commentators were received. The majority of them seemed to be in favour of a directive on the re-use of public sector data.[189] One of the main complaints

185. Commission of the European Communities, *eEurope 2002 Communication*, 10.

186. Commission of the European Communities, *Towards a European Union framework for the exploitation of public sector information. Working document*, 2002, <ftp://ftp.cordis.europa.eu/pub/econtent/docs/consultation_0102.pdf>, 22 Feb. 2010.

187. Corien Prins, 'Access to Public Sector Information; In Need of Constitutional Recognition?', in *Public Sector Information in the Digital Age, Between Markets, Public Management and Citizens' Rights*, ed. Georg Aichholzer & Herbert Burkert (Cheltenham: Edward Elgar Publishing Limited, 2004), 64.

188. Yvo Volman, 'Exploitation of Public Sector Information in the Context of the *eEurope* Action Plan', in *Public Sector Information in the Digital Age, Between Markets, Public Management and Citizens' Rights*, ed. Georg Aichholzer & Herbert Burkert (Cheltenham: Edward Elgar Publishing Limited, 2004), 102.

189. Nathalie Poupaert & Katleen Janssen, 'La directive du 17 Novembre 2003 concernant la réutilisation des informations du secteur public', *Revue du droit des technologies de l'information* no. 19 (2004): 32 [in French]; Commission of the European Communities, *Towards a*

originated from various cultural institutions, such as galleries, archives, and museums, which felt that their particular positions were not taken into account. They feared that their existence would be endangered by the introduction of a general right of re-use, as their funding is growingly based on a combination of public funds and profitable activities.[190]

B. The PSI Directive

Timeline – On 5 June 2002, thirteen years after the Synergy Guidelines, the European Commission finally presented its proposal on the re-use of public sector information.[191] The final version of the directive was adopted formally on 17 November 2003 and had to be implemented by 1 July 2005.[192]

European Union framework for the exploitation of public sector information. Working document, 2002, <ftp://ftp.cordis.europa.eu/pub/econtent/docs/consultation_0102.pdf>, 22 Feb. 2010. The replies to the Working Document can be found at <http://cordis.europa.eu/econtent/psi/psi_re-use_consultation_replies.htm>, 22 Feb. 2010.

190. Katleen Janssen, 'De exploitatie van overheidsinformatie: eindelijk een ontwerprichtlijn', Mediarecht no. 19 (2002): 114 [in Dutch].

191. Commission of the European Communities, *Proposal for a Directive of the European Parliament and of the Council on the re-use and commercial exploitation of public sector documents*, COM (2002) 207 final, 5 Jun. 2002.

192. Only four Member States made this deadline: Denmark, France, Ireland and the United Kingdom. Other Member States followed slowly, and in December 2006, the European Commission took the five remaining countries that had not transposed the directive to the ECJ – Austria, Belgium, Portugal, Spain and Luxembourg (Rapid Press Release, 'Commission takes 5 countries to Court of Justice for failure on EU rules to re-use public sector information – a resource worth up to €48 billion', 2006, <http://europa.eu/rapid/pressReleasesAction.do?reference=IP/06/1891&format=HTML&aged=0&language=EN&guiLanguage=en>, 22 Feb. 2010. Belgium, Spain, Luxemburg and Austria were convicted by the ECJ. By 8 May 2008, all twenty-seven Member States notified complete transposition of the directive (see <http://ec.europa.eu/information_society/policy/psi/news/index_en.htm>, 22 Feb. 2010). In October 2008, the European Commission launched new procedures against Sweden and Poland for failing to fully implement the directive, by sending a letter of formal notice. In March 2009, it did the same against Italy. In June 2009, the Commission issued a reasoned opinion against Poland. By January 2010, all three countries announced changes to their legislation in order to bring it in line with the PSI Directive (Rapid Press Release, 'Re-use of Public Sector Information: Commission Launches Infringement Proceedings against Poland and Sweden', 2008, <http://europa.eu/rapid/pressReleasesAction.do?reference=IP/08/1524&type=HTML&aged=0&language=EN&guiLanguage=en>, 22 Feb. 2010; Rapid Press Release, 'Re-use of Public Sector Information: Commission Starts Infringement against Italy', 2009, <http://europa.eu/rapid/pressReleasesAction.do?reference=IP/09/425&format=HTML&aged=0&language=EN&guiLanguage=en>, 22 Feb. 2010; Rapid Press Release, 'Commission Steps up Legal Action against Poland for Incorrect Transposition of EU Public Sector Information Rules', 2009, <http://europa.eu/rapid/pressReleasesAction.do?reference=IP/09/1011&format=HTML&aged=0&language=EN&guiLanguage=en>, 22 Feb. 2010; European Commission, 'Public Sector Information: What's New', <http://ec.europa.eu/information_society/policy/psi/news/index_en.htm>, 22 Feb. 2010).

Focus – While the PSI Directive recognizes the democratic value of PSI, its focus is on the economic aspects of this information, linked to the common European information market. It proposes a number of actions with a view to overcoming the commercial barriers that exist at the European level, in particular with regard to the development of the position of the information industry on the European and on the global markets.[193]

1. Objective

Stimulation of the information industry – While the PSI Directive addresses any type of re-use, whether commercial or non-commercial, by private individuals, by the non-profit sector, by commercial companies, and so on, its main objective is the development of a framework that will stimulate the use of PSI by the information industry: 'the creation of Community-wide information products and services based on public sector documents, to enhance an effective cross-border use of public sector documents by private companies for added-value information products and services and to limit distortions of competition on the Community market'.[194]

Barriers to re-use – In 2002, cross-border re-use of PSI was only reserved for a limited number of big companies, which had enough resources and connections across the different Member States to find all the necessary data. The considerable differences in the rules and practices in the Member States 'constitute barriers to bringing out the full economic potential of this key document resource'.[195] These barriers needed to be addressed by minimum harmonization of the national rules and practices on the re-use of public sector documents 'in cases where the differences in national regulations and practices or the absence of clarity hinder the smooth functioning of the internal market and the proper development of the information society in the Community'.[196]

Benefits to different target groups – According to the European Commission's proposal, '[b]etter conditions for the exploitation of public sector information will therefore boost economic activity and job creation. In addition, a better use of public sector information will lead to other benefits for the citizens in the form of a range of added-value information products that the public sector itself cannot provide'. Thus, while the PSI Directive's main target group is the information industry, it should not only bring economic benefits through the growth of the information industry and employment opportunities, but it could also entail

193. Nathalie Poupaert & Katleen Janssen, 'La directive du 17 Novembre 2003 concernant la réutilisation des informations du secteur public', *Revue du droit des technologies de l'information* no. 19 (2004): 30 [in French]; Anne Fitzgerald, '*Open Access Policies, Practices and Licensing: a Review of the Literature in Australia and Selected Jurisdictions*' (Brisbane: Queensland University of Technology, 2009), 174.

194. Recital 25 of the PSI Directive.

195. Recital 6 of the PSI Directive.

196. *Ibid.*

advantages for the informed participation of the citizens in the information society as a secondary effect.

2. Title

TITLE V. TEXT – The title of the PSI Directive deserves some attention in its own right, because it is the only part of the PSI Directive that refers to *public sector information.* The recitals and the articles all use the term *public sector documents.* This raises the question of whether there was a reason for this difference or whether this was merely an accident in the wording, unintentionally creating confusion or inconsistency.

PREPARATORY PROCESS: DOCUMENT V. INFORMATION – The original Commission proposal referred to the 're-use and commercial exploitation of public sector documents'. On the one hand, this indicated that re-use and commercial exploitation were considered two different things. On the other hand, the title seemed to indicate the choice for a document system as opposed to an information system.[197] In its 2002 opinion, the Committee of the Regions regretted the title of the Commission proposal 'because, as is clear from the title of the Communication [of 2001], it is the exploitation of public sector information and not the exploitation of public sector documents that needs regulating, the latter already being covered by the legislation of some Member States'.[198] The Committee of the Regions seemed to prefer the information system over the document system, the former system implying that it is not the document that is re-used but rather the

197. The distinction that is sometimes made between information and document can be linked to the theories on *information systems* and *document systems* in legislation on access to public sector data. The document system provides a right of access to documents, that is, information materialized on a carrier, while in the information system the public authority has to provide the citizen with information on a certain topic or matter. The duty of the authority is to inform the citizen, and if necessary look for the information, which can be spread over different carriers, itself. The authority can decide under what form it makes the information available. In practice, the two systems cannot be found in their pure form, but appear in different mixes in national and sub-national access legislation. Both systems have merged to a greater or lesser extent in the different access legislations: as *document* usually implies any carrier of information, the information itself has become the point of focus, while the authorities have to provide the citizen with the information in the form of his choice, and can no longer decide for themselves how they inform the citizen (Commissie Grondrechten in het Digitale Tijdperk, 'Rapport Grondrechten in het Digitale Tijdperk', 2000, 182, <www.ivir.nl/dossier/grondrechten/bronnen/rapport_gdt_5-00.pdf>, 22 Feb. 2010 [in Dutch]; Frankie Schram, *Begrenzingen aan de openbaarheid van bestuur: een analyse en vergelijking van de internrechtelijke openbaarheidsregelingen in het Belgische recht*, Doctoral Thesis K.U.Leuven (Leuven: K.U.Leuven, 2002), 442 [in Dutch].

198. Committee of the Regions, *Opinion of 21 November 2002 on the Communication from the Commission to the Council, the European Parliament, the Economic and Social Committee and the Committee of the Regions eEurope 2002: creating a EU framework for the exploitation of public sector information; and the Proposal for a Directive of the European Parliament and of the Council on the re-use and commercial exploitation of public sector documents, OJ C* 73, 26 Mar. 2003, 38.

information resting within that document. The European Parliament had the same opinion and proposed including *public sector information* in the title instead of *public sector documents* and replacing *document* with *information* throughout the entire directive because 'the term "information" corresponds better with the concept intended than the term "document" does'.[199] The Council opted to keep the word *documents* because it felt that 'the notion of "information" is too vast and that the word "document" better reflects the essential concept of this Directive and is a clearer term with regard to its implementation'.[200] The European Commission replied to this that, '[t]here is no material difference between the two versions as long as the broad definition of "document" or "information" proposed by the Commission is maintained'.[201] After the second reading, the institutions could agree on a compromise: the title would refer to public sector *information*, while the rest of the directive would address the re-use of public sector *documents*.[202]

No difference? – Even though the PSI Directive perhaps should have rather been named the *PSD* Directive, the broad definition of *document* ensures that it does not matter whether public sector *information* or public sector *documents* are the subject of re-use.[203] While *information* is usually considered to refer to content, and *document* makes the link to the carrier of the content or the medium, the PSI

199. Committee on Industry, External Trade, Research and Energy, *Report on the proposal for a European Parliament and Council directive on the re-use and commercial exploitation of public sector documents*, doc. nr. A5-0025/2003, 29 Jan. 2003, <www.europarl.europa.eu/sides/getDoc.do?pubRef=-//EP//NONSGML+REPORT+A5-2003-0025+0+DOC+PDF+V0//EN>, 22 Feb. 2010; European Parliament, *Legislative resolution on the proposal for a European Parliament and Council directive on the re-use and commercial exploitation of public sector documents*, *OJ C* 43E, 19 Feb. 2004, 222.
200. Council of the European Union, *Common position adopted with a view to the adoption of a Directive of the European Parliament and of the Council on the re-use and commercial exploitation of public sector documents. Statement of the Council's reasons*, doc. nr. 7946/1/03, 26 May 2003, 6, <http://register.consilium.europa.eu/pdf/en/03/st07/st07946-re01ad01.en03.pdf>, 22 Feb. 2010.
201. Commission of the European Communities, *Communication to the European Parliament pursuant to the second subparagraph of Art. 251 (2) of the EC Treaty concerning the common position of the Council on the adoption of a directive on the re-use and commercial exploitation of public sector information*, SEC(2003) 627 final, 4, 28 May 2003.
202. European Parliament, *Legislative resolution on the Council common position for adopting a European Parliament and Council directive on the re-use of public sector documents OJ C* 77E, 26 Mar. 2004, 314; Commission of the European Communities, *Opinion pursuant to Art. 251(2), third subparagraph, point (c) of the EC Treaty, on the European Parliament's amendments to the Council's common position regarding the proposal for a directive of the European Parliament and of the Council on the re-use of public sector information, amending the proposal of the Commission pursuant to Art. 250(2) of the EC Treaty*, COM (2003) 620 final, 15 Oct. 2003; Council of the European Union, *Projet de Procès-verbal de la 2536ème session du Conseil de l'Union européenne 'Environnement', tenue à Luxembourg le 27 octobre 2003*, doc. nr. 14140/03, 18 Nov. 2003, <http://register.consilium.eu.int/pdf/fr/03/st14/st14140.fr03.pdf>, 22 Feb. 2010.
203. Katleen Janssen, 'Hergebruik van overheidsinformatie – binnenkort ook bij u in de winkel?', *Privacy & Informatie* 9, no. 2 (2006): 60 [in Dutch].

Directive defines *document* as 'any content whatever its medium',[204] and hence reunites both ideas.

3. Subject Matter and Scope

a. Introduction

ONLY MINIMUM REQUIREMENTS – The PSI Directive 'establishes a minimum set of rules governing the re-use and the practical means of facilitating re-use of existing documents held by public sector bodies of the Member States'.[205] Such a general framework was deemed needed to ensure 'fair, proportionate and non-discriminatory conditions for the re-use of such information'.[206] By limiting itself to this *minimum* harmonization, the PSI Directive leaves a considerable margin to the Member States and does not go beyond what is necessary to achieve the objective of reducing the differences between the rules and practices concerning the re-use of PSI corresponding to the subsidiarity and proportionality principles of EU law.[207] The Member States can always allow for more extensive re-use than what is imposed in the PSI Directive. While the directive only sets out minimum requirements, it does encourage the Member States to go beyond these minimum rules and adopt 'open data policies, allowing widespread use of documents held by public sector bodies'.[208]

b. Re-use of Existing Documents

NO CREATION OF DOCUMENTS – The PSI Directive only applies to the re-use of *existing* public sector documents. It cannot be expected that the public sector bodies create documents in order to comply with a request for re-use.[209]

PREPARATORY PROCESS: GENERALLY ACCESSIBLE DOCUMENTS – In the original proposal, the Commission not only wanted the documents to exist, but also to be *generally accessible*. Re-use would be possible for documents 'to which a right of access is granted under the rules established in the Member State for access to

204. Article 2.3 of the PSI Directive. Cf. *supra*, Ch. 2, s. III.D.2.
205. Article 1.1 of the PSI Directive.
206. Recital 8 of the PSI Directive.
207. The subsidiarity principle of Art. 5, 2nd paragraph of the Treaty on the functioning of the EU states that: 'In areas which do not fall within its exclusive competence, the Union shall take action in accordance with the principle of subsidiarity, only if and insofar as the objectives of the proposed action cannot be sufficiently achieved by the Member States and can therefore, by reason of the scale or effects of the proposed action, be better achieved by the Union'. The principle of proportionality lies in Art. 5, para. 3: 'Any action by the Union shall not go beyond what is necessary to achieve the objectives of this Treaty'.
208. Micus Management Consulting Gmbh, 'Assessment of the Re-use of Public Sector Information (PSI) in the Geographical Information, Meteorological Information and Legal Information Sectors', 2008, 9, <http://ec.europa.eu/information_society/policy/psi/docs/pdfs /micus_report_december2008.pdf>, 22 Feb. 2010.
209. Cf. *infra*, s. III.B.7.

documents as well as any document used by public sector bodies as an input for information products or services which they commercialise'.[210]

PREPARATORY PROCESS: GENERALLY ACCESSIBLE BECAUSE OF ACCESS RULES – On the one hand, the notion of *generally accessible* documents was introduced to show that the PSI Directive intended to build on the existing access regimes of the Member States. If documents could not be made public under national access legislation, they would obviously not be available for re-use:

> The measures facilitating the re-use of public sector information throughout the European Union build upon the existing access rules applied in the Member States. [The proposal] does not propose changes to these rules. Since the proposal for a Directive builds upon existing access regimes, it will not apply to documents concerning public security, defence, state security and the activities of the state in court proceedings as far as they are not covered by special regimes. In addition, documents or the parts of documents that contain commercially sensitive information, for example business secrets, will not be addressed to the extent that they are covered by the national access regimes.[211]

The Council agreed with the point the Commission was trying to make but proposed a different solution by including a reference to national access legislation in the field of application.[212] It felt that including a definition of *generally accessible document* would 'cause unnecessary confusion in this Directive, which handles re-use and not access to documents'.[213]

PREPARATORY PROCESS: GENERALLY ACCESSIBLE BECAUSE OF PUBLIC SECTOR BODY RE-USE – On the other hand, if documents were used by public sector bodies for their own commercial products or services, they were also considered to be generally accessible: 'The mere fact that a public sector body commercialises its information cannot justify the exclusion of this information from the scope of this proposal'.[214] If a public sector body uses its own data for creating commercial products or services, it cannot refuse re-use by other parties. As the Council did not want to include a definition of generally accessible documents, Recital 9 was formulated as a compromise: 'This Directive should apply to documents that are made accessible for re-use when public sector bodies licence, sell, disseminate, exchange or give

---

210. Commission of the European Communities, *Proposal for a Directive of the European Parliament and of the Council on the re-use and commercial exploitation of public sector documents*, COM (2002) 207 final, 16, 5 Jun. 2002.
211. *Ibid.*, 9.
212. Cf. *infra*, s. III.B.3.e.
213. Council of the European Union, *Common position adopted with a view to the adoption of a Directive of the European Parliament and of the Council on the re-use and commercial exploitation of public sector documents. Statement of the Council's reasons*, doc. nr. 7946/1/03, 26 May 2003, 4, <http://register.consilium.europa.eu/pdf/en/03/st07/st07946-re01ad01.en03.pdf>, 22 Feb. 2010.
214. Commission of the European Communities, *Proposal for a Directive of the European Parliament and of the Council on the re-use and commercial exploitation of public sector documents*, COM (2002) 207 final, 9, 5 Jun. 2002.

out information'. In any case, the Commission's concern was also appeased by Article 10 on non-discrimination.[215]

c. *Re-use by any Natural or Legal Person*

NO NATIONALITY, RESIDENCE, OR REGISTRATION REQUIREMENT – The possibility of requesting re-use of public sector documents is open to any natural or legal person or group thereof, regardless of nationality, residence, or registration. In the proposal of the European Commission, the availability of public sector documents for re-use was limited to 'any citizen of the Union, and any natural or legal person residing or having its registered office in a Member State'.[216] During the discussions at the Council level, the Dutch delegation asked why the text was limited to Union citizens and questioned the legality of limiting the scope to persons based within the EU. Advice from the Council Legal Service was asked on the legal effects of the inclusion or deletion of this limitation, particularly in the context of international trade agreements.[217] In line with the Access Directive, the reference to citizenship or residence within the EU was removed.

d. *Practical Means of Facilitating Re-use*

NEED FOR PRACTICAL TOOLS TO LOCATE PUBLIC SECTOR DOCUMENTS – The PSI Directive did not intend to provide only rules governing re-use, but also rules on the practical means to facilitate this re-use. This was introduced by the European Parliament, because, '[a]s a citizen or company one should have the possibility to know which information is available for this purpose, where and how to get it'.[218] The Parliament also suggested combining this statement in the general subject matter with a new article on practical tools facilitating the search for data[219] and an explanatory Recital 23:

> Tools that help potential re-users to find documents available for re-use and the conditions for re-use can considerably facilitate the cross-border use of

215. Cf. *infra*, s. III.B.8.
216. Commission of the European Communities, *Proposal for a directive of the European Parliament and of the Council on the re-use and commercial exploitation of public sector documents*, COM (2002) 207 final, 15, 5 Jun. 2002.
217. Working Party on Telecommunications, *Proposal for a Directive of the European Parliament and of the Council on the re-use and commercial exploitation of public sector documents. Outcome of proceedings*, doc. nr. 13740/02, 31 Oct. 2002, <http://register.consilium.europa.eu/pdf/en/02/st13/st13740.en02.pdf>, 22 Feb. 2010.
218. Committee on Industry, External Trade, Research and Energy, *Report on the proposal for a European Parliament and Council directive on the re-use and commercial exploitation of public sector documents*, doc. nr. A5-0025/2003, 29 Jan. 2003, <www.europarl.europa.eu/sides/getDoc.do?pubRef=-//EP//NONSGML+REPORT+A5-2003-0025+0+DOC+PDF+V0//EN>, 22 Feb. 2010; European Parliament, *Legislative resolution on the proposal for a European Parliament and Council directive on the re-use and commercial exploitation of public sector documents*, *OJ C* 43E, 19 Feb. 2004, 222.
219. Cf. *infra*, s. III.B.7.

public sector documents. Therefore Member States should ensure that practical arrangements are in place that help re-users in their search for documents available for re-use. Assets lists, preferably accessible online, of main documents (documents that are extensively re-used or that have the potential to be extensively re-used), and portal sites that are linked to decentralized assets lists are examples of such practical arrangements.

e. *Limitations on the Field of Application*

ARTICLE 1.2 OF THE PSI DIRECTIVE – Article 1.2 of the PSI Directive excludes a considerable number of documents from the field of application. The directive is not applicable to the following documents:

(a) documents the supply of which is an activity falling outside the scope of the public task of the public sector bodies concerned as defined by law or other binding rules in the Member State, or in the absence of such rules as defined in line with common administrative practices in the Member State in question;
(b) documents for which third parties hold IPR;
(c) documents which are excluded from access by virtue of the access regimes in the Member States, including on the grounds of:
   - the protection of national security (i.e., State security), defence or public security;
   - statistical or commercial confidentiality;
(d) documents held by public service broadcasters and their subsidiaries, and by other bodies or their subsidiaries for the fulfilment of a public service broadcasting remit;
(e) documents held by educational and research establishments, such as schools, universities, archives, libraries and research facilities including, where relevant, organizations established for the transfer of research results;
(f) documents, held by cultural establishments, such as museums, libraries, archives, orchestras, operas, ballets and theatres.

TWO TYPES OF EXEMPTIONS – All these exemptions were included to protect a certain private or public interest, albeit with very different reasons. One can distinguish two kinds of exemptions: on the one hand, the first three categories concern certain types of documents, independent of the public sector body holding them, while on the other hand, the last three exemptions refer to all documents held by a particular type of public sector body. Notably, the last three types of documents were mainly included because of the lobbying efforts of the organizations involved. In what follows, each exemption is discussed in more detail.

EFFECT – It should be mentioned that the effect of the exemptions is different. The exemptions under (b) and (c) entail that the documents do not have to be made available at all. The other exemptions have a different starting point. For these types of documents, the public sector bodies involved are allowed to refuse re-use, but the main purpose of the exemption is to allow them to make their own rules for re-use, meaning that they may impose stronger conditions and higher charges, or that they may consider exclusive arrangements.

i. Documents the Supply of which is an Activity Falling outside of the Public Task

NO COMMERCIAL INFORMATION PRODUCTS OR SERVICES – It is not only the private information industry that offers information products or services with an added value on the market. The public sector bodies themselves can also use the documents they created in the course of the performance of their public task for later commercial activities on the information market, such as adding value to information for specific clients or developing added-value products for a broad consumer market. These information products and services are exempted from the PSI Directive because, '[w]hile the basic information gathered with public money and in the exercise of the public task should be available for re-use, the commercial information products and services derived from that information should not'.[220] This would be contrary to the interests of the public sector bodies, which in many cases have to provide for at least part of their own funding through the provision of these commercial products and services.

SAFEGUARDS – However, the public sector documents on which these commercial products are based have to be available for re-use under the non-discrimination provision of Article 10 of the directive.[221] In addition, the PSI Directive's inapplicability to the commercial products of the public sector bodies does not mean that those products are not subject to any rules at all: European and national competition law will be applicable.[222]

PUBLIC TASK – Hence, the definition of *public task* is not only important for the determination of whether a public body is re-using data or sharing, but also for the determination of whether the public sector spatial data or spatial information service that it has created is part of its public task. If this is the case, the PSI Directive applies. If not, these documents do not have to be made available for re-use.

ii. Documents for which Third Parties Hold Intellectual Property Rights

EXEMPTION FOR THIRD-PARTY INTELLECTUAL PROPERTY RIGHTS – Public sector bodies hold many documents that they did not create themselves, and for which third

---

220. Commission of the European Communities, *Proposal for a directive of the European Parliament and of the Council on the re-use and commercial exploitation of public sector documents*, COM (2002) 207 final, 7, 5 Jun. 2002.
221. Cf. *infra*, s. III.B.
222. Katleen Janssen, 'Hergebruik van overheidsinformatie – binnenkort ook bij u in de winkel?', *Privacy & Informatie*, no. 2 (2006): 60–61 [in Dutch]; Katleen Janssen, 'INSPIRE and the PSI Directive: Public Task Versus Commercial Activities?', paper given at the 11th EC-GIS Workshop. 'ESDI:Setting the Framework (Alghero, 2005), <www.ec-gis.org/Workshops/11ec-gis/papers/303janssen.pdf>, 22 Feb. 2010; Katleen Janssen & Jan J.C. Kabel, 'Commercialisering van overheidsinformatie door de overheid: rechtspraak en wetgeving in België en Nederland. De honden blaffen, maar de karavaan trekt voort', *Computerrecht* no. 3 (2005): 126 [in Dutch].

parties hold the IPR, such as reports of research institutions, commercial or industrial information provided by private companies in order to comply with national legislation, and documents created by public servants for which a public sector body has not obtained the rights. These documents are exempted from the PSI Directive. The Member State or the public sector body cannot decide to make these documents available for re-use; this can only be decided by the third-party right holder, according to the applicable IPR rules.

ONLY THIRD-PARTY INTELLECTUAL PROPERTY RIGHTS – The exemption of the PSI Directive, unlike the exceptions in the Access Directive and the INSPIRE Directive, only involves documents for which third parties hold the IPR, and not all documents on which IPR are held. Hence, documents on which the public sector bodies themselves hold IPR are not included under the exception.[223] Does this mean that the public sector bodies cannot refuse a request for re-use of their documents on the basis of IPR? I believe so.

SAFEGUARDS – However, safeguards are foreseen in the PSI Directive to make sure that, even if the public sector bodies cannot refuse re-use merely because of their IPR,[224] '[t]he obligations imposed by this Directive shall apply only insofar as they are compatible with the provisions of international agreements on the protection of IPR, in particular the Berne Convention and the TRIPS Agreement'.[225] Thus, even if public sector bodies could not refuse re-use because of their IPR, they could still impose conditions on this re-use, in alignment with existing IPR legislation. This existing legislation also includes the Community acquis on IPR, as mentioned in Recital 24 of the PSI Directive:

> This directive is without prejudice to Directive 2001/29/EC[226] of the European Parliament and of the Council of 22 May 2001 on the harmonisation of certain aspects of copyright and related rights in the information society and

---

223. There was considerable pressure by the Spanish, French and German delegations in the Working Party on Telecommunications of the Council regarding the exemption of all documents subject to IPR, but they did not succeed (Working Party on Telecommunications, *Proposal for a Directive of the European Parliament and of the Council on the re-use and commercial exploitation of public sector documents. Outcome of proceedings*, doc. nr. 11095/02, 31 Jul. 2002, <http://register.consilium.europa.eu/pdf/en/02/st11/st11095.en02.pdf>, 22 Feb. 2010; Working Party on Telecommunications, *Proposal for a Directive of the European Parliament and of the Council on the re-use and commercial exploitation of public sector documents. Outcome of proceedings*, doc. nr. 13740/02, 31 Oct. 2002, <http://register.consilium.europa.eu/pdf/en/02/st13/st13740.en02.pdf>, 22 Feb. 2010.
224. We will see later on how relative this is, as the PSI Directive leaves the decision to allow re-use to the Member States and the public sector bodies (cf. *infra*, s. III.B.5). Hence, if the Member States did not include a general right of re-use in their national legislation, but also left the decision to the public sector bodies, these could still exclude their documents from re-use for IPR reasons.
225. Article 1.5 of the PSI Directive.
226. Directive 2001/29/EC of the European Parliament and of the Council of 22 May 2001 on the harmonization of certain aspects of copyright and related rights in the information society, *OJ L* 167, 22 Jun. 2001, 10.

Directive 96/8/EC[227] of the European Parliament and of the Council of 11 March 1996 on the legal protection of databases. It spells out the conditions within which public sector bodies can exercise their [IPR] in the internal information market when allowing re-use of documents.

FACILITATING RE-USE – However, while the PSI Directive fully recognizes the right of the public sector bodies to exercise their IPR, it does encourage them to do it 'in a way that facilitates re-use'.[228]

iii. Documents Excluded from Access

DIFFERENT OBJECTIVES – The purpose of the PSI Directive is not to give citizens access to public sector documents, but mostly to enable the use of these documents by the private sector in order to gain economic benefits. As was mentioned in Chapter 2, access and re-use are two fundamentally different concepts with different purposes.[229] Access intends to enable citizens to fully exercise their democratic rights and obligations, while re-use has an economic slant.

CLOSE RELATIONSHIP – However, even though the legal regimes on access and on re-use of public sector documents have a different purpose and different slant,

227. Directive 96/9/EC of the European Parliament and of the Council of 11 Mar. 1996 on the legal protection of databases, *OJ L* 77, 27 Mar. 1996, 20.

228. Recital 22 of the PSI Directive.

229. During the course of the co-decision procedure on the PSI Directive, it was shown that this distinction was not always clear for all the EU bodies. The European Economic and Social Committee, for instance, regretted in its opinion on the proposal for the PSI Directive 'that the Commission does not take account of the idea of "active promotion", which is an integral part of freedom of information legislation, and is complementary to access on request of public sector documents, which can be regarded as the "passive" form of freedom of information' (European Economic and Social Committee, *Opinion on the Proposal for a Directive of the European Parliament and of the Council on the re-use and commercial exploitation of public sector documents*, *OJ C* 85, 8 Apr. 2003, 25). In addition, the European Parliament contributed to the confusion by proposing an amendment to the recitals, aiming to include that: 'Making public all generally accessible documents held by the public sector – concerning not only the political process but hence also the legal and administrative process – is a fundamental instrument for extending the right to knowledge, which is a basic principle of democracy . . . Public administrations, at every institutional level, should therefore promote and encourage the re-use of information made available by them' (European Parliament, *Legislative resolution on the proposal for a European Parliament and Council directive on the re-use and commercial exploitation of public sector documents*, *OJ C* 43E, 19 Feb. 2004, 222). According to the Commission, this amendment established a link between accessibility of information and the right to knowledge and it was more about access than re-use, therefore it should be removed (Commission of the European Communities, *Amended proposal for a directive of the European Parliament and of the Council on the re-use and commercial exploitation of public sector information*, COM (2003) 119 final, 17 Mar. 2003). Yet the Parliament maintained its amendment in a different form and it was included, despite protests of the Commission, in Recital 16 of the PSI Directive: 'Making public all generally available documents held by the public sector – concerning not only the political process but also the legal and administrative process – is a fundamental instrument for extending the right to knowledge, which is a basic principle of democracy. This objective is applicable to institutions at every level, be it local, national or international'.

they cannot be seen as two completely separate entities. They remain closely related, in that the limitations imposed on access to a certain document will reflect on the limitations for re-use. It seems only logical that documents to which an individual cannot gain access because of over-riding other interests cannot be re-used for commercial or non-commercial purposes and in this way offered to the general public.[230] This is reflected in the exemption from the PSI Directive of documents that are excluded from access by virtue of the access regimes in the Member States. As was mentioned above, the European Commission proposed reflecting this reasoning by only allowing the re-use of *generally accessible* public sector documents, but the Council preferred including an exemption to the field of application.[231]

iv. Documents held by Public Service Broadcasters

PARTICULAR POSITION OF PUBLIC SERVICE BROADCASTERS – The fourth exemption to the field of application of the PSI Directive concerns all documents that are held by public service broadcasters and their subsidiaries, and by other bodies or their subsidiaries for the fulfilment of the public service broadcasting remit. The European Commission excluded these 'to avoid any doubts about the question whether public service broadcasters should be considered as public sector bodies within the definition of this Directive'.[232] The exclusion reflects 'their particular position which has been recognized in the Protocol attached to the Amsterdam Treaty'.[233]

NO DISCUSSION ON THIS EXEMPTION – This exemption was never under much discussion between the three institutions. They believed it was necessary to maintain the particular positions of the public broadcasting companies, which had been and are still struggling with reconciling their public task of ensuring pluralism and

230. Katleen Janssen, 'INSPIRE and the PSI Directive: Public Task Versus Commercial Activities?', paper presented at the 11th EC-GIS Workshop. 'ESDI: Setting the Framework (Alghero, 2005), <www.ec-gis.org/Workshops/11ec-gis/papers/303janssen.pdf>, 22 Feb. 2010; Katleen Janssen, 'Hergebruik van overheidsinformatie – binnenkort ook bij u in de winkel?', *Privacy & Informatie*, no. 2 (2006): 61 [in Dutch].

231. Cf. *supra*, s. III.B.3.b.

232. Commission of the European Communities, *Proposal for a directive of the European Parliament and of the Council on the re-use and commercial exploitation of public sector documents*, COM (2002) 207 final, 9, 5 Jun. 2002.

233. In the Protocol to the Treaty of Amsterdam on the system of public broadcasting in the Member States (*OJ C* 340, 10 Nov. 1997, 109), the Member States agreed that, considering that the system of public broadcasting in the Member States is directly related to the democratic, social and cultural needs of each society and to the need to preserve media pluralism, 'the provisions of the Treaty establishing the European Community shall be without prejudice to the competence of Member States to provide for the funding of public service broadcasting insofar as such funding is granted to broadcasting organizations for the fulfilment of the public service remit as conferred, defined and organised by each Member State, and insofar as such funding does not affect trading conditions and competition in the Community to an extent which would be contrary to the common interest, while the realisation of the remit of that public service shall be taken into account'.

informing citizens with the increasing pressure of budget cuts and competition from the private networks.

v. Documents held by Educational, Research, and Cultural Establishments

TYPES OF ESTABLISHMENTS – The last two exemptions address all documents held by educational and research establishments, as wel as cultural establishments. Examples of the former are schools, universities, archives, libraries, and research facilities, including, where relevant, organizations established for the transfer of research results. Examples of the latter are museums, libraries, archives, orchestras, operas, ballets, and theatres. The lists in the directive are not exhaustive, so the Member States can include other public sector bodies in this exemption if they can be considered as educational, research, or cultural establishments. In addition, no definition of these establishments is given, so this is also left to the Member States to decide upon.

REASONS FOR DISCUSSING BOTH TYPES TOGETHER – The two exceptions can be discussed in the same section, as, on the one hand, their raison d'être is based on comparable arguments, and on the other hand, the distinction between educational/research and cultural establishments is not always clear. For instance, libraries and archives are mentioned in both categories, and many of these institutions can be considered both as institutions in the field of research or education and as institutions in the area of culture.

REASONS FOR EXEMPTION – According to the proposal of the European Commission, '[c]ertain public sector bodies in the cultural and educational area merit a special treatment in view of a combination of different factors'. The Commission stated three reasons for their exclusion: 'The application of the Directive may cause a relatively high administrative burden for them in comparison to the benefits to be gained. Much of their information would anyhow fall outside the scope of the Directive in view of third party copyrights. Finally, their function in society as carriers of culture and knowledge gives them a particular position'.[234] In addition to the reasons given by the Commission, there

234. Commission of the European Communities, *Proposal for a directive of the European Parliament and of the Council on the re-use and commercial exploitation of public sector documents*, COM (2002) 207 final, 9, 5 Jun. 2002. The European Economic and Social Committee once again showed that it did not fully grasp the economic purpose of the directive. It showed its regret about the scope of the exclusions of the PSI Directive, with a particular concern about documents that are part of the 'common cultural heritage' and that are often held by public museums and public and university libraries. According to the Committee, '[a]s soon as it is implemented or within a reasonable period the directive should include such documents within its scope. Once they become accessible their dissemination will increase knowledge of Europe's cultural heritage, making it a source of unity, pride and a sense of belonging to Europe. At the same time their dissemination will have a beneficial impact on cultural exchanges in Europe and artistic and educational activities, not forgetting the economic knock-on effect for local tourism' (European Economic and Social Committee, *Opinion on the Proposal for a Directive of the European Parliament and of the Council on the re-use and commercial exploitation of public sector documents*, *OJ C* 85, 8 Apr. 2003, 25).

is another rationale that lies behind these exemptions: money and the lobbying activities of the particular sectors. The establishments mentioned are increasingly required to provide for at least part of their own funding. They are required to find a balance between providing the public with access to their work or collections for democratic prices and finding enough resources to maintain that work. Often, such resources are obtained from the sale of information products or services, such as books, reports, posters, CD-ROMs, and so on.[235] If these institutions would have to make their documents available for re-use under the conditions of the directive, they would become more dependent on public funds.[236]

FOLLOW-UP – In May 2009, a study performed for the European Commission showed that benefits of the cultural institutions being included in the field of application of the PSI Directive were difficult to assess. While there was potential for commercial re-use, the cultural institutions were concerned about the consequences of their activities. The report recommended that the situation would be closely monitored over the next years.[237] The European Commission's review of the PSI Directive also showed divided opinions on the possible inclusion of cultural and educational institutions. The Member States and institutions involved were generally not inclined towards inclusion, while the other stakeholders (mostly the private sector) were in favour.[238]

f. Documents Containing Personal Data

NO EXEMPTION – Documents containing personal data are not per se excluded from the possibility of re-use under the PSI Directive. However, the PSI Directive 'leaves intact and in no way affects the level of protection of individuals with

235. This requirement of self-financing is not only applicable to cultural, educational or research establishments. Many other examples can be found in the sectors of spatial data, statistical information, business information, etc. These institutions were not exempted from the field of application of the directive, but their concern was met by the provision of Art. 6 of the PSI Directive on charging (cf. *infra*, s. III.B.G.7.b.).
236. Katleen Janssen, 'INSPIRE and the PSI Directive: Public Task Versus Commercial Activities?', paper given at the 11th EC-GIS Workshop. 'ESDI – Setting the framework (Alghero, 2005), <www.ec-gis.org/Workshops/11ec-gis/papers/303janssen.pdf>, 22 Feb. 2010; Katleen Janssen & Jan J.C. Kabel, 'Commercialisering van overheidsinformatie door de overheid: rechtspraak en wetgeving in België en Nederland. De honden blaffen, maar de karavaan trekt voort', *Computerrecht* no. 3 (2005): 126 [in Dutch]; Johan Pas & Bruno De Vuyst, 'Re-Establishing the Balance Between the Public and the Private Sector: Regulating Public Sector Information Commercialization in Europe', *Journal of Information, Law and Technology* no. 2 (2004): 11, <www2.warwick.ac.uk/fac/soc/law/elj/jilt/2004_2/pasanddevuyst> 22 Feb. 2010.
237. Rob Davies et al., 'Economic and Social Impact of the Public Domain: EU cultural institutions and the PSI Directive', 2009, <http://ec.europa.eu/information_society/policy/psi/docs/pdfs/report/comm_0022_psi_v6.0.pdf>, 22 Feb. 2010.
238. Commission of the European Communities, *Staff working document. Accompanying document to the Communication from the Commission to the European Parliament, the Council, the European Economic and Social Committee and the Committee of the Regions on the re-use of Public Sector Information – Review of Directive 2003/98/C*, SEC (2009) 597, 15–16, 7 May 2009. Cf. *infra*, s. III.C.

regard to the processing of personal data under the provisions of Community and national law, and in particular does not alter the obligations and rights set out in Directive 95/46/EC'.[239]

APPLICATION OF DIRECTIVE 95/46/EC – This means that, if a Member State or a public sector body has to decide on the possible re-use of public sector documents holding personal data, they do not have to automatically refuse the re-use, but in deciding whether or not re-use can be allowed, they have to apply the principles of Directive 95/46/EC on the processing of personal data.[240] This corroborates with the opinion of the Article 29 Working Party, which agreed that the disclosure of personal data for the purpose of re-use is not forbidden per se, but that, '[t]he question of whether the data protection Directive allows the re-use of public sector information that contains personal data requires a careful and case-by-case assessment in order to strike the balance between the right to privacy and the right to public access'.[241]

IMPORTANCE OF THE FINALITY PRINCIPLE – According to the Article 29 Working Party, the finality principle set out in Article 6 of Directive 95/46 will be crucial for determining the possibility of re-use: personal data must be collected for specified purposes and not processed further in a way that is incompatible with those purposes. Often, the public sector bodies that deal with a request for re-use will not know the purpose of the re-use, as there is no obligation under the PSI Directive to declare the purposes for which the documents are requested. However, Directive 95/46/EC may require such a declaration if the disclosure of personal data is requested, in order for the authority to be in a position to assess whether certain re-use is compatible with the original purpose and lawful in general.[242]

ONLY IF A DOCUMENT IS NOT EXCLUDED FROM ACCESS – One should also keep in mind that the requirement to act according to the principles on the processing of personal data of Directive 95/46 may only come into play in a second decision cycle. As many pieces of national access legislation contain the possibility or the duty to refuse access because it might harm the protection of individuals' privacy, some of the public sector documents holding personal data will already be outside of the scope of the PSI Directive because they are 'excluded from access by virtue of the access regimes in the Member States'.[243] Only if this is not the case will the public sector body have to worry about the principles of protection of personal data with regard to re-use.

---

239. Article 1.4 of the PSI Directive.
240. Directive 95/46/EC of the European Parliament and of the Council of 24 Oct. 1995 on the protection of individuals with regard to the processing of personal data and on the free movement of such data, *OJ L* 281, 23 Nov. 1995, 31.
241. Article 29 Data Protection Working Party, 'Opinion 7/2003 on the re-use of public sector information and the protection of personal data – Striking the Balance', 2003, 11, <http://ec.europa.eu/justice_home/fsj/privacy/docs/wpdocs/2003/wp83_en.pdf>, 22 Feb. 2010.
242. *Ibid.*, 8.
243. Cf. *supra*, s. III.B.3.e.

4. Definitions

OUTLINE – Article 2 of the PSI Directive contains the definitions of the most important terms of the directive, including *public sector body, body governed by public law, document*, and *re-use*. Each of these is discussed below. Next, the lack of definitions of *documents held by public sector bodies* and *public task* is also examined.

a. *Public Sector Body*

REFERENCE TO CHAPTER 2 – As the definitions of *public sector body* and *body governed by public law* have already been addressed extensively in Chapter 2,[244] I will only remind the reader of these definitions. *Public sector body* means 'the State, regional or local authorities, bodies governed by public law and associations formed by one or several of such authorities or one or several of such bodies governed by public law'.[245] In turn, a *body governed by public law* means any body:

(1) established for the specific purpose of meeting needs in the general interest, not having an industrial or commercial character; and
(2) having legal personality; and
(3) financed, for the most part by the State, or regional or local authorities, or other bodies governed by public law; or subject to management supervision by those bodies; or having an administrative, managerial or supervisory board, more than half of whose members are appointed by the State, regional or local authorities or by other bodies governed by public law.[246]

b. *Document*

BROAD DEFINITION – Earlier in the section describing the PSI Directive, it was stated that, while the directive's title refers to re-use of public sector *information*, the rest of the directive uses the term *document*. As was already mentioned in Chapter 2,[247] the PSI Directive defines *document* as 'any content whatever its medium (written on paper or stored in electronic form or as a sound, visual or audio-visual recording); [or] any part of such content'.[248] According to the European Commission, this definition was sufficiently broad to reflect the needs of the Information Society: 'It covers all types of content, varying from audiovisual material to databases, digitized or not. It would, for example, be awkward in the Internet age to make a

244. Cf. *supra*, Ch. 2, s. IV.C.
245. Article 2.1 of the PSI Directive.
246. Article 2.2 of the PSI Directive.
247. Cf. *supra*, s. III.D.2
248. Article 2.3 of the PSI Directive.

distinction between the proceedings of a governmental conference and the web-cast of that same conference'.[249]

c. Re-use

DEFINITION – As was discussed in Chapter 2,[250] re-use under Article 2.4 of the PSI Directive means 'the use by persons or legal entities of documents held by public sector bodies, for commercial or non-commercial purposes other than the initial purpose within the public task for which the documents were produced. Exchange of documents between public sector bodies purely in pursuit of their public tasks does not constitute re-use'.

USE OUTSIDE OF THE PUBLIC TASK – Re-use comprises any use of documents held by public sector bodies, whether for commercial or non-commercial purposes, except for two types of use: use for the initial purpose within the public task for which the document was created and the exchange of documents between public sector bodies purely for public task purposes. Thus, the main criterion for determining whether or not the use of a document constitutes *re-use* is the performance of the public task. No distinction is made between commercial or non-commercial activities. Hence, this is a very broad definition: all commercial and non-commercial use outside of the public task. This includes re-use by the information industry for the creation of information products and services; re-use by non governmental organizations or not-for-profit organizations for publications or publicity material; re-use by students for papers and research; re-use by ordinary citizens on blogs and websites or other user-generated content applications, and so on.

NO USE WITHIN THE PUBLIC TASK – The definition excludes the exchange of documents between public sector bodies, but it does not mention the further use of the documents within the public sector body where they were created for purposes other than the initial one for which the documents were produced, but still within the performance of the public task. However, if the lack of a public task purpose is the criterion to determine whether there is re-use or not, it seems only logical that this kind of use is also excluded from the definition of re-use.[251] If disseminating documents to other public sector bodies for the performance of their public task is not re-use, disseminating the same documents within the public sector body itself for public task purposes will not be either.

RE-USE BY PUBLIC SECTOR BODIES – Public sector bodies can also re-use data if they perform activities outside of the public task. The directive is therefore also applicable if documents are made available to other public sector bodies that use

249. Commission of the European Communities, *Proposal for a directive of the European Parliament and of the Council on the re-use and commercial exploitation of public sector documents*, COM (2002) 207 final, 9, 5 Jun. 2002.

250. Cf. *supra*, Ch. 2, s. VI.B.

251. Katleen Janssen, 'Hergebruik van overheidsinformatie – binnenkort ook bij u in de winkel?', *Privacy & Informatie* 9, no. 2 (2006): 60 [in Dutch].

them for activities other than their public task or if the public sector body itself uses its own documents for any purpose outside of its public task, such as creating commercial or value-added products. The conditions for re-use of the directive will also be applicable to this exchange or further use of documents. This is especially important when the documents are also re-used by the private sector, ensuring that the private sector will obtain the information under the same conditions as the public sector body.[252]

PREPARATORY PROCESS: DEFINITION – The initial definition proposed by the European Commission was much more concise: 'the use by persons or legal entities of documents held by public sector bodies, for commercial or non-commercial purposes'.[253] In the Working Party on Telecommunications, the Belgian delegation wondered whether the very notion of *re-use* 'implied prior and privileged use of the data by the public administration, which could undermine competition',[254] since the definition of the Commission proposal did not make any distinction between the types of use that could be made of the documents. It entailed *any use*, without any distinctions, including the further use for public task purposes. Germany and Austria requested the deletion of *non-commercial purposes,*[255] entailing that use of public sector documents for non-commercial purposes would not constitute re-use. This would mean, for example, that the use of public sector documents by non-profit organizations for their activities would not fall under the notion of re-use. This would be in line with the original focus of the EU policy on re-use, that is, the stimulation of the information industry. To end this discussion, the Presidency proposed to the Working Party on Telecommunications the definition that is currently in the directive and a new clarification in the recitals: 'Public sector bodies collect, produce, reproduce and disseminate documents to fulfil their public tasks. Use of such documents for other reasons constitutes a re-use'.[256]

252. Cf. *infra*, s. III.B.8.

253. Commission of the European Communities, *Proposal for a directive of the European Parliament and of the Council on the re-use and commercial exploitation of public sector documents*, COM (2002) 207 final, 9, 5 Jun. 2002.

254. Working Party on Telecommunications, *Proposal for a Directive of the European Parliament and of the Council on the re-use and commercial exploitation of public sector documents. Outcome of proceedings*, doc. nr. 11095/02, 31 Jul. 2002, <http://register.consilium.europa.eu/pdf/en/02/st11/st11095.en02.pdf>, 22 Feb. 2010.

255. Working Party on Telecommunications, *Proposal for a Directive of the European Parliament and of the Council on the re-use and commercial exploitation of public sector documents. Outcome of proceedings*, doc. nr. 12633/02, 27 Sep. 2002, <http://register.consilium.europa.eu/pdf/en/02/st12/st12633.en02.pdf>, 22 Feb. 2010.

256. Presidency of the Council of the European Union, *Note to the Working Party on Telecommunications on the Proposal for a Directive of the European Parliament and of the Council on the re-use and commercial exploitation of public sector documents*, doc. nr. 12761/02, 7 Oct. 2002, <http://register.consilium.europa.eu/pdf/en/02/st12/st12761.en02.pdf>, 22 Feb. 2010; Working Party on Telecommunications, *Preparation on transport/telecommunications/energy Council on 5/6 December 2002 on the Proposal for a Directive of the European Parliament and of the Council on the re-use and commercial exploitation of public sector documents. Outcome of proceedings*, doc. nr. 13225/02, 10 Oct. 2002, <http://register.consilium.europa.eu/pdf/en/02/st13/st13225.en02.pdf>, 22 Feb. 2010.

This would ensure that the public sector bodies could at least use their documents for the policy purposes they were created for, without automatically falling under the scope of the directive and subsequently being required to make their information available to the private sector. The Finnish delegation proposed also addressing the question of the exchange of data between public bodies in this context,[257] not only further use for the public task within the same public sector body, but also the exchange between public sector bodies for those purposes. This was added in the next meeting of the Working Party on Telecommunications[258] and in the political agreement, 'in order to avoid ambiguity concerning the exchange of documents between public sector bodies in pursuit of their public tasks'.[259]

d. Held by

No definition, but clarification – A definition that is not included in the PSI Directive, because it was not deemed necessary, was documents *held by* the public sector bodies. As it was already a point of discussion between the European Parliament, the Commission, and the Council, one can only assume that it might lead to difficulties in the national legislation transposing the PSI Directive, and especially for the national courts that have to interpret this legislation. However, Recital 11 of the PSI Directive gives some guidance: 'A document held by a public sector body is a document where the public sector body has the right to authorise re-use'.

e. Public Task

No definition – Another definition that is ostentatiously absent from the PSI Directive is *public task*.[260] The only reference that is made to it lies in Recital 9: 'Activities falling outside the public task will typically include supply of

257. Working Party on Telecommunications, *Preparation on transport/telecommunications/energy Council on 5/6 December 2002 on the Proposal for a Directive of the European Parliament and of the Council on the re-use and commercial exploitation of public sector documents. Outcome of proceedings*, doc. nr. 13225/02, 10 Oct. 2002, http://register.consilium.europa.eu/pdf/en/02/st13/st13225.en02.pdf>, 22 Feb. 2010.
258. Working Party on Telecommunications, *Proposal for a Directive of the European Parliament and of the Council on the re-use and commercial exploitation of public sector documents. Outcome of proceedings*, doc. nr. 13740/02, 31 Oct. 2002, <http://register.consilium.europa.eu/pdf/en/02/st13/st13740.en02.pdf>, 22 Feb. 2010.
259. Council of the European Union, *Common position adopted on 26 May 2003 with a view to the adoption of a Directive of the European Parliament and of the Council on the re-use and commercial exploitation of public sector documents. Statement of the Council's reasons*, doc. nr. 7946/1/03, 4, <http://register.consilium.europa.eu/pdf/en/03/st07/st07946-re01ad01.en03.pdf>, 22 Feb. 2010.
260. See also Michael Kamps, 'Thunder and Lightning: Public Sector Information Policy Experiences of Private Meteorological Service Providers', in *Public Sector Information in the Digital Age: Between Markets, Public Management and Citizens' Rights*, ed. Georg Aichholzer & Herbert Burkert (Cheltenham: Edward Elgar Publishing Limited, 2004), 183–184.

documents that are produced and charged for exclusively on a commercial basis and in competition with others in the market'.

COMPETENCE OF THE MEMBER STATES – As was already indicated in Chapter 2,[261] the notion of *public task* is pivotal in determining whether information is made available for *access*, *re-use* or *sharing*. However, in the original proposal for the PSI Directive, the European Commission already made it clear that it did not intend to harmonize the scope of the public tasks assigned by the Member States.[262] It left the scope of the public task to be defined by the legislation or the practices of the Member States, even though in the Council discussions, further clarification was repeatedly requested.[263]

5. General Principle

NO OBLIGATION TO ALLOW RE-USE – Article 3 of the PSI Directive contains the general principle for re-use of public sector documents: 'Member States shall ensure that, where the re-use of documents held by public sector bodies is allowed, these documents shall be re-usable for commercial or non-commercial purposes in accordance with the conditions set out in Chapters 3 and 4. Where possible, documents shall be made available through electronic means'. Hence, the PSI Directive does not impose any obligation on the Member States to make public sector documents available for re-use. Only if they, or the public sector bodies themselves, choose to do so, will they have to comply with the obligations of the directive.

PREPARATORY PROCESS: NO OBLIGATION TO ALLOW RE-USE – The Commission proposal started from the same viewpoint: the directive 'does not oblige the public sector bodies to allow the re-use of specific documents, but Member States are invited to stimulate public sector bodies to make the documents available for re-use'.[264] Most of the Member States' representatives in the Council expressed support for the principle, even though the Swedish, Dutch, and Finnish delegation considered that the impact of the directive would be greatly reduced by this

261. Cf. *supra*, Ch. 2, ss VI.B. and VI.C.
262. Commission of the European Communities, *Proposal for a directive of the European Parliament and of the Council on the re-use and commercial exploitation of public sector documents*, COM (2002) 207 final, 8, 5 Jun. 2002.
263. Working Party on Telecommunications, *Proposal for a Directive of the European Parliament and of the Council on the re-use and commercial exploitation of public sector documents. Outcome of proceedings*, doc. nr. 11095/02, 31 Jul. 2002, <http://register.consilium.europa.eu/pdf/en/02/st11/st11095.en02.pdf>, 22 Feb. 2010; Working Party on Telecommunications, *Proposal for a Directive of the European Parliament and of the Council on the re-use and commercial exploitation of public sector documents. Outcome of* proceedings, doc. nr. 12633/02, 27 Sep. 2002, <http://register.consilium.europa.eu/pdf/en/02/st12/st12633.en02.pdf>, 22 Feb. 2010.
264. Commission of the European Communities, *Proposal for a directive of the European Parliament and of the Council on the re-use and commercial exploitation of public sector documents*, COM (2002) 207 final, 9-10, 5 Jun. 2002.

reserved approach.[265] After realizing that a stronger general principle would not be possible, the Dutch delegation set its goal on a limited category of documents, proposing that, '[o]fficial texts of a legislative, administrative, and legal nature, and the official translations of such texts, will be made available for re-use'.[266] The European Parliament, of which the Rapporteur was also Dutch, proposed during the first reading a comparable distinction between *basic information of the democratic constitutional state* and other information, the former always being available for re-use free of charge and comprising at least 'texts of legislation and regulations, judicial decisions and the information of representative bodies (e.g., parliamentary information)'.[267] The Council was not happy with either suggestion, because they referred to 'access issues and transparency of access vis-à-vis the citizens in general, whereas the Directive concerns the re-use of documents'.[268] The Commission agreed to this, and the Parliament dropped its amendment during the second reading.[269]

---

265. Working Party on Telecommunications, *Proposal for a Directive of the European Parliament and of the Council on the re-use and commercial exploitation of public sector documents. Outcome of proceedings*, doc. nr. 11095/02, 31 Jul. 2002, <http://register.consilium.europa.eu/pdf/en/02/st11/st11095.en02.pdf>, 22 Feb. 2010; Working Party on Telecommunications, *Proposal for a Directive of the European Parliament and of the Council on the re-use and commercial exploitation of public sector documents. Outcome of proceedings*, doc. nr. 12633/02, 27 Sep. 2002, <http://register.consilium.europa.eu/pdf/en/02/st12/st12633.en02.pdf>, 22 Feb. 2010.
266. Working Party on Telecommunications, *Proposal for a Directive of the European Parliament and of the Council on the re-use and commercial exploitation of public sector documents. Outcome of proceedings*, doc. nr. 13740/02, 31 Oct. 2002, <http://register.consilium.europa.eu/pdf/en/02/st13/st13740.en02.pdf>, 22 Feb. 2010.
267. Committee on Industry, External Trade, Research and Energy, *Report on the proposal for a European Parliament and Council directive on the re-use and commercial exploitation of public sector documents*, doc. nr. A5-0025/2003, 29 Jan. 2003, <www.europarl.europa.eu/sides/getDoc.do?pubRef=-//EP//NONSGML+REPORT+A5-2003-0025+0+DOC+PDF+V0//EN>, 22 Feb. 2010; European Parliament, *Legislative resolution on the proposal for a European Parliament and Council directive on the re-use and commercial exploitation of public sector documents, OJ C* 43E, 19 Feb. 2004, 222.
268. Council of the European Union, *Common position adopted with a view to the adoption of a Directive of the European Parliament and of the Council on the re-use and commercial exploitation of public sector documents. Statement of the Council's reasons*, doc. nr. 7946/1/03, 26 May 2003, 7, <http://register.consilium.europa.eu/pdf/en/03/st07/st07946-re01ad01.en03.pdf>, 22 Feb. 2010.
269. The European Economic and Social Committee had also pleaded for the inclusion of a category of information that would be available at less stringent conditions than other information. It reiterated its opinion on the Green Paper, which stated that 'a distinction must be drawn between information essential to citizens, especially that which relates to the exercise of democratic rights – [. . .] – and information for commercial purposes [. . .]'. Essential information should include the following in the Committee's eyes:
    - constitutions, codes, laws and regulatory acts, treaties and jurisprudence emanating from the European Union;
    - statistical information, reports and studies of general interest;
    - legal, regulatory and political information, such as the minutes of decision making or advisory assemblies.

EVALUATION OF THE DEFINITION – The combination of the confined general principle and the extensive list of exemptions might lead one to believe that the impact of the PSI Directive will be limited to a gentle encouragement towards the availability of public sector documents, without any enforceable obligations for the Member States. However, upon closer examination, the PSI Directive does have some more far-reaching implications, which should cause changes in the activities of the public sector bodies. By making it explicit that documents should be available for commercial and non-commercial purposes, the PSI Directive ensures that no distinction can be made between both types of re-use as far as the availability itself of the documents is concerned.[270] From the moment a document is made available for one kind of re-use, it also has to be available for re-use with any other purpose.[271] Hence, the public sector body cannot decide to allow non-commercial re-use while excluding re-use for commercial purposes, or decide to use its own information for creating its own commercial products and refuse to provide the information to private competitors. Only the conditions or charges may differ, within the limits of the PSI Directive. As re-use includes any use for purposes outside of the public task, this entails that any information that is not used by a public sector body purely for the pursuit of its public task will also have to be available to the private sector for non-commercial and commercial re-use.[272]

The scope of essential information could be extended over time: '[i]nitially, "essential information" will be information from the public authorities in the Member States. Later, within a reasonable period, the Member States will be asked to make available similar documents from their public bodies on the same terms'. The category of essential information from the European Economic and Social Committee was wider than the basic information proposed by the Committee on Industry, External Trade, Research and Energy, and the official texts from the Dutch delegation in the Working Party on Telecommunications. Again, it seems that the mind of the Committee was more on the idea of access of the citizen, and the need of the citizen to be informed to be able to enjoy his democratic rights, than on the actual concept of re-use of PSI (European Economic and Social Committee, *Opinion on the Proposal for a Directive of the European Parliament and of the Council on the re-use and commercial exploitation of public sector documents*, *OJ C* 85, 8 Apr. 2003, 25; European Economic and Social Committee, *Opinion on Public Sector Information: a key resource for Europe – Green Paper on public sector information in the information society*, *OJ C* 169, 16 Jun. 1999, 30).

270. Katleen Janssen, 'INSPIRE and the PSI Directive: Public Task Versus Commercial Activities?', paper given at the 11th EC-GIS Workshop. 'ESDI – Setting the framework (Alghero, 2005), <www.ec-gis.org/Workshops/11ec-gis/papers/303janssen.pdf>, 22 Feb. 2010; Katleen Janssen, 'Hergebruik van overheidsinformatie – binnenkort ook bij u in de winkel?', *Privacy & Informatie*, no. 2 (2006): 59–63 [in Dutch].

271. Axelle Hovine, 'La directive du 17 Novembre 2003 concernant la réutilisation des informations du secteur public: présentation et préfiguration des incidences de transposition en droit français', *Gazette du Palais* no. 65/66 (2004): 756 [in French]; Katleen Janssen & Jan J.C. Kabel, 'Commercialisering van overheidsinformatie door de overheid: rechtspraak en wetgeving in België en Nederland. De honden blaffen, maar de karavaan trekt voort', *Computerrecht* no. 3 (2005): 126 [in Dutch].

272. Cf. *infra*, s. III.B.8.

6. Requirements for the Processing of Requests for Re-use

Outline – The second chapter of the PSI Directive contains one article, Article 4, which addresses a number of requirements for the processing of a request for re-use by the public sector bodies. These requirements concern time limits, communication of the grounds for refusal, and references to the possible means of redress. Article 4 does not apply to public service broadcasters, educational and research establishments, or cultural establishments.[273]

a. Time Limits

Consistency with time limits in national access legislation – According to Article 4.1 of the PSI Directive, '[p]ublic sector bodies shall, through electronic means where possible and appropriate, process requests for re-use and shall make the document available for re-use to the applicant or, if a licence is needed, finalize the licence offer to the applicant within a reasonable time that is consistent with the timeframes laid down for the processing of requests for access to documents'. If freedom of information legislation has been adopted, the Member States should look at this legislation to determine the time limits. However, the PSI Directive requires the interpretation of *reasonable time* to be *consistent* with freedom of information legislation, but not *identical*. What exactly is meant by *consistent*? Do the time limits need to be the same as those for access to public sector documents, or is it enough that they are proportional? Recital 12 of the PSI Directive adds that the 'time limit for replying to requests for re-use should be reasonable and *in line*[274] with the equivalent time for requests to access the document under the relevant access regimes'. In case of public sector spatial data, this means that the time limits of the Access Directive apply.

Time limits in the absence of national access legislation – Some Member States did not have fully developed freedom of information legislation at the time the PSI Directive was proclaimed. For example, the United Kingdom's Freedom of Information Act was only fully implemented in 2005, and the German Federal Freedom of Information Act only came into force in 2006. Luxembourg and the national level in Spain still do not have any access legislation. To ensure that this would not leave the possibility that the Member States or the public sector bodies could determine time limits long enough to discourage requests for re-use or make creating added-value products based on the public sector documents economically uninteresting due to the long waiting periods and the short-term relevance of the documents involved, the PSI Directive sets its own time limit: 'Where no time limits or other rules regulating the timely provision of documents have been

273. Presidency of the Council of the European Union, *Note to the Working Party on Telecommunications on the Proposal for a Directive of the European Parliament and of the Council on the re-use and commercial exploitation of public sector documents*, doc. nr. 12761/02, 7 Oct. 2002, <http://register.consilium.europa.eu/pdf/en/02/st12/st12761.en02.pdf>, 22 Feb. 2010.

274. Emphasis added.

established, public sector bodies shall process the request and shall deliver the documents for re-use to the applicant or, if a licence is needed, finalise the licence offer to the applicant within a timeframe of not more than 20 working days after its receipt. This timeframe may be extended by another 20 working days for extensive or complex requests. In such cases the applicant shall be notified within three weeks after the initial request that more time is needed to process it'.[275] Regarding the life span of certain types of information, such as traffic data or weather reports, Recital 12 adds that, '[o]nce a request for re-use has been granted, public sector bodies should make the documents available in a timeframe that allows their full economic potential to be exploited. This is particularly important for dynamic content (e.g. traffic data), the economic value of which depends on the immediate availability of the information and of regular updates'.

TIME LIMITS IN CASE OF LICENSES – In the case where a license is proposed to the applicant, the time limit of twenty working days only refers to the first license offer. The PSI Directive does not add a following time limit for the actual delivery of the documents after the license conditions have been accepted by the applicant. On the one hand, this is understandable, as it is also up to the applicant to respond to the license offer and accept it, or to try to negotiate it. If the applicant is not concerned with a speedy completion of the license, a timetable should not be forced upon the public sector body. On the other hand, the absence of a time limit for providing the documents in case they are subject to a license gives the public sector body the possibility of dragging out the negotiation process, in practice making it impossible for the applicant to react quickly to opportunities in the market. Moreover, it will not be easy for the applicant to search for redress in this situation, as he or she will have to prove that the public sector body's attitude can in practice be considered as a refusal.

PREPARATORY PROCESS: TIME LIMITS IN CASE OF LICENSES – The original Commission proposal took situations like this into account and suggested a time limit that was 'a total deadline covering both processing the request and dispatching the information',[276] but this was unacceptable for the Council. The Council Presidency proposed a compromise and set the timeframe for the delivery of the documents *or* the finalizing of the license agreement.[277] Note that the Presidency proposed three weeks to finalize the *agreement*, instead of the *offer*, as it appears in the final version of the directive. Thus, the burden on the public sector body is much lighter in the final version of the PSI Directive. It is only obliged to get the first offer for a license out within the time limit of the access legislation or twenty

275. Article 4.2 of the PSI Directive.

276. Commission of the European Communities, *Proposal for a directive of the European Parliament and of the Council on the re-use and commercial exploitation of public sector documents*, COM (2002) 207 final, 17, 5 Jun. 2002.

277. Presidency of the Council of the European Union, *Note to the Working Party on Telecommunications and Information Society Services on the proposal for a Directive of the European Parliament and of the Council on the re-use and commercial exploitation of public sector documents*, doc. nr. 5862/03, 24 Jan. 2003, <http://register.consilium.europa.eu/pdf/en/03/st05/st05862.en03.pdf>, 22 Feb. 2010.

working days. If it had to finalize the agreement, the public sector body would have also been dependent on the applicant to accept the offer in time.

b. *Communication of the Grounds for Refusal and the Means of Redress*

ARTICLE 4.3 OF THE PSI DIRECTIVE – In the event of a negative decision, the public sector body has to communicate the grounds for refusal to the applicant on the basis of the relevant provisions of the national access legislation[278] or of the national provisions that implement the PSI Directive, in particular regarding the exemptions and the general principle of the directive.[279] Any negative decision also has to contain a reference to the means of redress in case the applicant wishes to appeal the decision.

REFUSAL BECAUSE OF THIRD-PARTY INTELLECTUAL PROPERTY RIGHTS – If a request for re-use is turned down because the public sector document is protected by third-party IPR, the public sector body has to inform the applicant of the natural or legal person who is the right holder, when known, or alternatively, of the licensor from which the public sector body has obtained the relevant material. Contrary to the Access Directive, the PSI Directive does not give the public sector body the choice between referring the applicant to the right holder or transferring the request to the right holder itself. This solves the problem of determining how to apply the time limit: starting from the original request or from the moment the right holder receives the request for information.

EVALUATION OF ARTICLE 4.3 – The obligation to state the reasons for refusing a request for re-use and the possibilities of redress are among the most important merits of the PSI Directive. Before the directive, in many cases, the public sector body could refuse a request for re-use without informing the applicant of the reason, and the possibilities of appealing a decision from the public sector body were often unclear. Thanks to the PSI Directive, anyone faced with a refusal from a public sector body to deliver documents will now know whom to address.[280]

MEANING OF A NEGATIVE DECISION – One could ask the question of what exactly is meant by a negative decision. Obviously, a refusal to make public sector documents available constitutes a negative decision. But what happens when the license offer is not according to the rules of the PSI Directive or the national provisions based on the directive? And what happens if the public sector body does not reply at all within the time limit? Should this be considered a negative decision?[281]

278. This entails that any request for re-use will have to be checked against the national access legislation first. Only if the document can be made available for access, the question can be considered whether or not it is available for re-use.
279. Article 4.3 of the PSI Directive.
280. Nathalie Poupaert & Katleen Janssen, 'La directive du 17 Novembre 2003 concernant la réutilisation des informations du secteur public', *Revue du droit des technologies de l'information* no. 19 (2004): 37 [in French].
281. Katleen Janssen, 'De exploitatie van overheidsinformatie: eindelijk een ontwerprichtlijn', *Mediarecht* no. 19 (2002): 118 [in Dutch].

Availability of means of redress – As the directive requires a negative decision to mention the possible means of redress, this implies that the Member States have to make sure that appropriate means of redress are in place. How these means are organized is left to the Member States. The directive does not mention anything in this article about means of redress against unreasonable licensing conditions or charges, or about the lack of response within the time limit, or about the public sector body not delivering the information as agreed upon in the license. However, it does mention in Article 7 on transparency that the applicants have to be informed of the available means of redress.[282] Recital 15 adds that '[a]pplicants for re-use of documents should be informed of available means of redress relating to decisions or practices affecting them. This will be particularly important for small and medium enterprises, which may not be familiar with interactions with public sector bodies from other Member States and corresponding means of redress'.

7. Conditions for Re-use

Outline – If public sector bodies make their documents available for re-use, either based on an obligation required by the Member State or on their own decision, then not only the treatment of the request, but also the delivery of the documents is subject to a number of conditions. These conditions include the format of the documents, the charging principles, and obligations concerning transparency, licensing, and the practical arrangements for facilitating the search for documents.

a. Available Formats

Any pre-existing format – Public sector bodies should make their documents available 'in any pre-existing format or language, through electronic means where possible and appropriate'.[283] This is in line with the principle of many existing pieces of access legislation, whereby a citizen can choose the format in which he or she wants to obtain the documents he or she requested access to, if the document is available in different formats. The possibility of choosing the format is even more important for requests for re-use, as '[t]he possibilities for re-use can be improved by limiting the need to digitise paper-based documents or to process digital files to make them mutually compatible'. Moreover, the public sector body should, to the greatest extent possible, try to make its documents available in a format that 'is not dependent on the use of specific software'.[284]

282. Cf. *infra*, s. III.B.7.c.
283. Article 5.1 of the PSI Directive.
284. Recital 13 of the PSI Directive. The European Parliament adopted an amendment pleading for a more explicit reference to the use of open standards, and even open source software: 'The public authorities must make available to any citizen of the Union and any natural or legal person residing or having its registered office in a Member State their own documents in a format which as far as possible is not dependent on the use of specific software subject to payment, in particular by favouring systems with publicly accessible codes. Where possible

NO OBLIGATION TO CREATE OR ADAPT DOCUMENTS – However, the public sector bodies are only under the obligation to make their documents available in any pre-existing format. They cannot be required 'to create or adapt documents in order to comply with the request'.[285] Thus, public sector bodies do not have to create or adapt documents in order to answer concrete questions of applicants. They only have to make available what they already have, because they have created it during the performance of their public task.[286] It would be too great a burden on the public servants and on the functioning of the public sector bodies to demand of them the creation of new documents. In addition, there is no obligation to 'provide extracts from documents where this would involve disproportionate effort, going beyond a simple operation'.[287] *A contrario*, one could infer that extracts of which creation would not cause any disproportionate effort should be delivered upon request. In line with the Access Directive,[288] an extract could be delivered because the applicant is only interested in a part of the document or because other parts of the document cannot be made available, for example, if they contain confidential commercial information or information on national or public security.

NO OBLIGATION TO CONTINUE PRODUCTION WITH A VIEW TO RE-USE – The PSI Directive does not require the public sector bodies to 'continue the production of a certain type of documents with a view to the re-use of such documents by a private or public sector organization'.[289] Public sector bodies should not become

and appropriate, all the information must be presented not only in graphic form but also in textual mode, to enable people with sensory difficulties to use it' (European Parliament, *Legislative resolution on the proposal for a European Parliament and Council directive on the re-use and commercial exploitation of public sector documents*, *OJ C* 43E, 19 Feb. 2004, 222). This amendment was not accepted by the Council, and not repeated in the Parliament's second reading.

285. The European Economic and Social Committee believed that re-use should be promoted in a much broader way, because users would be inclined to 'develop new services if public sector bodies provide them with information which is adapted to changes in economic trends and published on a long-term basis'. Therefore it proposed to delete the provision that there is no obligation for public sector bodies to create or adapt documents. Instead it suggested that 'the directive should encourage public sector bodies to examine requests for documents to be created or adapted and to meet such requests where they are justified by a legitimate interest (public, economic, social, cultural or educational interest, or for scientific or university research purposes)' and to include in para. 2 of Art. 5 that '[p]ublic bodies must ensure that they do not stop the production of certain types of documents when these are used commonly and regularly by other public or private bodies, or else, where necessary, they must provide alternative resources' (European Economic and Social Committee, *Opinion on the Proposal for a Directive of the European Parliament and of the Council on the re-use and commercial exploitation of public sector documents*, *OJ C* 85, 8 Apr. 2003, 25). Not only would this considerably increase the burden on the public servants, but it would also go beyond the purpose of the PSI Directive, that is, making *existing* documents that were created by the public sector bodies in the course of the performance of their public task, available for re-use.
286. Cf. *supra*, s. III.B.3.e. on documents the supply of which is an activity falling outside the scope of the public task of the public sector bodies.
287. Article 4.1 of the PSI Directive.
288. Cf. *supra*, s. II.B.2.d.
289. Article 4.2 of the PSI Directive.

providers supplying documents upon the order of applicants. If a public sector body no longer needs to create the documents for the purposes of performing its public task, it cannot be expected to update the documents or create additional ones only because they will be re-used by the private sector or by another public sector body. If a public sector body would continue to produce documents purely because the private sector needs them, these documents would fall under the exception of Article 1.2(a) of the PSI Directive, documents the supply of which is an activity falling outside the scope of the public task.[290]

NO OBLIGATION TO INFORM THE RE-USERS OF DISCONTINUATION – The PSI Directive does not impose any obligation on a public sector body to inform the private organization that is using the public sector documents of the fact that it intends to stop their production.[291] It only contains a 'moral obligation' in this regard in Recital 18: 'If the competent authority decides to no longer make available certain documents for re-use, or to cease updating these documents, it should make these decisions publicly known, at the earliest opportunity, via electronic means where possible'.

b. Charges

POSSIBLE BARRIER FOR RE-USE – One of the factors that are frequently considered to be hampering re-use is the prices the public sector bodies charge for their data. As they are often the only producers of a certain type of data, the private sector has no choice but to turn to them and pay whatever price is required. The European Commission saw the danger in this:

> The control over information in a specific area puts public sector bodies in a position where they can have market power. In the same way as undertakings in a dominant position cannot abuse their market power, public sector bodies should not set their prices arbitrarily and should not charge excessive prices for information that has been established within the public tasks and with public money.[292]

290. Of course, if there is an obligation under any other law or regulation to continue the production of these documents, this obligation will continue and the supply of documents will be considered as part of the public task of the public sector body.

291. Nathalie Poupaert & Katleen Janssen, 'La directive du 17 Novembre 2003 concernant la réutilisation des informations du secteur public', *Revue du droit des technologies de l'information* no. 19 (2004): 38 [in French]; Katleen Janssen, 'INSPIRE and the PSI Directive: Public Task Versus Commercial Activities?', paper given at the 11th EC-GIS Workshop. 'ESDI – Setting the framework (Alghero, 2005), <www.ec-gis.org/Workshops/11ec-gis/papers/303janssen.pdf>, 22 Feb. 2010; Katleen Janssen & Jan J.C. Kabel, 'Commercialisering van overheidsinformatie door de overheid: rechtspraak en wetgeving in België en Nederland. De honden blaffen, maar de karavaan trekt voort', *Computerrecht* no. 3 (2005): 127 [in Dutch].

292. Commission of the European Communities, *Proposal for a directive of the European Parliament and of the Council on the re-use and commercial exploitation of public sector documents*, COM (2002) 207 final, 10, 5 Jun. 2002.

UPPER LIMIT – Therefore, the PSI Directive puts a limit on the charges for public sector documents: 'Where charges are made, the total income from supplying and allowing re-use of documents shall not exceed the cost of collection, production, reproduction and dissemination, together with a reasonable return on investment'.[293] Hence, the PSI Directive leaves a large margin of discretion to the Member States and the public sector bodies, 'having due regard to the self-financing requirements of the public sector body concerned, where applicable'.[294] It only poses an upper limit to the charges – reasonable return on investment – and no indications or criteria are given how to determine whether this return on investment is *reasonable* or not. This is left to the Member States, the public sector bodies, and ultimately, the courts to define.

COST-ORIENTED CHARGES – The charges have to be cost-oriented over the appropriate accounting period and calculated in line with the accounting principles applicable to the public sector bodies involved. This could mean that charging based on royalties is no longer possible under the directive, as this is not cost-oriented, but based on the income of the re-user.[295] The European Commission felt that the burden of proof that the charges are cost-oriented should be on the public sector body, unless there is such transparent accounting in place as to allow the prospective re-user to verify whether or not this principle is respected.[296]

WHAT COSTS CAN BE INCLUDED? – The costs of collection, production, reproduction, and dissemination can all be charged to the applicant. What can be included in these costs is suggested in Recital 14 of the PSI Directive: 'Production includes creation and collation, and dissemination may also include user support'. But what about the salary of the public servant sending the requested documents, or the maintenance of the copy machine or the computer network?[297] This is up to the Member State to determine. As the total income from supplying and allowing re-use cannot pass the level of a reasonable return on investment on the costs of collection, production, reproduction, and dissemination, it will be up to the public sector body to estimate the demand for the documents, as the charging limit is set on the total income and has to be divided over all the potential applicants.

ENCOURAGEMENT FOR MARGINAL COSTS – While the maximum limit is set, the PSI Directive does not require any minimum charge. The Member States and the public sector bodies are free to decide that they will make their documents available for

293. Article 6 of the PSI Directive.
294. Recital 14 of the PSI Directive.
295. *Ibid.*
296. Commission of the European Communities, *Proposal for a directive of the European Parliament and of the Council on the re-use and commercial exploitation of public sector documents*, COM (2002) 207 final, 10, 5 Jun. 2002.
297. Katleen Janssen, 'INSPIRE and the PSI Directive: Public Task Versus Commercial Activities?', paper given at the 11th EC-GIS Workshop. 'ESDI – Setting the framework (Alghero, 2005), <www.ec-gis.org/Workshops/11ec-gis/papers/303janssen.pdf>, 22 Feb. 2010; Katleen Janssen & Jan J.C. Kabel, 'Commercialisering van overheidsinformatie door de overheid: rechtspraak en wetgeving in België en Nederland. De honden blaffen, maar de karavaan trekt voort', *Computerrecht* no. 3 (2005): 127 [in Dutch].

re-use without any charge.[298] Realizing that this might be too challenging and not necessarily beneficial for the availability of information, considering the self-financing requirements of many public sector bodies, the PSI Directive intends for the Member States to 'encourage public sector bodies to make documents available at charges that do not exceed the marginal costs for reproducing and disseminating the documents'.[299]

PREPARATORY PROCESS: MARGINAL COSTS – The Committee on Industry, External Trade, Research and Energy suggested including an amendment in this regard in the text of the directive, stating that:

> Public sector information is a valuable resource for society as a whole. It should be as widely available for re-use as possible. The best way to achieve this is for only costs of reproduction and dissemination to be charged for this information. Moreover, a pricing model where only marginal costs of reproduction and dissemination are charged is the best way to ensure a level playing field without unfair competition and cross-subsidization from one public sector body to another. This is particularly true if one considers that, in many cases, a public sector body is the monopoly source of public sector information.[300]

However, the plenary session of Parliament did not accept this and maintained the spirit of the Commission proposal.

c. *Transparency*

ARTICLE 7 OF THE PSI DIRECTIVE – One of the main reasons why the European information market based on public sector documents did not reach its full potential was the lack of transparency about the availability of documents and the conditions under which the documents could be re-used. The PSI Directive aims to rectify this in Article 7:

> Any applicable conditions and standard charges for the re-use of documents held by public sector bodies shall be pre-established and published, through electronic means where possible and appropriate. On request, the public sector body shall indicate the calculation basis for the published charge. The public sector body in question shall also indicate which factors will be taken into account in the calculation of charges for atypical cases. Public sector bodies

298. Nathalie Poupaert & Katleen Janssen, 'La directive du 17 Novembre 2003 concernant la réutilisation des informations du secteur public', *Revue du droit des technologies de l'information* no. 19 (2004): 38–39 [in French].
299. Recital 14 of the PSI Directive.
300. Committee on Industry, External Trade, Research and Energy, *Report on the proposal for a European Parliament and Council directive on the re-use and commercial exploitation of public sector documents*, doc. nr. A5-0025/2003, 29 Jan. 2003, <www.europarl.europa.eu/sides/getDoc.do?pubRef=-//EP//NONSGML+REPORT+A5-2003-0025+0+DOC+PDF+V0//EN>, 22 Feb. 2010.

shall ensure that applicants for re-use of documents are informed of available means of redress relating to decisions or practices affecting them.

TWO ASPECTS OF TRANSPARENCY – The requirement of transparency covers two aspects. On the one hand, the applicant needs to be informed about the conditions for re-use, the charges, and the way in which charges are calculated, if he or she so requires. This will ensure that competition is not harmed and that the potential re-user can make informed investment decisions and develop business strategies.[301] Standard charges can be easily published on the Internet, and in cases that are atypical due to the type, amount, or complexity of the information, the public sector body has to at least be able to clarify the factors that it will take into account in setting the charges. On the other hand, the applicant also needs to know about the means of redress that are open to him or her if he or she disagrees with a decision or practice of the public sector body regarding his or her request for re-use. The original Commission proposal did not include this, but it was introduced by the European Parliament during its second reading.[302] It is complementary to the obligation of the public sector bodies to include in their negative decision a reference to the means of redress. In Article 7, not only are negative decisions involved, but also license terms or conditions that are not in line with the directive and any issues regarding the execution of the license agreement.

EVALUATION OF ARTICLE 7 OF THE PSI DIRECTIVE – According to De Vries, the obligations concerning transparency play an important role in the implementation and enforcement of the PSI Directive because they enable the re-users to verify whether the rules of the directive are being followed. Moreover, they provide a framework to increase the accessibility of public sector documents, which De Vries feels is the third pillar of a well-functioning system of dissemination of public sector documents, in addition to a right of access and a right of re-use.[303]

d. Licenses

NO OBLIGATION TO USE A LICENSE – If the public sector bodies impose charges or conditions on the re-use of their documents, they will often do so under the form of a license. The PSI Directive tries to bring some transparency and harmonization to the licensing activities of the public sector bodies. However, it does not impose the use of licenses: ‘Public sector bodies may allow for re-use of documents without conditions or may impose conditions, where appropriate through a licence, dealing

301. Commission of the European Communities, *Proposal for a directive of the European Parliament and of the Council on the re-use and commercial exploitation of public sector documents*, COM (2002) 207 final, 11, 5 Jun. 2002.

302. European Parliament, *Legislative resolution on the Council common position for adopting a European Parliament and Council directive on the re-use of public sector documents*, *OJ C* 77E, 26 Mar. 2004, 314.

303. Marc De Vries, ‘Implementatie van de EU-richtlijn hergebruik overheidsinformatie in de Wob. Beter ten halve gekeerd dan ten hele gedwaald’, *Nederlands Juristenblad* 79, no. 39 (2004): 2041 [in Dutch].

with relevant issues. These conditions shall not unnecessarily restrict possibilities for re-use and shall not be used to restrict competition'.[304] The public sector bodies can decide to use a license where they feel it is appropriate, but they can also impose conditions in another way or not impose any conditions at all.

STANDARD LICENSES – In cases in which public sector bodies would decide to use a license to impose conditions, the second paragraph of Article 8 comes into play: 'In Member States where licences are used, Member States shall ensure that standard licenses for the re-use of public sector documents, which can be adapted to meet particular licence applications, are available in digital format and can be processed electronically. Member States shall encourage all public sector bodies to use the standard licences'. Standard licenses have to be made available by the Member States, but that does not necessarily mean that the public sector bodies are under an obligation to use them. The PSI Directive only requires encouragement of the use of the standard license, but it does not demand a result. As the European Commission states it, '[i]n practice each public sector body licensing information could have its own online licence, although a certain standardisation would facilitate the re-use of information for re-users that have to deal with different public sector bodies'.[305]

CONTENT OF THE STANDARD LICENSES – Determining the content of such standard licenses is left to the Member States. The PSI Directive mainly refers in Recital 17 to 'issues such as liability, the proper use of documents, guaranteeing non-alteration and the acknowledgement of source'. In any case, the content of the license cannot be drafted or used in any way to unnecessarily restrict possibilities for re-use or to restrict competition.

e. *Practical Arrangements*

FACILITATING THE SEARCH FOR DOCUMENTS – Public sector documents will not be re-used much if nobody knows that they exist. That is why the PSI Directive intends for the Member States to take some measures to ensure that potential re-users are not just aware of the conditions and charges for re-use of the documents that they applied for and already know, but that they also can discover new documents and can find out where to get them in order to develop new products or services and to pursue new business opportunities: 'Member States shall ensure that practical arrangements are in place that facilitate the search for documents available for re-use, such as asset lists, accessible preferably online, of main documents and portal sites that are linked to decentralized assets lists'.[306] This requirement for practical arrangements can be linked to the issue of transparency and actually precedes it. First, a potential re-user has to know where he or she can find possible

304. Article 8.1 of the PSI Directive.
305. Commission of the European Communities, *Proposal for a directive of the European Parliament and of the Council on the re-use and commercial exploitation of public sector documents*, COM (2002) 207 final, 11, 5 Jun. 2002.
306. Article 9 of the PSI Directive.

documents to re-use. Once he or she has found them, he or she needs to know under what conditions he or she can re-use them.

TWO TYPES OF PRACTICAL MEASURES – The PSI Directive gives two possible types of practical measures, asset lists and portal sites, but these are only examples, and the Member States can choose to include other arrangements. The asset lists do not necessarily have to contain every document that is available for re-use, but only the *main* documents, that is, the documents that are often requested or that the public sector body can reasonably assume to be interesting for potential re-users, or, as Recital 23 describes it, 'documents that are extensively re-used or that have the potential to be extensively re-used'.

PREPARATORY PROCESS: ASSET LISTS – The original proposal of the Commission did not contain any provisions on such practical arrangements.[307] The European Parliament added an amendment during the first reading,[308] not only requiring asset lists, but also requiring that those lists show the different categories of information – *basic* information or other information. The distinction between the two types of information was not maintained, but the idea of creating asset lists was picked up by the Commission, which suggested adding that, '[s]uch lists shall be made available for free and shall indicate which are the public sector bodies responsible for making the information available for re-use'.[309] The Council did not agree with either suggestion, so a compromise was found in the wording of the current article. The asset lists are only an example, and it is no longer mentioned that they need to be accessible free of charge. The latter is to be regretted, as charging for access to the asset lists defies the purpose of having such lists. It would be asking a lot of potential re-users to pay a fee to find out if there might be public sector documents that could be interesting to use as a resource for their businesses.

PREPARATORY PROCESS: PORTAL SITES – While an obligation of the Member States to provide asset lists was included by the European Parliament during the first reading, in the second reading, the Parliament also introduced portal sites linked to de-centralized asset lists as a new example.[310] This was accepted by the Council and the European Commission without any comments.

307. While the Commission felt that asset lists could greatly facilitate the re-use of public sector information, it also believed that it would be hard to define a precise legal obligation that would bind the Member States in this area. It was therefore decided that this was one of the issues that should be tackled elsewhere (Yvo Volman, 'Exploitation of Public Sector Information in the Context of the *eEurope* Action Plan', in *Public Sector Information in the Digital Age: Between Markets, Public Management and Citizens' Rights*, ed. Georg Aichholzer & Herbert Burkert (Cheltenham: Edward Elgar Publishing Limited, 2004), 105.

308. European Parliament, *Legislative resolution on the proposal for a European Parliament and Council directive on the re-use and commercial exploitation of public sector documents*, *OJ C* 43E, 19 Feb. 2004, 222.

309. Commission of the European Communities, *Amended proposal for a directive of the European Parliament and of the Council on the re-use and commercial exploitation of public sector information*, COM (2003) 119 final, 17 Mar. 2003.

310. European Parliament, *Position adopted at second reading with a view to the adoption of Directive 2003/ . . . /EC of the European Parliament and of the Council on the re-use of public sector information*, *OJ C* 77E, 26 Mar. 2004, 314.

8. Non-discrimination

NO DISCRIMINATION BETWEEN COMPARABLE CATEGORIES OF RE-USE – Any conditions for the re-use of public sector documents have to be non-discriminatory for comparable categories of re-use.[311] Hence, the conditions may be different for re-use for commercial or non-commercial purposes. Recital 19 of the PSI Directive adds that the principle of non-discrimination should 'not prevent the exchange of information between public sector bodies free of charge for the exercise of public tasks, whilst other parties are charged for the re-use of the same documents'. While the idea should be applauded, this statement seems out of place here, as the PSI Directive has made it clear that the exchange of public sector documents between public sector bodies for the pursuit of their public task does not constitute re-use.[312] Thus, this is not relevant with regard to the non-discriminatory nature of conditions for re-use.

WHAT ARE COMPARABLE CATEGORIES? – The conditions – including the charges – have to be non-discriminatory for *comparable* categories of re-use. The only distinction that the PSI Directive suggests is between commercial and non-commercial re-use. But is all commercial re-use, or all non-commercial re-use, comparable? Is a company using census information to determine the location of its next branch comparable to a company providing commercial weather services based on meteorological data from the National Meteorological Service? Does an environmental non-governmental organization (NGO) publishing and criticizing results of government environmental studies find itself in a comparable category of re-use as that of a university student copying graphs or statistics from the website of his or her local authority into his or her essay? Considering the very wide definition of *re-use*, this will not be easy to determine. While the PSI Directive does not require the types of re-use to be the same but only comparable, it may still be a difficult task for the public sector bodies, re-users, and the courts to determine the comparable character.

PUBLIC SECTOR BODIES PERFORMING COMMERCIAL ACTIVITIES – The second paragraph of Article 10 addresses a specific situation wherein concerns for discrimination and distortion of competition could arise: a public sector body performing its own commercial activities. It states that '[i]f documents are re-used by a public sector body as input for its commercial activities which fall outside the scope of its public tasks, the same charges and other conditions shall apply to the supply of the documents for those activities as apply to other users'. As was mentioned before, in some Member States, some public sector bodies are required to provide for at least a part of their own funding. They do this by charging for the supply of their data, but also by creating their own products and services based on that information and selling them on the market. If they do this, they have to apply the same charges and conditions to themselves as they do to other users. If they do not, '[t]here is a risk that the public sector body involved could abuse its

311. Article 10.1 of the PSI Directive.

312. Article 2.4 of the PSI Directive. Cf. *supra*, s. III.B.4.c.

privileged position to gain a competitive advantage over other market players'.[313] The requirement of Article 10 may require the public sector bodies to have separate accounts or even separate their commercial activities from their public task activities.[314]

IMPORTANCE OF THE PUBLIC TASK – Article 10.2 shows the importance of the notion of *public task*. Whether or not an activity of a public sector body is defined as part of the public task determines the applicability of the conditions for re-use. As was mentioned before, the PSI Directive leaves it up to the Member States to determine the scope of the public task of their public sector bodies. This could cause problems for the harmonization of the rules for re-use at the level of the EU.

9. Prohibition of Exclusive Arrangements

RE-USE FOR ALL POTENTIAL ACTORS IN THE MARKET – Re-use of public sector documents has to be open to all potential actors in the market, even if one or more market players already exploit added-value products based on these documents. Contracts or other arrangements between public sector bodies and third parties should not grant exclusive rights.[315] This provision shows the PSI Directive's main target group: *market players that exploit value-added products*, that is, the information industry, as a subcategory of general re-use.

EXCEPTION TO THE PROHIBITION – Exclusive rights are no longer allowed. However, according to Article 10.2 of the PSI Directive, '[w]here an exclusive right is necessary for the provision of a service in the public interest, the validity of the reason for granting such an exclusive right shall be subject to regular review, and shall, in any event, be reviewed every three years'. Thus, exclusive agreements would still be possible if they are needed for a service in the public interest. What constitutes a service of public interest is not defined by the directive. Recital 20 actually refers to services 'of general economic interest', which, at first thought, does not have the same scope. *Service of general economic interest* is a term widely used in the field of competition law, mainly in application of Article 106 of the Treaty on the Functioning of the European Union (former Article 86 of the EC Treaty),[316] so one can look there for an interpretation, while *service of public interest* is not commonly used in EU law or policy. Can one assume from this that the Commission intended *services of public interest* to have the same meaning as *services of general economic interest*? In its proposal, the Commission itself referred to Article 86 of the EC Treaty, and hence to the services of general economic interest considered therein: 'Whether a situation justifies exclusivity,

313. Commission of the European Communities, *Proposal for a directive of the European Parliament and of the Council on the re-use and commercial exploitation of public sector documents*, COM (2002) 207 final, 11, 5 Jun. 2002.
314. Cf. *infra*, Ch. 8, s. IV.C.1.c.
315. Article 11.1 of the PSI Directive.
316. For the relationship between the PSI Directive and Art. 106 of the Treaty on the Functioning of the European Union, cf. *infra*, Ch. 8, s. IV.D.

and thus whether it does not unjustifiably restrict competition, would be eventually decided on a case by case basis and in application of Article 86 of the Treaty'.[317] In any case, Recital 20 gives an example of a situation where exclusive rights would be possible: 'If no commercial publisher would publish the information without such an exclusive right', for instance, because it would not be profitable or the investment for starting up the information service would be too costly. The exclusive arrangements that are established under the conditions of Article 11 have to be transparent and made public. Existing exclusive arrangements had to be examined against the conditions mentioned above, and if they did not qualify, they had to be terminated at the end of the contract, or in any case not later than 31 December 2008.

FOLLOW-UP – While in some Member States exclusive arrangements were being phased out before the end of 2008, only the United Kingdom and the Netherlands had undertaken a systematic identification of all exclusive arrangements in their territory. The European Commission questioned whether all exclusive arrangements had disappeared and launched a study for assessing the existence of possible exclusive arrangements within the Member States. A first set of nine Member States should be covered by mid-2010, after which the other Member States will be addressed.[318]

C. THE REVIEW OF THE PSI DIRECTIVE

ARTICLE 13 OF THE PSI DIRECTIVE – Under Article 13 of the PSI Directive, the European Commission has to carry out a review of its application before 1 July 2008, and communicate the results, together with any proposals for modifications of the directive, to the European Parliament and the Council.

TIMELINE – This review started in the spring of 2008 and consisted of two main parts. First, a questionnaire was sent to the Member States, which could respond until 15 September 2008. Second, a consultation was also started with all stakeholders in the PSI domain.

RESPONSES TO THE REVIEW: MEMBER STATES – Fourteen Member States responded to the Commission's questionnaire. According to the Commission's analysis, the consultation yielded 'substantial positive feedback on the impact of the PSI re-use among MS, which consider that from a qualitative point of view, the PSI Directive has been positive as it provides a legal framework for the re-use of the PSI in their respective countries, unregulated before'.[319] However, from a

317. Commission of the European Communities, *Proposal for a directive of the European Parliament and of the Council on the re-use and commercial exploitation of public sector documents*, COM (2002) 207 final, 11, 5 Jun. 2002.

318. European Commission, 'Public Sector Information – Exclusive Agreements', <http://ec.europa.eu/information_society/policy/psi/facilitating_re-use/exlusive_agreements/index_en.htm>, 22 Feb. 2010.

319. Commission of the European Communities, 'Results of the MS Consultation on the review of the application of Directive 2003/98/EEC on the re-use of public sector information', 2008, 1,

quantitative point of view, the Member States were more sceptical. Some Member States believed it was too early to determine the impact of the PSI Directive due to the late transposition into national legislation in some countries. Others believed that it was 'very difficult to provide factual economic figures for measuring the impact of re-use policies in their respective countries'.[320] Moreover, Member States signalled that the full potential of PSI re-use was 'far from being attained due to the lack of awareness by public sector bodies of the opportunities offered and of their responsibilities as regard PSI re-use' and the difficulties of the private sector in finding what type of PSI is available and in communicating with different levels of government.[321] In general, the 'overwhelming majority' of the Member States felt that it was too early to review the directive and that more time should be allowed for its full implementation.

Responses to the review: stakeholders – The European Commission also held an online consultation of stakeholders, the purpose of which was to 'gather information from as many sources as possible, including public sector content holders and commercial and non-commercial re-users (universities, NGOs) on their views on different aspects related to the implementation, impact and scope of the PSI Directive'.[322] Thirty-seven replies were received.[323] Overall, the stakeholders believed, as the Member States did, that the PSI Directive has had a positive effect on promoting PSI re-use by providing and establishing a legislative framework where there previously was none. However, the public sector body suppliers and the re-users had different views on the existing re-use environment. While the suppliers considered the re-use framework to be working well, the re-users felt that the implementation of the PSI Directive had been much too slow, or that the directive even was not implemented or applied properly in some Member States. According to the re-users, important barriers still remained, including lack of awareness and little effort among some of the public sector bodies, the non-mandatory character of re-use, strict licensing conditions, and limited transparency on re-use policies. In addition, questions arose about the limits of the public task when public sector bodies commercially compete with the private sector and about unfair competition practices by public sector bodies. The absence of efficient

---

<http://ec.europa.eu/information_society/policy/psi/docs/consultations/replies/members_states/report_ms_questionnaire.pdf>, 22 Feb. 2010. The individual replies can be retrieved from <http://ec.europa.eu/information_society/policy/psi/online_consultation/member_states/index_en.htm>, 22 Feb. 2010.

320. *Ibid.*

321. *Ibid.*, 2.

322. Commission of the European Communities, Results of the online consultation of stakeholders. 'Review of the PSI Directive', 2008, 1, <http://ec.europa.eu/information_society/policy/psi/docs/pdfs/online_consultation/report_psi_online_consultaion_stakeholders.pdf>, 22 Feb. 2010. The individual replies can be retrieved from <http://ec.europa.eu/information_society/policy/psi/online_consultation/stakeholders/index_en.htm>, 22 Feb. 2010).

323. This is in stark contrast with the 185 responses that were received to the Green Paper on Public Sector Information and the 79 responses to the consultation following the eEurope 2002 Communication.

means of redress in most countries was also felt as a deficiency.[324] Next, while the Member States stated in their consultation that it was too soon to amend the PSI Directive, the bulk of the re-users felt that legislative amendments were needed to make the directive 'sharper' and 'tighter' in its obligations for public sector bodies.[325] Suggestions included creating a general obligation to allow re-use, requiring the establishment of independent regulators or dispute resolution mechanisms, imposing marginal cost pricing and introducing greater clarification on the public tasks of public sector bodies. More guidance was requested from the Commission on terms such as *public task*, *document*, *marginal cost pricing*, and *reasonable return on investment*.

Reaction of the European Commission – After analysing the responses of the stakeholders and the Member States, the European Commission issued a Communication setting out its future plans. The Commission recognized that:

> Although re-use of PSI is growing, much of its potential is still not being tapped because of the way public sector bodies handle their information resources. One of the reasons for this is the focus on short-term cost recovery, as opposed to benefits in the wider economy . . . In addition, there are practical issues hindering the re-use of PSI, e.g. lack of information on available PSI. In other cases, public sector bodies are lukewarm to the basic idea of commercial re-use of their information or fail to realise its economic potential.[326]

The directive has led to the termination of exclusive arrangements, changes in charging policies, and the establishment of conflict resolution mechanisms.

Future actions – While the European Commission believes that the Member State practices on re-use can still be improved, it 'does not intend to propose amendments to the Directive at this point. The Commission will carry out a further review at the latest in 2012 when more evidence on the impact, effects and application of the Directive should be available . . . Legislative amendments will be considered at that stage if the potential of re-use is still being hampered'.[327] In the meantime, it encourages Member States to improve the way they deal with their public sector data resources in the market.

IV. SHARING

Definition – As was discussed in Chapter 2, sharing involves the availability of public sector spatial data for the purpose of the performance of public task

324. Commission of the European Communities, Results of the online consultation of stakeholders. 'Review of the PSI Directive', 2008, 2, <http://ec.europa.eu/information_society/policy/psi/docs/pdfs/online_consultation/report_psi_online_consultaion_stakeholders.pdf>, 22 Feb. 2010.
325. *Ibid.*, 3.
326. Commission of the European Communities, *Communication of 7 May 2009 to the European Parliament, the Council, the European Economic and Social Committee and the Committee of the Regions. Re-use of Public Sector Information – Review of Directive 2003/98/EC*, COM (2009) 212 final, 3, 7 May 2009.
327. *Ibid.*, 7.

activities. In the EU legal framework on the availability of public sector spatial data, sharing is regulated by the INSPIRE Directive.

OUTLINE – As was already indicated in the section on the INSPIRE Directive under access, this section addresses the history and background of the INSPIRE Directive, and then moves on to the general objective of the directive to stimulate sharing and the provisions containing the obligations of the Member States and the public authorities in this regard.

A. HISTORY

ATTEMPTS FOR A EUROPEAN SDI STRATEGY – During the 1990s, the European Commission had already attempted to set up a strategy for the implementation of a European SDI. However, these attempts all stalled before the Commission could adopt an official policy document under the form of a Communication, Green Paper, or proposal.

DRAFT COMMUNICATION GI2000 – In 1994, the European Commission initiated a consultation process with the goal of setting up a framework for a European Geographic Information Infrastructure. A wide consultation was held during 1995 and 1996, and a draft Communication, GI2000: Towards a European Policy Framework for GI, was produced in 1997.[328] The draft stated that: 'A European policy for geographic information is needed in the European information society. The policy will help to provide the basis for better and more efficient government, more effective management of scarce resources, as well as fostering new business and employment opportunities. The policy is necessary to ensure important trans-national initiatives are nurtured and managed cost-effectively'.[329]

NO FINAL VERSION – As Longhorn mentions, the draft Communication listed problems relating to regional geographic information and recommended actions that might improve wider use of pan-European geographic information. These included the identification, collection, and wide dissemination of pan-European

328. For a more detailed overview of the process towards the draft communication, see Roger A. Longhorn, 'Regional Geographic Information Policy: Fact or Fiction – The Case in Europe and Lesssons for GSDI', paper given at the GSDI-4 Conference (Capetown, 2000), <http://gsdidocs.org/docs2000/capetown/longhorn.rtf>, 22 Feb. 2010; Abbas Rajabifard, 'Diffusion of Regional Spatial Data Infrastructures: With Particular Reference to Asia and the Pacific', 2002, 41, <www.sli.unimelb.edu.au/research/publications/Rajabifard_thesis.pdf>, 22 Feb. 2010; Richard Groot, 'Spatial Data Infrastructure (SDI) for Sustainable Land Management', *ITC Journal* no. 3/4 (1997): 289. For more information on the initiatives and programmes regarding spatial data at the European level, see Arif Cagdas Aydinoglu, Philippe De Maeyer & Tahsin Yomralioglu, 'Examining European SDI Initiatives in Anticipation of a SDI Framework for Turkey', paper given at the GSDI-8 Conference (Cairo, 2005), <http://gsdidocs.org/gsdiconf/GSDI-8/papers/ts_31/ts31_04_aydinoglu_etal.pdf>, 22 Feb. 2010.

329. Commission of the European Communities, *GI2000: Towards a European Policy Framework for Geographic Information. A Communication from the Commission to the Council, to the European Parliament, to the Economic and Social Committee and to the Committee of the Regions. Draft*, 1997, <www.ec-gis.org/docs/F20680/MICROSOFT%20WORD%20-%20GI2000V10_7_98.PDF>, 22 Feb. 2010.

base data, strengthening the emerging national and pan-European metadata services and directories, removing barriers to wider access to geographic information, encouraging market growth for spatial data, and the application of standards and interoperability specifications.[330] However, the draft never reached the stage of completion, because, at least according to Longhorn, the Commission hierarchy itself was not convinced of the importance of an SDI. Rase formulates it less explicitly: 'The new Commission has decided to integrate this project in the overall context of the green paper on access to public sector information in Europe'.[331]

B. Preparatory Phase

E-ESDI initiative aka INSPIRE – Even though steps were being taken to promote the information society in general and the availability of public sector data in particular, a need for actions specifically dedicated to spatial data was still felt. This led to the Directorate General Environment, the Statistical Office of the European Communities (EUROSTAT), and the Joint Research Centre starting up a new initiative, initially referred to as European Environmental Spatial Data Infrastructure (E-ESDI) but soon renamed INSPIRE. The main principles of INSPIRE were set out in a Memorandum of Understanding, signed by the Commissioners responsible for Environment, Economy and Monetary Affairs, and Research in 2002.[332]

---

330. Roger A. Longhorn, 'Regional Geographic Information Policy: Fact or Fiction – the case in Europe and lesssons for GSDI', paper given at the GSDI-4 Conference (Capetown, 2000), <http://gsdidocs.org/docs2000/capetown/longhorn.rtf>, 22 Feb. 2010.

331. Daniel Rase, 'Initiatives taken by the Commission to Establish a European Geographic Information Infrastructure', *Statistical Journal of the United Nations ECE* (2000): 141–147, <www.unece.org/stats/documents/2000/04/gis/6.e.pdf>, 22 Feb. 2010. See also George Cho, *Geographic Information Science: Mastering the Legal Issues* (Chicester: John Wiley & Sons Ltd, 2005), 101.

332. Margot Wallstrom, Pedro Solbes Mira & Philippe Busquin, 'Memorandum of Understanding of 11 Apr. 2002. Infrastructure for Spatial Information in Europe (INSPIRE)', <http://inspire.jrc.ec.europa.eu/reports/MoU.pdf>, 22 Feb. 2010. The Memorandum divided the roles and responsibilities regarding INSPIRE between DG Environment, EUROSTAT and Joint Research Centre. For an overview of the preparation and implementation of INSPIRE, see Ian Masser, *Building European Spatial Data Infrastructures* (Redlands: ESRI Press, 2007), 62–78; Max Craglia & Alessandro Annoni, 'INSPIRE: An Innovative Approach to the Development of Spatial Data Infrastructures in Europe', in *Research and Theory in Advancing Spatial Data Infrastructure Concepts*, ed. Harlan Onsrud (Redlands: ESRI Press, 2007), 93–105; Alessandro Annoni & Massimo Craglia, 'Towards a Directive Establishing an Infrastructure for Spatial Information in Europe (INSPIRE)', paper given at the GSDI-8 Conference (Cairo, 2005), <http://gsdidocs.org/gsdiconf/GSDI-8/papers/ts_47/ts47_01_annoni_graglia.pdf>, 22 Feb. 2010; Danny Vandenbroucke, Katleen Janssen & Jos Van Orshoven, 'INSPIRE State of Play: Generic Approach to Assess the Status of NSDIs', in *A Multi-view Framework to Assess Spatial Data Infrastuctures*, ed. Joep Crompvoets et al. (Melbourne: The University of Melbourne, 2008), 146–148; Massimo Craglia & Andrew Johnston, 'Assessing the Impacts of Spatial Data Infrastructures: Methods and Gaps', paper given at the 7th AGILE Conference on

EXPERT GROUP – The first step in the process towards the INSPIRE Directive was setting up an expert group with representatives from the environmental and spatial data communities of all the Member States and representatives of the Commission and the European Environmental Agency. Its mandate comprised supporting the Commission in the preparation of the INSPIRE initiative by helping and following up the activities of the working groups and advising the Commission on the organization and action plan for the definition of the INSPIRE initiative.[333]

WORKING GROUP ON DATA POLICY AND LEGAL ISSUES – In a first phase, five working groups were created: Common Reference and Metadata, Data Policy and Legal Aspects, Architecture and Standards, Funding and Implementation Structures, and Impact Analysis.[334] In the scope of this book, the Position Paper on Data Policy and Legal Issues is of the most importance.[335] Its scope was to consider all relevant aspects related to data policy and to develop proposals on the issues to be addressed in the INSPIRE framework legislation. These included the following subjects:

- data policy initiatives at European level;
- open access for citizens to public sector information;
- coherence with the Aarhus Convention and the Access Directive, which was then still under preparation;
- consequences of applying environmental INSPIRE to other sectors, such as transport;
- dissemination modes;
- cost models for access to and re-use of data and information;
- existing experiences, including those of the US;
- IPR and copyright;
- freedom of information versus privacy issues;
- liability for data and information.

INSPIRE KEY PRINCIPLES – The Position Paper stated a number of key policy principles that were needed to set up a European SDI. INSPIRE data should be made available for access and view free of charge by citizens and other users, with delivery, downloading, and re-use on harmonized terms and conditions throughout the European Union. Sustainable funding, investment, and charging mechanisms should be put in place by the Member States and maintained in accordance with this idea. The fact that the vision and policy principles of INSPIRE were only

---

Geographic Information Science, (Heraklion, 2004), <http://plone.itc.nl/agile_old/Conference/greece2004/papers/1-1-1_Craglia.pdf>, 22 Feb. 2010; Danny Vandenbroucke, Katleen Janssen & Jos Van Orshoven, 'INSPIRE State of Play Development of the NSDI in 32 European countries between 2002 and 2007', paper given at the GSDI-10 Conference (St Augustine, 2008,), <www.gsdi.org/gsdiconf/gsdi10/papers/TS1.5paper.pdf>, 22 Feb. 2010.

333. See <www.ec-gis.org/inspire/about/expert.cfm>, 22 Feb. 2010.

334. The position papers of the Working Group can be found at <www.ec-gis.org/inspire/reports.cfm>, 22 Feb. 2010.

335. INSPIRE Data Policy and Legal Issues Working Group, 'Position Paper', 2003, <http://inspire.jrc.ec.europa.eu/index.cfm/pageid/541/downloadid/812>, 22 Feb. 2010.

possible to implement if funding would be guaranteed was considered fundamental. Also, harmonized licensing frameworks should be introduced to facilitate and optimize the sharing, trading, and extensive use of INSPIRE thematic data and information, and the unimpeded flow of data and information between the Commission, Member States, local authorities, and members of the public would have to be assured. Coordination bodies for the operation of the SDIs should be established at the European and national levels. Each Member State should mandate one national authority with the responsibility for relations with the Commission.

GROUP ON IMPLEMENTING STRUCTURES AND FUNDING – Another working group whose results are worth mentioning is the Working Group on Implementing Structures and Funding. In its position paper, it recommended that a European Spatial Data Committee be established, with a permanent role for the Commission in the SDI and one national authority that would serve as a contact point for the relations with the Commission.[336] The basic principle for funding should be a model combining grants and cost recovery in an efficient and acceptable way.

PROPOSAL FOR A DIRECTIVE – After an Extended Impact Assessment and an Internet consultation,[337] the Commission presented its proposal for a directive on establishing an infrastructure for spatial information in Europe in July 2004. After an extensive negotiation process, the final version of the directive was proclaimed on 14 March 2007.

C. THE INSPIRE DIRECTIVE

OUTLINE – The INSPIRE Directive contains provisions on many different aspects of an SDI, some more technical than others. As was mentioned above, this section looks at the provisions of the directive that are immediately relevant for the EU legal framework on the availability of public sector spatial data with regard to sharing. Hence, it does not address the provisions on public access to the INSPIRE network services.[338] Before the provisions on sharing are addressed, a short overview is given of the objective and general provisions of the directive.

1. Objective and General Provisions

ARTICLE 1 OF THE INSPIRE DIRECTIVE – The purpose of the INSPIRE Directive is to lay down general rules aimed at the establishment of the Infrastructure for Spatial

336. INSPIRE Implementing Structures and Funding Working Group, 'Position paper', 2003, <http://inspire.jrc.ec.europa.eu/reports/position_papers/inspire_isf_pp_v1_1_en.pdf>, 22 Feb. 2010.

337. For the reports on the Extended Impact Assessment and the Internet Consultation, see <www.ec-gis.org/inspire/reports.cfm>, 22 Feb. 2010.

338. Cf. *supra*, s. II.C.3.

Information in the EU for the purposes of Community environmental policies and policies or activities that may have an impact on the environment.[339] It focuses on environmental policy, but it is open for use by and future extension to other sectors such as agriculture, transport, and energy.[340] The European Commission intended the main beneficiaries of the proposal to be those involved in the formulation, implementation, monitoring, and evaluation of policies at the European, national, and local levels: 'These are public authorities, legislators and citizens and their organisations. However, other user groups are also expected to benefit, including the private sector, universities, researchers and the media'.[341]

EXISTING NATIONAL SDIS AS A BASIS – It is not the directive's ambition to create a new SDI from scratch. It builds upon the existing national SDIs that are established and operated by the Member States.[342] Hence, INSPIRE does 'not set off an extensive programme of new spatial data collection in the Member States. Instead, it is designed to optimize the scope for exploiting the data that are already available, by requiring the documentation of existing spatial data, the implementation of services aimed at rendering the spatial data more accessible and interoperable and by dealing with obstacles to the use of the spatial data'.[343]

NO PREJUDICE TO THE ACCESS DIRECTIVE AND THE PSI DIRECTIVE – As was mentioned before, Article 2 of the INSPIRE Directive makes clear that it is without prejudice to the Access Directive. In addition, it is without prejudice to the PSI Directive. This last provision was not necessary, as the field of application of both does not overlap: the INSPIRE Directive is applicable to data sharing and access, while the PSI Directive addresses re-use.

NO EFFECT ON EXISTENCE OR OWNERSHIP OF INTELLECTUAL PROPERTY RIGHTS – Article 2 also makes clear that the INSPIRE Directive does not affect the existence or the ownership of public authorities' IPR. All existing international, European, and national protection of IPR remains fully intact. In addition, Article 4.5 of the directive points out that for spatial data sets that are held by or on behalf of a public authority, but in respect of which a third party holds IPR, the public authority may only take action under the directive, that is, share or provide public access with the consent of that third party. In theory, this respect for third-party IPR should not even be made explicit, since it is a general rule, and the INSPIRE Directive does not contain any new rules on IPR.

---

339. Article 1.1 of the INSPIRE Directive.
340. Commission of the European Communities, *Proposal for a directive of the European Parliament and of the Council establishing an infrastructure for spatial information in the Community (INSPIRE)*, COM (2004) 516 final, 3, 23 Jul. 2004.
341. *Ibid.*
342. Article 1.2 of the INSPIRE Directive.
343. Commission of the European Communities, *Proposal for a directive of the European Parliament and of the Council establishing an infrastructure for spatial information in the Community (INSPIRE)*, COM (2004) 516 final, 3, 23 Jul. 2004.

2. Network Services

PUBLIC ACCESS AND SHARING – Public access to the network services was discussed in the section on access. However, it should be noted that these network services are also intended to be used by the public authorities for making data available and obtaining data for sharing purposes. This would actually be the preferred option once these services are fully operational.

LIMITATIONS AND CHARGES – However, the limitations and charges for access to the network services that are provided in Article 13 and 14 are only applicable to the public. When the public authorities share data through these network services, the main provisions on data sharing of Article 17 of the INSPIRE Directive remain applicable.

3. Data Sharing

OUTLINE – This section addresses the main elements that are included in the provisions on sharing in Article 17 of the INSPIRE Directive. This includes the main principle, restrictions at the point of use, the possibilities for charging and licensing, the availability of the arrangements to other public authorities outside of the Member State, and the limitations to sharing.

a. General Principle

ARTICLE 17.1 OF THE INSPIRE DIRECTIVE – According to Article 17.1 of the INSPIRE Directive, each Member State has to adopt measures for the sharing of spatial data sets and services between its public authorities referred to in (a) and (b) of the definition of *public authorities* in Article 3 of the directive.[344] These measures should enable the public authorities to gain access to spatial data sets and services, and to exchange and use those sets and services for the purposes of public tasks that may have an impact on the environment. This provision ensures that all the technical measures that are set up can actually be used by the public authorities, based on the agreements that are made within or between the Member States.

NOT ALL PUBLIC AUTHORITIES ARE INCLUDED – A number of remarks can be made with regard to this provision. First, the Member States only have to adopt measures for data sharing between some public authorities, but not all of them. The public authorities included are the government and other public administrations, and the natural or legal persons performing public administrative functions under national law. The public authorities mentioned under (c) of Article 3.9 are excluded. These include natural or legal persons 'having public responsibilities or functions, or providing public services relating to the environment under the control of a body or person' that is a public authority in the other categories mentioned above. As was mentioned before, this third category could include public bodies, but also

344. Cf. *supra*, Ch. 2, s. IV.B.

entities that are privately owned. However, if they are considered public authorities, it does not make sense to exclude them from the Member States' measures for sharing, which form one of the central elements of the directive.

USE OF THE TERM *ACCESS* – A second remark deals with the possibility that the public authorities can 'gain access to spatial data sets and services'. It should be made clear that the term *access* does not refer to access as it is defined in this book. It describes the physical access to the data for the purpose of sharing rather than the right of access.

b. Restrictions at the Point of Use

ARTICLE 17.2 OF THE INSPIRE DIRECTIVE – The measures that the Member States take to enable sharing between the public authorities must preclude any restrictions likely to create practical obstacles occurring at the point of use.

PREPARATORY PROCESS: TYPE OF RESTRICTIONS OR OBSTACLES – This preclusion of restrictions at the point of use was one of the points of debate in the negotiations on the INSPIRE Directive. The original Commission proposal stated that the measures for data sharing should 'preclude, at the point of use, any restrictions, in particular of a transactional, procedural, legal, institutional, or financial nature'.[345] The Working Party on the Environment of the Council proposed reformulating this section to 'any restrictions likely to create, at the point of use, obstacles to the sharing of spatial data sets and services' and refer to the possible types of restrictions – transactional, procedural, legal, institutional, or financial – in the recitals of the directive.[346] Based on this wording, having restrictions on sharing would not be a problem in itself, but only if they created obstacles at the point of use. What should be understood by *point of use* was questioned by many delegations in the Working Party.[347] In the Common Position, the Council referred to *practical* obstacles at the point of use, for example, 'a public authority employee using the data on [his or her] computer',[348] and removed any reference to the possible types

345. Commission of the European Communities, *Proposal for a directive of the European Parliament and of the Council establishing an infrastructure for spatial information in the Community (INSPIRE)*, COM (2004) 516 final, 3, 23 Jul. 2004.

346. General Secretariat of the Council of the European Union, *Note to the Working Party on the Environment. Proposal for a directive of the European Parliament and of the Council establishing an infrastructure for spatial information in the Community (INSPIRE)*, doc. nr. 8891/05, 13 May 2005, <http://register.consilium.europa.eu/pdf/en/05/st08/st08891-co01.en05.pdf>, 22 Feb. 2010.

347. General Secretariat of the Council of the European Union, *Note to the Permanent Representatives Committee. Proposal for a directive of the European Parliament and of the Council establishing an infrastructure for spatial information in the Community (INSPIRE)*, doc. nr. 9112/05, 20 May 2005, <http://register.consilium.europa.eu/pdf/en/05/st09/st09112.en05.pdf>, 22 Feb. 2010.

348. Council of the European Union, *Common Position adopted by the Council with a view to the adoption of a Directive of the European Parliament and of the Council establishing an Infrastructure for Spatial Information in the European Community (INSPIRE). Statement of the Council's reasons*, doc. nr. 12064/2/05, 23 Jan. 2006, <http://register.consilium.europa.eu/pdf/en/05/st12/st12064-re02ad01.en05.pdf>, 22 Feb. 2010.

of obstacles in the recitals.[349] However, the European Parliament felt in its second reading that not only practical obstacles at the point of use should be avoided, but any practical obstacle, wherever it would occur.[350] It failed to convince the Council and the final joint text referred to practical obstacles at the point of use. One might wonder what the scope of these practical obstacles is. Do they include financial obstacles, for instance? The fact that high charges are made for the use of data does not necessarily make it practically difficult to use the data at a computer station. A service may be very quick and user-friendly but also lead to a very high invoice. Yet, this price does not form a practical obstacle. However, it may do so if the charges are to be paid in advance or if they are too high to obtain any access at all.

c. *Licensing and Charging*

ARTICLE 17.3 OF THE INSPIRE DIRECTIVE – Under Article 17.3 of the INSPIRE Directive, public authorities that make spatial data sets and services available to other public authorities for sharing purposes can impose licenses or charges, if the Member State allows them to. However, there are limits. The charges and the licenses must be fully compatible with the general aim of facilitating sharing. The charges have to be 'kept to the minimum required to ensure the necessary quality and supply of spatial data sets and services together with a reasonable return on investment, while respecting the self-financing requirements of public authorities supplying spatial data sets and services, where applicable'.

PREPARATORY PROCESS: NO LICENSING OR CHARGING – The European Commission's proposal did not contain any possibility of licensing or charging. This did not mean that agreements on the financing of the gathering and sharing would never be possible, but such measures should not be financial obstacles to the data sharing. Therefore, charging and licensing per use of the data should be out of the question in the Commission's line of thinking.

PREPARATORY PROCESS: NEED FOR REVENUES – During the first meetings of the Council's Working Party on the Environment, the subject of charging or licensing for data sharing was not raised. Only in May 2005, after almost ten months of deliberation, the Council added a section on charging, stating that: 'Member States shall take appropriate measures to ensure that the implementation of this paragraph does not adversely affect the availability of spatial data sets and services, in particular where the data are supplied by public authorities upon whom Member States place a duty to recover all or a part of their costs. These measures may

---

349. Council of the European Union, *Common Position adopted by the Council on 23 January 2006 with a view to the adoption of a Directive of the European Parliament and of the Council establishing an Infrastructure for Spatial Information in the European Community (INSPIRE), OJ C* 126E, 30 May 2006, 16.

350. European Parliament, *Position adopted at second reading on 13 June 2006 with a view to the adoption of Directive 2006/ . . . /EC of the European Parliament and of the Council establishing an Infrastructure for Spatial Information in the European Community (INSPIRE), OJ C* 124E, 25 Jun. 2006, 116.

include, for example, licensing or financial arrangements, provided that this does not have the effect of creating restrictions at the point of use'.[351] In a later meeting, the Council made a shorter suggestion, leaving out any concerns for an adverse affect on the availability of spatial data sets, but referring directly to the need for making money. It proposed that 'licensing or financial arrangements may be applied, in particular where the spatial data sets are supplied by public authorities upon whom Member States place a duty to raise revenue'.[352]

PREPARATORY PROCESS: NEED FOR COST RECOVERY – Later on, the Common Position re-instated the more subtle approach and dropped any references to raising revenue, but only talked about recovering costs, which seems more modest. The Council merely stated that the fact that there should be no restrictions that may create practical obstacles should not 'prevent public authorities that supply spatial data sets and services from licensing them to, and requiring payment from, the public authorities or institutions and bodies of the Community that use these spatial data sets and services'.[353] According to the Council, allowing data providers to recover their costs would ensure that 'the quality and currency of the data are maintained'.[354] The Commission strongly disagreed, and stated that:

> The common position makes the obligation to avoid obstacles to data sharing, as well as the rules for ensuring harmonized conditions for Community institutions and bodies, subordinate to the right of public data providers to charge and license other authorities for their data. It is also vague about the obstacles to be avoided. It will therefore be ineffective in achieving one of the key aims of the proposal, and could even have the effect of increasing obstacles to the sharing of data.[355]

351. General Secretariat of the Council of the European Union, *Note to the Working Party on the Environment. Proposal for a directive of the European Parliament and of the Council establishing an infrastructure for spatial information in the Community (INSPIRE)*, doc. nr. 7985/05, 3 May 2005, <http://register.consilium.europa.eu/pdf/en/05/st07/st07985.en05.pdf>, 22 Feb. 2010.
352. General Secretariat of the Council of the European Union, *Note to the Working Party on the Environment. Proposal for a directive of the European Parliament and of the Council establishing an infrastructure for spatial information in the Community (INSPIRE)*, doc. nr. 8891/05, 13 May 2005, <http://register.consilium.europa.eu/pdf/en/05/st08/st08891-co01.en05.pdf>, 22 Feb. 2010.
353. Council of the European Union, *Common Position adopted by the Council on 23 January 2006 with a view to the adoption of a Directive of the European Parliament and of the Council establishing an Infrastructure for Spatial Information in the European Community (INSPIRE)*, *OJ C* 126E, 30 May 2006, 16.
354. Council of the European Union, *Common Position adopted by the Council on 23 January 2006 with a view to the adoption of a Directive of the European Parliament and of the Council establishing an Infrastructure for Spatial Information in the European Community (INSPIRE). Statement of the Council's reasons*, doc. nr. 12064/2/05, 23 Jan. 2006, <http://register.consilium.europa.eu/pdf/en/05/st12/st12064-re02ad01.en05.pdf>, 22 Feb. 2010.
355. Commission of the European Communities, *Communication to the European Parliament pursuant to the second subparagraph of Art. 251(2) of the EC Treaty concerning the common position of the Council on the adoption of a Directive of the European Parliament and of the*

Preparatory process: limits on the charges – In the meantime, the European Parliament had taken a more moderate approach in its first reading by requiring only that the implementation of the measures for data sharing did 'not aversely affect the availability of spatial data sets and services'.[356] In the second reading, the Parliament moved closer to the Council's position and accepted that 'Member States may allow public authorities that supply data sets and services to license them to, and/or require payment from, the public authorities or institutions and bodies of the Community that use these spatial data sets and services. In any event, where charges are made, the total income from supplying documents shall not exceed the cost of collection, production, reproduction and dissemination'.[357] The final conciliation procedure resulted in the rewording of a very large part of the section on charging and licensing to the final state of Article 17.3.

Reporting obligations – If the spatial data sets and services are provided by the Member States to the European Commission or other institutions and bodies of the EU in order to fulfil their reporting obligations under Community legislation relating to the environment, this is not subject to any charging. Such reporting obligations are laid down in EU rules, for example greenhouse gas emissions, pollutants, waste incineration, and so on.[358]

d. Authorities outside the Borders of the Member States

Articles 17.4, 17.5, and 17.6 of the INSPIRE Directive – Paragraph 4 of Article 17 makes it clear that the measures the Member States take to facilitate the data sharing between their public authorities are also applicable to the public authorities of other Member States and to the Community bodies and institutions,

Council establishing an infrastructure for spatial information in the Community (INSPIRE), COM (2006) 51 final, 3, 10 Feb. 2006.

356. European Parliament, *Position adopted at first reading on 7 June 2005 with a view to the adoption of European Parliament and Council Directive 2005/ . . . /EC establishing an infrastructure for spatial information in the Community (INSPIRE), OJ C* 124 E, 25 May 2006, 132.
357. European Parliament, *Position adopted at second reading on 13 June 2006 with a view to the adoption of Directive 2006/ . . . /EC of the European Parliament and of the Council establishing an Infrastructure for Spatial Information in the European Community (INSPIRE), OJ C* 124E, 25 Jun. 2006, 116.
358. See e.g., Directive 2003/87/EC of the European Parliament and of the Council of 13 Oct. 2003 establishing a scheme for greenhouse gas emission allowance trading within the Community and amending Council Directive 96/61/EC, *OJ L* 275, 25 Oct. 2003, 32; Directive 2008/1/EC of the European Parliament and of the Council of 15 Jan. 2008 concerning integrated pollution prevention and control, *OJ L* 24, 29 Jan. 2008, 8; Regulation 166/2006/EC of the European Parliament and of the Council of 18 Jan. 2006 concerning the establishment of a European Pollutant Release and Transfer Register and amending Council Directives 91/689/EEC and 96/61/EC, *OJ L* 33, 4 Feb. 2006, 1; Commission Decision 2000/479/EC of 17 Jul. 2000 on the implementation of a European pollutant emission register (EPER) according to Art. 15 of Council Directive 96/61/EC concerning integrated pollution prevention and control (IPPC), *OJ L* 192, 28 Jul. 2000, 36; Decision 280/2004/EC of the European Parliament and of the Council concerning a mechanism for monitoring Community greenhouse gas emissions and for implementing the Kyoto Protocol, *OJ L* 49, 19 Feb. 2004, 1.

should they choose to invoke these measures.[359] The arrangements for data sharing also have to be open to bodies established by international agreements to which the Community and the Member States are parties, on a reciprocal and equivalent basis (paragraph 5). The national conditions that are included in the arrangements that are made for the data sharing between the public authorities within the Member State can also be imposed on the public authorities of the other Member States, the Community institutions, and bodies and the bodies established by international agreements (paragraph 6 of Article 17).

ARTICLE 17.8 OF THE INSPIRE DIRECTIVE – SPECIAL POSITION OF EU INSTITUTIONS AND BODIES – The institutions and bodies of the European Community are in a unique position. For the availability of public sector spatial data from public authorities in the Member States to these institutions and bodies of the European Community, a separate Regulation will be adopted, holding harmonized conditions for data sharing.[360] Hence, the EU institutions and bodies will have their own set of rules for obtaining data. However, they can also choose to use the national measures from the Member States under Article 17.4 if this would be more advantageous to them. A draft regulation was approved by the INSPIRE Committee in June 2009 and sent to the European Parliament for scrutiny.[361]

e. *Limitations on Sharing*

ARTICLE 17.7 OF THE INSPIRE DIRECTIVE – Under Article 17.7, the Member States may limit data sharing when this would compromise the course of justice, public security, national defence, or international relations. The Council introduced this possibility based on the concern of some of the Member States about the sharing of classified information, particularly in relation to defence and financial issues.[362]

ADVERSE AFFECT V. COMPROMISE – Contrary to the limitations on public access, where access would have to adversely affect the protected interests, in the case of sharing the interests have to be *compromised*, that is endangered by the availability of the data. This seems a lighter condition than the adverse effect, as for the latter, there has to be an actual impact, while for sharing only a potential danger to the interests seems to be needed.

---

359. The Community institutions and bodies can obtain the data under the harmonized conditions that are laid out in the implementing rule that is created on the basis of Art. 17.8, so they do not need to invoke the national arrangements for data sharing between the public authorities, and hence, abide to the conditions and restrictions of these national arrangements to obtain the data sets or services. However, if they choose to use these national arrangements instead of the harmonized conditions organized by the implementing rule, they are free to do so.
360. Article 17.8 of the INSPIRE Directive.
361. See <http://inspire.jrc.ec.europa.eu/>, 22 Feb. 2010.
362. General Secretariat of the Council of the European Union, *Note to the Delegations. Proposal for a directive of the European Parliament and of the Council establishing an infrastructure for spatial information in the Community (INSPIRE)*, doc. nr. 5648/05, 1 Feb. 2005, <http://register.consilium.europa.eu/pdf/en/05/st05/st05648.en05.pdf>, 22 Feb. 2010.

f. References to Re-use?

PREPARATORY PROCESS: RE-USE IN THE INSPIRE DIRECTIVE – The proposal of the European Commission also contained references to the use of spatial data sets and services by the private sector for purposes outside of the public tasks relating to the environment. On the one hand, it wanted to ensure that distortion of competition would be prevented in cases where public authorities also carried out commercial activities unrelated to the performance of their public tasks. On the other hand, the development of an implementing rule was planned to increase the potential re-use of spatial data sets and services by third parties, possibly including the establishment of common licensing conditions. The Commission felt that 'this provision will complement Directive 2003/98 by allowing the adoption of rules that will further facilitate re-use of such data'.[363] The Council questioned the need for such provisions and deleted them from Article 17.[364] The European Parliament followed suit in the second reading, and the references to re-use were completely deleted from the directive.

EVALUATION – Even though the reason for the removal of this article probably has less to do with coherence than with concerns about financial resources and the autonomy of public authorities that have to provide for their own funding, the removal seems justified. As was mentioned above, any reference to re-use of data by third parties should be removed from the INSPIRE Directive and dealt with by the PSI Directive. The INSPIRE Directive should not be used as a means to resolve PSI-related issues. This would counterbalance all the attempts for transparency in both frameworks.[365]

V. SUMMARY

OUTLINE – In this section, a short summary is given of the main points of the EU legal framework on the availability of public sector spatial data for each purpose of use: access, re-use, and sharing. After these overviews, some remarks are made with regard to the evaluation of this legal framework.

---

363. General Secretariat of the Council of the European Union, *Note to the Delegations. Proposal for a directive of the European Parliament and of the Council establishing an infrastructure for spatial information in the Community (INSPIRE)*, doc. nr. 7488/05, 22 Mar. 2005, <http://register.consilium.europa.eu/pdf/en/05/st07/st07488.en05.pdf>, 22 Feb. 2010.

364. General Secretariat of the Council of the European Union, *Note to the Working Party on the Environment. Proposal for a directive of the European Parliament and of the Council establishing an infrastructure for spatial information in the Community (INSPIRE)*, doc. nr. 7985/05, 3 May 2005, <http://register.consilium.europa.eu/pdf/en/05/st07/st07985.en05.pdf>, 22 Feb. 2010.

365. Katleen Janssen, 'INSPIRE and the PSI Directive: Public Task Versus Commercial Activities?', paper given at the 11th EC-GIS Workshop. 'ESDI – Setting the framework (Alghero, 2005), <www.ec-gis.org/Workshops/11ec-gis/papers/303janssen.pdf>, 22 Feb. 2010.

A. Access

Definition – *Access* is described as 'obtaining public sector spatial data for the purpose of exercising one's democratic rights or obligations'. The availability of public sector spatial data for access is regulated by the Access Directive and the provisions on public access to the network services in the INSPIRE Directive.

Access Directive – The Access Directive includes an obligation for the public authorities to make their environmental information available upon request and to disseminate environmental information of their own motion.

Access Directive: access upon request – An applicant does not have to state an interest to obtain access. He or she has to be granted access to the information, in the form of consultation or a copy, as soon as possible and at the latest within a month after the request has been received by the public authority. The public authority should deliver the information in the specific format requested by the applicant, unless it is already available in another format that is easily accessible or it is reasonable to make the information available in another format. Limitations on access can be made to protect a number of interests, but in every case, the interests in disclosure and the protected interest have to be weighed. If the interest of the public in the availability of the environmental information for access is greater, the information should be provided. If access is refused, the reasons should be given to the applicant and the possible means of redress should be indicated. If access is granted in the form of consultation, this should be free of charge. If a copy of the information is supplied, charges may be made that cannot exceed a reasonable amount. A schedule of the charges should be made available. Other practical arrangements to facilitate the access of citizens can include the designation of information officers and registers of information.

Access Directive: dissemination by public authorities of their own motion – The public authorities also have to make environmental information available of their own motion. The Member States have to ensure that the public authorities organize their information with a view to its active and systematic dissemination to the public, preferably via means of information technology. A number of specific types of information have to be made available in electronic databases, including legislative texts, policies, and plans relating to the environment, as well as reports on the state of the environment. The Member States can add other information that they feel should be made available.

INSPIRE Directive – The INSPIRE Directive is without prejudice to the Access Directive, so all the obligations of the Access Directive remain intact. For the environmental information that is also spatial data that falls under the INSPIRE Directive, some additional measures are imposed. Under the INSPIRE Directive, the public has the right of access to the network services that have to be established by the Member States. While this is not a direct right of access to spatial data, it is inherently included. Limitations to the discovery services can only be made if access would adversely affect international relations, public security, or national defence. Access limitations to the other services can only be made for the same reasons as the ones that are included in the Access Directive, and, once again, the interests have to be weighed in each case. The charges for access are also subject to a differentiated arrangement. Access to the discovery services always has to be free of charge, while the download services, transformation services, and invoke services can always be charged for. The view services in principle have to be free of charge. However, charges are possible if they secure the maintenance of spatial data sets and services, especially in cases involving very large volumes of frequently updated data.

B. Re-use

Definition – *Re-use* was defined in Chapter 2 as 'any use for non-democratic commercial or non-commercial purposes outside of the public task'. In the EU legal framework on the availability of public sector spatial data, re-use is regulated by the PSI Directive.

Conditions for re-use – The PSI Directive does not include any obligation for the Member States to make their documents available for re-use. It only provides rules and conditions for making documents available for re-use if the Member States or the public sector bodies choose to do so. If this is the case, the documents have to be made available within a time limit that is consistent with national access legislation, or in the absence of such legislation, within twenty working days. If re-use is refused, the grounds for refusal and the possible means of redress have to be communicated to the applicant. If a request is accepted, the documents have to be made available in any pre-existing format. No documents should be created or adapted to comply with a request, nor should a public sector body continue the production of particular documents with a view to their re-use if it no longer needs them for its own tasks.

Charging and licensing – Charges can be made for re-use, but they have to be cost-oriented and the total income cannot exceed the cost of collection, production, reproduction, and dissemination, together with a reasonable return on investment. Any applicable standard charges and conditions have to be pre-established and published, and the public sector bodies have to indicate which factors will be taken into account in the calculation of charges for atypical cases. If the conditions and charges are imposed by the public sector bodies under the form of licenses, the Member States have to ensure that standard licenses are available and encourage the public sector bodies to use these licenses. In addition, practical arrangements have to be made to make it easier for potential re-users to find re-usable documents, such as asset lists or portal sites.

Level playing field – The PSI Directive also aims to ensure that the re-use of public sector documents takes place on a level playing field. Therefore, any conditions for re-use have to be non-discriminatory between comparable categories of re-use. If public sector bodies themselves use their documents as input for their commercial activities, they have to be subject to the same conditions and charges for obtaining the data as the private sector. Next, the public sector bodies are not allowed to grant exclusive rights for the re-use of their documents. The only exception to this is when such an exclusive arrangement is necessary for the provision of a service in the public interest.

C. Sharing

Definition – *Sharing* was defined in Chapter 2 as 'any use of public sector spatial data for the purpose of performing public task activities'. In the EU legal framework for the availability of public sector spatial data, sharing is regulated by the INSPIRE Directive.

Measures for data sharing – The INSPIRE Directive requires the Member States to adopt measures for the availability of spatial data sets and services to public authorities for the purpose of performing public tasks that may have an impact on the environment. The arrangements that are made for data sharing within a Member State also have to be open to public authorities from other Member States, the institutions and bodies of the EU, and bodies established by international agreements to which the EU and the Member States are parties. The institutions and bodies of the European Community have the choice, however, to use these arrangements or the regulation that will provide harmonized conditions specifically for their needs.

Charging and licensing – The measures that are taken for data sharing have to preclude any restrictions likely to create obstacles at the point of use. However, the Member States may allow public authorities to charge for and license their spatial data sets and services, under the condition that this is fully compatible with the general aim of facilitating sharing. In addition, the charges have to be limited to the minimum required to ensure the necessary quality and supply of spatial data sets and services, together with a reasonable return on investment, while respecting the self-financing requirements of public authorities that supply the spatial data sets and services.

Limitations on data sharing – Sharing may be limited only for the following reasons: if it would compromise the course of justice, public security, national defence, or international relations.

D. Differences between the Rules on Access, Re-use, and Sharing

Outline – As the provisions regulating access, re-use, and sharing all provide obligations for the public bodies for making their public sector spatial data available and impose conditions on how the public bodies should do this, it is natural that some of these provisions address the same topics. In this section, the main topics for which a comparison is relevant are addressed. This includes the obligatory character of the provisions, the limitations on the availability of data, the charges, and the time limits.

1. Obligatory Character

ACCESS AND SHARING V. RE-USE – The first reason why it is important to know whether an applicant is requesting public sector spatial data for access, re-use or sharing lies in the obligatory character of the rules. If data are requested for access or sharing – under the Access Directive or the INSPIRE Directive – the public bodies have an obligation to comply with this request, unless one of the limitations to access or sharing can be invoked. If the applicant is requesting re-use, the public body does not necessarily have to provide the data. The PSI Directive does not include an obligation for the public bodies to make their data available for re-use. The Member State may leave it to the public bodies to decide this themselves for their own data. Hence, if a potential re-user wants to obtain data, the public body may refuse such a request by stating that it does not wish to make its documents available for re-use. From the moment it decides to make its data available, the rules and conditions of the PSI Directive come into play.

2. Limitations on the Availability of Data

EXCEPTIONS – The rules for access, re-use, and sharing all contain a number of exceptions under which the public bodies can refuse to make their data available.

ACCESS: LIST OF LIMITATIONS – The limitations on access are included in Article 4 of the Access Directive and Article 13 of the INSPIRE Directive. The list of limitations is the same, so the question does not have to be asked of whether a particular data set falls under both the INSPIRE Directive and the Access Directive. The only exception to this are the limitations that can be made on access to the discovery services under the INSPIRE Directive. Access to these services can only be restricted if it would adversely affect international relations, public security, or national defence. As these discovery services only give access to the metadata and not to the data sets or services themselves, this does not cause any problems. Such metadata do not fall under the definition of environmental information, as they do not directly provide information on any of the topics mentioned in Article 2.1 of the Access Directive and described in Chapter 2 but only provide the characteristics of these data. In this way, the metadata are more related to the obligation of the Member States to create asset lists rather than the limitations to access.

RE-USE: LIST OF EXEMPTIONS – The availability of public sector spatial data for re-use can also be limited for a number of reasons. However, these reasons are not formulated as limitations or possibilities of refusal for the public bodies, but they are already taken into account in the field of application of the PSI Directive. First, it was already mentioned that the public bodies can decide for themselves whether or not to allow their public sector data to be re-used. Next, a number of documents are exempted from the field of application. The rules for access and re-use are coordinated well, in the sense that if access to public sector spatial data is limited under the Access Directive or the INSPIRE Directive, the documents are exempted from the PSI Directive. If data cannot be accessed, they also cannot be re-used. Hence, in this way, it is not important to define the demarcation line between access and re-use.

EXEMPTED ESTABLISHMENTS – Documents held by public sector broadcasters, cultural establishments, and research or educational establishments are exempted. It may be possible that such public bodies hold documents that can be considered spatial data or environmental information, particularly research and educational institutions. Hence, for these documents, access or sharing has to be allowed under the Access Directive or the INSPIRE Directive, but the rules for re-use in the PSI Directive do not have to be followed.

THIRD-PARTY INTELLECTUAL PROPERTY RIGHTS – The PSI Directive also exempts from re-use documents on which third parties hold IPR. The Access Directive and the INSPIRE Directive provide a possibility to limit access for all IPR, including the rights that are held by the public bodies themselves. This would mean that the public bodies could limit access to a wider set of data than they could limit for re-use purposes. However, some nuances can be made to this statement. First, one should keep in mind that the limitations to access have to be weighed. Hence, access can only be limited if the interest in the protection of IPR is higher than the public interest in disclosure of the data. In addition, the PSI Directive, while encouraging the public bodies to use their IPR in a way that facilitates re-use, still leaves the IPR intact, so the public bodies can impose conditions on the re-use to safeguard their rights.

PERSONAL DATA – The PSI Directive does not exempt documents that can be considered personal data, but it does ensure that the provisions of Directive 95/46/EC on the processing of personal data are complied with. However, this question of compliance only comes into play when the documents are not exempted because they cannot be accessed under the Access Directive or the INSPIRE Directive. Under these rules, access can be limited for the confidentiality of personal data. However, once again, it should be kept in mind that the interests in the protection of the personal data and the interests in disclosure have to be weighed.

SHARING – Finally, it should be noted that limiting sharing under the INSPIRE Directive is possible in much fewer cases than limiting access or re-use. Sharing can only be restricted if it would compromise the course of justice, public security, national defence, or international relations. Hence, it is important to know whether public sector spatial data are requested for sharing or for access or re-use.

3. Charges

DIFFERENCES – As the charges can differ considerably between access, re-use, and sharing, it is important for the public bodies and the applicants to know which rules they have to follow.

ACCESS – Once again, the charges for access have to take into account the provisions from the Access Directive and the INSPIRE Directive. Under the Access Directive, on-site consultation of information is free, while charges can be made for the supply of the information, not exceeding a reasonable amount. As a general rule, the charges may not exceed the costs of reproduction. Exceptionally, market-based charges can also be considered reasonable. Access to the registers of documents should be free. The INSPIRE Directive allows charges to the download services, the transformation services, and the invoke services without any restrictions. View services should in principle be free of charge, but they can be charged for where this is necessary to secure the maintenance of spatial data sets and services, especially in cases involving very large volumes of frequently updated data.

RELATIONSHIP BETWEEN ACCESS DIRECTIVE AND INSPIRE DIRECTIVE – The question can be asked how the charging provisions of the Access Directive and the INSPIRE Directive relate. The Access Directive seems to have been written with mostly traditional paper documents and services in mind, and it did not consider the possibility of remote consultation of information through Internet services. When this is combined with the obligations of the INSPIRE Directive on access to the network services, it remains unclear whether view services should be seen as a form of supply of information or rather as analogous to examination *in situ*. Consultation *in situ* entails that a citizen requesting information can see it but will not receive a copy, so he or she does not have any physical result of his or her request and will have to file a new request if he or she wants to see the information again. In the case of supply of information, the citizen obtains his or her own copy of the information – on paper or in another form– which he or she can consult later. If a technical argumentation would consider viewing services to be a supply of information, as technically a copy is made on the hard drive of a computer, in my opinion, a teleological and more common-sense interpretation of 'consultation' and 'supply' would lead to a different and more suitable conclusion. In the contemporary information society, making a difference between consultation of information on site or from a remote access point is outmoded. It is merely a matter of physical presence. In both cases, the citizen does not obtain a tangible copy of the information.

RE-USE – Hence, the charging issue may already cause confusion with regard to access. The combination with the charging requirements for re-use and sharing shows even more the importance of a clear distinction between the three types of use. For re-use, the total income cannot exceed the cost of collection, production, reproduction, and dissemination, together with a reasonable return on investment. While a reasonable charge under the Access Directive should generally be limited to the marginal cost, the reasonable return on investment limit under the PSI Directive shows that higher charges can be made for re-use.

SHARING – Next, the difference between re-use and sharing with regard to charging should also be considered. As was discussed before, the original proposal for the INSPIRE Directive did not contain any possibility for charging for data sharing. During the preparation of the final version of the directive, the possibility for charging was included, but without any limits. As this would lead to the strange situation that charges for re-use of public sector spatial data outside of the public task would be limited and charges for sharing would not, limits were introduced. Now the charges have to be kept to the minimum required to ensure the necessary quality and supply of spatial data sets and services together with a reasonable return on investment. Hence, the reasonable return on investment is still possible, but does the minimum required to ensure quality and supply include all costs of collection, production, reproduction, and dissemination, or might it include even more?

4. Time Limits

APPROPRIATE TIMEFRAME – Regardless of the purpose for which public sector spatial data are provided, it is important for all the purposes that the data are provided within a timeframe that is appropriate to the purpose and that time limits are foreseen.

ACCESS – For access, the Access Directive foresees a time limit of one month to reply to a request and to provide the requested data. However, the public body should have regard for any timeframe that the applicant has specified and make a reasonable effort to provide the data within that timeframe. Under the INSPIRE Directive, no time limits have been determined, but one could assume that the network services should be available on-line continuously, so that time limits are not necessary.

RE-USE – The time limits for re-use should follow the time limits foreseen in national access legislation. Thus, for public sector spatial data that is also environmental information, this entails that the maximum time limits of the Access Directive will also apply.

SHARING – With regard to sharing, the INSPIRE Directive does not specify a time limit but leaves it to the Member States to determine the arrangements. Once again, if the network services are used, the issue of time limits should not arise. However, the provision of data may occur in another form, at least for several more years or for some types of data for which other forms of delivery may be considered more suitable.

E. TOWARDS THE NEXT CHAPTER

OUTLINE – Now that the rules on access, re-use, and sharing have been discussed, and some differences in the rules have been indicated, it is time to determine why this legal framework for the availability of public sector spatial data as defined by the three directives discussed in this chapter does not work. The next chapter addresses the problems with this legal framework and explains the problem statement.

Chapter 4

The Problematic Character of the EU Legal Framework for the Availability of Public Sector Spatial Data

I. THE DISTINCTION BETWEEN ACCESS, RE-USE, AND SHARING

RECAPITULATION – In Chapter 2, it was stated that the EU legal framework for the availability of public sector spatial data examined in this book consists of the rules on access, re-use, and sharing. These rules are included in the Access Directive, the PSI Directive, and the INSPIRE Directive.

CRITERION FOR DISTINCTION – Chapter 2 explained that the distinction between the three types of use that are discerned in the EU legal framework for the availability of public sector spatial data are based on the purpose of the use. *Access* means 'obtaining public sector spatial data for the purpose of exercising one's democratic rights or obligations'. Next, *re-use* is defined as 'any use of public sector spatial data for commercial or non-commercial purposes outside of the public task, requiring the processing of the data beyond access needs'. Thirdly, *sharing* refers to 'any use of public sector spatial data for the purpose of performing public task activities'. Hence, the purpose for which the requested public sector spatial data will be used is the criterion for the application of the rules of the Access Directive, the PSI Directive, and the INSPIRE Directive.

II. IMPORTANCE OF THE DIFFERENCE BETWEEN ACCESS, RE-USE, AND SHARING

RECAPITULATION – Chapter 3 explained which rules are applicable to access, re-use, and sharing in the EU legal framework on the availability of public sector spatial data. The short overview at the end of the chapter showed that the rules with regard to availability, practical arrangements, limitations, charges, and so on are comparable, but not the same.

PRACTICAL IMPORTANCE – Hence, it is important for the public bodies and the users to know which rules apply to them and under what conditions public sector spatial data have to be made available. Moreover, some rules are only applicable to one or two of the three purposes and can be disregarded for the other(s). For instance, if a public body makes its public sector spatial data available for re-use, it has to make sure that it complies with the non-discrimination principles provided by Article 10 of the PSI Directive. This includes the rule that: 'If documents are re-used by a public sector body as input for its commercial activities which fall outside the scope of its public tasks, the same charges and other conditions shall apply to the supply of the documents for those activities as apply to other users'. This would mean that if a public body re-uses its own public sector spatial data, it has to make these data available under the same conditions to its private competitors. This is only relevant for re-use purposes. There is no comparable rule, for instance, stating that if a public body uses its own public sector spatial data for sharing purposes, that is, for the performance of its public tasks, that it has to apply the same conditions and charges to its own use as it would to other public bodies. It seems logical that a public body would not need to pay itself for further using its own data for performing its public tasks, even though it might charge other public bodies that request the same data under the INSPIRE Directive. Yet, one can see the importance of knowing whether this public body is actually using its own public sector spatial data for its public task or if it is re-using its own data.

CONCEPTUAL IMPORTANCE – A clear distinction between access, re-use, and sharing is indispensable for the proper functioning of the existing EU legal framework for the availability of public sector spatial data. The public bodies and the users both need legal certainty about what they can and cannot do when accessing, re-using, or sharing public sector spatial data. However, the distinction is not only important on a practical level with regard to the application of the proper rules, but it is also important for the development of the underlying theories and concepts supporting access, re-use, and sharing. Unclear distinctions and confusion between access, re-use, and sharing can confuse the debates that underpin the fundamental policy choices that form the basis for further development of the framework for the availability of public sector spatial data. If choices are made based on the value that is attached to the democratic purpose of access or the economic purpose of re-use, it is vital that these choices have a solid foundation in the demarcation between these types of use and purpose.

EXAMPLE TO DEMONSTRATE THE IMPORTANCE – I will demonstrate this importance of a clear delineation of the concepts of access, re-use, and sharing for both practical and conceptual reasons with the example of the discussion on charges for the availability of public sector spatial data. On the one hand, it is important to know the difference between access, sharing, and re-use because the rules in the Access Directive, the PSI Directive, and the INSPIRE Directive impose different limits on the charges that can be made. On the other hand, the underlying debate on the principle of whether charging for public sector spatial data should be allowed also needs clear concepts on access, re-use, and sharing as a starting point. We will see that this has not been the case for many participants in the debate and how this has clouded the arguments that have been used in favour of and against charging.

III. EXAMPLE: CHARGING FOR PUBLIC SECTOR SPATIAL DATA

A. Introduction

Value of the example – The discussion on charging for data is a good example to demonstrate the importance of clear concepts in the EU legal framework on the availability of public sector spatial data. First, the rules in the Access Directive, PSI Directive and INSPIRE Directive are different for access, re-use, and sharing. Second, the example also has a broader value. The charges for public sector spatial data have been the centre of attention in the relationship between the public bodies and the users of their spatial data for a number of years. Numerous debates have taken place between proponents of charging and supporters of free availability.[1] The arguments found in the literature very often mix the different purposes for the availability of public sector spatial data and in this way cloud the debate.

Outline – Before I look into the arguments against and in favour of charging for public sector spatial data, I will shortly recapitulate the existing rules in the Access Directive, the PSI Directive, and the INSPIRE Directive and then take a short look at the basis of the debate on charging: the dichotomy between open access and cost recovery charging regimes.

B. Existing Rules on Charges for Access, Re-use, and Sharing in the Access Directive, the PSI Directive, and the INSPIRE Directive

Access – Under the Access Directive, public authorities can charge for the supply of environmental information, as long as the charge does not exceed a reasonable amount.[2] This reasonable character of the charges implies that, as a general rule, charges may not exceed actual costs of producing the material in question. However, in particular cases where public authorities make available

1. The discussion on the charges for public access to the network services and for sharing during the negotiations on the INSPIRE Directive are a good example of these debates. Comparable discussions took place in the preparation of the PSI Directive, and in the responses to the Green Paper on Public Sector Information, eEurope 2002 Communication and the 2008 Review of the PSI Directive.
2. Article 5.2 of the Access Directive.

environmental information on a commercial basis, and where this is necessary in order to guarantee the continuation of collecting and publishing such information, a market-based charge is considered to be reasonable.[3] Additionally, Chapter IV of the INSPIRE Directive, which deals with access of the public to the INSPIRE network services, allows charges for download services and transformation services, without any limit to those charges. View services can only be charged for where such charges secure the maintenance of spatial data sets and corresponding spatial data services, especially in cases involving very large volumes of frequently updated data.[4] Hence, while the view services should generally be free, in the cases where public authorities are allowed to charge, no limit on this charge is imposed. However, the discovery services always have to be available free of charge.

RE-USE – With regard to re-use, the PSI Directive leaves a large margin for the possible charges: from no charge to a charge reflecting the cost of collection, production, reproduction, and dissemination, together with a reasonable return on investment.[5] The PSI Directive encourages marginal cost charging but also recognizes the self-financing requirements of some public sector bodies.[6] In addition, the PSI Directive makes it possible for the Member States to use licenses to restrict the conditions of re-use of the public sector spatial data that are made available. These conditions may not unnecessarily restrict possibilities for re-use or be used to restrict competition,[7] but within those limits, they may contain whatever conditions the Member States or public authorities choose to include.

SHARING – With respect to sharing, the INSPIRE Directive makes it possible for the Member States to allow their public authorities to provide spatial data sets and services under a license and to require payment. The charges have to be fully compatible with the general aim of facilitating the sharing of spatial data sets and services between public authorities, and they have to be kept to the minimum required to ensure the necessary quality and supply of spatial data sets and services together with a reasonable return on investment, while respecting the self-financing requirements of the supplying public authorities.[8]

MARGIN OF APPRECIATION FOR THE MEMBER STATES – It is clear that, while the three directives contain rules on the charges and conditions for making public sector spatial data available, they also leave a lot of room for the public bodies to determine their own prices and conditions. The public bodies could choose to make their data freely available to all applicants, or, on the contrary, to impose charges on all types of use that reflect a reasonable charge or a reasonable return on investment. However, the public bodies may also distinguish between different types of use and users, both regarding the charges and the conditions that are imposed.

3. Recital 18 of the Access Directive.
4. Article 14 of the INSPIRE Directive.
5. Article 6 of the PSI Directive.
6. Recital 14 of the PSI Directive.
7. Article 8 of the PSI Directive.
8. Article 17.3 of the INSPIRE Directive.

C. Basis of the Debate on Charging: Cost Recovery v. Open Access

1. Background of the Debate

Economic value v. social and policy value – The debate on whether or not a public body can charge for making public sector spatial data available can be traced back to the underlying tension between the potential economic value of public sector data in general and public sector spatial data in particular, and to the social and policy value of these data. On the one hand, increasing attention is given to making public sector spatial data available as widely as possible in order to encourage public participation and accountability, and on the other hand, public bodies are confronted with budget cuts and the rising demand of the national treasuries to provide, at least partly, for their own funding. From this latter perspective, charging for or *selling* public sector spatial data is seen as an excellent opportunity to raise revenue for the public bodies.[9]

Dichotomy between open access and cost recovery – The debate on whether or not a public body should charge for its data is often referred to as the *open access – cost recovery debate*.[10] However, these are not the only methods for funding public sector spatial data. Other possibilities that have been mentioned are the sale of public bonds, funds from outside grants, allocating a portion of each department of a public body's operating budget, allocating a portion of the increased savings from the better use of public sector spatial data for spatial planning, and so on.[11] However, in the end, many of these other models can be brought back to two possibilities: the collection of public sector spatial data is either funded by the central budget or by user contributions, that is, open access or cost recovery.

9. See François Salgé, 'Geographic Information Network in Europe (GINIE). Recommendations for action', in *Spatial Data Infrastructure and Policy Development in Europe and the United States*, ed. Bastiaan Van Loenen & Bas C. Kok (Delft: DUP Science, 2004), 40; Bastiaan Van Loenen & Bas C. Kok, 'Spatial Data Infrastructures: Legal and Economic Issues', in *Spatial Data Infrastructure and Policy Development in Europe and the United States*, ed. Bastiaan Van Loenen & Bas C. Kok (Delft: DUP Science, 2004), 7.
10. See for instance Bastiaan Van Loenen, *Developing Geographic Information Infrastructures: The Role of Information Policies* (Delft: DUP Science, 2006), 100–107; Harlan J. Onsrud, 'In Support of Cost Recovery for Publicly Held Geographic Information', *GIS Law* 1, no. 1 (1992): 1–7, <www.spatial.maine.edu/%7eonsrud/pubs/Cost_Recovery_for_GIS.html>, 22 Feb. 2010; Harlan J. Onsrud, 'In Support of Open Access for Publicly Held Geographic Information', *GIS Law* 1, no. 2 (1992): 3–6, <www.spatial.maine.edu/~onsrud/pubs/In_Support_OA.htm>, 22 Feb. 2010; Jeff P. Johnson & Harlan J. Onsrud, 'Is Cost Recovery Worthwhile?', 1995, <www.spatial.maine.edu/%7eonsrud/pubs/Cost_Recovery_Worthwhile.html>, 22 Feb. 2010.
11. Peter L. Croswell & Alex Wernher, *GIS Program Revenue Generation and Legal Issues in Public Sector Organizations* (Park Ridge: URISA, 2004), 6; Bruce Joffe, '10 Ways to Support your GIS Without Selling Data', <www.opendataconsortium.org/documents/10Ways2Support GIS-3.pdf>, 22 Feb. 2010.

2. Description of the Open Access and Cost Recovery Models

OPEN ACCESS – The open access model assumes that the public bodies responsible for collecting, creating, and making public sector spatial data available are fully financed with public funds to accomplish their public tasks. In the open access model, data are accessible for a price not exceeding the cost of reproduction and distribution (mostly referred to as 'marginal cost' or 'dissemination cost'), with the imposition of as few restrictions as possible.[12] Onsrud refers to this approach as the *free distribution* approach and states that it is based on the idea that public sector spatial data are 'acquired at taxpayer expense and should be free to anyone who has an interest in them and requests them'.[13] It is assumed under this approach that the well-being of the community as a whole is promoted through the practice of 'free distribution'.[14]

COST RECOVERY – On the contrary, the cost recovery model attempts to gain revenues from the sale of public sector spatial data to support the development and maintenance of data. The collection, maintenance, and dissemination of public sector spatial data are not fully financed by public funds, and the costs must be recovered through other means. Hence, the public bodies are forced to generate income from the sale of their data.[15] Cost recovery involves transferring some or all of the costs of a government activity from the general taxpayer to those who more directly receive a personal benefit from a product or service.[16]

SOMEONE HAS TO PAY – In any case, whether collecting and making available public sector spatial data is paid by the central budget or by user fees, Clark states very clearly that: 'What is inescapable is that, whichever way you cut it, it still comes down to the question *who pays*? – somebody has to, simply because there is a cost attached to producing it'.[17] Hence, the question is essentially whether the taxpayer pays or whether the user pays.

---

12. Bastiaan Van Loenen, *Developing Geographic Information Infrastructures: The Role of Information Policies* (Delft: DUP Science, 2006), 100–101; Bastiaan Van Loenen & Bas C. Kok, 'Spatial Data Infrastructures: Legal and Economic Issues', in *Spatial Data Infrastructure and Policy Development in Europe and the United States*, ed. Bastiaan Van Loenen & Bas C. Kok (Delft: DUP Science, 2004), 7.
13. Harlan J. Onsrud, 'In Support of Cost Recovery for Publicly Held Geographic Information', *GIS Law* 1, no. 1 (1992): 1–7, <www.spatial.maine.edu/%7eonsrud/pubs/Cost_Recovery_for_GIS.html>, 22 Feb. 2010.
14. *Ibid.*
15. Bastiaan Van Loenen, *Developing Geographic Information Infrastructures: The Role of Information Policies* (Delft: DUP Science, 2006), 102; Bastiaan Van Loenen & Bas C. Kok, 'Spatial Data Infrastructures: Legal and Economic Issues', in *Spatial Data Infrastructure and Policy Development in Europe and the United States*, ed. Bastiaan Van Loenen & Bas C. Kok (Delft: DUP Science, 2004), 8–9. See also William S. Holland, 'Copyright, Licensing and Cost Recovery for Geographic and Land Information System Data: A Legal, Economic and Policy Analysis', 1997, <www.spatial.maine.edu/~onsrud/tempe/holland.html>, 22 Feb. 2010.
16. Garry Sears, 'Geospatial Data Policy Study. Executive Summary', 2001, 3, <www.geoconnections.org/programsCommittees/proCom_policy/keyDocs/KPMG/KPMG_E.pdf>, 22 Feb. 2010.
17. Mike Clark, 'Fee or Free? The Hidden Costs of Free Public Sector Information', *Business Information Review* 24, no. 1 (2007): 50; See also Roger A. Longhorn & Michael Blakemore,

3. Demarcation between Open Access and Cost Recovery

CUT-OFF POINT: MARGINAL COST – In the debate on cost recovery versus open access, the question can be asked which costs are included in the 'recovery'. Considering that open access policies do not necessarily imply that the public sector spatial data are made available for free, but that a marginal cost can be charged for the data, one can deduct that *cost recovery* includes all the policies that impose charges above the marginal costs. This would involve not only full cost recovery, but also partial cost recovery above marginal costs, return on investment, or even the charging of market prices or profit-maximizing.[18] Thus, even though marginal cost charging *in se* is also a form of cost recovery, it seems to be the pivotal point between the two policies. Therefore, it may be interesting to have a first insight of what marginal costs for public sector spatial data actually include.

INCLUDED COSTS – In the context of data, marginal costs are often referred to as the costs of disseminating data. No charges are levied on the costs of collecting, editing, and organizing the data, because these costs are fixed and do not vary with the number of data users.[19] Onsrud stated that marginal costs for responding to citizen and business requests for public sector spatial data, records, and services might typically include the costs of staff time for selecting, downloading, and handling data, computer time, consumable supplies (such as disks, tapes, and printouts), hardware and software maintenance, and staff time to assist requesters in using the public sector spatial data.[20] The United Kingdom's Treasury report, Cross Cutting Review of the Knowledge Economy, recommended *marginal costs* to be defined as 'costs, including costs of staff time, reasonably incurred in locating and retrieving the information, and giving effect to the requesters preferred medium for the reply (which could be different to that in which the department currently held it); and also the disbursements directly incurred in communicating the information, for example, printing, postage etc'.[21] Rather than marginal cost, Van Eechoud and Kabel refer to dissemination cost as the costs that are related to a specific request for data: the selection of data, possible processing, reproduction on a carrier, and administrative and shipping costs.[22] The Berenschot study defines *marginal cost* as all costs related to the dissemination of PSI, including shipping,

---

Geographic Information: Value, Pricing, Production, and Consumption (Boca Raton: CRC Press, 2008), 63–94.

18. If these charges stay within the charging conditions that are imposed in the Access Directive, the PSI Directive and the INSPIRE Directive.
19. James Love, 'Pricing Government Information', *Journal of Government Information* 22, no. 5 (1995): 365.
20. Harlan J. Onsrud, 'In Support of Cost Recovery for Publicly Held Geographic Information', *GIS Law* 1, no. 1 (1992): 1–7, <www.spatial.maine.edu/%7eonsrud/pubs/Cost_Recovery_ for_GIS. html>, 22 Feb. 2010.
21. HM Treasury, 'Cross Cutting Review of the Knowledge Economy. Review of Government Information', 2000, <www.hm-treasury.gov.uk/spend_sr00_ccr.htm>, 22 Feb. 2010.
22. Mireille Van Eechoud & Jan Kabel, *Prijsbepaling voor elektronische overheidsinformatie* (Deventer: Kluwer, 1998), 6–7 [in Dutch].

promotional costs, personnel, and information technology costs.[23] The National Research Council distinguishes marginal cost from incremental cost pricing, which includes the cost of recompiling the data, perhaps maintaining a computer site for downloading, purchasing CD-ROM blanks, recording the data, shipping it to the user, and customer support, but not the cost for the core service.[24]

SUMMARY – From these definitions it follows that marginal costs include the staff time needed for processing a request, the reproduction of the public sector spatial data on the medium that is used for the transfer, and the administrative costs in getting the data to the applicant. In any case, as marginal costs are seen as the expenditure incurred in providing an extra copy of a piece of public sector spatial data to a user, production costs are not part of marginal costs, as they do not vary with the number of users. Also, Newbery et al. feel that many aspects of distribution display economies of scale and involve fixed costs that cannot be included in the marginal cost estimation, such as buying an additional server to handle downloads and the purchase of bandwidth, even though, strictly, they may be costs involved in the dissemination of public sector spatial data.[25]

OUTLINE – Now that a first insight has been given into the concept of *marginal cost*, which forms the demarcation line between open access and cost recovery, and the reader now has an idea about what value may be at stake when the decision of making public sector spatial data available is made, the arguments in favour of and against open access and cost recovery can be lined up.

D. ARGUMENTS FOR OPEN ACCESS AND COST RECOVERY

1. Introduction

ARGUMENTS ARE SUBJECTIVE – The debate about the funding and pricing of public sector data has often been said to be more about personal feelings and opinions than about facts, or as Litman puts it:

> The debate has seemed to be fuelled by some combination of almost religious faith (that privatization, or nationalization, or commercialization, or liberation is good; that markets are best; that facts should be free; that commons are tragic; and so forth) with self interest: those who have invested in public sector spatial data collection argue that the public interest is best served by the enforcement of strong property rights in collected data, while those with an

23. Berenschot and NEI, 'Welvaartseffecten van verschillende financieringsmethoden van elektronische gegevensbestanden', 2001, 11, <www.epsiplus.net/media/files/final_eindrapport_onderzoek_welvaartseffecten_feb2001>, 22 Feb. 2010 [in Dutch].
24. National Research Council, *Bits of Power. Issues in Global Access to Scientific Data* (Washington: National Academies Press, 1997), 125–126.
25. David Newbery, Lionel Bently & Rufus Pollock, 'Models of Public Sector Information Provision via Trading Funds', 2008, 50, <www.berr.gov.uk/files/file45136.pdf>, 22 Feb. 2010.

> interest in mounting competing products incorporating the data assert that the public interest lies in unfettered movement of information.[26]

Litman believes that especially this faith that guides people in their policy choices is particularly hard to dislodge.[27]

ARGUMENTS ARE DIFFICULT TO PROVE – Furthermore, the empirical evidence is often said to be non-conclusive.[28] Case studies and anecdotal evidence sometimes point in the direction of open access, while others find that cost recovery would be the best option for making public sector spatial data available.[29] Blakemore and Singh already came to this conclusion in a 1992 study. They looked at various data financing and charging policies around the world and could not find that one was better than the other.[30] Blakemore later repeated that: 'There is no certainty that a centrally funded . . . model will work, and neither is it certain that effective and nationally beneficial usage will not occur if user-charging strategies are put in place'.[31]

DECISION MAKERS' OPINION DETERMINES THE CHOICE – Thus, while the arguments defending or rejecting cost recovery or open access are important for public bodies to take into account in determining which charging policy they adopt, one should also keep in mind that the choice of the argumentation that is followed may often depend on the personal considerations and values of the people that have to decide. As there is no conclusive evidence in favour of either of the approaches, decision makers have no choice but to fall back on their views of what is most beneficial to

26. Jessica Litman, 'Rights in Government-generated Data', paper given at the Conference on Law and Information Policy for Spatial Databases (Tempe, 1994), <www.spatial.maine.edu/~onsrud/tempe/litman.html>, 22 Feb. 2010. See also Roger A. Longhorn & Michael Blakemore, *Geographic Information: Value, Pricing, Production, and Consumption* (Boca Raton: CRC Press, 2008), 12; Krysia Rybaczuk & Michael Blakemore, 'Selling Government Information. a Comparative Perspective on UK and USA Developments', paper given at the Conference on Law and Information Policy for Spatial Databases (Tempe, 1994), <www.spatial.maine.edu/%7eonsrud/tempe/rybaczuk.html>, 22 Feb. 2010.
27. *Ibid.*
28. See Paul F. Uhlir et al., *The Socioeconomic Effects of Public Sector Information on Digital Networks* (Washington: National Academies Press, 2009), 25, 45, 61–67.
29. See e.g., Michael Blakemore, 'Financing the NGDI', *GIS Development*, s.d., <www.gisdevelopment.net/policy/international/interna012pf.htm>, 22 Feb. 2010; Peter N. Weiss, 'Borders in Cyberspace: Conflicting Public Sector Information Policies and their Economic Impacts', in *Public Sector Information in the Digital Age: Between Markets, Public Management and Citizens' Rights*, ed. Georg Aichholzer & Herbert Burkert (Cheltenham: Edward Elgar Publishing Limited, 2004), 137–159; Jeff P. Johnson & Harlan J. Onsrud, 'Is Cost Recovery Worthwhile?', 1995, <www.spatial.maine.edu/%7eonsrud/pubs/Cost_Recovery_Worthwhile.html>, 22 Feb. 2010; Bruce Joffe, '10 Ways to Support your GIS Without Selling Data', s.d., <www.opendataconsortium.org/documents/10Ways2SupportGIS-3.pdf>, 22 Feb. 2010; Economic Development and Infrastructure Committee, 'Inquiry into Improving Access to Victorian Public Sector Information and Data', 2009, 10–11, <www.parliament.vic.gov.au/edic/inquiries/access_to_PSI/final_report.html>, 22 Feb. 2010.
30. See Michael Blakemore, 'Financing the NGDI', *GIS Development*, s.d., <www.gisdevelopment.net/policy/international/interna012pf.htm>, 22 Feb. 2010.
31. *Ibid.*

society or to their own organization. Or as Love formulates it: 'Often the decisions to sell or disseminate for free depend upon the whims of agency officials or lower technicians, who may or may not have considered the broader consequences of their actions'.[32]

VALUE OF CHARGING DEBATE – Before one starts worrying too much about all the difficulties that exist in selecting a charging policy, Longhorn and Blakemore put these concerns into perspective: 'The crucial debate is not about price or charging regimes per se; it is about consistent resources for reinvestment and maintenance of information that is fit for a wide range of purposes, while at the same time maximizing the ability of information providers to respond to the widest possible constituency or market'.[33] Newbery et al. also warn that 'charging policy is not the central issue'. It is best seen as 'secondary, and dependent upon, the primary matter of the regulatory/governance structures under which data provision (and collection) by trading funds occurs'.[34] Without such adequate structures in place, 'all charging regimes can result in poor incentives, inefficiency and overall poor performance'.[35] One should also realize that charging and pricing are together only one of the possible barriers to the availability of public sector spatial data. These barriers may also include difficulties in finding public sector spatial data or even knowing that it exists, constraining licensing terms, wrong formats, unhelpful officials lacking knowledge, confusion about where to apply for a license, delays in the delivery in the data, and so on.[36] However, charging remains one of the essential elements of the policies on the availability of public sector spatial data.

OUTLINE – Below are the arguments that regularly occur in the literature on cost recovery and open access policies. They can be divided into five types of arguments. The first type can be described as democratic arguments relating to the empowerment and participation of the citizen, while the second type relates more to the fact that the public sector spatial data may be paid for twice. The third category of arguments looks at the economic impact of the policies at the level of society, and the fourth takes the economic approach at the level of the public bodies. In the fifth type, some political issues may also play a role. As was mentioned before, these arguments in favour of and against open access or cost recovery are often just as much based on personal opinions and ideologies as on facts.

---

32. James Love, 'Pricing Government Information', *Journal of Government Information* 22, no. 5 (1995): 385.
33. Roger A. Longhorn & Michael Blakemore, *Geographic Information: Value, Pricing, Production, and Consumption* (Boca Raton: CRC Press, 2008), 208.
34. David Newbery, Lionel Bently & Rufus Pollock, 'Models of Public Sector Information Provision via Trading Funds', 2008, 114, <www.berr.gov.uk/files/file45136.pdf>, 22 Feb. 2010.
35. *Ibid.*, 116.
36. See Ed Mayo & Tom Steinberg, 'The Power of Information: An independent Review', 2007, 19–20, <www.cabinetoffice.gov.uk/media/cabinetoffice/strategy/assets/power_information.pdf>, 22 Feb. 2010; Office of Fair Trading, 'The Commercial Use of Public Information (CUPI)', 2006, 78–119, <www.oft.gov.uk/shared_oft/reports/consumer_protection/oft861.pdf>, 22 Feb. 2010.

Moreover, considering all the conflicting interests that are in play in these arguments, it is not even clear how *better* should be defined. How *good* or *bad* a charging policy is depends not only on the type of public sector spatial data that is being made available, but also on the users and especially on the purpose that the data are used for. That is why it is unfortunate that most arguments dealing with open access and cost recovery do not consider whether these policies should be the same for access, re-use, and sharing, or whether a differentiation is possible between these purposes. For every argument, I try to examine whether a distinction between access, re-use, and sharing is needed.

2. 'Democratic Arguments': Open Access Increases Democracy and Participation of the Citizen

ARGUMENT FOR OPEN ACCESS: PARTICIPATION – According to many scholars, greater availability of public sector data in general, and public sector spatial data in particular, for free or at the cost of dissemination would increase democracy and the accountability of the public sector towards the citizens.[37] In addition, it would facilitate the participation of the citizens in decision making processes. Onsrud states strongly that: 'All other rights in a democratic society extend from our ability to access information. Democracy can't function effectively unless people have ready access to government information in order to keep government accountable'.[38] Uhlir adds that one of the greatest non-economic values associated with public sector data is '*transparency* of governance and the promotion of democratic ideals: equality, democracy, openness. The more information that is openly available from the government and about the government, the less likely will it be that government is able to hide illegal acts, corruption and misrule'.[39]

ARGUMENT FOR OPEN ACCESS: DIGITAL DIVIDE – As an additional barrier to democracy, charging for access would cause a division between those who can pay for data and those who cannot, often referred to as the *information haves* and the *information have-nots*. As Love puts it: 'If access is too expensive, it will deter

---

37. See Harlan J. Onsrud, 'In Support of Open Access for Publicly Held Geographic Information', *GIS Law* 1, no. 2 (1992): 3–6, <www.spatial.maine.edu/~onsrud/pubs/In_Support_OA.htm>, 22 Feb. 2010. See also Bruce Joffe, '10 Ways to Support your GIS Without Selling Data', s.d., <www.opendataconsortium.org/documents/10Ways2SupportGIS-3.pdf>, 22 Feb. 2010; Garry Sears, 'Geospatial Data Policy Study. Executive Summary', 2001, 3, <www.geoconnections.org/programsCommittees/proCom_policy/keyDocs/KPMG/KPMG_E.pdf>, 22 Feb. 2010; Renate Mason, 'Developing Australian Spatial Data Policies – Existing Practices and Future Strategies', 2000, 3, <http://handle.unsw.edu.au/1959.4/18646>, 22 Feb. 2010; OECD Working Party on the Information Economy, 'OECD Workshop on Public Sector Information held on 31 May 2006: Summary', 2006, 4–5, <www.oecd.org/dataoecd/54/35/36854745.pdf>, 22 Feb. 2010.
38. Harlan J. Onsrud, 'In Support of Open Access for Publicly Held Geographic Information', *GIS Law* 1, no. 2 (1992): 3–6, <www.spatial.maine.edu/~onsrud/pubs/In_Support_OA.htm>, 22 Feb. 2010.
39. UNESCO, 'Policy Guidelines for the Development and Promotion of Governmental Public Domain Information' (Paris: UNESCO, 2004), 6–7, <http://unesdoc.unesco.org/images/0013/001373/137363eo.pdf>, 22 Feb. 2010.

low-income persons or disinterested citizens from taking an active role in governance'.[40] However, this *digital divide* issue is not solved by just making data available to the public, but it also needs 'empowering of the citizen' to make sensible use of the data.[41] This entails that public sector spatial data not only have to be delivered, but they also have to be made accessible and understandable for the target group they are addressed to.[42] In this respect, Longhorn and Blakemore warn that: 'Data can both empower and disenfranchise citizens, depending upon their particular circumstances in being able to make the best use of the information'.[43]

ARGUMENT FOR OPEN ACCESS: INTERNATIONAL RECOGNITION – As will be described in Chapter 5, many international and European legislative and policy texts recognize the importance of access to public sector spatial data for participation of the citizens in democracy.[44] Therefore, it seems only logical that access is as widely spread as possible by the free (or marginal cost) dissemination of public sector spatial data.

ARGUMENT FOR COST RECOVERY: HISTORY – Peterson Dando is very critical of the argument that access to public sector data is essential to democracy. She argues that, while forms of democracy have existed for hundreds of years, access to public sector data is a right created by legislation during the last fifty years. Moreover, freedom of information legislation is not a global phenomenon in all democracies.[45] For instance, the United Kingdom's Freedom of Information Act only entered into force in 2005, and the German access legislation dates from 2006. Longhorn and Blakemore have comparable concerns: they wonder if the availability of data really has an impact on the participation of the citizen, referring to the low number of people who vote in the United States (in which federal public sector data are available for free) and the higher number in the United Kingdom, where charging for public sector data is much more common.[46]

ARGUMENT FOR COST RECOVERY: PURPOSE OF REQUESTS FOR DATA – As another counter-argument for the statement that open access increases democracy, Holland states that the requests for '[t]he bulk of public records are not motivated by the need of the public to know what their government is doing. To the contrary, a large

---

40. James Love, 'Pricing Government Information', *Journal of Government Information* 22, no. 5 (1995): 385.
41. See Roger Longhorn & Michael Blakemore, 'Re-visiting the Valuing and Pricing of Digital Geographic Information', *Journal of Digital Information* 4, no. 2 (2004): 7, <http://journals.tdl.org/jodi/article./view/103/102>, 22 Feb. 2010.
42. Cf. *infra*, Ch. 8, s. III.B.4.a.
43. Roger A. Longhorn & Michael Blakemore, *Geographic Information: Value, Pricing, Production, and Consumption* (Boca Raton: CRC Press, 2008), 13.
44. Cf. *infra*, Ch. 5, s. I.C.
45. Lori Peterson Dando, 'A Case for the Commercialization of Public Information', paper given at the Conference on Law and Information Policy for Spatial Databases (Tempe, 1994), <www.spatial.maine.edu/%7eonsrud/tempe/dando.html>, 22 Feb. 2010. See also Renate Mason, 'Developing Australian Spatial Data Policies – Existing Practices and Future Strategies', 2000, 112, <http://handle.unsw.edu.au/1959.4/18646>, 22 Feb. 2010.
46. Roger A. Longhorn & Michael Blakemore, *Geographic Information: Value, Pricing, Production, and Consumption* (Boca Raton: CRC Press, 2008), 72.

proportion (if not the vast majority) of requests are made to agencies who have custodial responsibilities for information that depict or may affect property'.[47] Peterson Dando also feels that the data that are requested mostly have little to do with democracy, as they are mainly data wanted by the information industry for creating their own products and services: 'Obtaining data for commercial purposes has nothing to do with holding government accountable'.[48]

EVALUATION: ACCESS – It seems clear that the democratic argument has more value for the situation where the public sector has to make public sector spatial data available for access than for re-use. As the purpose of making public sector spatial data available for access is to enable the citizen to check up on the activities of the government that he or she selected in an election process and to fully exercise his or her rights and obligations in today's society, it seems only logical that the availability of data for free or at marginal cost will increase this ability of the citizen. While Peterson Dando may argue that democracy has existed for hundreds of years, and that the recent emergence of access legislation shows its relative unrelated character to democracy, one should also take into account the enormous changes that our society has gone through in the past fifty years. The evolution towards the information society of today has changed the position of the citizen, in that he or she is bombarded with data from many different sources and that he or she needs to determine which of these data are reliable to base decisions on. The public sector has an important role to play in making data available that the citizen can rely on as accurate and trustworthy. Hence, as the size, composition, and characteristics of today's information society are fundamentally different from what they were fifty years ago, the argument that democracy has existed without access for so long is no longer valid in my opinion.

EVALUATION: RE-USE – There is some value to the argument that a considerable part of public sector spatial data is requested for other purposes than ensuring democracy or participation of the citizen. According to Peterson Dando, for instance, this entails that access to these data should not necessarily be free of charge or charged at marginal cost, at least not for the reason of safeguarding democracy. The requests that Peterson Dando is referring to are in most cases made by people interested in data on property or the information industry. Hence, they are made for economic purposes and involve re-use of public sector spatial data rather than access for democratic purposes. Therefore, the democratic argument is not fully appropriate for requests for re-use. This does not mean that one should automatically assume that these requests should always be subject to cost recovery. It just entails that the free availability of public sector spatial data for re-use cannot be justified solely on the basis of such *democratic* arguments.

47. William S. Holland, 'Copryight, Licensing and Cost Recovery for Geographic and Land Information System Data: A Legal, Economic and Policy Analysis', 1997, <www.spatial.maine.edu/~onsrud/tempe/holland.html>, 22 Feb. 2010.
48. Lori Peterson Dando, 'A Case for the Commercialization of Public Information', paper given at the Conference on Law and Information Policy for Spatial Databases (Tempe, 1994), <www.spatial.maine.edu/%7eonsrud/tempe/dando.html>, 22 Feb. 2010.

Evaluation: re-use – However, one should also keep in mind that the free of charge or marginal cost availability of public sector spatial data for re-use may still have an indirect impact on democracy, as the information industry has public sector spatial data at its disposal for creating information services with an added value that can be marketed to the citizen. Hence, the citizen can obtain public sector spatial data from different sources, in a value-added and user-friendly way. Whether such information services or products would not be created if the public sector spatial data was charged for under a cost recovery policy for re-use unfortunately remains without empirical evidence.

Evaluation: sharing – With regard to sharing, that is, the availability of public sector spatial data for the performance of public tasks, the democracy argument also only has an indirect value. As sharing is mainly for the purpose of preparing, implementing, and evaluating policy by the government or by public authorities, the free availability of public sector spatial data for sharing does not have a direct impact on the democratic rights and participation of the citizen. However, one could state that there is an indirect effect, and this effect would be larger or more important than the possible indirect effect regarding re-use. As the public task of the public authorities entails making policy and providing services to the citizen, the citizen will also benefit if the public authorities have the public sector spatial data they need at their disposal. After all, if the public sector has all the data it needs, this should entail that better policies are made and that the services to the citizens are optimized.

3. The 'Why Pay Twice' Arguments

Outline – The second set of arguments relates to the question of whether the collection and dissemination of public sector spatial data is being paid for twice ('The taxpayer has already paid'), or to the shifting of payment from one group to another ('The investments have been made for the public task'; 'Charges between public bodies merely shift the budget').

a. 'The Investments Have Been Made for the Public Task'

Argument for open access: data are collected for the public bodies' own needs – Reinsma and Van Der Sluijs feel that the primary goal of the data sets should be to support the performance of the public task, so therefore, public sector spatial data should be available at the maximum fee of the cost of dissemination.[49] Public bodies collect their public sector spatial data in order to support and improve the preparation, implementation, and evaluation of their policy-making and in order to provide better public services to the citizens. One should assume that if the public bodies do not need public sector spatial data for their own purposes, they would not collect them. Hence, since the data are already paid for by the public body, whether

49. Marijke Reinsma & H. Van Der Sluijs, 'Naar ruimere openbaarheid en een vrij gebruik van bestuurlijke informatie', 2002, 101, <www.minbzk.nl/aspx/download.aspx?file=/contents/pages/5006/RapportReinsmaenVanderSluijs.pdf>, 22 Feb. 2010 [in Dutch].

or not they are made available for others afterwards, the public body should not charge more than the extra costs it incurs for making the public sector spatial data available, that is, the marginal cost of dissemination.[50]

ARGUMENT FOR OPEN ACCESS: PUBLIC AGENCIES MAKE RATIONAL INVESTMENTS – As Holland puts it, budgets and resource allocations are made to meet mandates, not to accommodate the potential demand for public sector data by the public: 'One must assume that these investments would be made to support the business functions of government irrespective of whether there is an opportunity for cost recovery or not'.[51] This entails that rational public agencies would seek to maximize those preferences and data needs that meet their mandates and not necessarily those of their potential customers.

EVALUATION: SHARING – This argument is mostly relevant for the availability of public sector spatial data for access and sharing, but not so much for re-use. With regard to sharing, it does not make a difference whether the public sector spatial data are collected for the public task of one public body and used again for another public task of that same body, or whether they are passed on to other public bodies for their public tasks. The data are still needed for the public task, so they still need to be paid for with tax money. This is related to the argument below that under cost recovery, tax budgets are merely shifted from one public body to another.

EVALUATION: ACCESS – One might also consider that making public sector spatial data available for access is a public task: the European and national legislations on access to public sector data oblige the public bodies to make data available to the citizens. Hence, if public sector spatial data are used for one part of the public task that is financed by tax money, for instance, the preparation of a better environmental policy, then the use of these data for another part of the public task, that is, giving information to the public, should also be financed by this tax money.

EVALUATION: RE-USE – It is more difficult to apply this argument to re-use, as it is harder to see making public sector spatial data available for re-use as a public task. However, if one believes, like the European Commission[52], that the availability of public sector data for re-use at marginal cost stimulates economic growth and employment, it may still be considered a public task to make public sector spatial data available for re-use.

b. 'The Taxpayer Has Already Paid'

ARGUMENT FOR OPEN ACCESS: DATA ARE FUNDED BY TAXPAYER – One of the arguments that is most often made in favour of making public sector spatial data available for

50. See James Love, 'Pricing Government Information', *Journal of Government Information* 22, no. 5 (1995): 385.
51. William S. Holland, 'Copryight, Licensing and Cost Recovery for Geographic and Land Information System Data: A Legal, Economic and Policy Analysis', 1997, <www.spatial.maine.edu/~onsrud/tempe/holland.html>, 22 Feb. 2010.
52. See Commission of the European Communities, *Proposal for a Directive of the European Parliament and of the Council on the re-use and commercial exploitation of public sector documents*, COM (2002) 207 final, 5–6, 5 Jun. 2002.

free or at marginal cost is that the taxpayer has already paid for their collection.[53] Supporters of this argument feel that where data are funded by the taxpayer, through government funding, these should be available for access and use for no more than the costs of dissemination.[54]

ARGUMENT FOR OPEN ACCESS: COOPERATION OF TAXPAYER – Rybaczuk and Blakemore add another argument in favour of open access: the cooperation of the taxpayers may be reduced if they find they are being charged large amounts of money for data they contributed in the first place. This cooperation is 'essential to ensure that information is relevant, timely and of a high quality'.[55]

ARGUMENT FOR COST RECOVERY: UNKNOWN INCOME FROM TAXATION – Longhorn and Blakemore raise some practical concerns against the funding of the availability of public sector spatial data by the central budget. First, they warn that a funding mechanism that would only involve state financing, with no onward user costs for data access and usage, assumes that: 'There will be enough taxation income to cover the costs of production and dissemination, no matter what the demand ... Second, it assumes that those who make decisions on the allocation of taxation income have full cognisance of the strategic needs of the data production process ... Third, direct taxation models of funding critically assume a correspondence in the trends of data production costs, data consumption demands, and taxation income'. Models that would support such assumptions would be hard to find.[56]

53. Monika Bargmann, Gerald Pfeifer & Boris Piwinger, 'A Citizen's Perspective on Public Sector Information', in *Public Sector Information in the Digital Age: Between Markets, Public Management and Citizens' Rights*, ed. Georg Aichholzer & Herbert Burkert (Cheltenham: Edward Elgar Publishing Limited, 2004), 261. See also Roger Longhorn & Michael Blakemore, 'Re-visiting the Valuing and Pricing of Digital Geographic Information', *Journal of Digital Information* 4, no. 2 (2004): 8, <http://journals.tdl.org/jodi/article/view/103/102>, 22 Feb. 2010; Renate Mason, 'Developing Australian Spatial Data Policies – Existing Practices and Future Strategies', 2000, 120–121, <http://handle.unsw.edu.au/1959.4/18646>, 22 Feb. 2010; UNESCO, 'Policy Guidelines for the Development and Promotion of Governmental Public Domain Information' (Paris: UNESCO, 2004), 13, <http://unesdoc.unesco.org/images/0013/001373/137363eo.pdf>, 22 Feb. 2010; OECD Working Party on the Information Economy, 'OECD Workshop on Public Sector Information held on 31 May 2006: Summary', 4–5, <www.oecd.org/dataoecd/54/35/36854745.pdf>, 22 Feb. 2010; Jean-Michel Bruguière, *Les données publiques et le droit* (Paris: Editions Litec, 2002), 179 [in French]; Inigo Baten and Gijs Van Der Starre, *Elektronische toegankelijkheid van overheidsinformatie* (The Hague: Rathenau Instituut, 1996), 40 [in Dutch]; Herbert Burkert, 'The Legitimacy of Commercialization of Public Sector Information as a Research Topic', in *Towards a Legal Framework for a Diffusion Policy for Data Held by the Public Sector*, ed. Cécile De Terwangne, Herbert Burkert & Yves Poullet (Deventer: Kluwer Law and Taxation, 1995), 87.

54. Office of Fair Trading, 'The Commercial Use of Public Information (CUPI)', 2006, 60, <www.oft.gov.uk/shared_oft/reports/consumer_protection/oft861.pdf>, 22 Feb. 2010.

55. Krysia Rybaczuk & Michael Blakemore, 'Selling Government Information: A Comparative Perspective on UK and USA Developments', paper given at the Conference on Law and Information Policy for Spatial Databases (Tempe, 1994), <www.spatial.maine.edu/%7eonsrud/tempe/rybaczuk.html>, 22 Feb. 2010.

56. Roger Longhorn & Michael Blakemore, 'Re-visiting the Valuing and Pricing of Digital Geographic Information', *Journal of Digital Information* 4, no. 2 (2004): 10, <http://journals.tdl.org/jodi/article/view/103/102>, 22 Feb. 2010. See also Edeltraud Hanappi-Egger, 'Cultural

ARGUMENT FOR COST RECOVERY: TAXPAYERS AS THE BOARD OF DIRECTORS – Rybaczuk and Blakemore also consider another view on the taxpayers: they may be 'viewed as being the equivalent of a board of directors that expects a return on investment. Selling data is one way of helping to achieve this'.[57] Litman also mentions this argument: 'Because taxpayers have paid for the government's gathering and analysis of the data, however, arguments that they should be able to recoup their costs, by selling access to the data in the marketplace, also seem attractive'.[58] Mason reasons along the same lines: 'Governmental departments can be held accountable if they fail to protect the public information in their care. This information can also be referred to as public trust. If a government department gives public trust away or sells it below market value, then it can be viewed that they are neglecting their duty'.[59] Hence, all taxpayers would benefit from public bodies charging more than marginal costs for making public sector spatial data available, as these public bodies are seen as custodians of the data and are considered responsible for making these public assets generate as much revenue as possible.

EVALUATION: RE-USE – This last argument assumes that all taxpayers benefit from the maximization of the revenues of the public bodies. The other side of the coin is that all taxpayers would pay for the public sector spatial data that are only disseminated to very few interested parties. This argument is mainly made with reference to commercial re-use by the information industry, suggesting that making public sector spatial data available for commercial purposes at marginal cost unfairly subsidizes private profit at taxpayer expense.[60] Volman formulates the argument as follows: 'Only specific user-groups benefit from the information

---

Heritage: The Conflict between Commercialisation and Public Ownership', in *Public Sector Information in the Digital Age, Between Markets, Public Management and Citizens' Rights*, ed. Georg Aichholzer & Herbert Burkert (Cheltenham: Edward Elgar Publishing Limited, 2004), 249; Roger A. Longhorn & Michael Blakemore, *Geographic Information: Value, Pricing, Production, and Consumption* (Boca Raton: CRC Press, 2008), 65, 74.

57. Krysia Rybaczuk & Michael Blakemore, 'Selling Government Information: A Comparative Perspective on UK and USA Developments', paper given at the Conference on Law and Information Policy for Spatial Databases (Tempe, 1994), <www.spatial.maine.edu/%7eonsrud/tempe/rybaczuk.html>, 22 Feb. 2010.
58. Jessica Litman, 'Rights in Government-generated Data', paper given at the Conference on Law and Information Policy for Spatial Databases (Tempe, 1994), <www.spatial.maine.edu/~onsrud/tempe/litman.html>, 22 Feb. 2010.
59. Renate Mason, 'Developing Australian Spatial Data Policies – Existing Practices and Future Strategies', 2000, 104, <http://handle.unsw.edu.au/1959.4/18646>, 22 Feb. 2010.
60. William S. Holland, 'Copyright, Licensing and Cost Recovery for Geographic and Land Information System Data: A Legal, Economic and Policy Analysis', 1997, <www.geoanalytics.com/library/publications/copyrightAndLicensing.pdf>, 22 Feb. 2010. See also George Cho, *Geographic Information Science: Mastering the Legal Issues* (Chicester: John Wiley & Sons Ltd, 2005), 62; Renate Mason, 'Developing Australian Spatial Data Policies – Existing Practices and Future Strategies', 2000, 110–111, <http://handle.unsw.edu.au/1959.4/18646>, 22 Feb. 2010; Jean-Michel Bruguière, *Les données publiques et le droit* (Paris: Editions Litec, 2002), 2002, 175 [in French]; Economic Development and Infrastructure Committee, 'Inquiry into Improving Access to Victorian Public Sector Information and Data', 2009, 97–98, <www.parliament.vic.gov.au/edic/inquiries/access_to_PSI/final_report.html>, 22 Feb. 2010.

resources, so it is only right to ask for a substantial contribution if they want to use these resources for commercial activities'.[61] The Office of Fair Trading report states it more directly: 'Some proportion of taxation used to fund [public sector data] will be providing profits for private businesses'.[62] This is even more true if we consider one of the underlying purposes of the stimulation of commercial re-use of public sector spatial data in particular: an increase in the development of cross-border or even pan-European information services. This entails not only that the taxpayer is contributing to the activities of the commercial companies that are subject to the same national tax regime and will pay taxes on their sales, but also that he or she contributes to the activities of foreign companies, which are taxed elsewhere and do not contribute to the national treasury.[63]

Evaluation: re-use – However, if one follows the argument mentioned above, that the investments of the public body have been made for the performance of a public task, they would also be made even if there are no companies trying to make a private profit from the data. The only extra cost would be the dissemination to these private companies, which should be covered by the marginal cost pricing. Thus, if cost recovery is applied, the taxpayer would indeed gain a benefit and would have to pay less taxes, but this is a more indirect effect and does not allow the full statement that the taxpayer is partly paying the profit-raising activities of commercial companies. After all, there is no guarantee that the revenues of the public bodies will actually be used to lower taxes. Moreover, there is no guarantee that the money gained by cost recovery will be used for data collection, so the sustainability of the public body making public sector spatial data available is still not ensured. The final destination of the revenues will depend on the structure of the public sector and the place of the public body that makes the public sector spatial data available. If there are no guarantees that the money obtained from the re-users is going directly towards the creation of data, this argument is not very convincing. If the money would be used for other purposes than data collection by a national treasury, one could even consider the argument from the other side: for example, why should the re-users of public sector spatial data pay for an increase in the budget of the national railways, for the subsidization of a theatre company, etc?

Evaluation: sharing – The argument that the taxpayer has already paid is easier to understand with relation to access and sharing than to re-use. With regard

61. Yvo Volman, 'Exploitation of Public Sector Information in the Context of the *eEurope* Action Plan', in *Public Sector Information in the Digital Age: Between Markets, Public Management and Citizens' Rights*, ed. Georg Aichholzer & Herbert Burkert (Cheltenham: Edward Elgar Publishing Limited, 2004), 98.
62. Office of Fair Trading, 'The Commercial Use of Public Information (CUPI)', 2006, 61, <www.oft.gov.uk/shared_oft/reports/consumer_protection/oft861.pdf>, 22 Feb. 2010.
63. Lori Peterson Dando, 'A Case for the Commercialization of Public Information', paper given at the Conference on Law and Information Policy for Spatial Databases (Tempe, 1994), <www.spatial.maine.edu/%7eonsrud/tempe/dando.html>,22 Feb. 2010. See also Roger A. Longhorn & Michael Blakemore, *Geographic Information: Value, Pricing, Production, and Consumption* (Boca Raton: CRC Press, 2008), 70–73.

to sharing, one could assume that if other bodies need the same public sector spatial data, the taxpayer might have paid for them twice if the other bodies decided to collect their own data. In this way, the taxpayer saves money.

EVALUATION: ACCESS – In the case of access, the argument that all the taxpayers are funding the needs of only a few interested parties is certainly not valid. If we consider that access involves obtaining data for the purpose of exercizing democratic rights and obligations, this does not involve only a small group of people, but every single taxpayer in the national tax system. Everyone may be in need of data at a certain point in time, and while this may be not be true to the same extent for all the taxpayers, it is impossible to state that there are only a few interested parties. All citizens are stakeholders in this case. As public sector spatial data are collected with taxpayers' money, charging any price higher than the cost of dissemination for access would make the taxpayers pay twice.

c. *'Charges Between Public Bodies Merely Shift the Budget'*

ARGUMENT FOR OPEN ACCESS: NO SOLUTION FOR INSUFFICIENT BUDGET – This argument only relates to public bodies charging other bodies for making available public sector spatial data. Some argue that payments between public bodies are unnecessary and that these payments should be replaced with taxation funding.[64] After all, as Weiss states: 'Charging other government users merely shifts the expenses from one agency to another rather than actually saving the national treasury any money'.[65] He warns that 'cost recovery policies often mean that budgetary constraints prevent some government agencies from acquiring information that has already been created or collected by another part of government, resulting in agencies either doing without or using inferior alternatives'.[66] Andrews agrees: 'For one government body to charge another government body for information may be useful for accounting and budgeting purposes, but it certainly does not address the government's overall shortage of money'.[67]

ARGUMENT FOR OPEN ACCESS: DIVISION OF THE BUDGET IS NOT THE TAXPAYERS' PROBLEM – For the taxpayer, it makes no difference which part of the public sector

64. Office of Fair Trading, 'The Commercial Use of Public Information (CUPI)', 2006, 60, <www.oft.gov.uk/shared_oft/reports/consumer_protection/oft861.pdf>, 22 Feb. 2010.
65. Peter N. Weiss, 'Borders in Cyberspace: Conflicting Public Sector Information Policies and their Economic Impacts', in *Public Sector Information in the Digital Age: Between Markets, Public Management and Citizens' Rights*, ed. Georg Aichholzer & Herbert Burkert (Cheltenham: Edward Elgar Publishing Limited, 2004), 157. See also Ravi Bedrijvenplatform, *Commercialisering van geo-informatie. Een drietal rapporten ter ondersteuning van de Ravi-discussie 1997* (Amersfoort: Ravi, 1997), 102 [in Dutch].
66. Peter N. Weiss, 'Borders in Cyberspace: Conflicting Public Sector Information Policies and their Economic Impacts', in *Public Sector Information in the Digital Age, Between Markets, Public Management and Citizens' Rights*, ed. Georg Aichholzer & Herbert Burkert (Cheltenham: Edward Elgar Publishing Limited, 2004), 138.
67. William J. Andrews, 'Nurturing the Global Information Commons: Public Access, Public Infrastructure', *Government Information in Canada* 2, no. 3 (1996), <http://library2.usask.ca/gic/v2n3/andrews/andrews.html>, 22 Feb. 2010.

holds his or her money, or whether one public body should pay the other for its public sector spatial data or not. It is not the citizen's concern whether his or her tax contribution goes to a national, regional, or local authority. Furthermore, it is not always straightforward which public authority is competent for performing a public task. This may change over time and depends on the type of state structure, and especially on political agreements and negotiations. These are not the tax-payers' problem. No matter how much the public authorities charge each other, in the end, the taxpayers are presented with the same bill.

EVALUATION: PRACTICAL PROBLEM – Of course, while this argument makes sense on the level of the national treasury, it may have a very big impact on the public bodies themselves.[68] It assumes that the relationships between public bodies involve a mutual exchange of public sector spatial data of comparable proportions. However, some public bodies, such as National Mapping Agencies or National Meteorological Offices, make public sector spatial data available to many other public bodies, but they do not receive any data from those bodies in return. In that case, there is no trade-off between the gains or losses of either public bodies. Hence, this argument only works if a national treasury awards the public data providers with enough resources to perform their tasks.

EVALUATION: CONCEPTUAL PROBLEM – A more conceptual problem with this argument is that it makes no distinction between public bodies performing their public tasks and public bodies performing commercial activities, that is, re-using public sector spatial data. The reasoning can only work for sharing, as in the case of re-use, the public bodies will have to obtain the public sector spatial data under the same conditions as their competitors from the private sector.

4. 'The Macro-Economic Arguments': Open Access Is an Incentive for Economic Activity and Economic Growth

ARGUMENT FOR OPEN ACCESS: MORE INCOME FROM TAXES – Probably, the objective that is mentioned most in the literature to justify open access is that low-cost access supposedly serves as an incentive for economic activity that will generate more tax income than the supply of public sector data at prices above marginal costs could.[69]

68. Ravi Bedrijvenplatform, *Commercialisering van geo-informatie. Een drietal rapporten ter ondersteuning van de Ravi-discussie 1997* (Amersfoort: Ravi, 1997), 62 [in Dutch].
69. Mireille Van Eechoud, 'The Commercialization of Public Sector Information: Delineating the Issues', in *The Future of the Public Domain: Identifying the Commons in Information law*, ed. Lucie Guibault & P. Bernt Hugenholtz (Alphen a/d Rijn: Kluwer Law International, 2006), 299. See also Herbert Kubicek, 'Third-Generation Freedom of Information in the Context of E-Government: the Case of Bremen, Germany', in *Public Sector Information in the Digital Age: Between Markets, Public Management and Citizens' Rights*, ed. Georg Aichholzer & Herbert Burkert (Cheltenham: Edward Elgar Publishing Limited, 2004), 284; Herbert Burkert & Peter N. Weiss, 'Towards a Blueprint for a Policy on Public Sector Information', in *Public Sector Information in the Digital Age: Between Markets, Public Management and Citizens' Rights*, ed. Georg Aichholzer & Herbert Burkert (Cheltenham: Edward Elgar Publishing Limited, 2004), 329; Roger A. Longhorn & Michael Blakemore, Geographic Information: Value, Pricing,

In a much discussed and cited paper, Weiss states that, at least for the United States, the benefits 'that accrue from the corporate and individual taxes from the secondary publishing and service activities stimulated by open access policies far exceed any revenues that might be generated through cost recovery policies'.[70]

ARGUMENT FOR OPEN ACCESS: BENEFITS FOR SOCIETY – Weiss also feels that: 'The consensus of recent research is that charging marginal cost of dissemination for [public sector data] will lead to optimal economic growth in society and will far outweigh the immediate perceived benefits of aggressive cost recovery'.[71] Volman mentions two components of this indirect benefit for society: the immaterial component (such as the availability of better products for citizens), and the material component, which 'looks at the activity that can be generated by companies that want to re-use the [public sector data] for value added products and services. This activity leads to additional income and employment, which in turn will lead to more tax income and other benefits for the state finances ... Better possibilities

---

Production, and Consumption (Boca Raton: CRC Press, 2008), 71; Lori Peterson Dando, 'A Case for the Commercialization of Public Information', paper given at the Conference on Law and Information Policy for Spatial Databases (Tempe, 1994), <www.spatial.maine.edu/%7eonsrud/tempe/dando.html>,22 Feb. 2010; Mireille Van Eechoud, 'Het hergebruik regime voor overheidsinformatie in de Wob, een tussenstand', *Mediaforum* 20, no. 1 (2008): 8 [in Dutch]; Brian Klinkenberg, 'The True Cost of Spatial Data in Canada', *The Canadian Geographer* 47, no. 1 (2003): 41; Harlan J. Onsrud, 'Legal Access to Geographic Information: Measuring Losses or Developing Responses?', in *Information, Place and Cyberspace: Issues in Accessibility*, ed. Donald G. Janelle & David Charles Hodge (Berlin: Springer-Verlag, 2000), <www.spatial.maine.edu/%7Eonsrud/pubs/LegalAccessToGI.pdf>, 22 Feb. 2010; Acil Tasman, 'The Value of Spatial Information: The Impact of Modern Spatial Information Technologies on the Australian Economy', 2008, 154–155, <www.crcsi.com.au/uploads/publications/PUBLICATION_324.pdf>, 22 Feb. 2010; Roger Longhorn & Michael Blakemore, 'Re-visiting the Valuing and Pricing of Digital Geographic Information', *Journal of Digital Information* 4, no. 2 (2004): 10, <http://journals.tdl.org/jodi/article/view/103/102>, 22 Feb. 2010; Office of Fair Trading, 'The Commercial Use of Public Information (CUPI)', 2006, 61, <www.oft.gov.uk/shared_oft/reports/consumer_protection/oft861.pdf>, 22 Feb. 2010; OECD Working Party on the Information Economy, 'OECD Workshop on Public Sector Information held on 31 May 2006: Summary', 4–5, <www.oecd.org/dataoecd/54/35/36854745.pdf>, 22 Feb. 2010; Ravi Bedrijvenplatform, *Economische effecten van laagdrempelige beschikbaarstelling van overheidsinformatie* (Amersfoort: Ravi, 2000), 29 [in Dutch]; Estelle Derclaye, 'Does the Directive on the Re-use of Public Sector Information affect the State's database sui generis right?', in *Knowledge rights – Legal, Societal and Related Technological Aspect. Knowright Symposium, 18–19 September 2008*, ed. Jens Gaster, Erich Schweighofer & Peter Sint (Austrian Computer Society, 2008), 143; Richard Maven, 'Interview: Online mapping guru Ed Parsons', *E-consultancy* 2007, <www.e-consultancy.com/news-blog/362454/interview-online-mapping-guru-ed-parsons.html>, 22 Feb. 2010; Economic Development and Infrastructure Committee, 'Inquiry into Improving Access to Victorian Public Sector Information and Data', 2009, 8, <www.parliament.vic.gov.au/edic/inquiries/access_to_PSI/final_report.html>, 22 Feb. 2010.

70. Peter N. Weiss, 'Borders in Cyberspace: Conflicting Public Sector Information Policies and their Economic Impacts', in *Public Sector Information in the Digital Age: Between Markets, Public Management and Citizens' Rights*, ed. Georg Aichholzer & Herbert Burkert (Cheltenham: Edward Elgar Publishing Limited, 2004), 138.

71. *Ibid.*

for re-use by market players will lead to higher tax incomes that can be invested to keep up the high quality of the data collections concerned'.[72]

Argument for cost recovery: no conclusive evidence – However, there is no consensus that marginal cost pricing would necessarily increase social benefits.[73] For instance, Clark argues that: 'Increased use does not automatically mean either increased tax revenue or decreased societal cost'.[74] Longhorn and Blakemore also 'do not infer that there is an automatic, direct, and immutable link between free-of-cost (to the end user) access to [spatial data] and increased usage or societal impact'.[75] In any case, so far, there is no conclusive evidence that economic growth is truly stimulated by open access. Even Weiss admits that while he strongly believes in this argument, it is difficult to find easily quantifiable economic benefits to society.[76]

Argument for cost recovery: assignment of the tax income – If there were evidence that the tax revenues gained under an open access approach would be higher than the revenues raised by charging for the public sector spatial data, the question remains to what public body or public service the tax income will be assigned. Hence, from the point of view of the public sector as a whole, the increase in tax income may be beneficial. However, the extra money may not be spent on collecting public sector spatial data, but on an entirely different public service. Thus, once again, the public bodies whose main task is to make public sector spatial data available may not benefit from the economic growth, while they would directly benefit from the revenues they raised themselves under cost recovery.

Evaluation: access – Some proponents also mention the benefit to the citizens, because the citizens will have more information services at their disposal on the market, so they will be better informed. This is related to the democratic purpose discussed previously. According to this democratic argument, the citizens needs public sector data to check up on the government that they elected and to exercise their democratic rights and obligations. From a socio-economic point of view, citizens also need data beyond their *democratic* rights in order to interact with all the other parties in society and on the market. This will stimulate the efficient operation of the market. However, one should keep in mind, again, that there is no conclusive evidence that open access will have such a beneficial effect,

72. Yvo Volman, 'Exploitation of Public Sector Information the Context of the *eEurope* Action Plan', in *Public Sector Information in the Digital Age, Between Markets, Public Management and Citizens' Rights*, ed. Georg Aichholzer & Herbert Burkert (Cheltenham: Edward Elgar Publishing Limited, 2004), 99.
73. Office of Fair Trading, 'The Commercial Use of Public Information (CUPI)', 2006, 61, <www.oft.gov.uk/shared_oft/reports/consumer_protection/oft861.pdf>, 22 Feb. 2010.
74. Mike Clark, 'Fee or Free? The Hidden Costs of Free Public Sector Information', *Business Information Review* 24, no. 1 (2007): 50.
75. Roger A. Longhorn & Michael Blakemore, Geographic Information: Value, Pricing, Production, and Consumption (Boca Raton: CRC Press, 2008), 65.
76. Peter N. Weiss, 'Borders in Cyberspace: Conflicting Public Sector Information Policies and their Economic Impacts', in *Public Sector Information in the Digital Age: Between Markets, Public Management and Citizens' Rights*, ed. Georg Aichholzer & Herbert Burkert (Cheltenham: Edward Elgar Publishing Limited, 2004), 157.

and that such information services would not be created if a price was charged that exceeds the marginal cost of dissemination of the data.

EVALUATION: SHARING – With regard to sharing, open access might also have an indirect effect on market growth, as the availability of public sector spatial data to public bodies for the purpose of performing their public tasks would be expected to lead to better policy making and better services to the public. These better policies and services should lead to better regulation (or de-regulation) of the market and society, stimulating economic activities.

5. 'The Micro-Economic Argument': Better Performance of the Public Bodies

OUTLINE – The arguments relating to the performance of public bodies are mostly in favour of cost recovery, because they estimate that this raises the resources to improve quality, efficiency and fitness for purpose. However, an important counterargument in favour of open access claims that cost recovery costs more than it produces.

a. Quality of the Data

ARGUMENT FOR COST RECOVERY: QUALITY AND UPDATES – An argument often heard by public bodies who charge for their public sector spatial data is that a data policy with lower charges could lead to lower income and could thereby endanger the quality of the public service.[77] They often refer to the United States (US) model: the general reliance on government funding for public sector data at the US federal level and the lack of revenue that can be invested means 'that the quality of some of their [public sector data] may be inferior to countries that charge directly for [public sector data]'.[78] The private sector has overtaken the public sector in terms of quality in some sectors, such as mapping. For instance, the United States Geological Survey has admitted that much of its topographic data has not been updated for more than fifty years, due to a lack of financial resources, which are only provided from central government funding.[79] OFT even feels that this raises the

77. Yvo Volman, 'Exploitation of Public Sector Information in the Context of the *eEurope* Action Plan', in *Public Sector Information in the Digital Age: Between Markets, Public Management and Citizens' Rights*, ed. Georg Aichholzer & Herbert Burkert (Cheltenham: Edward Elgar Publishing Limited, 2004), 98. See also Micus Management Consulting Gmbh, 'Assessment of the Re-use of Public Sector Information (PSI) in the Geographical Information, Meteorological Information and Legal Information Sectors', 2008, 25, <http://ec.europa.eu/information_society/policy/psi/docs/pdfs /micus_report_december2008.pdf>, 22 Feb. 2010.

78. Office of Fair Trading, 'The Commercial Use of Public Information (CUPI)', 2006, 58, <www.oft.gov.uk/shared_oft/reports/consumer_protection/oft861.pdf>, 22 Feb. 2010.

79. Kathryn Brown, 'Mapping the Future', *Science* 298, no. 5600 (2002): 1874–1875; Roger A. Longhorn & Michael Blakemore, *Geographic Information: Value, Pricing, Production, and Consumption* (Boca Raton: CRC Press, 2008), 12, 72–73.

question of whether the public bodies should continue to make public sector spatial data available at all.[80]

Argument for open access: no conclusive evidence – However, Van Loenen did not find any evidence for a relationship between open access policies and lower quality of the data.[81] Longhorn and Blakemore also find that other examples show that: 'There is not a direct relationship between central government funding and poor data. The USGS [United States Geological Survey] mapping example must not, therefore, be taken as indicating a general rule that a reliance on government funding produces bad or incomplete data'. Nor does the quality of the public sector spatial data produced by the Ordnance Survey Great Britain – which works under a cost recovery policy – indicate any general rule that cost recovery is always needed to produce excellent data.[82]

Argument for open access: over-investment in quality – While under-investment may lead to a lack in quality, one could also question whether cost recovery might not lead to an over-investment in quality, where this quality would not be necessary for the public body to fulfil its own public task.[83] This would entail that the public body focuses on producing public sector spatial data with high quality, regardless of the need for such quality. This raises the issue of production costs, which in their turn ask for a higher cost recovery charge. This does not provide any incentive for the public body to limit itself to the level of quality that it needs to perform its own tasks, but it makes it spend more money on the production than is actually necessary.

Argument for open access: standardization – The Ravi Bedrijvenplatform believes that quality will actually increase with the low-level availability of public sector spatial data.[84] This is also linked to another benefit that was observed by the Geospatial Information and Technology Association: the process of making public sector spatial data freely available has forced public bodies and industry organizations to develop data standards, which ultimately fosters more widespread sharing and use of these data.[85]

Summary – In short, many public bodies whose main task is to collect public sector spatial data with the purpose of making it available to other parties within or outside the public sector argue that they have to charge prices for their data under a

80. Office of Fair Trading, 'The Commercial Use of Public Information (CUPI)', 2006, 59, <www.oft.gov.uk/shared_oft/reports/consumer_protection/oft861.pdf>, 22 Feb. 2010.

81. Bastiaan Van Loenen, *Developing Geographic Information Infrastructures: The Role of Information Policies* (Delft: DUP Science, 2006), 303–308.

82. Roger A. Longhorn & Michael Blakemore, *Geographic Information: Value, Pricing, Production, and Consumption* (Boca Raton: CRC Press, 2008), 77.

83. See David Newbery, Lionel Bently & Rufus Pollock, 'Models of Public Sector Information Provision via Trading Funds', 2008, 117, <www.berr.gov.uk/files/file45136.pdf>, 22 Feb. 2010.

84. Ravi Bedrijvenplatform, *Economische effecten van laagdrempelige beschikbaarstelling van overheidsinformatie* (Amersfoort: Ravi, 2000), 24 [in Dutch].

85. Geospatial Information & Technology Association, 'Free or Fee: The Governmental Data Ownership Debate. GITA White Paper', 2005, 5, <www.gita.org/resources/whitepapers/free_or_fee.pdf>, 22 Feb. 2010.

cost recovery scheme, because not doing so would endanger the sustainable availability of the public sector spatial data and would lead to a decrease in quality of the data. However, this may not be a valid argument for the charging for all purposes.

EVALUATION: ACCESS, RE-USE AND SHARING – First, relating to the argument above that the public sector spatial data were created for the public task, if it is felt that data of a certain quality is needed for the performance of a public task, one can assume that these data will be collected, whether or not they are subject to a cost recovery regime. And if these public sector spatial data would not be necessary for the public sector's own public task purposes, then one might ask the question of whether or not the public sector should even engage in collecting these data, or at least data with such a quality level. In addition, the quality requirements will differ according to the purposes of a request for public sector spatial data access, re-use, or sharing. A citizen who wants access to public sector spatial data may not need the same quality of data as a re-user would or that a public body would for sharing, or vice versa.

b. Quality and Efficiency of the Service

ARGUMENT FOR COST RECOVERY: BETTER DELIVERY OF THE DATA – Not only is the improved quality of the public sector spatial data a perceived benefit of cost recovery policies, but also, so is the better delivery of these data, because of the higher efficiency of the public bodies' processes and better insight into the needs of the users. Peterson Dando argues that: 'A policy allowing cost recovery provides incentives to the government to provide the type and quality of information and services that the public wants'.[86] Van Eechoud agrees that by applying a cost recovery policy, the data provider has better insight into the value of particular data for consumers, which in turn may lead to a more efficient organization of the data provision.[87] Cho states that: 'User charges will gauge market demands as well as eliminate frivolous requests at the same time empowering consumers – "user pays, user says" '.[88] Furthermore,

86. Lori Peterson Dando, 'A Case for the Commercialization of Public Information', paper given at the Conference on Law and Information Policy for Spatial Databases (Tempe, 1994), <www.spatial.maine.edu/%7eonsrud/tempe/dando.html>, 22 Feb. 2010. See also Krysia Rybaczuk & Michael Blakemore, 'Selling Government Information: A Comparative Perspective on UK and USA Developments', paper given at the Conference on Law and Information Policy for Spatial Databases (Tempe, 1994), <www.spatial.maine.edu/%7eonsrud/tempe/rybaczuk.html>, 22 Feb. 2010.

87. Mireille Van Eechoud, 'Het hergebruik regime voor overheidsinformatie in de Wob, een tussenstand', *Mediaforum* 20, no. 1 (2008): 8 [in Dutch]. See also Economic Development and Infrastructure Committee, 'Inquiry into Improving Access to Victorian Public Sector Information and Data', 2009, 99, <www.parliament.vic.gov.au/edic/inquiries/access_to_PSI/final_report.html>, 22 Feb. 2010.

88. George Cho, *Geographic Information Science: Mastering the Legal Issues* (Chicester: John Wiley & Sons Ltd., 2005), 62.

Newbery refers to the responsiveness of the public bodies to complaints and their concern about general service quality.[89]

ARGUMENT FOR OPEN ACCESS: REQUIREMENTS FOR DELIVERY MAY DIFFER – Weiss is sceptical about this argument: 'No one supplier, public or private, can design all information products required to meet the needs of all users'.[90] Moreover, he states that the 'private user base that can be charged is not large enough to support recovery of the full costs of a comprehensive unsubsidized information service'.[91] Bargmann et al. agree: 'The personal situation and requirements of individual citizens may differ significantly, which makes it virtually impossible to define which information citizens actually need'.[92]

ARGUMENT FOR OPEN ACCESS: PUBLIC BODIES ARE NEVER AS EFFICIENT AS THE PRIVATE SECTOR – Onsrud believes that even under a cost recovery policy, a public body will not deliver services as efficiently as the private sector can: 'After a government department has gained an exclusive right to provide a specific service which generates income, there is little incentive for the government bureaucracy to improve that service'.[93] The public body does not have an incentive to minimize the costs, but may rather seek to match the costs to the revenue. Furthermore, given the market of some public bodies, they can always raise prices and increase revenue to address overspending.[94]

ARGUMENT FOR COST RECOVERY: FUNDING BASED ON EXPENDITURE – However, public bodies that operate under an open access policy might also not be inclined to reduce their costs and to start performing more efficiently, as their funding is based on the expenditures of the previous years. Onsrud does not agree with this: if the collection and dissemination of public sector spatial data 'is a burden on tax revenues, it is likely to be more efficient over time, because it needs continual justification in the budgeting process'.[95]

---

89. David Newbery, Lionel Bently & Rufus Pollock, 'Models of Public Sector Information Provision via Trading Funds, 2008, 112, <www.berr.gov.uk/files/file45136.pdf>, 22 Feb. 2010.
90. Peter N. Weiss, 'Borders in Cyberspace: Conflicting Public Sector Information Policies and their Economic Impacts', in *Public Sector Information in the Digital Age: Between Markets, Public Management and Citizens' Rights*, ed. Georg Aichholzer & Herbert Burkert (Cheltenham: Edward Elgar Publishing Limited, 2004), 138.
91. *Ibid.*, 157.
92. Monika Bargmann, Gerald Pfeifer & Boris Piwinger, 'A Citizen's Perspective on Public Sector Information', in Georg Aichholzer & Herbert Burkert (ed.), *Public Sector Information in the Digital Age: Between Markets, Public Management and Citizens' Rights* (Cheltenham: Edward Elgar Publishing Limited, 2004), 257.
93. Harlan J. Onsrud, 'In Support of Open Access for Publicly Held Geographic Information', *GIS Law* 1, no. 2 (1992): 3–6, <www.spatial.maine.edu/~onsrud/pubs/In_Support_OA.htm>, 22 Feb. 2010.
94. David Newbery, Lionel Bently & Rufus Pollock, 'Models of Public Sector Information Provision via Trading Funds', 2008, 117, <www.berr.gov.uk/files/file45136.pdf>, 22 Feb. 2010.
95. Harlan J. Onsrud, 'In Support of Open Access for Publicly Held Geographic Information', *GIS Law* 1, no. 2 (1992): 3–6, <www.spatial.maine.edu/~onsrud/pubs/In_Support_OA.htm>, 22 Feb. 2010.

ARGUMENT FOR OPEN ACCESS: OTHER MEANS TO DETERMINE USER NEEDS – Hence, according to the proponents of cost recovery charging policies, public bodies will have a better insight into the demands and needs of the users of the public sector spatial data, which enables them to provide a better service in delivering the data. However, charging for making data available is not the only way to discover what the user wants. There are other means to discover this, such as customer surveys or market research.

EVALUATION: ACCESS, RE-USE, AND SHARING – Moreover, making public sector spatial data available in a way that the user wants will entail a different approach to different user groups, that is, to access, re-use, and sharing. As making data available for access is a public task, the public should find out what level of service towards the citizens is required, independent of whether this delivery service could be charged for. The same statement can be made with regard to sharing for public task purposes.

c. *Liability*

ARGUMENT FOR COST RECOVERY: LESS RISK FOR LIABILITY – Some public bodies feel that making their public sector spatial data widely available subjects them to considerable liability risks, so they try to discourage potentially interested parties from obtaining their data by imposing stringent conditions or charging high prices. They are not only worried that their public sector spatial data might be inaccurate or of a below-standard quality, but also their concern is that the data they make available will be altered or misused by the users, while they are being presented as coming from the public body. They fear that the public will rely on the faulty data, believing that the data are made available in such a faulty form by the public bodies. The public bodies feel that charging a price that is higher than the cost of dissemination is a disincentive to use the public sector spatial data.

ARGUMENT FOR OPEN ACCESS: LOWER LIABILITY BURDEN – Johnson and Onsrud actually feel that liability is a considerable disincentive to cost recovery policy: 'At first glance, it might appear that a cost recovery policy (or a restricted access policy) might reduce liability exposure by reducing the number of people who use the GIS data'.[96] However, '[c]harging for information or restricting secondary uses may actually increase liability exposure for GIS agencies through reduced opportunities to claim sovereign immunity protection and greater exposure to specific theories of liability'.[97]

ARGUMENT FOR OPEN ACCESS: INCREASE OF RELIABILITY – While the concern of the public bodies is understandable, one could also consider that making public sector spatial data easily available would rather encourage the users to use these data in a

96. Jeff P. Johnson & Harlan J. Onsrud, 'Is Cost Recovery Worthwhile?', 1995, <www.spatial.maine.edu/%7eonsrud/pubs/Cost_Recovery_Worthwhile.html>, 22 Feb. 2010.
97. *Ibid.* See also Harlan J. Onsrud, 'In Support of Open Access for Publicly Held Geographic Information', *GIS Law* 1, no. 2 (1992): 3–6, <www.spatial.maine.edu/~onsrud/pubs/In_Support_OA.htm>, 22 Feb. 2010.

correct way, as this would only enhance the reliability of their service or product. It is against the user's own interests to change or misrepresent the data, as he or she wants to promote his or her service or product as reliable. This reliability is due to the incorporation of the public sector spatial data, as the potential customer can always check these data by requesting access from the public body that holds the data.

EVALUATION: ACCESS, RE-USE, AND SHARING – As was already implied in the previous paragraphs, the main concern of the public bodies regards re-use of their public sector spatial data rather than access or sharing. In the case of access or sharing, the concern for liability relates to the applicant making decisions based on the public sector spatial data made available by the public body. When the data are made available for re-use, the public bodies are more worried about the re-users misusing or misrepresenting their data, whether accidentally or not. While liability issues may be a legitimate concern of the public bodies, they should not be solved by charging higher prices and discouraging use, but by providing sound liability and warranty clauses that make clear that the public sector spatial data are made available 'as is' without any guarantees. Moreover, if we go back to the basic idea that the public bodies originally collected the data for their own needs, one should assume that they tried to make their data as accurate as possible.

d. Transaction Costs of Charging for Public Sector Spatial Data

ARGUMENT FOR OPEN ACCESS: CHARGING NOT WORTHWHILE – Kingma argues that charging fees for making public sector spatial data available 'requires recording and collecting fees that, for individual transactions, can add a significant cost in addition to the cost of producing and providing the information' and because data have low marginal costs of production, 'the cost of pricing may not make it worthwhile to implement user fees'.[98] Bargmann et al. agree that charging will cause 'bureaucratic overhead'.[99] Onsrud refers to the 'administrative burden' of developing written contractual relationships.[100] According to Andrews, these costs are

98. Bruce R. Kingma, *The Economics of Information: A Guide to Economic and Cost-Benefit Analysis for Information Professionals*, 2nd edn (Englewood: Libraries Unlimited, 2001), 117. See also Kirsti Nilsen, 'Economic Theory as It Applies to Statistics Canada: A Review of the Literature', 2007, 34, <www.chass.utoronto.ca/datalib/misc/Nilsen%20Economics%20Paper%202007%20final%20version.pdf>, 22 Feb. 2010.

99. Monika Bargmann, Gerald Pfeifer & Boris Piwinger, 'A Citizen's Perspective on Public Sector Information', in *Public Sector Information in the Digital Age: Between Markets, Public Management and Citizens' Rights*, ed. Georg Aichholzer & Herbert Burkert (Cheltenham: Edward Elgar Publishing Limited, 2004), 271. See also Herbert Burkert, 'Offentliche Informationen – Privater Markt – Thesen zur staatlichen Informationsverantwortung', paper given at the International Symposium 'Public Data on the Private Market – New Regulations on the Re-use of Public Sector Information' (Potsdam, 2007), 6, <www.lda.brandenburg.de/media/lbm1.a.2628.de/Burkert_070730.pdf>, 22 Feb. 2010 [in German].

100. Harlan J. Onsrud, 'In Support of Open Access for Publicly Held Geographic Information', *GIS Law* 1, no. 2 (1992): 3–6, www.spatial.maine.edu/~onsrud/pubs/In_Support_OA.htm> 22 Feb. 2010.

typically underestimated by public bodies.[101] Moreover, Johnson and Onsrud warn that even 'the time and monetary cost of pursuing cost recovery enabling legislation can only be characterized as disincentives to cost recovery'.[102]

ARGUMENT FOR COST RECOVERY: BURDEN ON PUBLIC BODIES – On the contrary, it is also argued that in an open access policy, the demand would put a strain on government departments, distracting them from the tasks of government in order to respond to requests for public sector spatial data and misdirecting government resources to benefit a select group of users.[103] As Longhorn and Blakemore state it: 'Assigning zero end-user cost to information can lead end-users to denigrate the value of the data or demonstrate lack of discipline in requesting data. Since all data has a cost, this can lead to significant inefficiencies in data production and over supply, especially for [public sector spatial data]'.[104] Moreover, in the case where marginal costs are charged, there will still be some administrative costs of setting up and maintaining a payment system, and so on. One might also wonder if the income from the marginal costs – which in theory corresponds perfectly with the charges made, but in practice still is not easy to determine – will be enough to cover this overhead.

EVALUATION: ACCESS – However, the argument that the demand for public sector spatial data would distract the public bodies from performing their public tasks is not valid if one considers that making data available for access is also a public task of the public bodies. Hence, this task does not *misdirect* resources from the tasks of the public bodies. It is also not correct to state that it would only benefit a select group of users, as any citizen is a potential candidate for filing a request for access – any citizen is a stakeholder. In addition, one could argue that taking care of the administrative requirements for cost recovery would distract the public body even more from its core tasks.

EVALUATION: ACCESS, RE-USE, AND SHARING – An additional element should be considered regarding the transaction costs. The more distinctions the public bodies make between the prices and conditions under which they make their public sector spatial data available, the higher the transaction costs will be. Therefore, making a distinction in charging between access, re-use, and sharing may not always be beneficial for the public body. From the point of view of transaction costs, preference should go to the same charges for the three types of use. Moreover, this

---

101. William J. Andrews, 'Nurturing the Global Information Commons: Public Access, Public Infrastructure', *Government Information in Canada* 2, no. 3.1 (1996), <http://library2.usask.ca/gic/v2n3/andrews/andrews.html>, 22 Feb. 2010.
102. Jeff P. Johnson and Harlan J. Onsrud, 'Is Cost Recovery Worthwhile?', 1995, <www.spatial.maine.edu/%7eonsrud/pubs/Cost_Recovery_Worthwhile.html>, 22 Feb. 2010.
103. Lori Peterson Dando, 'A Case for the Commercialization of Public Information', paper given at the Conference on Law and Information Policy for Spatial Databases (Tempe, 1994), <www.spatial.maine.edu/%7eonsrud/tempe/dando.html>, 22 Feb. 2010.
104. Roger Longhorn & Michael Blakemore, 'Re-visiting the Valuing and Pricing of Digital Geographic Information', *Journal of Digital Information* 4, no. 2 (2004): 9, <http://journals.tdl.org/jodi/article/view/103/102>, 22 Feb. 2010.

prevents the public body from having to decide whether a certain request deals with access, re-use, or sharing.

6. 'The Political Argument': Cost Recovery Improves the Negotiation Position of the Public Body

ARGUMENT FOR COST RECOVERY: POSITION TOWARDS POSSIBLE PARTNERS – If a public body operates under a cost recovery policy, it may improve its position towards possible partners in cost-sharing partnerships and towards the national treasury. According to Onsrud et al., one of the strongest practical incentives for following a revenue-generating approach is the 'improved negotiating position for data or cost-sharing partnerships with other government agencies that can result from imposing a revenue-generating policy'.[105] If a public body maintains an open access policy, the private sector will not be inclined to enter into a partnership, as on the one hand it can obtain the public sector spatial data for free, and on the other hand, it has to worry that its data may accidentally be released into the public domain through the open access policy of the public body.[106]

ARGUMENT FOR OPEN ACCESS: PARTNERSHIPS CAN ALSO SUCCEED – However, Johnson and Onsrud admit that: 'In spite of the difficulties of developing partnerships for an open access agency, partnerships based on exchange of services and expertise, or coordination of data quality and types could succeed in an open access environment'.[107]

ARGUMENT FOR COST RECOVERY: NO DEPENDENCE ON CENTRAL BUDGET – Next, the open access policy implies that the availability of public sector spatial data is funded by the central budget and assumes 'an ability for government to commit to payments both now and in the future. Such an ability cannot be taken for granted. Governments . . . have frequently demonstrated the difficulty of making such commitments and the impact of political considerations on infrastructure investment'.[108] In cost recovery models, the public bodies are, in principle, 'in control of their own budget, making them independent of fluctuating budgets in national government'.[109] The appearance of bringing in revenue may also improve the position of the public body with the highest-level budget approvers. According to Joffe, for

105. Harlan J. Onsrud, Jeffrey Johnson and Judy Winnecki, 'GIS Dissemination Policy: Two Surveys and a Suggested Approach', *Journal of the Urban and Regional Information Systems Association* 8 (1996): 8–23, <www.spatial.maine.edu/~onsrud/pubs/GIS_Dissemination/GIS_Diss_Policy.html>, 22 Feb. 2010.

106. Jeff P. Johnson & Harlan J. Onsrud, 'Is Cost Recovery Worthwhile?', 1995, <www.spatial.maine.edu/%7eonsrud/pubs/Cost_Recovery_Worthwhile.html>, 22 Feb. 2010.

107. *Ibid.*

108. David Newbery, Lionel Bently and Rufus Pollock, 'Models of Public Sector Information Provision via Trading Funds', 2008, 110, <www.berr.gov.uk/files/file45136.pdf>, 22 Feb. 2010. See also Richard Maven, 'Interview: Online mapping guru Ed Parsons', *E-consultancy* 2007, <www.e-consultancy.com/news-blog/362454/interview-online-mapping-guru-ed-parsons.html>, 22 Feb. 2010.

109. Bastiaan Van Loenen, *Developing Geographic Information Infrastructures: The Role of Information Policies* (Delft: DUP Science, 2006), 105.

some public bodies, '[a] revenue stream, even perhaps at the expense of more valuable staff time, fostered credibility and protection from The Axe'.[110]

7. Evaluation of the Arguments

DIFFERENT VALUE OF ARGUMENTS FOR ACCESS, RE-USE, AND SHARING – As has been discussed above, the arguments that are usually involved in the debate on open access versus cost recovery are used for access, re-use, and sharing interchangeably. However, some of the arguments are not entirely valid for all three types of use, but relate more to either access, re-use, or sharing separately. For instance, the argument that open access is needed for democracy and accountability of government is more suited for access than for re-use or sharing, while the arguments related to the taxpayer paying for the benefits of a limited group of users has more value regarding re-use.

DIFFERENT CONSEQUENCES FOR THE ENTIRE SOCIETY OR INDIVIDUAL STAKEHOLDERS – One should also take into account the different consequences of both policies for either the entire society or the individual stakeholders. For instance, the arguments that were in favour of cost recovery for re-use mostly addressed the individual public bodies, while the arguments in favour of open access referred to the interests of the entire society and market. However, one could also state that the argument of open access for re-use was a benefit for the individual re-users, while it could increase costs for the entire taxpaying community. The same could be argued for sharing: the free of charge or marginal cost charging policies would be beneficial for the public sector as a whole, while cost recovery would serve the individual public bodies that deliver public sector spatial data. One could imagine, however, that a marginal cost policy is adhered to regarding to access or sharing, while re-use is charged for under a cost recovery policy. Once again, one is reminded of the need for clear distinction between access, re-use, and sharing. However, I believe that this distinction is problematic.

IV. PROBLEMATIC CHARACTER OF THE DISTINCTION BETWEEN ACCESS, RE-USE, AND SHARING

RECAPITULATION – The previous discussion on charging indicates that a clear legal framework is needed for the availability of public sector spatial data. As this framework is based on the distinction between access, re-use, and sharing, these three types of use need to be clearly defined and demarcated.

HYPOTHESIS – However, I question whether this distinction is clear, and whether the legal framework for the availability of public sector spatial data can properly ensure this availability. My hypothesis is that this is not the case. This hypothesis can be formulated as follows: the distinction between access, re-use,

110. Bruce Joffe, '10 Ways to Support your GIS Without Selling Data', s.d., <www.opendataconsortium.org/documents/10Ways2SupportGIS-3.pdf>, 22 Feb. 2010.

and sharing on which the EU legal framework for the availability of public sector spatial data is built is problematic and hinders the availability of these data. This causes uncertainty for the public bodies and the users of public sector spatial data and does not promote the availability of these data. This problem occurs on two levels.

PUBLIC BODIES AS DATA PROVIDERS – On the first level, the public bodies operate as the provider of public sector spatial data to the users. They will encounter two main problems when making their data available. First, it will not be possible to find a perfectly fitting demarcation between access and re-use, because it is not always possible to distinguish between the democratic purpose of access and the economic purpose of re-use. In addition, it cannot not always be properly controlled whether an applicant is requesting access or re-use. As no interest has to be stated when access is requested, the public body should look for other indications that re-use might be the purpose instead of access. The second problem regards the distinction between sharing and re-use. This distinction is based on the public task. As most public task activities are performed by public bodies, the question will mostly occur when a public body has to make public sector spatial data available to other public bodies, or when it intends to use its own data for the creation of information services. The distinction between re-use and sharing by these public bodies will be difficult to make, as there is no definition of public task and the scope of the public task seems impossible to determine with legal certainty.

PUBLIC BODIES AS SERVICE PROVIDERS – This second problem also has its effect on the second level where the problem occurs. This second level refers to the public bodies that are using public sector spatial data to perform their activities. On the one hand, public bodies use data to prepare, implement, and evaluate their policies. On the other hand, it is also part of the public task of public bodies to deliver services to the public, for example, public transport, waste disposal, and so on. This also includes the service to inform the public about matters that are important to them. This cannot only be done by making public sector spatial data available to the citizens for access upon request, but also by actively disseminating data to the public. Public bodies are under increasing pressure to make their data accessible to the citizens. This often will include processing the documents or adding some value to it. Hence, the public bodies can use public sector spatial data to create information services that are part of their public task to inform the public. However, they might also create such services outside of their public task, as a market service possibly in competition with the private sector. In the first case, the INSPIRE Directive applies; in the second case, the PSI Directive is applicable. Hence, the question is whether providing such information services is part of the public task or not. Here we encounter the same problem as above: the public task is not defined and its scope is not clear.

OUTLINE: CHAPTERS 5 AND 6 – The following chapters will attempt to assess how far our perception of the problems in the EU legal framework for the availability of public sector spatial data is valid. The first two chapters, Chapter 5 and Chapter 6, deal with the first level of the availability of public sector spatial data. First, Chapter 5 examines other legislative or policy documents that apply to all the

Member States of the EU and that relate to the availability of public sector spatial data. This should provide a view on what the scope of the three types of use is and how these can be distinguished. This may shed more light on the distinction between access, re-use, and sharing. Next, Chapter 6 takes a look at possible indicators that could help determine the distinction between the three types of use and takes a deeper look into the criterion of the purpose of use itself.

OUTLINE: CHAPTERS 7 AND 8 – The following two chapters, Chapter 7 and Chapter 8, deal with the second level, that is, the provision of information services by the public bodies. As was stated earlier, the public bodies are sharing public sector spatial data if they are providing these services as part of their public task. If this is not the case, they are re-using the data. First, in Chapter 7, I examine whether the international and European legislative and policy documents that were discussed in Chapter 5 determine the scope of the public task of the public bodies to provide information services. In Chapter 8, I try to determine whether defining the scope of the public task is actually possible and whether my perception that the public task is not a good criterion for the distinction between re-use and sharing is valid.

Chapter 5

Obligations to Make Public Sector Spatial Data Available in International and European Legislation and Policy Documents

I. INTRODUCTION

PURPOSE OF THE CHAPTER – Public bodies hold an enormous amount of data that may be of interest to users for the purposes of access, re-use, and sharing. In many cases they are obliged to make these data available to other parties. The three parts of the obligation to make public sector spatial data available are guaranteed in different sets of legislation. As we have seen in Chapter 3, it mainly concerns the Access Directive, the PSI Directive, and the INSPIRE Directive. However, the scope of such obligations is also formed by other European and national legislation and policy documents. These are addressed in this chapter in order to examine whether they can give more indications on the precise scope of the obligations of the public bodies to make public sector spatial data available for access, sharing, and re-use. This can also help determine the impact of the problem of distinguishing between the different purposes for which public sector spatial data are made available.

BASIS FOR THE ORDER OF THE CHAPTERS – I refer to the obligation of public bodies to make public sector spatial data available, but I must admit that in some cases this might also indicate the provision of information, that is, data that are put in a context. The distinction is not always clearly indicated in the legal texts and policy documents. However, the criterion that is considered is whether the public body has to perform any additional processing of the data (apart from creating extracts or the mere physical compilation of data) that makes the data understandable for the user, and that goes beyond what the public body needs for performing its own tasks. If such additional processing has to take place, the public body is providing an information service rather than making public sector spatial data available. While this distinction may relate to the distinction that is often made between raw data and added-value information, this does not mean that it is related to the roles of government and the market that are usually attached to it.[1] On the contrary, it is chosen to avoid such raw data – value added information dichotomy. As the reader will notice, in the different rules and texts, the distinction between making data available and providing information services is not always clearly represented and the two tasks are often combined. Admittedly, the distinction between these two types of obligations may be artificial, but it is maintained to demonstrate the different issues that come into play for both tasks.

DISTINCTION IS NOT ALWAYS CLEAR – In theory, it is possible to make the distinction between the obligation to make public sector spatial data available for access, re-use, or sharing, but in many policy documents and literature, the distinction is not so clear, and the reader is left to decide whether the provision of data is mainly meant for access, for re-use, or for sharing, and whether a distinction can be made between making public sector spatial data available or providing spatial information services. This will be indicated when the documents are discussed, and in some cases this will lead to documents being addressed multiple times, for example, when they address both access and re-use.

OUTLINE – The first part of this section looks at the obligation of providing spatial data for access, the second addresses the obligation of making spatial data available for re-use, and the third part deals with the availability of public sector spatial data for sharing purposes.

II. THE OBLIGATION OF MAKING PUBLIC SECTOR SPATIAL DATA AVAILABLE FOR ACCESS

A. INTRODUCTION

OUTLINE – While there are very few documents that specifically refer to the right of the citizens to access public sector spatial data – and the equivalent task of the public bodies to make these data available – there are numerous international and European documents on access to data held by the public bodies in general, which are also applicable to public sector spatial data and may therefore give an indication on what spatial data should be made available by the public bodies. Before these documents are addressed, the first question that can be asked is whether the fundamental right to freedom of information entails an obligation of the public bodies to make public sector data available. This would entail that access to public sector data would be part of a human right. Next, we will address the obligation of making public sector spatial data available in the European and international policy documents and legislation, including a short revisitation of the Access Directive and the INSPIRE Directive.

1. Cf. *infra*, Ch. 8, s. III.B.4.a.

SCOPE – The documents that are discussed are chosen because they are adopted on an international or European level and they are applicable to all the Member States of the EU. In this way, they have an influence on the EU legal framework for the availability of public sector spatial data. As this book only deals with such data, I will only address documents that might be applicable to the three fields that determine this common denominator, that is, public sector data, environmental data, and spatial data. Documents regarding other obligations of public bodies to provide data, for example in the field of cultural data, will not be addressed.

DATA V. INFORMATION – While in Chapter 2 it was stated that in this book, I choose to use the term *public sector data* rather than *public sector information*, in this chapter I have chosen to use the term of the particular legislation or document that is addressed. Only when providing my own remarks and summaries do I attempt to consistently use the term *public sector data.* Therefore, *information, data*, and *documents* are all used in this chapter.

B. THE OBLIGATION OF MAKING PUBLIC SECTOR DATA AVAILABLE FOR ACCESS AS A PART OF FREEDOM OF INFORMATION

OUTLINE – Can access to public sector data be considered as a human right or a fundamental right? To determine this, we have to look at the main international and European organizations that have acknowledged human rights, notably the United Nations and the Council of Europe. Both organizations have shaped the way human rights or fundamental rights are treated in today's Western civilization. The treaties on human rights that were concluded under their auspices do not contain any references to access to public sector data, but only to freedom of information in general. Hence, it has been up to the institutions of both international organizations to determine whether access could form a part of freedom of information.

1. United Nations

a. Universal Declaration on Human Rights

ARTICLE 19 UDHR – The Universal Declaration of Human Rights (UDHR) was adopted by the General Assembly of the United Nations in 1948. Its Article 19 proclaims that 'everyone has the right to freedom of opinion and expression; this right includes freedom to hold opinions without interference and to seek, receive and impart information and ideas through any media and regardless of frontiers'.[2] Freedom of information was not considered as an independent right, but was considered to be part of the freedom of expression.

2. United Nations, *Universal Declaration on Human Rights*, 10 Dec. 1948, <www.un.org/Overview/rights.html>, 22 Feb. 2010. For more information and background, see Henry J. Steiner & Philip Alston, *International Human Rights in Context: Law, Politics, Morals*, 2nd edn (Oxford: Oxford University Press, 2000), 1497; Uhani Kortteinen, Kristian Myntti & Laura Hannikainen,

No binding effect – As it is a declaration rather than a convention or a treaty, the UDHR does not have a binding effect on the Member States of the United Nations, but its power lies in its moral authority and influence due to the fact that it has been accepted on a global level.[3]

b. The International Covenant on Civil and Political Rights

Timeline – In 1966, the United Nations adopted two treaties on human rights, the International Covenant on Civil and Political Rights (ICCPR)[4] and the International Covenant on Economic, Social and Cultural Rights.[5] By 1976, both Covenants had received sufficient ratifications to enter into force.[6]

Article 19 of the ICCPR – Article 19 of the ICCPR recognizes freedom of information, once again as part of freedom of expression:

> (1) Everyone shall have the right to hold opinions without interference.
> (2) Everyone shall have the right to freedom of expression; this right shall include freedom to seek, receive and impart information and ideas of all kinds, regardless of frontiers, either orally, in writing or in print, in the form of art, or through any other media of his choice.
> (3) The exercise of the rights provided for in paragraph 2 of this article carries with it special duties and responsibilities. It may therefore be subject to certain restrictions, but these shall only be such as are provided by law and are necessary:
>
> (1) For respect of the rights or reputations of others;
> (2) For the protection of national security or of public order (ordre public), or of public health or morals.

Obligation for public bodies to make data available for access? – Article 19 of the ICCPR protects the freedom to seek, receive, and impart information, but this does not immediately entail an obligation on public bodies to make information available. Mason believes, that:

> Although paragraph 2 [of Article 19] does not expressly impose any obligation on state agencies or private persons to make available information not otherwise accessible, it cannot be assumed that the interpretation of the article

'Article 19', in *The Universal Declaration of Human Rights: A Common Standard of Achievement*, ed. Gudmundur Alfredsson & Asbjorn Eide (The Hague: Martinus Nijhoff Publishers, 1999), 393–415; M. Glen Johnson & Janusz Symonides, *The Universal Declaration of Human Rights: a History of Its Creation and Implementation 1948–1998* (Paris: UNESCO, 1998), 166; Barend Van Der Heijden & Bahia Tahzib-Lie, *Reflections on the Universal Declaration of Human Rights: a Fiftieth Anniversary Anthology* (The Hague: Nijhoff, 1998), 344.

3. Steiner & Alston, *supra* n. 2, 142.
4. United Nations, *International Covenant on Civil and Political Rights*, 16 Dec. 1966, <www2.ohchr.org/english/law/ccpr.htm>, 22 Feb. 2010.
5. United Nations, *International Covenant on Economic, Social and Cultural Rights*, 16 Dec. 1966, <www2.ohchr.org/english/law/cescr.htm>, 22 Feb. 2010.
6. Steiner & Alston, *supra* n. 2, 139.

will stop short of recognizing such an obligation. The essentiality of freedom of expression and information to modern liberal democratic government, as well as the emergence of the modern information society, as indicated by the growing number of statutory enactments requiring disclosure of information, notably in the area of public administration, may lead to the implication in Article 19 of such an obligation.[7]

SPECIAL RAPPORTEUR ON FREEDOM OF OPINION AND EXPRESSION – In 1998, the Special Rapporteur on Freedom of Opinion and Expression followed the same reasoning. He stated in his annual report that freedom of information includes the right to access information held by the State: 'The right to seek, receive and impart information imposes a positive obligation on States to ensure access to information, particularly with regard to information held by Government in all types of storage and retrieval systems'.[8] In 1999, the Special Rapporteur, the OSCE Representative on Freedom of the Media, and the OAS Special Rapporteur on Freedom of Expression adopted a Joint Declaration stating that: 'Implicit in freedom of expression is the public's right to open access to information and to know what governments are doing on their behalf, without which truth would languish and people's participation in government would remain fragmented'.[9]

UNITED NATIONS HUMAN RIGHTS COMMITTEE – However, the United Nations Human Rights Committee,[10] which monitors the implementation of the ICCPR by considering how nations are enforcing the Covenant and by hearing individual complaints, has not found that freedom of information includes a positive duty for the Member States to ensure access to their data.[11]

---

7. Anthony Mason, 'The Relationship between Freedom of Expression and Freedom of Information', in *Freedom of Expression and Freedom of Information: Essays in Honour of Sir David Williams*, ed. Jack Beatson & Yvonne Cripps (Oxford: Oxford University Press, 2000), 226–227.
8. United Nations Special Rapporteur on Freedom of Opinion and Expression, *Promotion and protection of the right to freedom of opinion and expression*, UN Doc. E/CN.4/1998/40, 28 Jan. 1998, <http://daccess-ods.un.org/access.nsf/Get?Open&DS=E/CN.4/1998/40&Lang=E>, 22 Feb. 2010. See also Toby Mendel, 'Freedom of Information: An Internationally protected Human Right', *Comparative Media Law Journal* no. 1 (2003): 44–45.
9. United Nations Special Rapporteur on Freedom of Opinion and Expression, OSCE Representative on Freedom of the Media and OAS Special Rapporteur on Freedom of Expression, *International Mechanisms for Promoting Freedom of Expression. Joint Declaration*, 26 Nov. 1999, <http://portal.unesco.org/ci/en/files/26152/12047264643UNFEX1999.pdf/UNFEX1999.pdf>, 22 Feb. 2010.
10. While the ICCPR does have a binding effect on the Member States, its importance in Europe is considerable smaller than the ECHR of 1950. Though the content of the treaties overlap to a great extent, the ECHR contains a much stronger enforcement procedure in the shape of a European Court of Human Rights, to which the citizen can direct its complaints about a Member States disrespecting its human rights. The ICCPR only installs a Human Rights Committee, which can make reports on suspected infractions and submit them to the Member States involved, and make an annual report for the General Assembly of the United Nations.
11. Cheryl Ann Bishop, 'Access to Information as a Human Right: Analysis of the United Nations Human Rights Committee Documents', paper given at the annual meeting of the International

2. Council of Europe

a. *The European Convention on the Protection of Human Rights and Fundamental Freedoms*

TIMELINE – The European Convention on the Protection of Human Rights and Fundamental Freedoms (ECHR) was adopted on 4 November 1950 by the Member States of the Council of Europe. It obtained the ten ratifications necessary for entry into force in 1953 and has been ratified by thirty-seven other European states since then.[12]

ARTICLE 10 OF THE ECHR – The ECHR also protects freedom of information as part of freedom of expression in Article 10:

(1) Everyone has the right to freedom of expression. This right shall include freedom to hold opinions and to receive and impart information and ideas without interference by public authority and regardless of frontiers. This article shall not prevent States from requiring the licensing of broadcasting, television or cinema enterprises.
(2) The exercise of these freedoms, since it carries with it duties and responsibilities, may be subject to such formalities, conditions, restrictions or penalties as are prescribed by law and are necessary in a democratic society, in the interests of national security, territorial integrity or public safety, for the prevention of disorder or crime, for the protection of health or morals, for the protection of the reputation or rights of others, for preventing the disclosure of information received in confidence, or for maintaining the authority and impartiality of the judiciary.

ELEMENTS OF FREEDOM OF EXPRESSION – Article 10 of the ECHR conceives freedom of expression as including both freedom of opinion and freedom to receive and impart information and ideas. Unlike the ICCPR, it does not protect the right actively to seek information.[13] However, according to Mason, such a right may be implied from the terms of the article.[14]

Communication Association, Dresden International Congress Centre (Dresden, 2006), 20–21, <www.allacademic.com/meta/p91640_index.html>, 22 Feb. 2010.

12. <http://conventions.coe.int/Treaty/Commun/ChercheSig.asp?NT=005&CM=7&DF=7/14/2007&CL=ENG>, 22 Feb. 2010.
13. Anthony Mason, 'The relationship between freedom of expression and freedom of information', in *Freedom of expression and freedom of information. Essays in honour of Sir David Williams*, ed. Jack Beatson and Yvonne Cripps (Oxford: Oxford University Press, 2000), 228; Dirk Voorhoof, 'Artikel 10 - Vrijheid van meningsuiting', in *Handboek EVRM. Deel 2. Artikelsgewijze commentaar*, ed. Johan Vande Lanotte & Yves Haeck (Antwerp: Intersentia, 2004, I), 850 [in Dutch]; Frédéric Sudre, *Droit Européen et International des Droits de l'Homme*, 2nd edn (Paris: Presses Universitaires de France, 2006,) 493 [in French].
14. Anthony Mason, 'The relationship between freedom of expression and freedom of information', in *Freedom of expression and freedom of information. Essays in honour of Sir David Williams*, ed. Jack Beatson and Yvonne Cripps (Oxford: Oxford University Press, 2000), 228.

No obligation for public bodies to make data available? – Similar to the ICCPR, Article 10 of the ECHR seems to assume that the right to receive information relates only to sources of information that are otherwise generally accessible, and it does not recognize a general right of the public to access to public sector data.[15] However, according to Hins and Voorhoof, '[o]ne could argue that a positive obligation for the State to supply relevant information and to give access to official documents regarding matters of public interest is inherent in that article'.[16] The European Court of Human Rights was reluctant to recognize this for many years, but recently there have been indications that this might be changing.[17]

b. Documents of the Council of Ministers and the Parliamentary Assembly[18]

1970: Recommendation of the Parliamentary Assembly – In a 1970 Recommendation on mass communication media and human rights, the Parliamentary Assembly of the Council of Europe stated that the right to freedom of expression includes the freedom to seek, receive, impart, publish, and distribute information and ideas, and that there is a corresponding duty for public authorities to make available information on matters of public interest within reasonable limits.[19]

1982: Declaration of the Committee of Ministers – The 1982 Declaration on the freedom of expression and information of the Committee of Ministers of the Council of Europe also went further than the ECHR. According to the Declaration, freedom of expression and information was 'a fundamental element' of the principles

15. Anthony Mason, 'The relationship between freedom of expression and freedom of information', in *Freedom of expression and freedom of information. Essays in honour of Sir David Williams*, ed. Jack Beatson and Yvonne Cripps (Oxford: Oxford University Press, 2000), 228; François Dubuisson, 'Quand le "Droit" du Public à l'Information Rencontre les "Droits" sur l'Information: Vrai ou Faux Débat?', in *Droit d'Auteur et Liberté d'Expression. Regards Francophones, d'Europe et d'Ailleurs*, ed. Alain Strowel & François Tulkens (Brussels: Larcier, 2006), 91–92 [in French].
16. Wouter Hins & Dirk Voorhoof, 'Access to State-held Information as a Fundamental Right under the European Convention on Human Rights', *European Constitutional Law Review* 3, no.1 (2007): 114.
17. Cf. *infra*, s. I.B.2.c.
18. See <www.coe.int/t/cm/home_en.asp>, 22 Feb. 2010, and <http://assembly.coe.int/default.asp>, 22 Feb. 2010. The Council of Ministers and the Parliamentary Assembly are two of the institutions of the Council of Europe. The others are the Congress for Local and Regional Authorities, the European Court of Human Rights, the Commissioner for Human Rights, and the Conference of INGOs. The institutions are supported by a Secretariat General of circa 1800 people. The Committee of Ministers is the Council of Europe's decision making body, composed of the Foreign Ministers of the Member States or their deputies. The Parliamentary Assembly groups 636 members (318 representatives and 318 substitutes) from the forty-seven national parliaments.
19. Council of Europe Parliamentary Assembly, *Recommendation 582 (1970) on mass communication media and human rights*, 1970, <http://assembly.coe.int/main.asp?Link=/documents/adoptedtext/ta70/erec582.htm>, 22 Feb. 2010.

of 'genuine democracy, the rule of law and respect for human rights', and it was necessary for 'the social, economic, cultural and political development of every human being, and constitutes a condition for the harmonious progress of social and cultural groups, nations and the international community'.[20] The goals of the Declaration included 'the pursuit of an open information policy in the public sector, including access to information, in order to enhance the individual's understanding of, and his ability to discuss freely political, social, economic and cultural matters'. According to Mason, to see the Declaration as doing more than enunciating policies that are consistent with the general thrust of Article 10 might be going too far. At the same time, the Declaration identified values that may influence future interpretation of the article.[21] Beers suggests paying attention to the rather strict distinction between the right to freedom of expression on the one hand, and access to data on the other. Freedom of expression is qualified as a matter of right and access to data as a matter of policy. He feels that: 'This seems to imply that the opinion of the Committee of Ministers is that public access to government information forms no part of the right to freedom of expression as guaranteed by Article 10'.[22]

RECENT DEVELOPMENTS – During more recent years, the Committee of Ministers repeatedly recognized a duty of the Member States and the public bodies to make public sector data available to the public, mostly in the framework of the development of new technologies. For instance, in its 1999 Declaration on a European policy for new information technologies, the Committee of Ministers urged the governments of Member States to use information technologies to 'provide easy access to information about and direct links to local, regional and national administrative and judicial services' and to 'make available official texts of local, regional and national laws and regulations, of international agreements and of the jurisprudence of national and international jurisdictions'.[23] In 2007, the Committee of Ministers added that Member States should ensure that 'public authorities increase the provision and transparency of their online services to citizens and

---

20. Council of Europe Committee of Ministers, *Declaration on the freedom of expression and information*, 1982, <https://wcd.coe.int/com.instranet.InstraServlet?command=com.instranet.CmdBlobGet&InstranetImage=601273&SecMode=1&DocId=675536&Usage=2>, 22 Feb. 2010. See also Ton A.L. Beers, 'Public Access to Government Information towards the 21st Century', in *Information Law towards the 21st Century*, ed. Willem F. Korthals Altes (Deventer: Kluwer, 1992), 192; Frankie Schram, 'Openbaarheid van bestuur: een voorbeeld van interactie tussen het Europees en nationaal publiekrecht', *Tijdschrift voor Bestuurswetenschappen en Publiekrecht* no.3 (1999): 157 [in Dutch].
21. Anthony Mason, 'The relationship between freedom of expression and freedom of information', in *Freedom of expression and freedom of information. Essays in honour of Sir David Williams*, ed. Jack Beatson and Yvonne Cripps (Oxford: Oxford University Press, 2000), 229.
22. Ton A.L. Beers, 'Public Access to Government Information towards the 21st Century', in *Information Law towards the 21st Century*, ed. Willem F. Korthals Altes (Deventer: Kluwer, 1992), 192.
23. Council of Europe Committee of Ministers, *Declaration on a European policy for new information technologies*, 1999, <https://wcd.coe.int/ViewDoc.jsp?id=448133>, 22 Feb. 2010.

businesses so that they allow every individual access to public information'[24] and that the Member States should promote public domain information accessibility via the Internet, including government documents.[25]

c. *The Case Law of the European Court of Human Rights*

NO OBLIGATION FOR PUBLIC BODIES TO MAKE DATA AVAILABLE? – The European Court of Human Rights and the European Commission of Human Rights[26] repeatedly recognized the right of the public to be properly informed.[27] Yet, the European Court of Human Rights never accepted a duty for the public authorities to actively provide data to the public. However, in the 2006 *Sdruzeni Johoceske Matky*[28] decision, this might have changed. Before we address this decision, we take a short look at some important cases in the evolution towards it.

NO BASIS IN ARTICLE 10 OF THE ECHR FOR AN OBLIGATION TO MAKE DATA AVAILABLE – In *Leander v. Sweden* and *Gaskin v. the United Kingdom*, the European Court of Human Rights did not find a basis in Article 10 of the ECHR for a positive obligation of the Member State to make data available to citizens upon request. The former case concerned the access of an individual to his personal file held by the Swedish Secret Service, and the latter regarded a personal file held by the English Child Protection Services. In both cases, the Court found that: 'Freedom to receive information basically prohibits a government from restricting a person from receiving information that others wish or may be willing to impart to him. That freedom cannot be construed as imposing on a State, in circumstances such as those of the present case,

24. Council of Europe Committee of Ministers, *Recommendation CM/Rec(2007)11 on promoting freedom of expression and information in the new information and communications environment*, 2007, <https://wcd.coe.int/ViewDoc.jsp?id=1188541>, 22 Feb. 2010.
25. Council of Europe Committee of Ministers, *Recommendation CM/Rec(2007)16 on measures to promote the public service value of the Internet*, 2007, <https://wcd.coe.int/ViewDoc.jsp?id=1207291>, 22 Feb. 2010.
26. The European Commission of Human Rights was abolished in 1998. Previously, it performed the first treatment of the complaints made to the European Court of Human Rights. A citizen could not go directly to the European Court of Human Rights, but had to launch a complaint to the Human Rights Commission, which would do a first validity check and then take the case to the European Court of Human Rights on the individual's behalf. From 1998 onwards, the Court, now enlarged with a number of additional judges, could treat complaints of the citizens directly.
27. European Court of Human Rights, *Sunday Times v. United Kingdom*, 24 Apr. 1979, paras 64–66; European Court of Human Rights, *Open Door and Dublin Well Women v. Ireland*, 29 Oct. 1992, para. 55; European Court of Human Rights, *Lingens v. Austria*, 8 Jul. 1986, para. 41; European Court of Human Rights, *Thorgeir Thorgeirson v. Iceland*, 25 Jun. 1992, para. 63; European Court of Human Rights, *Jersild v. Denmark*, 23 Sep. 1994, para. 31; European Court of Human Rights, *Hertel v. Switzerland*, 25 Aug. 1998, paras 47–49; European Court of Human Rights, *Colombani and others v. France*, 25 Jun. 2002, para. 55; European Court of Human Rights, *Cetin and others v. Turkey*, 13 Feb. 2003, para. 64; European Court of Human Rights, *Ukrainian Media Group v. Ukraine*, 29 Mar. 2005, para. 38. All cases were retrieved from <http://cmiskp.echr.coe.int/tkp197/search.asp?skin=hudoc-en>, 22 Feb. 2010.
28. European Court of Human Rights, *Sdruzeni Johoceske Matky v. Czech Republic*, 10 Jul. 2006, <http://cmiskp.echr.coe.int/tkp197/search.asp?skin=hudoc-en>, 22 Feb. 2010.

positive obligations to collect and to disseminate information of its own motion'.[29] This approach was repeated in several other cases.[30]

POSSIBLE BASIS IN ARTICLE 8 OF THE ECHR FOR AN OBLIGATION TO MAKE DATA AVAILABLE – In the *Guerra* case, forty Italian citizens living in the proximity of a high-risk chemical factory filed a complaint because the authorities had neglected to take measures to inform them about the risks the population was running and the measures they had to take in case of a major accident.[31] The claimants considered this a breach of Article 10 of the ECHR. The European Court of Human Rights refused to decide this case on the basis of Article 10 of the ECHR, and turned to Article 8 of the ECHR protecting the right to respect for private and family life, stating that this awarded a right to receive information that is essential to enjoy this right.[32] Once again, the Court reiterated what was stated in the *Leander* and *Gaskin* cases. Interesting to note, however, is that, more so than in the two cases above, it seems that a decision in favour of the right to data could have been reached in a different context from the case at hand: nine out of the twenty judges were ready to recognize, considering the circumstances, that Article 10 can impose on the state positive obligations to communicate data to the public.[33]

AN OBLIGATION TO MAKE DATA AVAILABLE 'DEPENDING ON THE CIRCUMSTANCES?' – In addition, the fact that the ECHR felt the need to refer to 'the current circumstances' in each of these cases seems to indicate that it had not taken a definite decision to either recognize or deny an obligation to the State.[34] According to Hins

29. European Court of Human Rights, *Leander v. Sweden*, 26 Mar. 1987, para. 74; European Court of Human Rights, *Gaskin v. United Kingdom*, 7 Jul. 1989, para. 52, <http://cmiskp.echr.coe.int/tkp197/search.asp?skin=hudoc-en>, 22 Feb. 2010.
30. For example, European Court of Human Rights, *Sirbu and others v. Moldova*, 15 Jun. 2004, para. 18; European Court of Human Rights, *Roche v. United Kingdom*, 23 May 2002, <http://cmiskp.echr.coe.int/tkp197/search.asp?skin=hudoc-en>, 22 Feb. 2010.
31. European Court of Human Rights, *Guerra and others v. Italy*, 19 Feb. 1998, <http://cmiskp.echr.coe.int/tkp197/search.asp?skin=hudoc-en>, 22 Feb. 2010.
32. See Jean-Pierre Marguenaud, 'La Convention d'Aarhus et la Convention européenne des droits de l'homme', *Revue Juridique de l'Environnement* (1999): 83–85 [in French]; Maria Gavouneli, 'Access to Environmental Information: Delimitation of a Right', *Tulane Environmental Law Journal* 13 (2000): 324–325; Maguelonne Déjaent-Pons, 'Le Conseil de l'Europe et l'Accès à l'Information Environnementale: la Convention Européenne des Droits de l'Homme', in *Ten Years of Access to Environmental Information in International, European and Belgian Law: Stock-Taking and Perspectives*, ed. Christine Larssen (Brussels: Bruylant, 2003), 72–74 [in French]; Maguelonne Déjaent-Pons, 'La Convention européenne des droits de l'homme et le droit à l'information en matière d'environnement', in *La Convention européenne des droits de l'homme: développements récents et nouveaux défis*, ed. Jean-François Flauss & Michel De Salvia (Brussels: Bruylant, 1997), 154–166 [in French].
33. Cécile De Terwangne, 'La Convention Européenne des Droits de l'Homme et le Droit de Recevoir des Informations de la Part des Autorités Publiques', *Aménagement - Environnement* (1998): 267 [in French].
34. Ton A.L. Beers, 'Public Access to Government Information towards the 21st Century', in *Information Law towards the 21st Century*, ed. Willem F. Korthals Altes (Deventer: Kluwer, 1992), 196; Dirk Voorhoof, 'Artikel 10 – Vrijheid van meningsuiting', in *Handboek EVRM. Deel 2. Artikelsgewijze commentaar*, ed. Johan Vande Lanotte and Yves Haeck (Antwerp: Intersentia, 2004, I), 914–916 [in Dutch]; Dirk Voorhoof, 'Europees Mensenrechtenhof maakt

and Voorhoof, the words 'in circumstances such as those of the present case' suggest that there might be situations in which a positive obligation for the State does exist, yet the Court never established such a situation.[35] However, in the *Eccleston v. United Kingdom* decision, regarding the rejection of a request for access to a personal file by a local authority, the European Court of Human Rights seemed to move further away than ever from a right to obtain data from the public bodies. The Court, referring *mutatis mutandis* to *Guerra*, stated that: 'The right to receive information contained in Article 10 refers to information which others wish to communicate. It does not impose an obligation on others to provide information'.[36] It did not add its usual phrase 'in the circumstances such as those in the present case', but formulated the refusal in a general manner, suggesting a definite refusal of such an obligation. However, according to Voorhoof, because the case concerned a personal file of the applicant, one can assume that the rejection had to be interpreted in the context or the circumstances of the case.[37]

Change in the scope of article 10? – The decision of 10 July 2006 in the *Sdruzeni Johoceske Matky v. Czech Republic* case indicates a change in the European Court of Human Rights' approach. Sdruzeni Johoceske Matky was a Czech association with the principal objectives of protecting nature, improving the state of the environment, and supporting sustainable development. One of its main activities was protesting against the production and use of nuclear energy. It had requested access to data on the technology behind a new nuclear power plant that was being built in the Czech Republic. The Czech authorities refused to make these data available. The European Court of Human Rights stated that the arguments for refusal of the Czech authorities to make these documents available were in agreement with Article 10 of the ECHR and the limits it allowed to freedom of expression and freedom of information. The Court, referring to *Leander* and *Guerra*, still stated that Article 10 of the ECHR 'cannot be construed as imposing on a State, in circumstances such as those of the present case, positive obligations to collect and to disseminate information of its own motion',[38] and even added that 'it is difficult to deduct from the Convention a general right of access to documents of an

(eindelijk) toepassing van artikel 10 EVRM in het kader van openbaarheid van bestuur en toegang tot bestuursdocumenten', *Mediaforum* 18, no. 10 (2006): 290–294 [in Dutch].

35. Wouter Hins & Dirk Voorhoof, 'Access to State-Held Information as a Fundamental Right under the European Convention on Human Rights', *European Constitutional Law Review* 3, no.1 (2007): 118.
36. European Court of Human Rights, *Eccleston v. United Kingdom*, 18 May 2004, para. 3, <http://cmiskp.echr.coe.int/tkp197/search.asp?skin=hudoc-en>, 22 Feb. 2010.
37. Dirk Voorhoof, 'Europees Mensenrechtenhof maakt (eindelijk) toepassing van artikel 10 EVRM in het kader van openbaarheid van bestuur en toegang tot bestuursdocumenten', *Mediaforum* 18, no. 10 (2006): 293 [in French].
38. European Court of Human Rights, *Sdruzeni Johoceske Matky v. Czech Republic*, 10 Jul. 2006, <http://cmiskp.echr.coe.int/tkp197/search.asp?skin=hudoc-en>, 22 Feb. 2010. The above text is a translation of 'ne saurait se comprendre comme imposant à un Etat, dans des circonstances telles que celles de l'espèce, des obligations positives de collecte et de diffusion, *motu proprio*, des informations'.

administrative nature'.[39] However, at the same time, the Court recognized that the refusal to give access to administrative documents that fall under the field of application of the national access legislation is *'une ingérence au droit de la requérante de recevoir des informations'* (an interference with the right of the applicant to receive information).[40] The Court did not find any infraction of Article 10 of the ECHR and therefore declared the applicant's complaint inadmissible.

Reasons for the application of article 10 – The European Court of Human Rights mentioned three characteristics of the case that made it decide that Article 10 of the ECHR was applicable. First, the applicant had filed a request, so there was no failure of the authorities to spread data of its own motion. To the Court, this made an important difference. However, as Hins and Voorhoof mention, '[i]n other judgments and decisions of the Court, Article 10 has been declared inapplicable also in cases where the applicant had requested information or had sought access to personal records. In the Sruzeni Johoceske Matky decision, the Court, by clearly emphasizing the circumstance that the applicant had filed a request to have access to administrative documents, upgraded this circumstance to a relevant, if not decisive, element in order to make Article 10 applicable'.[41] The second characteristic is that the data were contained in administrative documents already held by the authorities, so they did not need to collect or create the data. Finally, unlike the applicant, other citizens had or could have access to the data, under the conditions of the Czech law on access to documents held by public authorities.

Enforceable obligations under article 10 – According to Voorhoof, the decision holds an important impulse to assess the refusal of requests for accessing or obtaining a copy of documents held by public authorities on the basis of Article 10 of the ECHR. Such a refusal not only has to be based on a clearly determined legal restriction imposed by law, but its application also has to be 'necessary in a democratic society' from the perspective of Article 10 of the ECHR.[42] Hence, Article 10 actually imposes enforceable obligations and direct claims to a right to access public sector data. Hins and Voorhoof summarize this as follows: 'The crucial point . . . is that Article 10 was considered to be applicable in the first place'.[43]

39. *Ibid.* The text is a translation of 'il est difficile de déduire de la Convention un droit général d'accès aux données et documents de caractère administratif'.

40. *Ibid.*

41. Wouter Hins & Dirk Voorhoof, 'Access to State-Held Information as a Fundamental Right under the European Convention on Human Rights', *European Constitutional Law Review* 3, no. 1 (2007): 118, 125. See also Jean-Jacques Paradissis, *The Right to Access Environmental Information* (Saarbrucken: VDM Verlag, 2009), 267.

42. Dirk Voorhoof, 'Europees Mensenrechtenhof maakt (eindelijk) toepassing van artikel 10 EVRM in het kader van openbaarheid van bestuur en toegang tot bestuursdocumenten', *Mediaforum* 18, no. 10 (2006): 293–294 [in Dutch].

43. Wouter Hins and Dirk Voorhoof, 'Access to State-Held Information as a Fundamental Right under the European Convention on Human Rights', *European Constitutional Law Review* 3, no. 1 (2007): 123.

FOLLOW-UP – In 2009, the European Court of Human Rights admitted itself that it was moving towards the recognition of an obligation to make public sector data available. In the *Társaság a Szabadságjogokért v. Hungary* case, the Court found that the case 'essentially concerns an interference by virtue of the censorial power of an information monopoly – with the exercise of the functions of a social watchdog like the press, rather than a denial of a general right of access to official documents', but it did admit that 'the Court has recently advanced towards a broader interpretation of the notion of "freedom to receive information". . . . and thereby towards the recognition of a right of access to information'.[44] A next step was taken in the *Kenedi v. Hungary* case, in which the European Court of Human Rights stated that: 'Access to original documentary sources for legitimate historical research was an essential element of the exercise of the applicant's right to freedom of expression'.[45] However, the decision of the Court was made fairly straightforward as the national courts had already decided that the government should give access and an enforcement procedure had been started, but the government had never provided the documents.

3. Interim Summary

NO OBLIGATION TO MAKE PUBLIC SECTOR DATA AVAILABLE (YET) – Neither the ICCPR nor the ECHR recognize a right to access public sector data as a part of freedom of information. However, there are some signs indicating an evolving interpretation of the institutions of the UN and the Council of Europe towards the acceptance of such a right to access. The UN Special Rapporteur believes access should be included in freedom of information (although the Human Rights Committee does not); and the Parliamentary Assembly and Committee of Ministers of the Council of Europe also indicate, albeit very carefully, the importance of an open information policy in the public sector for freedom of information. As for the European Court of Human Rights, it held on to its viewpoint that access to public sector data is not part of Article 10 of the ECHR for a very long time, but in the 2006 *Sdruzeni* case, it showed some first signs of acceptance of enforceable obligations on the Member States to make public sector data available, which were confirmed in 2009. However, the Court made it clear that it still found it difficult to deduct from the Convention a general right of access, and that it is only moving towards it. Therefore, it is premature to fully acknowledge access as a part of Article 10 of the ECHR without any further case law of the European Court of Human Rights.

---

44. European Court of Human Rights, *Társaság a Szabadságjogokért v. Hungary*, 14 Apr. 2009, paras 35–36, <http://cmiskp.echr.coe.int/tkp197/search.asp?skin=hudoc-en>, 22 Feb. 2010. See also Jean-Jacques Paradissis, *The Right to Access Environmental Information* (Saarbrucken: VDM Verlag, 2009), 268–269.
45. European Court of Human Rights, *Kenedi v. Hungary*, 26 May 2009, para. 43, <http://cmiskp.echr.coe.int/tkp197/search.asp?skin=hudoc-en>, 22 Feb. 2010. See also Jean-Jacques Paradissis, *The Right to Access Environmental Information* (Saarbrucken: VDM Verlag, 2009), 269–270.

Hence, there is currently no obligation for public bodies to provide access to their data based on human rights.

C. International and European Documents

General recognition of an obligation to make public sector data available – Even if access to public sector data may not have received full recognition as a human right, this does not mean that making public sector data available for access cannot be seen as a task of the public bodies. Access received considerable attention in the second part of the twentieth century from the international and European organizations and the national legislators. They all consider that there is an obligation of the public bodies to provide information on their activities and to give access to the data they hold. As this book is focused on the European level, it will look at the most important documents from the international institutions but not address the national access legislation of the Member States.

Outline – This section will deal with public sector data in general, as this also includes public sector spatial data, and then move on to look at environmental data and spatial data in particular. First, introductory remarks are needed on the difference between passive access to data upon request and the active dissemination of public sector data.

1. Access: From Passive to Active?

No clear distinction between active and passive access – The documents that are discussed in this section often make a distinction between the task of the public bodies to make data available upon request and the obligation to disseminate data of their own motion. This distinction is often referred to as the difference between *passive* and *active* access.[46] However, this distinction is no longer as clear as it used to be. Van Eechoud and Van Der Wal state that: 'The use of ICT increasingly blurs the boundary between active and passive dissemination. The more government publishes metadata (registers) on the documents it holds, which can then be requested on-line for example, the less meaningful the distinction passive/active becomes'.[47] De Terwangne wonders if disseminating a database online is not simply the same as allowing access to that database. There is an act of publication of the public body, but on the other hand the user still has to query and search the database.[48]

46. The distinction is sometimes also referred to as pull and push models, see e.g., Economic Development and Infrastructure Committee, Inquiry into improving access to Victorian public sector information and data. Discussion Paper, 2008, 5, <www.parliament.vic.gov.au/edic/inquiries/access_to_PSI/call_for_submissions.html>, 22 Feb. 2010.
47. Mireille Van Eechoud & Brenda Van Der Wal, 'Creative commons licensing for public sector information. Opportunities and pitfalls', 2008, 49, <www.ivir.nl/publications/eechoud/CC_PublicSectorInformation_report_v3.pdf>, 22 Feb. 2010.
48. Cécile De Terwangne, *Société de l'information et mission publique d'information*, Doctoral Thesis FUNDP (Namur: Faculté Universitaire de Notre Dame de Paix, 2001), 309 [in French].

RELEVANCE OF THE DISTINCTION – Baten and Van Der Starre agree that for electronic public sector data, the distinction between active and passive access seems to fade, and especially when data are accessed on-line, because the effort that the public servant needs to make is reduced to a few mouse-clicks. However, they feel that the distinction is still relevant, because the initiative lies in a different place.[49] This entails that for the active dissemination of data, the public bodies have to consciously decide which data they think should be disseminated to the public, and they cannot wait for the citizens to give a signal about which data they need. The public bodies need to develop a vision about the scope of their obligations to make public sector data available and how they will fulfil that task, for example through which dissemination channels, in which formats, or at which time intervals they disseminate data.[50] In addition, the need of the public for certain data can change. Sometimes the public bodies will have to adapt their activities in order to actively disseminate data that previously was only available upon request.

2. United Nations

a. United Nations Special Rapporteur on Freedom of Opinion and Expression

2000 ANNUAL REPORT – The United Nations (UN) has not adopted any specific treaties or covenants relating to access to public sector data. In 2000, the Special Rapporteur stated in his Annual Report that: 'Public bodies have an obligation to disclose information and every member of the public has a corresponding right to receive information; "information" includes all records held by a public body, regardless of the form in which it is stored'. In addition, '[f]reedom of information implies that public bodies publish and disseminate widely documents of significant public interest, for example, operational information about how the public body functions and the content of any decision or policy affecting the public'. Next to a number of requirements for the procedures of making information available, the Rapporteur also felt that: 'The cost of gaining access to information held by public bodies should not be so high as to deter potential applicants and negate the intent of the law itself'.[51]

49. Inigo Baten and Gijs Van Der Starre, *Elektronische toegankelijkheid van overheidsinformatie* (The Hague: Rathenau Instituut, 1996), 20 [in Dutch].
50. Katleen Janssen, Jo Steyaert and Roland Van Gompel, *Transparantie van overheidsinformatie in Vlaanderen. Beschouwingen en aanbevelingen* (Bruges: Die Keure, 2003), 172–173 [in Dutch].
51. United Nations Special Rapporteur on Freedom of Opinion and Expression, *Promotion and protection of the right to freedom of opinion and expression*, UN Doc. E/CN.4/2000/63, 18 Jan. 2000, 15–16, <http://daccess-ods.un.org/access.nsf/Get?Open&DS=E/CN.4/2000/63&Lang=E>, 22 Feb. 2010.

2004 ANNUAL REPORT – In 2004, the Special Rapporteur affirmed that:

> All information held by public bodies shall be publicly available unless it is subject to a legitimate exemption, and all bodies performing public functions, including governmental, legislative and judicial bodies, should be obliged to respond to requests for information. The expression 'bodies performing public functions' also pertains to enterprises, societies and associations performing a unique role and/or receiving public funds'. In addition access should be to information rather than to documents, and its cost to the requestor should be limited to the supply costs and should not be so high as to prove an obstacle to access.[52]

No new declarations were made regarding the active dissemination of information.

JOINT DECLARATION ON ACCESS TO INFORMATION – Also in 2004, the UN Special Rapporteur on Freedom of Opinion and Expression, the OCSE Representative on Freedom of the Media, and the OAS Special Rapporteur on Freedom of Expression signed a joint declaration on access to information, calling on all countries to enact freedom of information laws based on the principle of maximum disclosure. Exceptions to the right to access should be clearly and precisely delimited and countries should set up independent bodies authorized to hear appeals against refusals to make information available.[53]

b. United Nations Educational, Scientific and Cultural Organization (UNESCO)

BACKGROUND – In line with 29 C/Resolution 28 of the UNESCO General Conference of 1997, which invited the Director-General to undertake action to 'facilitate access to information in the public domain with the ultimate aim of building up a general electronic repository of all information of a public nature relevant to UNESCO's fields of competence',[54] UNESCO put the preparation of guidelines on its agenda to advise Member States on policies for the development and promotion of public domain information, taking account of both national needs and international practices. The Policy Guidelines for the

---

52. United Nations Special Rapporteur on Freedom of Opinion and Expression, *Civil and political rights, including the question of freedom of expression. The right to freedom of expression and opinion*, UN Doc. E/CN.4/2005/64, 17 Dec. 2004, 10–11, <http://daccess-ods.un.org/access.nsf/Get?Open&DS=E/CN.4/2005/64&Lang=E>, 22 Feb. 2010.
53. United Nations Special Rapporteur on Freedom of Opinion and Expression, OSCE Representative of the Media and OAS Special Rapporteur on Freedom of Expression and Information, *International Mechanisms for Promoting Freedom of Expression. Joint Declaration*, 6 Dec. 2004, <www.cidh.org/Relatoria/showarticle.asp?artID=319&lID=1>, 22 Feb. 2010. See also Thorsten Ader & Max Schoental, 'Access to Information on Government Action, especially from the Media Point of View', *IRIS plus* no. 2 (2005): 2.
54. UNESCO, *Records of the General Conference. Twenty-ninth session, Paris, 21 October to 12 November 1997. Volume 1. Resolutions*, 58, 1997, <http://ulis2.unesco.org/images/0011/001102/110220e.pdf>, 22 Feb. 2010.

Development and Promotion of Governmental Public Domain Information were presented in 2004.[55]

NEED FOR A COMPREHENSIVE NATIONAL INFORMATION POLICY – The UNESCO policy guidelines make it clear that, in order to fulfil their critical leadership role in expanding access to and use of governmental public domain information,[56] governments need to develop an integrated and comprehensive national information policy to promote the production and the availability of governmental information in the public domain. Three main elements need to be included:

- Scope of information to be made available: as a guiding principle, information produced by public entities in all branches and at all levels should be presumed to be available to the public, and any formal exceptions preventing citizens from accessing public information should be specifically justified and formulated as narrowly as possible.
- Access to and use of public information as a legal principle: countries that do not yet have access legislation in place should adopt it, following a comparative analysis of such similar laws in other countries, while those countries that already do have such a law may wish to revise their existing legislation. Any exceptions to the principle of availability, such as national security restrictions, and the protection of personal privacy and of trade secrets, should be carefully balanced.
- Comprehensive governmental information policy framework: adopting access legislation is not sufficient. In addition, a policy framework for the management and active dissemination of government information is needed, focusing on producing and disseminating information that meets the needs of citizens as openly and inexpensively as possible, with special attention to multicultural or disadvantaged communities.[57]

PRINCIPLES OF ACCESS TO PUBLIC SECTOR INFORMATION – According to the Guidelines, public bodies should make their information broadly and openly available because it is 'an important part of building participatory democracy, fostering open debate, and promoting effective government processes. It also provides citizens with a means to learn about their country, their fellow citizens, and their government that

---

55. UNESCO, 'Policy Guidelines for the Development and Promotion of Governmental Public Domain Information' (Paris: UNESCO, 2004), <http://unesdoc.unesco.org/images/0013/001373/137363eo.pdf>, 22 Feb. 2010.
56. The UNESCO policy guidelines give two possible descriptions of 'public domain information'. On the one hand, it can be defined as what is left outside the scope of copyright or other forms of statutory protection. On the other hand, it also refers to information of an intrinsically public nature; that is, certain types of information that are produced by public authorities ('government' in the broad sense) in the course of their duties, and that are seen as a public good. The latter category is referred to as 'governmental public domain information', which is considered as a subcategory of public sector information, as some public sector information may also be protected by intellectual property rights (*Ibid.*, iv-v). Cf. *supra*, Ch. 2.
57. *Ibid.*, v–vii.

in many cases will not be available from any other source'.[58] With regard to access or freedom of information legislation, the following principles should be adhered to:

- The right of citizens to access public sector information should be recognized in the national constitution, and implemented by statute.
- The type of information that should be actively disseminated should be defined; and the legal value and authenticity of electronic formats should be recognized.
- The interest of the applicant in the information should not be controlled by the public bodies.
- Exceptions to the availability of data should be carefully balanced.
- Time limits should be specified, and the reasons for refusal of a request should be sufficiently detailed, so that the applicant can determine the basis for an appeal of the decision.
- An independent office needs to be established to handle appeals of decisions denying access to the information, often referred to as an 'ombudsman'.[59]

NEED FOR A COMPREHENSIVE POLICY FRAMEWORK – While these principles are important, the Guidelines admit that they are not sufficient to increase the availability of public sector information and transparency in governance. A comprehensive policy framework for the management and active dissemination of public sector information is needed. Three main areas of action are identified.

CREATION OF AN INFORMATION MANAGEMENT STRUCTURE – First, an appropriate information management structure should be created. This includes establishing a high-level executive office for national public information policy providing a comprehensive vision, high-level arbitration, and strong leadership. Next, each public body should assign an Information Officer, who has the prime responsibility for the organization's information management and policy. To create such an information management structure, an effective work force and appropriate budgets are needed.

DEFINITION OF INFORMATION MANAGEMENT POLICY REQUIREMENTS – The second element is the definition of the public information management policy requirements. This includes providing information to the public, such as information that describes the organization, activities, programmes, meetings, systems of records, and any other information that is necessary or appropriate for the proper performance of the public body's functions. When disseminating data, the public bodies should take care to do this in a manner that achieves the best balance between maximizing the usefulness of the data and minimizing the cost to the government and the public. They should disseminate the data on equitable and timely terms and take advantage of all possible and appropriate dissemination channels. Metadata and registers should be created to support these activities. Special attention should

58. *Ibid.*, 9.
59. *Ibid.*, 14–16.

be given to multilingual or disadvantaged communities. The data should be produced and disseminated in appropriate formats. In addition, the public bodies should work with experts to develop information products and services that can meet the special needs of people with physical disabilities. Generally, user charges should be no higher than the costs of dissemination, and all public bodies should develop a comprehensive information resource management plan. Electronic media and formats should be used to make data more easily available and useful to the public.

MANAGEMENT SYSTEMS FOR INFORMATION SYSTEMS AND TECHNOLOGY – Third, strategies on information systems and information technology management should be adopted. These include technical management, strategic and operational management, and monitoring of the performance of the public bodies.[60]

ACCESS OR RE-USE? – In addition to the value of public sector data for democracy, the Guidelines also refer to the importance of public sector data for the growth of the private sector, but it is not entirely clear from the text whether they are referring to the benefits of access or the benefits of re-use. The critical leadership role of the government is recognized in expanding access to and use of public domain information.[61] Throughout the Guidelines, the combination of access and use of data is often mentioned, yet the recommendations only seem to directly address access possibilities. Considering that the document is written for UNESCO by a United States (US) author, it might be that he applied the US approach that the right to access public sector data automatically entails the right to use these data for further purposes.[62] However, in the legal frameworks of many Member States of the EU, this is not necessarily the case, as a distinction is made between applications for access and applications for re-use.

3. Organization for Economic Co-operation and Development

TIMELINE – In March 2006, the Working Party on the Information Economy of the Information, Computer and Communication Policy Committee (ICCP Committee) of the Organization for Economic Co-Operation and Development (OECD) presented a report on Digital Broadband Content: Public Sector Information and Content,[63] which was discussed in a follow-up workshop on 31 May 2006.[64] Based

60. *Ibid.*, 17–27.
61. *Ibid.*, 10.
62. See Johan Pas, 'The Use and Re-use of Public Sector Information (PSI): Some Legal and Policy Reflections', paper given at the Conference of the Canadian Association for Information Science (Winnipeg, 2004), 9, <www.cais-acsi.ca/proceedings/2004/pas_2004.pdf>, 22 Feb. 2010.
63. OECD Working Party on the Information Economy, 'Digital Broadband Content: Public Sector Information and Content', 2006, 82 p., <www.oecd.org/dataoecd/10/22/36481524.pdf>, 22 Feb. 2010.
64. OECD Working Party on the Information Economy, 'OECD Workshop on Public Sector Information held on 31 May 2006: Summary', 84 p., <www.oecd.org/dataoecd/54/35/36854745.pdf>, 22 Feb. 2010.

on this report, the Working Party started drafting principles relating to public sector data in 2007. This work led to a Recommendation for Enhanced Access and More Effective Use of Public Sector Information, adopted by the OECD Council on 30 April 2008.[65]

a. Report on Digital Broadband Content: Public Sector Information and Content

OUTLINE – The 2006 report on Digital Broadband Content: Public Sector Information and Content also addressed access. The report made a distinction between making public sector information available for re-use and the diffusion of public sector content. The former will be addressed in the appropriate section on re-use, while the latter relates more to the availability of public sector data for access, so it will be discussed in this section.

FOCUS ON PUBLIC SECTOR CULTURAL CONTENT – Public sector content was described as static information that is held by the public bodies rather than created by it (e.g., cultural archives and artistic works), not directly associated with the functioning of government, and not necessarily associated with commercial uses but having other public good purposes (such as culture and education). The term referred mostly to cultural, educational, and scientific public knowledge, for which wide public diffusion and long-term preservation are major governmental objectives.[66] Democratic societies are committed to make public sector content available to everyone, including reaching out to socially and economically disadvantaged groups. Additionally, many public bodies also want to stimulate interest and use of cultural and educational resources. While public sector spatial data can in some cases also be of a scientific or educational nature, perhaps even related to cultural content (e.g., the location of cultural heritage), the focus of the report on cultural content makes it less relevant regarding access to public sector spatial data.

ATTENTION FOR PUBLIC SECTOR CONTENT IN GENERAL – However, the report also listed a number of policy issues regarding public sector content in general and presented a number of best practices in making such content available. Measures for increasing the diffusion of public sector content included the use of institutional networks for collaboration across institutions and sectors, the development of portals, metadata and standardization, and the development of technology and human capital. The report also paid attention to the development of communities through which citizens can interact and contribute their own content. This highlights a possible difficulty in distinguishing access from re-use. On-line

---

65. OECD, *Recommendation C(2008)36 of the Council for Enhanced Access and More Effective Use of Public Sector Information*, 2008, <www.oecd.org/dataoecd/0/27/40826024.pdf>, 22 Feb. 2010.

66. OECD Working Party on the Information Economy, 'Digital Broadband Content: Public Sector Information and Content', 2006, 8, <www.oecd.org/dataoecd/10/22/36481524.pdf>, 22 Feb. 2010.

communities with participation from the users often imply that these users add their own contribution to the content provided, which may lead to re-use rather than mere access to the content. The same issue came up when the report discussed the financing of online cultural projects. It referred to business opportunities in online education markets, public private partnerships, and commercialization. All of these topics blur the distinction between access to public sector content and re-use of public sector data.

b. Recommendation for Enhanced Access and More Effective Use of Public Sector Information

PRINCIPLES – The OECD's Seoul Declaration for the future of the Internet economy of 2008 recommended that public sector data and content be made more widely accessible in digital format.[67] The Recommendation for Enhanced Access and More Effective Use of Public Sector Information served as a background for this Declaration.

PRINCIPLES RELATING TO ACCESS – As the Recommendation is mostly directed towards re-use of public sector data, it will be addressed more extensively in that section of this chapter. However, a number of principles are also applicable to access and deserve to be mentioned here. A first reference that is made to access lies in the first principle on openness: the recommendation states that a regime of access principles or assuming openness in public sector information should be developed as a default rule whenever possible, no matter what the level of funding is for the development and maintenance of the information.[68] A next reference can be found in the second principle on access and transparent conditions for re-use, which includes that access to information over the Internet and in electronic form should be improved. However, this seems to be part of the call for making data available for re-use and not just for access, so it is not entirely clear what the value of this principle is for access purposes. On the contrary, the third principle that asset lists need to be created is clearly directed towards both access and re-use. Regarding pricing, the recommendations state that if the data are not free of charge, they should generally be charged at a marginal cost of maintenance and distribution. Pricing policies should be transparent and consistent within and between public bodies. Redress mechanisms should be available, and greater consistency in cross-border access regimes should be sought.

---

67. OECD, *The Seoul Declaration for the future of the internet economy*, 2008, <www.oecd.org/dataoecd/49/28/40839436.pdf>, 22 Feb. 2010. See also Anne Fitzgerald, *Open access policies, practices and licensing: a review of the literature in Australia and selected jurisdictions* (Brisbane: Queensland University of Technology, 2009), 154.
68. OECD Working Party on the Information Economy, 'Digital Broadband Content: Public Sector Information and Content', 2006, 5, <www.oecd.org/dataoecd/10/22/36481524.pdf>, 22 Feb. 2010.

4. Council of Europe

OUTLINE – Up to 2008, the Council of Europe only addressed access to public sector data under the form of non-binding documents from the Council of Ministers and the Parliamentary Assembly. The most important documents were a 1979 recommendation of the Parliamentary Assembly and recommendations of the Committee of Ministers from 1981 and 2002.[69] While these documents were not binding, they did have a large moral authority and, as the European Court of Human Rights has admitted, they can be relevant for the interpretation of the ECHR.[70] In November 2008, the Committee of Ministers adopted the binding Convention on Access to Official Documents, which was opened for signing in June 2009.

a. *Recommendation 854 (1979) of the Parliamentary Assembly on Access by the Public to Government Records and Freedom of Information*[71]

GROUNDS FOR ACCESS – The Parliamentary Assembly of the Council of Europe pointed out that a parliamentary democracy can function only if the people in general and their elected representatives are fully informed,[72] and that: 'In today's society public life has become so complicated and technical that government departments and agencies frequently generate and possess information which cannot be obtained from other sources'. Therefore, it was desirable that, subject to certain inevitable exceptions, the public should have access to government records.[73]

INVITATION TO THE MEMBER STATES TO ALLOW ACCESS – Hence, the Assembly recommended that the Committee of Ministers invite the Member States that had not already done so to introduce a system of freedom of information, that is, access to government files. It also wanted a Committee of Experts on Public Authorities

69. For a more extensive overview of relevant Council of Europe recommendations, see Maguelonne Déjaent-Pons, 'Le Conseil de l'Europe et l'Accès à l'Information Environnementale: la Convention Européenne des Droits de l'Homme', in *Ten Years of Access to Environmental Information in International, European and Belgian Law: Stock-Taking and Perspectives*, ed. Christine Larssen (Brussels: Bruylant, 2003), 65–83 [in French].
70. *Ibid.*, referring to European Court of Human Rights, 9 May 2003, *Tepe* v. *Turkey*, para. 18; European Court of Human Rights, 10 Nov. 2005, *Leyla Sahin v. Turkey*, para. 136; European Court of Human Rights, 9 Nov. 2006, *Leempoel & SA Cine Revue* v. *Belgium*, para. 78, retrieved from <http://cmiskp.echr.coe.int/tkp197/search.asp?skin=hudoc-en>, 22 Feb. 2010. See also Peter Seipel, 'The Technology of Insight: Computers and Informed Citizens', *Chicago-Kent Law Review* 69, no. 2 (1993): 443.
71. Council of Europe Parliamentary Assembly, *Recommendation 854 (1979) of 1 February 1979 on access by the public to government records and freedom of information*, <http://assembly.coe.int/main.asp?Link=/documents/adoptedtext/ta79/erec854.htm>, 22 Feb. 2010.
72. Ton A.L. Beers, 'Public Access to Government Information towards the 21st Century', in *Information Law towards the 21st Century*, ed. Willem F. Korthals Altes (Deventer: Kluwer, 1992), 192.
73. Council of Europe Parliamentary Assembly, *Recommendation 854 (1979) of 1 February 1979 on access by the public to government records and freedom of information*, <http://assembly.coe.int/main.asp?Link=/documents/adoptedtext/ta79/erec854.htm>, 22 Feb. 2010.

and Access to Information to make a full study on the question of access to government files.[74] The Recommendation referred to the active dissemination of information only in relation to the activities of the Committee of Ministers itself, but did not address this issue for the Member States.

b. Recommendation No. R(81)19 of the Committee of Ministers on the Access to Information Held by Public Authorities[75]

GROUNDS FOR ACCESS – The Committee of Ministers considered that adequate information on public issues is important for the public in a democratic society, or as Gavouneli formulates it, 'democratic decision-making cannot be achieved without adequate information being available to the public',[76] and access to information held by public bodies is likely to strengthen the confidence of the public in the administration. Therefore, '[t]he utmost endeavour should be made to ensure the fullest possible availability to the public of information held by public authorities'.[77]

GENERAL PRINCIPLES – R(81)19 recommended that the governments of Member States be guided in their law and practices by a number of basic principles recognizing and organizing the right to access public information. Every legal or natural person within the jurisdiction of a Member State should have the right to obtain, upon request and without having to state an interest, information held by public bodies other than legislative bodies and judicial authorities. Limitations and restrictions to this right of access had to be necessary in a democratic society for the protection of legitimate public or private interests. Refusals had to be motivated and subject to review on request.[78] The Recommendation did not address the active dissemination of public sector information.

IMPACT OF R(81)19 – Even though the recommendation had considerable moral authority, Seipel felt that the Member States did not appear to have been in haste to turn it into national law. Italy and Luxembourg even explicitly reserved the right of their governments to comply with it or not.[79]

---

74. From the text of Recommendation 854(1979) it seems that the Parliamentary Assembly identifies freedom of information with access to public sector information ('a system of freedom of information, i.e., access to government files'). While narrowing the meaning of freedom of information considerably, this also shows that the Assembly takes a different position regarding the scope of Art. 10 ECHR than the European Court of Human Rights. The Assembly and the Committee of Ministers are more inclined to interpret Art. 10 ECHR as including an obligation of the public authorities to make their information available.
75. Council of Europe Committee of Ministers, *Recommendation No. R(81)19 of 25 November 1981 to Member States on the Access to Information held by public authorities*, <www.coe.int/t/e/legal_affairs/legal_co-operation/administrative_law_and_justice/Texts_&_Documents/Conv_Rec_Res/Recommendation(81)19.asp>, 22 Feb. 2010.
76. Maria Gavouneli, 'Access to Environmental Information: Delimitation of a Right', *Tulane Environmental Law Journal* 13 (2000): 323.
77. *Ibid.*
78. *Ibid.*
79. Peter Seipel, 'The Technology of Insight: Computers and Informed Citizens', *Chicago-Kent Law Review* 69, no. 2 (1993): 442.

c. *Recommendation Rec(2002)2 of the Committee of Ministers to Member States on Access to Official Documents*[80]

ACTIVE AND PASSIVE ACCESS – On 21 February 2002, the Committee of Ministers adopted a new recommendation on access to public sector documents. Rec(2002)2 recognized both an enforceable subjective right of citizens to have access upon request to official documents and a positive obligation of the authorities to provide, on their own initiative, the public with relevant documents in matters of public interest.[81]

GROUNDS FOR ACCESS – The Recommendation considered that wide access to official documents on a basis of equality and in accordance with clear rules 'allows the public to have an adequate view of, and to form a critical opinion on, the state of the society in which they live and on the authorities that govern them, whilst encouraging informed participation by the public in matters of common interest', 'fosters the efficiency and effectiveness of administrations and helps maintain their integrity by avoiding the risk of corruption', and 'contributes to affirming the legitimacy of administrations as public services and to strengthening the public's confidence in public authorities'.[82] The Recommendation repeated the statement of R(81)19 that: 'The utmost endeavour should be made to ensure the fullest possible availability to the public of information held by public authorities'.[83]

GENERAL PRINCIPLES: ACCESS ON REQUEST – The principles of this recommendation were intended as a minimum standard, without prejudice to national legislation recognising a wider right of access.[84] The main principle of the recommendation recognized a general right of access to public sector documents: the Member States should guarantee the right of everyone to have access upon request to official documents held by public authorities, without discrimination on any ground, including national origin. An applicant should not be obliged to give reasons for requesting access and formalities should be kept to a minimum. Limitations on access should be set down precisely in law, be necessary in a democratic society, and be proportionate to the aim of protecting a number of interests mentioned in the recommendation, such as national security, public safety, nature, and so on. Once again, the recommendation made it clear that any refusal has to be motivated and mention the means of redress.

---

80. Council of Europe Committee of Ministers, *Recommendation Rec(2002)2 of 21 February 2002 on access to official documents*, <www.coe.int/T/E/Human_rights/rec(2002)2_eng.pdf>, 22 Feb. 2010.
81. Wouter Hins and Dirk Voorhoof, 'Access to State-Held Information as a Fundamental Right under the European Convention on Human Rights', *European Constitutional Law Review* 3, no. 1 (2007): 116–117.
82. Council of Europe Committee of Ministers, *Recommendation Rec(2002)2 of 21 February 2002 on access to official documents*, <www.coe.int/T/E/Human_rights/rec(2002)2_eng.pdf>, 22 Feb. 2010.
83. *Ibid.*
84. Katleen Janssen, Jo Steyaert and Roland Van Gompel, *Transparantie van overheidsinformatie in Vlaanderen. Beschouwingen en aanbevelingen* (Bruges: Die Keure, 2003), 69 [in Dutch].

GENERAL PRINCIPLES: ACTIVE DISSEMINATION – With regard to active dissemination of public sector documents, Rec(2002)2 stated in the preamble that: 'Public authorities should commit themselves to conducting an active communication policy, with the aim of making available to the public any information which is deemed useful in a transparent democratic society'.[85] The eleventh and final recommendation of the document dealt with documents made public at the initiative of the public authorities: 'A public authority should, at its own initiative and where appropriate, take the necessary measures to make public information which it holds available when the provision of such information is in the interest of promoting the transparency of public administration and efficiency within administrations or will encourage informed participation by the public in matters of public interest'.[86] The public authorities were free to choose the most appropriate means of publishing information (billboards, official publications, websites, or any other medium easily accessible to the public). As far as possible, the public authorities should make available information on the matters or activities for which they were responsible, for example by drawing up lists or registers of the documents they hold.

d. Convention on Access to Official Documents[87]

TIMELINE – Building on Recommendation Rec(2002)2, the Committee of Ministers of the Council of Europe instructed the Steering Committee for Human Rights to draft a convention on access to official documents.[88] The Convention will be the first international binding instrument that recognizes a general right of access to official documents held by public authorities. The final version of the Convention was adopted by the Council of Ministers on 27 November 2008. It was opened for signature by the Member States in June 2009.[89] By February 2010, the Convention was signed by twelve countries and ratified by two.[90]

---

85. Council of Europe Committee of Ministers, *Recommendation Rec(2002)2 of 21 February 2002 on access to official documents*, <www.coe.int/T/E/Human_rights/rec(2002)2_eng.pdf>, 22 Feb. 2010.
86. *Ibid.*
87. Council of Europe, *Convention on Access to Official Documents of 27 November 2008*, <https://wcd.coe.int/ViewDoc.jsp?id=1377737&Site=CM>, 22 Feb. 2010.
88. Steering Committee for Human Rights, *Draft Convention of the Council of Europe on Access to Official Documents and Draft Explanatory Report*, doc. nr. CDDH(2008)008 Add. III Bil., 3 Apr. 2008, 4, <http://gyla.ge/foi/doc_upload/capmaign/3.%20evropis%20sabchos%20konvenciis%20proeqti%20inglisur%20enaze.pdf>, 22 Feb. 2010.
89. See Council of Europe, '29th Conference of Council of Europe Ministers of Justice', <www.coe.int/t/dghl/standardsetting/minjust/mju29/>, 22 Feb. 2010.
90. Belgium, Estonia, Finland, Georgia, Hungary, Lithuania, Montenegro, Norway, Serbia, Slovenia, Sweden and the former Yugoslav Republic of Macedonia have signed the Convention. Norway and Hungary have ratified it. The United Kingdom has indicated that it currently does not intend to sign the Convention (see Council of Europe, 'Council of Europe Convention on Access to Official Documents. Signatures and ratifications', <http://conventions.coe.int/Treaty/Commun/ChercheSig.asp?NT=205&CM=8&DF=10/02/2010&CL=ENG>, 22 Feb. 2010.

Background for the drafting procedure – The Steering Committee was guided by the concern to identify among the various national legal systems a core of basic obligatory provisions reflecting what was already accepted in the legislation of a number of countries and that at the same time could be accepted by States that did not have such legislation.[91] Rather than bringing together the best practices in the field of access to public sector data, the Committee chose to elaborate an instrument that was acceptable to the greatest number of Member States of the Council of Europe.

General principle: access upon request – The Parties to the Convention have to guarantee the right of everyone, without discrimination on any ground, to have access upon request to official documents held by public authorities. They also have to take the necessary measures in domestic law to give effect to the provisions set out in the Convention, including adopting or changing access laws. However, the principles of the Convention should remain without prejudice to wider rights of access recognized by national or international legislation.

Limitations on access – The Member States may limit the right to access official documents, and the limitations have to be set down precisely in law, be necessary in a democratic society, and be proportionate to the aim of protecting a list of eleven different interests, such as national security, defence and international relations; public safety; privacy, and other legitimate private interests; or the environment. Access can only be refused if disclosure would harm or would be likely to harm any of the protected interests, unless there is an overriding interest in disclosure.[92] The list of protected interests is exhaustive: Member States cannot add to it, but they can choose to reduce the limitations or formulate them more narrowly. They should also consider time limits beyond which the limitations would no longer apply.

Procedure – The public authority has to help the applicant, as far as reasonably possible, to identify the requested official document. Requests have to be dealt with promptly, and the decision should be reached, communicated, and executed as soon as possible or within a reasonable time limit that has been specified beforehand.[93] If a request is refused, the public authority has to give the reasons for the refusal and, upon request from the applicant, a written justification. Where the applicant who received the document is unable to obtain a basic understanding of its content, the public authority should help him or her as far as it is possible and reasonable. However, help with comprehension does not imply an obligation to translate the documents or to provide highly specialized technical or legal explanations.[94] Next, the Member States to the Convention have to inform the public

---

91. Steering Committee for Human Rights, *Draft Convention of the Council of Europe on Access to Official Documents and Draft Explanatory Report*, Doc. nr. CDDH(2008)008 Add. III Bil., 3 Apr. 2008, 4, <http://gyla.ge/foi/doc_upload/capmaign/3.%20evropis%20sabchos%20konvenciis%20proeqti%20inglisur%20enaze.pdf> 22 Feb. 2010.
92. Article 3 of the Convention on Access to Official Documents.
93. Article 5 of the Convention on Access to Official Documents.
94. Steering Committee for Human Rights, *Draft Convention of the Council of Europe on Access to Official Documents and Draft Explanatory Report*, doc. nr. CDDH(2008)008 Add. III

about its right of access and how to exercise it. They have to take appropriate measures to educate the public authorities on how to apply the right of access; to provide information on the matters or activities for which they are responsible; to manage their documents efficiently so that they are easily accessible; and to apply clear and established rules for the preservation and destruction of their documents.[95]

Charges – Inspection of the document on the premises of the public authority has to be free of charge. However, archives and museums are not prevented from establishing charges. A fee can be charged for the provision of a copy of the document, which should be reasonable and not exceed the actual costs of reproduction and delivery of the document. The tariffs for the charges shall be published.[96]

General principle: active dissemination – Public authorities also have to make available official documents they hold at their own initiative, in the interest of promoting the transparency and efficiency of public administration and to encourage informed participation by the public in matters of general interest.[97] While the explanatory report to Article 9 states that national rules on proactive publication are encouraged, the obligations imposed on the Member States and the public authorities are very limited.

Criticism on the Convention – While the Convention was set up to be acceptable for as many Member States as possible, it was still subject to a lot of criticism from non-governmental organizations and other action groups defending freedom of information and the right of access. The comments were mainly directed towards the scope of the Convention, the means of redress, and the absence of time limits for the response to requests for access. For instance, the Office of the Representative on Freedom of the Media from the Organization for Security and Co-operation in Europe (OSCE) stated that the Convention should require all public bodies to affirmatively publish data about their structures, personnel, activities, rules, guidance, decisions, procurement, and other data of public interest on a regular basis in formats, including the use of information and communication technologies and in public reading rooms or libraries to ensure easy and widespread access.[98] In September 2008, the Committee on Legal Affairs and Human Rights of the Parliamentary Assembly also gave its opinion on the draft Convention. The Committee also found a number of shortcomings: 'The definition of "public authorities" is too narrow, allowing some public bodies to continue

---

Bil., 3 Apr. 2008, 4, <http://gyla.ge/foi/doc_upload/capmaign/3.%20evropis%20sabchos%20 konvenciis%20proeqti%20inglisur%20enaze.pdf>, 22 Feb. 2010.

95. Article 9 of the Convention on Access to Official Documents.
96. Article 7 of the Convention on Access to Official Documents.
97. Article 10 of the Convention on Access to Official Documents.
98. OSCE Office of the Representative on Freedom of the Media, *Comments on the draft European Convention on access to official documents of the Council of Europe, as prepared by the Group of Specialists on Access to Official Documents*, 11 Sep. 2007, <www.access-info.org/data/File/ OSCE%20Comments%20to%20CoE%20access%20convention%2011SEP07.doc>, 22 Feb. 2010.

operating in the shadows, no time-limits are laid down, which would allow public bodies to delay publication of awkward data until it had lost its value, and review bodies are not given the power to order disclosure of a requested official document where access has initially been refused. Finally, Member States can opt out of parts of the treaty by entering wide-ranging reservations when they ratify it'.[99] Therefore, the Committee recommended sending the draft back to the experts for further consideration. However, the Committee of Ministers approved the Convention without any changes.

5. European Union

PUBLIC SECTOR DATA HELD BY MEMBER STATES AND PUBLIC SECTOR DATA HELD BY EU INSTITUTIONS – In the analysis of legal and policy sources on the level of the EU on the obligations of public bodies to make public sector data available for access, one has to make a distinction according to the target group of the rule. On the one hand, the institutions and bodies of the EU might impose obligations regarding access on the Member States and their public bodies in directives, regulations, or decisions, while on the other hand, these institutions are themselves the holders of an enormous amount of public sector data. Access to EU institutions' own data also has to be regulated, especially considering the growing importance of the competences of this supranational level of decision making.

OUTLINE – The obligations of the Member States to make public sector spatial data available for access under the Access Directive and the INSPIRE Directive have already been addressed in Chapter 3, so this section pays more attention to other EU documents that provide indications on how far the obligation of the public bodies to make public sector spatial data available should go. But first, I examine the obligations of the EU institutions to make their data available for access. This can also provide us with a view on the scope of access to public sector data.

a. Access to Public Sector Data Held by the European Union Institutions and Bodies

i. Article 15 of the Treaty on the Functioning of the European Union (Former Article 255 of the EC Treaty)

GENERAL RIGHT OF ACCESS – Before the Treaty of Lisbon, a right of access to documents held by the European Parliament, the European Council, and the European Commission was recognized in Article 255 of the EC Treaty: 'Any citizen of the Union, and any natural or legal person residing or having its registered office in a Member State, shall have a right of access to European Parliament, Council and

99. Committee on Legal Affairs and Human Rights of the Parliamentary Assembly, *Draft Council of Europe Convention on Access to Official Documents. Report*, 12 Sep. 2008, <http://assembly.coe.int/Main.asp?link=/Documents/WorkingDocs/Doc08/EDOC11698.pdf>, 22 Feb. 2010.

Commission documents'.[100] The Council had to lay down general principles and limits on grounds of public or private interest governing this right of access within two years of entry into force of the Treaty of Amsterdam, when Article 255 was introduced.[101] According to Schram, due to the need for legislation to implement Article 255, the right to access may be a fundamental right, but only an indirect one.[102]

BACKGROUND – The introduction of Article 255 of the EC Treaty was part of the more general move of the EU towards more openness and transparency. This process was laid down in Article 1 of the EU Treaty: 'This Treaty marks a new stage in the process of creating an ever closer union among the peoples of Europe, in which decisions are taken as openly as possible and as closely as possible to the citizen'. According to Curtin, the evolution of this principle of open decision making should be seen within the context of the evolution of the democratic principle at the EU level.[103]

---

100. The introduction of Art. 255 in the EC-Treaty in 1998 does not mean that the public could not get access to the information of the European Parliament, the Council and the European Commission before. The Council and the European Commission already signed a joint Code of Conduct on 6 Dec. 1993 regarding the access to their documents (Council and Commission of the European Communities, *Code of conduct concerning public access to Council and Commission documents, OJ L* 340, 31 Dec. 1993, 41–42). This Code of Conduct contained the conditions for the citizen to obtain access to documents, the possible means of redress against refusals, and a number of exceptions to the right of access. It was implemented by the Council in Council Decision 93/731/EC of 20 Dec. 1993 on public access to Council documents (*OJ L* 340, 31 Dec. 1993, 43–44), and by the European Commission in Commission Decision 94/90/ECSC, EC, Euratom of 8 Feb. 1994 on public access to Commission documents, *OJ L* 46, 18 Feb. 1994, 58–59. The European Parliament only introduced its own formal procedure 4 years later, with European Parliament Decision 97/632/EC, ECSC, Euratom of 10 Jul. 1997 on public access to European Parliament documents (*OJ L* 263, 25 Sep. 1997, 27). The other institutions and bodies of the EU did not have any formal policy on public access to their documents (see Bernard Vanheusden, 'Vrije toegang tot milieu-informatie: een stand van zaken en de implicaties van het verdrag van Aarhus voor Europa en voor België', *Milieu- en Energierecht* (1999): 272–275 [in Dutch]; Ralph E. Hallo, 'The EU Institutions', in *Access to environmental information in Europe: the implementation and implications of directive 90/313/EEC*, ed. Ralph E. Hallo (London: Kluwer Law International, 1996), 409–422; Maria Gavouneli, 'Access to Environmental Information: Delimitation of a Right', *Tulane Environmental Law Journal* 13 (2000): 307–312; Mireille Van Eechoud, 'Nader tot U – Brussel verordent openbaarheid EU-documenten', *Mediaforum* 12, no. 3 (2000): 73–79 [in Dutch]; Geert Van Calster, 'Openbaarheid van bestuur in de Europese Unie: recente ontwikkelingen', *Rechtskundig Weekblad* 58, no. 36 (1994–1995): 1209–1219 [in Dutch]; Rik De Baere, 'Openbaarheid van milieu-informatie', *Tijdschrift voor Milieurecht* 4, no. 2 (1995): 90–98 [in Dutch]; Jean-Jacques Paradissis, *The Right to Access Environmental Information* (Saarbrucken: VDM Verlag, 2009), 44).
101. The Treaty of Amsterdam entered into force on 1 May 1999, so the Council had to act by 31 Apr. 2001.
102. Frankie Schram, 'Openbaarheid van Europese bestuursdocumenten', *Nieuw Juridisch Weekblad* no. 32 (2003): 583 [in Dutch].
103. Deirdre M. Curtin, 'Citizens' Fundamental Right of Access to EU Information: an Evolving Digital Passepartout?', *Common Market Law Review* 37, no. 1 (2000): 14.

Extension in Treaty of Lisbon – When the Treaty of Lisbon entered into force, the right of access was extended. Article 255 of the EC Treaty was replaced by Article 15 of the Treaty on the Functioning of the European Union, which extends the right of access to documents of all the institutions and bodies of the European Union, and not just the European Parliament, the Council and the Commission. In addition, the European Parliament and the Council have to ensure the publication of documents relating to the legislative procedures.

ii. Article 42 of the Charter of Fundamental Rights of the EU

Originally non-binding general right – This article recognizes a general right of access to documents: 'Any citizen of the Union, and any natural or legal person residing or having its registered office in a Member State, has a right of access to European Parliament, Council and Commission documents'.[104] The Charter was not formally binding – according to Birkinshaw largely as a result of British opposition[105] – so it was of less importance than Article 255 of the EC Treaty. However, with the Treaty of Lisbon, the new Article 6 of the Treaty on European Union recognizes the rights, freedoms, and principles set out in the Charter and gives the Charter the same legal value as the Treaties themselves.

iii. Regulation 1049/2001 Regarding Public Access to European Parliament, Council, and Commission Documents[106]

Timeline – On 30 May 2001, the Council and the European Parliament adopted Regulation 1049/2001 to implement Article 255 of the EC Treaty (current Article 15 of the Treaty on the Functioning of the European Union).[107] The deadline imposed by Article 255 forced the institutions to act quickly. The European Parliament, the European Commission, and the Council set up a 'trialogue' to sidestep the normal co-decision procedure, entailing that there was no conciliation procedure, despite misgivings by the Parliament about parts of the Regulation.[108]

---

104. Charter of Fundamental Rights of the European Union, *OJ C* 364, 18 Dec. 2000, 1.
105. Patrick Birkinshaw, 'Freedom of Information in the UK and Europe: Further Progress?', *Government Information Quarterly* 19, no. 1 (2002): 82.
106. Regulation 1049/2001/EC of the European Parliament and the Council of 30 May 2001 regarding public access to European Parliament, Council and Commission documents, *OJ L* 145, 31 May 2001, 43.
107. For more information on Regulation 1049/2001, see Rhiannon Williams, 'Access to Documents Held by European Institutions with Particular Regard to Environmental Infringement Procedures, or the Emperor's New Clothes, and the Chimera of "Amicable Dispute Resolution" ', in *Ten Years of Access to Environmental Information in International, European and Belgian Law: Stock-Taking and Perspectives*, ed. Christine Larssen (Brussels: Bruylant, 2003), 85–109; Frankie Schram, 'Openbaarheid van Europese bestuursdocumenten', *Nieuw Juridisch Weekblad* no. 32 (2003): 582–592 [in Dutch]; Herke Kranenborg and Wim Voermans, *Access to Information in the European Union. A Comparative Analysis of EC and Member State Legislation* (Groningen: Europa Law Publishing, 2005), 129 p.
108. Rhiannon Williams, 'Access to Documents Held by European Institutions with Particular Regard to Environmental Infringement Procedures, or the Emperor's New Clothes, and the

FIELD OF APPLICATION – The Regulation only covers documents[109] held by the Commission, the European Parliament, and the Council. At the moment, there is no other legal obligation in EU law to guarantee access to public sector data held by the other institutions and bodies of the EU (e.g., the European Central Bank, the ECJ, and the European Environment Agency), and they accord access to data on a merely discretionary basis.[110] However, the Parliament, Council, and Commission issued a joint declaration stating that 'the agencies and similar bodies created by the legislator should have rules on access to their documents which conform to those of this Regulation'.[111] Therefore, the Commission intended to propose amendments to the acts establishing the existing agencies and bodies and introduce provisions in acts establishing new agencies and bodies. The existing institutions and bodies were encouraged to adopt internal rules on public access to documents which take account of the principles and limits in Regulation 1049/2001. Some of the institutions and bodies answered by adopting their own rules of procedure based on the Regulation.[112]

LISBON TREATY – Now that the Lisbon Treaty has entered into force and Article 16 of the Treaty on the Functioning of the European Union has extended the right of access to documents of all Union institutions, bodies, offices, and agencies,[113] the Regulation will have to be extended accordingly.[114]

GENERAL PRINCIPLE – Any citizen of the Union, and any natural or legal person residing or having its registered office in a Member State, has a right of access to documents falling under Regulation 1049/2001, without having to state his or her interest. The institutions may also grant access to any other natural or legal persons. The Regulation applies to all documents held by the European Parliament, the

Chimera of "Amicable Dispute Resolution" ', in *Ten Years of Access to Environmental Information in International, European and Belgian Law: Stock-Taking and Perspectives*, ed. Christine Larssen (Brussels: Bruylant, 2003), 86–87.

109. Document is defined as 'any content whatever its medium (written on paper or stored in electronic form or as a sound, visual or audiovisual recording) concerning a matter relating to the policies, activities and decisions falling within the institution's sphere of responsibility (Art. 3(a) of Regulation 1049/2001).

110. Vera Rodenhoff, 'The Aarhus Convention and its Implications for the "Institutions" of the European Community', *RECIEL* 11, no. 3 (2002): 351–352.

111. European Parliament and Council, *Joint declaration relating to Regulation 1049/2001/EC of the European Parliament and of the Council of 30 May 2001 regarding public access to European Parliament, Council and Commission documents*, *OJ L* 173, 27 Jun. 2001, 5.

112. For example, the Committee of Regions reacted to the joint declaration by adopting Committee of the Regions Decision 64/2003 of 11 Feb. 2003 on public access to Committee of the Regions documents (*OJ L* 160, 28 Jun. 2003, 96). The European Economic and Social Committee adopted Decision 2003/603/EC of the European Economic and Social Committee of 1 Jul. 2003 on public access to European Economic and Social Committee (EESC) documents (*OJ L* 205, 14 Aug. 2003, 19). The European Central Bank followed in 2004, with Decision of the European Central Bank ECB/2004/3 of 4 Mar. 2004 on public access to European Central Bank documents (*OJ L* 80, 18 Mar. 2004, 82).

113. Cf. *supra*, s. II.C.5.a.i.

114. Cf. *infra*, s. II.C.5.a.iv.

Council, or the European Commission, not only the documents they created, but also the documents received by them or in their possession, in all areas of activity of the European Union.

LIMITATIONS – Article 4 of Regulation 1049/2001 contains the exceptions that the institutions may call upon to refuse access. The Regulation does not mention that the exceptions have to be interpreted restrictively, but this is a well established principle of EU law, so it does not need an explicit reference.[115] First, the institutions covered in the Regulation must refuse access to a document where disclosure would undermine the protection of a number of public interests (public security, defence and military matters, international relations, and the financial, monetary, or economic policy of the Community or a Member State). This also applies to the protection of privacy and the integrity of the individual. The second group of exceptions requires a weighing of interests. The institutions must refuse access to a document where disclosure would undermine the protection of particular interests (commercial interests of a natural or legal person, including intellectual property, court proceedings, and legal advice, and the purpose of inspections, investigations, and audits), unless there is an overriding public interest in disclosure.

TIME LIMIT ON LIMITATIONS – An original element of the Regulation is the time limit imposed on the exceptions. They only apply for the period during which protection is justified on the basis of the content of the document, and for a maximum period of thirty years. If necessary, the exceptions may apply for longer in the case of documents covered by the exceptions relating to privacy or commercial interests and in the case of sensitive documents.[116]

DOCUMENTS ORIGINATING FROM THIRD PARTIES – Since the Regulation applies to all documents held by the European Parliament, the Commission, and the Council, it also includes documents that have not been created by these institutions themselves, but by third parties, including the Member States, other Community or non-Community institutions and bodies, and third countries.[117] In cases when such third-party documents are requested, the institutions have to consult that third party with a view to assessing whether one of the exceptions is applicable, unless it is clear that the document shall or shall not be disclosed. The wording 'consult with a view to assessing' appears to indicate that the ultimate decision as to whether to disclose remains with the institution.[118]

---

115. Pedro Cabral, 'Access to Member State Documents in EC Law', *European Law Review* 31, no. 3 (2006): 381.
116. Article 4.7 of Regulation 1049/2001.
117. Article 3(b) of Regulation 1049/2001.
118. Rhiannon Williams, 'Access to Documents Held by European Institutions with Particular Regard to Environmental Infringement Procedures, or the Emperor's New Clothes, and the Chimera of "Amicable Dispute Resolution" ', in *Ten Years of Access to Environmental Information in International, European and Belgian Law: Stock-Taking and Perspectives*, ed. Christine Larssen (Brussels: Bruylant, 2003), 94. See also Frankie Schram, 'Openbaarheid van Europese bestuursdocumenten', *Nieuw Juridisch Weekblad* no. 32 (2003): 585 [in Dutch].

DOCUMENTS ORIGINATING FROM MEMBER STATES – Even if no exception could be invoked, the Member States[119] can still request the institutions not to disclose a document originating from them without their prior agreement.[120] Under Article 5, they do not explicitly have to justify this request.[121] The Court of First Instance (CFI) decided in *IFAW Internationaler Tierschutz-Fonds v. Commission of the European Communities* that the Member States had a right of veto: 'Thus, a request made by a Member State under Article 4(5) does constitute an instruction to the institution not to disclose the document in question'.[122] The CFI's decision was annulled by the ECJ in December 2007. The ECJ stated that to interpret Article 4(5) as 'conferring on the Member State a general and unconditional right of veto, so that it can oppose, in an entirely discretionary manner and without having to give reasons for its decision, the disclosure of any document held by a Community institution simply because it originates from that Member State, is not compatible with the objectives' of the Regulation.[123] The Member State can only refuse access for one of the exceptions that are listed in the regulation.[124]

PROCEDURE – Applications for access have to be handled promptly, and within fifteen working days from registration of the application, the institution shall grant and provide access or, in a written reply, state the reasons for the refusal and the means of redress. The time limit may be extended by another fifteen working days if the application relates to a very long document or a very large number of documents. If a request is approved, access shall be granted either by providing a copy of the document, including an electronic copy if it is available, or by enabling consultation on the spot, according to the applicant's preference.

---

119. The Court of First Instance has had the opportunity to address the scope of the term Member State in three cases: CFI, Case T-76/02, *Mara Messina v. Commission of the European Communities*, 17 Sep. 2003, *European Court Reports* 2003, II-3203; CFI, Case T-168/02, *IFAW Internationaler Tierschutz-Fonds v. Commission of the European Communities*, 30 Nov. 2004, *European Court reports* 2004, II-4135; CFI, Case T-187/03, *Isabella Scippacercola v. Commission of the European Communities*, 17 Mar. 2005, *European Court reports* 2005, II-1063. For a critical comment on the interpretation by the CFI, see Pedro Cabral, 'Access to Member State Documents in EC Law', *European Law Review* 31, no. 3 (2006): 378–389.
120. Article 4.5 of Regulation 1049/2001.
121. Herke Kranenborg and Wim Voermans, *Access to Information in the European Union. A Comparative Analysis of EC and Member State Legislation* (Groningen: Europa Law Publishing, 2005), 34. See also Rhiannon Williams, 'Access to Documents Held by European Institutions with Particular Regard to Environmental Infringement Procedures, or the Emperor's New Clothes, and the Chimera of "Amicable Dispute Resolution" ', in *Ten Years of Access to Environmental Information in International, European and Belgian Law: Stock-Taking and Perspectives*, ed. Christine Larssen (Brussels: Bruylant, 2003), 94.
122. CFI, Case T-168/02, *IFAW Internationaler Tierschutz-Fonds v. Commission of the European Communities*, 30 Nov. 2004, para. 58, *European Court reports* 2004, II-4135. See also Pedro Cabral, 'Access to Member State Documents in EC Law', *European Law Review* 31, no. 3 (2006): 378–389.
123. ECJ, Case C-64/05 P, *Sweden v. Commission*, 18 Dec. 2007, para. 58, *European Court Reports* 2007, I-11389.
124. *Ibid.*, para. 88.

CHARGES – The cost of producing and sending the copies may be charged for, but the charge cannot exceed the real cost of producing and sending the copies. Consultation on the spot, copies of less than twenty A4 pages and direct access in electronic form or through the register has to be free of charge. Documents have to be supplied in an existing version and format, with regard to the applicant's preference. No new versions or formats have to be created.[125]

ACTIVE DISSEMINATION – The Regulation also contains obligations for the institutions to make their data available of their own motion – often referred to as active publicity. They have to provide public access to a register of documents; as far as possible make documents directly accessible to the public in electronic form or through a register, in particular legislative documents and, where possible, other documents such as documents relating to the development of policy or strategy; and take the requisite measures to inform the public of the rights they enjoy under the Regulation.[126] Dissemination of data of the institutions is also foreseen via the *Official Journal.*

RULES OF PROCEDURE – The European Parliament, the Council, and the European Commission had to adapt their own rules of procedure to the provisions of the Regulation. The Council did this in a new decision of 29 November 2001,[127] the Parliament Bureau on 28 November 2001,[128] and the Commission on 3 December 2001.[129]

iv. Towards a Review of Regulation 1049/2001: the Green Paper on Public Access to Documents Held by Institutions of the European Community[130]

TIMELINE – On 18 April 2007, the European Commission issued a Green Paper as part of a review process of the impact of Regulation 1049/2001. In turn, the review of this Regulation is part of the European Transparency Initiative, launched by the Commission on 9 November 2005.[131]

ASSESSMENT BY THE COMMISSION – Despite the Commission's impression that Regulation 1049/2001 was having the desired effect,[132] the European Parliament

125. Article 10 of Regulation 1049/2001.
126. Article 11-14 of Regulation 1049/2001.
127. Council Decision of 29 Nov. 2001 amending the Council's Rules of Procedure, *OJ L* 313, 30 Nov. 2001, 40.
128. European Parliament Bureau Decision of 28 Nov. 2001 on Public Access to European Parliament Documents, *OJ C* 374, 29 Dec. 2001, 1.
129. Commission Decision of 5 Dec. 2001 amending its rules of procedure, *OJ L* 345, 29 Dec. 2001, 94.
130. Commission of the European Communities, *Green Paper on Public Access to Documents Held by Institutions of the European Community*, COM (2007) 185 final, 18 Apr. 2007.
131. The President of the European Commission, Margot Wallstrom, Siim Kallas, Danuta Hübner and Mariann Fischer Boel, *Communication to the Commission proposing the launch of a European Transparency Initiative*, <http://ec.europa.eu/commission_barroso/kallas/doc/etik-communication_en.pdf>, 22 Feb. 2010. See also Commission of the European Communities, *Green Paper on European Transparency Initiative*, COM (2006) 194 final, 3 May 2006.
132. The number of access requests has been increasing with ca. 50% yearly since the introduction of Regulation 1049/2001, mainly due to the enormous increase in the requests to the

asked the Commission to come forward with proposals for amending the Regulation.[133] After a new assessment of the Regulation, the Commission felt that there was room for improvement in several areas, such as the scope of the registers, in particular those of the Commission; the number of documents or the amount of data made directly accessible to the public; and the user-friendliness of the electronic information systems.[134] Moreover, the relationship with Regulation 1367/2006 on the application of the Aarhus Convention to the Community Bodies and Institutions[135] should be clarified.

RESPONSES TO CONSULTATION – A public consultation was launched that was open until mid July 2007. The Commission received eighty-one contributions from civil society, public bodies, the corporate world, and individual citizens.[136] A large majority of all categories of respondents considered that information from the institutions and bodies was difficult to find on the registers and websites. The scope of the registers and websites was also considered insufficient by half of the respondents. Regarding the active dissemination of information, opinions were divided. The vast majority of civil society organizations were in favour of more active dissemination, and public bodies unanimously supported more proactive transparency. The corporate sector, on the other hand, did not see the need for such a policy, and citizen respondents were divided on the issue.

COMMISSION PROPOSAL – The European Commission issued its proposal for a recast of the Regulation in April 2008. The proposal intended to update Regulation 1049/2001 following the entry into force of Regulation 1367/2006, taking into account recent case law on transparency and access from the ECJ and the CFI; and to address problems with regard to the handling of requests.[137] The proposed amendments address the purpose and beneficiaries of the regulation, the scope and

---

Commission (from 991 in 2002 to 6712 in 2006). Overall, the Council and the Commission grant two out of three applications for access, while Parliament grants more than 80% of the requests it receives. Globally, about 4% of all initial requests lead to confirmatory requests (i.e., applications asking the institution to reconsider a refusal to grant access). Applications to the Court and complaints to the European Ombudsman represent even smaller margins of the total number of requests handled (See Commission of the European Communities, *Green Paper on European Transparency Initiative*, COM (2006) 194 final, 3 May 2006).

133. European Parliament, *Resolution with recommendations to the Commission on access to the institutions' texts*, *OJ C* 293E, 2 Dec. 2006, 151.

134. Commission of the European Communities, *Green Paper on European Transparency Initiative*, COM (2006) 194 final, 6, 3 May 2006.

135. Regulation 1367/2006/EC of the European Parliament and of the Council of 6 Sep. 2006 on the application of the provisions of the Aarhus Convention to Community institutions and bodies, *OJ L* 264, 25 Sep. 2006, 13. Cf. *infra*, s. I.D.4.a.

136. Commission of the European Communities, *Commission Staff Working Document. Report on the Outcome of the Public Consultation on the Review of Regulation (EC) No 1049/2001 regarding public access to European Parliament, Council and Commission documents*, SEC (2008) 29/2, 11 Jan. 2008.

137. General Secretariat of the Council of the European Union, *Report to the Delegations. Proposal for a Regulation of the European Parliament and of the Council regarding public access to European Parliament, Council and Commission documents (recast)*, doc. nr. 5671/09, 26 Jan. 2009, <http://register.consilium.europa.eu/pdf/en/09/st05/st05671.en09.pdf>, 22 Feb. 2010.

definition, the exceptions, consultations with third parties, the procedural rules, and active dissemination.[138] After discussing the proposal in a first reading in March 2009, the European Parliament decided to postpone its final conclusion of this reading to the next legislative term.[139] At the time of writing in February, 2010, the European Parliament had not yet delivered its final opinion in the first reading. In the meantime, the Council kept working on the document, and the European Commission issued an explanatory note on the rewording of the proposal towards the changes introduced by the Lisbon Treaty, that is, Article 15 of the Treaty on the Functioning of the European Union as a new legal base and the application of the right of access to all institutions, bodies, offices, and agencies of the EU rather than just the Council, the Commission, and the Parliament. As this is only a formal change, the Commission requested the Parliament and the Council to change the wording of their documents accordingly.[140]

b. Access to Public Sector Data Held by the Member States

NO EU LEGISLATION, EXCEPT FOR ENVIRONMENTAL DATA – As discussed in Chapter 3, the EU has addressed public access to environmental data as part of the implementation of the Aarhus Convention, but it has never had any intention of extending this to a directive or regulation concerning all types of public sector data. The European Commission stated this explicitly in the eEurope 2002 Communication: 'However, with the exception of the environmental area arrangements for access are first and foremost a national, regional and local responsibility. Therefore no harmonization on the specific aspects of access to public sector information is envisaged'.[141] Access to environmental information was considered closely related to environmental matters, for which the EU does have competence, so the Commission felt that action could be taken there.[142]

138. Commission of the European Communities, *Proposal for a Regulation of the European Parliament and of the Council regarding public access to European Parliament, Council and Commission documents*, COM (2008) 229 final, 30 Apr. 2008.
139. General Secretariat of the Council of the European Union, *Note to the Permanent Representatives Committee/Council. Proposal for a Regulation of the European Parliament and of the Council regarding public access to European Parliament, Council and Commission documents. Outcome of the European Parliament's First Reading*, doc. nr. 7394/09, 17 Mar. 2009, <http://register.consilium.europa.eu/pdf/en/09/st07/st07394.en09.pdf>, 22 Feb. 2010; General Secretariat of the Council of the European Union, *Report to the Delegations. Proposal for a Regulation of the European Parliament and of the Council regarding public access to European Parliament, Council and Commission documents (recast)*, doc. nr. 9234/09, 8 May 2009, <http://register.consilium.europa.eu/pdf/en/09/st09/st09234.en09.pdf>, 22 Feb. 2010.
140. General Secretariat of the Council of the European Union, *Note to the delegations. Proposal for a Regulation of the European Parliament and of the Council regarding public access to European Parliament, Council and Commission documents (recast)*, doc. nr. 5461/10, 19 Jan. 2010, <http://register.consilium.europa.eu/pdf/en/10/st05/st05461.en10.pdf>, 22 Feb. 2010.
141. Commission of the European Communities, *eEurope 2002 Communication*, 11.
142. Katleen Janssen & Jos Dumortier, 'The Legal Framework for a European Spatial Data Infrastructure – Uncrossing the Wires', in *Research and Theory in Advancing Spatial Data Infrastructure Concepts*, ed. Harlan Onsrud (Redlands California: ESRI Press, 2007), 236.

EUROPEAN COMMISSION POLICY DOCUMENTS – The European Commission has addressed access to public sector data in a number of policy documents. As this has mostly been in relation to the question of re-use of public sector data, the documents will be discussed in the section on re-use. Yet some elements regarding access deserve to be mentioned here.

i. Publaw Reports

STARTING POINT – The Publaw reports made it clear that they only looked at access legislation or freedom of information legislation from the point of view of what this legislation could do to improve the European market for information products and services, that is, the market for re-use, particularly with commercial purposes. Therefore, they will be discussed in the appropriate section on the obligation of public bodies to make public sector data available for re-use.

UNIVERSAL INFORMATION SERVICE – However, one comment can already be made about the reports. The third Publaw report proposed a universal information service to be developed, along the lines of the universal service model from the field of telecommunications. However, the concept that was presented can relate to the availability of public sector data for access, the availability of data for re-use, or the provision of information services, depending on how the scope is filled in. The report stated that it is 'impossible to provide a single definition of the universal public information service, either in terms of content (raw data, information with added value) or in terms of technical means of access (paper medium, electronic media etc). Only a definition in terms of objects and characteristics of universal service is possible'.[143] The report believed the public information service should go 'beyond the simple right of access referred to by the legislation on access' and 'perceives the obligation held by the public authority here to be a positive obligation to make information services available using the appropriate technologies for their wide dissemination'.[144] In addition, the public bodies would be required to make data available to citizens using the same techniques as those used internally for their own public servants. While the latter task could refer to making data available for access as it is considered in this book, the former obligation of the public bodies is more directed towards information services. Therefore, the scope of the universal information service is not clear.

ii. Green Paper on Public Sector Information

BACKGROUND – In 1999, the European Commission presented its Green Paper on Public Sector Information. While the Green Paper was part of the process towards the PSI Directive, and hence mainly involved re-use, it also discussed some elements on access.

GROUNDS FOR ACCESS – According to the Green Paper on Public Sector Information, access to public sector data is essential for the mobility of both workers and

143. Commission of the European Communities, *Publaw 3. Final report* (Luxembourg: CEC Legal Advisory Board, 1995), 86.
144. *Ibid.*, 88.

categories such as students and retired people within the EU. Next, the participation of the citizens in the European integration process requires that they are well informed on issues related to the functioning of the EU and its activities. Therefore, the Green Paper also included some questions on access in its consultation. From the replies to the consultation, it was clear that the respondents felt that access to public sector data was fundamental for the participation of the citizens in the European democratic society. Many of the responses stressed the importance of metadata and search facilities, and in general helping the citizens to find, understand, and use the public sector data.

D. The Obligation of Making Environmental Data Available for Access in European and International Documents

Outline – While the previous section addressed public sector data in general, this part of the chapter deals with the task of making environmental data available for access. As a very large part of environmental data has a spatial component and can also be categorized as spatial data, this category is very important to determine the obligations of the public bodies to make public sector spatial data available. This section will show that access to environmental data held by public bodies is considered very important by the international and European legislators and policymakers.

1. United Nations

Outline – Protecting the integrity of the global environmental and developmental system has been high on the United Nations agenda since the early 1990s. Access to environmental data is considered to be essential for the full deployment of adequate protection of the environment. This was recognized for the first time in the Rio Declaration, and it has led to what globally can be considered as one of the most important legal instruments on access to environmental data, the Aarhus Convention. This Convention in its turn led to the adoption of the Access Directive in the EU, so it may be of importance for the interpretation of the scope of the obligations in the Access Directive.

a. Rio Declaration on Environment and Development

Grounds for access – In 1992, the members of the United Nations met in Rio de Janeiro for the Conference on Environment and Development. At the end of the conference, they adopted the Rio Declaration on Environment and Development.[145] Principle 10 of the Declaration states that: 'Environmental issues are best handled

145. United Nations Conference on Environment and Development, *Rio Declaration on Environment and Development*, 13 Jun. 1992, <www.unep.org/Documents.Multilingual/Default.asp?DocumentID=78&articleID=1163>, 22 Feb. 2010.

with participation of all concerned citizens, at the relevant level. At the national level, each individual shall have appropriate access to information concerning the environment that is held by public authorities, including information on hazardous materials and activities in their communities, and the opportunity to participate in decision-making processes. States shall facilitate and encourage public awareness and participation by making information widely available'.[146] Hence, the Rio Declaration acknowledged that information on the environment is essential for the citizens to participate in decision making regarding the environment and to enjoy a healthy and productive life in harmony with nature (Principle 1). The Rio Declaration did not include any examples of types of information that the citizen needed, nor did it state whether the public sector should disseminate information to the citizens of if it would only need to make information available upon request.

AGENDA 21 ACTION PLAN – Agenda 21 is the action plan to implement the Rio Declaration. It contains a chapter on information for decision making, which holds planned actions regarding the collection and sharing of information. However, the availability of information does not seem to be directed towards the public in this section, but rather towards the authorities that use the information in their policies regarding the sustainability of the environment. Therefore, this will be addressed in the section on sharing. However, Chapter 23 of Agenda 21 does address access and proclaims that: 'Individuals, groups and organizations should have access to information relevant to environment and development held by national authorities, including information on products and activities that have or are likely to have a significant impact on the environment, and information on environmental protection measures'.[147]

b. Guidelines for Access to Environmental Information

BACKGROUND – The Ministers of Environment of the United Nations Economic Commission for Europe (UNECE)[148] adopted guidelines for access to environmental information and public participation to decision making regarding environmental matters during the Third Ministerial Conference of Environment for Europe, held in Sofia in 1995.[149] These guidelines drew significantly from Directive 90/313/EEC, but in some aspects they went further than the directive.

146. *Ibid.* See also Patricia Salkin, 'GIS in an age of homeland security: accessing public information to ensure a sustainable environment', *William and Mary Environmental Law and Policy Review* 33 (2005): 60.

147. United Nations Commission on Sustainable Development, *Agenda 21*, 1992, <www.un.org/esa/sustdev/documents/agenda21/english/agenda21toc.htm>, 22 Feb. 2010.

148. Not only European countries can be a member of the ECE. In 2007, the ECE included 57 countries from all across Europe to the former republics of the Soviet Union and the United States and Canada.

149. United Nations Economic Commission for Europe, *Guidelines on Access to Environmental Information and Public Participation in Environmental Decision-Making*, 1995, <www.eea.europa.eu/publications/92-9167-020-0/page017.html>, 22 Feb. 2010.

In their turn, the Guidelines served as one of the cornerstones for the development of the Aarhus Convention.[150] The Guidelines were not binding and did not include sanctions for non-compliance, but they did contain very clear instructions on the implementation of the Guidelines into national law.[151]

General principle: access upon request – According to the Guidelines, '[a]ny natural or legal person should have free access to environmental information at their request, subject to the terms and conditions contained in these Guidelines, without regard to citizenship, nationality or domicile and without having to prove a legal or other interest. Public authorities (at national, regional and local level) and bodies having public responsibilities for the environment, with the exception of bodies which are acting in a judicial or legislative capacity, should supply environmental information subject to the terms and conditions contained in these Guidelines'.[152] Transparency was encouraged by specifying the information that is available and the conditions under which the citizens can obtain access to it, and by creating registers and appointing information officers.

Exceptions – The Guidelines contained a list of exceptions to the right of access. A request could be refused if it affected a number of public and private interests, such as public security, commercial and industrial confidentiality, including IPR, or the confidentiality of personal data, or if it involved the supply of any material that is not complete. The exceptions had to be interpreted restrictively and weighed against the public interest in each occasion.

Procedure – Public authorities should respond to a request as soon as possible and at the latest within six weeks. The information should be available to the public for inspection free of charge, and copies should be available on payment of cost of reproduction and dissemination, if appropriate. Means of redress should be available and made known to the citizens.

General principle: active dissemination – The Guidelines also addressed the dissemination of environmental information by public bodies of their own motion. They did not directly impose obligations on the public authorities but required the States to make sure that information was published and disseminated. The States should regularly publish up-to-date information on the state of the environment and actively publicize the availability of important national and

---

150. Vera Rodenhoff, 'The Aarhus Convention and its Implications for the "Institutions" of the European Community', *RECIEL* 11, no. 3 (2002): 351–352; Frankie Schram, 'Openbaarheid van bestuur: een voorbeeld van interactie tussen het Europees en nationaal publiekrecht', *Tijdschrift voor Bestuurswetenschappen en Publiekrecht* no. 3 (1999): 172 [in Dutch]; Ralph E. Hallo, *Public Access to Environmental Information. European Environment Agency's Experts Corner Number 1997/1* (Luxembourg: Office for Official Publications of the European Communities, 1997), 16.
151. Frankie Schram, 'Openbaarheid van bestuur: een voorbeeld van interactie tussen het Europees en nationaal publiekrecht', *Tijdschrift voor Bestuurswetenschappen en Publiekrecht* no. 3 (1999): 171 [in Dutch].
152. United Nations Economic Commission for Europe, *Guidelines on Access to Environmental Information and Public Participation in Environmental Decision-Making*, 1995, <www.eea.europa.eu/publications/92-9167-020-0/page017.html>, 22 Feb. 2010.

international documents on the environment and the texts of international legal instruments to which they were a party, and which established procedures for public access to environmental information or public participation rights.

CRITICISM FROM NGOs – Even though the environmental NGOs were involved in the creation of the Guidelines, they still found a number of reasons to criticize them, such as the wide range of exceptions, the blanket exemption of legislative bodies from the obligation to supply data, and the lengthy timeframe for responding to requests for data.[153] In general, also with regards to the sections that dealt with public participation, the NGOs felt that the Guidelines did not go far enough and were a missed opportunity to establish new norms of democratic accountability in environmental matters.[154]

FOLLOW-UP – In addition to adopting the Guidelines, the Environment Ministers also agreed that the possibility of a Convention on Public Participation should be considered, with 'appropriate involvement' of NGOs.[155] The NGOs had already pointed out during the preparatory procedure that the power and use of the Guidelines was limited and they should be considered as a first step towards a legally binding convention. They were supported in this position by the European Parliament.[156] The ECE's Committee on Environmental Policy started preparing the Aarhus Convention in January 1996.

c. *The Aarhus Convention*

i. Background

TIMELINE – The Aarhus Convention was adopted on 25 June 1998 at the Fourth Ministerial Conference in the 'Environment for Europe' programme, with the

153. Frankie Schram, 'Openbaarheid van bestuur: een voorbeeld van interactie tussen het Europees en nationaal publiekrecht', *Tijdschrift voor Bestuurswetenschappen en Publiekrecht* no. 3 (1999): 172 [in Dutch]; Ralph E. Hallo, *Public Access to Environmental Information. European Environment Agency's Experts Corner Number 1997/1* (Luxembourg: Office for Official Publications of the European Communities, 1997), 17.
154. Jeremy Wates, 'The UN ECE Guidelines and Draft Convention on Access to Environmental Information and Public Participation in Environmental Decisionmaking', in *Access to environmental information in Europe: the implementation and implications of directive 90/313/EEC*, ed. Ralph E. Hallo (London: Kluwer Law International, 1996), 427–429.
155. *Ibid.*, 423. See also Kurt Deketelaere & Frankie Schram, *Milieurechten van het publiek in de participatiedemocratie* (Bruges: Vanden Broele, 2002), 23 [in Dutch]; Michel Prieur, 'La Convention d'Aarhus, Instrument Universel de la Démocratie Environnementale', *Revue Juridique de l'Environnement* 1999: 19 [in French]; Bernard Vanheusden, 'Vrije toegang tot milieu-informatie: een stand van zaken en de implicaties van het verdrag van Aarhus voor Europa en voor België', *Milieu- en Energierecht* 1999: 276 [in Dutch]; Tom De Gendt, 'Het Verdrag van Aarhus in Vlaanderen: Actieve Openbaarheid van Milieu-Informatie, Participatie van de Burger aan het Milieubeleid en Toegang tot de Rechter in Milieuzaken', *Milieu- en Energierecht* no. 4 (2001): 234 [in Dutch]; Ralph E. Hallo, *Public Access to Environmental Information. European Environment Agency's Experts Corner Number 1997/* (Luxembourg: Office for Official Publications of the European Communities, 1997), 8–9.
156. Jeremy Wates, 'The UN ECE Guidelines and Draft Convention on Access to Environmental Information and Public Participation in Environmental Decisionmaking', in *Access to*

signature of thirty-six countries. In 2010, it has forty signatories and forty-one Parties who have ratified or accepted or acceded to the Convention.[157] It entered into force on 30 October 2001.

VALUE OF THE CONVENTION – Even though it was adopted under the auspices of the Economic Commission for Europe, the Convention is open for all members of the United Nations and has the ambition of being a universal instrument for democratization of environmental decision making.[158] The importance of the Convention was underlined by Mr Annan, former Secretary-General of the United Nations: 'Although regional in scope, the significance of the Aarhus Convention is global. It is by far the most impressive elaboration of principle 10 of the Rio Declaration, which stresses the need for citizen's participation in environmental issues and for access to data on the environment held by public authorities. As such it is the most ambitious venture in the area of environmental . . . democracy so far undertaken under the auspices of the United Nations'.[159]

ACCOMPANYING DOCUMENTS – Along with the signing of the Convention, the Conference adopted a Resolution on Access to Information, Public Participation in Decision-Making and Access to Justice in Environmental Matters, which called for the Convention to be implemented as soon as possible and 'to seek to apply the Convention to the maximum extent possible pending its entry into force'.[160] In addition, an Implementation Guide to the Convention was developed, which has no legal value.[161]

Environmental Information in Europe: the Implementation and Implications of Directive 90/313/EEC, ed. Ralph E. Hallo (London: Kluwer Law International, 1996), 431.

157. See <www.unece.org/env/pp/ratification.htm> (last visited 22 Feb. 2010).

158. Michel Prieur, 'La Convention d'Aarhus, Instrument Universel de la Démocratie Environnementale', *Revue Juridique de l'Environnement* (1999): 20 [in French]; Carrie Noteboom, 'Addressing the External Effects of Internal Environmental decisions: Public Access to Environmental Information in the International Law Commission's Draft Articles on the Prevention of Transboundary Harm', *New York University Environmental Law Journal* 12 (2003): 261.

159. See <www.unece.org/env/pp/welcome.html>, 22 Feb. 2010. Unlike much of international law, which creates obligations between the Member States, the Aarhus Convention imposes obligations on the Member States towards the public. The fact that the European union also a party to the Convention has complicated the implementation of the Treaty in the EU Member States. On the one hand, they have to implement the Aarhus Convention directly, while on the other hand they also have to implement the Access Directive, which is a part of the EU's measures to implement the Convention. However, the Convention and the Directive are not completely identical, possibly causing the Member States difficulties in complying with both. The reason why the Convention was signed by the European union is that the Member States felt that, unlike Directive 90/313/EEC, it should also be applicable to the EU bodies and institutions (Bernard Vanheusden, 'Vrije toegang tot milieu-informatie: een stand van zaken en de implicaties van het verdrag van Aarhus voor Europa en voor België', *Milieu- en Energierecht* 1999: 276 [in Dutch]).

160. United Nations Economic Commission for Europe, *Resolution on Access to Information, Public Participation in Decision-Making and Access to Justice in Environmental Matters*, adopted at the Fourth Ministerial Conference 'Environment for Europe', Aarhus, 23–25 Jun. 1998, <www.unece.org/env/documents/1998/cep/cep.43.add.1.rev.1.e.pdf>, 22 Feb. 2010.

161. United Nations Economic Commission for Europe, *Aarhus Implementation Guide*, 156 p. See also Tom De Gendt, 'Het Verdrag van Aarhus in Vlaanderen: Actieve Openbaarheid

THREE PILLARS OF THE CONVENTION – The Aarhus Convention is divided into three large *pillars*. These pillars successively deal with access to information, public participation, and access to justice. They can be considered as instrumental to 'contribute to the protection of the right of every person of present and future generations to live in an environment adequate to his or her health and well-being', the ultimate objective of the Aarhus Convention.[162] Hence, the three types of rights that are recognized in the Convention are merely functional rights that serve a higher goal.[163] In the scope of this book, the only pillar that is of importance is the first one, on access to environmental information. Therefore, the other two pillars will not be addressed except where necessary to clarify any matters under the first pillar, and the general provisions of the Convention will only be discussed as relevant to access to environmental information.[164]

ii. Article 3 of the Aarhus Convention: General Provisions

SUPPORTING MEASURES TO THE THREE PILLARS – Before addressing access to environmental information, the Aarhus Convention provides a number of obligations for the signatories to the Convention that are not specifically linked to any of the pillars but could be considered as prerequisites or supporting measures for the rights of the three pillars to be fully implemented and enjoyed by the public. They clarify the Convention's other provisions and provide guidance for their interpretation.[165]

---

van Milieu-Informatie, Participatie van de Burger aan het Milieubeleid en Toegang tot de Rechter in Milieuzaken', *Milieu- en Energierecht* no. 4 (2001): 235 [in Dutch]. Although this Implementation Guide has an enormous value for the interpretation and implementation of the Convention, one must keep in mind that it was drawn up by the NGOs, who were actively involved in the drafting process of the Convention. So, while the information in the Implementation Guide comes from a 'privileged source', it might also be coloured or perhaps even biased by the point of view of the NGOs. Yet, one should not underestimate its moral authority.

162. Article 1 of the Aarhus Convention. For more information on the underlying purposes of access to information, public participation and access to justice, see Michel Prieur, 'La Convention d'Aarhus, Instrument Universel de la Démocratie Environnementale', *Revue Juridique de l'Environnement* 1999: 9–29 [in French]; Jean-Jacques Paradissis, *The Right to Access Environmental Information* (Saarbrucken: VDM Verlag, 2009), 52–54.

163. Frankie Schram, 'Openbaarheid van bestuur: een voorbeeld van interactie tussen het Europees en nationaal publiekrecht', *Tijdschrift voor Bestuurswetenschappen en Publiekrecht* no. 3 (1999): 172 [in Dutch]; Kurt Deketelaere & Frankie Schram, *Milieurechten van het publiek in de participatiedemocratie* (Bruges: Vanden Broele, 2002), 25 [in Dutch].

164. For more information on the second and third pillar of the Aarhus Convention, see Kurt Deketelaere & Frankie Schram, *Milieurechten van het publiek in de participatiedemocratie* (Bruges: Vanden Broele, 2002), 384 p. [in Dutch]; Vera Rodenhoff, 'The Aarhus Convention and its Implications for the "Institutions" of the European Community', *RECIEL* 11, no. 3 (2002): 343–357; Sean T. McAllister, 'The Convention on Access to Information, Public Participation in Decisionmaking and Access to Justice in Environmental Matters', *Colorado Journal of International Environmental Law and Policy* 187 (1998): 187–198; Benoit Jadot & Christine Larssen, 'L'accès à la justice en matière d'environnement au regard de la convention d'Aarhus', in *L'accès à la justice en matière d'environnement*, ed. Christine Larssen & Marc Pallemaerts (Brussels: Bruylant, 2005), 195–261 [in French].

165. Vera Rodenhoff, 'The Aarhus Convention and its Implications for the "Institutions" of the European Community', *RECIEL* 11, no. 3 (2002): 345.

Measures for implementing the Convention – The Members of the Convention have to 'take the necessary legislative, regulatory and other measures, including measures to achieve compatibility between the provisions implementing the information, public participation and access to justice provisions of the Convention, as well as proper enforcement measures, to establish and maintain a clear, transparent and consistent framework to implement the provisions of this Convention'.[166] Other measures include ensuring that officials and authorities assist and provide guidance to the public, and promoting environmental education and awareness among the public.[167]

Minimum standard – The signatories to the Convention can always maintain or introduce measures that broaden the rights given to the public regarding access to information, participation in decision making, and access to justice in environmental matters. On the other side of the equation, the Convention does not require any derogation from existing rights to access to information, public participation in decision making and access to justice.[168] The Implementation Guide expresses this as the Convention providing 'a floor, not a ceiling', in that it only sets a minimum standard for the States.[169]

iii. Article 4 of the Aarhus Convention: Access to Information upon Request

General principle – The Parties to the Convention have to ensure that public authorities make information available upon request, without the applicant having to state an interest. There is no need for any formalities for the requests under the Convention, such as a request in writing,[170] so the request may even be made orally. The form of access may include copies of the information requested, but also other forms, such as consultation on-site or lending out the carriers of the information.[171] The environmental information has to be made available as soon as possible and at the latest within one month after the request has been submitted, unless the volume and the complexity of the information justify an extension of this period up to two months.

Exceptions – The Aarhus Convention contains two different types of exceptions to the right of public access to environmental information. On the one hand, these exceptions are limitative. The signatories to the Convention cannot include

166. Article 3.1 of the Aarhus Convention.
167. *Ibid.* See also Kurt Deketelaere & Frankie Schram, *Milieurechten van het publiek in de participatiedemocratie* (Bruges: Vanden Broele, 2002), 32, 44 [in Dutch].
168. Articles 3.5–3.6 of the Aarhus Convention.
169. United Nations Economic Commission for Europe, *Aarhus Implementation Guide*, 45. See also Christine Larssen, 'Het Verdrag van Aarhus en de toepassing ervan in het Belgisch recht – deel 1', *Tijdschrift voor Milieurecht* 14 (2005): 251 [in Dutch].
170. Christine Larssen, 'L'Accès aux Informations sur L'Environnement en Droit International: la Convention d'Aarhus', in *Ten Years of Access to Environmental Information in International, European and Belgian Law: Stock-Taking and Perspectives*, ed. Christine Larssen (Brussels: Bruylant, 2003), 34 [in French].
171. *Ibid.*

any other reasons for refusal of a request for access to environmental information in their national legislation.[172] On the other hand, the list is not binding: the States do not have to include all the exceptions; they can decide that some of the interests that are mentioned in the list are not in need of protection.

FORMAL EXCEPTIONS – The first type concerns so-called *formal* exceptions. These refer to formal conditions, such as the fact that the public authority to which the request is addressed does not hold the data. Next, the request may be manifestly unreasonable or formulated in too general a manner. Thirdly, a request may be refused if it concerns material in the course of completion or concerns internal communications, taking into account the public interest served by disclosure.[173]

EXCEPTIONS BASED ON CONTENT – The second type of exceptions refers to the content of the environmental information that is requested. The Convention contains a list of another eight reasons for refusal of access, which the States can but do not have to implement. However, as was mentioned above, these are the only reasons they can incorporate in their national legislation. In any case, disclosure of the environmental information has to *adversely affect* the interests that are mentioned in the Convention, meaning that they would have a negative impact on these interests. The list of exceptions includes confidentiality of the proceedings of public authorities; international relations, national defence or public security; the course of justice, the ability of a person to receive a fair trial or the ability of a public authority to conduct an enquiry of a criminal or disciplinary nature; confidentiality of commercial and industrial information; IPR; confidentiality of personal data and/or files relating to a natural person; the interest of a third party that has voluntary supplied the information; and the environment to which the information relates.[174]

REQUIREMENTS FOR INVOKING THE EXCEPTIONS – Whatever grounds for refusal are included in the national legislation, they all have to be 'interpreted in a restrictive way, taking into account the public interest served by disclosure and taking into account whether the information requested relates to emissions into the environment'.[175] In every case, an *in concreto* judgment has to be made whether or not it is permitted to give access to the information.[176] This judgment has to be made by weighing the protected interest against the *public* interest, not just the interest of the applicant.[177] Refusals of a request have to state the reasons why and give

172. Christine Larssen, 'Het Verdrag van Aarhus en de toepassing ervan in het Belgisch recht – deel 1', *Tijdschrift voor Milieurecht* 14 (2005): 258 [in Dutch].
173. Article 4.3 of the Aarhus Convention.
174. Article 4.4 of the Aarhus Convention.
175. Article 4.4 in fine of the Aarhus Convention.
176. Christine Larssen, 'Het Verdrag van Aarhus en de toepassing ervan in het Belgisch recht – deel 1', *Tijdschrift voor Milieurecht* 14 (2005): 260 [in Dutch].
177. Christine Larssen, 'L'Accès aux Informations sur L'Environnement en Droit International: la Convention d'Aarhus', in *Ten Years of Access to Environmental Information in International, European and Belgian Law: Stock-Taking and Perspectives*, ed. Christine Larssen (Brussels: Bruylant, 2003), 36 [in French].

information on access to means of redress against the decision of the public authority.

CHARGES – With regard to charges, the Aarhus Convention embraces the concept that if information is to be truly accessible, it must also be affordable.[178] Public authorities can be allowed to charge for supplying information, but these charges may not exceed a reasonable amount, so as to ensure the effectiveness of the right of access to environmental information. The charges may be levied beforehand. Charging for on-site consultation of the environmental information is not allowed under the Aarhus Convention, nor is any charging in case of the refusal of a request for access. The public authorities have to make available to applicants a schedule of charges that may be levied, indicating the circumstances in which they may be levied or waived and when the supply of information is conditional on the advance payment of such a charge.[179]

iv. Article 5 of the Aarhus Convention: Dissemination of Environmental Information

WIDE RANGE OF OBLIGATIONS – In addition to the obligations of the public authorities to make environmental information available for access upon request, the Aarhus Convention also contains provisions on the active dissemination of information by the public authorities. It covers a wide range of information and an equally broad range of implementation requirements. Some of the provisions require the states or the public authorities to take certain specific steps for implementation, while others give some guidance on the end result but leave the choice of process and the implementation methods open.[180] The Convention goes beyond the recognition of the right of access to data: it also refers to accessibility and requires the public authorities to collect certain information. However, accessibility is only seen from a practical point of view. Larssen criticizes the Convention for disregarding another aspect of accessibility, that is, intellectual accessibility: 'What is the point for the public to have material access to the information when it cannot decipher or understand its contextual meaning?'[181]

TRANSPARENCY AND EFFECTIVE ACCESSIBILITY – The parties to the Aarhus Convention have to ensure that: 'Within the framework of national legislation, the way in which public authorities make environmental information available to the public

178. United Nations Economic Commission for Europe, *Aarhus Implementation Guide*, 65.
179. Article 4.8 of the Aarhus Convention.
180. United Nations Economic Commission for Europe, *Aarhus Implementation Guide*, 66.
181. Larssen sees the collection and possession of information and accessibility matters as part of the right of access to environmental information by way of active dissemination by the public authorities. She feels that a complete system of dissemination should encompass at least four aspects: regular production and collection of information; dissemination of information by the public authorities; measures to guarantee the accessibility of the information; and dissemination of information by private parties at the advice of the public authorities. Article 5 of the Aarhus Convention addresses all these items. (Christine Larssen, 'Het Verdrag van Aarhus en de toepassing ervan in het Belgisch recht – deel 1', *Tijdschrift voor Milieurecht* 2005: 265, 269 [in Dutch] (translation by the author).

is transparent and that environmental information is effectively accessible'. The Implementation Guide describes *transparency* as follows: 'The public can clearly follow the path of environmental information, understanding its origin, the criteria that govern its collection, holding and dissemination, and how it can be obtained'. *Effectively accessible* is taken to mean 'that the established information systems should go beyond simply making the information systems available to the public. Records, databases and documents can be considered effectively accessible when, for example, the public can search for specific information, or when the public can easily obtain access because of convenient opening hours of the public authorities, locations, equipment such as copy machines, etc'.[182]

MEASURES FOR ENSURING TRANSPARENCY AND EFFECTIVE ACCESSIBILITY – Article 5.2 gives a non-limitative list of measures with which the public authorities can ensure this transparency and effective accessibility. First, the public authorities can provide sufficient information to the public about the type and scope of environmental information held by the relevant public authorities, the basic terms and conditions under which such information is made available and accessible, and the process by which it can be obtained. Next, they can establish and maintain practical arrangements, such as publicly accessible lists, registers or files making it easy to find information, the identification of points of contact, or requiring public servants to support the public in seeking access to information.

INFORMATION TO BE MADE AVAILABLE IN ELECTRONIC DATABASES – Another aspect of the obligation of the public authorities to actively make available their environmental information comes into play in Article 5.3 of the Aarhus Convention. The signatories to the Convention have to ensure that environmental information progressively becomes available in electronic databases that are easily accessible to the public through public telecommunications networks. However, they can be flexible in the timeframe for meeting the obligation and in determining who will manage the process.[183] Article 5.3 lists four types of information that should at least be available[184]:

- Reports on the state of the environment. These national reports have to be published and disseminated by the States at regular intervals not exceeding three or four years. They have to include information on the quality of the environment and information on pressures on the environment;[185]
- Texts of legislation on or relating to the environment;
- As appropriate, policies, plans and programmes on or relating to the environment, and environmental agreements;

182. United Nations Economic Commission for Europe, *Aarhus Implementation Guide*, 71–72. See also Jean-Jacques Paradissis, *The Right to Access Environmental Information* (Saarbrucken: VDM Verlag, 2009), 105–107.
183. *Ibid.*, 75.
184. According to Deketelaere and Schram, this list was the result of very difficult negotiations that led to a laborious compromise (Kurt Deketelaere & Frankie Schram, *Milieurechten van het publiek in de participatiedemocratie* (Bruges: Vanden Broele, 2002), 161 [in Dutch]).
185. Article 5.4 of the Aarhus Convention.

– Other information, to the extent that the availability of such information would facilitate the application of national law implementing the Convention.

ONLY EXISTING INFORMATION – The information only has to be made available if it already exists in electronic form. The Implementation Guide warns that: 'It is important that the electronic versions do not replace other forms of the same information, as computers and public telecommunications networks are not readily accessible to all members of the public in every country'.[186]

OTHER INFORMATION TO BE MADE AVAILABLE – The states also have to take measures, within the framework of their legislation for the purpose of making available, inter alia, legislation and policy documents[187]; international treaties, conventions, and agreements on environmental issues; and other significant international documents on environmental issues as appropriate.[188] This list of documents is only a preliminary list, and the states may add any other relevant types of information that will help to implement the Convention. The documents have to be disseminated to the audience by means of publications, mailings or e-mails. Notifying the public that certain information is available and informing them where and when it is accessible could also be sufficient. In this case, the states have to make sure that the information is made available freely or at low cost.[189] In addition to the previous documents, the states also have to publish the facts and the analyses of facts that they consider relevant and important in framing major environmental policy proposals and publish, or otherwise make accessible, available explanatory material on their dealings with the public in matters falling within the scope of the Convention.[190] They also have to provide information, in an appropriate form, on the performance of public functions or the provision of public services relating to the environment by government at all levels.[191] The main purpose of the obligation to publish the documents mentioned in this paragraph is to provide the public, the public authorities, and the other parties to the Convention with some insight into the manner in which government decisions are made in order to allow them to inspect the way in which these

186. United Nations Economic Commission for Europe, *Aarhus Implementation Guide*, 75. See also Christine Larssen, 'Het Verdrag van Aarhus en de toepassing ervan in het Belgisch recht – deel 1', *Tijdschrift voor Milieurecht* 2005: 269 [in Dutch].

187. Such as documents on strategies, policies, programmes and action plans relating to the environment, and progress reports on their implementation, prepared at various levels of government.

188. Article 5.5 of the Aarhus Convention.

189. Kurt Deketelaere & Frankie Schram, *Milieurechten van het publiek in de participatiedemocratie* (Bruges: Vanden Broele, 2002), 168 [in Dutch]; Tom De Gendt, 'Het Verdrag van Aarhus in Vlaanderen: Actieve Openbaarheid van Milieu-Informatie, Participatie van de Burger aan het Milieubeleid en Toegang tot de Rechter in Milieuzaken', *Milieu- en Energierecht* no. 4 (2001): 245–246 [in Dutch].

190. Article 5.7 of the Aarhus Convention.

191. *Ibid.*

decisions are implemented. In this way they have the possibility to contribute to policymaking.[192]

EXCEPTIONS – The final section of Article 5 ensures that the right of the states to refuse to disclose certain environmental information in accordance with the exceptions listed in the Convention is not prejudiced.[193] Hence, the states can use the same exceptions to refuse to actively disseminate information as they can to refuse to make it available on request. However, one should keep in mind that in the case of imminent threats to human health or the environment,[194] the environmental information should always be made available, as there will always be an overriding public interest.[195]

2. Organization for Economic Cooperation and Development: Recommendation of the Council on Environmental Information

GENERAL PRINCIPLE – The OECD has adopted several recommendations that deal, directly or indirectly, with environmental information,[196] but the most relevant one is the Recommendation on Environmental Information, adopted by the Council on 3 April 1998.[197] It encourages all Member States to take all necessary actions within the framework of their national laws to increase the availability to the public of environmental information held by public authorities. In doing so, each Member State can determine which information needs to be widely distributed and under which form it should be disseminated. Information should be available to the citizens without their having to prove an interest, without unreasonable charges, and as soon as possible.

192. United Nations Economic Commission for Europe, *Aarhus Implementation Guide*, 79; Tom De Gendt, 'Het Verdrag van Aarhus in Vlaanderen: Actieve Openbaarheid van Milieu-Informatie, Participatie van de Burger aan het Milieubeleid en Toegang tot de Rechter in Milieuzaken', *Milieu- en Energierecht* no. 4 (2001): 250 [in Dutch]; Kurt Deketelaere & Frankie Schram, *Milieurechten van het publiek in de participatiedemocratie* (Bruges: Vanden Broele, 2002), 176–177 [in Dutch].
193. Article 5.10 of the Aarhus Convention.
194. See Art. 5.1(c) of the Aarhus Convention.
195. United Nations Economic Commission for Europe, *Aarhus Implementation Guide*, 84. See also Kurt Deketelaere & Frankie Schram, *Milieurechten van het publiek in de participatiedemocratie* (Bruges: Vanden Broele, 2002), 185 [in Dutch]; Tom De Gendt, 'Het Verdrag van Aarhus in Vlaanderen: Actieve Openbaarheid van Milieu-Informatie, Participatie van de Burger aan het Milieubeleid en Toegang tot de Rechter in Milieuzaken', *Milieu- en Energierecht* no. 4 (2001): 254 [in Dutch].
196. See OECD Secretariat, 'OECD: The OECD Recommendation and Other International Acts on Environmental Information', paper presented at the OECD Seminar 'Public Access to Environmental Information. Proceedings' (Athens, 2000), 65, <www.olis.oecd.org/olis/2000doc.nsf/LinkTo/NT00000F96/$FILE/00080429.PDF>, 22 Feb. 2010.
197. OECD, *Recommendation of the Council on Environmental Information*, 3 Apr. 1998, <http://webdomino1.oecd.org/horizontal/oecdacts.nsf/linkto/C(98)67>, 22 Feb. 2010.

REPORTS ON THE STATE OF THE ENVIRONMENT – Member States should promote adequate dissemination of environmental information, such as periodic reports on the state of the environment and its changes over time, or the publication of environmental indicators. Such periodic reports could possibly be seen as the provision of an information service to the citizens, as they go beyond the processing of data for the public body's own needs.[198]

EFFECTIVE MECHANISMS FOR MAKING DATA AVAILABLE – The Council also states that the Member States should establish effective mechanisms to better inform the public, decisionmakers, and the authorities on environmental and sustainable development conditions and issues. While the Council mentions decisionmakers and public authorities as receiving parties of the data and therefore may also indicate sharing, the first suggestions that it makes to elaborate on this recommendation are more directed towards access:

- encourage appropriate levels of government to make publicly available reports on the results of public policies and related actions;
- use modern effective information communication methods to enable timely, easy and inexpensive access to large volumes of information;
- promote co-operation on dissemination of environmental information among different administrations and government levels as well as non-governmental organizations concerned.

The OECD Recommendation requires assistance by public authorities to inform the public of its rights and the establishment of educational measures to help the public to effectively use environmental information.[199]

3. Council of Europe: The Lugano Convention

GENERAL PRINCIPLE – The Council of Europe has until now not addressed access to environmental data as an issue needing separate attention from access to public sector data in general. However, it did consider access to environmental data as a part of the 1993 Convention on Civil Liability for Damage Resulting from Activities Dangerous to the Environment.[200] Article 14 of the Convention contains the

198. Cf. *infra*, Ch. 7, s. II.D.2.
199. OECD Secretariat, 'The OECD Recommendation and other international acts on environmental information', in paper presented at the OECD Seminar Public Access to Environmental Information. Proceedings (Athens, 2000), 67, <www.olis.oecd.org/olis/2000doc.nsf/LinkTo/NT00000F96/$FILE/00080429.PDF>, 22 Feb. 2010.
200. Council of Europe, *Convention of 21 June 1993 on Civil Liability for Damage Resulting from Activities Dangerous to the Environment*, <http://conventions.coe.int/Treaty/EN/Treaties/Html/150.htm>, 22 Feb. 2010. See also Maguelonne Déjaent-Pons, 'Le Conseil de l'Europe et l'Accès à l'Information Environnementale: la Convention Européenne des Droits de l'Homme', in *Ten Years of Access to Environmental Information in International, European and Belgian Law: Stock-Taking and Perspectives*, ed. Christine Larssen (Brussels: Bruylant, 2003), 71 [in French].

right of any person to have access to information relating to the environment held by public authorities, at his or her request and without his or her having to prove an interest. It is based almost literally on Directive 90/313/EEC, regarding the definitions, exceptions, and procedural guarantees.[201]

INFORMATION HELD BY PERSONS OPERATING A DANGEROUS ACTIVITY – Article 16 of the Convention contains a right to specific information held by operators, that is, any person operating a dangerous activity. Any person who suffered damage can request the court to order an operator to provide him or her with specific information, insofar as this is necessary to establish the existence of a claim for compensation under the Lugano Convention.

VALUE OF THE CONVENTION – However, the value of this Convention is negligible, as seventeen years after the adoption of the Convention, none of the Member States of the Council of Europe have ratified it,[202] and it has not entered into force.

4. European Union

OUTLINE – While the EU institutions made it clear that they considered access to public sector data as a matter to be addressed by the Member States, an exception was made with regard to the task of making environmental information available, as was mentioned in Chapter 3.[203] Hence, in 1990, Directive 90/313/EEC was introduced. It was replaced by the Access Directive in 2003. As these directives were discussed in Chapter 3, only their main points will be repeated here for the sake of completeness. A short introduction will also be given to Shared Environmental Information System (SEIS), a European policy on environmental information that is currently under development. First, the legislation of the EU regarding the obligation on its own institutions to make environmental information available for access needs to be mentioned, considering the exemplary position of these institutions. This may give indications on the scope of the obligations of the public bodies to make public sector spatial data available for access purposes.

a. Access to Documents Held by the Community Institutions and Bodies: Regulation 1367/2006

BACKGROUND – As the Access Directive only implements the provisions of the Aarhus Convention for the Member States of the EU, a separate legal instrument was needed to make the institutions and bodies of the Community adapt to the Aarhus Convention.[204] In September 2006, the European Parliament and the council

201. Frankie Schram, 'Openbaarheid van bestuur: een voorbeeld van interactie tussen het Europees en nationaal publiekrecht', *Tijdschrift voor Bestuurswetenschappen en Publiekrecht* no.3 (1999): 169 [in Dutch].
202. Cf. <http://conventions.coe.int/Treaty/Commun/ChercheSig.asp?NT=150&CM=8&DF=8/9/2007&CL=ENG>, 22 Feb. 2010.
203. Cf. *supra*, Ch. 3, s. II.A.
204. Jean-Jacques Paradissis, *The Right to Access Environmental Information* (Saarbrucken: VDM Verlag, 2009), 48.

adopted Regulation 1367/2006 to take care of this.[205] In addition, the Regulation takes into account the Sixth Community Environment Action Programme,[206] which 'stresses the importance of providing adequate environmental information and effective opportunities for public participation in environmental decision making'.[207]

OBJECTIVE – The Regulation's objective is to contribute to the implementations of the Aarhus Convention, by:

(a) guaranteeing the right of public access to environmental information received or produced by Community institutions or bodies and held by them, and by setting out the basic terms and conditions of, and practical arrangements for, the exercise of that right;
(b) ensuring that environmental information is progressively made available and disseminated to the public in order to achieve its widest possible systematic availability and dissemination. To that end, the use, in particular, of computer telecommunication and/or electronic technology, where available, shall be promoted;
(c) providing for public participation concerning plans and programmes relating to the environment;
(d) granting access to justice in environmental matters at Community level under the conditions laid down by this Regulation.[208]

CONFORMITY WITH REGULATION 1049/2001 – The Regulation makes sure that there is no distinction in the procedure for obtaining access to environmental information held by Community bodies and institutions and other public sector data held by them by applying the rules of Regulation 1049/2001 to the requests for environmental information, 'without discrimination as to citizenship, nationality or domicile and, in the case of a legal person, without discrimination as to where it has its registered seat or an effective centre of its activities'.[209] This broadens the field of application, as Regulation 1049/2001 was only applicable to documents held by the European Parliament, the Council, and the Commission, and could only be used by citizens of the European Union and any natural or legal person residing or having its registered office in a Member State.[210]

EXCEPTIONS: DIFFERENCES WITH REGULATION 1049/2001 – In its Article 6, Regulation 1367/2006 slightly adapts the exceptions of Regulation 1049/2001. In the cases where the public interest should be weighed, an overriding public

205. Regulation 1367/2006/EC of the European Parliament and of the Council of 6 Sep. 2006 on the application of the provisions of the Aarhus Convention on Access to Information, Public Participation in Decision-Making and Access to Justice in Environmental Matters to Community institutions and bodies, *OJ L* 264, 25 Sep. 2006, 13. Cf. *supra*, Ch. 5, s. I.D.4.a.
206. Decision 1600/2002/EC of the European Parliament and of the Council of 22 Jul. 2002 laying down the Sixth Community Environment Action Programme, *OJ L* 242, 10 Sep. 2002, 1.
207. Recital 2 of Regulation 1367/2006.
208. Article 1(a) of Regulation 1367/2006.
209. Article 3 of Regulation 1367/2006.
210. Article 2 of Regulation 1049/2001. See Jean-Jacques Paradissis, *The Right to Access Environmental Information* (Saarbrucken: VDM Verlag, 2009), 48–49.

interest shall be deemed to exist where the requested information relates to emissions into the environment (except in the case of court proceedings and legal advice, and investigations, in particular those concerning possible infringements of Community law). An additional exception was added for situations where disclosure of the information would adversely affect the protection of the environment to which the information relates, such as the breeding sites of rare species.[211] The provisions on the exceptions to the right of access caused a great deal of discussion between the Commission, the Council, and the Parliament, and they have been subject to considerable criticism since then because they are not consistent with the Aarhus Convention, for example regarding the inclusion of an exception for the financial, monetary, or economic policy of the Community or a Member State, the broader protection of commercial interests, or the exception for legal advice.[212]

ACTIVE DISSEMINATION – With regards to the active dissemination of environmental information, Article 4 of the Regulation requires the Community bodies and institutions to organize the environmental information that is relevant to their functions and that is held by them, with a view to its active and systematic dissemination to the public, in particular by means of computer telecommunication and/or electronic technology, in accordance with Regulation 1049/2001. They have to make this environmental information progressively available in electronic databases that are easily accessible to the public through public telecommunications networks.

INFORMATION TO BE INCLUDED IN DATABASES – The databases or registers that have to be set up under Regulation 1049/2001 must include environmental information, for example:

- Texts of international treaties, conventions or agreements, and of Community legislation on the environment or relating to it, and of policies, plans and programmes relating to the environment;
- Progress reports on the implementation of these texts, if they are prepared or held in electronic form by the Community bodies and institutions;
- Steps taken in proceedings for infringements of Community law;
- Authorizations with a significant impact on the environment, and environmental agreements, or a reference to the place where such information can be requested or accessed;
- Environmental impact studies and risk assessments concerning environmental elements, or a reference to the place where such information can be requested or accessed;[213]

211. Article 6.2 of Regulation 1367/2006.
212. European Centre on Sustainable Politics for Human and Environmental Rights, 'The Aarhus Regulation: New Opportunities for Citizens in the EU and Beyond? Report of the ECOSPHERE Forum held in Brussels on 27 October 2006', 2006, 3–4, 7–10, <www.ecosphere.be/pdfflyers/ReportForumOct2006.pdf>, 22 Feb. 2010.
213. Article 4 of Regulation 1367/2006.

- Reports on the state of the environment. The Commission has to ensure that such reports, including information on the quality of, and pressures on the environment, are published at regular intervals not exceeding four years.
- Data or summaries of data derived from the monitoring of activities affecting, or likely to affect the environment. Such summaries might only entail aggregation of data, but it is also possible that these data need to be processed further and commented on, in order to make a useful summary for the citizens.

b. *Access to Documents Held by Member States*

OUTLINE – As the rules for access to public sector spatial data from the Access Directive and its predecessor Directive 90/313/EEC have already been discussed in Chapter 3, this section will only remind the reader of the main principles of these directives very briefly.

i. Directive 90/313/EEC

GENERAL PRINCIPLE: ACCESS ON REQUEST – Under Directive 90/313/EEC, any natural or legal person could make a request for information without having to prove an interest. The Member States had to define the practical arrangements under which the information was effectively made available.[214]

EXCEPTIONS – The public authorities could refuse a request for access where it would involve the supply of unfinished documents or data or internal communications, or where the request was manifestly unreasonable or formulated in a too general manner.[215] In addition, a request for information could be refused for reasons of confidentiality of proceedings, public security, judicial or other enquiries, commercial and industrial confidentiality, third-party materials, or damage to the environment.[216] These interests had to be affected, but they did not have to be weighed against the public interest in disclosure of the information.

PROCEDURE – The public authorities had to respond to a request for information as soon as possible and at the latest within two months.[217] However, Directive 90/313/EEC did not include a time limit for the delivery of the information if the request was accepted, nor any provisions on the form in which the access could be given or in which the information should be supplied. If the request was denied, the reasons for refusal must be given. A charge could be made for the supplying the information, not exceeding a reasonable cost.[218]

GENERAL PRINCIPLE: ACTIVE DISSEMINATION – Directive 90/313/EEC had very limited ambitions regarding the active dissemination of information relating to the environment by the public authorities, for which it was heavily criticized.

214. Article 3.1 of Directive 90/313/EEC.
215. Article 3.3 of Directive 90/313/EEC.
216. Article 3.2. of Directive 90/313/EEC.
217. Article 3.4 of Directive 90/313/EEC.
218. Article 5 of Directive 90/313/EEC.

In its Article 7, it only required the Member States to 'take the necessary steps to provide general information to the public on the state of the environment by such means as the periodic publication of descriptive reports'.

ii. Access Directive

GENERAL PRINCIPLE: ACCESS UPON REQUEST – Under the Access Directive, the public authorities have to make environmental information available for access upon request without the applicant having to state an interest. The requested information has to be made available as soon as possible or at the latest within one month after the receipt by the public authority.[219]

EXCEPTIONS – Article 4 of the directive contains a limitative number of exceptions under which the public authority can refuse to provide access to the requested environmental information. These exceptions include the situation where the information requested is not held by or for the public authority to which the request is addressed; the request is manifestly unreasonable or formulated in too general a manner; or the request concerns unfinished material or internal communications. Other interests that may be protected by a refusal to make information available include the confidentiality of the proceedings of public authorities; international relations, public security, or national defence; the course of justice, the ability of any person to receive a fair trial, or the ability of a public authority to conduct an enquiry of a criminal or disciplinary nature; confidentiality of commercial or industrial information; IPR; the confidentiality of personal data; the interests or protection of a person who has voluntarily supplied information; and the protection of the environment.

PRACTICAL ARRANGEMENTS – The Member States have to provide practical arrangements to support the right of access of the citizens. They can do this by ensuring that information officers are designated, and that registers or lists are made of the environmental information held by public authorities, with clear indications of where this information can be found.

GENERAL PRINCIPLE: ACTIVE DISSEMINATION – With regards to the active dissemination of environmental information, Member States have to ensure that: 'Environmental information is progressively made available and disseminated to the public in order to achieve the widest possible systematic availability and dissemination to the public of environmental information'.[220] Information that has to be made available and disseminated includes texts of international treaties, conventions, or agreements, and of Community, national, regional, or local legislation on the environment or relating to it; policies, plans, and programmes relating to the environment; progress reports on the implementation of these rules; environmental impact studies and risk assessments concerning the elements of the environment; and information in the event of an imminent threat to human health or the environment.[221]

219. Article 3 of the Access Directive.
220. Article 1 of the Access Directive.
221. Article 7 of the Access Directive.

iii. Access to Documents Held by Member States and the European Union: Shared Environmental Information System (SEIS)

General principle – On 1 February 2008, the European Commission published the Communication Towards a Shared Environmental Information System (SEIS). The Communication sets out an approach to 'simplify the collection, exchange and use of the data and information required for the design and implementation of environmental policy'.[222] As this is more related to the topic of sharing, the SEIS Communication will be addressed more extensively in that section, but it deserves to be mentioned that one of the principles underpinning SEIS is that 'information should be fully available to the general public, after due consideration of the appropriate level of aggregation and subject to appropriate confidentiality restraints, and at national level in the relevant national languages'.[223] This availability will 'help to empower European citizens by making relevant, timely information available to them, enabling them to take informed decisions concerning their environment, including taking appropriate action in cases of emergency, and to influence public policy'.[224]

E. The Obligation of Making Spatial Data Available for Access in European and International Documents

Outline – While access to public sector data has been given a lot of attention in European and international legislative and policy documents, and the duties of public bodies regarding environmental data have also been moving to the forefront, the attention for access to spatial data in particular is of a much more recent nature. To a certain extent, such special attention is not necessary, as most spatial data can be categorized as public sector data and/or environmental data. However, the European and international organizations seemed to believe that spatial data should be seen as a separate category of data, with characteristics that need their own approach.

1. United Nations

a. *United Nations Spatial Data Infrastructure*

General principle – In February 2007, the United Nations Geospatial Information Working Group published its strategy for developing and implementing a United Nations Spatial Data Infrastructure (UNSDI).[225] While this UNSDI is directed more

222. Commission of the European Communities, *Communication to the Council, the European Parliament, the European Economic and Social Committee and the Committee of the Regions. Towards a Shared Environmental Information System (SEIS)*, COM (2008) 46 final, 2, 1 Feb. 2008.
223. *Ibid.*, 3.
224. *Ibid.*, 5.
225. United Nations Geographic Information Working Group, 'Strategy for developing and implementing a United Nations Spatial Data Infrastructure in support of Humanitarian Response,

towards data sharing rather than making spatial data available for access, it also includes the objective to 'ensure that current, quality assured geospatial data and information can be easily discovered, and are immediately and openly available via the Internet from within a distributed matrix of interoperable data resources resident in UN bodies, regional organizations, national governments, academia, industry, the NGO network and the community at large'.[226]

FOCUS ON SHARING AND RE-USE – The reference to the community at large may indicate that the UN is also considering access to spatial data. However, as the focus is on sharing and possibly re-use, we will address the UNSDI in those sections.

b. Global Earth Observation System of Systems

BACKGROUND – The United Nations 2002 World Summit on Sustainable Development and the G8 countries – the eight most important industrial countries in the world – called for action on international collaboration for exploiting the growing potential of Earth observations to support decision making in an increasingly complex and environmentally stressed world. In response to this call, the Group on Earth Observations (GEO) was launched.[227] GEO is a voluntary partnership. As of February 2010, its Members include eighty governments and the European Commission. In addition, fifty-eight intergovernmental, international, and regional organizations with a mandate in Earth observation or related issues are recognized as Participating Organizations. In 2005, the GEO started with a ten-year plan to develop a Global Earth Observation System of Systems (GEOSS).

OBJECTIVE OF GEOSS – The purpose of GEOSS is 'to achieve comprehensive, coordinated and sustained observations of the Earth system, in order to improve monitoring of the state of the Earth, increase understanding of Earth processes, and enhance prediction of the behaviour of the Earth system. GEOSS will meet the need for timely, quality long-term global data as a basis for sound decision making and will enhance delivery of benefits to society' relating to the environment.[228]

Economic Development, Environmental Protection, Peace and Safety', 2007, 32 p., <www.ungiwg.org/docs/unsdi/UNSDI_Strategy_Implementation_Paper.pdf>, 22 Feb. 2010.

226. *Ibid.*, 15.
227. See GEO, 'About GEO', <www.earthobservations.org/about_geo.shtml>, 22 Feb. 2010.
228. Group on Earth Observations, 'The Global Earth Observation System of Systems (GEOSS) 10-Year Implementation Plan', 2005, <www.earthobservations.org/documents/10-Year%20Implementation%20Plan.pdf>, 22 Feb. 2010. GEOSS initially focuses on the following areas:
    (1) Reducing loss of life and property from natural and human-induced disasters;
    (2) Understanding environmental factors affecting human health and well-being;
    (3) Improving management of energy resources;
    (4) Understanding, assessing, predicting, mitigating, and adapting to climate variability and change;
    (5) Improving water resource management through better understanding of the water cycle;
    (6) Improving weather information, forecasting, and warning;
    (7) Improving the management and protection of terrestrial, coastal, and marine ecosystems;

Focus on sharing – AS GEOSS' main focus is sharing, as well as some elements of re-use, its main characteristics will be treated in those sections. However, as GEOSS is encouraging all the Member States to make their data, metadata, and products available without restriction for the widest possible array of users all over the world, it has the potential of making a significant contribution to access to spatial data as well.

General principles – GEOSS has three main data sharing principles:

- There will be full and open exchange of data, metadata, and products shared within GEOSS, recognizing relevant international instruments and national policies and legislation.
- All shared data, metadata, and products will be made available with minimum time delay and at minimum cost.
- All shared data, metadata, and products free of charge or no more than cost of reproduction will be encouraged for research and education.

Data sharing guidelines – One of the tasks scheduled in the GEO Workplan for 2007–2009 was the development of GEOSS data sharing guidelines, based on the three main principles.[229] These Implementation Guidelines for the GEOSS data sharing principles are currently under development and should be adopted in their final version by the GEO plenary meeting and Ministerial Summit in November 2010.[230] The draft Implementation Guidelines will be addressed in the sections on re-use and sharing.

2. European Union

Outline – In addition to the INSPIRE Directive, some attention should also be given to the Global Monitoring for Environment and Security Initiative. This will be discussed in the second part of this section.

a. The INSPIRE Directive

General principle: access to the network services – As the INSPIRE Directive was addressed in Chapter 3, this section will only shortly revisit the tasks of the Member States and the public authorities to make data available for access. The Member States have to establish and operate a network of services, including discovery services, view services, download services, and transformation services.[231] Hence, the public only has an indirect right of access to the spatial

---

(8) Supporting sustainable agriculture and combating desertification;
(9) Understanding, monitoring, and conserving biodiversity.

229. Group on Earth Observations, 'GEO 2007–2009 Work Plan. Toward Convergence', 2007, 25, <www.earthobservations.org/documents/wp0709_v6.pdf>, 22 Feb. 2010.
230. See GEO, 'GEO Data Sharing Principles Implementation', <www.earthobservations.org/geoss_dsp.shtml>, 22 Feb. 2010.
231. Article 11 of the INSPIRE Directive. Cf. *supra*, Ch. 3, s. II.C.3.

data, but access to the services implies access to the data, so the end result is the same.

EXCEPTIONS – Access to these services can be limited for a limitative number of reasons, which are the same as in the Access Directive. The discovery services have to be available free of charge, while view services should in principle be free of charge, but could be charged for if those charges would 'secure the maintenance of spatial data sets and corresponding data services, especially in cases involving very large volumes of frequently updated data'.[232] Download services and transformation services can be charged for without restrictions. If the Member States choose to levy charges, they have to make sure that e-commerce services are available.[233]

TYPES OF DATA THAT NEED TO BE MADE AVAILABLE – The data sets that need to be available through the network services are grouped under thirty-four data themes mentioned in the Annexes to the INSPIRE Directive.[234] Hence, for those data themes, it is clear that the European institutions see an obligation for the Member States and the public authorities to make these data available for access to the public.

b. Global Monitoring for Environment and Security (Kopernikus)

BACKGROUND – Global Monitoring for Environment and Security (GMES) is a European initiative for the implementation of information services dealing with environment and security.[235] It is a joint initiative of the European Commission, the European Space Agency, and the Member States. It is based on observation data received from Earth observation satellites and ground-based data. In this way, it is the EU's contribution to GEOSS. It will enter the operational phase in 2011.[236] In September 2008, the change of GMES' name to Kopernikus was announced, in honour of the famous astronomer who greatly advanced the science of his time.[237] However, later documents from the European institutions still keep referring to GMES.

FOCUS ON SHARING – The possible GMES users can be very diverse: a national or European policymaker, a researcher who requires large climate change data sets for modeling purposes, a service provider who needs access to various observing

232. Article 14 of the INSPIRE Directive. Cf. *supra*, Ch. 3, s. II.C.3.c.
233. Cf. *supra*, Ch. 3, s. II.C.3.c.
234. Annex I, II and III of the INSPIRE Directive. Cf. *supra*, Ch. 2, s. III.B.
235. See GMES, 'Overview', <http://ec.europa.eu/gmes/overview.htm>, 22 Feb. 2010.
236. Commission of the European Communities, *Communication to the European Parliament, the Council, the European Economic and Social Committee and the Committee of the Regions. Global Monitoring for Environment and Security (GMES): we care for a safer planet*, COM (2008) 748 final, 12 Nov. 2008.
237. Rapid Press Release, 'Kopernikus: Observing our planet for a safer world' 2008, <http://europa.eu/rapid/pressReleasesAction.do?reference=IP/08/1330&format=HTML&aged=0&language=EN&guiLanguage=fr>, 22 Feb. 2010.

system data and model output in order to produce an information product for an end-user, a commercial user who requires an actual information product, an NGO, or a member of the general public.[238] Hence, GMES will make spatial data available for the purposes of access, sharing, and re-use. However, as the main focus in the beginning stages of GMES appears to lie with the availability of data for policy and decision making, we will discuss its main elements in the section on the obligation of public bodies to make public sector spatial data for sharing. Yet, it deserves to be mentioned that the European Commission acknowledges, in light of the objectives of the Aarhus Convention, that 'a significant part of GMES information should be freely available' to the citizens.[239] This is also acknowledged in the European Commission's proposal for a Regulation on GMES, published in May 2009. The Regulation should provide the legal basis for the GMES programme, enabling it to contribute to the production and availability to the public of environmental information, 'in line, in particular, with the principles of the Aarhus Convention, the INSPIRE Directive and the Shared Environmental Information System (SEIS)'.[240] Article 8 of the proposal requires the data and information policy of GMES actions to have as an objective the 'full and open access to information produced by GMES services and data collected through GMES infrastructure, subject to relevant security restrictions'. While the European Parliament insists on such a free and open access policy, some of the Member States have expressed their concerns in the Council.[241] This will be addressed further in the sections on re-use and sharing.

F. Interim Summary

Objective – As was stated in the introduction, the purpose of this chapter is to have an overview of what the international and European legislation and policy

238. GMES Support Team, 'GMES Vision and targets 2008 and Work progress of the Initial Period of the GMES Action Plan (2002–2003)', 2003, 12, <www.gmes.info/pages-principales/library/reference-documents/?no_cache=1&download=GMES_Vision_and_progress.pdf&did=61>, 22 Feb. 2010.

239. Commission of the European Communities, *Communication to the European Parliament and the Council. Global Monitoring for Environment and Security (GMES): Establishing a GMES capacity by 2008 – (Action Plan (2004–2008))*, COM (2004) 65 final, 13, 3 Feb. 2004.

240. Commission of the European Communities, *Proposal for a Regulation of the European Parliament and of the Council on the European Earth observation programme (GMES) and its initial operations (2011–2013)*, COM (2009) 223 final, 5, 20 May 2009. See also Commission of the European Communities, *Communication to the European Parliament, the Council, the European Economic and Social Committee and the Committee of the Regions*, COM (2009) 589 final, 28 Oct. 2009.

241. Presidency of the Council of the European Union, *Note to the Research Working Party. Proposal for a Regulation of the European Parliament and of the Council on the European Earth observation programme (GMES) and its initial operations (2011–2013 – key points*, doc. nr. 16176/09, 19 Nov. 2009, <http://register.consilium.europa.eu/pdf/en/09/st16/st16176.en09.pdf>, 22 Feb. 2010.

documents find to be obligations for public bodies to make their public sector data available for the different purposes. First, this may indicate the scope and importance of the different purposes for which public sector spatial data are used. Second, this may give an idea about the impact of an unclear distinction between access, re-use, and sharing.

MORE ATTENTION FOR PUBLIC SECTOR DATA IN GENERAL AND FOR ENVIRONMENTAL DATA THEN SPATIAL DATA – This section has looked at the obligations of the public bodies to make public sector spatial data available for access as they are perceived by international and European legislation and policy documents. Generally, access to public sector data and environmental data have received a lot of attention from the international and especially the European organizations. The availability of public sector spatial data for public access has only been on the agenda for a much shorter period. This may be explained by the fact that public sector spatial data are already covered by the documents on general access, and by the fact that a lot of the spatial data that will interest the citizens is either covered by environmental data or by specific legislation, for example cadastre and land registry legislation. Moreover, a lot of public sector spatial data is often used as a background to show other thematic data, so the citizens may not have a direct need for the spatial data themselves. For such reasons, the data are more valuable for re-use and sharing.

PUBLIC SECTOR DATA – From the overview that was given in this section, it is clear that making data available upon request is clearly seen as a task for the public bodies, with extensively elaborated obligations on timing, format, exceptions, and so on. The active dissemination of public sector data for public access is regulated much more loosely. For instance, the Special Rapporteur of the UN feels that 'documents of significant public interest' need to be made available, while the UNESCO guidelines refer to 'information that meets the needs of citizens'. In its Recommendation Rec(2002)2 and in the Convention on Access to Official Documents, the Council of Europe spoke about dissemination of information 'in the interest of promoting the transparency of public administration and efficiency within administrations' or that will 'encourage informed participation by the public in matters of public interest'. The EU sees access as important for participation in the European integration process and for mobility, but it did not take any initiatives on access for subsidiarity reasons.

ENVIRONMENTAL DATA – With regard to environmental data, the rules have been made more concrete: lists or examples of documents that need to be disseminated are included in the Aarhus Convention, the Access Directive, and the OECD Recommendation on environmental information. These documents include national and international legislation and agreements regarding the environment; environmental policies, plans, and programmes; environmental impact studies and risk assessments; and other significant international documents. The Aarhus Convention and the Access Directive emphasize that these documents have to be available electronically. Other documents that need to be made available are, for instance, reports on the state of the environment, explanatory material on dealings with the public in environmental matters, or summaries of data derived from monitoring activities.

Spatial data – With regard to spatial data, the obligation of the public bodies to make data available for access purposes is only explicitly mentioned in the INSPIRE Directive, while even in that directive, the main focus is on sharing between public bodies for the purpose of performing the public task. However, the categories of data sets that need to be made available through the network of services that needs to be set up under INSPIRE include a broad range of spatial data.

Practical arrangements – For all types of data, there are a number of obligations, often referred to as practical arrangements, that need to be fulfilled to make sure that the public has easy access. These measures include setting up registers or access lists; providing metadata, assistance, and education; the appointment of information officers in the public bodies; and providing data in an electronic format. In the INSPIRE Directive, one could say that such practical arrangements are part of the main elements of the directive: it only concerns data in an electronic form, and the network services and metadata are an essential part of the rules.

Summary – Hence, there are a number of obligations with regard to making public sector spatial data available for access that are common to all these texts. In essence, the background of all texts lies in the exercise of democratic rights and obligations and participation in democracy. Although most of them focus on the availability of data for access on request, the dissemination of data by public bodies of their own motion is growing increasingly important.

III. THE OBLIGATION OF MAKING PUBLIC SECTOR SPATIAL DATA AVAILABLE FOR RE-USE

A. Introduction

Recent interest in making data available for re-use – The previous section showed that the obligation of public bodies to make data available for access purposes has gained increasing attention from international and European legislators and policy makers. The interest in the value of these data for any use beyond access for democratic purposes is more recent. The task of making data available for re-use has not taken on such a solid form as the obligation of making data available for access. During the 1980s and the 1990s, governments and public bodies began to understand that public sector data have a considerable value for increasing economic growth and employment, particularly in the information market. The idea began to grow that public bodies should make their data available for all types of use by third parties, including commercial use. The European Commission has taken on a front-runner's role in promoting this idea.

Outline – In the next part of this section, the international and European legislative and policy documents on re-use are examined in order to have an indication of the scope of the obligation to make public sector spatial data available for re-use. In line with the structure of the first part of this chapter, we will first question whether the fundamental right to freedom of information includes the right to use the data for re-use purposes, that is, any commercial or non-commercial use outside of the public tasks that requires the processing of the data beyond access needs.

B. Does Freedom of Information Include a Right to Re-use the Information?

No evolution towards recognition under article 10 of the ECHR – In the previous part of the chapter, it was stated that access to public sector data was at the verge of being recognized as part of the fundamental right of freedom of information and freedom of expression in Article 10 of the ECHR. The next step would be the acknowledgement that this right to access data would also include the right to re-use it for any purpose. So far, the European Court of Human Rights has not acknowledged such an evolution.

First indication – A first indication that the right to access public sector data, if it were ever fully recognized by the Court, could include the right to re-use the data for any purpose can be found in the *Casado Coca* case. In this case, the Court stated that Article 10 applies to 'everyone', whether the type of aim pursued is profit-making or not.[242] In addition, 'Article 10 does not apply solely to certain types of information or ideas or forms of expression, in particular those of political nature; it also encompasses ... information of a commercial nature ... and even light music and commercials transmitted by cable'. However, this case addressed freedom of expression rather than freedom of information, so it is a wide stretch to apply this to the access to public sector data debate.[243]

C. International and European Documents

Recent developments – The obligation of public bodies to make public sector data available for re-use has been on the agenda of the European and international organizations mainly since the 1990s. The biggest promoter of re-use has been the European Commission. Therefore, most of this section will deal with the policy documents adopted by the Commission, which ultimately culminated in the PSI Directive. While these documents have already been mentioned in Chapter 3, that chapter only addressed them as part of the process towards the PSI Directive and did not look at their specific views on the re-use of public sector data. This will be examined in this section.

---

242. European Court of Human Rights, *Casado Coca v. Spain*, 24 Feb. 1994, para. 35, <http://cmiskp.echr.coe.int/tkp197/search.asp?skin=hudoc-en>, 22 Feb. 2010.
243. *Ibid.*, See also Dirk Voorhoof, 'Artikel 10 - Vrijheid van meningsuiting', in *Handboek EVRM. Deel 2. Artikelsgewijze commentaar*, ed. Johan Vande Lanotte and Yves Haeck (Antwerp: Intersentia, 2004, I), 940 [in Dutch].

NO DISTINCTION BETWEEN DOCUMENTS ON PUBLIC SECTOR DATA, ENVIRONMENTAL DATA AND SPATIAL DATA – Contrary to the previous section, no distinction is made in this part of the chapter between documents on public sector data, environmental data, and spatial data because most documents discuss public sector data in general and they do not foresee any different treatment for environmental data or spatial data (although they do emphasize the importance of spatial data for re-use). The only exceptions to this are UNSDI, GEOSS, and GMES.

OBJECTIVE – The examination of the documents involved should give us an insight into what the international and European legislators and policy makers feel should be part of the obligations to make public sector spatial data available for re-use and what the scope of such re-use should be.

1. United Nations

a. UNESCO Policy Guidelines

VALUE OF THE POLICY GUIDELINES FOR RE-USE – The UNESCO Policy Guidelines for the Development and Promotion of Governmental Public Domain Information were addressed in the section on the task of making public sector data available for access purposes because they are mainly focused on access.[244] However, as the guidelines are directed towards the promotion of access and *use* of governmental public domain information, they deserve to be mentioned here for the sake of completeness. The Guidelines refer to the economic value of public sector information for the information industry, yet they do not address any issues that could be considered to go beyond access into the re-use debate. As was mentioned above, the fact that access and use are named in one breath may be due to a federal American point of view: access to public sector data automatically entails the right to use those data for any further purpose.

b. United Nations Spatial Data Infrastructure

VALUE OF THE UNSDI INITIATIVE FOR RE-USE – The importance of this initiative for re-use is limited, but it has the potential to grow. While the UN Spatial Data Infrastructure mentions the benefits of the availability of spatial data for academia, industry, NGOs, and the community at large, the emphasis currently lies on the sharing of spatial data for the performance of the UN activities for peacekeeping, sustainability, food and environment programmes, and so on.[245] The UNSDI is

244. UNESCO, 'Policy Guidelines for the Development and Promotion of Governmental Public Domain Information' (Paris: UNESCO, 2004), 2–3, <http://unesdoc.unesco.org/images/0013/001373/137363eo.pdf>, 22 Feb. 2010.

245. United Nations Geographic Information Working Group, 'Strategy for developing and implementing a United Nations Spatial Data Infrastructure in support of Humanitarian Response, Economic Development, Environmental Protection, Peace and Safety', 2007, 5–6, <www.ungiwg.org/docs/unsdi/UNSDI_Strategy_Implementation_Paper.pdf>, 22 Feb. 2010.

counting on cooperation with the private sector, NGOs, and interest groups, but these relationships will all be in light of the performance of the tasks of the UN, which could be seen as public tasks, so generally, there will be no re-use. However, as the UNSDI vision includes the availability of spatial data outside of the UN, it may be possible that in some cases there will be re-use, for example, when academia obtains the data for research that is not commissioned by the UN or that is not immediately related to UN activities.

c. *Global Earth Observation System of Systems*

RE-USE FOR RESEARCH AND EDUCATION PURPOSES – The main focus of the development of GEOSS is the availability of spatial Earth observation data for the purpose of performing public tasks. Therefore a further description can be found in the section on sharing. However, GEOSS also intends to promote the availability of data for research and education purposes. Depending on the type or purpose of the research or education, this may fall under the category of sharing (if it is a public task), or it can be categorized as re-use.

IMPLEMENTATION GUIDELINES – The Guidelines implementing the GEOSS data sharing principles can shed more light on this. As was mentioned above, these Guidelines are currently under development and will not be definitive until November 2010. Therefore, we will briefly discuss the draft Guidelines. These Guidelines urge the members of GEO to define research and education in order to comply with the promotion of the availability of data for research and education purposes. Data, metadata, and information products should be made available free of charge or at no more than the cost of reproduction.[246] The users should be informed of the costs and possible reductions, so that they know 'about the costs they should expect when they move from educational and research applications to other operational applications'.[247] Any other type of re-use is not considered directly in the draft Guidelines.

2. Organization for Economic Cooperation and Development

OUTLINE – The OECD Recommendation for Enhanced Access and More Effective Use of Public Sector Information (PSI) was announced in the previous section on access. However, as the principles of this Recommendation are almost solely related to re-use rather than access, they are addressed in this section.

GENERAL PRINCIPLES – The Recommendations has thirteen principles, 'which provide a general framework for the wider and more effective use of public sector

246. Group on Earth Observations, 'Draft Implementation Guidelines for the GEOSS Data Sharing Principles', 9–10, <www.earthobservations.org/geoss_dsp.shtml>, 22 Feb. 2010.
247. *Ibid.*, 10.

information and content and the generation of new uses from it'.[248] These principles include:

- Openness: the availability of PSI for use and re-use based upon the presumption of openness should be the default rule to facilitate access and re-use;
- Access and transparent conditions for re-use: broad non-discriminatory competitive access and conditions for re-use should be encouraged and unnecessary restrictions should be removed, so that in principle all accessible information would be open to re-use by all;
- Asset lists: awareness of what PSI is available for re-use should be strengthened. This could be done by providing information asset lists and inventories, preferably published online, and by a clear presentation of the conditions for access and re-use;
- Quality: methodical data collection and curation practices need to be ensured to enhance quality and reliability;
- Integrity: the integrity and availability of information should be ensured by implementing appropriate safeguards and best practices in information management;
- New technologies and long-term preservation;
- Copyright: IPR should be respected;
- Pricing: when PSI is not free of charge, it should be priced transparently and consistently across public bodies. Where possible, charges should not be higher than marginal costs for maintenance and distribution, and in special case extra costs, for example for digitization. Any higher charges should be based on clearly expressed policy ground. It is not clear from this recommendation whether such costs should be charged for both access and re-use, or if re-use should be treated differently from access;
- Competition: unfair competition should be avoided in situations where both the public and the private sector promote value added services. Particular attention should be paid to information resources where the public body is the only source;[249]
- Redress mechanisms: transparent complaints and appeals processes should be available;
- Public private partnerships: where appropriate and feasible, public private partnerships should be promoted to make PSI available;
- International access and use: cross-border re-use should be facilitated;
- Best practices: information on best practices and on enhanced implementation should be shared, users and re-users should be educated and institutional capacity should be built.

248. OECD, *Recommendation C(2008)36 of the Council for Enhanced Access and More Effective Use of Public Sector Information*, 2008, 5, <www.oecd.org/dataoecd/0/27/40826024.pdf>, 22 Feb. 2010.

249. The issue of competition will be addressed in Ch. VIII, s. 4.

OBJECTIVE – The purpose of the Recommendation is to encourage member countries to realize the potential of PSI. OECD will review the implementation of these principles in the members countries in three years' time and rethink how they can be potentially modified according to the countries' experience.[250]

REPORT ON PUBLIC SECTOR INFORMATION AND CONTENT – The Recommendation partly builds on the findings of the 2006 report on public sector information and content. The part of the report on making public sector content available for access was already addressed in the previous section of the chapter, so this section only mentions the re-use of PSI.[251] The report lists four main challenges for PSI: the definition of roles between the public and the private sector; operational management and licensing agreements; data quality; and lack of statistics on the economic value of PSI.[252] It warns that in many countries access does not necessarily imply rights to commercial re-use and that IPR may have a role to play in some countries.[253] Moreover, pricing policies range from free dissemination and marginal cost pricing to various levels of production cost recovery plus profit pricing.[254] These issues might restrict the re-use of PSI.

3. Council of Europe

ATTENTION FOR NON-COMMERCIAL RE-USE ONLY – The Council of Europe has not paid much attention to the issue of re-use and to the obligation of public bodies of making public sector data available for re-use purposes. The only reference to re-use can be found in a 2007 Recommendation of the Committee of Ministers, which encourages the Member States to promote 'an open offer of services and accessible, usable and exploitable content via the Internet which caters to the different needs of users and social groups, in particular by . . . encouraging, where appropriate, the re-use of public data by non-commercial users, so as to allow every individual access to public information, facilitating their participation in public life and democratic processes'.[255] Hence, the Committee of Ministers was only concerned with non-commercial re-use. Previously, it was made clear that only allowing non-commercial re-use is not possible in the Member States of the European Community, so the Committee of Ministers' Recommendation adds no extra value to the re-use debate.

---

250. Andrej Osterman, 'Minutes of the 11th Meeting of the Public Sector Information Group', 2008, 6, <http://ec.europa.eu/information_society/policy/psi/docs/pdfs/minutes_psi_group_meetings/11th.pdf>, 22 Feb. 2010.
251. OECD Working Party on the Information Economy, 'Digital Broadband Content: Public Sector Information and Content', 2006, 82 p., <www.oecd.org/dataoecd/10/22/36481524.pdf>, 22 Feb. 2010.
252. *Ibid.*, 30–31.
253. *Ibid.*, 43.
254. *Ibid.*, 45.
255. Council of Europe Committee of Ministers, *Recommendation CM/Rec(2007)16 on measures to promote the public service value of the Internet*, 2007, <https://wcd.coe.int/ViewDoc.jsp?id=1207291>, 22 Feb. 2010.

4. European Union

BACKGROUND – The EU has issued binding legislation (at least with regards to the objectives that need to be reached) in the form of the PSI Directive, but this directive was the result of a long process involving the European institutions, the Member States, and the stakeholders. This process is still ongoing, as the European Commission launched its review procedure in the summer of 2008 and will perform a new review in 2012.

a. Synergy Guidelines

OUTLINE – The Synergy Guidelines were addressed in Chapter 3.[256] In this section, we briefly review what these Guidelines recommended on the task of the public bodies to make public sector data available for re-use. The position of the Synergy Guidelines on the provision of information services by public bodies will be discussed later.[257]

GENERAL PRINCIPLES – The Synergy Guidelines indicated that public bodies should allow re-use of the data they had collected by the private sector and the information industry through electronic information services. The public bodies should regularly review their data with regard to the possibility of their re-use. The availability of data, the conditions for their re-use, and the procedures by which they can be obtained should be publicized. The conditions for re-use should be developed by consultation with the appropriate bodies and publicized by the Member States, and the negotiation procedures and charging principles should be as harmonized as possible. This could be supported by the establishment of an advisory body and model contracts. Charges should reflect the costs of preparing the data for dissemination and disseminating the data, but not necessarily the full costs of collecting and processing the data.

IMPACT OF THE SYNERGY GUIDELINES – Hence, the Guidelines promoted the general availability of public sector data for re-use. When access to the data was not restricted for reasons of privacy, security, and so on, the data should also be available for re-use. However, these Guidelines did not have much influence in the Member States, due to their non-binding character and the differences in practices between and within the Member States.

b. The Publaw Reports

RE-USE AS A SUPPLEMENT TO ACCESS? – While the first Publaw study addressed access to data held by public bodies and thus at first thought should be discussed in that section, the study only looked at access legislation in light of the information market. Hence, it is discussed in this section on re-use. According to the study, '[i]n general the private sector use of access legislation may be seen as a use

256. Cf. *supra*, Ch. 3, s. III.A.1.
257. Cf. *infra*, Ch. 7, s. III.B.2.a.

complementary to the use by individuals. The private sector may offer additional services based on public sector information . . . The private sector, however, can only play this role if the principle of public access as such is not put into question. Private sector involvement in public sector information should therefore not be seen as a contradiction but as a supplement to the principle of public access'.[258] The Publaw report was right in stating that there can be no re-use if access is not allowed. However, the right to obtain access to public sector documents does not necessarily imply a right to re-use the documents that were obtained.

RE-USE AS A REQUIREMENT FOR ACCESS? – The study also indicated that an exclusion of re-use would jeopardise the right to access because:

> The effectiveness of access rights very often demands that organizations specialize in obtaining information from the public sector on the basis of access rights, upgrade this information and make it accessible to the citizen in more 'digestible' formats. Such organizations, very often public interest groups, would have to demand some sort of compensation for these services. Also one has to take into account that the public sector might cease to collect or to hold available certain kinds of information. In such circumstances it is useful if private organization[s] have already established themselves in providing such information. Finally, such restrictions on re-use would also have to be analyzed in view of their compatibility with EEC competition law.[259]

However, as making public sector data available for access purposes can be seen as an obligation in its own right with a lot of international and European support, this will most likely remain standing independent of any decisions on re-use. This line of reasoning indicates once again that the relationship between access and re-use is not clear.

PUBLIC OR PRIVATE SECTOR? – According to the Publaw 2 study, '[g]overnment departments collect information in order to perform their primary role of government and administration. The subsequent release and exploitation of the information they collect is secondary to their main purpose'.[260] Therefore, the exploitation of public information should best be left to the private sector. The report seems to confuse making data available and creating information services. The former is a part of the task of the public bodies under access legislation, so it cannot be left to the private sector to disseminate the data that were collected by public bodies. The latter deals with the relationship between the public and the private sector for the provision of information services on the market.[261]

---

258. Commission of the European Communities, *Publaw. Subject Report: General Access to Information Legislation* (Luxembourg: CEC Legal Advisory Board, 1991), s.p.
259. *Ibid.*
260. Commission of the European Communities, *Publaw 2. A Report to the Commission of the European Communities on an Evaluation of the Implementation of the Commission's Guidelines for Improving the Synergy between the Public and Private Sectors of the Information Market* (Luxembourg: Cec Legal Advisory Board, 1993), 14.
261. Cf. *infra*, Ch. 7, s. II.C.5.b.

Recommendation to allow re-use – The third Publaw study looked at the evolution of the European information market since the first Publaw report in 1991 up to 1995. As most of its findings and recommendations were more related to the provision of information services, this study will be discussed in that chapter. We only indicate the following recommendations of the report. Any prohibitions on the re-use of public sector data should be removed from national access legislation. Charging should be possible, depending on the type and quantity of the requested data. However, some basic data should be made available free of charge. Regarding this last statement, one could wonder whether the report was still looking at access from the point of view of the information market, so that these data should be free to re-use, or whether it was taking the point of view that essential data should be freely available for access.[262] Government was expected in the report to play an active role in increasing the availability of data by indicating which data were available in what location and under which conditions, and by regularly updating the data that were suitable for commercial re-use.[263] Government should maximise the use of its data and stimulate variety in information products and services. Even if active dissemination were part of the task of a public body, it should still consider involving the private sector.[264]

c. *Green Paper on Public Sector Information*

Importance of public sector information – The European Commission's Green Paper on Public Sector Information stressed the importance of PSI for access and for re-use, and listed a number of issues that might create barriers for the development of the information industry in the European Community.

Consultation – The Green Paper launched a consultation on several matters relating to access and re-use, which was received with high interest and got 185 responses. The main questions of the consultation addressed the definition and categories of PSI, the conditions for access to PSI, and the barriers presented by copyright, privacy, liability, and so on.[265]

Responses to the consultation – While most of the responses to the Green Paper addressed the importance of access to PSI for guaranteeing democracy, accountability, and transparency, many of the replies also indicated the economic value of PSI, and the private sector strongly pleaded for a harmonized policy for re-use. The need for metadata and the importance of information technology was also highlighted in almost every response. Throughout the responses, the confusion between access and re-use often reappeared, and the need for a clear description of both concepts became clear.

262. Commission of the European Communities, *Publaw 3. Final report* (Luxembourg: CEC Legal Advisory Board, 1995), 81.
263. Simone Van Der Hof, *Overheidsinformatie in de etalage. Belangen rondom de toegang tot overheidsinformatie* (Alphen a/d Rijn: Samsom Bedrijfsinformatie, 1997), 22 [in Dutch].
264. *Ibid.*, 24.
265. Commission of the European Communities, *eEurope 2002 Communication. Annex*, 4.

d. *eEurope 2002 Communication*

GENERAL PRINCIPLE – In the eEurope 2002 Communication, the Commission once again stressed the importance of PSI for citizens and for economic growth, but it focused its attention on re-use. The Communication's leading principle was a general right of re-use: whenever PSI was generally accessible under access legislation, it should be available for re-use.

CONSULTATION – The Communication launched a new consultation round of stakeholders and received almost eighty responses. The Working Document that was prepared for the Consultation contained a first draft of a legal instrument that could be used for regulating the re-use of PSI. The main questions of the working document regarded the adequate measures that needed to be taken with regard to re-use, the scope of any legal instrument that would be introduced, IPR, and transparency. The first draft of a legal instrument contained a general principle that all documents that are generally accessible should be re-usable for commercial or non-commercial purposes.

RESPONSES TO THE CONSULTATION – Responses were received from Member States and public bodies, business, academia, and citizens. The sectors that were most represented were the cultural sector (with archives, libraries, and museums), the sector of spatial data, and newspaper and magazine publishers. Opinions were divided on whether re-use should be a general right or not. Many of the respondents felt that it was impossible to make a legal instrument that would apply to all types of PSI and that a sector-specific approach would be preferable. The cultural institutions were very concerned that making their data available for re-use would endanger their revenues from selling information products and services and, as a consequence, would jeopardize their activities and existence. The private sector was very much in favour of the idea of a directive and applauded the preference for data in electronic formats, the establishment of regulatory or arbitrary bodies, search facilities, the prohibition of exclusive agreements, and the introduction of a marginal cost pricing policy. Throughout the responses, access to PSI was often mentioned, even though the Communication and the working document explicitly announced that they would not address access issues. While some pleaded for clarification of the difference between access and re-use, others failed to see the distinction between both. This once again led to confusion about the demarcation between both and about the obligations of the public bodies regarding either issue.

e. *The PSI Directive*

OUTLINE – The PSI Directive was discussed extensively in Chapter 3, so in this section we will only address the main elements of the obligation of public bodies to make their data available for re-use.

GENERAL PRINCIPLE – As the PSI Directive's main purpose is to stimulate the availability of public sector documents for re-use for commercial and non-commercial purposes, one would expect the directive to include a clean-cut obligation for the public bodies to make their documents available. However, this is not the case. The general principle of the directive does not impose any obligation on the Member States or the public bodies to make their documents available for re-use. Only if they choose to do so should they follow the rules of the PSI Directive and its implementing legislation.[266]

MAKING DATA AVAILABLE FOR RE-USE NOT A TASK FOR THE PUBLIC BODIES? – Does this mean that the European legislator did not see making public sector data available for re-use as a task for the public bodies? Some Member States' delegations in the Council felt that there should be an obligation to allow re-use, and the Parliament tried to install an obligation to make at least basic information of the constitutional state available for re-use. However, these aspirations were not fulfilled, and no general obligation to allow re-use was agreed upon.[267] With regard to the intention of the European Parliament to make basic information of the constitutional state available, once again, there is a possibility for confusion between access and re-use. If information is regarded as being of a basic importance for the constitutional state, it is important that it is available for everyone in order to enable them to participate in decision making, the democratic process, and society in general. Hence, it is important for access. This information does not necessarily have to be available for re-use more than any other type of information. One of the reasons for this confusion may be the extensive scope of re-use, that is, use for commercial and non-commercial purposes.[268]

FREEDOM OF DECISION FOR MEMBER STATES OR PUBLIC BODIES – Hence, it remains a decision of the Member State or even the public body itself whether it considers making data available for re-use obligatory. If they do not, then they do not have to allow re-use. If they do, they have to allow re-use for commercial and non-commercial purposes and follow the rules of the directive regarding time limits, format of delivery, means of redress, charging, non-discrimination, and so on.

EXEMPTIONS – The directive excluded some public bodies from its field of application, namely cultural institutions, education and research institutions, and public broadcasting organizations. One might wonder if these public bodies were left out because the European institutions did not feel they had an obligation at all to make their data available for re-use, or because they believed that a task might be assumed but a separate approach would be needed for these particular types of data.

TRANSPARENCY – According to Article 9 of the directive, the Member States should also take measures to ensure that potential re-users can discover new documents that are available and find out where to get them. This entails that the Member States and the public bodies should consciously consider for which public sector documents they see an obligation to allow re-use. Then they should make sure that practical arrangements are in place to facilitate the search for available documents, such as asset lists and portal sites.[269]

---

266. Article. 3 of the PSI Directive.
267. Cf. *supra*, Ch. 3, s. III.B.5.
268. Cf. *infra*, Ch. 6, s. V.
269. Article 9 of the PSI Directive.

IMPACT OF THE PSI DIRECTIVE – A study conducted by Micus Management Consulting in 2008 for the domains of geographical information, meteorological information, and legal and administrative information showed that the PSI Directive has had a positive impact on the re-use environment. Growth rates were reported in every market segment that was considered by the study, but mainly on the technical level: 'In particular, technical oriented changes . . . such as delivery formats, delivery speed and number of products have been pursued at high intensity. [Public sector information] is increasingly offered on Internet portals or by web services, thus providing instant access to a large amount of regularly updated data'.[270] However, other areas needed more improvement, such as transparency of pricing and licensing policies, and the definition of the obligation of the public bodies to make public sector documents available. Moreover, a significant lack of awareness of the PSI Directive among both the public bodies and the private sector was noted. The 2008 review also found that the potential of PSI 'is still not being tapped because of the way public sector bodies handle their information resources' and 'there are practical issues hindering the re-use of PSI, for example, lack of information on available PSI. In other cases, public sector bodies are lukewarm to the basic idea of commercial re-use of their information or fail to realize its economic potential'.[271]

f. The Commission Decision on the Re-use of Commission Information

GENERAL PRINCIPLE – On 7 April 2006, the European Commission adopted a decision to allow the re-use of its own data for commercial and non-commercial purposes.[272] As a general principle, all Commission documents are re-usable for commercial or non-commercial purposes in accordance with the conditions set out in the Decision. The following documents are kept outside of the field of application of the Decision:

- software or documents covered by industrial property rights such as patents, trademarks, registered designs, logos and names;
- documents for which the Commission is not in a position to allow re-use in view of IPR of third parties;
- the research results of the Joint Research Centre;

270. Micus Management Consulting Gmbh, 'Assessment of the Re-use of Public Sector Information (PSI) in the Geographical Information, Meteorological Information and Legal Information Sectors', 2008, 98, <http://ec.europa.eu/information_society/policy/psi/docs/pdfs/micus_report_december2008.pdf>, 22 Feb. 2010.
271. Commission of the European Communities, *Communication to the European Parliament, the Council, the European Economic and Social Committee and the Committee of the Regions. Re-use of Public Sector Information – Review of Directive 2003/98/EC*, COM (2009) 212 final, 3, 7 May 2009.
272. Commission Decision 2006/291/EC, Euratom of 7 Apr. 2006 on the re-use of Commission information, *OJ L* 107, 20 Apr. 2006, 38.

- documents made accessible to a party under specific rules governing privileged access to documents.[273]

TRANSPARENCY – To make things easier for potential re-users and limit the need to make individual re-use applications, the Commission services have to indicate, where appropriate, whether documents are re-usable, for example, by a notice on a web page or a 're-usable' icon next to the title of a document.[274] Conditions and standard charges have to be pre-established and published, and practical arrangements will be made to facilitate the search for re-use, such as asset lists of main documents. If conditions are imposed, this may be done through a license or a disclaimer.

TIME LIMIT – A time limit of fifteen working days has been decided to handle a request and either allow re-use and provide a copy of the document or refuse the request. Such a refusal has to be made in writing and state the reasons and the possibility for the applicant to file a complaint with the CFI or the European Ombudsman.

CHARGES – In principle, the re-use is free of charge, but in specific cases, marginal costs incurred for the reproduction and dissemination of documents may be charged. If the Commission finds that adaptations are needed in order to satisfy a specific application, the costs involved in that adaptation may also be charged to the applicant.[275]

OTHER EU INSTITUTIONS AND BODIES – The other institutions of the EU have not taken any general measures with regard to allowing the re-use of their data.

g. Global Monitoring for Environment and Security (Kopernikus)

VALUE OF THE GMES INITIATIVE – As was mentioned above, the GMES initiative is mainly focused on making spatial data available for the purpose of sharing; therefore, it is discussed in that section. However, in 2001, the European Commission stated that GMES is also intended to develop the market for European industry: 'Operational GMES services may be provided by industry on commercial basis giving the opportunity to develop a strong and innovative industrial pole both for information services and space/terrestrial infrastructure. Revenues have been estimated for data suppliers, value adding companies and other business activities'.[276] Yet, until now, the developments in the implementation phase of GMES have shown more attention for the development of the services for policy makers.

In the 2009 proposal for a Regulation, the European Commission also pays attention to re-use: the GMES data policy should promote the use and sharing of

273. Article 2 of Commission Decision 2006/291.
274. Article 5 of Commission Decision 2006/291.
275. Article 7 of Commission Decision 2006/291.
276. Commission of the European Communities, *Communication to the Council and the European Parliament. Global Monitoring for Environment and Security(GMES) – Outline GMES EC Action Plan (InitialPeriod: 2001–2003)*, COM (2001) 609 final, 23 Oct. 2001.

GMES information and data; strengthen Earth observation markets in Europe, in particular the downstream sector, with a view to enabling growth and job creation; and support the European research communities.[277] This promotion should be done by lowering the cost to re-use information.[278] The Council was concerned about the impact of the proposed 'open access' policy, and while it agreed with the free of charge availability of data – at least for the Community-funded GMES services and data – it found that registration and acceptance of user licenses should be required, subject to relevant security and licensing restrictions.[279]

D. Interim Summary

Objective – This section has looked at the obligations to make public sector spatial data available for re-use in the view of the international and European legislator and policy makers. This should give us some insight in the scope of re-use and the impact of an unclear distinction between access, re-use, and sharing.

Limited acceptance of an obligation to make data available for re-use – The obligation of public bodies to make their public sector spatial data available for re-use is not nearly as well recognized or accepted as the obligation of making data available for access purposes. This is partly due to the more recent nature of the discussion on this part of the public bodies' information activities, which in turn can be brought back to the developments in information technology that make it easier to select, combine, process, and disseminate data to a large audience. Another possible cause could be that public bodies are not inclined to make their data available for re-use for several reasons, such as security, liability, or protection of personal data, or because they realize the economic value of their data and they want to exploit this economic value for themselves.

United Nations – As was set out above, the attention of the United Nations is not focused on the obligation of public bodies to make public sector data available for re-use purposes, but rather on access or sharing.

OECD – However, the OECD and the EU are more interested in re-use. For the OECD, this interest is fairly recent and has culminated in its Recommendation for enhanced access and more effective use of public sector information. The Recommendation promotes the general availability of public sector data for re-use and the presumption of openness as the default rule, so it sees a general task for the public bodies to make their data available for re-use. Unfortunately, this recommendation is

277. Commission of the European Communities, *Proposal for a Regulation of the European Parliament and of the Council on the European Earth observation programme (GMES) and its initial operations (2011–2013)*, COM (2009) 223 final, 18, 22 Feb. 2010.

278. *Ibid.*, 5.

279. Presidency of the Council of the European Union, *Note to the Delegations. Proposal for a Regulation of the European Parliament and of the Council on the European Earth observation programme (GMES) and its initial operations (2011–2013) – Presidency Compromise Text*, doc. nr. 5233/10, 15 Jan. 2010, <http://register.consilium.europa.eu/pdf/en/10/st05/st05233.en10.pdf>, 22 Feb. 2010.

very recent, and it is not binding, so it is impossible to assess its impact on the availability of public sector data for re-use.

EUROPEAN UNION – The EU has been a big advocate of an obligation to make public sector data available for re-use since the 1980s. Since the Synergy Guidelines, it has been promoting a task of the public bodies to make their data available for re-use. However, this has never led to the true acknowledgement of an obligation in a legally binding text, except for the documents held by the European Commission itself. The Member States and the public bodies only have certain obligations to fulfil when they decide to make their data available for re-use, but the decision of whether they see it as an obligation to allow re-use in the first place is left to their discretion. Considering the efforts from the European Commission for over ten years, one could wonder why the obligation of making public sector data available for re-use was not fully endorsed in the directive. Did some Member States contest the existence of such a task, or could they not agree on a scope?[280] In any case, the EU was not ready to officially recognize it. Moreover, it even felt the need to exclude documents from research and educational institutions, cultural institutions, and public broadcasters from the PSI Directive. What it did acknowledge, however, were some secondary obligations that come into play when the public bodies allow their data to be re-used. These obligations do not relate to the scope of the obligation to make data available or the question of what data should be provided, but rather to *how* this obligation should be performed. Data should be provided in an electronic format and in a timely manner, preferably under a charge that does not exceed the marginal costs of dissemination. The conditions for re-use should be transparent, and means for redress should be available on a national level. Exclusive agreements should be avoided, and fair competition should be ensured.

NO SPECIFIC ATTENTION FOR SPATIAL DATA OR ENVIRONMENTAL DATA – Some final points of interest with regard to the obligation of making data available for re-use can be mentioned. First, until now, no specific attention has been paid to the re-use of spatial data, while this type of data is almost always mentioned as one of the types of data that are most suitable for re-use. From this, one could expect at least some policy papers or guidelines on the re-use of spatial data, but these were never created. The general rules on public sector data were apparently considered sufficient to deal with spatial data as well. The same remark can be made for environmental data.

CONFUSION BETWEEN ACCESS AND RE-USE – Second, during the development of the policies on re-use, confusion with the obligations to make data available for access seems to appear on several occasions. For example, the replies to the Green

280. Leith and McCullagh feel that with the introduction of the PSI Directive 'the EU is trying to run before it learned to walk. One tries to regulate the use of public sector information on a European level without a total agreement about some essential elements, which should be articulated first'. (Philip Leith and Karen Mcullagh, 'Developing European Legal Information Markets based on Government Information', *International Journal of Law and Information Technology* 12, no. 3 (2004): 249.

Paper on Public Sector Information and the 2001 consultation showed that many of the respondents confused the concepts, making it difficult to define what the actual task regarding re-use is. The Publaw reports also showed the difficulty in distinguishing between re-use and access. Determining the scope of the obligations for either access or re-use is complicated even more by this confusion.

IV. THE OBLIGATION OF MAKING PUBLIC SECTOR SPATIAL DATA AVAILABLE FOR SHARING

A. Introduction

Outline – From what was stated above, it is clear that the obligation of making public sector spatial data available for access is much more clearly described in international and European policy documents than the obligation of making re-use of public sector spatial data possible. This section looks at the third part of the information obligations of the public bodies, that is, the obligation to make data available for the purpose of sharing, or performing other public tasks. This sharing of public sector spatial data will mostly take place between public bodies, but this may not always be the case. It may also involve private bodies that are entrusted with a public task.

Aspects of the public task – The public task of the body receiving the public sector spatial data may consist of policy making, but it may also consist of providing services to the public (such as social security, education, public transport, and so on) or making data or information available to the public. The former could be considered as an *internal public task*, while the latter can be seen an *external public task*. In both cases, the public is the beneficiary of the actions in the end, but in the case of policy making, the data that are received are not used directly to provide a particular service to the public.

External public task: information services – In the case of sharing as it is defined in this book, the body receiving the data may in turn provide these data (possibly combined with other data or included in an information product or service) to the public because it is under an obligation to do so. Hence, it is performing its own public task of providing access to data or disseminating an information service to the public. However, it is also possible that the receiving public body uses the data it received to sell information products or services outside of its public task, in competition with private players on the market. In that case, it is re-using the data rather than performing its public task of providing an information service. This means that if the receiving body is providing an information service to the public as part of its public information task – that is, access – the supplying body has to apply the rules on sharing to the exchange of data with the receiving body. If the receiving body is performing market activities, the supplying body has to look towards the rules on re-use for guidance. Therefore, the distinction between access and re-use also has an impact on the sharing framework. This makes it even more crucial to clarify the distinction between the types of use.

No distinction between public sector data, environmental data, and spatial data – In comparison to access and re-use, the sharing of public sector spatial data has also not received a lot of attention from the international and European institutions. While with the increase of environmental and sustainability issues on the political agenda, the need for sharing spatial data has risen over the past years, the material that is relevant for public sector spatial data remains limited. This is also the case for the obligation of sharing environmental data and public sector data in general. Therefore, this section does not make a distinction between the types of data, but merely lists the most relevant documents by organization. It should be mentioned that we have not looked at any initiatives regarding sharing for specific types of data beyond the scope of our analysis, for example, the exchange of passenger name records, DNA data exchange, or cultural heritage.

B. International and European Documents

1. United Nations

a. Rio Declaration and Agenda 21

General principle – The Rio Declaration of 1992 was already mentioned in the section on access, but it has a more important role with regard to sharing. Two principles on sharing in concrete situations are included in the Declaration. First, Principle 18 states that: 'States shall immediately notify other States of any natural disasters or other emergencies that are likely to produce sudden harmful effects on the environment of those States'. Second, Principle 19 adds that: 'States shall provide prior and timely notification and relevant information to potentially affected States on activities that may have a significant adverse transboundary environmental effect and shall consult with those States at an early stage and in good faith'.[281] While these two principles relate to specific situations, the Rio Declaration concerns cooperation between the Member States on many different levels, so they will need to share environmental information for many public task purposes. In addition, the Agenda 21 working plan for implementing the Rio Declaration contains obligations for the Member States to share environmental information.

Agenda 21 implementation plan – Chapter 40 of the Agenda 21 work plan deals with 'information for decision making'. It includes measures for bridging the gaps in collecting data and for improving the availability of information. According to the work plan, '[e]xisting national and international mechanisms of information processing and exchange, and of related technical assistance, should be

281. United Nations Conference on Environment and Development, *Rio Declaration on Environment and Development*, 13 Jun. 1992, <www.unep.org/Documents.Multilingual/Default.asp?DocumentID=78&ArticleID=1163>, 22 Feb. 2010.

strengthened to ensure effective and equitable availability of information generated at the local, provincial, national and international levels, subject to national sovereignty and relevant [IPR]'. While this availability is not strictly intended for public bodies only, its main purpose is to improve decision making on sustainability issues, which in most cases will be considered a public task. Therefore, Agenda 21 is more relevant for the sharing of spatial data than for re-use or access.

TYPES OF DATA – The information that the Member States are required to make available for sharing involves data and information that could be used for the management of sustainable development. While it is not indicated which data sets this would include, an indication is given as to which information should be collected in order for such management to be feasible. The list includes the areas of urban air, freshwater, land resources, desertification, other habitats, soil degradation, biodiversity, the high seas, and the upper atmosphere.[282]

b. Framework Convention on Climate Change

BACKGROUND – Also in 1992, the United Nations Framework Convention on Climate Change (UNFCC) was adopted as a global response to the issue of climate change.[283] Currently, 192 countries have signed the Convention. It is complemented by the Kyoto Protocol.[284] The Convention and the Protocol promote the sharing of data on climate change between public bodies to meet the needs of the Convention.

OBLIGATIONS FOR SHARING – Article 4 of the Convention puts an obligation on the Parties for 'full, open and prompt exchange of relevant scientific, technological, technical, socio-economic and legal information related to the climate system and climate change, and to the economic and social consequences of various response strategies'.

c. United Nations Spatial Data Infrastructure

BACKGROUND – The United Nations Geographic Information Working Group (UNGIWG) was set up in March 2000. It coordinates activities and formulates policies concerning spatial data that affect the work of UN Organizations and Member States. The Working Group approved the concept of a UN-specific SDI

282. United Nations Commission on Sustainable Development, *Agenda 21*, 1992, <www.un.org/esa/sustdev/documents/agenda21/english/agenda21toc.htm>, 22 Feb. 2010.
283. United Nations, *United Nations Convention on Climate Change*, 1992, <http://unfccc.int/resource/docs/convkp/conveng.pdf>, 22 Feb. 2010
284. See UNFFC, 'An introduction to the United Nations Framework Convention on Climate Change (UNFCCC) and its Kyoto Protocol', <http://unfccc.int/press/fact_sheets/items/4978.php>, 22 Feb. 2010. See also also Anne Fitzgerald, *Open access policies, practices and licensing: a review of the literature in Australia and selected jurisdictions* (Brisbane: Queensland University of Technology, 2009), 146–148.

in 2005 at its Sixth Plenary Meeting.[285] Historically, the production and use of spatial data within the United Nations was done by its component organizations for their own particular purposes. This resulted into considerable duplication of effort, reduced opportunities for sharing data and unnecessary costs for the United Nations as a whole.[286]

UNSDI COMPONENTS – At its 2005 meeting, the UNGIWG found that it could deliver the essential components of an operational UNSDI, including the following generic elements:

- data, standards, data sharing and metadata;
- development of common services including infrastructure, charter, and spatial data web services providing data and metadata;
- the organizational and institutional environment, that includes a policy structure, financial, human resources and legal issues, and an interface for information management, and how these would be structured and linked to other parties;
- technical capacity building, interfaces with the outside world and partners (including advocacy and outreach components).[287]

OBJECTIVE – The UNSDI should support the UN's achievement of the Millennium Development Goals.[288] The main mission statement of UNSDI is that it is 'a mechanism to establish system coherence for the applications and exchange of geospatial data for UN activities'.[289] The Frascati Statement of March 2007 added to this that, '[h]aving a UNSDI is substantially conducive to improving decision making for social, economic, and environmental development enhanced at all levels of societies, in peace and safety'.[290] The supported UN activities include food and environmental security, sustainable development, and humanitarian and peacekeeping operations. It should benefit stakeholders such as UN organizations

285. United Nations Geographic Information Working Group, 'UNSDI Compendium. A UNSDI Vision, Implementation Strategy and Reference Architecture', 2007, 53, <www.ungiwg.org/docs/unsdi/UNSDI_Compendium_13_02_2007.pdf>, 22 Feb. 2010.
286. *Ibid.*, 38.
287. *Ibid.*, 54.
288. In September 2000, the General Assembly of the United Nations adopted the United Nations Millennium Declaration, and committed to achieving eight Millennium Development Goals by 2015: eradicate extreme poverty and hunger; achieve universal primary education; promote gender equality and empower women; reduce child mortality; improve maternal health; combat HIV/AIDS, malaria and other diseases; ensure environmental sustainability; and develop a global partnership for development (United Nations General Assembly, *United Nations Millennium Declaration*, 18 Sep. 2000, <www.un.org/millennium/declaration/ares552e.pdf>, 22 Feb. 2010).
289. United Nations Geographic Information Working Group, 'Strategy for developing and implementing a United Nations Spatial Data Infrastructure in support of Humanitarian Response, Economic Development, Environmental Protection, Peace and Safety', 2007, 4, <www.ungiwg.org/docs/unsdi/UNSDI_Strategy_Implementation_Paper.pdf>, 22 Feb. 2010.
290. United Nations Geographic Information Working Group, *Frascati Statement*, 2007, <www.unsdi.nl/downloads/frascatistatement.pdf>, 22 Feb. 2010.

and their staff, Member States and their citizens, regional organizations, strategic partners, and other organizations and institutions involved in social, economic, and environmental development.

TASKS – The priority tasks in the development of the UNSDI relating to the availability of spatial data include the development of UNSDI sharing agreements, metadata standards and a metadata policy, and the progressive availability of data sets that are identified as core geographic data sets. These data sets will include cartographic data, satellite imagery, and GIS services system-wide; thematic data to support sustainability (e.g., climate, water resources, topography, soil, land use, environmental overlays, and demographic patterns); data from and for global environmental observation and assessment; and data to support emergency response and disaster preparedness (e.g., local infrastructure, population, and satellite data).

d. Global Earth Observation System of Systems

OBJECTIVE – As was mentioned before, the purpose of GEOSS is 'to achieve comprehensive, coordinated and sustained observations of the Earth system, in order to improve monitoring of the state of the Earth, increase understanding of Earth processes, and enhance prediction of the behaviour of the Earth system. GEOSS will meet the need for timely, quality long-term global information as a basis for sound decision making and will enhance delivery of benefits to society' relating to the environment.[291] It will provide the overall conceptual and organizational framework to build towards integrated global Earth observations to meet user needs. GEOSS will be a 'system of systems' built on existing and future Earth observation systems, supplementing but not supplanting their own mandates and governance arrangements.[292]

TARGET AREAS – The main areas or public tasks for which GEOSS wants to enhance the availability of spatial data as a basis for sound decision making and the delivery of benefits to society are:

- reducing loss of life and property from natural and human induced disasters;
- understanding environmental factors affecting human health and well-being;
- improving management of energy resources;
- understanding, assessing, predicting, mitigating, and adapting to climate variability and change;
- improving water resource management through better understanding of the water cycle;
- improving weather information, forecasting and warning;

291. Group on Earth Observations, 'The Global Earth Observation System of Systems (GEOSS) 10-Year Implementation Plan', 2005, 1, <www.earthobservations.org/documents/10-Year%20Implementation%20Plan.pdf>, 22 Feb. 2010.
292. *Ibid.*, 2.

- improving the management and protection of terrestrial, coastal, and marine ecosystems;
- supporting sustainable agriculture and combating desertification;
- understanding, monitoring, and conserving biodiversity.[293]

TYPES OF DATA – Data that are needed for decision making in these areas include land use, energy resources, hydrography, meteorological data, species distribution, satellite imagery, climate, natural resources, ecosystems, and so on. The GEONETCast will be established as a distribution system for GEOSS related data, information, and products by 2011.[294]

PRINCIPLES ON DATA SHARING – As was indicated before, the GEO Work Plan for 2007–2009 contained a task to develop guidelines for implementing the three main GEO principles on data sharing:

- There will be full and open exchange of data, metadata, and products shared within GEOSS, recognizing relevant international instruments and national policies and legislation.
- All shared data, metadata, and products will be made available with minimum time delay and at minimum cost.
- All shared data, metadata, and products free of charge or no more than cost of reproduction will be encouraged for research and education.

IMPLEMENTATION GUIDELINES – It was also mentioned earlier that Implementation Guidelines for the GEOSS Data Sharing Principles are currently under development and will not be finalized until November 2010. Yet, some more information can already be given about the draft Guidelines. These Guidelines urge the members of GEOSS to promote the implementation of the principle of full and open exchange because it is deemed essential for GEOSS to realize its vision and potential.[295] Public bodies receiving data should be able to integrate, use, and further disseminate these data with minimal restrictions, but with recognition of all data sources. Yet, the Guidelines recognize that there may be national or international policies or legislation that could limit the availability. These restrictions should of course be respected. In order for the applicants to know about these instruments, GEO should collect them in a compendium. With regard to charging, the Implementation Guidelines contain some elaboration on the second sharing principle of minimum cost. The costs of data collection and system development should be considered a previously incurred cost and an unallowable part of cost recovery. Metadata should generally be available at no cost, and the GEO members should encourage the development of flexible, online cost recovery mechanisms that allow different types of users to understand their costs.[296]

---

293. *Ibid.*, 1–2.
294. Group on Earth Observations, 'GEO 2009–2011 Work Plan', 2009, 7, <www.earthobservations.org/documents/work%20plan/geo_wp0911_rev2_091210.pdf>, 22 Feb. 2010.
295. Group on Earth Observations, 'Draft Implementation Guidelines for the GEOSS Data Sharing Principles', 2008, 3, <www.earthobservations.org/geoss_dsp.shtml>, 22 Feb. 2010.
296. *Ibid.* 7.

Follow-up – The 2009–2011 Work Plan organizes the follow-up of the Data Sharing Principles. It requires GEO to '[u]nder the guidance of the Data Sharing Principles Task Force, define near-term milestones needed to come to consensus on *Implementation Guidelines for the GEOSS Data Sharing Principles* and move expeditiously towards the development of working data sharing procedures for GEOSS'.[297]

2. European Union

a. INSPIRE

Outline – The INSPIRE Directive was discussed extensively in Chapter 3, so in this section I only revisit the obligation of making spatial data available for sharing as it is envisaged by the directive.

General principle – The obligation of public bodies to make spatial data available for sharing is incorporated in Article 17 of the INSPIRE Directive. The Member States have to adopt measures to facilitate sharing of the spatial data sets and services that fall under the thirty-four data themes for the purposes of performing public tasks that may have an impact on the environment. The measures of the Member States have to enable the public authorities to gain access to the spatial data sets and services and to exchange and use those sets and services. They should facilitate the preparation, implementation, and evaluation of environmental policies. The arrangements for sharing must be open to public authorities of other Member States, the institutions and bodies of the EU, and bodies established by international agreements.

Limitations – Any restrictions likely to create practical obstacles occurring at the point of use have to be precluded. The policy makers and decision makers should be able to obtain and use the data when they need it, without technical difficulties, but also without delay due to negotiating licensing conditions, prices, and so on.

Charges – If charges are made or licenses are imposed, these must be fully compatible with the general aim of facilitating sharing. Charges have to be kept to the minimum required to ensure the necessary quality and supply of the data sets and services together with a reasonable return on investment, while respecting the self-financing requirements of the public authorities. This is to ensure the sustainability of the production and dissemination of spatial data, while still emphasizing that the charges should be as low as possible.

Types of data – The list of spatial data themes included in the Annexes to the INSPIRE Directive is considered to hold the data themes that are needed to perform the public tasks relating to the environment addressed in INSPIRE.[298] However, as was mentioned in Chapter 2, this list of data has undergone a

297. Group on Earth Observations, 'GEO 2009–2011 Work Plan', 2009, 8, <www.earthobservations.org/documents/work%20plan/geo_wp0911_rev2_091210.pdf>, 22 Feb. 2010.

298. Cf. *supra*, Ch. 2, s. III.B.

considerable number of changes since the INSPIRE process was started in 2002.[299] The precise scope of the data themes is currently being worked out by Thematic Working Groups.

b. *Shared Environmental Information System (SEIS)*

OBJECTIVE – As was mentioned before, the European Commission's Communication Towards a Shared Environmental Information System addresses the sharing of environmental data for the purpose of preparing and implementing environmental policy. The aim of SEIS is 'to maintain and improve the quality and availability of information required for environmental policy, in line with better regulation, while keeping the associated administrative burdens to a minimum'.[300]

GENERAL PRINCIPLE – The following principles underpin SEIS:

- information should be managed as close as possible to its source;
- information should be collected once, and shared with others for many purposes;
- information should be readily available to public authorities and enable them to easily fulfil their legal reporting obligations;
- information should be readily accessible to end-users, primarily public authorities at all levels from local to European, to enable them to assess in a timely fashion the state of the environment and the effectiveness of their policies, and to design new policy;
- information should also be accessible to enable end-users, both public authorities and citizens, to make comparisons at the appropriate geographical scale (e.g., countries, cities, catchment areas) and to participate meaningfully in the development and implementation of environmental policy;
- information should be fully available to the general public, after due consideration of the appropriate level of aggregation and subject to appropriate confidentiality constraints, and at national level in the relevant national language(s); and
- information sharing and processing should be supported through common, free open-source software tools.

SIMPLIFICATION – SEIS is intended to modernize the legal provisions related to the way the reporting requirements in European environmental legislation are sculpted. It should lead to 'further simplification benefits in relation to (i) the content of information requirements in thematic environmental legislation, (ii) the

299. For a short overview of the history of the INSPIRE data specifications, see INSPIRE Drafting Team on Data Specifications, 'Deliverable D2.3: Definition of Annex Themes and Scope', 2008, 132 p., <http://inspire.jrc.ec.europa.eu/reports/ImplementingRules/DataSpecifications/D2.3_Definition_of_Annex _Themes_and_scope_v3.0.pdf>, 22 Feb. 2010.

300. Commission of the European Communities, *Communication to the Council, the European Parliament, the European Economic and Social Committee and the Committee of the Regions. Towards a Shared Environmental Information System (SEIS)*, COM (2008) 46 final, 2, 1 Feb. 2008.

content and procedure for reporting at international level, and (iii) more efficient organization of data-gathering activities within Member States'.[301]

IMPROVEMENT OF INFORMATION COLLECTION AND AVAILABILITY – SEIS is not only about simplification, but it also recognizes the importance of relevant and timely information of a high quality for better regulation and better policy: 'Improving the mechanisms for collecting, exchanging and using the data can be expected to significantly increase the use that is made of environmentally-relevant data, together with a significant reduction in cost for the users. This will improve the effectiveness of environmental policy across the entire agenda, including for example adaptation to climate change, protection of biodiversity, management of water resources, and prevention and management of environmental crises such as floods and forest fires'.[302]

RELATIONSHIP WITH EXISTING INITIATIVES – SEIS should fit in with existing initiatives on the streamlining of reporting requirements from thematic environmental legislation, the INSPIRE Directive, the Access Directive, the GMES initiative, and GEOSS. A Commission proposal for a new directive was announced for the second half of 2008, but it has not been published by February 2010.

c. Global Monitoring for Environment and Security

BACKGROUND – Global Monitoring for Environment and Security (GMES) is a European initiative for the implementation of information services dealing with environment and security.[303] It is a joint initiative of the European Commission, the European Space Agency, and the Member States of the EU. It is based on observation data received from Earth observation satellites and ground-based data. In this way, it is the EU's contribution to GEOSS.

OBJECTIVE – As the European Commission formulates it, '[t]he GMES initiative seeks to bring together the needs of society related to the issue of environment and security with the advanced technical and operational capability offered by terrestrial and space borne observation systems. It is a direct response to the growing concerns amongst policy makers to ensure in a timely manner access to data on the environment at global, regional and local scales without sacrificing independence'.[304] GMES aims at coordinating existing and new technologies and systems to better meet a structured demand for data on the part of European, national, regional, and local decision makers and users.[305] Under the

---

301. *Ibid.*, 3–5.
302. *Ibid.*, 5.
303. See GMES, 'Overview', <http://ec.europa.eu/gmes/overview.htm>, 22 Feb. 2010.
304. Commission of the European Communities, *Communication to the Council and the European Parliament. Global Monitoring for Environment and Security (GMES) – Outline GMES EC Action Plan (Initial Period: 2001–2003*, COM (2001) 609 final, 2, 23 Oct. 2001.
305. Commission of the European Communities, *Communication to the European Parliament and the Council. Global Monitoring for Environment and Security (GMES): Establishing a GMES capacity by 2008 (Action Plan 2004–2008)*, COM (2004) 65 final, 3, 3 Feb. 2004.

initiative, services are developed that will support making informed decisions on trade-offs that are required to meet environmental and security objectives at acceptable economic costs and that will allow launching efficient response actions at local, regional, and global levels. These services will make it possible to manage natural resources and biodiversity more efficiently, to monitor the key factors in climate change, to respond to disasters, and to ensure more effective border surveillance.[306] GMES will also develop a market for the European industry. This was addressed in the previous section on re-use.

THREE MAIN OBJECTIVES – The first years of the GMES programme were dedicated to exploration and development of the concept. Three main aims of GMES were identified:

- Establishing a GMES Partnership of the key European Actors, who have agreed on the division of tasks as well as on the ownership of data and information. This Partnership should agree on a strategy and programme for a GMES Shared Information System; agree on the repartition of tasks between the partners; and set up a GMES Memorandum of Understanding, or other agreed mechanisms, regarding the shared use of assets, data and information within GMES.
- Creating a GMES Shared Information System. This System will consist of a large number of distributed facilities brought together into a coordinated system oriented towards meeting user needs and will include the following functionality: operational observations; operational production and dissemination of information; storage, cataloguing and library access to information.
- Securing a Mechanism for Dialogue between stakeholders ensuring that user needs are met.[307]

FOCUS OF GMES – From 2004 onwards, the focus was on establishing the GMES capacity and implementation of the concept. A number of EU objectives and policy domains were identified to which GMES should provide support, such as Europe's environmental commitments; other EU policy areas such as agriculture, regional development, fisheries, transport, and external relations with respect to the environmental dimension; and the Common Foreign and Security Policy and other policies relevant to European citizens' security.[308]

306. Commission of the European Communities, *Proposal for a Regulation of the European Parliament and of the Council on the European Earth observation programme (GMES) and its initial operations (2011–2013)*, COM (2009) 223 final, 2, 22 Feb. 2010.
307. GMES Support Team, 'GMES Vision and targets 2008 and Work progress of the Initial Period of the GMES Action Plan (2002–2003)', 2003, 9, <www.gmes.info/pages-principales/library/reference-documents/?no_cache=1&download=GMES_Vision_and_progress.pdf&did=61>, 22 Feb. 2010.
308. Commission of the European Communities, *Communication to the European Parliament and the Council. Global Monitoring for Environment and Security (GMES): Establishing a GMES capacity by 2008 (Action Plan 2004–2008)*, COM (2004) 65 final, 5, 3 Feb. 2004.

INCREMENTAL DEVELOPMENT – GMES was developed in steps through the introduction of pilot phase services, starting with the areas of land monitoring, marine services, and emergency response at the end of 2008.[309]

LEGAL BASIS – The European Commission's 2009 proposal for a Regulation on GMES is intended to provide a legal basis for the GMES initiative and to regulate its funding. The financial cost was determined to be 107 million Euros (EUR).[310] A data policy has to be developed that promotes use and sharing of GMES data and information. According to Article 8 of the proposal, the data and information policy for actions financed under the GMES programme has the objective of promoting the sharing of GMES information and data, and full and open access to information, subject to relevant security restrictions. As was mentioned before, the Council suggested including a requirement for registration and user licenses, with the possibly of additional licensing restrictions.[311] At the time of this writing in February 2010, the proposal was still under discussion.

C. INTERIM SUMMARY

RECENT DEVELOPMENTS – The obligation of public bodies to make public sector spatial data available for sharing has recently received increasing attention from the international and European organizations. The public tasks for which the policies are developed are centred around environmental and sustainability issues (except from GMES, which has a broader scope), and the data that should be provided in order to enable the performance of those public tasks is indicated more or less concretely for all the policies.

TYPES OF DATA – This entails that the obligation of making public sector spatial data available for sharing seems quite clear-cut. It includes data on land use, soil, satellite imagery, meteorological data, hydrography, biodiversity, habitats, topography, agricultural data, infrastructure and resources, species distribution, health statistics, and so on. However, while many data sets are mentioned in all of the policies on sharing that have been described above, many of them are not. Yet, these data sets all deal with environmental and sustainability issues and might often deal with the same situations. Hence, what data should be made available and what the obligation of sharing exactly contains depends on how the policies are defined

309. Commission of the European Communities, *Communication to the Council and the European Parliament. Global Monitoring for Environment and Security (GMES): From Concept to Reality*, COM (2005) 565 final, 8, 10 Nov. 2005.
310. Commission of the European Communities, *Proposal for a Regulation of the European Parliament and of the Council on the European Earth observation programme (GMES) and its initial operations (2011–2013)*, COM (2009) 223 final, 18, 20 May 2009.
311. Presidency of the Council of the European Union, *Note to the Delegations. Proposal for a Regulation of the European Parliament and of the Council on the European Earth observation programme (GMES) and its initial operations (2011–2013) – Presidency Compromise Text*, doc. nr. 5233/10, 15 Jan. 2010, <http://register.consilium.europa.eu/pdf/en/10/st05/st05233.en10.pdf>, 22 Feb. 2010. Cf. *supra*, s. III.C.4.g.

by the stakeholders that are involved. In addition, the changes in the scope of the INSPIRE data themes that have been discussed in Chapter 2 show that the scope of the obligation of sharing public sector spatial data was not so easily agreed upon by the different Member States and the European institutions, and that views changed over time. Therefore, while it would most likely be easier to define the obligation of sharing public sector spatial data than it is for access and re-use, the scope of this task is still contingent and will vary according to the specific needs of the stakeholders.

V. CONCLUSION AND OUTLOOK TOWARDS THE NEXT CHAPTER: WHAT IS THE SCOPE OF THE OBLIGATIONS OF THE PUBLIC BODIES TO MAKE PUBLIC SECTOR SPATIAL DATA AVAILABLE?

ACCESS – Section I of this chapter found that the obligations for public bodies to make public sector spatial data available have evolved very strongly over the last few years and may continue to do so in the future. For instance, the international and European institutions have only recently started to impose heavier obligations on the Member States and the public bodies to make data available of their own motion, rather than just providing data upon request. Thus, the legislator has evolved to a point where it felt that the citizens should be given data without asking for it. However, since the attacks of 11 September 2001, some governments have also moved in the other direction and made less data available, even upon request. This indicates that the scope of the obligation of public bodies to make their data available is highly dependent on time, place, and circumstances. The attention for the availability of environmental data is another example of this. With the increasing concerns about climate change and the influence of environmental factors on human health and safety, the emphasis on the active dissemination of environmental data to the public has been much greater than for other types of data. However, with the financial crisis of 2009, the public may be more interested in obtaining financial data. Does this mean that there is a new task for the government to disseminate such data?

RE-USE – The obligation to make data available for re-use is also of a very recent nature and is mostly based on the development of information and communication technologies and the possibilities they bring for creating information products and services. Hence, the task has arisen due to technical circumstances rather than a conscious decision of the public sector that it has a task of making its data available for re-use. Moreover, until the PSI Directive of 2003, some Member States of the EU even did not allow re-use of their data for commercial purposes, apart from some particular databases, for example, company registers or vehicle registration databases.[312] Hence, the view of the public sector on its obligations

312. See e.g., Commission of the European Communities, *Publaw. Subject report: general access to information legislation* (Luxembourg: CEC Legal Advisory Board, 1991), s.p.

depended on the demand of the private sector. However, this demand would depend on the size, the organization, and the lobbying power of the sector, shutting out small and medium-sized enterprises (SMEs) and certain types of niche data. In 2003, the PSI Directive made it clear that across Europe, there was no consensus on the scope of the task of making data available for re-use, as no obligation to allow re-use could be agreed upon.

SHARING – With regard to sharing, the explicit recognition to share public sector spatial data is also of a very recent nature, and it is subject to many discussions on the conditions against which this obligation should be fulfilled. As from the layperson's point of view, it seems self-evident that public bodies share the data they need for performing their tasks, it is difficult to understand why, for example, it was so difficult for the Member States and the European Commission to come to an agreement about sharing public sector spatial data sets and services under INSPIRE.

UNCLEAR SCOPE – Thus, the scope of the obligations for the EU Member States and the public bodies to make public sector spatial data available for access, re-use, or sharing is not yet completely clear. In addition, the review of the documents involved has shown that many of them mix the availability of data for the three purposes, in this way creating confusion between the rules and conditions that should be applicable to the types of use. Moreover, the overview has shown that the views on the obligations of the public bodies can change.

EVALUATION – Hence, the overview of the international and European legislation and policy documents has not made the legal framework for the availability of public sector spatial data any easier, and it has not given us more insight into the distinction between access, re-use, and sharing, or into the impact of these differences.

NEXT CHAPTER – Is it not possible to determine some criteria for the distinction between access, re-use, and sharing? If such a clear view is not developed, the danger arises that arguments are taken out of context and cloud the vision on the obligations of the public bodies to make data available. In Chapter 2, we stated that the purpose of access, sharing, and re-use determined the difference. The next chapter tries to determine if there are other possible criteria for making the distinction and examines whether the criterion of purpose itself is actually suitable.

Chapter 6

Defining the Distinctions between Access, Re-use, and Sharing

I. INTRODUCTION

Recapitulation – The previous chapter gave an overview of the obligations of public bodies in the EU Member States to make public sector spatial data available for access, re-use, and sharing. It concluded that the scope of these obligations is not yet completely clear and that the legislator and policy makers are not always well aware that there is a distinction and sometimes mix the different types of use. Hence, the overview of the international and European legislation and policy documents has not given us any more insight into the distinctions between access, re-use, and sharing.

Objective of this chapter – In Chapter 2, I stated that the criterion on which the distinction is based according to the EU legislator is the purpose of the use. This criterion can be deducted from the texts of the Access Directive, the PSI Directive, and the INSPIRE Directive. In this chapter I examine this criterion and indicate whether it is sufficient for making the EU legal framework for the availability of public sector spatial data clear for the public bodies holding the data and for the users of the data. My starting point is that this is not the case, and that the criterion of purpose is not suitable for a number of reasons.

Outline: chapter – I approach this statement in an indirect way. First I look at a number of factors that could indicate whether access, re-use, or sharing is the purpose for which public sector spatial data are needed. I will argue that these indicators cannot clearly indicate which purpose is present. Next I take an in-depth look at these purposes and why the demarcation is not possible in the EU legal framework on the availability of public sector spatial data as it is set up under the Access Directive, the PSI Directive, and the INSPIRE Directive.

OUTLINE: FIRST INDICATOR – The first indicator that is examined is the type of user. Maybe access, re-use, and sharing involve different types of persons or organizations. It could be assumed that citizens need data in order to check up on government, that is for democratic purposes, and that the information industry needs data for re-use. Sharing would then be limited to the public bodies, as they would be the ones to perform a public task. Hence, I will determine whether we can deduct from the type of user whether access, re-use, or sharing is the objective for the use of the public sector spatial data.

OUTLINE: SECOND INDICATOR – The second indicator that could possibly determine whether access, re-use, or sharing is present is the type of data that are needed. First, I address the type of data from a content point of view. Could we assume that because the Access Directive addresses environmental data and the INSPIRE Directive involves spatial data that the former type of data will be needed for access and the latter type for sharing? And which data falling under the common denominator of public sector spatial data are most likely being requested for re-use purposes? Second, the type of data based on the level of processing is examined. Is raw data only requested for re-use, and does access always involve a further processing of the data? Third, the volume and format of the data are considered: do re-use and sharing always entail requests for an entire database rather than a small set of data? And is the reverse also true for access?

OUTLINE: PURPOSE – After these indications are addressed, the purposes or objectives for making available public sector spatial data will be examined themselves. We assume that the indications are not sufficient to show whether access, sharing, or re-use is present, so we need to examine whether the purpose criterion in itself can work for making the demarcation.

II. TYPE OF USER TO WHOM PUBLIC SECTOR SPATIAL DATA IS MADE AVAILABLE

ASSUMPTION – At first thought, it might seem logical to assume that the availability of public sector spatial data for each type of use is aimed at a certain category of users. Roughly, we can divide them into three categories: the citizens, the undertakings, and the public bodies. The citizens are assumed to have a democratic purpose in requesting data, wanting to know what the government they have selected is doing with their tax money. Citizens need to *access* public sector spatial data. Next, undertakings have economic purposes and want to use the data they obtain to gather commercial gains, so they are *re-using* public sector spatial data. Finally, public bodies want to obtain public sector spatial data for the performance of their public task, so the form of use that is most fitting for them would be *sharing*.

EVALUATION OF ASSUMPTION – As was shown before, this distinction does not reflect reality. Access, re-use, or sharing cannot be linked to a particular type of user. All three user categories (which are, admittedly, very generalized) can find themselves in a situation where they would need public sector spatial data for each type of use.

A. Users Requesting Public Sector Spatial Data for Access

Citizens – Access entails a democratic purpose: it is pre-eminently directed towards the citizens in order to improve transparency and participation. This generates the assumption that the beneficiaries of access are the individuals that make up the electorate and decide on the legitimacy of their government.

Undertakings – However, access legislation is not just applicable to natural persons; it also applies to legal persons, such as undertakings, NGOs, and so on. For instance, this is stated clearly in Article 2 of Regulation 1049/2001, Article 2 of the Aarhus Convention, and Article 2.5 of the Access Directive. Hence, if an undertaking wants to know the status of its building permit or its application for a government grant to make sure that the public body involved acted in accordance with the applicable legislation, it will request *access* to data held by the public body.

Public bodies – A public body might also find itself in a comparable situation: for example, if the National Health Service wants to build a new hospital, it may also have to pass through the process of obtaining all the proper permits, with the possible need for checking up on the authorities involved. This means that *access* to data held by another public body is requested. The case law of the ECJ also shows that access legislation is sometimes used by the Member States or public bodies themselves against the EU institutions and bodies.[1]

B. Users Requesting Public Sector Spatial Data for Re-use

Undertakings – Re-use is easily associated with the information industry: for instance, an undertaking that requests data on coordinates, geographical names, addresses, or transport networks from the public bodies to create added-value products such as GPS or location-based services. Undertakings may also re-use public sector spatial data in order to determine their business policies: for instance, an undertaking may request land use data, possibly in combination with statistical units, school services, and public transport to determine where it will open its next branch.

1. CFI, Case T-380/08, *Kingdom of the Netherlands v. Commission*, 9 Sep. 2008, *OJ C* 285, 8 Nov. 2008, 51; CFI, Case T-236/06, *Landtag Schleswig-Holstein v. Commission*, 1 Sep. 2006, *European Court Reports* 2008, II-461.

CITIZENS – However, not only undertakings re-use public sector spatial data. Considering the very broad definition of *re-use,* citizens will also very often be re-using data, for instance, by using a local map for writing a book on the history of the area or publishing an article in the local newspaper. The development of the information society and the Internet has exponentially increased the possibilities for the individual to re-use public sector spatial data. As Mayo and Steinberg found in their Power of Information Review, '[c]itizens have increasingly changed their role from passive recipients of information provided by experts, to active producers of information themselves'.[2] They distinguish two groups of citizens. The first group comprises people who take part in user-generated websites, such as Wikipedia. Such websites often include public sector data, and their purpose goes beyond access for democratic purposes. Hence, they are the result of re-use. The second category are data mashers, described by Mayo and Steinberg as 'businesses, non-profit organizations and normal Internet users who want to mix and combine information to generate valuable new forms of information and new services'.[3] They believe that public sector spatial data is one of the most desirable types of data for these data mashers to be used as a background for other thematic data. While such data mashing initiatives will in many cases be taken by commercial undertakings, they may also be developed by ordinary citizens who want to create a service that is a benefit to society without the intention of gaining revenues. An example is OpenStreetMap, which uses maps that are no longer protected by copyright and other existing databases as a basis to edit its maps.[4]

PUBLIC BODIES – Next, re-use is not just a private sector matter. Many public bodies are forced to resort to creating added-value products based on the data they hold, such as tourist maps, walking or cycling guides, museum guides, postcards, CD-ROMs, DVDs, and so on, because government requires them to provide at least part of their own funding and sometimes to be even completely self-sufficient. In this case, whether they are using data held by another public body or data they created themselves, these public bodies are re-using data, based on the wide definition of re-use that is given in the PSI Directive.

C. USERS REQUESTING PUBLIC SECTOR SPATIAL DATA FOR SHARING

PUBLIC BODIES – The third purpose for which public sector spatial data are made available is sharing. This includes any use of public sector spatial data for the purpose of performing a public task. One can assume that such public tasks will be

2. Ed Mayo & Tom Steinberg, 'The Power of Information: An Independent Review', 2007, 11, <www.cabinetoffice.gov.uk/media/cabinetoffice/strategy/assets/power_information.pdf>, 22 Feb. 2010.
3. *Ibid.*, 13. See also Power of Information Taskforce, 'Power of Information Taskforce Report. February 2009', 2009, 14–16, <www.cabinetoffice.gov.uk/reports/power_of_information.aspx>, 22 Feb. 2010.
4. See <www.openstreetmap.org>, 22 Feb. 2010.

performed by public bodies, which prepare, implement, and evaluate policies and deliver public services to the citizens.

UNDERTAKINGS – However, the fact that the public bodies are responsible for performing their public tasks does not mean that they have to execute them themselves. In Chapters 3 and 4, we have seen that under the Access Directive and the Aarhus Convention, private persons can also be charged with the provision of a public service, in which case they are considered a public authority. In general, public bodies may outsource some parts of their activities under different types of contracts to undertakings in the private sector. This means that requests from such undertakings to obtain public sector spatial data could be made for the performance of the public task that has been commissioned to them. Hence, such situations will be considered as sharing.

CITIZENS – While the performance of a public task by individual citizens might be rare, the possibility exists that such a task is outsourced to a private individual. For instance, the Access Directive and the INSPIRE Directive have taken natural persons into account in their definition of *public authority*.[5]

III. TYPE OF PUBLIC SECTOR SPATIAL DATA THAT IS MADE AVAILABLE

RECAPITULATION – The previous section showed that the type of user cannot be used as a criterion to determine whether the rules for access, sharing, and re-use should be applied to the public bodies making available public sector spatial data. This section will try to determine whether the type of data can give an indication on whether the Access Directive, the INSPIRE Directive, or the PSI Directive should be applicable.

ASSUMPTION – As was explained in Chapter 2, this book addresses public sector data that falls under the definition of environmental information of the Access Directive and under one of the spatial data themes listed in the Annexes of the INSPIRE Directive. One might wonder whether this means that environmental information is more suitable for access rather than for re-use or sharing, and whether spatial data is particularly relevant for sharing rather than for access or re-use.

ARGUMENTS FOR ASSUMPTION – What is clear from the documents discussed in the previous chapter is that the public bodies should make their data available for access upon request. This obligation relates to all data held by public bodies that are not subject to specific legislation and that do not fall under any of the exceptions that can be invoked by a public body in order not to allow public access. Public sector spatial data are included under these data, insofar as they are not subject to specific legislation. This is often the case for cadastral data, for instance. As for making data available upon request for re-use, the public task is not nearly as

5. Cf. *supra*, Ch. 2, ss IV.B.1. and IV.B.2.

clear as with regard to access. While such a task is recognized in several policy documents, the one binding document that has been introduced on re-use, the PSI Directive, does not explicitly decide that there is a public task to allow re-use but leaves it up to the Member States to make the definite decision. This entails that there is no direct indication on the European level on what spatial data should be made available for re-use. With regard to sharing, the INSPIRE Directive gives a detailed overview of the spatial data themes that need to be made available on request. The other policy documents are not as specific in determining the types of data to be made available.

OUTLINE – While there is no true certainty about what data should be made available for access, re-use, or sharing, there are some types of data that are considered important or valuable by policy documents or in the literature. First, the type of data based on the content is examined. These types of data include spatial data and environmental data, but they could also include research data, statistical data, legal data, tourism data, cultural data, and so on. For all these types of data, it could be examined whether their type is an indicator in determining whether access, sharing, or re-use is intended with a request. While this would be an interesting exercise, it would be beyond the scope of our analysis. Therefore, the discussion is limited to spatial data and environmental data. Next, the type of data based on the level of processing is considered, and finally we look at the format and volume of the data that are requested. However, the section starts with an overview of existing typologies of data to show the possibilities for dividing data into different categories. I would argue that for these categories, the examination of whether they could be a factor to determine the type of use would lead to the same results as for the categories that are examined.

A. TYPOLOGIES OF PUBLIC SECTOR DATA

OECD TYPOLOGY – While there are many different distinctions made between types of public sector data,[6] the OECD typology is one of the most interesting because it is very extensive and gives an overview of several different types of data. It is also interesting because it makes an assumption that certain types of data will be more interesting for re-users, while others are more linked to access of the citizens for democratic reasons and social growth. The typology is represented in Figure 5.

6. See e.g., Commission of the European Communities, *Green Paper on Public Sector Information*, 11; PIRA International, 'Commercial Exploitation of Europe's Public Sector Information. Final Report', 2000, 10, <http://ec.europa.eu/information_society/policy/psi/docs/pdfs/pira_study/commercial_final_report.pdf>, 22 Feb. 2010; Makx Dekkers et al., 'Measuring European Public Sector Information Resources (MEPSIR). Final report', 2006, 14–16, <http://ec.europa.eu/information_society/policy/psi/docs/pdfs/mepsir/final_report.pdf>, 22 Feb. 2010.

Figure 5. Public Sector Information and Public Sector Content Domains with Examples

	Domain	Examples
Commercial re-use of PSI	Geographic Information	cartographic information
		land use info (cadastral data)
		spatial data/geographical coordinates
		administrative and political boundaries
		topographical information
		elevation data
	Meterological and Environmental Information	oceanographic data
		hydrographic data
		environmental (quality) data
		atmospheric data
		meteorological (weather) data
	Economic and Business Information	financial information
		company information
		economic and statistics
		industry and trade information
	Social Information	demographic information
		attitude surveys
		data on health/illness
		education and labour statistics
	Traffic and Transport Information	transport network information
		traffic information
		transport statistics
		car registration data
	Tourist and Leisure Information	hotel information
		tourism statistics
		entertainment (local and national)
	Agricultural, Farming, Forestry and Fisheries information	cropping/land use data
		farm incomes/use of resources
		fish farming/harvest information
		live stock data
	Natural Resource Information	biologic and ecologic information
		energy resource/consumption information
		geological and geophysical information
	Legal System Information	crime/conviction data
		laws
		information on rights and duties
		information on legislation
		information on judicial decisions
		patent and trademark information
	Scientific Information and Research data	university research
		publicly-funded research institutes
		governmental research
	Educational Content	academic papers and studies
		lecture material
	Political Content	governmental press releases
		local and national proceedings of governments
		green papers
Making available PSC	Cultural Content	museum material
		gallery material
		archeological sites
		library resources
		public service broadcast archives
		other public archives

(Source: OECD Working Party on the Information Economy, 2006, 12)

Evaluation of OECD typology – The higher the type of data is on the list, the more the OECD feels that it is valuable for re-use purposes. The lower part of the list includes data for which, according to the OECD, it is more important for the public bodies to make these data accessible to the public. One might wonder if this assumption is valid. Scientific or cultural data are, for instance, also often mentioned as a resource for the creation of information products and services on the market. And the citizens may be just as interested in information services from the public bodies regarding the weather as in cultural content. In addition, the OECD acknowledges that the list is not collectively exhaustive or mutually exclusive.[7] For example, the category *natural resource information* contains data that can also be a part of the categories *scientific information* and *geographic information*.[8] Therefore, the scheme is a good example of the difficulties in making categories of data to determine their purpose.

PIRA STUDY – Several examples of other typologies exist. Pas derives eight categories from Europe's Public Sector Information (PIRA) study:

- Economic information (financial information, information concerning undertakings and economic statistics);
- Environmental information (of a hydrographic nature, information on land use, information regarding the quality of the environment, geographical and meteorological information);
- Agricultural and fishing information (information on harvests, use of resources, and fisheries);
- Social information (demographic information, behavioural information, information concerning health and sickness);
- Legal information (information regarding crimes, but also on laws and jurisprudence);
- Scientific information (results of research by universities and government subsidized research institutions),
- Cultural information (objects situated in museums and art galleries, as well as library services); and
- Political information (press releases by governments, proposals and consultations).[9]

GREEN PAPER ON PUBLIC SECTOR INFORMATION – The European Commission's Green Paper on Public Sector Information also suggested a number of classifications. First, it proposed making a distinction between administrative and non-administrative information. The first category related to the function of government and the administration itself, and the second category to information on the outside world that is gathered during the execution of public tasks, such as spatial data, information on businesses, and so on.[10] Within this administrative information, a further

7. OECD Working Party on the Information Economy, 'Digital Broadband Content: Public Sector Information and Content', 2006, 11, <www.oecd.org/dataoecd/10/22/36481524.pdf>, 22 Feb. 2010.
8. *Ibid.*
9. Johan Pas, 'The Commercialization of Government Information and the Proposal for a Directive COM (2002) 207 by the European Commission', *Murdoch University Electronic Journal of Law* 9, no. 4 (2002), <www.murdoch.edu.au/elaw/issues/v9n4/pas94_text.html>, 22 Feb. 2010.
10. Commission of the European Communities, *Green Paper on Public Sector Information*, 11.

distinction could be made between information that is fundamental for the functioning of democracy (such as laws, court cases, and parliamentary information) and information that does not have such a fundamental character. A second possible distinction that the Green Paper mentioned drew a line between information that is relevant for a general public (such as parliamentary information) and information that is only relevant for a very limited set of persons that have a direct interest. For the third distinction, the Green Paper took a market perspective: information can be divided according to its (potential) economic value. Both administrative and non-administrative information can have a considerable market value.[11]

TYPOLOGY BASED ON ACTIVITIES – De La Presle makes a distinction between types of data according to the activities of the state: the publication of official texts; the collection of information for the purpose of providing a public service (e.g., the national census, maintenance of criminal records); and the collection of data for policy purposes.[12]

TYPOLOGY BASED ON PURPOSE: BRUGUIÈRE – Bruguière also does not believe in distinctions based on the type of data. He believes that what should be looked at is the finality of the information. A teleological approach is needed: data should be divided into data produced as a goal of the public service, and data created as a means used in the performance of the public service. He indicates these categories as *information objet* (information as an objective) and *information moyen* (information as a means).[13] The first can be described as the data that are collected, processed, or disseminated by a public body in order to perform its public task of making public sector data available.[14] *Information moyen* can itself be divided in three categories:

- *Données de gestion*, collected by the public bodies for their managerial or administrative tasks, such as personnel reports, procedural documents on building permits, etc;
- *Données d'information*, which are produced for the benefit of the public, in order to facilitate formalities, or to inform the public about public transport time tables;
- *Données de decision*, attached to all decisions adopted by the public bodies in the course of performing their public tasks. These data relate to the decision-making process of the public bodies.

11. *Ibid.*
12. Anne De La Presle, 'Vers un statut des données publiques?', *Revue Brises* no. 12 (1998): 40 [in French]. See also Jean-Michel Bruguière, *Les données publiques et le droit* (Paris: Editions Litec, 2002), 20 [in French].
13. Jean-Michel Bruguière, *Les données publiques et le droit* (Paris: Editions Litec, 2002), 22 [in French].
14. Some public bodies, such as IGN France, can have the task of creating and disseminating data, while other public bodies, e.g., La Documentation Française, only have a task to disseminate data (Jean-Michel Bruguière, *Les données publiques et le droit* (Paris: Editions Litec, 2002), 23–24 [in French]).

Once again, the distinction is relative, as Bruguière indicates that the three categories are not completely separate.[15]

TYPOLOGY BASED ON PURPOSE: BDO CONSULTANTS – Another typology that looks at the purpose for which the data are used was made by BDO Consultants, based on criteria originally set out by the Dutch Ministry of the Interior. BDO divides the electronic public sector data into four groups:

- Research data: data created by specialized research institutions such as statistical agencies or meteorological services. Collecting and analysing data is part of these agencies' public task. These data are not necessarily directly aimed at policy making, but only get such a relationship with policy when they are used by other public bodies.
- Registry data: data that are collected and maintained based on legal obligations, in order to ensure the reliability of social intercourse, for example, the civil registry, the company registry and the cadastre. Disseminating data from these registries is a public task described by the law.
- Administrative data: electronic data that are used for administration of the performance of the tasks of the public bodies. Collecting such administrative data is not a legal obligation in itself, but it is a consequence of the public tasks of the public bodies.
- Supporting data: the public bodies hold data sets that are used to support the development or the implementation of policy. In this way, they are related to the public tasks of the public bodies.[16]

The study found that making available administrative data was considered the least important for all purposes, while research data and supporting data were used by different types of users for different purposes. Registry data were found to be used mostly for sharing with other public bodies.[17]

TYPOLOGY BASED ON PURPOSE: WELLS BRANSCOMB – Wells Branscomb makes a distinction between at least four types of government information:

- Information which is necessary for citizens acting in their roles as voters engaging responsibly in the electoral process;
- Information which is necessary for law-abiding residents in order to comply with the legislative enactments and judicial decisions that are the law of the land;
- Information which is mandated by the purpose for which the agency is established, for example, to provide medical, environmental, commercial, technical, or educational information; and

15. Jean-Michel Bruguière, *Les données publiques et le droit* (Paris: Editions Litec, 2002), 22 [in French].
16. BDO Consultants Bestuur en Management, *Elektronische bestanden van het bestuur* (Groningen: BDO, 1998), 17–18 [in Dutch].
17. *Ibid.*, 64–65.

- Information upon which the very essence of the deliberative process rests, and which cannot be collected reliably and accurately in the private sector, such as census data and sensitive economic data – necessary in the aggregate but damaging or invasive of privacy if disclosed with identifying attributes.[18]

The overview of typologies shows that there are many ways to determine categories of data. I have chosen to look at data types from the point of view of three criteria: the content, the level of processing, and the volume and format.

B. Types of Public Sector Data Based on the Content

Outline – As was mentioned in the introduction, the first classification of public sector data that I examine in order to determine if the type of data can be an indicator for access, re-use, or sharing is based on the content. The OECD typology described above shows that a very detailed distinction can be made between different types of public sector data. While it would be interesting to examine for each type of data whether it is requested for access, re-use, or sharing, this section only addresses spatial data and environmental data, as the common denominator between these is the subject of this book. An exhaustive examination of all types of data would be beyond the scope of my analysis, and the point that I intend to make can already be seen from the discussion on spatial data and environmental data.

1. Spatial Data

Sharing – According to Onsrud, the sharing of spatial data is 'highly valuable in better dealing with a broad range of critical social problems . . . and environmental problems from local levels to global scales'.[19] As was found earlier, this importance of spatial data for sharing has been widely recognized by the international and European legislator and policy makers. The INSPIRE initiative was set up to meet the needs of public bodies for spatial data to perform their public tasks. Sharing is also one of the main objectives of the UNSDI, GEOSS, and GMES. Purposes that can be mentioned are environmental policy, national defence, emergency measures and services, planning, decision making and resource allocation, transport coordination, public utilities, and legislative and policy development.[20]

18. Anne Wells Branscomb, *Who Owns Information? From Privacy to Public Access* (New York: Basic Books, 1994), 164–165.
19. Harlan J. Onsrud, 'Role of Law in Impeding and Facilitating the Sharing of Geographic Information', in *Sharing Geographic Information*, ed. Harlan J. Onsrud & Gerard Rushton (New Brunswick: Rutgers CUPR Press, 1995), 292–306, <www.spatial.maine.edu/%7eonsrud/pubs/roleoflaw36.pdf>, 22 Feb. 2010.
20. See Andrew U. Franck, *The Surveying Activities at the Austrian Federal Office for Metrology and Surveying: an Economic Analysis* (Vienna: Austrian Federal Ministry of Economics and Labour, 2003), 21 <ftp://ftp.geoinfo.tuwien.ac.at/frank/reportbevfinalv31prn.pdf>, 22 Feb. 2010.

Re-use – From this, one might assume that spatial data are most relevant for sharing. However, spatial data are also seen as an important category of data for re-use. According to Rybaczuk and Blakemore, it is a general trend that spatial data are 'being viewed by many governments as a potential commodity that can be traded'.[21] The European Commission's eEurope 2002 Communication stated that spatial data are one of the areas of particular interest for the development of rules for re-use: they are 'for example essential for wireless applications like car navigation systems, a segment where considerable market growth can be expected'.[22] This was also based on the PIRA study, which found that spatial data accounted for over half the value of the European information market.[23] Other examples of possible re-use of spatial data include telecommunication, engineering and consultancy companies, geomarketing, transportation, utilities, health, education and entertainment, tourism, research, and so on.[24] For instance, Sears found that especially reference data and topographic data should be available for re-use, as they are often used as an underlay for thematic data.[25]

Access – Thirdly, spatial data 'that are collected or created by any level of government should be made as readily available ... to the public as possible'.[26] When one considers access, it is hard to miss the overlaps between environmental data and spatial data. Hence, the obligations of the Aarhus Convention and the Access Directive are also applicable to many spatial data sets. As was indicated in Chapter 3, this is also why the INSPIRE Directive refers to the Access Directive and tries to ensure conformity between both. De Terwangne also links the availability of spatial data for citizens to the issue of security. She believes that road and traffic data, meteorological data, maritime maps, and geological maps should be made available because the citizens need these for their safety on the road and in their own homes.[27]

---

21. Krysia Rybaczuk & Michael Blakemore, 'Selling Government Information: A Comparative Perspective on UK and USA Developments', paper given at the Conference on Law and Information Policy for Spatial Databases (Tempe, 1994), <www.spatial.maine.edu/%7eonsrud/tempe/rybaczuk.html>, 22 Feb. 2010. See also William S. Holland, 'Copyright, Licensing and Cost Recovery for Geographic and Land Information Systems Data: A Legal, Economic and Policy Analysis', 1997, <www.spatial.maine.edu/~onsrud/tempe/holland.html>, 22 Feb. 2010.
22. Commission of the European Communities, *eEurope 2002 Communication*, 4.
23. PIRA International, 'Commercial Exploitation of Europe's Public Sector Information. Final Report', 2000, 16, 15, <http://ec.europa.eu/information_society/policy/psi/docs/ pdfs/pira_study/commercial_final_report.pdf>, 22 Feb. 2010.
24. Andrew U. Franck, *The Surveying Activities at the Austrian Federal Office for Metrology and Surveying: an Economic Analysis* (Vienna: Austrian Federal Ministry of Economics and Labour, 2003), 21 <ftp://ftp.geoinfo.tuwien.ac.at/frank/reportbevfinalv31prn.pdf>, 22 Feb. 2010.
25. Garry Sears, 'Geospatial Data Policy Study. Executive Summary', 2001, 23, <www.geoconnections.org/programsCommittees/proCom_policy/keyDocs/KPMG/KPMG_E.pdf>, 22 Feb. 2010.
26. *Ibid.*, 22.
27. Cécile De Terwangne, *Société de l'information et mission publique d'information*, Doctoral Thesis FUNDP (Namur: Faculté Universitaire de Notre Dame de Paix, 2001), 487–490 [in French].

SUMMARY – Hence, spatial data held by public bodies are seen as very valuable for access, for re-use, and for sharing, so it is not possible to assume that requests for spatial data will mostly concern either of these purposes. The exact purpose of the request will still have to be determined for each case.

2. Environmental Data

ACCESS – As was made clear in the previous chapter, the availability of environmental data has received a lot of attention over the past years, particularly with regard to access: 'Public access to environmental information is essential to increase public participation and awareness and to develop greater accountability of all stakeholders concerned'.[28] The Aarhus Convention, the OECD Recommendation on environmental information, and the Access Directive all require that reports on the state of the environment should be available to the public. The OECD Recommendation also wants environmental indicators to be published, and the Access Directive calls for the dissemination of environmental impact studies and risk assessments, as well as data or summaries of data derived from the monitoring of activities affecting or likely to affect the environment. In addition, information in the event of an imminent threat to human health or the environment should be made available. According to the European Commissioner for the Environment, '[c]itizens are entitled to know of the quality of the air and water where they live or if floods, droughts or pollution is risking their property and livelihood'.[29]

RE-USE – Environmental data are also interesting for re-use. The PIRA study indicated that the environmental domain was one of the most promising areas: 'Each month, there are several new information companies entering the environmental domain . . . More environmental [public sector data] is being made available (albeit slower than the market would like) and people are prepared to pay for good quality weather and mapping services and products'.[30] One should take note that PIRA included meteorological and spatial data under the environmental domain. Examples of re-use of environmental data include property information services, insurance, real estate markets, environmental investigations, decontamination services, research on sustainable development and energy resources, and so on.

SHARING – While the specific attention for re-use of environmental data has been limited and recent, sharing of environmental data has been on the agenda of the international organizations and the national governments for a longer time, as

---

28. Jacqueline Burkhardt, 'Conclusions from Session 4. Public Access to Environmental Information: Current Practices' paper given at the OECD Seminar Public Access to Environmental Information (Athens, 2000), 19, <www.olis.oecd.org/olis/2000doc.nsf/LinkTo/NT00000F96/ $FILE/ 00080429.PDF>, 22 Feb. 2010.
29. Rapid Press Release, 'Commission proposes to improve and streamline the European system for collecting, analyzing and reporting environmental information', 2008, <http://europa.eu/rapid/ pressReleasesAction.do?reference=IP/08/185>, 22 Feb. 2010.
30. PIRA International, 'Commercial Exploitation of Europe's Public Sector Information. Final Report', 2000, 71, 15, <http://ec.europa.eu/information_society/policy/psi/docs/pdfs/pira_study/ commercial_final_report.pdf>, 22 Feb. 2010.

'[e]nvironmental governance is plagued by uncertainty, with regard both to bio-geophysical processes and to socio-economic costs and benefits'.[31] According to Sands, the obligation of one state to make data available on one or more matters on an ad hoc basis to another state is found, 'in one form or another, in virtually every international environmental agreement'.[32] Agenda 21 of the United Nations calls for the strengthening of data collection activities, 'especially in the areas of urban air, freshwater, land resources (including forests and rangelands), desertification, other habitats, soil degradation, biodiversity, the high seas and the upper atmosphere'.[33] In addition, according to the OECD Secretariat, '[i]nitially the demand for environmental information was closely related to the definition and implementation of environmental policies and their effects on the state of the environment . . . In the 1980s, new issues and commitments on the political agenda created new demands for information regarding the economic implications of environmental policy decisions . . . environment-health issues including risks to specific groups in the society, and environmental implications of sectoral policy decisions'. In the 1990s, more information was needed about sustainability, environmental consequences of economic growth, climate change, and biodiversity. During recent years, the OECD Secretariat finds that there has been greater interest in the social implications of environmental policies and more demands for data from outside the environmental community (e.g., foreign aid, finance administration, development planning).[34] In addition, '[e]nvironmental information is now considered an essential instrument to carry out more responsive and cost-effective environmental policies'.[35]

Summary – Hence, environmental data are considered important for access, sharing, and re-use. Once again, it is not possible to assume that a request for environmental data will automatically involve access, re-use, or sharing.

C. Types of Public Sector Spatial Data Based on the Level of Processing

Assumption – Can the level of processing of the requested data be a factor to determine whether they are requested for access, re-use, or sharing? Such an

31. Peter H. Sand, 'The Right to Know: Environmental Information Disclosure by Government and Industry', 2002, 2, <www.inece.org/forumspublicaccess_sand.pdf>, 22 Feb. 2010.
32. Philippe Sands, *Principles of International Environmental Law,* 2nd edn (Cambridge: Cambridge University Press, 2003), 827.
33. United Nations Commission on Sustainable Development, *Agenda 21*, 1992, <www.un.org/esa/sustdev/documents/agenda21/english/agenda21toc.htm>, 22 Feb. 2010.
34. OECD Secretariat, 'OECD: Overview and Main Issues', paper given at the OECD Seminar Public Access to Environmental Information (Athens, 2000), 26, <www.olis.oecd.org/olis/2000doc.nsf/LinkTo/NT00000F96/$FILE/00080429.PDF>, 22 Feb. 2010.
35. Anne Teller, 'Conclusions of the Chair', paper given at the OECD Seminar Public Access to Environmental Information (Athens, 2000), 16, <www.olis.oecd.org/olis/2000doc.nsf/LinkTo/NT00000F96/$FILE/00080429.PDF>, 22 Feb. 2010.

assumption is based on the difference between raw data and value-added data. One could assume that access would be requested to value-added data, because the citizens are not interested in obtaining data sets for participating in the democratic society, but they rather need information, that is, data that are put into a context or to which value is added. On the contrary, one would assume that re-users would like to obtain raw, unprocessed data in order to do this processing themselves. With regard to sharing, it would seem that the decision makers would also benefit more from information that has been put in a context rather than the raw data.

EVALUATION OF THE ASSUMPTION – However, once again, such an assumption is not valid for a number of reasons. First, due to the very broad definition of *re-use,* this not only involves the information industry using data held by the public bodies to create information products or services. It also involves businesses needing value-added data in order to make informed decisions about their plans and practices, or students needing information for writing their theses or papers. Also, the increase of user-generated content was already mentioned earlier in this chapter. Such users are in many cases not interested in raw data sets, but they are looking for ready-to-use information they can incorporate in their websites, blogs, forums, and so on. In addition, the assumption that sharing would mainly involve value-added data is also not necessarily correct. While the decision makers on the political level will require value-added data, the public bodies that implement these decisions will in many cases need to combine their own data with data from other public bodies. Hence, raw data are needed for sharing just as much as value-added data.

DISTINCTION BETWEEN RAW DATA AND VALUE-ADDED DATA IS PROBLEMATIC – One should also keep in mind that the level of processing can be relative. The distinction between raw data and value-added data has been subject to many discussions and many definitions. What are value-added data for one organization may be raw data for another. In addition, several types of added value can be distinguished, which can make it difficult to determine whether particular data could still be considered raw or not. For instance, the value chain suggested by the OECD in its 2006 report on public sector information and content includes a number of levels, which may be different for different types of data. What is considered value creation in this data chain may still be considered raw data from the perspective of the user needs. The first step is the creation of the data itself as single data items. The second step is the aggregation and organization of numerous single items of data, in order to produce a complete or at least more comprehensive set of data. From the point of view of the user, such data will most likely be considered as unprocessed and will be seen as raw data. The third element of the value chain involves processing, editing, re-packaging, or re-modelling, and the fourth function is marketing, distribution, and delivery.[36]

TYPES OF ADDED VALUE – Bruguière makes a distinction between three types of added value. First, technical added value consists of presenting the data on a

36. OECD Working Party on the Information Economy, 'Digital Broadband Content: Public Sector Information and Content', 2006, 13, <www.oecd.org/dataoecd/10/22/36481524.pdf>, 22 Feb. 2010.

different carrier than the original one for the purpose of adapting, communicating, or preserving the data. For example, digitization is a form of adding technical value.[37] This would mean that re-users often use value-added data rather than raw data, as their request for data will in many cases entail that the data are transferred to another carrier, such as a DVD or an ftp-server. The second type of added value is intellectual value: this includes comments, classification, or systematization. Third, documentary added value entails that the data can be selected, compiled, and correlated. Hence, the mere compilation of data into one data set already means that value is added. Once again, re-users may also request value-added data from that perspective.

SUMMARY – Hence, the level of processing of the public sector spatial data requested does not necessarily indicate whether these data are requested for access, sharing, or re-use. In addition, the distinction between raw data and value-added data is not always clear. Therefore, as was mentioned before, I have chosen to make the distinction between the availability of public sector spatial data and the provision of spatial information services based on the processing of the data that is needed for the public bodies' own activities, and I have not linked this distinction to the debate on raw data and value-added data.

D. Types of Public Sector Spatial Data Based on Format and Volume

ASSUMPTION: ELECTRONIC DATA INDICATE RE-USE – Does the electronic format of the requested data indicate whether public sector spatial data is required for access, re-use, or sharing? According to De Terwangne, re-use definitely requires electronic data, as this is the only way that data can be immediately processed to information products.[38] Pas agrees that: 'In particular technology has created incentives for the private sector to create value-added products from the raw data produced by government agencies. With records being increasingly computerized, entire record systems, rather than individual records can be easily searched, copied and transferred'[39] Therefore, the PSI Directive shows a preference for the provision of data in a digital format: 'Where possible, documents shall be made available through electronic means'.[40] The OECD Recommendation for Enhanced Access and More

37. Jean-Michel Bruguière, *Les données publiques et le droit* (Paris: Editions Litec, 2002), 40–42 [in French].
38. Cécile De Terwangne et al., *L'accès à l'information administrative et la commercialisation des données publiques: analyse théorique et pratique du problème d'accès, pour le citoyen en général, aux informations détenues par le service public* (Brussels: Story-Scientia, 1994), 94 [in French].
39. Johan Pas, 'The Use and Re-use of Public Sector Information (PSI): Some Legal and Policy Reflections', paper given at the Conference of the Canadian Association for Information Science (Manitoba, 2004), 9, <www.cais-acsi.ca/proceedings/2004/pas_2004.pdf>, 22 Feb. 2010.
40. Article 3 of the PSI Directive.

Effective Use of Public Sector Information also lists new technologies and long-term preservation as one of the principles that should underpin the framework for re-use.[41]

EVALUATION: SHARING – However, this does not distinguish re-use from sharing or access. First, electronic data are also considered essential for sharing. For instance, the INSPIRE Directive is clear: it is only applicable to spatial data in an electronic format. The SEIS initiative is also particularly intended to do away with paper environmental reporting data and move towards harmonized electronic reporting on environmental issues from the Member States to the European Commission. Taking into account the differences in the availability of information technology throughout the world, Agenda 21 does not go as far as the European initiatives, but it does encourage the establishment and strengthening of electronic networking capabilities and the proliferation of technologies to improve data collection and dissemination.[42]

EVALUATION: ACCESS – In addition, the importance of electronic data for wider access has been recognized for a long time. The Council of Europe already started this process in 1979 by stating in its Recommendation 854 (1979) that '[t]he time has come to recognize this principle [of access] with regard to any records, whether electronic or manual'.[43] It is generally accepted that the applicant can choose the format in which he or she wants to obtain the data (e.g., on paper or electronically, via e-mail or on CD-ROM), if this format is available and within reasonable limits.[44] Next, it is also encouraged to actively disseminate public sector data through 'the media and . . . electronic or other, future forms of communication'[45]; to use 'modern effective information communication methods'[46] or to use 'computer telecommunication and/or electronic technology, where available'.[47] However, with regard to access, the traditional means of communication are still considered valuable, as 'computers and public telecommunications networks are not readily accessible to all members of the public in every country'[48] and '[d]issemination of public sector information on the Internet does not automatically imply that all citizens have an equal access to it'.[49]

---

41. OECD, *Recommendation C(2008)36 of the Council for Enhanced Access and More Effective Use of Public Sector Information*, 2008, 5, <www.oecd.org/dataoecd/0/27/40826024.pdf>, 22 Feb. 2010.
42. United Nations Commission on Sustainable Development, *Agenda 21*, 1992, <www.un.org/esa/sustdev/documents/agenda21/english/agenda21toc.htm>, 22 Feb. 2010.
43. Council of Europe Parliamentary Assembly, *Recommendation 854 (1979) of 1 February 1979 on access by the public to government records and freedom of information*, <http://assembly.coe.int/main.asp?Link=/documents/adoptedtext/ta79/erec854.htm>, 22 Feb. 2010.
44. See e.g., Art. 6 of the Convention on Access to Official Documents, Art. 4.1. of the Aarhus Convention.
45. Preamble of the Aarhus Convention.
46. OECD, *Recommendation of the Council on Environmental Information*, 3 Apr. 1998, <http://webdomino1.oecd.org/horizontal/oecdacts.nsf/linkto/C(98)67>, 22 Feb. 2010.
47. Article 1 of the Access Directive.
48. United Nations Economic Commission for Europe, *Aarhus Implementation Guide*, 75.
49. Commission of the European Communities, *Green Paper on Public Sector Information*, 10.

SUMMARY – Hence, the electronic format of the data does not show whether access, re-use, or sharing is intended. Considering the wide scope of potential activities that can be undertaken based on public sector spatial data, 'a one-size-fits-all format does not work for everyone. Data needs to be available in a number of formats to suit the varying needs of users'.[50]

ASSUMPTION: LARGE VOLUMES FOR RE-USE – A related question regards the volume of data that is requested. Do re-use and sharing automatically entail that a large volume is needed, for example an entire database or multiple data sets? And can the contrary be believed for access? It has been argued that companies or public bodies would need large volumes of data for re-use or sharing purposes, while the individual citizen could suffice with access to a small subset of those data.[51]

EVALUATION – This assumption is also not always valid. Considering the wide definition of *re-use,* it is very well possible that a very small data set is re-used, either for commercial or non-commercial purposes, for example, a restaurant showing a small piece of a map on its website to indicate its location, a publisher using a photograph held by a public body on the cover of a book, or a blogger showing the weather information of his or her holiday location. Moreover, the development of web services, such as the network services that are required under INSPIRE, makes it possible to query for small pieces of data sets for any purpose.

SUMMARY – Hence, the volume of data that is requested does not indicate whether access, re-use, or sharing is intended. All three purposes may need different volumes of data, depending on their concrete needs.

E. INTERIM SUMMARY

RECAPITULATION – After giving an overview of existing typologies, this section has attempted to demonstrate that the type of data that is requested from a public body cannot always be an indicator of the type of use that is intended with those data. Three examples of categories of data were examined.

---

50. Joint Information Systems Committee and Office of Public Sector Information, 'The Power of Information Review and Unlocking UK Government Information – A Consultative Forum', 2008, 4, <www.opsi.gov.uk/advice/poi/poi-consultative-forum.pdf>, 22 Feb. 2010.
51. Lori Peterson Dando, 'A Case for the Commercialization of Public Information', paper given at the Conference on Law and Information Policy for Spatial Databases (Tempe, 1994), <www.spatial.maine.edu/%7eonsrud/tempe/dando.html>, 22 Feb. 2010; Micus Management Consulting Gmbh, 'Assessment of the Re-use of Public Sector Information (PSI) in the Geographical Information, Meteorological Information and Legal Information Sectors', 2008, 76, <http://ec.europa.eu/information_society/policy/psi/docs/pdfs/micus_report_december2008.pdf>, 22 Feb. 2010; Mireille Van Eechoud & Jan Kabel, *Prijsbepaling voor elektronische overheidsinformatie* (Deventer: Kluwer, 1998), 51 [in Dutch]; Marijke Reinsma & H. Van Der Sluijs, 'Naar ruimere openbaarheid en een vrij gebruik van bestuurlijke informatie', 2002, 119, <www.minbzk.nl/aspx/download.aspx?file=/contents/pages/5006/RapportReinsmaenVanderSluijs.pdf>, 22 Feb. 2010 [in Dutch]; Oxera, 'Public Information, Private Profit: How Should Government Agencies Compete?', 2005, 2, <www.oxera.com/cmsDocuments/Agenda_Oct%2005/Public%20information%20private%20profit.pdf>, 22 Feb. 2010.

TYPE OF DATA BASED ON CONTENT – First, it was questioned whether one could assume that environmental data would mostly be requested for access, and spatial data for sharing, considering the purposes of the Access Directive and the INSPIRE Directive. It was found that environmental data are important for access, re-use, and sharing, and that spatial data are also valuable for all three purposes of use. While only two data categories were examined, I believe that the same result would be found for most other types of data, and that these types also could be useful for all three types of use that are distinguished in this book.

TYPE OF DATA BASED ON LEVEL OF PROCESSING – Next, the level of processing was examined as an indicator for the purpose for which public sector spatial data could be requested. It was found that the distinction between raw or unprocessed data and value-added data could not be used to determine whether a request for public sector spatial data concerned access, re-use, or sharing. On the one hand, this was due to the fact that all purposes might require different levels of processing. On the other hand, it was caused by the relative character of the distinction between raw data and value-added data: what are value-added data for one person may be raw data for another.

TYPE OF DATA BASED ON VOLUME AND FORMAT – The third indicator that was considered was the volume and the format of the public sector spatial data that are requested. This showed that neither volume nor format can guarantee whether public sector spatial data are requested for access, re-use, or sharing. All types of use may prefer electronic data, and while a request for bulk data may give a hint that either re-use or sharing is intended, a request for smaller parts of a data set or single data items does not necessarily involve access.

IV. THE PURPOSE FOR WHICH PUBLIC SECTOR SPATIAL DATA SHOULD BE MADE AVAILABLE

RECAPITULATION – In the previous sections, it was found that the type of user or the type of data cannot be used as an indicator to determine whether public sector spatial data are requested for the purpose of access, re-use, or sharing. Hence, there are no indirect ways to determine whether an applicant requesting data has an intention of access, re-use, or sharing. Therefore, we have to take a closer look at the purposes or objectives of the availability of public sector spatial data themselves in order to determine whether this distinction can be maintained at all.

OUTLINE – This section examines the main objectives offered in the literature for which public sector spatial data should be made available in order to determine whether these objectives are really the best way to determine the demarcation between access, re-use, and sharing. First, we look at the purposes of making public sector spatial data available that could be seen as democratic. Next, socio-economic purposes are examined. Third, we take a look at the purpose of better decision making and better policy making.

A. Democratic Objectives

Outline – The obligation of public bodies to make public sector spatial data available has traditionally been based on democratic reasons and objectives. These objectives include transparency, public participation in decision making, legitimacy and accountability of government and the public bodies towards the public, and the possibility of the citizens to exercise their rights and obligations. As was mentioned in Chapter 2, democratic purposes are linked to the concept of access and not to re-use or sharing. However, we will see later on that this distinction is not always easy to maintain.

1. Transparency

Transparency as an umbrella term – According to De Terwangne, making public sector data available is an instrument for transparency. What she refers to as transparency can include different elements in the relationship between the public bodies and the citizens, such as the right to have a voice in the decision-making process; the right to make sure that the public bodies are spending the public budget well; or the need of the citizens to know what rights and obligations they have towards the State and towards other citizens. These elements will be discussed briefly.

2. Public Participation

Recognition in legislation and policy documents – Citizens should be informed at every stage of the decision-making process, so that they are able to make informed political choices and participate in the democratic process. This is a clear objective in the international and European texts on the availability of public sector data. As was mentioned before, according to the Council of Europe Recommendation Rec(2002) 2, access to official documents encourages 'informed participation by the public in matters of common interest'.[52] The earlier Recommendation 854 (1979) had already expressly linked access to participation: 'Parliamentary democracy can function adequately only if the people in general and their elected representatives are fully informed'.[53] Participation of the public is found especially important in environmental matters, for example, the Rio Declaration, the Aarhus Convention, the Access Directive, and the OECD Recommendation on environmental information.

52. Council of Europe Committee of Ministers, *Recommendation Rec(2002)2 of 21 February 2002 on access to official documents*, <www.coe.int/T/E/Human_rights/rec(2002)2_eng.pdf>, 22 Feb. 2010.

53. Council of Europe Parliamentary Assembly, *Recommendation 854 (1979) of 1 February 1979 on access by the public to government records and freedom of information*, <http://assembly.coe.int/main.asp?Link=/documents/adoptedtext/ta79/erec854.htm>, 22 Feb. 2010.

IMPORTANCE FOR DIRECT AND INDIRECT PARTICIPATION – Bovens states that: 'The right to vote can only be exercised in a meaningful fashion when all deliberations and decision making on the part of the legislative body are open and accessible. Over the course of time, this has been expanded to include the activities of the executive power. In most Western countries, it is acknowledged that a high degree of open government forms an extremely important condition for the effective democratic control of government action'.[54] As '[g]overnment is the largest repository of information, information that it holds on its own activities and deliberations, information that it holds on others in its many regulatory responsibilities, and information that it holds on us as citizens', access to this information is vital for the citizen to participate in decision making.[55] In the case of indirect participation, the public needs to have sufficient information in order to know whether it is choosing the right person to represent them. In moments of direct participation, for example referenda, the public also needs to know and understand the goal at stake and what measures are being proposed.[56]

INFLUENCE OF INFORMATION TECHNOLOGY – The development of the information society and the growing possibilities for citizens to make their opinion heard via the Internet, for example via e-mail, blogs, discussion forums, and so on, have increased the call for public sector data throughout the entire political process from the very start: 'political debate is no longer restricted to results, conclusions and recommendations, but extends to assumptions, starting points and presuppositions . . . All this means that open government in the information society cannot remain limited to access by the public to official, definitive documents, but that data sets, models and background studies – even when unused – may also be relevant . . . Hence information rights . . . can serve as a tool in helping to expand the reflexive nature of democracy'.[57]

---

54. Mark Bovens, 'Information Rights: Citizenship in the Information Society', *The Journal of Political Philosophy* 10, no. 3 (2002): 324.
55. Patrick Birkinshaw, 'Freedom of Information and Openness: Fundamental Human Rights?', *Administrative Law Review* 58 (2006): 183. See also Dag Wiese Schartum, 'Access to Government-held Information: Challenges and Possibilities', *Journal of Information, Law and Technology* no. 1 (1998): 3–4, <www2.warwick.ac.uk/fac/soc/law/elj/jilt/1998_1/schartum/>, 22 Feb. 2010; see also Elizabeth Longworth, 'The Role of Public Authorities in Access to Information: the Broader and More Efficient Provision of Public Content'. Study Prepared for the Third UNESCO International Congress on Ethical, Legal and Societal Challenges of Cyberspace INFOethics, 2000, 23, <http://unesdoc.unesco.org/images/0012/001210/121051e.pdf>, 22 Feb. 2010; Estelle Derclaye, 'Does the Directive on the Re-use of Public Sector Information Affect the State's Database Sui Generis Right?', in *Knowledge rights – Legal, Societal and Related Technological Aspect. Knowright Symposium, 18–19 September 2008*, ed. Jens Gaster, Erich Schweighofer & Peter Sint (Austrian Computer Society, 2008), 143; Economic Development and Infrastructure Committee, 'Inquiry into Improving Access to Victorian Public Sector Information and Data', 2009, 16, <www.parliament.vic.gov.au/edic/inquiries/access_to_PSI/final_report.html>, 22 Feb. 2010.
56. Frankie Schram, *Begrenzingen aan de openbaarheid van bestuur: een analyse en vergelijking van de internrechtelijke openbaarheidsregelingen in het Belgische recht,* Doctoral Thesis K.U.Leuven (Leuven: K.U.Leuven, 2002), 96–97 [in Dutch].
57. Mark Bovens, 'Information Rights: Citizenship in the Information Society', *The Journal of Political Philosophy* 10, no. 3 (2002): 325.

CRITICISM – Peterson Dando does not agree that access to public sector data is essential to democracy. As an example, she states that the US democracy has existed for well over 200 years, while the Freedom of Information legislation only dates back to 1966: 'Access to government information is a right created by statute. *A fortiori* it can be changed by statute'.[58] In addition, there are numerous democracies throughout the world that have flourished without access legislation.[59]

3. Legitimacy and Accountability

PUBLIC BODIES AS SUBJECTS TO THE LAW – Democracy and the rule of law entail that the actions of government and of the public bodies are not without rules or boundaries. They are also subject to the law and can only be performed in accordance with it. This entails that the citizen is not subject to arbitrary rules or decisions of the public bodies, but that there are norms that the public bodies need to comply with in their relationship with the citizen.[60]

ACCESS AS SAFEGUARD AGAINST ARBITRARINESS – In order for the citizen to ensure that the public bodies are staying within the limits of the law and within the limits of their remit – that is, to exercise *a posteriori* political control over the government activities[61] – he or she needs to know about those limits and about the actions of the public bodies. As Kranenborg and Voermans state, access ensures legal certainty and acts as a safeguard against arbitrariness.[62] According to Beers, the need of the citizen for information in order to be able to approve or disapprove of its

58. Lori Peterson Dando, 'A Case for the Commercialization of Public Information', paper given at the Conference on Law and Information Policy for Spatial Databases (Tempe, 1994), <www.spatial.maine.edu/%7eonsrud/tempe/dando.html>, 22 Feb. 2010.
59. *Ibid.*
60. Frankie Schram, *Begrenzingen aan de openbaarheid van bestuur: een analyse en vergelijking van de internrechtelijke openbaarheidsregelingen in het Belgische recht*, Doctoral Thesis K.U.Leuven (Leuven: K.U.Leuven, 2002), 93–95 [in Dutch].
61. *Ibid.*, 96–97 [in Dutch]. See also Dag Wiese Schartum, 'Access to Government-held Information: Challenges and Possibilities', *Journal of Information, Law and Technology* no. 1 (1998): 12–13, <www2.warwick.ac.uk/fac/soc/law/elj/jilt/1998_1/schartum/>, 22 Feb. 2010; Anthony Mason, 'The Relationship between Freedom of Expression and Freedom of Information', in *Freedom of Expression and Freedom of Information. Essays in honour of Sir David Williams*, ed. Jack Beatson & Yvonne Cripps (Oxford: Oxford University Press, 2000), 237; Keene Matsunaga & Jack Dangermond, 'Promoting a Free Access or Minimal Cost of Dissemination Arrangement for Government-Held Geographic Information Systems Data', paper given at the Conference on Law and Information Policy for Spatial Databases (Tempe, 1994), <www.spatial.maine.edu/~onsrud/tempe/matsunaga.html>, 22 Feb. 2010; Bruce Joffe, 'The GIS Data Sales Dilemma – Finding a Middle Ground', s.d., <www.opendataconsortium.org/article_gis_data_sales_dilemma.htm>, 22 Feb. 2010; Economic Development and Infrastructure Committee, 'Inquiry into improving access to Victorian public sector information and data. Discussion Paper', 2008, 2–3, <www.parliament.vic.gov.au/edic/inquiries/access_to_PSI/call_for_submissions.htm>, 22 Feb. 2010.
62. Herke Kranenborg & Wim Voermans, *Access to Information in the European Union: A Comparative Analysis of EC and Member State Legislation* (Groningen: Europa Law Publishing, 2005), 129.

government is answered 'by the information service of the government about the current and planned policies. Because the objective is for the citizen to form an opinion, this type of information service is usually offered to the mass media, or at least stated in such a way that opinions can be formed via the mass media'.[63] However, the citizen should also always be able to directly obtain the information he or she wants from the public bodies.

RECOGNITION IN LEGISLATION AND POLICY DOCUMENTS – The Council of Europe recognized this in its Recommendation Rec(2002) 2: 'The wide access to official documents . . . helps maintain the integrity [of administrations] by avoiding the risk of corruption'.[64] Therefore, the citizens need to have access to the information that is held by the public bodies about their decisions and activities, and to the information on which they base these decisions and activities. As the UNESCO Policy Guidelines for the Development and Promotion of Governmental Public Domain Information formulate it, '[t]he more information that is openly available from the government and about the government, the less likely it will be that government is able to hide illegal acts, corruption and misrule'.[65] The legitimacy of public bodies increases when the public knows what they are doing.[66]

GOOD GOVERNANCE – The citizen needs to be able to verify that government and the public bodies are not putting themselves above their remit or even above the law, but he or she may also be interested in how his or her tax money is being spent. Accountability also contains aspects of good governance and cost-effectiveness.[67]

---

63. Anton A.L. Beers, *Informatica Publica: publieke toegankelijkheid van elektronische overheidsinformatie* (The Hague: Rathenau Instituut, 1996), 10 [in Dutch]; Anton A.L. Beers, 'Openbaarheid van overheidsinformatie', in *Elektronische toegankelijkheid van overheidsinformatie*, ed. Inigo Baten & Gijs Van Der Starre (The Hague: Rathenau Instituut, 1996), 64–65 [in Dutch].
64. Council of Europe Committee of Ministers, *Recommendation Rec(2002)2 of 21 February 2002 on access to official documents*, <www.coe.int/T/E/Human_rights/rec(2002)2_eng.pdf>, 22 Feb. 2010. See also Toby Mendel, 'Freedom of Information as an Internationally Protected Right', *Comparative Media Law Journal* no. 1 (2003): 39–70.
65. UNESCO, *Policy Guidelines for the Development and Promotion of Governmental Public Domain Information* (Paris: UNESCO, 2004), 6–7, <http://unesdoc.unesco.org/images/0013/001373/137363eo.pdf>, 22 Feb. 2010. See also Estelle Derclaye, 'Does the Directive on the Re-use of Public Sector Information Affect the State's Database Sui Generis Right?', in *Knowledge Rights – Legal, Societal and Related Technological Aspect. Knowright Symposium, 18–19 September 2008*, ed. Jens Gaster, Erich Schweighofer & Peter Sint (Austrian Computer Society, 2008), 143.
66. Henry H. Perrit Jr. & Christopher J. Lhulier, 'Information Access Rights Based on International Human Rights Law', *Buffalo Law Review* 45 (2007): 899. See also Margot Wallstrom, '10 Years of Access to Environmental Information in European, International and Internal Law: Evaluation and Perspectives', in *Ten Years of Access to Environmental Information in International, European and Belgian Law: Stock-Taking and Perspectives*, ed. Christine Larssen (Brussels: Bruylant, 2003), 113.
67. See UNESCO, *Policy Guidelines for the Development and Promotion of Governmental Public Domain Information* (Paris: UNESCO, 2004), 7, <http://unesdoc.unesco.org/images/0013/001373/137363eo.pdf>, 22 Feb. 2010; OECD, *Recommendation of the Council on Environmental Information,* 3 Apr. 1998, <http://webdomino1.oecd.org/horizontal/oecdacts.nsf/linkto/C(98)67>, 22 Feb. 2010.

Kranenborg and Voermans link access to public sector data to the boosting of efficiency, the promoting of legitimacy, and assistance in smooth implementation.[68] The Council of Europe makes it clear that access to public sector data 'constitutes an adequate check on . . . waste of public funds' [69] and it 'fosters the efficiency and effectiveness of administrations'.[70] Hence, availability of public sector data should also lead to more rational decision making by the public bodies.[71] Derclaye even links this rationality to the issue of charging: 'If the government can charge for its information and holds a monopoly it can of course abuse the latter and charge excessive prices'.[72]

4. Fulfilling One's Rights and Obligations

GROWING NEED FOR DATA – As our society is increasingly governed by formal rules laid down in laws and regulations, access to public sector data is necessary 'in order to obtain an overview, to foresee one's own legal position, and to take initiatives to fulfil duties and realize rights'.[73] Bovens believes that the need for data has grown with the development of the information society: 'To a far greater extent public institutions will be expected to support individual subjects in determining their legal position by, for example, actively providing information in various forms specifically tailored to specific categories or situations'.[74] There are so many regulations that the citizen needs more elaborate information to find his or her

68. Herke Kranenborg & Wim Voermans, *Access to Information in the European Union. A Comparative Analysis of EC and Member State Legislation* (Groningen: Europa Law Publishing, 2005), 129.
69. Council of Europe Parliamentary Assembly, *Recommendation 854 (1979) of 1 February 1979 on access by the public to government records and freedom of information*, <http://assembly.coe.int/main.asp?Link=/documents/adoptedtext/ta79/erec854.htm>, 22 Feb. 2010.
70. Council of Europe Committee of Ministers, *Recommendation Rec(2002)2 of 21 February 2002 on access to official documents*, <www.coe.int/T/E/Human_rights/rec(2002)2_eng.pdf>, 22 Feb. 2010.
71. Frankie Schram, *Begrenzingen aan de openbaarheid van bestuur: een analyse en vergelijking van de internrechtelijke openbaarheidsregelingen in het Belgische recht*, Doctoral Thesis K.U.Leuven (Leuven: K.U.Leuven, 2002), 99 [in Dutch]. See also Estelle Derclaye, 'Does the Directive on the Re-use of Public Sector Information Affect the State's Database Sui Generis Right?', in *Knowledge Rights – Legal, Societal and Related Technological Aspect. Knowright Symposium, 18–19 September 2008*, ed. Jens Gaster, Erich Schweighofer & Peter Sint (Austrian Computer Society, 2008), 143.
72. Estelle Derclaye, 'Does the Directive on the Re-use of Public Sector Information Affect the State's Database Sui Generis Right?', in *Knowledge Rights – Legal, Societal and Related Technological Aspect. Knowright Symposium, 18–19 September 2008*, ed. Jens Gaster, Erich Schweighofer and Peter Sint (Austrian Computer Society, 2008), 143.
73. Dag Wiese Schartum, 'Access to Government-Held Information: Challenges and Possibilities', *Journal of Information, Law and Technology* no. 1 (1998): 6, <www2.warwick.ac.uk/fac/soc/law/elj/jilt/1998_1/schartum/>, 22 Feb. 2010; See also Secretary of State for Communities and Local Government, 'Communities in control: real people, real power', 2008, 48–49, <www.communities.gov.uk/documents/communities/pdf/886045.pdf>, 22 Feb. 2010.
74. Mark Bovens, 'Information Rights: Citizenship in the Information Society', *The Journal of Political Philosophy* 10, no. 3 (2002): 324.

way. According to Beers, this means that the public bodies should provide an information service to the citizens: this information service includes, in addition to a 'manual' on how to use the regulations, also guidance services, which involve the stimulating of certain behaviour in light of the general interest of individual well-being, without any reference to a certain regulation.[75]

LINK BETWEEN DEMOCRATIC AND SOCIO-ECONOMIC OBJECTIVES – The need for data to exercise one's rights and obligations shows on the one hand that the democratic reasons for the availability of public sector data and the socio-economic reasons are not always that far apart, as exercising one's rights and obligations is a part of participating in socio-economic life. On the other hand, it touches upon the question how far the public service should provide an information service for the citizen to enjoy his or her rights. The former will be addressed in the next paragraphs, and the latter issue will be discussed in Chapter 7 and Chapter 8.

B. SOCIO-ECONOMIC OBJECTIVES

PARTICIPATION IN ECONOMIC AND SOCIAL LIFE – The availability of public sector data is also an instrument of socio-economic development.[76] Citizens need information to participate in economic and social life, to develop their relationships with other people and with the community and on the market, and to enjoy a good standard of living. The socio-economic objectives for the availability of information can be divided into the more socially oriented objectives of societal growth and welfare on the one hand and economic growth on the other hand. A newly emerging purpose that lies between social welfare and economic growth is the development of communities.

OUTLINE – This section briefly discusses the objectives of societal growth, the development of communities, and economic growth. Next, I address the theory of public sector data as a public good. This theory can be used to clarify the relationship between these socio-economic objectives and the availability of public sector data.

1. Participation in Society: Societal Growth

ENABLING THE CITIZENS – In European welfare societies, information activities constitute an important part of public bodies' tasks. Public information often addresses citizens' duties and rights as well as their standard of living, health and safety,[77]

75. Anton A.L. Beers, *Informatica Publica: publieke toegankelijkheid van elektronische overheidsinformatie* (The Hague: Rathenau Instituut, 1996), 10 [in Dutch]; Anton A.L. Beers, 'Openbaarheid van overheidsinformatie', in *Elektronische toegankelijkheid van overheidsinformatie*, ed. Inigo Baten & Gijs Van Der Starre (The Hague: Rathenau Instituut, 1996), 64–65 [in Dutch].
76. Cécile De Terwangne, *Société de l'information et mission publique d'information*, Doctoral Thesis FUNDP (Namur: Faculté Universitaire de Notre Dame de Paix, 2001), 282 [in French].
77. Dag Wiese Schartum, 'Access to Government-held Information: Challenges and Possibilities', *Journal of Information, Law and Technology* no. 1 (1998): 3, <www2.warwick.ac.uk/fac/soc/

and the general social welfare. In this way, citizens become better able to make informed decisions about their daily life, their environment, and their future.

Wide range of objectives – There is a wide range of social objectives underlying the provision of public sector data. The UNESCO Policy Guidelines for the Development and Promotion of Governmental Public Domain Information link these objectives to 'circumstances where the public's welfare will be better served through access to or disclosure of information, rather than a paternalistic approach where decisions are made on behalf of the people without informing or consulting them'. Examples are health issues and environmental pollution, but also the recognition of the importance of social capital: there are many social benefits to be derived from a more knowledgeable population. Public funding of libraries, archives, museums, educational institutions and research institutes are all manifestations of this recognition.[78] Another key argument in favour of making public sector data available is that it will enhance the potential for innovation and creativity throughout society.[79]

Environmental information – The importance of environmental information for the enjoyment of a healthy life and environment has especially been stressed over the last few years. As the Access Directive states, '[i]ncreased public access to environmental information and the dissemination of such information contribute to a greater awareness of environmental matters, a free exchange of views, more effective participation by the public in environmental decision making and, eventually, to a better environment'.[80] The Aarhus Convention also has the purpose of starting an awareness or behavioural change in order to ensure a healthy environment for everyone now and in the future. The Convention has even made the citizen co-responsible as an upholder of environmental law. To realize this, the citizen has to have access to environmental information.[81] As the environment is a public good that belongs to everyone, it is essential for this information to be widely shared and available. Recent experiences of 'forest fires, floods and droughts underline the demand for accurate environmental information to be made quickly and easily available'.[82]

---

law/elj/jilt/1998_1/schartum/>, 22 Feb. 2010. See also Economic Development and Infrastructure Committee, 'Inquiry into Improving Access to Victorian Public Sector Information and Data. Discussion Paper', 2008, 2–3, <www.parliament.vic.gov.au/edic/inquiries/access_to_PSI/call_for_submissions.html>, 22 Feb. 2010; Patrick Birkinshaw, 'Freedom of Information and Openness: Fundamental Human Rights?', *Administrative Law Review* 58 (2006): 203.

78. UNESCO, *Policy Guidelines for the Development and Promotion of Governmental Public Domain Information* (Paris: UNESCO, 2004), 7, <http://unesdoc.unesco.org/images/0013/001373/137363eo.pdf>, 22 Feb. 2010.
79. Economic Development and Infrastructure Committee, 'Inquiry into Improving Access to Victorian Public Sector Information and Data. Discussion Paper', 2008, 10, <www.parliament.vic.gov.au/edic/inquiries/access_to_PSI/call_for_submissions.html>, 22 Feb. 2010.
80. Recital 1 of the Access Directive.
81. Frankie Schram, *De publieke toegang tot milieu-informatie in internationaal en internrechtelijk perspectief*, (Brussels: De Boeck & Larcier, 2008), 3 [in Dutch].
82. Commission of the European Communities, *Communication to the Council, the European Parliament, the European Economic and Social Committee and the Committee of the Regions.*

INFLUENCE OF THE INFORMATION SOCIETY – In the current information and knowledge society, the availability of digital information via the Internet and digital information networks is essential for the empowerment of the citizens.[83] In addition, the active role of the government and the public bodies is increasingly emphasized. According to Bovens, the 'basic issue is whether the nature of the information in question is crucial to the social functioning of citizens. The more affirmative the answer to this question, the more active a stance will the government have to assume'.[84]

RELATIONSHIP BETWEEN DEMOCRATIC AND SOCIO-ECONOMIC OBJECTIVES – As was mentioned before, the distinction with the democratic objective of the availability of public sector data for the citizens to fulfil their rights and obligations is not always clear. Full enjoyment of rights and obligations implies the benefits of health care, a clean environment, safety, and social welfare in general. Hence, the democratic and socio-economic reasons are not always easy to separate. De Terwangne also shows that the distinction with transparency is not always clear: for instance, information about the environment is important for the people that live there, but it may also be important for checking the decision making of the government regarding the environment, public health, or economic planning.[85]

SHARING – With regard to societal growth, public sector data should not only be available to the citizen to improve their quality of life, but also to other public bodies in order for them to create public policy that ensures this quality of life, for example, the protection of the environment, sustainability, health, safety, and so on. Hence, societal growth is not only an objective of making data available for access or re-use, but also for sharing. For example, this is clear for spatial data in the GEOSS initiative: 'Understanding the Earth system – its weather, climate, oceans, atmosphere, water, land, geodynamics, natural resources, ecosystems, and natural and human-induced hazards – is crucial to enhancing human health, safety and welfare, alleviating human suffering including poverty, protecting the global environment, reducing disaster losses, and achieving sustainable development. Observations of the Earth system constitute critical input for advancing this understanding'.[86]

Towards a Shared Environmental Information System (SEIS), COM (2008) 46 final, 3, 1 Feb. 2008.

83. UNESCO, *Policy Guidelines for the Development and Promotion of Governmental Public Domain Information* (Paris: UNESCO, 2004), 1, <http://unesdoc.unesco.org/images/0013/001373/137363eo.pdf>, 22 Feb. 2010.

84. Mark Bovens, 'Information Rights: Citizenship in the Information Society', *The Journal of Political Philosophy* 10, no. 3 (2002): 331.

85. Cécile De Terwangne, *Société de l'information et mission publique d'information*, Doctoral Thesis FUNDP (Namur: Faculté Universitaire de Notre Dame de Paix, 2001), 282–283 [in French].

86. Group on Earth Observations, 'The Global Earth Observation System of Systems (GEOSS) 10-Year Implementation Plan, 2005', 1, <www.earthobservations.org/documents/10-Year%20Implementation%20Plan.pdf>, 22 Feb. 2010.

2. Development of Communities

INFLUENCE OF THE INFORMATION SOCIETY – The development of information and communication technologies, and especially the Internet, has led to new purposes for public sector data that are linked to societal growth, but that may deserve some particular attention. These purposes are related to the development of communities, with new patterns of participative information creation and use.[87]

INVOLVEMENT OF THE CITIZEN – According to Longworth, making public sector data available is a very powerful incentive for citizens to become content producers, for example, as they start to communicate with each other to build a collective consciousness and to make their views known through the Internet.[88]

DATA MASHERS – As was mentioned above, in their Power of Information Review, Mayo and Steinberg state that there are two groups of citizens: people who create information on the Internet and a second group composed of people who take information from various sources, including government, and mix it together to make new tools and services, widely known as data mashers. Some of the most desirable information for this new group is data generated by government, especially public sector spatial data, 'which can often be used like a glue to bind together disparate information'.[89] According to the Review, '[t]he government should support the development of new and innovative services that provide tailored advice to specific groups. These are outside government's direct influence, but government has a role to play in supporting them – for example by ensuring that they are not undermined by government programmes or websites with similar objectives, and have easy access to publicly available information'.[90] While these new and innovative services may be developed by commercial companies (and rather relate to economic growth), there are also many initiatives with user-generated content in pursuit of common social and economic objectives, such as OpenStreetMap,[91] on-line fora about tax-returns, support groups on health issues or parenting, and so on.

87. Office of Public Sector Information, 'The United Kingdom Implementation of the European Directive on the Re-use of Public Sector Information – the first two years', 2007, 21–22, <www.opsi.gov.uk/advice/psi-regulations/uk-implementation-first-years.pdf>, 22 Feb. 2010.
88. Elizabeth Longworth, 'The role of public authorities in access to information: the broader and more efficient provision of public content. Study prepared for the Third UNESCO International Congress on Ethical, Legal and Societal Challenges of Cyberspace INFOethics', 2000, 23, <http://unesdoc.unesco.org/images/0012/001210/121051e.pdf>, 22 Feb. 2010.
89. Ed Mayo & Tom Steinberg, 'The Power of Information: An independent Review', 2007, 12–13, <www.cabinetoffice.gov.uk/media/cabinetoffice/strategy/assets/power_information.pdf>, 22 Feb. 2010; Economic Development and Infrastructure Committee, 'Inquiry into improving access to Victorian public sector information and data. Discussion paper', 2008, 12, <www.parliament.vic.gov.au/edic/inquiries/access_to_PSI/call_for_submissions.html>, 22 Feb. 2010.
90. Ed Mayo & Tom Steinberg, 'The Power of Information: An independent Review', 2007, 7, <www.cabinetoffice.gov.uk/media/cabinetoffice/strategy/assets/power_information.pdf>, 22 Feb. 2010. See also Secretary of State for Communities and Local Government, 'Communities in control: real people, real power', 2008, 49, <www.communities.gov.uk/documents/communities/pdf/886045.pdf>, 22 Feb. 2010.
91. See <www.openstreetmap.org>, 22 Feb. 2010.

3. Economic Growth

RECOGNITION IN LEGISLATION AND POLICY DOCUMENTS – Public sector data has an important value for economic growth. This economic value was recognized by the European Commission: '[public sector data] has a considerable economic potential. It is an essential basis for many digital information products and could become an important raw material for new services'.[92] The availability of public sector data 'could considerably boost economic activity and job creation. In turn the new added-value information products could have a positive economic impact, increasing efficiency and leading to better informed decisions'.[93] The UNESCO Policy Guidelines for the Development and Promotion of Governmental Public Domain Information emphasize the importance of public sector data for SMEs: 'Easy access to public information supports the growth of the private sector, especially small businesses, for which information costs can represent a real difficulty'.[94] Moreover, innovative use may be made of the data that may not have been anticipated by the creators of the original data.[95]

ECONOMIC VALUE FOR THE PRIVATE SECTOR – According to Papapavlou, '[a]ccess to public sector information is of great importance for the information content industry. The information content industry is a big generator of revenue and jobs'.[96] However, public sector data has two economic objectives: it can serve as a resource for the development of value-added information products and services, but it can also be of essential value to businesses for the development of their business strategies or marketing policies.[97] Hence, it is not only the information industry creating information products and services that benefits from the availability of data, but also the wider economy. This is for instance reflected in the concept of re-use of the PSI Directive: *any* use outside of the public task.

SPATIAL DATA – With regard to the economic importance of spatial data, the White Paper of the Geospatial Information & Technology Association stated that '[m]ost GIS professionals are generally in consensus on one point: giving away map data has helped society at large and the commercial industry in particular.

---

92. Commission of the European Communities, *eEurope 2002 Communication*, 3. See also Economic Development and Infrastructure Committee, 'Inquiry into Improving Access to Victorian Public Sector Information and Data', 2009, 10–11, <www.parliament.vic.gov.au/edic/inquiries/access_to_PSI/final_report.html>, 22 Feb. 2010.
93. Commission of the European Communities, *eEurope 2002 Communication*, 5.
94. UNESCO, *Policy Guidelines for the Development and Promotion of Governmental Public Domain Information* (Paris: UNESCO, 2004), 7, <http://unesdoc.unesco.org/images/0013/001373/137363eo.pdf>, 22 Feb. 2010.
95. Economic Development and Infrastructure Committee, Inquiry into Improving Access to Victorian Public Sector Information and Data, 2009, 13, <www.parliament.vic.gov.au/edic/inquiries/access_to_PSI/final_report.html>, 22 Feb. 2010.
96. George Papapavlou, 'Public Sector Information Initiatives in the European Union', 1999, 2, <http://webworld.unesco.org/infoethics2000/documents/paper_papapavlou.rtf>, 22 Feb. 2010.
97. Katleen Janssen & Jos Dumortier, 'Towards a European Framework for the Re-use of Public Sector Information: A Long and Winding Road', *International Journal of Law and Information Technology* 11, no. 2 (2003): 185.

Specifically, access to basic map data has cultivated an entire industry of private sector value-added firms. They use the data as the basis to create products and services that not only allow them to make money, but their products often enable end users to do their jobs faster, better, and cheaper. This creates a win-win situation for everyone'.[98]

ECONOMIC VALUE FOR THE PUBLIC SECTOR – On the other hand, public sector spatial data may also have economic value for the public bodies themselves. 'Many government institutions recognize the economic value of government information in electronic form and also recognize that monopolists can extract more revenue by maintaining control and discouraging competition. Accordingly, they set up government-run or government-sponsored monopolies to sell access to their information resources while blocking access by others'.[99] 'Public authorities are afraid to lose revenue by freeing up their information for which they used to charge'.[100] However, the argument is often made that by giving the information away, returns from taxation will far exceed the revenue public authorities expect from the sale of their information.

4. Public Sector Spatial Data as a Public Good

OUTLINE – A rationale that actually may be considered to lie at the basis of the socio-economic reasons for the availability of public sector spatial data is the perception of these data as a public good. If public sector spatial data can be seen as a public good, the public bodies may have a duty to make them available to the public. Therefore, this section addresses the concept of public goods and examines whether information in general and public sector spatial data in particular could qualify as being public goods.

a. What Is a Public Good?

MAIN CHARACTERISTICS – The theory of public goods was mainly developed by Samuelson.[101] A public good is characterized by two attributes: it is non-excludable and non-rivalrous.[102] Non-excludability entails that it is difficult, if

98. Geospatial Information & Technology Association, 'Free of Fee: the Governmental Data Ownership Debate. GITA White Paper', 2005, 3, <www.gita.org/resources/whitepapers/free_or_fee.pdf>, 22 Feb. 2010.
99. Henry Perrit Jr. & Christopher J. Lhulier, 'Information Access Rights Based on International Human Rights Law', Buffalo Law Review 45 (1997): 900.
100. Estelle Derclaye, 'Does the Directive on the Re-use of Public Sector Information Affect the State's Database Sui Generis Right?', in *Knowledge Rights – Legal, Societal and Related Technological Aspect. Knowright Symposium, 18–19 September 2008*, ed. Jens Gaster, Erich Schweighofer & Peter Sint (Austrian Computer Society, 2008), 143.
101. Paul A. Samuelson, 'The Pure Theory of Public Expenditure', *Annual Review of Information Science and Technology* 36, no. 4 (1954): 387–389.
102. Inge Kaul, Isabelle Grunberg & Marc A. Stern, 'Defining Global Public Goods', in *Global Public Goods*, ed. Inge Kaul, Isabelle Grunberg & Marc A. Stern (Oxford: Oxford University Press, 1999), 3.

not impossible, to exclude an individual from enjoying the good, while non-rivalrousness means that the consumption of the good by one individual does not detract another person from consuming it.[103] According to the National Research Council, non-depletability is the main reason economists conclude that free use of public goods can be justified – there is no social cost when another person uses them, and there is no justification for the disincentive of a substantial price for that use. In addition, absence of excludability makes it very difficult for the provider to collect reimbursement for the cost of supplying the public good. Therefore, it is often difficult for any other provider than government to provide public goods, and the state will have to play some role in the provision of these goods.[104] Classic examples of public goods include national defence, environmental protection, and investments in new technical information.[105]

Impure public goods – Goods that only partly meet either or both of the defining criteria are called impure public goods. These goods can be divided into

---

103. Joseph E. Stiglitz, 'Knowledge as a Global Public Good', in *Global Public Goods*, ed. Inge Kaul, Isabelle Grunberg & Marc A. Stern (Oxford: Oxford University Press, 1999), 309. See also National Research Council, *Bits of Power: Issues in Global Access to Scientific Data* (Washington: National Academies Press, 1997), 112; Harlan J. Onsrud, 'Legal Access to Geographic Information: Measuring Losses or Developing Responses?', in *Information, Place and Cyberspace: Issues in Accessibility*, ed. Donald G. Janelle & David Charles Hodge (Berlin: Springer-Verlag, 2000), 303–316, <www.spatial.maine.edu/%7Eonsrud/pubs/LegalAccessToGI.pdf>, 22 Feb. 2010; George Cho, *Geographic Information Science: Mastering the Legal Issues* (Chicester: John Wiley & Sons Ltd., 2005), 58; Patrick McNutt, 'Public Goods and Club Goods', in *Encyclopedia of Law and Economics*, ed. Boudewijn Bouckaert & Gerrit De Geest (Cheltenham: Edward Elgar Publishing Limited, 2000, vol. 1), 928–929; William Mock, 'On the Centrality of Information Law: a Rational Choice Discussion of Information Law and Transparency', *John Marshall Journal of Computer & Information Law* 17 (1999): 1085; Vito Tanzi, 'IMF Working Paper. The role of the state and the quality of the public sector', 2000, 4, <www.imf.org/external/pubs/ft/wp/2000/wp0036.pdf>, 22 Feb. 2010.
104. National Research Council, *Bits of Power. Issues in Global Access to Scientific Data* (Washington: National Academies Press, 1997), 112; Inge Kaul, Isabelle Grunberg & Marc A. Stern, 'Defining Global Public Goods', in *Global Public Goods*, ed. Inge Kaul, Isabelle Grunberg & Marc A. Stern (Oxford: Oxford University Press, 1999), 6–7; Joseph E. Stiglitz, 'Knowledge as a Global Public Good', in *Global Public Goods*, ed. Inge Kaul, Isabelle Grunberg & Marc A. Stern (Oxford: Oxford University Press, 1999), 311; Egbert J. Dommering, P. Bernt Hugenholtz & Jan C. Kabel, 'De overheid en het publiek domein van informatie voor wetenschappelijk onderzoek', in *De publieke dimensie van kennis*, ed. Huub Dijstelbloem & Cornelius J.M. Schuyt (The Hague: WRR, 2002), 259–260 [in Dutch]; Egbert J. Dommering, 'Informatie tussen economie en cultuur', in *Informatierecht. Fundamentele rechten voor de informatiesamenleving*, ed. Egbert J. Dommering et al. (Amsterdam: Otto Cramwinckel Uitgever, 2000), 476 [in Dutch]; William Mock, 'On the Centrality of Information Law: a Rational Choice Discussion of Information Law and Transparency', *John Marshall Journal of Computer & Information Law* 17 (1999): 1087; Kirsti Nilsen, 'Economic Theory as It Applies to Statistics Canada: A Review of the Literature', 2007, 5, <www.chass.utoronto.ca/datalib/misc/Nilsen%20Economics%20Paper%202007%20final%20version.pdf>, 22 Feb. 2010.
105. Keith Maskus & Jerome H. Reichman, 'The Globalization of Private Knowledge Goods and the Privatization of Global Public Goods', *Journal of International Economic Law* 7 (2004): 284.

two categories. First, goods that are non-rivalrous in consumption but excludable are club goods (e.g., golf courses or satellite television). Such goods can be made artificially scarce. Second, goods that are mostly non-excludable but rivalrous in consumption are common pool resources (e.g., fishing or hunting game).[106]

Merit goods – Another term that deserves some attention is *merit goods*: goods that are subsidized by the public sector because their existence or their consumption is highly valued by the community.[107] Society believes that, '[e]very-body should have access to certain goods and services, regardless of their ability and willingness to pay the market price'.[108] As Walsh puts it: 'Government needs to act to provide merit goods because individuals are not necessarily the best judges of what is in their own or the public interest'.[109]

Role of the public bodies – The public bodies should play some role in the provision of such goods; otherwise they might be undersupplied.[110] However, for different services in differing national contexts, there will be different *reasons* for the public bodies to be involved, and these imply that the intensity and form of involvement may be different.[111] For instance, the government may be providing the merit good itself, or it may lease or concession the task of creating the good; it may supervise the provision of the good or it may provide financial support; and so on. Contrary to public goods – for which it is impossible to demand payment, for example street lighting – merit goods are offered on the market, but not in a sufficient volume or quality, because market players only take their own costs and revenues into account, and not the public value of offering the good or service.[112] Hence, they will determine their prices based on their costs and their desired profit margin, and they will not lower their prices just because their product would be beneficial for the entire society.

Criticism – According to Georgiadou and Groot, the basic premises of public goods have recently been criticised for their lack of relevance and practicality in

106. Inge Kaul, Isabelle Grunberg & Marc A. Stern, 'Defining Global Public Goods', in *Global Public Goods*, ed. Inge Kaul, Isabelle Grunberg & Marc A. Stern (Oxford: Oxford University Press, 1999), 4–5.
107. *Ibid.*, 5. See also Egbert J. Dommering, 'Informatie tussen economie en cultuur', in *Informatierecht. Fundamentele rechten voor de informatiesamenleving*, ed. Egbert J. Dommering et al. (Amsterdam: Otto Cramwinckel Uitgever, 2000), 478 [in Dutch]; HM Treasury, 'Cross Cutting Review of the Knowledge Economy. Review of Government Information', 2000, <www.hm-treasury.gov.uk/spend_sr00_ccr.htm>, 22 Feb. 2010.
108. Richard Batley, 'Public-Private Relationships and Performance in Service Provision', *Urban Studies* 33, no. 4–5 (1996): 728. See also John M. Donahue, *The Privatization Decision: Public Ends, Private Means* (New York: Basic Books, 1989), 21.
109. Kieron Walsh, *Public Services and Market Mechanisms: Competition, Contracting and the New Public Management* (Basingstoke: Palgrave Macmillan, 1995), 10.
110. Joseph E. Stiglitz, 'Knowledge as a Global Public Good', in *Global Public Goods*, ed. Inge Kaul, Isabelle Grunberg & Marc A. Stern (Oxford: Oxford University Press, 1999), 311.
111. Richard Batley, 'Public-Private Relationships and Performance in Service Provision', *Urban Studies* 33, no. 4–5 (1996): 728.
112. Jan J.C. Kabel et al., *Kennisinstellingen en informatiebeleid. Lusten en lasten van de publieke taak* (Amsterdam: Universiteit van Amsterdam, 2001), 16 [in Dutch].

today's globalized society.[113] It is argued that the lines between public and private are blurred and constantly changing: 'The public or private nature of a good is not a given but a matter of policy choice'.[114] In addition, the role of non-state actors in the provision of public goods is increasing.

b. Is Information a Public Good?

PROPONENTS – Nilsen attempts to describe information as a public good by using an example: 'If one individual buys a newspaper and benefits from that purchase through reading it, the paper itself or the news within it can be shared with others without detracting from the purchaser's benefit. While the newspaper is produced by a private firm, and there is a small costs to the original purchaser, the private producer cannot exclude non-paying individuals from enjoying its content, and the use of it by the original purchaser does not keep others from purchasing it'.[115] From this, it seems clear that information is a public good. Onsrud agrees: 'Many people may use information and yet it remains available for many others to use as well'.[116] In addition, he specifies that public sector data is 'primarily nonexcludable. Such records must be made available and open to citizens so the electorate has the opportunity to scrutinize in detail the functioning of government, the actions of government officials, and the documents upon which government decisions have been made'.[117]

CRITICISM ON NON-EXCLUDABLE CHARACTER – However, many scholars have a more nuanced point of view. Love does not believe information is always a public good, as it is 'often possible to exclude consumption and even meter the use of information', for instance by IPR, or technologies to control access.[118] This is the

113. Yola Georgiadou & Richard Groot, 'Policy Development and Capacity Building for Geo-Information Provision: a Global Public Goods Perspective', *GIS Development* 6, no. 7 (2002): 33–40, <www.gisdevelopment.net/policy/international/interna014.htm>, 22 Feb. 2010.
114. *Ibid.* See also Kirsti Nilsen, 'Economic Theory as It Applies to Statistics Canada: A Review of the Literature', 2007, 6, <www.chass.utoronto.ca/datalib/misc/Nilsen%20Economics%20Paper%202007%20final%20version.pdf>, 22 Feb. 2010.
115. Kirsti Nilsen, 'Economic Theory as It Applies to Statistics Canada: A Review of the Literature', 2007, 13–14, <www.chass.utoronto.ca/datalib/misc/Nilsen%20Economics%20Paper%202007%20final%20version.pdf>, 22 Feb. 2010. See also Bastiaan Van Loenen & Bas C. Kok, 'Spatial Data Infrastructures: Legal and Economic Issues', in *Spatial Data Infrastructure and Policy Development in Europe and the United States*, ed. Bastiaan Van Loenen & Bas C. Kok (Delft: DUP Science, 2004), 6; Dieudonné Mandelkern & Bertrand Du Marais, 'Diffusion des données publiques et révolution numérique: rapport', 1999, 44, <www.ladocumentationfrancaise.fr/rapports-publics/994001620/index.shtml#>, 22 Feb. 2010 [in French].
116. Harlan J. Onsrud, 'The tragedy of the information commons', in *Policy issues in modern cartography*, ed. D.R. Fraser Taylor (Oxford: Elsevier Science Ltd., 1998), 145.
117. *Ibid.*, 145–146.
118. James Love, 'Pricing Government Information', *Journal of Government Information* 22, no. 5 (1995): 364. See also Egbert J. Dommering, P. Bernt Hugenholtz & Jan C. Kabel, 'De overheid en het publiek domein van informatie voor wetenschappelijk onderzoek', in *De publieke dimensie van kennis*, ed. Huub Dijstelbloem & Cornelius J.M. Schuyt (The Hague: WRR, 2002), 263 [in Dutch].

reason that there is an information market. He states that it is useful to distinguish between the information, which is 'nonrival' in consumption, and the system of dissemination of the information, which may require resources either to facilitate or to exclude consumption.[119] The information may be non-rivalrous, but the carrier cannot always be held by multiple persons at the same time, or access to the website holding the information may be limited by passwords, and so on. Cahir has comparable doubts regarding the inherent non-excludability of information: 'The non-rival nature of information inheres in the quantity itself – information has always been and will always be non-rival; non-excludability in contrast is an attribute that depends on both technology and social choice'.[120] He states that: 'Inasmuch as technology gives rise to non-excludability so too does it take it away'.[121] For instance, modern technology makes it easy to reproduce and disseminate information, which makes information approach the status of a public good. However, encryption technologies can make access to information conditional on payment or other conditions and in this way revive excludability. The assertion of IPR can also have such an effect. While Stiglitz feels that IPR give private producers the incentive to produce knowledge,[122] laws that grant exclusive rights to an individual to control the dissemination of information exclude the availability of this information. 'In sum, while the non-excludability attribute of information is in theory a reality, it is nonetheless susceptible to technological and legal excludability'.[123]

CRITICISM ON LACK OF EMPIRICAL EVIDENCE – Bruguière looks at the issue from a more economic point of view. He believes that if information is a public good by nature, it should show from empirical evidence that no public or private operator is interested in making a particular data set available as a product on the market. This may change over time: due to budgetary constraints, information that is now considered to be public, so within the task of the State, could be privatised later. For instance, road data are important for safety, so it could be considered to be within the task of the public bodies to make these data available. Yet, the collection of these data is increasingly expensive, so one can imagine the privatization of these data.[124] For this type of situations, Love feels that some information products

119. James Love, 'Pricing Government Information', *Journal of Government Information* 22, no. 5 (1995): 364.
120. John Cahir, 'Understanding information laws: a sociological approach', *Journal of Information, Law and Technology* 2002, <www2.warwick.ac.uk/fac/soc/law/elj/jilt/2002_3/cahir/>, 22 Feb. 2010.
121. *Ibid.* See also Egbert J. Dommering, P. Bernt Hugenholtz & Jan C. Kabel, 'De overheid en het publiek domein van informatie voor wetenschappelijk onderzoek', in *De publieke dimensie van kennis*, ed. Huub Dijstelbloem & Cornelius J.M. Schuyt (The Hague: WRR, 2002), 263 [in Dutch].
122. Joseph E. Stiglitz, 'Knowledge as a Global Public Good', in *Global Public Goods*, ed. Inge Kaul, Isabelle Grunberg & Marc A. Stern (Oxford: Oxford University Press, 1999), 311–312.
123. John Cahir, 'Understanding information laws: a sociological approach', *Journal of Information, Law and Technology* 2002, <www2.warwick.ac.uk/fac/soc/law/elj/jilt/2002_3/cahir/>, 22 Feb. 2010.
124. Jean-Michel Bruguière, *Les données publiques et le droit* (Paris: Editions Litec, 2002), 284 [in French].

and services created by public bodies can be thought of as *merit goods* – goods for which 'consumption should be encouraged based upon nonmarket value judgments by society', such as the awareness and participation of the citizen.[125]

c. *Is Public Sector Spatial Data a Public Good?*

PROPONENTS – According to Onsrud, '[spatial] data, like many other forms of scientific and technical data, possess the classic characteristics of public goods. That is, [spatial] data are typically nonrival and nonexcludable ... By example, the use of digital data for finding one's way does not make the same data any less useful to others for finding their way'. In addition, '[spatial] data, as well as many other forms of factual data, are additionally nonexcludable in that others may recollect the same or substitutable data'.[126] Georgiadou and Groot look at public sector spatial data as an intermediate public good: 'Ownership of and unlimited access to fundamental [spatial data] allows national governments to govern, i.e., to provide final public goods, notoriously underprovided if left to market forces. An example is defence, the purest public good of all'.[127]

REASONS – Corey explains why public sector spatial data can be regarded as a public good in Canada:

> There are compelling arguments related to the "public good" aspects of mapping programs, and the economic, social, environmental and public safety advantages that flow from readily available and nationally consistent map information. Such information is an essential part of governing, providing for the defence of the nation and demonstrating national sovereignty. Mapping is an essential tool in education in helping young and not-so-young Canadians to understand the dimensions and possibilities of their country ... In a country like Canada, with a relatively small population and a huge landmass, it would not be commercially viable for a private sector company to undertake the responsibility of developing and maintaining high quality map information for the whole country, on a commercial basis ... Such factors support the case

---

125. James Love, 'Pricing Government Information', *Journal of Government Information* 22, no. 5 (1995): 364.
126. Harlan J. Onsrud, 'Legal Access to Geographic Information: Measuring Losses or Developing Responses?', in *Information, Place and Cyberspace: Issues in Accessibility*, ed. Donald G. Janelle & David Charles Hodge (Berlin: Springer-Verlag, 2000), 303–316, <www.spatial.maine.edu/%7Eonsrud/pubs/LegalAccessToGI.pdf>, 22 Feb. 2010. See also Michael F. Goodchild & Robert P. Haining, 'GIS and Spatial Data Analysis: Converging Perspectives', *Papers in Regional Science* no. 83 (2004): 369; Andrew U. Franck, *The surveying activities at the Austrian Federal Office for Metrology and Surveying: an economic analysis* (Vienna: Austrian Federal Ministry of Economics and Labour, 2003), 22 <ftp://ftp.geoinfo.tuwien.ac.at/frank/reportbevfinalv31prn.pdf>, 22 Feb. 2010.
127. Yola Georgiadou and Richard Groot, 'Policy development and capacity building for geoinformation provision: a global public goods perspective', *GIS Development* 6, no. 7 (2002): 33–40, <www.gisdevelopment.net/policy/international/interna014.htm>, 22 Feb. 2010.

that mapping information should be provided by government, funded through general tax revenues.[128]

While the situation in Canada is specific due to the size and population of the country, the reasoning may be transposed to comparable situations in different countries.

CRITICISM – However, the same remark can be made as for information in general: if public sector spatial data is a public good, no private provider could be found to create spatial data sets. While this may have been true for some types of spatial data, such as topography, meteorological data, or cadastral data, evolutions on the spatial data market have shown that data that were collected by the public bodies are now also collected by private companies and vice versa.

MERIT GOOD? – Hence, it is difficult to state that public sector spatial data are *in se* a public good. Public sector spatial data could be considered a merit good, of which society or government values the existence or consumption and therefore the government needs to play a role in providing this good. However, the concept of merit goods does not bring us much further in defining the obligations of the public bodies to make public sector spatial data available and the distinctions between access, re-use, and sharing. It only asks the question which data should be considered so important or valuable that it should be made available for access, re-use, or sharing. Once again, we still have to determine what the distinction is. On that basis we could determine what would constitute a merit good for that purpose.

C. BETTER DECISION MAKING

DECISION MAKING IS NOT AN OBJECTIVE IN ITSELF – It has been mentioned above that the availability of public sector spatial data can also lead to more rational decision making. Public bodies prepare, implement, and evaluate policies for ensuring the quality of life of the citizens, and they need data in order to do so. Therefore, it was stated that societal growth also was an objective of sharing: ensuring such growth is exactly the public task of the public bodies. The same could be stated for economic growth. It is the Member States' and the public bodies' task to promote economic growth by measures relating to employment, stability of business, the promotion of entrepreneurship, and so on. In this way, one could state that decision making in itself is not an objective of the availability of public sector spatial data, but such decision making contributes to one of the other objectives.

EXAMPLES – Many examples can be found with regard to public sector spatial data: such data are needed for sustainable development, natural resource planning and management, environmental management, emergency planning and disaster

128. Mark Corey, 'Digital data, copyright, commercialization and related issues – a Canadian perspective', in *Policy issues in modern cartography*, ed. D.R. Fraser Taylor (Oxford: Elsevier Science Ltd., 1998), 30.

relief, the battle against poverty, crime monitoring, transport, agriculture, security, and so on.[129]

RECOGNITION IN LEGISLATION AND POLICY DOCUMENTS – Legislation and policy have mostly paid attention to the need for public sector data for environmental purposes. As was discussed in the previous chapter, GEOSS promotes sharing for the development of policies with regard to, among others, disaster management, energy resources, climate change, water resource management, and biodiversity.[130] Next, Agenda 21's main purpose regarding information is to improve decision making on sustainability issues. The INSPIRE Directive and the SEIS Communication also believe that environmental and spatial data are needed for the formulation and implementation of environmental policies.

D. INTERIM SUMMARY

RECAPITULATION – In this section, the three purposes of the availability of public sector spatial data have been considered in some more detail. The democratic objectives for the availability of public sector spatial data were divided in

129. See e.g., Ian Williamson, Abbas Rajabifard & Andrew Binns, 'The Role of Spatial Data Infrastructures in Establishing an Enabling Platform for Decision Making in Australia', in *Research and Theory in Advancing Spatial Data Infrastructure Concepts*, ed. Harlan J. Onsrud (Redlands: ESRI Press, 2007), 121–132; National Research Council, *Licensing geographic data and services* (Washington: National Academies Press, 2004), 33; Acil Tasman, 'The Value of Spatial Information: The Impact of Modern Spatial Information Technologies on the Australian Economy', 2008, 93, 143, <www.crcsi.com.au/uploads/publications/PUBLICATION_324.pdf>, 22 Feb. 2010; Richard E. Klosterman, 'The appropriateness of geographic information systems for regional planning in the developing world', *Computers, Environment and Urban Systems* 19, no. 1 (1995): 3; Kwabena O. Asante et al., 'Spatial Data Infrastructures in Management of Natural Disasters', in *Research and Theory in Advancing Spatial Data Infrastructure Concepts*, ed. Harlan J. Onsrud (Redlands: ESRI Press, 2007), 279–293; Charles M. Croner, 'Public Health, GIS, and the Internet', *Annual Review of Public Health* 24 (2003): 57–82; Felicia O. Akinyemi, 'Spatial Data Needs for Poverty Management', in *Research and Theory in Advancing Spatial Data Infrastructure Concepts*, ed. Harlan J. Onsrud (Redlands: ESRI Press, 2007), 261–277; Michael Blakemore, 'Financing the NGDI', <www.gisdevelopment.net/policy/international/interna012pf.htm>, 22 Feb. 2010; Jeffrey P. Johnson, 'Case studies of dissemination policy in local government GIS agencies', *Computer Environment and Urban Systems* 19, no. 5–6 (1996): 373; Arthur Getis et al., 'Geographic Information Science and Crime Analysis', *Journal of the Urban and Regional Information Systems Association* 12, no. 2 (2000): 7–14; François Salgé, 'Geographic information network in Europe (GINIE). Recommendations for action', in *Spatial data infrastructure and policy development in Europe and the United States*, ed. Bastiaan Van Loenen & Bas C. Kok (Delft: DUP Science, 2004), 41; Economic Development and Infrastructure Committee, 'Inquiry into Improving Access to Victorian Public Sector Information and Data', 2009, 11–12, <www.parliament.vic.gov.au/edic/inquiries/access_to_PSI/final_report.html>, 22 Feb. 2010.
130. Group on Earth Observations, 'The Global Earth Observation System of Systems (GEOSS) 10-Year Implementation Plan, 2005', 1–2, <www.earthobservations.org/documents/10-Year%20Implementation%20Plan.pdf>, 22 Feb. 2010.

transparency, public participation, accountability, and legitimacy, and the fulfillment of the rights and obligations of the citizen. The socio-economic objectives included societal growth, the development of communities, and economic growth. These were put against the background of the question of whether public sector spatial data are a public good. Thirdly, decision making was also considered an objective of the availability of public sector spatial data.

EVALUATION – While it can be stated that the first three democratic reasons are clearly linked to access and do not immediately call for making public sector spatial data available for re-use, the last reason starts to show that the distinction between democratic reasons and economic reasons is not always that easy to maintain. The problem lies with what one could refer to as the more 'social objectives' that are part of the socio-economic reasons that have been mentioned. For instance, in order to participate in society, one could state that knowing about one's rights and obligations towards the government and towards the other citizens is essential. Hence, the socio-economic objective also includes a democratic objective. This is also the case for the development of communities: these communities are considered to be of socio-economic importance, yet they are also important to increase public participation in decision making and they also provide a forum to assess the decisions of government and to hold government accountable for those decisions.

DISTINCTION REMAINS A PROBLEM – Hence, an examination of the objectives that should clearly be linked to either access, re-use, or sharing does not make it clear for which objective public sector spatial data are requested. Therefore, the distinction between access, re-use, and sharing remains a problem. The next section discusses the scope of this problem.

V. THE PROBLEMATIC CHARACTER OF THE DISTINCTIONS BETWEEN ACCESS, RE-USE, AND SHARING

OUTLINE – In the previous sections I have looked a little closer at indicators that could determine whether access, re-use, or sharing is the purpose of requests for public sector spatial data and at the objectives underlying the availability of public sector spatial data. It became clear that the indicators were not sufficient, and that the objectives also did not make the distinction much clearer. Based on this I take a closer look at the problematic character of the distinction.

ACCESS V. RE-USE V. SHARING – While the demarcation between access and sharing does not seem to cause much trouble according to the literature or policy documents, the difference between access and re-use on the one hand, and re-use and sharing on the other hand, has been more contested. Therefore, these two distinctions are described below. One might wonder why the difference between access and sharing is not considered a problem. It may be because they are both considered as non-economic purposes and therefore the relationship between the public and the private sector does not come into play. This relationship causes re-use to be the centre of the discussion, and this branches out into the debate on the difference with access on the one hand and sharing on the other hand.

A. Access and Re-use

Democratic v. economic purpose – In Chapter 2, the distinctions between access, re-use, and sharing were made on the basis of the objectives as they appeared from the three directives that determine the availability of public sector spatial data (Access Directive, PSI Directive and INSPIRE Directive). Access was related to democratic purposes and re-use to economic purposes. The United Kingdom's Advisory Panel on Crown Copyright gave a nice description of the difference between access and re-use: 'Although UK officials are keen to maintain a distinction between access and re-use, the distinction is far from clear . . . Although the subject matter (public information) and the broad scope (public bodies) of these instruments are similar, the underpinning policies are quite different. The [Freedom of Information] Act seeks to promote greater transparency and openness in the conduct of public affairs, while the PSI Directive recognizes the value, and aims to encourage the commercial exploitation, of public information. The focus of the FOI Act is enhancing the rights of individuals in a democratic society. At the heart of the PSI Directive is the smoother running of the internal market; the stimulation of the European information industry so it can compete more effectively in the global marketplace'.[131]

Problematic distinction – However, in practice it is hard to distinguish between access and re-use based on the objective. Hence, the model that has been installed by the three directives on the availability of public sector spatial data can be problematic to follow through in practice. This can be demonstrated by two examples.

Example: journalists as democratic watchdogs – The first example is the use of public sector spatial data by journalists. Traditionally, journalists have always made use of national access legislation to obtain data from the public bodies in order to publish articles about national and international policies, and to hold government accountable for those policies.[132] Journalists are seen to have a fundamental task of informing the public. This importance is reflected in the freedom of the press, the freedom of information, and the freedom of speech. Hence, their purpose can easily be described as democratic: they are an indispensable link in the process of democratic control of the citizens over their government. In the Convention on Access to Official Documents of the Council of Europe, journalists are mentioned especially as target group of the Convention, as they have a

131. Advisory Panel on Crown Copyright, 'Aligning the FOI and PSI initiatives in the UK', 2004, 2, <www.appsi.gov.uk/reports/aligning-FOI-and-PSI.pdf>, 22 Feb. 2010.

132. See James T. O'Reilly, 'Access to records versus access to evil': should disclosure laws consider motives as a barrier to records release?', *Kansas Journal of Law and Public Policy* 12 (2003): 565.

fundamental role 'as watchdog over the proper functioning of public institutions – a crucial role in a democratic society which has been fully recognized by the case-law of the European Court of Human Rights'.[133]

TRYING TO SELL NEWS ITEMS – However, with the increasing merging of publishers, broadcasting companies, and other information channels, there is a growing concern that commercial interests are replacing bringing information to the public as the main purpose of journalists and their employers. For instance, this has led to many discussions about safeguarding pluralism in the converging media and telecommunications environment. In short, journalists and newspapers want to increase the circulation of their products and make a profit. This could also be considered as an economic purpose. From the model presented in Chapter 2, this means that journalists would have to obtain data held by the public bodies under the rules of re-use. This is supported by the wide definition of *re-use:* any use for commercial or non-commercial purposes outside of the public task. This would entail that requests for public sector data would be subject to additional conditions that are not part of access legislation, such as concluding a license, with possibly time-consuming negotiations, or a charge that includes a reasonable return on investment. Van Eechoud and Van Der Wal seem to indicate that the media should still fall under access legislation, irrespective of the re-use definition. They state that interest groups and 'particularly of course the media as "public watchdog", play a vital role in informing citizens about decision-making processes, thus helping them to make informed political choices and participate in the democratic process. That they are profit-based entities should not be – and is not – relevant from the viewpoint of freedom of information law'.[134] Van Eechoud believes that the possibility alone of newspapers facing high charges after filing a request for public sector data is in itself unwelcome.[135]

EXAMPLE: LAWYERS – A second example is a lawyer filing a request to obtain public sector data in order to argue a case for his or her client. On the one hand, he or she is using the public sector data in order to do the job that the client has entrusted him or her with and to get his or her lawyer fee. Hence, there is an economic purpose involved. As was mentioned, the definition of re-use includes *any use* for commercial or non-commercial purposes outside of the public task, so the lawyer is re-using the public sector data. On the other hand, there is also a tradition that these data can be obtained under access legislation. In addition, one could argue that the lawyer is essentially helping his or her client to enjoy one of his or her human rights, that is the right to defend himself or herself in a fair trial. This

133. Steering Committee for Human Rights, *Draft Convention of the Council of Europe on Access to Official Documents and Draft Explanatory Report*, doc. nr. CDDH(2008)008 Add. III Bil., 3 Apr. 2008, 10–11, <http://gyla.ge/foi/doc_upload/capmaign/3.%20evropis%20sabchos%20konvenciis%20proeqti%20inglisur%20enaze.pdf> 22 Feb. 2010.

134. Mireille Van Eechoud & Brenda Van Der Wal, 'Creative commons licensing for public sector information. Opportunities and pitfalls', 2008, 54, <www.ivir.nl/publications/eechoud/CC_PublicSectorInformation_report_v3.pdf>, 22 Feb. 2010.

135. Mireille Van Eechoud, 'Het hergebruik regime voor overheidsinformatie in de Wob, een tussenstand', *Mediaforum* 20, no. 1 (2008): 9–10 [in Dutch].

can be considered an essential part of a democratic state where the rule of law is maintained. According to Marquand, lawyers are 'market traders, selling their wares in a highly competitive marketplace ... But that is not all they are. They also have duties to the court. Their primary duty is to ensure that justice is fairly and impartially administered; and that duty is supposed to override their economic interests. It is not confined to barristers whose fees are paid by the public purse. It applies to wealthy silks with lucrative commercial practices and annual earnings in seven figures as well as to the Rumpoles of the profession who depend on legal aid'.[136] Hence, once again, the wide definition of re-use creates confusion.

ACCESS INCLUDING RE-USE? – The relationship between access and re-use is seen as an interesting challenge by many authors. Peterson Dando asks the question of whether access legislation should be interpreted in a strict and traditional sense allowing an inquiring citizen to obtain a requested document or whether access legislation also includes the 'ability to manipulate data into entirely new documents'.[137] Van der Hof states that access actually often includes the possibility for the private sector to obtain public sector data for the creation of information products and services. These products in their turn can, due to their added value, improve the accessibility of public sector data for the citizen.[138] Perritt and Lhulier follow a comparable reasoning: 'Freedom of information and open government policies must make it clear that a commercial publishing motive does not disqualify anyone from the right to obtain basic government information'.[139] However, while Van der Hof and Perritt and Lhulier feel that access includes re-use, the separate attention that has been given to both topics indicates otherwise. In addition, before the transposition of the PSI Directive, the access legislation of a number of European countries included a provision prohibiting the use of the data that was obtained under such legislation to be used for commercial purposes, for example, Belgium, France, Greece, and the Netherlands.[140]

RE-USE AS A REQUIREMENT FOR ACCESS? – The first Publaw study also explicitly linked access and re-use. According to the study, '[i]n general the private sector use of access legislation may be seen as a use complimentary to the use by individuals. The private sector may offer additional services based on public sector [data] ... The private sector, however, can only play this role if the principle of public access

136. David Marquand, 'False Friend: the State and the Public Domain', in *Restating the State? Special issue of The Political Quarterly*, ed. Andrew Gamble & Tony Wright (Malden, Blackwell Publishing, 2004), 54.

137. Lori Peterson Dando, 'A Case for the Commercialization of Public Information', paper given at the Conference on Law and Information Policy for Spatial Databases (Tempe, 1994), <www.spatial.maine.edu/%7eonsrud/tempe/dando.html>, 22 Feb. 2010.

138. Simone Van Der Hof, *Overheidsinformatie in de etalage. Belangen rondom de toegang tot overheidsinformatie* (Alphen a/d Rijn: Samsom Bedrijfsinformatie, 1997), 5 [in Dutch].

139. Henry H. Perrit Jr. & Christopher J. Lhulier, 'Information Access Rights Based on International Human Rights Law', *Buffalo Law Review* 45 (1997): 927.

140. For examples, see Commission of the European Communities, *Publaw. Subject report: general access to information legislation* (Luxembourg: CEC Legal Advisory Board, 1991), s.p.

as such is not put into question. Private sector involvement in public sector information should therefore not be seen as a contradiction but as a supplement to the principle of public access'.[141] The study believed that not allowing re-use would jeopardize the right to access because 'the effectiveness of access rights very often demands that organizations specialize in obtaining information from the public sector on the basis of access rights, upgrade this information and make it accessible to the citizen in more "digestible" formats. Such organizations, very often public interest groups, would have to demand some sort of compensation for these services. Also one has to take into account that the public sector might cease to collect or to hold available certain kinds of information. In such circumstances it is useful if private organization[s] have already established themselves in providing such information'.[142] Schartum also believes that the possibility for private publication of public case documents and the competition for context that private publication could bring about, is likely to have clear positive effects on access to government-held information. The amount of information will increase as will the number of different statements about and interpretations of documents that apply to public activities. As a result, '[t]he individual, even more so than today, will be able to choose whether to act on the basis of the original documents or in accordance with the interpretations and explanations that the administration, professional and industrial bodies, and the press make available'.[143]

EVALUATION – A number of remarks can be made on this line of reasoning. First, the Publaw study assumes that access legislation includes the right to re-use, which we have seen was not automatically the case before the PSI Directive, and has been definitely ruled out by the introduction of the PSI Directive. Second, the fact that access legislation would not allow re-use of the obtained data would not necessarily put the access right of the citizens in danger. They can still obtain the data from the public bodies for their own access purposes. It is true that private organizations may create products that make public sector data more accessible and understandable, but the obligations regarding accessibility of access legislation are very limited, and they are directed towards the public bodies themselves. While such accessibility is very important, it cannot be said that without accessible private sector information-products based on public sector data, the right of access itself is being limited. Next, the study uses the argument that the public body might cease to collect or produce information, and that in such cases the private sector could help. However, if the private sector is re-using the public sector data to create information products or services, then the discontinuation of the collection of data

141. Commission of the European Communities, *Publaw. Subject report: general access to information legislation* (Luxembourg: CEC Legal Advisory Board, 1991), s.p. See also HM Treasury, 'Cross Cutting Review of the Knowledge Economy. Review of Government Information', 2000, <www.hm-treasury.gov.uk/spend_sr00_ccr.htm>, 22 Feb. 2010.

142. Commission of the European Communities, *Publaw. Subject report: general access to information legislation* (Luxembourg: CEC Legal Advisory Board, 1991), s.p.

143. Dag Wiese Schartum, 'Access to Government-Held Information: Challenges and Possibilities', *Journal of Information, Law and Technology* no. 1 (1998): 4, <www2.warwick.ac.uk/fac/soc/law/elj/jilt/1998_1/schartum/>, 22 Feb. 2010.

by the public sector will cause the private sector companies to look for other sources of that information, and this will significantly increase their costs – and consequently their prices. Moreover, the private sector will often use public sector data because they are the only source, so if the public body ceases to collect the data, both access and re-use will become impossible.

ACCESS NOT A TASK OF THE PRIVATE SECTOR? – According to the Publaw 2 study, '[g]overnment departments collect information in order to perform their primary role of government and administration. The subsequent release and exploitation of the information they collect is secondary to their main purpose'.[144] Therefore, the exploitation of public information should best be left to the private sector.

EVALUATION – The report seems to confuse making available data and creating information services. The former is a part of the public task of the public bodies under access legislation, so it cannot be left to the private sector to disseminate the information that was collected by the public bodies. This is not the same as the provision of added value information services, for which the Synergy Guidelines stated that it should be left to the private sector.[145]

POLICY DOCUMENTS – The two previous examples and literature show that it may not always be possible to make a distinction between access and re-use. Moreover, we have seen in the previous chapters that the European and international legislative and policy documents are not always clear whether they address access or re-use. This is also the case on the national level. For instance, the Dutch Policy Paper *Naar toegankelijkheid van overheidsinformatie* speaks about accessibility of PSI and making available basic information of the democratic constitutional state to maximize public participation, and in the same breath mentions access to public sector databases by business for creating information products.[146] Another example can be found in Papapavlou's opinion that '[a]ccess to public sector information is of great importance for the information content industry. The information content industry is a big generator of revenue and jobs'.[147]

PRACTICAL DIFFICULTIES – This difficulty in demarcating access and re-use can also be felt on the practical level, in that it will be hard to ensure the enforceability of the different rules that apply to access and re-use. Considering the broad definition of re-use and the confusion there might exist between some re-use activities and access (e.g., journalists, lawyers, but also user-generated content such as blogs, wikis, and so on), the applicant may sometimes not even know that he or she is

144. Commission of the European Communities, *Publaw 2. A report to the Commission of the European Communities on an evaluation of the implementation of the Commission's Guidelines for improving the synergy between the public and private sectors of the information market* (Luxembourg: CEC Legal Advisory Board, 1993), 14.
145. Cf. *infra*, Ch. 7, s. III.B.2.a.
146. Ministry of the Interior, 'Naar toegankelijkheid van overheidsinformatie', 1997, 13, <www.minbzk.nl/aspx/download.aspx?file=/contents/pages/302/nota_ntvo_nl.pdf>, 22 Feb. 2010 [in Dutch].
147. George Papapavlou, 'Public Sector Information Initiatives in the European Union', 1999, 2, <http://webworld.unesco.org/infoethics2000/documents/paper_papapavlou.rtf>, 22 Feb. 2010.

re-using public sector spatial data and apply for access in good faith. Next, even if this is not the case, and the applicant is well aware that he or she is requesting data for re-use, it may not be so easy for the public body to determine this, as under the Access Directive, and any access legislation in general, the applicant is not under any obligation to state his or her interest. Hence, the public body would have to follow up on every request for public sector spatial data and try to find out if the applicant uses the data for re-use rather than access. On the one hand, this would be too much of a burden on the public bodies, and on the other hand this would entail a far-reaching interference in the citizens' lives.[148]

CAUSE OF PROBLEMS: BROAD DEFINITION OF RE-USE – The broad definition of re-use plays an important role in this issue. Since it involves 'any use for commercial or non-commercial purposes' outside of the public task, it involves any activity beyond the mere looking at the data. So if an individual undertakes any action based on the public sector spatial data that he or she has obtained, it could possibly be qualified as re-use. This could include very basic democratic actions, such as filing a complaint against a wrongly refused or granted building permit, or using information on flooding risks for deciding whether to buy a certain house or piece of land. In almost all requests for public sector spatial data, the applicant will have a purpose beyond just knowing the content. He or she will almost always have a personal interest, which entails that, based on the objective, access will very often constitute re-use. This will especially be the case for non-commercial re-use but may also be true for commercial re-use.

CATEGORIES OF RE-USE – The difference between commercial and non-commercial re-use is also not clarified in the EU legal framework on the availability of public sector spatial data. The question can be asked how a distinction between different categories of re-use can be made. As was mentioned in Chapter 2, the public bodies can impose different conditions on different types of re-use under the PSI Directive, but the conditions have to be non-discriminatory for comparable categories of re-use. However, as was discussed in Chapter 2, no indication was given on which re-use constituted comparable categories. Is all commercial use comparable and all non-commercial use? Or are there categories within either type of use? This will be hard to determine, as it is even not always easy to make a difference between commercial and non-commercial re-use. For instance, Corey sees a significant problem with separating academic use from commercial use: 'Many university professors, under the current education spending squeezes and with the active blessing and support of their institutions, are also consultants. They do contract work, provide services and do studies, in competition with the private sector. For government to make inexpensive or free data available to them, with no restrictions, might give them an unfair advantage in the marketplace when bidding

148. See Commission of the European Communities, *Publaw. Subject report: general access to information legislation* (Luxembourg: CEC Legal Advisory Board, 1991), s.p.; Ravi Bedrijvenplatform, *Commercialisering van geo-informatie. Een drietal rapporten ter ondersteuning van de Ravi-discussie 1997* (Amersfoort: Ravi, 1997), 104 [in Dutch].

against companies who have to pay full price for this data.'[149] While academic re-use may be seen as non-commercial, such contract work could just as easily be considered commercial.

EVALUATION OF THE BROAD DEFINITION OF RE-USE – One can question whether the broad impact of re-use was the purpose of the PSI Directive. The main concern of the European Commission when it started the process towards the PSI Directive was with the information industry, which could not fully benefit from the availability of public sector spatial data to create information products and services.[150] Ronai actually saw this restrictive approach as the reason why any PSI initiative before the PSI Directive only had a limited impact.[151] If re-use should only be considered as the use made of public sector spatial data by the information industry to create information products and services, changes in the definition of re-use would need to be considered. However, it is not clear which criterion could be used in that case to determine what re-use is. As we have seen before, neither the type of user nor the type of data can provide an answer to this.

B. RE-USE AND SHARING

ECONOMIC V. PUBLIC TASK PURPOSE – Above it was stated that the purpose does not make it clear what the difference between access and re-use is. The same question now can be asked for the distinction between re-use and sharing. As was explained before, the demarcation between re-use and sharing is based on the public task. While this public task can be performed by private parties, in most cases they will do this under the assignment or supervision of a public body. The main concern with regard to the distinction between re-use and sharing therefore lies with the activities of the public bodies. As was mentioned before, they may perform public tasks, but they may also go outside of their public task and enter the market to provide economic services, possibly in competition with the private sector. This categorization of the activities of the public bodies, and subsequently the role of public bodies on the market, has been a recurring topic on the agenda with regard to the availability of public sector spatial data. Therefore, the definition of the scope of the public task of the public bodies is very important.

CONSEQUENCES – While there may be some activities for which it is clear that the public bodies should perform them as part of their public task, for other activities this is not so clear. These activities might be classified as market activities outside of the public task. This has consequences for the rules under which public

149. Mark Corey, 'Digital data, copyright, commercialization and related issues – a Canadian perspective', in *Policy issues in modern cartography*, ed. D.R. Fraser Taylor (Oxford: Elsevier Science Ltd., 1998), 45.

150. Office of Public Sector Information, 'The United Kingdom Implementation of the European Directive on the Re-use of Public Sector Information – the first two years', 2007, 3, <www.opsi.gov.uk/advice/psi-regulations/uk-implementation-first-years.pdf>, 22 Feb. 2010.

151. Maurice Ronai, 'L'Etat comme machine informationnelle', *Revue Française d'Administration Publique* no. 72 (1994): 571–580 [in French].

sector spatial data are made available in order to perform that activity. If the activity is part of the public task, the rules for sharing under the INSPIRE Directive will be applicable. If the activity is not part of the public task but a market activity instead, the public sector spatial data will have to be obtained under the rules of the PSI Directive. Therefore, it is important to know what the scope of the public task is.

ELEMENTS OF THE PUBLIC TASK – The public task consists of different elements. As was mentioned in Chapter 5, the public task of public bodies consists of preparing, implementing, and evaluating policy. This could be considered as the internal public task. As was indicated before, making public sector spatial data available is needed for better decision making. This decision making is not necessarily an objective in itself, but it helps in fulfilling the democratic and socio-economic objectives for which public sector spatial data should also be available. Next, the public task of the public bodies not only involves making policy or making rules that apply to the public, but it can also include the offering of services to the public. This makes up the external public task of the public bodies. Such services may be linked more or less directly to the policy choices made by the public bodies, such as health services, waste disposal, utilities, public transport, public broadcasting, and so on. However, for some of these services, the question can be asked whether these still belong to the public task or whether they are services that can be performed on the market.

BASIS FOR THE DEFINITION OF THE PUBLIC TASK – This leads us to the question of what the public task exactly contains. Attempts at defining this have started from different points of view. On the one hand, the public task is defined as the tasks that have been expressly acknowledged as such by law or that can be derived from there by interpretation.[152] On the other hand, the public task is also related to the safeguarding of the public interest or to the remedy of market failure. As Abraham Lincoln stated, 'The legitimate object of government is to do for a community of people whatever they need to have done, but cannot do in their separate and individual capacities'.[153]

EXAMPLES – Based on these points of view, a number of activities appear to be generally considered as part of the public task. For instance, Merusk believes that the substantial tasks of the state include, above all, defence of the state, police, foreign relations, financial administration, tax administration, customs, compulsory fulfilment of legal decisions, and administration of law.[154] Vermeulen thinks

152. Wetenschappelijke Raad voor het Regeringsbeleid, *Het borgen van publiek belang* (The Hague: Sdu Uitgevers, 2000), 56, <www.wrr.nl/content.jsp?objectid=2545>, 22 Feb. 2010 [in Dutch]. See also Udo Di Fabio, 'Privatisierung und Staatsvorbehalt', *Juristen Zeitung* 54 no. 12 (1999): 586 [in German]; Johannes Hengstschlager, 'Privatisierung von Verwaltungsaufgaben', in *Veröffentlichungen der Vereinigung Deutscher Staatsrechtlehrer 54*, ed. X. (Berlin: De Gruyter, 1995), 173 [in German].

153. See John M. Donahue, *The privatization decision. Public Ends, Private Means* (New York: Basic Books, 1989), 17.

154. Kalle Merusk, 'Assignment of public tasks to private legal persons. Legal bases and limits', *Juridica International* 5 (2000): 42, <www.juridica.ee/get_doc.php?id=327>, 22 Feb. 2010.

along the same lines: some tasks are so essential that the government needs to carry the final responsibility for them, such as national peace and safety, the defence of the nation, the administration of justice, and the development of democratic decision-making procedures.[155] According to Lim A Po, the fundamental public tasks of the state are to ensure law and order, safety, and security, as well as guarantee social justice for all. The responsibility of the state for law and order is based on the notion that the rule of law and an independent judiciary are guarantees for freedom and equality. Both of these basic values are the cornerstones of democracy. Social justice as a responsibility of the state ensues from 'human solidarity', the third basic value of a social-democratic system. He reminds us, however, that the public tasks are not restricted to these fundamental tasks. The state also has a public responsibility for the course and the result of development; however, it is no longer considered solely and fully responsible for the public tasks related to development.[156] Schaefer mentions setting and enforcing laws, securing domestic security and stability, defending against external threats, providing public goods and services, building and maintaining infrastructure, providing the means for impartial arbitration and justice, and undertaking policies to facilitate domestic tranquillity.[157] Tanzi has a more economic point of view: he sees a fundamental role for the state in the allocation of resources, the redistribution of income, the stabilization of economic activity, and the promotion of growth and employment.[158] Sopers warns that these goals may sometimes be contrary to each other, for example, economic growth and employment and protection of the environment.[159]

Contingency of public task – However, whether a particular activity is considered as a public task or not may change with time and depend on national situations and cultures. In addition, it remains a political decision. The role of employment agencies is a good example: the scope of the public task has changed considerably in this area; it used to be a monopoly of the state in many EU Member States, but gradually the market has started to play a bigger role and the public sector has withdrawn from many of the activities related to employment-finding.[160]

155. Ben P. Vermeulen, 'De publieke taak in het algemeen', in *De publieke taak. Staatsrechtconferentie 2002, Vrije Universiteit Amsterdam*, ed. Jan W. Sap, Ben P. Vermeulen & Carla M. Zoethout (Amsterdam: Kluwer, 2003), 22–24 [in Dutch].

156. Hans R. Lim A Po, 'Good governance in the shared public domain'. Speech held at the symposium on 'Good governance, a basic condition for economic development, towards a Surinamese agenda (Paramaribo, 2003), <www.fhrinstitute.com/good_governance.htm>, 22 Feb. 2010.

157. Brett D. Schaefer, 'How the scope of government shapes the wealth of nations', <www.heritage.org/research/tradeandforeignaid/hl925.cfm>, 22 Feb. 2010.

158. Vito Tanzi, 'IMF Working Paper. The role of the state and the quality of the public sector', 2000, 5, <www.imf.org/external/pubs/ft/wp/2000/wp0036.pdf>, 22 Feb. 2010; Herman Matthijs, Frank Naert & Jef Vuchelen, *Handboek Openbare Financiën* (Antwerp: Intersentia, 2004), 10–12 [in Dutch].

159. Jan M.M. Sopers, *De bakens verzet. Publieke taakorganisaties in verandering* (Assen: Uitgeverij Van Gorcum, 1999), 10 [in Dutch].

160. Ulrich Walwei, 'Arbeitsvermittlung als öffentliche Aufgabe under privatwirtschaftliche Dienstleistung', *Mitteillungen aus der Arbeitsmarkt- und Berufsforschung (MittAB)* no. 1 (1996):

Another example can be found in the telecommunications domain. According to De Wit, the role of the government with regard to telephony changed continuously between the end of the nineteenth and the end of the twentieth century.[161] When the telephone was first introduced, the government did not see a task for itself to promote connectivity, but during the course of the twentieth century, it realized that it was in the public interest that the public had access to a telephone and started to increase its involvement. The last years of the twentieth century saw the balance move in the other direction, with the limitation of state intervention and the growing privatization movement.

EVALUATION – This contingent character of the public task at first sight forms a considerable problem for determining the distinction between re-use and sharing. In the next chapters, we examine how far the concept of the public task may be a problem for the proper functioning of the EU legal framework for the availability of public sector spatial data. Can the scope of the public task be determined in a way that is sufficient for the concept to work as a criterion for determining the application of the legal framework?

OUTLINE – The next two chapters will examine this question with regard to one particular kind of services that is offered by the public bodies, that is, information services based on public sector spatial data. As was mentioned before, the provision of information services entails in my view that the public bodies process their spatial data beyond their own needs in order to provide them to their users. These information services may need to be provided under the public task, for example, because the public bodies have an obligation under national law to provide access to information services for the public. On the other hand, these information services may also be commercial products that are offered on the market in competition with the private sector. The demarcation line between the two types of information services is not clear. It is a good example to show the discussion that has been and could also be held with regard to other services. In addition, it shows that the availability of public sector spatial data and the provision of spatial information services are part of a chain that is aimed at increasing the use of spatial data. Therefore, Chapter 7 first examines the existing obligations for public bodies to provide spatial information services. While Chapter 5 was directed towards the availability of data, Chapter 7 looks at the same legislation and policy documents, but with specific attention to the provision of spatial information services rather than data. Next, Chapter 8 questions the concept of the public task and the possibility of defining it. This is done by looking at the arguments that are used to justify the assignment of a public task and how this is handled in the related area of competition law.

54–72 [in German]. See also Katleen Janssen & Jan C. Kabel, 'Commercialisering van overheidsinformatie door de overheid: rechtspraak en wetgeving in België en Nederland. De honden blaffen, maar de karavaan trekt voort', *Computerrecht* no. 3 (2005): 119–121 [in Dutch].

161. Onno De Wit, 'Langs lijnen van geleidelijkheid: de telefonie in Nederland als object van staatszorg, 1877–1989', in *Toeval of noodzaak? Geschiedenis van de overheidsbemoeienis met de informatievoorziening*, ed. Rathenau Instituut (The Hague, Rathenau Instituut, 1995), 72–110 [in Dutch].

Chapter 7

The Public Task of Providing Spatial Information Services in International and European Legislation and Policy Documents

I. INTRODUCTION

RECAPITULATION – The previous two chapters addressed the problematic character of the basis for the EU legal framework for the availability of public sector spatial data and the differences this legal framework makes between access, re-use, and sharing. On the one hand, the difference between access and re-use is problematic due to the wide definition of re-use. On the other hand, the difference between re-use and sharing is based on the public task, while no definition has been given of this public task.

INFORMATION SERVICES WITHIN THE SCOPE OF THE PUBLIC TASK – When the public sector obtains public sector spatial data in order to perform its public tasks, it falls under the rules for sharing. As was mentioned above, this public task includes the *internal* public task of performing the public bodies' own functions of preparing, implementing and evaluating policy, and the *external* public task of providing services to the public. Our attention is focused on one type of services, that is, information services. In today's information society, the demands on the information that should be provided to the public are continuously increasing. Not only does the public have the right to obtain information, it should also receive it in an accessible and understandable way. This entails that the public task of the public bodies may sometimes go beyond merely making available public sector spatial data as they were collected for the public bodies' own needs, and extend to providing information services. In light of my analysis, I consider information services to involve the further processing of the public sector spatial data beyond the public bodies' own requirements. Such information services may be needed to provide the users with *information* or *knowledge* rather than *data* to ensure that they do not only get access to a data set, but that the data are placed in a context so that they can be understood by non-experts. If the public bodies are providing such information services under their public task, they can obtain the public sector spatial data on which the information service is based under the rules for sharing.

INFORMATION SERVICES OUTSIDE OF THE SCOPE OF THE PUBLIC TASK – The public sector may also create information services outside of its public task on the information market, possibly in competition with the private sector. In this case, it is acting as a re-user and has to obtain the public sector spatial data it needs under the rules of the PSI Directive. The public body is once again adding value, but it is acting outside of its public task. This raises the question of when the public body is adding value as a re-user in order to provide the market with an information service, and when the public body is adding value with the purpose of fulfilling its public task of informing the public. How far does this public task of the public sector go?

ORIGIN OF THE PUBLIC SECTOR SPATIAL DATA UNDERLYING THE INFORMATION SERVICES – Both in the case where the information services are provided as a public task and the case where the information services are offered on the market, two situations are possible for the underlying public sector spatial data. First, the public body that provides the information service may have obtained these data from another public body. This means that the relationship between these two bodies is determined by the limits of the public task. If the receiving public body is providing an information service under its public task, the supplying public body will have to make the underlying data available under the conditions of the INSPIRE Directive. If the receiving body creates an information service on the market, the supplying body has to follow the rules of the PSI Directive. Second, in many cases the public body will not obtain public sector spatial data from another public body, but use data that it created or collected itself. In this situation, drawing the boundary between the types of information service that are provided may be even more challenging because it may impact the internal procedures of the public body. For instance, the PSI Directive has made it very clear that if documents are re-used by a public sector body as input for its commercial activities which fall outside the scope of its public tasks, the same charges and conditions shall apply to the supply of the documents for those activities as apply to other users.[1] This will most likely require the public body to separate its re-use activities from its other functions. While there is no such provision in the INSPIRE Directive regarding sharing, one may wonder if the public body should impose the same conditions and charges on itself for making public information services as it would on other public bodies.

OUTLINE – This chapter examines whether the international and European legislation and policy documents that have been discussed in Chapter 5 can give indications on the scope of the public task to provide information services. On the

1. Article 10.2 of the PSI Directive.

one hand, it is investigated what information services the public bodies *should* provide, that is, the public task from a *positive* point of view. While there are a number of texts that give some hints towards an obligation to provide information services, such obligation is very limited. Mostly, we will see that the texts are not clear on the scope of the public task. On the other hand, some legislative or policy texts address the scope of the public task of providing information services from a *negative* point of view. They try to determine what the public bodies should not do and which information services they should not provide, but rather leave up to the private sector on the market. As we will see, these legislative and policy texts have rarely succeeded in defining the public task in a negative way. They are mostly limited to describing what the consequences are of a public body performing activities outside of its public task, without discussing what this public task actually is.

II. THE SCOPE OF THE PUBLIC TASK OF PROVIDING INFORMATION SERVICES FROM A POSITIVE POINT OF VIEW

A. Introduction

Outline – This section looks at the positive obligations that are imposed on Member States and public bodies to provide information services. In order to keep the focus on public sector spatial data, only the documents are addressed that would apply to that category of data. Hence, we look at the obligations of public bodies to provide public sector information services, environmental information services, and spatial information services. While there may exist other obligations to provide a certain type of information service to the public, for example, cultural information, news events and so on, these are not considered in this chapter. Before the legislation and policy documents are addressed, the same question is asked as in Chapter 5: Is the public task of public bodies to provide spatial information services so important that it can be seen as included in the fundamental right of freedom of information as it is acknowledged in international treaties on human rights?

B. The Public Task of Providing Spatial Information Services as Part of Freedom of Information?

Outline – In Chapter 5, it was described that including a public task of the public bodies to make public sector spatial data available to the public in the fundamental right of freedom of information was problematic. Such an obligation could not be clearly determined from the text of the Treaties or from the jurisprudence of the European Court of Human Rights. Hence, one can assume that deriving an obligation to provide information services to the public from Article 19 of the ICCPR or Article 10 of the ECHR would be even more far-reaching.

1. United Nations

a. Universal Declaration on Human Rights

No recognition of a public task to provide information services – The question of whether freedom of information under the Declaration includes an obligation for the public bodies to provide public sector data or information services has never been raised. More attention was given to the binding International Covenant on Civil and Political Rights.

b. The International Covenant on Civil and Political Rights

ICCPR – The 1966 International Covenant on Civil and Political Rights (ICCPR) deals with freedom of information, as a part of freedom of expression, in Article 19.[2] As was mentioned in Chapter 5, Article 19 of the ICCPR does not include an obligation of public bodies to make public sector data available, so it definitely cannot be extended to the provision of information services.

Special Rapporteur on Freedom of Opinion and Expression – While Mason believes that the interpretation should include the recognition of such an obligation,[3] the UN Special Rapporteur on Freedom of Opinion and Expression did not feel that information services should be included under Article 19. While he stated in 1998 that 'the right to see, receive and impart information imposes a positive obligation on States to ensure access to information',[4] he did not extend this to the provision of information services. Next, the 2004 Joint Declaration of the UN Special Rapporteur, the OSCE Representative on Freedom of the Media, and the OAS Special Rapporteur on Freedom of Expression asked to promote awareness of freedom of information legislation and the need to make public sector data available, but it did not require that any specific information service be set up to do this.[5]

2. United Nations, *International Covenant on Civil and Political Rights*, 16 Dec. 1966, <www.unhchr.ch/html/menu3/b/a_ccpr.htm>, 22 Feb. 2010.
3. Anthony Mason, 'The relationship between freedom of expression and freedom of information', in *Freedom of expression and freedom of information. Essays in honour of Sir David Williams*, ed. Jack Beatson & Yvonne Cripps (Oxford: Oxford University Press, 2000), 226–227. Cf. *supra*, Ch. 5, s. I.B.1.a.
4. United Nations Special Rapporteur on Freedom of Opinion and Expression, *Promotion and protection of the right to freedom of opinion and expression*, UN Doc. E/CN.4/1998/40, 28 Jan. 1998, <http://daccess-ods.un.org/access.nsf/Get?Open&DS=E/CN.4/1998/40&Lang=E>, 22 Feb. 2010.
5. United Nations Special Rapporteur on Freedom of Opinion and Expression, OSCE Representative on Freedom of the Media & OAS Special Rapporteur on Freedom of Expression, *International Mechanisms for Promoting Freedom of Expression. Joint Declaration*, 6 Dec. 2004, <www.cidh.org/Relatoria/showarticle.asp?artID=319&lID=1>, 22 Feb. 2010.

SUMMARY – The United Nations did not believe that a public task to make public sector data available could be deducted from Article 19 of the ICCPR.[6] *A fortiori*, the provision of information services cannot be deemed to be a part of freedom of information in their opinion.

2. Council of Europe

a. The European Convention on the Protection of Human Rights and Fundamental Freedoms

NO RECOGNITION OF A PUBLIC TASK TO PROVIDE INFORMATION SERVICES – Like Article 19 of the ICCPR, Article 10 of the ECHR does not recognize a general obligation of the Member States or public bodies to make public sector data available to the public,[7] let alone to create information services for the public. While Voorhoof and Hins argue that the European Court of Human Rights is moving into the direction of acknowledging a public task to make public sector data available, such an acknowledgement is only in its infancy, and it can definitely not yet be extended to include the public task of providing information services.[8]

b. Documents of the Committee of Ministers and the Parliamentary Assembly

PARLIAMENTARY ASSEMBLY – In 1970, the Parliamentary Assembly felt that in the framework of freedom of information there was a public task of the Member States and the public bodies 'to make information available on request, subject to appropriate limitations', but it did not feel that the public bodies should make information available on their own initiative or create information services directed towards the citizens.[9]

---

6. Cheryl Ann Bishop, 'Access to Information as a Human Right: Analysis of the United Nations Human Rights Committee Documents', Paper given at the annual meeting of the International Communication Association, Dresden International Congress Centre (Dresden, 2006), 20–21, <www.allacademic.com/meta/p91640_index.html>, 22 Feb. 2010. See Ch. V, s. 1.2.A.b.
7. Anthony Mason, 'The relationship between freedom of expression and freedom of information', in *Freedom of expression and freedom of information. Essays in honour of Sir David Williams*, ed. Jack Beatson & Yvonne Cripps (Oxford: Oxford University Press, 2000), 228; François Dubuisson, 'Quand le "Droit" du Public à l'Information Rencontre les "Droits" sur l'Information: Vrai ou Faux Débat?', in *Droit d'Auteur et Liberté d'Expression. Regards Francophones, d'Europe et d'Ailleurs*, ed. Alain Strowel & François Tulkens (Brussels: Larcier, 2006), 91–92 [in French]. See Ch. V, s. 1.2.B.a.
8. Wouter Hins & Dirk Voorhoof, 'Access to State-Held Information as a Fundamental Right under the European Convention on Human Rights', *European Constitutional Law Review* 3, no.1 (2007): 123. See Ch. V, s. 1.2.B.c.
9. Council of Europe Parliamentary Assembly, *Recommendation 582 (1970) on mass communication media and human rights*, 1970, <http://assembly.coe.int/main.asp?Link=/ documents/adoptedtext/ta70/erec582.htm>, 22 Feb. 2010.

Committee of Ministers – In the 1982 Declaration on Freedom of Expression and Information, the Committee of Ministers sought an open information policy in the public sector, but it limited the active duties of the Member States to ensure plurality among the information sources. No obligation for the public bodies to act as an information source or provide information services was included.[10] While later texts did encourage the public bodies to make information available of their own motion,[11] the link from freedom of information to the availability of public sector data and especially to the provision of information services by the public bodies still seems a bridge too far. Only a first hint was given in this direction in a 2007 Recommendation of the Committee of Ministers: the Member States should 'ensure that public authorities increase the provision and transparency of their online services to citizens and businesses so that they allow every individual access to public information'.[12]

c. *The Case Law of the European Court of Human Rights*

No recognition of a public task to provide information services – As was mentioned above and in Chapter 5, the European Court of Human Rights is slowly moving in the direction of acknowledging an obligation of public bodies to make public sector data available under Article 10 of the ECHR, but it may still be too soon to accept a general task for the public bodies. A fortiori, one can assume that Article 10 of the ECHR does not include any public task to create and provide information services directed towards the public.

3. Interim Summary

No recognition of a public task to provide information services – Neither the ICCPR nor the ECHR recognize a positive obligation of the Member States or the public bodies to provide information services to the public. The European Court of Human Rights is slowly evolving towards acknowledging a public task to make public sector data available, and one may assume that the evolution towards information services will take a while longer. The other United Nations and Council of Europe entities also seem to be valuing an obligation for public bodies to make public sector data available, but none of them have taken the next step towards information services. Hence, it is currently not possible to deduct a public task to provide information services from the fundamental right of freedom of information.

---

10. Council of Europe Committee of Ministers, *Declaration on the freedom of expression and information*, 1982, <https://wcd.coe.int/com.instranet.InstraServlet?command=com.instranet.CmdBlobGet&InstranetImage=601273&SecMode=1&DocId=675536&Usage=2>, 22 Feb. 2010.
11. Cf. *supra*, Ch. 5, s. I.B.2.b.
12. Council of Europe Committee of Ministers, *Recommendation CM/Rec(2007)11 on promoting freedom of expression and information in the new information and communications environment*, 2007, <https://wcd.coe.int/ViewDoc.jsp?id=1188541>, 22 Feb. 2010.

C. The Public Task of Providing Public Sector Information Services in International and European Documents

1. Introduction

No distinction between access, re-use, and sharing – In Chapter 5, we have seen that there are many international and European documents that impose obligations on public bodies to make public sector data available. This section examines whether these documents also contain a public task to provide information services to the public. However, while Chapter 5 made a distinction between the availability of public sector data for access, re-use, and sharing, such a distinction does not need to be maintained in this chapter. It does not matter whether the public bodies have to provide information services to the users for their access, sharing or, re-use purposes. The focus is on the public task of the public body that provides the service. Hence, from the moment this public body has an obligation to provide an information service, it is performing a public task. This means that the public sector data on which the information service is based should be obtained under the rules for sharing. If there is no such obligation to provide an information service, the underlying data should be obtained under the rules for re-use under the PSI Directive. Hence, it is sufficient to examine whether there is such a public task or not, in order to determine whether the public sector data on which the information service is created are being re-used or shared.

Outline – As not all documents that were discussed in Chapter 5 contain obligations to provide information services, they are not all repeated in this section. The documents that clearly contain no provisions on obligations for public bodies to provide information services are not addressed. Only the ones for which a possible interpretation in the sense of a public task of providing information services could be found are included.

2. United Nations

a. *UN Special Rapporteur on Freedom of Opinion and Expression*

Information about the functioning of public bodies – As was mentioned previously in Chapter 5, the UN Special Rapporteur found that public bodies had a duty to disclose information, that is, all records held by them. They should also 'publish and disseminate widely documents of significant public interest, for example, operational information about how the public body functions and the content of any decision or policy affecting the public'.[13] While this mostly refers to the

13. United Nations Special Rapporteur on Freedom of Opinion and Expression, *Civil and political rights, including the question of freedom of expression. The right to freedom of expression and opinion, UN Doc. E/CN.4/2005/64*, 2004, 15–16, <http://daccess-ods.un.org/access.nsf/Get?Open&DS=E/CN.4/2005/64&Lang=E>, 22 Feb. 2010.

availability of public sector data to the public, the suggestion to make operational information available about the functioning of the public bodies requires further processing of public sector data than the public body would need for its own activities. Hence, it would constitute providing an information service as it is defined in the context of this analysis.

PUBLIC EDUCATION – In addition, the Special Rapporteur felt that the public bodies should 'make provision for public education and the dissemination of information regarding the right to have access to information'.[14] Educating the public also entails that information is given that requires processing beyond the public bodies' own needs, so once again the Rapporteur asked for a specific information service to be created.

b. United Nations Economic, Social and Cultural Organization (UNESCO)

LIMITED RECOGNITION OF A PUBLIC TASK TO PROVIDE INFORMATION SERVICES – The main elements of the 2004 UNESCO Policy Guidelines for the Development and Promotion of Governmental Public Domain Information were already addressed in Chapter 5, so this section focuses on the elements of the Guidelines that specifically deal with the provision of information services by public bodies. From what follows, it shows that the Guidelines do not make any clear statement on whether public bodies should create information services or should stick with making available public sector data, except for a few specific cases.

MAXIMIZING USEFULNESS OF PUBLIC SECTOR DATA – According to the Guidelines, it is 'essential for the government and other public bodies whose duties involve creating and making available information . . . to maximize the usefulness of its information'.[15] Does this mean that the public bodies have to process the data they hold for the purpose of performing their own tasks in order to create information services for the public? Or does this simply entail that the public bodies should ensure that their data are optimally available for the public to use for any purpose they like, without any further processing being necessary? The latter option is supported by the Guidelines asking the government and the public bodies to define the types of public sector data that should be actively disseminated, but not to process this information any further for accessibility purposes.[16]

MEETING THE NEEDS OF THE CITIZENS – This is contradicted, however, by the Guidelines' statement that: 'The focus should always be on producing and disseminating public information that meets the needs of citizens as openly and inexpensively as possible, with special attention to multicultural or disadvantaged communities'.[17] Such special attention might bring along accessibility issues that

14. *Ibid.*
15. UNESCO, *Policy Guidelines for the Development and Promotion of Governmental Public Domain Information* (Paris: UNESCO, 2004), 11, <http://unesdoc.unesco.org/images/0013/001373/137363eo.pdf>, 22 Feb. 2010.
16. *Ibid.*, 14.
17. *Ibid.*, 16.

call for data activities from the public bodies that go beyond their own needs. The public bodies should disseminate information to the public 'in a manner that achieves the best balance between the goals of maximizing the usefulness of the information and minimizing the cost to the government and the public'.[18] One could interpret this statement as only referring to format or pricing issues, but one could also imagine the public bodies having to adapt or revise the information beyond their own needs in order to make it useful to the public. Considering that the document is mostly concerned with issues of access to information rather than accessibility, the former interpretation might be more plausible. However, the attention and demand for '[i]nformation products and services that can meet the special needs of people with physical disabilities', albeit very specific, might suggest otherwise.[19]

INFORMATION ABOUT THE FUNCTIONING OF PUBLIC BODIES – For certain types of information, the Guidelines state the need for information services more clearly: public bodies have to provide information that describes their organization, activities, programmes, meetings, systems of records, and other information holdings, and how the public may obtain access to them.[20] While this information is specific and straightforward, and may fall under active dissemination of public sector data, it may also require the public bodies to create additional information especially directed towards the public. Therefore, one could categorize this as providing information services.

SUMMARY – In summary, the UNESCO Guidelines promote attention for the usefulness of public sector data and meeting the needs of citizens and different target groups, but they only contain some more concrete proposals with regard to providing information services on the public bodies' activities and procedures and the way of getting access to their data. Hence, the public task to provide information services only has a limited scope.

3. Organization for Economic Co-operation and Development

RECOMMENDATION FOR ENHANCED ACCESS AND MORE EFFECTIVE USE OF PUBLIC SECTOR INFORMATION – The OECD Recommendation of the Council for Enhanced Access and More Effective Use of Public Sector Information was discussed extensively in Chapter 5.[21] The Recommendation hardly mentions what information services should be created for the purpose of informing the public. Only a small reference is made to the need for multilingualism: Member States are encouraged to do research on 'improving access and availability of public sector information in multiple languages, and ensuring development of the necessary related skills'.[22]

18. *Ibid.*
19. *Ibid.*, 20.
20. *Ibid.*, 19.
21. Cf. *supra*, Ch. 5, ss II.C.3.b. and III.C.2.
22. OECD, *Recommendation C(2008)36 of the Council for Enhanced Access and More Effective Use of Public Sector Information*, 2008, 5, <www.oecd.org/dataoecd/0/27/40826024.pdf>, 22 Feb. 2010.

Report on public sector information and content – More attention was given to the provision of information services in the OECD Working Party on the Information Economy's 2006 report on Digital Broadband Content: Public Sector Information and Content. As was mentioned in Chapter 5, the report made a distinction between making public sector data available and the diffusion of public content. This public content referred mostly to cultural, educational, and scientific public knowledge. Therefore, it is of less interest with regard to the question of the provision of spatial information services. Yet, many of the initiatives given as examples of the diffusion of public content can be considered as information services, so the report is still of some importance. According to the report, '[t]he aim is to diffuse information rather than merely to provide it on line. This means that cultural content programmes should aim to attract users and to provide useful services'.[23] Therefore, user needs and capabilities should be examined and taken into account when disseminating cultural content online. This entails that further actions should be undertaken to make the cultural data useful and accessible for the public, beyond any of the data processing that might be needed for the public bodies' own purposes. One might even question if making such cultural content available on the Internet at all, regardless of the additional actions, does not go beyond the own needs of the public bodies.

Summary – In summary, while the OECD had paid attention earlier to the provision of information services, especially with regard to cultural content, none of these concerns were transferred to the 2008 Recommendation. The Recommendation does not contain a public task to provide information services.

4. Council of Europe

Outline – The Council of Europe has adopted a number of documents relating to the availability of public sector data, all related to access rather than sharing or re-use. This section examines the documents that might hold any public tasks for the Member States and the public bodies to provide information services.

a. Recommendation No. R(81)19 of the Committee of Ministers on the Access to Information Held by Public Authorities

No recognition of a public task to provide information services – As was mentioned in Chapter 5, Recommendation R(81)19 did not hold any provisions on the active dissemination of public sector data, let alone on the provision of information services to the public. The Recommendation limited itself to stating that 'the utmost endeavour should be made to ensure the fullest possible availability to the public of information held by public authorities'.[24]

23. OECD Working Party on the Information Economy, 'Digital Broadband Content: Public Sector Information and Content', 2006, 40, <www.oecd.org/dataoecd/10/22/36481524.pdf>, 22 Feb. 2010.
24. Council of Europe Committee of Ministers, *Recommendation No. R (81) 19 of 25 November 1981 to Member States on the Access to Information held by public authorities,*

b. Recommendation Rec(2002)2 of the Committee of Ministers to Member States on Access to Official Documents

NO RECOGNITION OF A PUBLIC TASK TO PROVIDE INFORMATION SERVICES – It was explained in Chapter 5 that Recommendation Rec(2002)2 held public tasks for the Member States both for making public sector data available upon request and of their own motion. Public bodies should make available 'any information which is deemed useful in a transparent democratic society'.[25] One could question whether this means that the public bodies should make public sector data available if they consider that they might be useful to the public, or whether they should take a more pro-active approach and ensure that the information they disseminate is useful, by processing it beyond their own needs and making it accessible for the appropriate target groups. Recommendation XI seems to indicate the former. It states that a public body should take the necessary measures to make available 'public information which it holds', when such information is in the interest of promoting transparency and encouraging informed participation in matters of public interest.

c. Convention on Access to Official Documents

LIMITED RECOGNITION OF A PUBLIC TASK TO PROVIDE INFORMATION SERVICES – The Convention repeats in its Article 10 the provision of Recommendation Rec(2002)2 that public bodies should make available the official documents they hold, without any requirement of making these documents more accessible to the public by processing them beyond the public body's own purposes for which the documents were created. The Steering Committee does clarify, however, that: 'Any policy which seeks to make official documents of general interest public without the need for individual requests must ensure that citizens are able to form an opinion on the authorities that govern them and to become involved in the decision-making process'.[26] To ensure this, it might be necessary for the public bodies to process the official documents, but there is no enforceable obligation to guarantee this. The only hint at an information service obligation can be found in Article 9 of the Convention, which requires the Member States to inform the public about its right of access and how to exercise it, and to provide information on the matters or activities for which they are responsible. This can entail a very specific and limited public task to provide information services.

SUMMARY – Hence, the Council of Europe only has a very limited view on the public task of public bodies to provide information services.

<www.coe.int/t/e/legal_affairs/legal_co-operation/administrative_law_and_justice/Texts_&_Documents/Conv_Rec_Res/Recommendation(81)19.asp>, 22 Feb. 2010. Cf. *supra*, Ch. 5, s. I.C.4.b.

25. Council of Europe Committee of Ministers, *Recommendation Rec(2002)2 of 21 February 2002 on access to official documents*, <www.coe.int/T/E/Human_rights/rec(2002)2_eng.pdf> 22 Feb. 2010>. See Ch. 5, s. I.B.4.c.
26. Steering Committee for Human Rights, *Draft Convention of the Council of Europe on Access to Official Document and Draft Explanatory Report*, doc.nr CDDH (2008)008 Add. III Bil., 3 Apr. 2008, 28, <http://gyla.ge/foi/doc_upload/capmaign/3.%20evropis%20sabchos%20konvenciis%20proeqti%20inglisur%20enaze.pdf>. 28.

5. European Union

Outline – The first part of this section looks at the obligations of the institutions and bodies of the EU to provide information services, and the second part deals with the documents that are addressed to the Member States.

a. *The Provision of Information Services by the European Union Institutions and Bodies*

i. Regulation 1049/2001 Regarding Public Access to European Parliament, Council, and Commission Documents

Limited recognition of a public task to provide information services – For an extensive discussion of Regulation 1049/2001, the reader is referred to the appropriate section of Chapter 5.[27] While the Regulation contains a number of measures for promoting the availability of public sector data held by the EU institutions, it does not include any obligations with regard to making available information services for the European citizens. Apparently, the EU institutions did not deem it part of their public task to create information services beyond their own needs, directed towards the public. The only reference in the direction of an obligation to create or process public sector data in a way that goes beyond the needs of the public bodies themselves, is the obligation that the institutions take the requisite measures to inform the public of their right of access. This is a very specific obligation, which is in line with the UNESCO guidelines and the European Convention on Access to Official Documents.

ii. Green Paper on Public Access to Documents held by Institutions of the European Community

European Commission: No recognition of a public task to provide information services – The Green Paper on Public Access to Documents held by Institutions of the European Community started a consultation on the possible need for review of Regulation 1049/2001.[28] The Commission suggested that the institutions could 'develop a more systematic policy of making documents directly accessible to the public in the most user-friendly way'.[29] However, this was mostly directed at improvements of the workflow, the databases, and websites, rather than processing the information itself to an extent beyond the institution's own purposes in order to make it more accessible for the public. No public task to provide information services was foreseen.

---

27. Regulation (EC) No 1049/2001 of the European Parliament and the Council of 30 May 2001 regarding public access to European Parliament, Council and Commission documents, *OJ L* 145, 31 May 2001, 43. Cf. *supra*, Ch. 5, s. I.C.5.a.
28. Commission of the European Communities, *Green Paper on Public Access to Documents Held by Institutions of the European Community*, COM (2007) 185 final, 6, 18 Apr. 2007.
29. *Ibid.*, 12.

CONSULTATION: NO RECOGNITION OF A PUBLIC TASK TO PROVIDE INFORMATION SERVICES – None of the respondents to the consultation requested the provision of information services that entailed processing of the public sector data beyond the public bodies' own needs. Consequently, the European Commission's proposal for a new regulation on access to documents held by the EU institutions, which followed the consultation, did not contain any obligations for the public bodies to provide information services to the public for access purposes.[30]

SUMMARY – Hence, the EU institutions and bodies currently only have a very limited public task to provide information services under the existing legislation. It does not seem likely that these obligations will be extended under the new regulation that will adapt Regulation 1049/2001. The Commission's proposal for a recast of Regulation 1049/2001 does not contain any changes to the provisions that could be interpreted as containing an obligation to provide information services, that is, the obligations for the institutions to take the requisite measures to inform the public of their right of access.[31]

b. Documents Held by the Member States

OUTLINE – As was mentioned in Chapter 5, the European institutions considered the availability of public sector data for access in general to be outside of their area of competence. They only addressed public access to the particular category of environmental information. However, the provision of information services based on public sector data in general has been discussed in some documents adopted by the European Commission. These documents will be discussed below, insofar as they are relevant for this section.

i. Publaw Reports

FOCUS ON THE PUBLIC TASK FROM A NEGATIVE POINT OF VIEW – The Publaw reports looked at access legislation from the point of view of the information market. Hence, they were mostly concerned with the development of information services by the public bodies on the market, in competition with private parties. With regard to this topic, the reports looked at the limits of the remit of the public bodies, that is, whether or not they were allowed to provide information services on the market. The main concern of the reports was to determine what the public bodies *could not* do, rather than the information services they *should* provide to the public for access purposes. Therefore, the Publaw reports will be discussed in more detail in the appropriate section.[32]

RECOGNITION OF A PUBLIC TASK TO PROVIDE (UNIVERSAL) INFORMATION SERVICES – However, a few short remarks with regard to the provision of information services

30. Commission of the European Communities, *Proposal for a Regulation of the European Parliament and of the Council regarding public access to European Parliament, Council and Commission documents*, COM (2008) 229 final, 30 Apr. 2008.
31. *Ibid.*
32. Cf. *infra*, s. III.B.2.b.

for access purposes can be made here. As was mentioned in Chapter 5, the Publaw 3 report introduced the concept of a universal information service to the image of the universal service of the telecommunications sector. While the report questioned whether a single definition of such a universal information service was possible, it stated that in any case, such a public service should 'comprise a certain level of access to information of a certain quality, reliability, continuity and, where possible, neutrality – therefore providing a certain added value'.[33] Thus, universal service should no longer be understood as only access to the technical means of communication, but also as the requirement to set up and make available the content of the information.[34]

NO DEFINITION OF UNIVERSAL INFORMATION SERVICES – The Publaw 3 report found it impossible to provide a single definition of the universal public information service, 'either in terms of content (raw information, information with added value) or in terms of technical means of access (paper medium, electronic media, etc.)'.[35] It could be sufficient to make public sector data available in some cases, while in other situations a true information service with added value should be provided in order to fulfil the tasks of the universal information service. Hence, the Publaw 3 study did not have a clear view on which information services should be provided by the public bodies as part of their public task. It only stated that such choices should be made following a public debate with the stakeholders: 'The universal public information service is a concept of variable geometry developing in time whose definition calls for an open and transparent debate, placed under the responsibility of the public authority'.[36] Beyond making public sector data available under national access legislation and an obligation to make data available in the same manner as the public body holds it for its own purposes, the Publaw study felt that it is up to the public bodies themselves to determine what the content of the universal information service should be.

ii. Green Paper on Public Sector Information

EUROPEAN COMMISSION: RECOGNITION OF A PUBLIC TASK TO PROVIDE INFORMATION SERVICES – The Green Paper mostly limited itself to the availability of public sector data for access purposes, but it did acknowledge that: 'In the execution of its primary tasks, the public sector produces a vast amount of raw data. The format of this public sector information may often be unattractive and difficult to access. Having taken this issue into account, many public sector bodies are engaged in tailoring raw data to customer needs'.[37] It was not questioned whether or not this was appropriate.

33. Commission of the European Communities, *Publaw 3. Final report* (Luxembourg: CEC Legal Advisory Board, 1995), 80.
34. *Ibid.*, 85.
35. *Ibid.*, 86.
36. *Ibid.*, 88.
37. Commission of the European Communities, *Green Paper on Public Sector Information*, 12.

CONSULTATION: LIMITED RECOGNITION OF A PUBLIC TASK TO PROVIDE INFORMATION SERVICES – The responses to the consultation that was launched by the Green Paper showed that several organizations advocated the availability of more user-friendly, transparent, and accessible information services, while others – mostly the private sector – found that the public bodies should stay away from processing public sector data beyond their own needs and provide information services.

SUMMARY – Hence, on the EU level, the opinions were divided on whether the Member States and the public bodies should have a public task to provide information services, and it was left to the Member States to decide this, as was done with regard to the availability of public sector data for access.

D. THE PUBLIC TASK OF PROVIDING ENVIRONMENTAL INFORMATION SERVICES IN EUROPEAN AND INTERNATIONAL DOCUMENTS

OUTLINE – The previous section addressed the public task of public bodies to provide information services based on all types of public sector data, while this section specifically examines the obligations of the public bodies to provide environmental information services. It already became clear in Chapter 5 that environmental data have received considerable attention in the international and European legislation and policy documents. Therefore, one could assume that more information services would also be required under these legislative and policy documents. This section will examine whether this assumption is valid.

RELATIONSHIP WITH SPATIAL INFORMATION SERVICES – As a considerable part of environmental data can also be categorized as spatial data and vice versa, this category is very relevant to determine the public tasks of the public bodies with regard to providing spatial information services.

1. United Nations

a. Rio Declaration on Environment and Development and Agenda 21

RIO DECLARATION – The Rio Declaration on Environment and Development recognizes the importance of access to environmental information, but it does not suggest any information services beyond 'making information widely available'.[38]

AGENDA 21: INFORMATION SERVICES FOR ACCESS – Chapter 23 of Agenda 21, which addresses the availability of information for the citizens, does not impose that information services are created to ensure the availability and accessibility of this information.[39]

---

38. United Nations Conference on Environment and Development, *Rio Declaration on Environment and Development*, 13 Jun. 1992, <www.unep.org/Documents.Multilingual/Default.asp?DocumentID=78&articleID=1163>, 22 Feb. 2010.
39. United Nations Commission on Sustainable Development, *Agenda 21*, 1992, <www.un.org/esa/sustdev/documents/agenda21/english/agenda21toc.htm>, 22 Feb. 2010.

AGENDA 21: INFORMATION SERVICES FOR SHARING – Chapter 40 of Agenda 21 on information for decision making states that in sustainable development, information should be considered in the broad sense, including 'data, information, appropriately packaged experience and knowledge'.[40] This might imply that information services are needed for decision-making purposes with regard to environment and sustainability issues, mostly done by government and public bodies. One of the concrete information services that are suggested is the development of indicators of sustainable development. Next, countries should establish supporting mechanisms to provide local communities and resource users with the information and know-how they need to manage their environment and resources sustainably. Existing information should be transformed into forms more useful for decision making and targeting information at different user groups. This last requirement would entail the provision of information services beyond the internal needs of the public body that collected the environmental data, even if the recipients of the information service are still part of the public sector.

SUMMARY – Hence, while the Rio Declaration remains silent, Agenda 21 contains an obligation to provide information services – maybe not towards the public, but at least towards other public bodies.

b. *Guidelines for Access to Environmental Information*

LIMITED RECOGNITION OF A PUBLIC TASK TO PROVIDE INFORMATION SERVICES – The 1995 UN ECE Guidelines for access to environmental information hardly addressed the provision of information services. The preamble of the Guidelines stated the UN ECE's conviction that '[e]nvironmental authorities should raise public awareness in order to promote greater public understanding and support for environmental policies and enforcement'.[41] While this could be interpreted as calling for accessible environmental information services directed towards the public, the Guidelines primarily focused on the availability of environmental data to the public.

ENSURING AN ADEQUATE FLOW OF INFORMATION – Some provisions could actually be considered to go a little further than the mere availability of environmental data without any further processing. For example, the Guidelines required the Member States to ensure that there was 'an adequate flow of information' about activities significantly affecting the environment to the public authorities.[42] The question can be asked what the use of the word *adequate* entails. Does this mean that the Guidelines deemed it sufficient that there was a flow of information, or did they

40. *Ibid.*
41. United Nations Economic Commission for Europe, *Guidelines on Access to Environmental Information and Public Participation in Environmental Decision-Making*, 1995, <www.eea.europa.eu/publications/92-9167-020-0/page017.html>, 22 Feb. 2010. Cf. *supra*, Ch. 5, s. I.D.1.c.
42. *Ibid.* See also Frankie Schram, 'Openbaarheid van bestuur: een voorbeeld van interactie tussen het Europees en nationaal publiekrecht', *Tijdschrift voor Bestuurswetenschappen en Publiekrecht* no.3 (1999): 172 [in Dutch].

require this flow of information to actually be meaningful to the user? The second option may require that the information is processed beyond the public bodies' own needs.

STATE OF THE ENVIRONMENT – Next, the Guidelines required the Member States to regularly publish up-to-date information on the state of the environment, for example, in a report. Such reports are specifically intended towards the public and are not needed for the functioning of the public bodies themselves, so they involve the provision of an information service.

SUMMARY – Hence, the Guidelines for access to environmental information contained a first indication that there is a public task for the public bodies to provide environmental information services to the public, even though this task only had a limited scope.

c. *The Aarhus Convention*

OUTLINE – The background and the general provisions of the Aarhus Convention were already discussed in Chapter 5.[43] Therefore, this section focuses on the public tasks of the public bodies to provide information services that can be deducted from the articles of the Convention.

PROMOTING ENVIRONMENTAL EDUCATION AND ENVIRONMENTAL AWARENESS – These public tasks are particularly embodied in Article 5 of the Convention. As was indicated in Chapter 5, Article 5 mixes obligations of the Member States and the public bodies that can clearly be considered as making environmental data available for access, with obligations that arguably entail the provision of information services. However, one should also take into account that Article 3 of the Aarhus Convention, which contains some general provisions, requires the Member States to 'promote environmental education and environmental awareness among the public, especially on how to obtain access to information, to participate in decision making and to obtain access to justice in environmental matters'.[44] Environmental education involves general education at all levels, while environmental awareness-raising is more topic-oriented.[45] Such educational and awareness-raising activities require activities from the Member States and the public bodies that go beyond merely making environmental data available to the public, as dissemination activities targeted to specific audiences need to be undertaken. This entails that the provision of information services with regard to education and awareness is a public task under the Aarhus Convention. However, the generic formulation of the task makes it difficult to determine how far the Member States should go to fulfil this task. In any case, they should inform the citizens on how to exercise their rights under the Convention.

ENSURING AN ADEQUATE FLOW OF INFORMATION – Article 5.1(b) requires the Member States to ensure that: 'Mandatory systems are established so that there

43. Cf. *supra*, Ch. 5, s. I.D.1.c.
44. Article 3.3 of the Aarhus Convention.
45. United Nations Economic Commission for Europe, *Aarhus Implementation Guide*, 44.

is an adequate flow of information to public authorities about proposed and existing activities which may significantly affect the environment'. Such an obligation could already be found in the 1995 Guidelines on access to environmental information. The Implementation Guide to the Aarhus Convention gives a few examples to demonstrate what should be understood under an 'adequate' flow of information, for example, 'through mandatory monitoring and research programmes', 'systems of self-monitoring and record-keeping by facilities on data such as air and water emissions and waste disposal'.[46]

Ensuring transparency and effective accessibility – The Aarhus Convention requires the Member States to 'ensure that, within the framework of national legislation, the way in which public authorities make environmental information available to the public is transparent and that environmental information is effectively accessible'.[47] The Implementation Guide to the Convention clarifies that transparency entails that 'the public can clearly follow the path of environmental information, understanding its origin, the criteria that govern its collection, holding and dissemination and how it can be obtained'.[48] 'Effectively accessible' means that 'the established information systems should go beyond simply making the information available to the public'. Hence, one can assume that environmental information services are required of which the content is targeted towards the public. Yet, the measures that the Implementation Guide proposes to implement this provision all refer to technical and practical assistance rather than a larger intellectual accessibility of the content, for example, search engines, convenient office hours and locations, the availability of copy machines, and so on.[49]

Providing specific information services – The Convention also lists certain types of information that should in any case be made available. The information that needs to be provided to the public includes (1) information about the type and scope of environmental information held by the public authorities and the conditions and process to obtain these data; (2) publicly accessible lists, registers or files; and (3) points of contact. While the first could be considered as an information service, as the public authorities have to create this information specifically for the public, the second and third types of information only increase the practical accessibility but do not require any processing of the environmental information itself.

46. *Ibid.*, 69.
47. Article 5.2 of the Aarhus Convention. See also Tom De Gendt, 'Het Verdrag van Aarhus in Vlaanderen: Actieve Openbaarheid van Milieu-Informatie, Participatie van de Burger aan het Milieubeleid en Toegang tot de Rechter in Milieuzaken', *Milieu- en Energierecht* no. 4 (2001): 241 [in Dutch]; Kurt Deketelaere & Frankie Schram, *Milieurechten van het publiek in de participatiedemocratie* (Bruges: Vanden Broele, 2002), 153 [in Dutch]; Christine Larssen, 'L'Accès aux Informations sur L'Environnement en Droit International: la Convention d'Aarhus', in *Ten Years of Access to Environmental Information in International, European and Belgian Law: Stock-Taking and Perspectives*, ed. Christine Larssen (Brussels: Bruylant, 2003), 40 [in French].
48. United Nations Economic Commission for Europe, *Aarhus Implementation Guide*, 71.
49. *Ibid.*, 72. See also Tom De Gendt, 'Het Verdrag van Aarhus in Vlaanderen: Actieve Openbaarheid van Milieu-Informatie, Participatie van de Burger aan het Milieubeleid en Toegang tot de Rechter in Milieuzaken', *Milieu- en Energierecht* no. 4 (2001): 241–242 [in Dutch].

MAKING REPORTS ON THE STATE OF THE ENVIRONMENT – In addition to the previous types of information, the Member States also have to ensure that certain categories of information are progressively made available via electronic means,[50] such as reports on the state of the environment. These reports have to be published and disseminated under Article 5.4. of the Convention, at least once every three or four years. They have to pay attention to the state of the environment and to the pressure that is being put on the environment due to different factors.[51] If these reports have to be made specifically to inform the public, they can be considered as information services. However, the Implementation Guide indicates that these reports already exist electronically in many countries, based on other legislation, so the Aarhus Convention does not necessarily introduce a new task to provide an information service. De Gendt adds that the Convention does not demand that the public bodies transpose information from a non-electronic format into an electronic format.[52]

PUBLISHING FACTS AND ANALYSES OF FACTS ON ENVIRONMENTAL POLICY PROPOSALS – Finally, under Article 5.7 of the Convention, the Member States have to publish the facts and analyses of facts that they consider relevant and important in framing major environmental policy proposals.[53] Hence, the Member States have to provide background information to the public, which in most cases will entail the provision of an information service. This is also the case with regard to the obligation to publish available explanatory material on the Member States' dealings with the public, and to provide information on the performance of public functions or the provision of public services relating to the environment.

SUMMARY – The Aarhus Convention contains some obligations for the public bodies to provide information services, but the scope of such a public task remains limited.

2. OECD: Recommendation of the Council on Environmental Information

LIMITED RECOGNITION OF A PUBLIC TASK TO PROVIDE INFORMATION SERVICES – The OECD Recommendation of the Council on Environmental Information was adopted in 1998, while the Aarhus Convention was still under preparation. The Recommendation holds a number of provisions relating to the creation of reports and other

50. Article 5.3 of the Aarhus Convention.
51. Tom De Gendt, 'Het Verdrag van Aarhus in Vlaanderen: Actieve Openbaarheid van Milieu-Informatie, Participatie van de Burger aan het Milieubeleid en Toegang tot de Rechter in Milieuzaken', *Milieu- en Energierecht* no. 4 (2001): 245 [in Dutch]; Kurt Deketelaere & Frankie Schram, *Milieurechten van het publiek in de participatiedemocratie* (Bruges: Vanden Broele, 2002), 165–166 [in Dutch].
52. Tom De Gendt, 'Het Verdrag van Aarhus in Vlaanderen: Actieve Openbaarheid van Milieu-Informatie, Participatie van de Burger aan het Milieubeleid en Toegang tot de Rechter in Milieuzaken', *Milieu- en Energierecht* no. 4 (2001): 244 [in Dutch].
53. See also Kurt Deketelaere & Frankie Schram, *Milieurechten van het publiek in de participatiedemocratie* (Bruges: Vanden Broele, 2002), 176–178 [in Dutch]; Christine Larssen, 'Het Verdrag van Aarhus en de toepassing ervan in het Belgisch recht – deel 1', *Tijdschrift voor Milieurecht* 2005, 268 [in Dutch].

information that could be interpreted as a public task for the public bodies to provide information services. Such an interpretation is supported by the Council's consideration that 'public awareness of environmental conditions and risks is essential to protection of human health and the environment', and that there is a need for intensified efforts to 'upgrade the extent and quality of environmental data, indicators and information dissemination systems'.[54]

Adequate dissemination and effective mechanisms – The Council stated that the Member States of the OECD should 'promote adequate dissemination of environmental information (e.g., periodic reports on the state of the environment and its changes over time, environmental indicators publication) and establish effective mechanisms to better inform the public, decision-makers and the authorities on environmental and sustainable development conditions and issues'.[55] In order to fulfil these tasks properly, information services would need to be provided to the public that require the processing of environmental data beyond the Member States' or the public bodies' own needs. Hence, a public task to provide information services is present in the Recommendation.

3. European Union

Outline – With regard to environmental information services, the legislative activities of the European institutions can also be divided into two tracks, as was the case for public sector information services in general. First, the Aarhus Convention had to be transposed for environmental information held by the European Community institutions and bodies. Second, access to environmental information was also seen as an issue that required harmonization between the Member States. Therefore, Directive 90/313/EEC and its replacement, the Access Directive, were issued.

a. Access to Documents Held by the European Union Institutions and Bodies: Regulation 1367/2006

Limited recognition of a public task to provide information services – Regulation 1367/2006 ensures the application of the Aarhus Convention to the institutions and bodies of the European Community.[56] The Regulation contains a number of obligations regarding the active dissemination of environmental information. Some of these obligations could be interpreted as requiring the Member States or the public bodies to set up an information service rather than merely making their environmental data available. First, Article 4 of the Regulation requires regular reports on the state of the environment to be published and disseminated, including via

54. OECD, *Recommendation of the Council on Environmental Information*, 3 Apr. 1998, <http://webdomino1.oecd.org/horizontal/oecdacts.nsf/linkto/C(98)67>, 22 Feb. 2010.
55. *Ibid.*
56. Regulation 1367/2006/EC of the European Parliament and of the Council of 6 Sep. 2006 on the application of the provisions of the Aarhus Convention on Access to Information, Public Participation in Decision-Making and Access to Justice in Environmental Matters to Community institutions and bodies, *OJ L* 264, 25 Sep. 2006, 13. Cf. *supra*, Ch. 5, s. I.D.4.a.

electronic means, at least once every four years. Next, data or summaries of data derived from the monitoring of activities affecting or likely to affect the environment have to be made available via electronic means. In order for these summaries to be useful, the environmental data may need to be processed beyond the public bodies' own needs and prepared for its target audience, that is the general public.

b. Access to Documents Held by Member States

i. Directive 90/313/EEC

LIMITED RECOGNITION OF A PUBLIC TASK TO PROVIDE INFORMATION SERVICES – Directive 90/313/EEC did not include many obligations for the Member States or the public bodies to actively disseminate environmental information or to provide information services.[57] With regard to the latter, it only stated in Article 7 that the Member States should 'take the necessary steps to provide general information to the public on the state of environment by such means as the periodic publication of descriptive reports'. As was mentioned in Chapter 5, one could assume that the fact that these reports should be descriptive entails that the public bodies have to perform additional activities to their environmental data in order to 'describe' the state of the environment. Like Vanheusden, one can regret that the directive did not suggest the content of those reports.[58]

ii. Access Directive

OUTLINE – The Access Directive was addressed extensively in Chapter 3 and shortly revisited in Chapter 5. Therefore, this section will only address the provisions of the directive that are directly related to the provision of environmental information services.

ACCESSIBLE ENVIRONMENTAL INFORMATION – The Preamble of the Access Directive emphasizes the importance of effectively and easily accessible environmental information[59] and the need for public bodies to make sure that when 'environmental information is compiled by them or on their behalf, the information is comprehensible, accurate and comparable'.[60] In order to achieve such a comprehensible character, further processing of the environmental data beyond the public bodies' own needs will be required in many cases.

---

57. Gavouneli feels this was justified 'both by the format of the Directive itself and the administrative peculiarities of each Member State' (Maria Gavouneli, 'Access to Environmental Information: Delimitation of a Right', *Tulane Environmental Law Journal* 13 (2000): 314). Heyvaert disagrees and states that the directive did not go far enough (Veerle Heyvaert, 'Wie wordt er wijzer van? Enkele bedenkingen bij het recht op toegang tot milieu-informatie', in *Ten Years of Access to Environmental Information in International, European and Belgian Law: Stock-Taking and Perspectives*, ed. Christine Larssen (Brussels: Bruylant, 2003), 59–60 [in Dutch]).
58. Bernard Vanheusden, 'Vrije toegang tot milieu-informatie: een stand van zaken en de implicaties van het verdrag van Aarhus voor Europa en voor België', *Milieu- en Energierecht* 1999: 270 [in Dutch].
59. Recital 15 of the Access Directive.
60. Recital 20 of the Access Directive.

REPORTS ON THE STATE OF THE ENVIRONMENT AND ACTIVITIES AFFECTING THE ENVIRONMENT – Article 7 of the Access Directive includes the obligations of the Aarhus Convention concerning the active dissemination of environmental information. As in Regulation 1367/2006, the public bodies are required to publish regular reports on the state of the environment at least once every four years; and to publish data or summaries of data derived from the monitoring of activities affecting, or likely to affect, the environment. Once again, this will entail the provision of information services rather than merely making environmental data available.

SUMMARY – Hence, the Access Directive recognizes a public task for the public bodies to provide information services based on environmental data.

c. *Access to Documents Held by the Member States and the Institutions and Bodies of the European Union: Shared Environmental Information System (SEIS)*

BACKGROUND – The European Commission's communication on a Shared Environmental Information System (SEIS) addresses the 'collection, exchange and use of the data and information required for the design and implementation of environmental policy'.[61] The European Commission felt that, despite several European initiatives being underway, 'a major challenge in Europe and globally remains to organize the vast array of already collected environmental data and information, to integrate these, where desirable, with existing data and information from the social and economic realms; to make them available together with tools that allow experts to do their own analyses, and to communicate them in ways which the public policy makers and the public can readily understand and use as a basis for their own actions'.[62] The European Commission was expected to address this challenge in a proposal for a European directive on the Shared Environmental Information System in the course of 2009. However, at the time of this writing, no proposal has been made public.

RECOGNITION OF A PUBLIC TASK TO PROVIDE INFORMATION SERVICES – The intention of SEIS is to provide an open, shared environmental information system that will enable more efficient use to be made of available environmental data and the provision of information services to both public authorities and citizens. The information should be readily accessible to enable the public authorities 'to assess in a timely fashion the state of the environment and the effectiveness of their policies,

61. Commission of the European Communities, *Communication to the Council, the European Parliament, the European Economic and Social Committee and the Committee of the Regions. Towards a Shared Environmental Information System (SEIS)*, COM (2008) 46 final, 2, 1 Feb. 2008. Cf. *supra*, Ch. 5, s. I.D.4.b.

62. Commission of the European Communities, *Staff working document. Accompanying document to the Communication from the Commission to the Council and the European Parliament. Towards a Shared Environmental Information System (SEIS) – Executive summary of the impact assessment*, SEC (2008) 111, 22, 1 Feb. 2008.

and to design new policy'.[63] Hence, the provision of information services forms a considerable part of the public task as it is conceived under SEIS.

E. The Public Task of Providing Spatial Information Services in European and International Documents

Outline – This section addresses the obligation of public bodies to provide spatial information services. While a considerable amount of spatial data and services could also be categorized as environmental and vice versa, the provision of spatial data has also received separate attention from the European and international policy makers and legislator. As this attention is fairly recent, many initiatives are still in a starting phase, and only very few include obligations for the public bodies to provide spatial information services.

1. United Nations: Global Earth Observation System of Systems

Recognition of a public task to provide information services – While GEOSS' main intention is to promote the availability of observation data, its implementation plan also states that it will facilitate the development of 'shared data, metadata, and products'.[64] One of the functional components of the GEOSS ten-year implementation plan is to 'process data into useful information, that is, geo-products that are part of GEOSS, recognizing the value of modeling, integration and assimilation techniques'.[65] This indicates that GEOSS sees it as its task to create such information services that process the spatial data beyond the needs of the public body that holds them or has collected them, for the domains of interest that it has selected. It 'will place a high priority on data and information products commonly required across diverse societal benefit areas'.[66] Products would include current status assessments and descriptions, weather and air quality forecasts, seasonal and climate predictions, and so on.

Implementation Guidelines for the data sharing principles – It was already mentioned in Chapter 5 that Implementation Guidelines for the GEOSS data sharing principles are expected to be finalized by November 2010. The draft Implementation Guidelines state that the full and open exchange of products should be ensured, without restrictions on further use or dissemination. However, they do not

63. Commission of the European Communities, *Communication to the Council, the European Parliament, the European Economic and Social Committee and the Committee of the Regions. Towards a Shared Environmental Information System (SEIS)*, COM (2008) 46 final, 3, 1 Feb. 2008.
64. *Ibid.*, 6.
65. Group on Earth Observations, 'Global Earth Observation System of Systems (GEOSS) 10-Year Implementation Plan Reference Document', 2005, 127, <www.earthobservations.org/documents.shtml>, 22 Feb. 2010.
66. *Ibid.*, 130.

give any further guidance on which information products or services should be provided. A public task to provide an information service is only put on GEO itself: it should establish an online service that holds a compendium of the relevant international instruments and national policies and legislation.[67]

SUMMARY – Hence, GEOSS recognizes that there is a public task for the public bodies to provide information services based on Earth observation data.

2. European Union

a. *The INSPIRE Directive*

OUTLINE – The INSPIRE Directive was discussed extensively in Chapter 3 and briefly revisited in Chapter 5,[68] so this section only addresses the provisions of the directive that directly relate to the public task of public bodies to provide information services as they were defined at the beginning of the analysis.

NO RECOGNITION OF A PUBLIC TASK TO PROVIDE INFORMATION SERVICES – The public bodies are expected to provide an extensive range of network services with a view to making spatial data more accessible and easy to use. These services might indicate the presence of a public task of the public bodies to provide spatial information services. However, they only make it easier for the public on a technical and practical level to obtain the spatial data and to perform their own processing activities on the data. They do not entail any further processing of the spatial data by the public bodies themselves. Hence, no provision of spatial information services is required.

SPATIAL DATA SERVICES? – With regard to sharing, the INSPIRE Directive requires the Member States and the public bodies to make available spatial data sets and services to other Member States and public authorities for the purpose of performing their public tasks. While the directive uses the term 'spatial data services' and the reader might assume that the provision of information services is required, one should keep in mind that spatial data services are defined as 'the operations which may be performed, by invoking a computer application, on the spatial data contained in spatial data sets or on the related metadata'.[69] This would imply that the Member States and public bodies have to provide the technical services that enable the user to combine and interpret the spatial data, but not to do this combination and interpretation themselves. Hence, they do not have to provide information services, but only the technical network services that have been mentioned above. The INSPIRE Directive rather lays down the obligation to share spatial data so that the receivers of this spatial data can create their own information services, either as part of their public task or outside of it.

---

67. Group on Earth Observations, 'Draft Implementation Guidelines for the GEOSS Data Sharing Principles', 2008, 6, <www.earthobservations.org/geoss_dsp.shtml>, 22 Feb. 2010.
68. Cf. *supra*, Ch. 3, s. II.C. and IV.C; Ch. 5, ss I.E.2.a. and III.B.2.a.
69. Article 3.4 of the INSPIRE Directive.

SUMMARY – Hence, the public task of providing information services is not part of the INSPIRE Directive, but the directive contains a number of obligations on the establishment of technical services that may enable the provision of information services.

b. Global Monitoring for Environment and Security: (Kopernikus)

RECOGNITION OF A PUBLIC TASK TO PROVIDE INFORMATION SERVICES – At the 2008 INSPIRE conference in Maribor, Kaschl stated that GMES is 'all about services'.[70] It will 'deliver services of public interest. Examples include maps for emergency operations, monitoring of climate change parameters of chemical composition of the atmosphere'.[71] In 2008, marine and atmosphere services were launched, including services for producing information for monitoring and understanding climate change or contributing to improvement in the transport sector or marine knowledge. Next, services were also started for land monitoring, emergency response, and security applications at national, regional, European, and global levels. The Commission believes that such services should be provided under the control of national and regional authorities. However, the services are incomplete and their sustainability is not yet fully guaranteed.

FUNDING FOR THE INFORMATION SERVICES – The European Commission foresees that in time the market uptake of activities and services will entail that these become more mature and will be financed by the private sector, so public investment can be reduced or directed towards less mature sectors, depending on the policy priorities. However, as GMES delivers services of public interest, its financing is expected to be mainly public for the first years.

SUMMARY – Hence, the public bodies that cooperate in GMES should provide the services mentioned above as part of their public task. However, it might be possible that the scope of their public task will change if the private sector becomes more involved in the delivery of services based on GMES.

F. INTERIM SUMMARY

RECAPITULATION – This section examined the obligations in international and European documents to provide public sector information services, environmental information services, and spatial information services. Such a public task was not often recognized.

---

70. Arno Kaschl, 'GMES Global Monitoring for Environment and Security: recent developments', paper given at the INSPIRE Conference 2008 (Maribor, 2008), <www.ec-gis.org/Workshops/inspire_2008/presentations/05_03_kaschl.pdf>, 22 Feb. 2010.
71. Commission of the European Communities, *Communication to the European Parliament, the Council, the European Economic and Social Committee and the Committee of the Regions. Global Monitoring for Environment and Security (GMES): we care for a safer planet*, COM (2008) 748 final, 4, 12 Nov. 2008.

Public sector information services: United Nations – The United Nations Rapporteur and the UNESCO Guidelines, while referring to the need for education and useful information, only made concrete suggestions towards the provision of information services with regard to the activities and the functioning of the public bodies.

Public sector information services: Council of Europe – The aspirations of the Council of Europe to have an impact on the issue of the availability of public sector data have grown over the years, but they are still not much greater than the UN institutions' objectives. The Council of Europe does not commit itself to installing any obligation on the public bodies to provide information services beyond requiring that citizens are able to form an opinion on the public authorities and can become involved in the decision-making process. The only exception is the obligation to inform the public about the right of access and the procedure to obtain it.

Public sector information services: European Community – The EU view on information services is also limited. The institutions and bodies of the EU should only inform the public of their right of access and the procedure on how to exercise it. With regard to the public task in the Member States, the question was asked whether an information service should be provided, but the EU institutions have not risked giving a definite answer to this.

Environmental information services – The provision of environmental information services has received more attention. While the Rio Declaration and Agenda 21 were mainly focused on the provision of information services between public authorities for policy-making purposes, the other European and international documents addressed the provision of information services to the general public. However, this growing call for a public task to provide environmental information services has mostly been restricted to requests for 'adequate' or 'accessible' information. For instance, the Aarhus Convention requires information to be transparent and effectively accessible, and the OECD Recommendation refers to adequate and effective information. The Access Directive mentions effectively and easily accessible information, and states the need for comprehensible, accurate, and comparable environmental information. The very few concrete requirements for the provision of information services that can be found involve the provision of reports on the state of the environment and analyses or summaries of facts and activities that could possibly harm the environment. However, with the introduction of SEIS, the European Commission has indicated that the provision of environmental information services is gaining importance in the European Community.

Spatial information services – As the initiatives in the spatial data sector are relatively recent, the focus of most of these initiatives is still on the availability of spatial data rather than spatial information services. However, GEOSS is already moving towards the provision of information services by placing a high priority on the creation of products. This is also the case for the EU initiative GMES. One of GMES's tasks is to provide services of public interest in the sectors of environment, sustainability, and security.

SUMMARY AND OUTLINE – In conclusion, the scope of the public task is difficult to define from a positive point of view, as the obligations of the governments and the public authorities are generally formulated in a very high-level manner without attaching any concrete measures. If it is difficult to determine what the public sector *should* do, can it be a solution to start from the other end, that is, defining what the public sector *should not* do and leave to the private sector? The next section looks at the attempts of the European and international legislator and policy maker to address the public task of providing information services from such a negative point of view.

III. THE SCOPE OF THE PUBLIC TASK OF PROVIDING INFORMATION SERVICES FROM A NEGATIVE POINT OF VIEW

A. INTRODUCTION

OBJECTIVE – The documents that were discussed in the previous section addressed the public task of providing information services to the public from the point of view of what the public bodies *should* do, that is, which services they are under an obligation to provide. This section addresses the scope of the public task to provide information services from the other side, namely what the public bodies should not do and leave to the private sector. This is also important to determine whether the public bodies are performing their public tasks to provide information services to the public, or whether they are performing activities outside of their public task, possibly in competition with private parties on the market.

OUTLINE – This section shows that, while the European and international legislator and policy makers were very brief when determining the public task of the public bodies to provide information services in a positive way, most of the legislative and policy documents do not even address the negative point of view on the public task. Therefore, what follows will only discuss the documents that actually refer to this topic. Only documents referring to public sector data and information services in general were found in this regard, and none referring to environmental or spatial information services in particular. However, one should keep in mind that these documents on public sector data and information services are also applicable to environmental and spatial information services.

B. THE LIMITS ON THE PUBLIC TASK IN EUROPEAN AND INTERNATIONAL DOCUMENTS

1. Organization for Economic Cooperation and Development

RECOMMENDATION FOR ENHANCED ACCESS AND MORE EFFECTIVE USE OF PUBLIC SECTOR INFORMATION – The OECD Recommendation for Enhanced Access and More

Effective Use of Public Sector Information recognized an aim to promote 'the development of new information products and services particularly through market-based competition among re-users of information',[72] but it did not see a need to limit the public bodies' participation in the market. It only recommended that if the public bodies chose to enter the market with information services, pricing considerations should take into account considerations of unfair competition. This included '[p]ursuing competitive neutrality, equality and timeliness of access where there is potential for cross-subsidization from other government monopoly activities or reduced charges on government activities', and '[r]equiring public bodies to treat their own downstream-value-added activities on the same basis as their competitors for comparable purposes, including pricing'.[73]

REPORT ON PUBLIC SECTOR INFORMATION AND CONTENT – In the 2006 report on 'Digital broadband Content: Public Sector Information and Content', the Working Party on the Information Economy had already named the definition of roles between the public and private sector as one of the main challenges concerning the re-use of public sector information.[74] One of the solutions the report proposed to create fair conditions of competition between them was 'to define and divide functions and activities of public and private sectors. This excludes competition between both as they produce different products'.[75] The Royal Dutch Meteorological Service (hereafter referred to as KNMI) was quoted as an example where the functions had been divided and where discussion on the role of the public sector had been settled.[76] However, the report did not give any suggestions for criteria based on which the distinction between the activities of the public and the private sector could or should be made.

WORKSHOP ON PUBLIC SECTOR INFORMATION AND CONTENT – The distinction between the public task of public bodies to provide information services and their provision of information services on the market was also discussed in an OECD workshop in May 2006, in which it was argued that it was almost impossible to classify a specific kind of information service as a public task or not.[77]

SUMMARY – In summary, the OECD acknowledged the need for a negative definition of the public task to provide information services, but it did not attempt to define any criteria for limiting this public task and even questioned whether this was possible.

---

72. OECD, *Recommendation C(2008)36 of the Council for Enhanced Access and More Effective Use of Public Sector Information*, 2008, 4, <www.oecd.org/dataoecd/0/27/40826024.pdf>, 22 Feb. 2010.
73. *Ibid.*, 6.
74. OECD Working Party on the Information Economy, 'Digital Broadband Content: Public Sector Information and Content', 2006, 30, <www.oecd.org/dataoecd/10/22/36481524.pdf>, 22 Feb. 2010.
75. *Ibid.*, 42.
76. However, in Ch. 8 it will be shown that the discussion on the public task of the Dutch National Meteorological Service has restarted (cf. *infra*, Ch. 8, s. III.B.1.)
77. OECD Working Party on the Information Economy, 'OECD Workshop on Public Sector Information held on 31 May 2006: Summary', 2006, 7, < www.oecd.org/dataoecd/54/35/36854745.pdf>, 22 Feb. 2010.

2. European Union

a. Guidelines for Improving the Synergy between the Public and the Private Sectors in the Information Market

BACKGROUND – The 1989 Synergy Guidelines were the result of the European Commission's first attempt to draw attention to the value of public sector data for the private sector. They addressed both the availability of public sector data and the provision of public sector information services.

RECOGNITION OF A LIMIT TO THE PROVISION OF INFORMATION SERVICES – With regard to the provision of information services, the Synergy Guidelines promoted a limited role of the public sector. When a public body provided electronic information services directly, it should avoid any practice that would lead to the distortion of competition. Next, '[b]efore establishing a new electronic service or continuing an existing one, public administration should consider whether an existing private sector service can be used or adapted to meet their requirements'.[78] So the public bodies should stay away from providing information services and leave it up to the private sector to provide these services, unless there was a reason for the public bodies to provide them. Such reasons could include the following:

- the service was deemed to be essential to the public interest, but the private sector was unwilling or unable to offer it on reasonable terms;
- the service was an inseparable part of public sector tasks;
- a visibly neutral service, independent of the private information industry, was required.[79]

PERIODIC REVIEW OF THE INFORMATION SERVICES – If the public bodies did provide an information service, they should review regularly whether it was most appropriate for them to keep providing it, or whether the involvement of the private sector in their production or distribution was desirable, or possibly even the replacement of the information services by appropriate commercial services.[80] Hence, the role of the public sector was only to react to the situation on the market developments and look at the intentions of the private sector before acting.[81]

CRITICISM – Bruguière criticized this approach because it presented the public sector as a producer of raw data and only a subsidiary supplier of information services. The division was based on the notion of added value, which is difficult to base a division of the roles of the public and private players on the market on. He felt that such a policy of limiting the role of the public sector was dangerous for the

78. Commission of the European Communities, *Synergy Guidelines*, 10.
79. *Ibid.*
80. *Ibid.*
81. Cécile De Terwangne, *Société de l'information et mission publique d'information*, Doctoral Thesis FUNDP (Namur: Faculté Universitaire de Notre Dame de Paix, 2001), 433 [in French].

social preservation of the public data patrimony.[82] Poullet added that the requirements of the Synergy Guidelines called into question the whole idea of a public task to provide information services or would reduce this task to the strictest minimum.[83]

SUMMARY – Hence, the Synergy Guidelines took a very minimalistic approach to the public task of the public bodies to provide information services. However, the Guidelines did not determine what that public task should be, but only stated that if the public bodies felt it to be their public task to provide an information service, they should have a reason for it. The public task seemed to consist of anything that the market could or would not do.

b. Publaw Reports

BACKGROUND – The Publaw studies looked at access issues, but only from the point of view of the information market. In that respect, they looked at the role of the public sector and asked the question whether public bodies could intervene on the market as a competitor.[84] According to the studies, public bodies should take into account competition law when they entered into competition with private parties already on the market, or if they delivered an information service that could be provided by private parties.

PUBLAW 1: NO RECOGNITION OF A LIMIT TO THE PROVISION OF INFORMATION SERVICES – The first Publaw study was performed in 1992. With regard to the role of the public sector on the information market, the study saw two possibilities. On the one hand, the Synergy Guidelines could be followed, implying that the public bodies should not enter the market if there was a private party offering or willing to offer the information. On the other hand, it was also possible to let the public bodies enter the market if they pleased, under the condition that they did so on an equal basis with their private competitors.[85] This meant that they would have to comply with European and national competition law, unless the service could be considered as a service of general economic interest and compliance with competition law would hinder their activities. Of course, this still did not answer the question on when the public bodies were actually leaving the realm of their public task and entering the market. In any case, the Publaw 1 study found that more research was needed into the role of the public bodies on the information market and the rules that were applicable to the market activities of the public bodies.[86]

UNIVERSAL INFORMATION SERVICE – While the Publaw 2 study only addressed access legislation, Publaw 3 looked deeper into the roles of the public and the

82. Jean-Michel Bruguière, *Les données publiques et le droit* (Paris: Editions Litec, 2002), 120 [in French].
83. Yves Poullet, 'The Commercialization of Data Held by the Public Sector', *The Computer Law and Security Report* 9 (1993): 230.
84. Commission of the European Communities, *Publaw. Subject report: general access to information legislation* (Luxembourg: CEC Legal Advisory Board, 1991), s.p.
85. *Ibid.*
86. The topic of competition law will be addressed in Ch. VIII, s. 4.

private sector on the information market. The report admitted that there was confusion about what the respective roles of the public and the private sector should be: 'Should the public sector sell any information or should it all be channelled through the private sector?'[87] The study recommended that government and private sector should work together in adding value to public sector information, rather than leaving adding value completely to the private sector. Government had to set up a universal service and take measures to guarantee a healthy and balanced information market. If public bodies performed commercial activities on the market, they had to comply with competition law. To ensure this, it was important that activities of collecting information and exploiting this information were separated.[88]

PUBLAW 3: RECOGNITION OF A LIMIT TO THE PROVISION OF INFORMATION SERVICES – Hence, the Publaw 3 study admitted that there should be a distinction between the information the public bodies provide on the basis of their public task, and the information services they offer on the market. However, as was mentioned before,[89] what exactly this public task – what Publaw refers to as the universal information service – should contain and what its limits are, was not made clear. The study found that making a definition of the scope of the universal information service was not possible.

c. Green Paper on Public Sector Information

EUROPEAN COMMISSION: QUESTION ON THE LIMIT TO THE PROVISION OF INFORMATION SERVICES – The 1999 Green Paper on Public Sector Information recognized that the 'commercial re-use of public sector information may however raise questions as to the boundaries and limitations on the roles of the different actors . . . If the public sector adds value to its own information, launching commercial information products on a hitherto private information market, the issue of fair competition may be raised'.[90] This is partly due to the fact that public bodies have different cost structures than private businesses and it might be possible that they offer services at less than the market price in some circumstances, or cross-subsidies other activities through the sale of information.[91] To make the stakeholders aware of this issue, the Commission asked in its consultation to what extent and under which conditions activities of public sector bodies on the market could distort the information market.[92]

87. Commission of the European Communities, *Publaw 2. A report to the Commission of the European Communities on an evaluation of the implementation of the Commission's Guidelines for improving the synergy between the public and private sectors of the information market* (Luxembourg: CEC Legal Advisory Board, 1993), 75.
88. Simone Van Der Hof, *Overheidsinformatie in de etalage. Belangen rondom de toegang tot overheidsinformatie* (Alphen a/d Rijn: Samsom Bedrijfsinformatie, 1997), 17 [in Dutch].
89. Cf. *supra*, s. II.C.5.b. and Ch. 5, s. I.C.5.b.
90. Commission of the European Communities, *Green Paper on Public Sector Information*, 9.
91. *Ibid.*, 14–15.
92. *Ibid.*, 15.

CONSULTATION RESPONSES: RECOGNITION OF A LIMIT TO THE PROVISION OF INFORMATION SERVICES – The responses to the consultation included considerable criticism on the public sector bodies for abuse of their dominant position when providing their own information services on the market. However, many of the respondents felt that the answer to such issues should be sought in existing competition law rather than in new specific rules for public sector information. Yet many stakeholders also found that the public bodies should adhere more strictly to their public tasks as they were prescribed by the law.[93]

d. *The eEurope 2002 Communication*

NO QUESTION ON THE LIMIT TO THE PROVISION OF INFORMATION SERVICES – While the responses to the consultation launched by the Green Paper on Public Sector Information had shown that there was serious concern among the stakeholders about the roles of the public and the private sector on the market and the demarcation of the public task, the topic was not touched upon in the European Commission's eEurope 2002 Communication.

DISCUSSION ON THE LIMIT TO THE PROVISION OF INFORMATION SERVICES – In the consultation process that followed the Communication, it became clear that opinions still differed on whether public bodies should be allowed to provide information services. According to many private sector respondents, this should not be the case, and the public bodies should leave it up to the private sector to provide what they perceived as value-added services. On the other hand, many of the responding public bodies felt that they should still be able to offer their information services on the market, as long as they did so under the existing rules of competition law. In many replies, the need was felt to make a distinction between the public tasks of the public bodies and commercial activities. Where exactly this distinction would have to lie, once again remained unclear.[94]

e. *PSI Directive*

NO RECOGNITION OF A LIMIT TO THE PROVISION OF INFORMATION SERVICES – The PSI Directive does not contain any provisions limiting the scope of the public task, apart from the definition of re-use. It only states that if public bodies perform activities on the market, they should avoid cross-subsidies,[95] by ensuring that '[i]f documents are re-used by a public sector body as input for its commercial activities which fall outside the scope of its public tasks, the same charges and other conditions shall apply to the supply of the documents for those activities as apply to other users'.[96] First, this assumes that there may be commercial activities that fall

---

93. Commission of the European Communities, *eEurope 2002 Communication. Annex*, 4.
94. The replies to the consultation can be found at <http://cordis.europa.eu/econtent/psi/psi_re-use_consultation_replies.htm>, 22 Feb. 2010.
95. Recital 9 of the PSI Directive.
96. Article 10.2 of the PSI Directive.

within the scope of the public tasks of the public bodies. For instance, this is the case for some public bodies in the United Kingdom. The so-called Trading Funds have the obligation of making a profit on their activities. Next, the PSI Directive uses the public task as a criterion to determine the applicability of the directive and the role of the public bodies, but it does not define this public task or indicate what the scope of this public task is.

CRITICISM – Prins states that '[u]nfortunately, the directive remains silent on the crucial borderline between the public tasks of [public sector bodies] on the one hand, and the commercial activities of these bodies on the other hand. Thus, the first European regulatory framework on the re-use of PSI does not offer an instrument for dealing with the highly important and often debated question as to what type of activities on the information market belong to the public sector and what is to be left to the private sector . . . the absence of further details on this issue could hinder the proper functioning of the information market'.[97]

f. *Commission Decision on the Re-use of Commission Information*

RECOGNITION OF A LIMIT TO THE PROVISION OF INFORMATION SERVICES? – The Commission Decision on the re-use of Commission information applies comparable rules to its own documents as the rules that were included in the PSI Directive.[98] However, the absence of any provision on the use by the Commission of its own documents for the purpose of commercial activities outside its public task indicates the Commission's position that it does not or should not perform any activities outside of its public task. On the one hand, this implies that the Commission feels that it has no business entering the market, possibly in competition with the private sector. On the other hand, it could also indicate that the Commission has a very broad view of its own public task, and that any information service it provides, is considered to be within its public task, regardless of whether there would be similar services on the market. It is not immediately clear which of these arguments weighed the most in the drafting of the Commission's text.

g. *Review of the PSI Directive*

QUESTIONS ON THE LIMIT TO THE PROVISION OF INFORMATION SERVICES: MEMBER STATES – The review of the PSI Directive started in 2008 with a consultation of the Member States and the stakeholders on the scope and impact of the directive. While the Member States indicated that it was too early for a full review of the impact of the PSI Directive, some of them did signal that problems arose when public bodies

---

97. Corien Prins, 'Directive 2003/98/EC of the European Parliament and of the Council', in *Concise European IT law*, ed. Corien Prins (Alphen a/d Rijn: Kluwer Law International, 2006), 414.
98. Commission Decision 2006/291/EC, Euratom of 7 Apr. 2006 on the re-use of Commission information, *OJ L* 107, 20 Apr. 2006, 38. Cf. *supra*, Ch. 5, s. II.C.4.f.

competed on the market with the private sector. Their view was that there is a 'lack of clarity as to what constitutes a public sector body's public task'[99] and guidance from the Commission was desired.

QUESTIONS ON THE LIMIT TO THE PROVISION OF INFORMATION SERVICES: STAKEHOLDERS – Comparable comments were received from the stakeholders, among which were public sector content holders and commercial and non-commercial re-users. They had questions with regard to the limits of the public tasks when public bodies commercially compete with private firms and unfair competition practices by public sector bodies. Many of the respondents felt that limits and boundaries should be established on the role of public bodies operating commercially in the market. They also strongly recommended the issuance of guidance on ambiguous terms such as *public task.*[100]

QUESTIONS ON THE LIMIT TO THE PROVISION OF INFORMATION SERVICES: MICUS REPORT – Around the same time as the consultation results, a report from Micus Management Consulting GmbH ordered by the Commission was published on the re-use of PSI in the sectors of spatial, meteorological, and legal information.[101] The study also reported considerable problems with regard to the scope of the public task of the public bodies and the activities of these public bodies on the market. It found that '[R]e-users advocated a strict separation from all business operations from PSI holders as is the case in several Member States. In their view, PSI holders should reduce their activities to the production of raw data, while re-users offer value added products and client-focused services'.[102] With regard to spatial data, the study stated that '[R]egarding the public task matter, the biggest problem seems to be that specific laws in certain Member States explicitly define the production of PSI originating material as a public task for certain PSI holders. This leads them to carry out commercial activity in competition with re-users. The fact that, for example, private publishers are eager to enter a part of the GI map market indicates that there might not be a need for general activity of a public body in this area'.[103]

COMMUNICATION – In its 2009 Communication on the PSI review, the European Commission indicated that it will not take any legislative initiatives until a new

99. Commission of the European Communities, 'Results of the MS Consultation on the review of the application of Directive 2003/98/EEC on the re-use of public sector information', 2008, 4, <http://ec.europa.eu/information_society/policy/psi/docs/consultations/replies/members_states/report_ms_questionnaire.pdf>, 22 Feb. 2010.

100. Commission of the European Communities, Results of the online consultation of stakeholders 'Review of the PSI Directive', 2008, 3–4, <http://ec.europa.eu/information_society/policy/psi/docs/pdfs/online_consultation/report_psi_online_consultaion_stakeholders.pdf>, 22 Feb. 2010.

101. Micus Management Consulting Gmbh, 'Assessment of the Re-use of Public Sector Information (PSI) in the Geographical Information, Meteorological Information and Legal Information Sectors', 2008, <http://ec.europa.eu/information_society/policy/psi/docs/pdfs/micus_report_december2008.pdf>, 22 Feb. 2010.

102. *Ibid.*, 60.

103. *Ibid.*, 98.

review has been performed in 2012. In the meantime, it also does not intend to give guidance on the role of the public bodies and when they should remain off the market.[104]

C. Interim Summary

No clarity on the limit to the provision of information services – The overview of the documents that deal with the scope of the public task from a negative point of view shows the uncertainty of the European and international legislators and policy makers on how to define this public task. There is a general consensus that the public bodies have to comply with competition law if they enter the market, and some argue that the commercial and public task activities should be separated. However, at which point the public task stops and there is a market entry remains unclear. For example, the Synergy Guidelines only saw a public task to provide information services where the private sector was not able or not willing to provide the service. The Publaw 3 report pleaded on the contrary for a universal information service that should be delivered as a service of public interest, regardless of market failure. Such a universal information service would contain a number of obligations for the Member States or the public bodies to provide information or data that could be perceived as a minimum obligation agreed upon by all the stakeholders. Unfortunately, the report admitted at the same time that defining this universal information service was not possible. Later, the lack of clarity about the scope of the public task was often denounced, yet no attempts were made to determine this scope.

IV. TOWARDS THE NEXT CHAPTER: WHAT IS THE SCOPE OF THE PUBLIC TASK OF PROVIDING INFORMATION SERVICES?

Conclusion – From what has been discussed above, it is clear that the European and international legislators and policy makers have not formed many explicit opinions about the scope of the public task, either from a positive or negative point of view. From the positive side, the main obligations that can be traced are requirements for the public bodies to provide 'adequate' information and make information 'effectively accessible'. Very few of these obligations are defined into detailed obligations. Only a few exceptions can be found with regard to information on the functioning of the public bodies, how to exercise the right of access, and some environmental information obligations. From the negative side, even less guidance

104. Commission of the European Communities, *Communication to the European Parliament, the Council, the European Economic and Social Committee and the Committee of the Regions. Re-use of Public Sector Information – Review of Directive 2003/98/EC*, COM (2009) 212 final, 9, 7 May 2009.

is given by the European and international documents. Most of them merely state that the separation of the public task and the other activities of the public bodies is a difficult matter, and that competition law should be complied with if a public body enters the market.

Outline of next chapter – Hence, as the European and international legislative and policy documents remain silent on the scope of the public task, it is left to the Member States to determine this scope. This forms a considerable risk for the development of a European information society and information market, as it hinders the development of pan-European information services. In the next chapter, I examine the possibility of determining the scope of the public task, both from the positive and the negative point of view, and look for inspiration in the field of competition law, where the same problem has been encountered many times.

Chapter 8
The Possibility of Defining the Public Task

I. INTRODUCTION

Recapitulation – Chapter 7 showed that the international and European legislators and policy makers have given very little guidance about the content or the scope of the public task, both from the point of view of what the public bodies *should* do and from what the public bodies *should not* do. And yet, this public task is one of the main points of demarcation in the legal framework that has been developed for the availability of public sector spatial data. Understanding the scope of the public task is vital to know when to apply the rules for sharing in the INSPIRE Directive or the rules for re-use in the PSI Directive. As was explained in the previous chapters, if a public body provides spatial information services that are part of its public task, it can obtain the data it uses for these spatial information services under the rules for sharing. If it provides these spatial information services on the market as an activity outside of its public task, it should obtain the underlying data under the provisions of the PSI Directive.

Objective – This chapter examines the possibility of defining the public task to provide spatial information services in a sufficiently clear manner, so that the distinction between re-use and sharing in the EU legal framework for the availability of public sector spatial data could be clear. In order to define whether this is possible, a number of steps need to be taken.

Outline – First, I examine the question of the scope of the public task as to its importance for the domain of spatial information services. The need for clarification is discussed on the basis of the policy documents and literature in the domain. After indicating the problem, I look at the criteria that are offered to determine the scope of the public task. These criteria are approached from two sides. First, the possible criteria to determine the public task are approached from the positive point of view, that is, criteria that determine what the public bodies *should* do and what is part of their public task. Second, the negative approach tries to determine what activities the public bodies should not undertake under their public task. Next, I look to another legal domain for inspiration to solve the question of the scope of the public task, that is, the domain of competition law. This domain has also had to deal with the distinction between activities performed by public bodies that have a public character and activities that have an economic character on the market. The rules of competition law are also applicable to public sector spatial data, in addition to the EU legal framework for the availability of these data that has been described in Chapter 3. Therefore, it is important to determine how this domain would deal with the scope of the public task and if it can help in demarcating this scope.

II. THE QUESTION OF THE SCOPE OF THE PUBLIC TASK

A. Introduction

Recurring issue – The need for a definition of the public task, or for a clear boundary between the public task activities and other activities of the public bodies, has often been recognized, but attempts have only rarely been made to actually formulate such a definition or lay down such a boundary. However, it is interesting to see that the problem is widely recognized and that the struggle with this criterion is a recurring issue. The next overview will show that the demand for a demarcation of the scope of the public task has repeatedly been made by policy makers and scholars from both civil and common law countries within the EU.

B. Recognition of the Problem

Recognition in the Member States – Questions on the limit between the public task and market or commercial activities of the public bodies did not just arise with the introduction of the PSI Directive. For instance, the Dutch Minister of Internal Affairs already wondered in 1997 in how far providing information was a market activity.[1] Also in 1997, a Dutch report was published on the relationship between the market and government, followed by a bill that was based on the distinction between the public task activities of the public bodies and their other activities. This bill described the conditions that the public bodies had to comply with to participate in the market.[2] However, it did not define the cut-off point between the

1. Ministry of the Interior, 'Naar toegankelijkheid van overheidsinformatie', 1997, 3, <www.minbzk.nl/aspx/download.aspx?file=/contents/pages/302/nota_ntvo_nl.pdf>, 22 Feb. 2010 [in Dutch].
2. Werkgroep Markt en Overheid, 'Eindrapport', 1997, 114, <www.ez.nl/dsresource?objectid=148229&type=PDF>, 22 Feb. 2010 [in Dutch]; Minister of Economics and Minister of Justice,

public task and the market activities. Possibly, this was one of the reasons why the bill was repealed.[3] In the United Kingdom, the 2000 Cross-Cutting Review of the Knowledge Economy concluded that there was a need for a clearer explanation and definition of the role of government in the information market.[4]

RECOGNITION ON THE LEVEL OF THE EUROPEAN UNION – With respect to the level of the EU, the previous chapter already showed that the European Commission had been addressing the role of the public bodies on the information market since the 1989 Synergy Guidelines, yet it had not addressed specifically what the public task of the public bodies with regard to the provision of information services should be. During the 1996 Stockholm conference that followed the publication of the Publaw reports, Gellman indicated that, also on the European level, '[t]he debate over the proper role of the government and the private sector in supplying government information products and services is ongoing. To date, no one has been able to define an appropriate boundary for government dissemination activities in a networked environment'.[5]

IMPORTANCE OF THE DEFINITION – Hence, how can we distinguish the activities 'in which the government should be providing the product to the end user from those where we ought to rely upon the private sector?'[6] A definition of the public

---

Voorstel van Wet – Regels omtrent marktactiviteiten van overheidsorganisaties en omtrent ondernemingen die van overheidswege over een bijzondere positie beschikken (Wet markt en overheid), Kamerdocumenten 2001–2002, 28 050, no. 1–2, <www.ez.nl/dsresource?objectid=148231&type=PDF>, 22 Feb. 2010 [in Dutch].

3. See letter from the Minister of the Economic Affairs repealing the bill: Laurens J. Brinkhorst, 'Intrekking voorstel van wet Markt en Overheid', 9 Apr. 2004, <www.ez.nl/dsresource? objectid=47703&type=PDF>, 22 Feb. 2010 [in Dutch].
4. HM Treasury, 'Cross Cutting Review of the Knowledge Economy. Review of Government Information', 2000, <www.hm-treasury.gov.uk/spend_sr00_ccr.htm>, 22 Feb. 2010.
5. Robert Gellman, 'The American Model of Access to and Dissemination of Public Information', paper given at the European Commission Conference: Access to Public Information. A Key to Commercial Growth and Electronic Democracy (Stockholm, 1996): 11.
6. Anne Wells Branscomb, *Who owns information? From privacy to public access* (New York: Basic Books, 1994), 171. See also Herbert Burkert, 'The Mechanics of Public Sector Information', in *Public Sector Information in the Digital Age: Between Markets, Public Management and Citizens' Rights*, ed. Georg Aichholzer & Herbert Burkert (Cheltenham: Edward Elgar Publishing Limited, 2004), 4; Krysia Rybaczuk & Michael Blakemore, 'Selling Government Information: A Comparative Perspective on UK and USA Developments', paper given at the Conference on Law and Information Policy for Spatial Databases (Tempe, 1994), <www.spatial.maine.edu/%7eonsrud/tempe/rybaczuk.html>, 22 Feb. 2010; Katleen Janssen, Jo Steyaert & Roland Van Gompel, *Transparantie van overheidsinformatie in Vlaanderen. Beschouwingen en aanbevelingen* (Bruges: Die Keure, 2003), 215–217 [in Dutch]; University of Sheffield et al., 'MADAME – Methods for Access to Data and Metadata in Europe. Increasing Access to Public Sector Information: Guidelines of Best Practice' (Sheffield: University of Sheffield, 2000); William S. Holland, 'Copyright, Licensing and Cost Recovery for Geographic and Land Information System Data: A Legal, Economic and Policy Analysis', 1997, <www.spatial.maine.edu/~onsrud/tempe/holland.html>, 22 Feb. 2010; Bastiaan Van Loenen & Jaap Zevenbergen, 'One Access Policy for High-quality Geographic Information: Results from a US-EU Comparative Study', paper given at the GSDI-9 conference (Santiago de Chili, 2006), 13, <http://gsdidocs.org/gsdiconf/GSDI-9/papers/TS35.4paper.pdf>, 22 Feb. 2010; Intelligent

task is important. Kabel et al. believe that without a vision on the public tasks of the government, it is not possible to determine the roles of the public bodies and other parties in making public sector data available and accessible.[7] As Baten and Van Der Starre indicate, sometimes this might lead to the public bodies making the information accessible, and sometimes to the press or commercial publishers creating the service.[8] The Advisory Panel on Public Sector Information (APPSI) saw this lack of a definition as one of the main obstacles to progress of the re-use of PSI in the UK: 'It is not clear, as a matter of policy, where the government's priorities lie. Possible priorities pull in different directions. Who should be the main beneficiary of re-use – public bodies, citizens, businesses, or not-for-profit organisations? Should this asset, PSI, be exploited to help fund public services, or to strengthen the competitive positioning of UK industry? There are no authoritative answers to either of these sets of questions'.[9] Especially for the Trading Funds, it is necessary to 'distinguish more clearly what is required by Government for public tasks'.[10]

COMPLICATING FACTORS – Some additional factors complicate the issue even more. First, the term *public task* is often used next to other terms such as *statutory function*, *public mission*, *core task*, and so on. While these terms are often found to be used interchangeably,[11] the lack of one term that can be used as the defining criterion may cloud the discussion by giving the impression that people are

---

Addressing, 'Press Release. Public Sector Information Debate', 2008, <www.iahub.net/docs/1221131589821.pdf>, 22 Feb. 2010; Jan J.C. Kabel et al., *Kennisinstellingen en informatiebeleid. Lusten en lasten van de publieke taak* (Amsterdam: Universiteit van Amsterdam, 2001), 123 [in Dutch]; Marc R. Bokkerink, 'De uitoefening van publiekrechtelijke taken door privaatrechtelijke rechtspersonen', in *Overheid en onderneming. Tussen publiek en privaat ondernemen*, ed. Sebastianus C.J.J. Kortmann et al. (Deventer: W.E.J. Tjeenk Willink, 1998), 293 [in Dutch]; Andrew Gamble & Tony Wright, 'Introduction', in *Restating the State? Special issue of The Political Quarterly*, ed. Andrew Gamble & Tony Wright (Malden: Blackwell Publishing, 2004), 1; Chris Corbin, 'Thematic Meeting Report – PSI Pricing 1,' 2007, 14, <www.epsiplus.net/reports/epsiplus_thematic_meeting_reports/fin_mtg_1>, 22 Feb. 2010; Paul F. Uhlir et al., *The Socioeconomic Effects of Public Sector Information on Digital Networks* (Washington: National Academies Press, 2009), 8.

7. Jan J.C. Kabel et al., *Kennisinstellingen en informatiebeleid. Lusten en lasten van de publieke taak* (Amsterdam: Universiteit van Amsterdam, 2001), 14 [in Dutch].
8. Inigo Baten & Gijs Van Der Starre, *Elektronische toegankelijkheid van overheidsinformatie* (The Hague: Rathenau Instituut, 1996), 23 [in Dutch].
9. Advisory Panel on Public Sector Information, 'The need for a government strategy on the re-use of public sector information', 2007, 2, <www.appsi.gov.uk/ministers/gov-strategy-psi.pdf>, 22 Feb. 2010.
10. HM Treasury, 'Stability and opportunity: building a strong, sustainable future. Budget 2008', 2008, 58, <www.hm-treasury.gov.uk/d/bud08_completereport.pdf>, 22 Feb. 2010. See also Stephen Saxby, 'Public sector information and re-use policy – where is the UK now?', paper given at the Second International Conference on Business, Law and Technology (New York, 2008), 13, <http://eprints.soton.ac.uk/52523/01/PSI_Re-use_Paper_28_April_2008_pdf.pdf>, 22 Feb. 2010; Advisory Panel on Public Sector Information, 'Response to the consultation on the Power of Information Report', 2009, 3, <www.appsi.gov.uk/2009/02/16/APPSIsResponseToTheDraftPowerOf InformationTaskForceReport>, 22 Feb. 2010.
11. See Office of Fair Trading, 'The Commercial Use of Public Information (CUPI)', 2006, 58, <www.oft.gov.uk/shared_oft/reports/consumer_protection/oft861.pdf>, 22 Feb. 2010.

discussing different things. Second, while some public bodies have actually defined and published their main activities and missions, they may not always determine their public task well enough to be operative or useful for applying the legal framework.[12] In addition, where the public tasks have been listed for a specific organization, these organizations may interpret them their own way, which may differ over time. Third, such laws or regulations listing the public tasks may not be the only aspect that needs to be taken into account. As APPSI stated, '[i]t is not clear whether the question is to be determined primarily by reference to (i) the public (including social or economic) need for the information; or to (ii) the duties of the public body as set by local law and practice'.[13] Fourth, the issue of using the public task as a criterion is 'further complicated when re-using PSI across borders as the public bodies either side of a border may well take diametric positions with respect to their public task. That is one public body does comply with the Directive as they deem their public holdings to be part of their public task whilst others state they are exempt as some or all of their public sector information holdings are not part of their public task'.[14] According to APPSI, the PSI Directive suggests that local law and practice should be the first reference for the public task, but this would militate against the harmonizing intent of the directive.[15]

Re-ignition of the debate – While the public task debate has lingered in the background for a long time, it became more acute with the introduction of the PSI Directive, as the directive directly referred to the public task as a criterion for the field of application. The INSPIRE Directive intensified this discussion by referring to 'public tasks with an impact on the environment'. Prins stated it clearly for the PSI Directive in 2006: 'For the rule to work, it will be crucial that clear guidance is given on what constitutes use as part of a public task and what must be qualified as part of [the public bodies'] commercial ambitions and activities'.[16] In its internal review of the first two years of the PSI Regulations that transposed the PSI Directive, the United Kingdom's Office of Public Sector Information (hereafter referred to as OPSI) found that it was not always apparent whether an activity carried out by a public body is a public task or a re-use activity. A more explicit definition was felt to be needed.[17] Therefore, OPSI undertook to produce a clearer

12. *Ibid.*
13. Advisory Panel on Public Sector Information, 'The impact and success of legislation on the re-use of public sector information,' 2008, 26, <www.appsi.gov.uk/reports/ec-submission-07-2008.doc>, 22 Feb. 2010.
14. Chris Corbin, 'Is there a need to define "public task"', 2007, <www.epsiplus.net/forum/forum/is_there_a_need_to_define_public_task>, 22 Feb. 2010. See also Geographic Information Network in Europe, 'New Issues for the European GI Strategy: Public Sector Information', 2003, 8, <www.ec-gis.org/ginie/doc/d293_issues_pse_fv3.pdf>, 22 Feb. 2010.
15. Advisory Panel on Public Sector Information, 'The impact and success of legislation on the re-use of public sector information, 2008', 26, <www.appsi.gov.uk/reports/ec-submission-07-2008.doc>, 22 Feb. 2010.
16. Corien Prins, 'Directive 2003/98/EC of the European Parliament and of the Council', in *Concise European IT law*, ed. Corien Prins (Alphen a/d Rijn: Kluwer Law International, 2006), 430.
17. Office of Public Sector Information, 'The United Kingdom Implementation of the European Directive on the Re-use of Public Sector Information – the first two years', 2007, 6–7,

interpretation of the term *public task* and to include this in its Guide to the Regulations and Best Practice. So far, this has not been done. OPSI's view on the situation in the United Kingdom was extrapolated by the Advisory Panel on Public Sector Information (APPSI) in its response to the consultation organized by the European Commission in the framework of the review of the PSI Directive. APPSI felt that '[t]he term "public task" plays a crucial role in the Directive since documents whose supply falls outside a public body's public task are excluded from its application'.[18] In its role as complaint authority, APPSI had been confronted firsthand with the lack of clarity of the criterion of the public task. APPSI was not the only one that requested clarification on the public task issue from the European Commission; so did other Member State representatives, such as Ireland and Slovenia, and a number of other stakeholders. For instance, the Locus Association stated that it would 'welcome a clarification of the role of Public Sector Information Holders (PSIHs) which it believes can be subject to undesirable conflicts of interest . . . the private sector needs real guidance over what PSIHs' public task is'.[19]

Benefits of a clear definition – Clarification of the term *public task* and the demarcation between the public tasks of the public bodies and their other activities would have a number of benefits. It would 'reinforce the message that if public sector organizations re-use information they should be subject to the same terms and conditions as any private sector re-user. In this way the principle of fairness and non-discrimination will be maintained'.[20] In addition, it will be easier for the administrative or judicial authorities that have to decide on any complaints launched against public bodies with regard to this issue to determine whether these complaints are justified. This is not only true on the national level, but also on the level of the EU. For instance, APPSI feels that an adjustment to the PSI Directive, or at least guidance on the principles, would assist in promoting a harmonized approach, 'which at present is impossible if the critical concept of "public task" is simply left to be interpreted in accordance with local law, binding rules, "common

<www.opsi.gov.uk/advice/psi-regulations/uk-implementation-first-years.pdf>, 22 Feb. 2010. See also Stephen Saxby, 'Public sector information and re-use policy – where is the UK now?', paper given at the Second International Conference on Business, Law and Technology (New York, 2008), 14, <http://eprints.soton.ac.uk/52523/01/PSI_Re-use_Paper_28_April_2008_pdf.pdf>, 22 Feb. 2010.

18. Advisory Panel on Public Sector Information, 'The impact and success of legislation on the re-use of public sector information', 2008, 25, <www.appsi.gov.uk/reports/ec-submission-07-2008.doc>, 22 Feb. 2010.
19. Locus Association, 'Submission to the European Commission on the Re-use of Public Sector Information Directive', 2008, 5, <http://ec.europa.eu/information_society/policy/psi/online_consultation/stakeholders/index_en.htm>, 22 Feb. 2010.
20. Office of Public Sector Information, 'The United Kingdom Implementation of the European Directive on the Re-use of Public Sector Information – the first two years', 2007, 6, <www.opsi.gov.uk/advice/psi-regulations/uk-implementation-first-years.pdf>, 22 Feb. 2010; Advisory Panel on Public Sector Information, 'The impact and success of legislation on the re-use of public sector information', 2008, 26–27, <www.appsi.gov.uk/reports/ec-submission-07-2008.doc>, 22 Feb. 2010.

administrative practice" (accepting that there should be some scope for variation between Member States on what is defined as "public task") or even departmental preference'.[21] Within such EU guidance, the Member States could then establish their own *public task* definitions.

C. IMPACT OF THE LACK OF DEFINITION OF *THE PUBLIC TASK*

RE-USE V. SHARING – The consequences of the categorization of a particular activity of a public body as part of the public task appear on different levels. The first level where concerns may rise has already been discussed. As explained before, if there is doubt that an information service created by a public body is delivered as part of its public task or not, this causes uncertainty about the application of the PSI Directive or the INSPIRE Directive. As these directives contain different rules for re-use and sharing, the wrong rules might be applied, possibly causing inequality between the different stakeholders, such as competing companies, non-profit organizations, public bodies, or citizens.[22] APPSI believes that: 'As a result of the lack of clarity in the process for defining an organization's "Public Task", there is much mistrust between the various parties'.[23]

LEVEL PLAYING FIELD – On a second level, the distinction is also important with regard to questions of fair competition and a level playing field.[24] The topic of

21. Advisory Panel on Public Sector Information, 'The impact and success of legislation on the re-use of public sector information', 2008, 27, <www.appsi.gov.uk/reports/ec-submission-07-2008.doc>, 22 Feb. 2010. See also Marc De Vries, 'Redress of PSI re-use disputes: the missing teeth?- ePSIplus Thematic Meeting Report', 2008, 5, 13, <www.epsiplus.net/reports/epsiplus_thematic_meeting_reports/law_and_regulation_meeting_3_report>, 22 Feb. 2010.
22. See also Commission of the European Communities, *Staff working document. Accompanying document to the Communication from the Commission to the European Parliament, the Council, the European Economic and Social Committee and the Committee of the Regions on the re-use of Public Sector Information – Review of Directive 2003/98/C*, SEC (2009) 597, 28, 7 May 2009.
23. David Rhind, 'Letter to Michael Wills MP, Minister of Justice – APPSI's recommendations to progress the re-use of public sector information', 2008, 2, <www.epsiplus.net/content/download/14491/181308/file/minister-letter-ec-submission-07-2008.pdf>, 22 Feb. 2010.
24. Yvo Volman, 'Exploitation of Public Sector Information in the Context of the *eEurope* Action Plan', in *Public Sector Information in the Digital Age, Between Markets, Public Management and Citizens' Rights*, ed. Georg Aichholzer & Herbert Burkert (Cheltenham: Edward Elgar Publishing Limited, 2004), 97. See also Roger Longhorn & Michael Blakemore, 'Re-visiting the Valuing and Pricing of Digital Geographic Information', *Journal of Digital Information* 4, no. 2 (2004): 14, <http://journals.tdl.org/jodi/article/view/103/102>, 22 Feb. 2010; Bertrand Nouel, 'Commercialisation des données publiques et concurrence', paper given at the European Commission Conference: Access to Public Information. A Key to Commercial Growth and Electronic Democracy (Stockholm, 1996): 8 [in French]; Jean-Michel Bruguière, 'Droit des données publiques', s.d., 37, <www.educnet.education.fr/chrgt/droitddp.pdf>, 22 Feb. 2010 [in French]; Her Majesty's Stationery Office, 'Future Management of Crown Copyright', 1999, 21, <www.opsi.gov.uk/advice/crown-copyright/future-management-of-crown-copyright.pdf>, 22 Feb. 2010; Jan J.C. Kabel et al., *Kennisinstellingen en informatiebeleid. Lusten en lasten van de publieke taak* (Amsterdam: Universiteit van Amsterdam, 2001), 53 [in Dutch];

competition law is addressed later on in this chapter, but some remarks can already be made here. Most of the data that are held by the public bodies were created in the framework of their public task and cannot be obtained elsewhere. Insofar as a public body would offer these data on the market or create information services or products with them, it is in principle a monopolist. As public bodies often have a natural monopoly over their information resources, their behaviour can have a considerable impact on the market.[25]

MISUSE OF MONOPOLY POSITION – Monopolists are often tempted to use their monopolies to impede healthy competition. If they enter the market, there are several risks for them to 'give in to the temptation to use this monopolistic situation to maximize the short-term revenues of certain data sets, without taking into account the need for fair competition and the optimal use of the data for society as a whole'.[26] They might 'perceive that they can increase their total revenue stream by setting prices higher than they would in a competitive market' or 'extend their monopolies into downstream markets' by exclusive partnerships.[27] Such exclusive deals might have the short-term advantage of a guaranteed stable income for a certain period of time, but they also prevent others from getting the public sector spatial data under the same conditions.[28]

CROSS-SUBSIDIZATION – Another possible concern regards the cross-subsidization of activities. There is no problem if the public body finances losses that occur in the performance of one public task with funds for another public task

---

Maeve McDonagh, 'Access to public sector information: the Australian experience', paper given at the European Commission Conference: Access to Public Information. A Key to Commercial Growth and Electronic Democracy (Stockholm, 1996): 2; Johan Pas & Bruno De Vuyst, 'Re-establishing the Balance between the Public and the Private Sector: Regulating Public Sector Information Commercialization in Europe' *The Journal of Information, Law and Technology* no. 2 (2004): 6–8, <www2.warwick.ac.uk/fac/soc/law/elj/jilt/2004_2/pasanddevuyst>, 22 Feb. 2010.

25. Yvo Volman, 'Exploitation of Public Sector Information in the Context of the *eEurope* Action Plan', in *Public Sector Information in the Digital Age, Between Markets, Public Management and Citizens' Rights*, ed. Georg Aichholzer & Herbert Burkert (Cheltenham: Edward Elgar Publishing Limited, 2004), 97. See also Henry Perrit Jr., 'Should local governments sell local spatial databases through state monopolies?', *Jurimetrics Journal* 35 (1994–1995): 454; Peter N. Weiss, 'Borders in Cyberspace: Conflicting Public Sector Information Policies and their Economic Impacts', in *Public Sector Information in the Digital Age, Between Markets, Public Management and Citizens' Rights*, ed. Georg Aichholzer & Herbert Burkert (Cheltenham: Edward Elgar Publishing Limited, 2004), 149.
26. Yvo Volman, 'Exploitation of Public Sector Information in the Context of the *eEurope* Action Plan', in *Public Sector Information in the Digital Age, Between Markets, Public Management and Citizens' Rights*, ed. Georg Aichholzer & Herbert Burkert (Cheltenham: Edward Elgar Publishing Limited, 2004), 97.
27. Henry Perrit Jr., 'Should local governments sell local spatial databases through state monopolies?', *Jurimetrics Journal* 35 (1994–1995): 454. See also Henry H. Perrit Jr., 'Sources of rights to access public information', *William and Mary Bill of Rights Journal* 4 (1995): 205.
28. Yvo Volman, 'Exploitation of Public Sector Information in the Context of the *eEurope* Action Plan', in *Public Sector Information in the Digital Age, Between Markets, Public Management and Citizens' Rights*, ed. Georg Aichholzer & Herbert Burkert (Cheltenham: Edward Elgar Publishing Limited, 2004), 97.

that were not completely depleted. But when 'public money' is used to make up for losses in the market activities of the public body, this could have a negative effect on the level playing field.[29] In these cases, the commercial activities are 'sponsored with information produced with public funding and within the public task'.[30] Peters and Vriesendorp refer to this situation as government money being employed for the 'sidelines'.[31] If other parties cannot obtain the public sector spatial data under the same conditions as the public bodies using their own data, competition on the market is distorted. The dissemination of information services by the public bodies is subsidized by their public funding, so they can keep the prices low (possibly even below the level of commercial profit[32]) making it virtually impossible for the private sector to develop similar services.[33]

SUMMARY – Such possible distortions of competition may have as a result that '[t]he diversity provided by commercial and non-commercial information providers will diminish ... and raise the spectre of monopoly control' of the public bodies.[34] Hence, the definition of the *public task* appears to be vital to safeguard the offer of a variety of information products that can respond to the social and economic needs of the society.

D. THE WHAT-QUESTION AND THE HOW-QUESTION

CONFUSION BETWEEN TWO QUESTIONS – When discussing the public task and the role of the public bodies on the market, two questions are often asked and mixed together. The Wetenschappelijke Raad voor het Regeringsbeleid refers to these

---

29. See Commission of the European Communities, *Staff working document. Accompanying document to the Communication from the Commission to the European Parliament, the Council, the European Economic and Social Committee and the Committee of the Regions on the re-use of Public Sector Information – Review of Directive 2003/98/C*, SEC (2009) 597, 28, 7 May 2009.
30. Yvo Volman, 'Exploitation of Public Sector Information in the Context of the *eEurope* Action Plan', in *Public Sector Information in the Digital Age, Between Markets, Public Management and Citizens' Rights*, ed. Georg Aichholzer & Herbert Burkert (Cheltenham: Edward Elgar Publishing Limited, 2004), 97.
31. Johannes A.F. Peters & Reinout D. Vriesendorp, 'Insolvency of public entities other than the State under Dutch law', *Electronic Journal of Comparative Law* 6 (2002): 412, <www.ejcl.org/64/art64-23.pdf>, 22 Feb. 2010.
32. Krysia Rybaczuk & Michael Blakemore, 'Selling Government Information: A Comparative Perspective on UK and USA Developments', paper given at the Conference on Law and Information Policy for Spatial Databases (Tempe, 1994), <www.spatial.maine.edu/%7eonsrud/tempe/rybaczuk.html>, 22 Feb. 2010. See also Henry Perrit Jr., 'Should local governments sell local spatial databases through state monopolies?', *Jurimetrics Journal* 35 (1994–1995): 450.
33. Krysia Rybaczuk & Michael Blakemore, 'Selling Government Information: A Comparative Perspective on UK and USA Developments', paper given at the Conference on Law and Information Policy for Spatial Databases (Tempe, 1994), <www.spatial.maine.edu/%7eonsrud/tempe/rybaczuk.html>, 22 Feb. 2010.
34. Robert Gellman, 'The American Model of Access to and Dissemination of Public Information', paper given at the European Commission Conference: Access to Public Information. A Key to Commercial Growth and Electronic Democracy (Stockholm, 1996): 10.

issues as the *what-question* and the *how-question.*[35] The distinction between the two questions is not often made, which has led to many misunderstandings.

WHAT? – The what-question determines whether there is a task for the government or the public bodies to provide a particular service or to respond to particular needs in society. For instance, should the government ensure that there is sufficient public transportation, that the mail is delivered every day or twice a day, or that spatial information services are made available to the public?

HOW? – The how-question relates to the execution of the public task. Should this be achieved by the public bodies themselves or can they assign the task to a contractor or a concession-holder? Hence, the fact that the provision of a particular spatial information service is considered as a public task does not mean that the public body has to perform every part of it from start to finish. It might outsource the creation of a web service or the dissemination of the spatial information service through a procurement contract. However, the public body remains responsible for the performance of the task, and it has to ensure that all citizens have access to the spatial information service on a fair and equal basis. The Wetenschappelijke Raad voor het Regeringsbeleid considered the what-question to be of a much more political nature than the how-question.[36]

FOCUS ON THE WHAT-QUESTION – As both questions are often on the table together, it is not surprising that in practice they get muddled. However, as I tackle the question on the scope of the public task, my analysis in principle only addresses the what-question. Yet, some elements may refer to the how-question in order to demonstrate a point on the what-level. The how-question itself addresses issues such as privatization, procurement, concessions, and so on.

III. ATTEMPTS AT CRITERIA TO DETERMINE THE PUBLIC TASK

A. INTRODUCTION

NEED FOR A DEFINITION – As the public task is the main criterion for determining the applicability of the PSI Directive, the INSPIRE Directive, and the Access Directive, it is important to have a good overview of the scope of this public task. This provides legal certainty for all the stakeholders involved: the citizens know what services they can expect from the public bodies, the public bodies know which conditions to follow when they deliver public sector spatial data to other public bodies or what conditions they have to comply with if they use public sector spatial data for the provision of their services.

OUTLINE – However, there is no true definition of *the public task* to provide spatial information services under the PSI Directive or the INSPIRE Directive.

35. Wetenschappelijke Raad voor het Regeringsbeleid, *Het borgen van publiek belang* (The Hague: Sdu Uitgevers, 2000), 43, <www.wrr.nl/content.jsp?objectid=2545>, 22 Feb. 2010 [in Dutch].
36. *Ibid.*, 21.

In this section, I try to determine what criteria can be used as input for this definition. First I look at some criteria that could be considered as positive in the sense that these criteria could determine the positive obligation of a public body to provide a spatial information service. Next, the scope of the public task is discussed from a negative point of view: are there criteria that determine what the public bodies are not allowed to do? From the description of both types of criteria, I then try to deduce whether the scope of the public task to provide spatial information services can be demarcated.

B. The Positive Approach: What the Public Bodies Should Do

Open character of criteria – While there may be no ready-to-use definition of the *public task*, some criteria can be found in literature and policy documents to determine whether the public bodies should provide spatial information services or not. While some of these criteria are reasonably concrete, most of them remain rather vague and open for interpretation. Often this is intentional to make sure that the decision of whether a particular activity of a public body falls under the public task can be adapted to the circumstances. However, this vague or general character of the criteria also leads to confusion and difficulty in applying the criteria, leaving a lot of room for personal interpretation.

Outline – The main criteria that can be found include the presence of a legal basis, the relationship with the core tasks of the public body, market failure, and the presence of a public interest. These criteria are discussed in this section.

1. Legal Basis

Assigned by law – The most straightforward criterion to determine whether an activity of a public body forms part of the public task or remains outside of its scope is the presence of a legal basis.[37] This would mean that a public body's public task can be described as the tasks that are assigned to this public body by the law. This legal basis does not necessarily have to be a law that is enacted by Parliament; it can also be an administrative decision.

37. Bertrand Nouel, 'Commercialisation des données publiques et concurrence', paper given at the European Commission Conference: Access to Public Information. A Key to Commercial Growth and Electronic Democracy (Stockholm, 1996): 8 [in French]; Chris Corbin, 'Is there a need to define "public task" ', 2007, <www.epsiplus.net/forum/forum/is_there_a_need_to_define_public_task>, 22 Feb. 2010; Udo Di Fabio, 'Privatisierung und Staatsvorbehalt', *Juristen Zeitung* no. 12 (1999): 586 [in German]; Wetenschappelijke Raad voor het Regeringsbeleid, *Het borgen van publiek belang* (The Hague: Sdu Uitgevers, 2000), 56, <www.wrr.nl/content.jsp?objectid=2545>, 22 Feb. 2010 [in Dutch]; Johannes Hengstschlager, 'Privatisierung von Verwaltungsaufgaben', in *Veröffentlichungen der Vereinigung Deutscher Staatsrechtlehrer 54*, ed. X. (Berlin: De Gruyter, 1995),173 [in German].

EXAMPLES: POLICY DOCUMENTS – This criterion was, for instance, considered by the Dutch proposal for a law on the relationship between market and government[38] and by the Swedish government.[39] The United Kingdom's OPSI also stated that one of the characteristics of a public task to provide information services was that there is 'a statutory requirement to produce or issue such information',[40] but it also admitted that a statute was not always necessary. A public task could be 'any activity of a public sector body which is described by law or in which it engages as a result of common administrative practice'.[41]

EXAMPLE: KNMI – This criterion has been applied in specific cases. For example, the Dutch KPMG report reviewing the Law on the Royal National Meteorological Service (KNMI) assessed it as positive that the KNMI's public task was laid down in legislation and felt that it should stay that way. The main elements of KNMI's public task should be laid down in a law and set out in more detail in administrative decisions that implement the law.[42]

EVALUATION: PUBLIC TASK IN A LAW – However, there are several problems with the description of a public task of a public body as the activities of the public body as they are prescribed by a law or by administrative decisions or even 'common administrative practices'. First, while laying down the public task in a law in the formal sense should provide for legal certainty and ensures the intervention of democratically elected parliamentarians, the extensive and time-consuming procedure for changing such a law makes this public task rigorous and unable to be adapted swiftly to changing circumstances. In order to avoid the description of the public task to be outdated before it has even gone through Parliament and entered into force, this description should be kept at a very high-level, only setting out the very main elements of the public task. However, having such a very general description does not add any clarity and could leave much room for narrow or

38. Minister of Economics and Minister of Justice, *Voorstel van Wet – Regels omtrent marktactiviteiten van overheidsorganisaties en omtrent ondernemingen die van overheidswege over een bijzondere positie beschikken (Wet markt en overheid)*, Kamerdocumenten 2001–2002, 28 050, no. 1–2, <www.ez.nl/dsresource?objectid=148231&type=PDF>, 22 Feb. 2010 [in Dutch].
39. See Marijke Reinsma & H. Van Der Sluijs, 'Naar ruimere openbaarheid en een vrij gebruik van bestuurlijke informatie', 2002, 107, <www.minbzk.nl/aspx/download.aspx?file=/contents/pages/5006/RapportReinsmaenVanderSluijs.pdf>, 22 Feb. 2010 [in Dutch].
40. Office of Public Sector Information, 'The Re-use of Public Sector Information: A Guide to the Regulations and Best Practice', 2005, 12, <www.opsi.gov.uk/advice/psi-regulations/advice-and-guidance/guide-to-psi-regulations-and-best-practice.doc>, 22 Feb. 2010.
41. Cabinet Office and Department of Trade and Industry, 'Consultation document – Proposal for Regulations implementing a Directive of the European Parliament and Council on the re-use of public sector information', 2005, 3, <www.opsi.gov.uk/advice/psi-regulations/consultations/regulation/consultation-document-2004-12.pdf>, 22 Feb. 2010, referred to in Chris Corbin, 'Is there a need to define "public task" ' 2007, <www.epsiplus.net/forum/forum/is_there_a_need_to_define_public_task>, 22 Feb. 2010; Jan J.C. Kabel et al., *Kennisinstellingen en informatiebeleid. Lusten en lasten van de publieke taak* (Amsterdam: Universiteit van Amsterdam, 2001), 14 [in Dutch].
42. KPMG, 'Ministerie van Waterstaat. Evaluatie van de wet op het KNMI', 2007, 71, <www.verkeerenwaterstaat.nl/Images/2008651%20bijlage_tcm195-231492.pdf>, 22 Feb. 2010 [in Dutch].

broad interpretation, causing just as much discussion as not giving a definition of the public task at all.[43] For instance, if the task of a national mapping agency is described as 'providing topographic mapping for the national territory', this can cause just as much discussion as there would be without such a law, as there is no clarity about the way this mapping should be done, how the data can be presented, which services can be provided based on these data, and so on. Hence, a detailed description in a law is too strict, and a general description in a law has little to no added value.

EVALUATION: PUBLIC TASK IN AN ADMINISTRATIVE DECISION – This dilemma might be helped by setting out the public task or the details thereof in an administrative decision. Such a decision is easier to adapt, so it is more suitable to react to circumstances that require the public task to be adapted. However, this approach also has a number of drawbacks. First, the easy adaptability of such administrative decisions also makes them vulnerable for changes purely based on political or personal convictions from the person responsible for taking the decisions. While it is true that the public task in general, including the public task to provide spatial information services, is continuously evolving, the risk of major changes to this task when a new political party holds power is not inconceivable. One could question whether this may discourage the private sector from offering information services on the market. They would not know whether the service they offer might later on be considered a public task or not. If this were the case, the public bodies might then offer these services without having to comply with competition law.

EVALUATION: LEVEL OF ADMINISTRATIVE DECISION – In addition, the level at which the administrative decision is made to determine the public task of a public body should be taken into account. In some cases, the public bodies are requested to create a document that sets out their main tasks and responsibilities. This means that a public body can decide by itself what its own task is, without any immediate control of a supervisory authority. This could cause concern that the public body will determine its public task based on its own preferences, and when questioned on this task, it can just refer to its own list.[44] In the United Kingdom, APPSI suggested that PSI holders should reach agreement with OPSI as to which products and services constitute their public task.[45] The Locus Association found that while this may provide a way of defining the coverage of the PSI Regulations, it could also be construed as PSI Holders 'marking their own exam papers'.[46] This problem

43. See Bertrand Nouel, 'Commercialisation des données publiques et concurrence', paper given at the European Commission Conference: Access to Public Information. A Key to Commercial Growth and Electronic Democracy (Stockholm, 1996): 9 [in French].

44. See Office of Public Sector Information, 'Unlocking PSI potential. The United Kingdom Report on the Re-use of Public Sector Information', 2008, 59, <www.opsi.gov.uk/advice/psi-regulations/uk-report-re-use-psi-2008.pdf>, 22 Feb. 2010.

45. Communities and Local Government Committee, 'Ordnance Survey, Fifth Report of Session 2007-08, HC 268', <www.publications.parliament.uk/pa/cm200708/cmselect/cmcomloc/268/26802.htm>, 22 Feb. 2010.

46. *Ibid.*

is even more obvious when we consider the possibility provided by OPSI to define the public task as tasks that a public body engages in as a result of common administrative practice.

EXAMPLE: ORDNANCE SURVEY – An example of this problem that has been discussed extensively is the United Kingdom's Ordnance Survey. Founded in 1791 as England's National Mapping Agency,[47] the Ordnance Survey took the form of a Trading Fund in 1999.[48] The scope of Ordnance Survey's public task has been the subject of many debates, not just in the SDI community, but also in the British Parliament. The Communities and Local Government Committee of the Parliament carried out an inquiry into Ordnance Survey's public task during the course of 2007 and 2008.[49] The Guardian's Free Our Data Campaign also made an attempt in May 2008. The Campaign's blog stated: 'So we wondered; where can the OS's public task be found? Not on the [Ordnance Survey] website, where you simply find "For the purposes of the PSI Regulations, Ordnance Survey's Public Task is defined as embracing everything we do from time to time to fulfil our obligations under the Ordnance Survey Trading Fund Order 1999 (SI 99/965) and the Ordnance Survey Framework Document 2004" '.[50] According to Arthur, '[t]his is akin to saying "whatever we do, it's a public task, because we do it" '.[51] According to Parsons, this is because nobody else within government was either qualified or interested to do so. He claims that it is not illogical that Ordnance Survey defines its own task: 'If you were the CEO of a large commercial organization would you not want to write your own mission statement?'[52] McDonald replied to this that 'OS is still a public agency. We expect public agencies and government in general, to consult when it sets out its remit'. Moreover, in the field of PSI, 'OPSI and APPSI have roles to safeguard public interests in PSI, and they haven't been asked for views on [the Ordnance Survey's] new statement'.[53]

EVALUATION: CIRCULAR ARGUMENT – This last issue demonstrates the main difficulty in defining the public task as the tasks laid down in legislation or administrative decisions. It leads to a circular argument. An activity of the public body is

---

47. See <www.ordnancesurvey.gov.uk>, 22 Feb. 2010.
48. Government Trading Funds Act 1973 of 25 Oct. 1973, amended by Government Trading Funds Act 1990. Trading Funds are established by a Trading Fund Order, and more than 50% of their revenue has to be gained from the provision of goods and services.
49. See Communities and Local Government Committee, 'Ordnance Survey, Fifth Report of Session 2007–08, HC 268', <www.publications.parliament.uk/pa/cm200708/cmselect/cmcomloc/268/26802.htm>, 22 Feb. 2010.
50. Charles Arthur, 'Where is Ordnance Survey's public task set out, exactly? And why is it paying an external PR company?', 2008, <www.freeourdata.org.uk/blog/?p=195>, 22 Feb. 2010. See also Ordnance Survey, 'Re-use of Public Sector Information', <www.ordnancesurvey.co.uk/oswebsite/aboutus/yourinforights/ropsi.html>, 22 Feb. 2010.
51. Charles Arthur, 'Where is Ordnance Survey's public task set out, exactly? And why is it paying an external PR company?', 2008, <www.freeourdata.org.uk/blog/?p=195>, 22 Feb. 2010.
52. Ed Parsons, 'Your starter for ten, Define public task . . . ', 2008, <www.edparsons.com/2008/05/your-starter-for-ten-define-public-task/>, 22 Feb. 2010.
53. Dan McDonald, Reply to 'Your starter for ten, Define public task . . . ', 2008, <www.edparsons.com/2008/05/your-starter-for-ten-define-public-task/>, 22 Feb. 2010.

considered a public task, so it is laid down in the law as such.[54] And these legally recognized tasks are considered public tasks because they are laid down in the law. Hence, while it would be an elegant solution to define the public task as the task that is laid down in legislation or administrative decisions, it only moves the problem instead of solving it. It still leaves the question of why the legislator or the administrative body considered a particular task to be a public task that should be laid down in the law in the first place. This might be because of other criteria, such as public interest reasons, market failure, and so on.

2. Link with the Core Responsibility of the Public Body

SPATIAL INFORMATION SERVICES AS AN EXTENSION OF THE CORE ACTIVITIES – Some do not consider the provision of information services as a separate activity within or outside of the public task, but they link these services to the other tasks of the public bodies. For instance, OPSI believes that information falls within the public task if it is essential to the business of the public bodies, and that the production of information services falls outside of the scope of a public body's public task if they are not related to its core responsibility.[55] Nouel also believes that information services can only be provided as a normal extension of the public body's public service activities,[56] and Burkert states that '[g]overnments will have to show a close link between their public sector missions and their information activities in order to legitimize these information activities'.[57]

EVALUATION – However, this approach only moves the problem along. OPSI admits that '[i]t may not always be a straightforward matter to identify with precision what a public sector body's core responsibilities are' and hence 'where the boundary lies between documents within the public task and those that fall outside of it'.[58]

3. Market Failure

REASONS FOR MARKET FAILURE – A third criterion for the public bodies to create spatial information services is often found in the concept of market failure, that is,

54. See KPMG, 'Ministerie van Waterstaat. Evaluatie van de wet op het KNMI', 2007, 16, <www.verkeerenwaterstaat.nl/Images/2008651%20bijlage_tcm195-231492.pdf>, 22 Feb. 2010 [in Dutch].
55. Office of Public Sector Information, 'The Re-use of Public Sector Information: A Guide to the Regulations and Best Practice', 2005, 12, <www.opsi.gov.uk/advice/psi-regulations/advice-and-guidance/guide-to-psi-regulations-and-best-practice.doc>, 22 Feb. 2010.
56. Bertrand Nouel, 'Commercialisation des données publiques et concurrence', paper given at the European Commission Conference: Access to Public Information. A Key to Commercial Growth and Electronic Democracy (Stockholm, 1996), 9 [in French].
57. Herbert Burkert, 'Conclusions', paper given at the workshop on commercial and citizen's access to government information (Luxembourg, 1995), 81.
58. Office of Public Sector Information, 'The Re-use of Public Sector Information: A Guide to the Regulations and Best Practice', 2005, 12, <www.opsi.gov.uk/advice/psi-regulations/advice-and-guidance/guide-to-psi-regulations-and-best-practice.doc>, 22 Feb. 2010.

the market is not providing spatial information services of which the policy maker, the government, or whoever decides the content of the public task[59] feels that they should be provided. There may be several reasons why these spatial information services are not provided by the market. For instance, the private sector may not be willing to supply information services because information has the characteristics of a public good.[60] Stiglitz et al. argue that if it is impossible to exclude individuals from the public good and no additional costs are involved in providing the good to an extra person, the private sector will not supply such goods at all or in any case not in sufficient quantities.[61] The private sector may also refuse to provide an information service because market demand is insufficient: market demand is so limited that it is not worth the private sector's time and effort.[62] Other reasons for market failure may be that there is failure of competition. This does not necessarily mean that the spatial information service will not be offered, but it might only be offered at a higher cost and lower productivity than the government would want,[63] or endanger pluriformity.[64] Dijstelbloem and De Beer consider market failure to be present also when the socially preferred form of quality control cannot be installed via the market, for example, reliable and objective medical

59. Cf. *infra*, s. V.C.
60. Cf. *supra*, Ch. 6, s. IV.B.4.
61. Joseph E. Stiglitz, Peter R. Orszag & Jonathan M. Orszag, 'The role of government in a digital age', 2000, 32, <http://unpan1.un.org/intradoc/groups/public/documents/apcity/unpan002055.pdf>, 22 Feb. 2010. See also Kirsti Nilsen, 'Economic Theory as It Applies to Statistics Canada: A Review of the Literature', 2007, 16, <www.chass.utoronto.ca/datalib/misc/Nilsen%20Economics%20Paper%202007%20final%20version.pdf>, 22 Feb. 2010; Ravi Bedrijvenplatform, *Commercialisering van geo-informatie. Een drietal rapporten ter ondersteuning van de Ravi-discussie 1997* (Amersfoort: Ravi, 1997), 92 [in Dutch]; Colin Scott, 'Services of general interest in EC law: matching values to regulatory technique in the public and privatised sectors', *European Law Journal* 6, no. 4 (2000): 313; Kees Huisman & Jan Van De Lei, 'Wet FIDO en publieke taak onder Wet Dualisering', *Tijdschrift B&G* 2003: 34, [in Dutch]. Cf. *supra*, Ch. 6, s. IV.B.4.
62. See Bertrand Nouel, 'Commercialisation des données publiques et concurrence', paper given at the European Commission Conference: Access to Public Information. A Key to Commercial Growth and Electronic Democracy (Stockholm, 1996): 8 [in French]; Kees Huisman & Jan Van De Lei, 'Wet FIDO en publieke taak onder Wet Dualisering', *Tijdschrift B&G* 2003: 34, <www.bng.nl/bng/pdf/200302Huisman_34-35.pdf>, 22 Feb. 2010 [in Dutch]; Eric Van Damme, 'Concurrentievervalsing door de overheid?', *Openbare Uitgaven* 30, no. 4 (1999): 194–203 [in Dutch].
63. Joseph E. Stiglitz, Peter R. Orszag & Jonathan M. Orszag, 'The role of government in a digital age', 2000, 32, <http://unpan1.un.org/intradoc/groups/public/documents/apcity/unpan002055.pdf>, 22 Feb. 2010.
64. Huub Dijstelbloem & Paul De Beer, 'Een kwestie van selectie: de overheid en de informatievoorziening via internet', *Beleid en Maatschappij* 30, no. 4 (2003): 242–250, <www.pauldebeer.nl/documenten/artikelen/Een%20kwestie%20van%20selectie%201.7.doc>, 22 Feb. 2010 [in Dutch]; G. Heijne, 'Perspectieven op informatiepolitiek. Een verslag gebaseerd op vier workshops', in *Een kwestie van toegang. Bijdragen aan het debat over het publieke domein van de informatievoorziening*, ed. Inigo Baten & Jolien Ubacht (The Hague: Rathenau Instituut, 1995), 58 [in Dutch].

information without commercial purposes.[65] Whatever the reason for the market failure, it entails that the private sector is not willing or not able to offer particular spatial information services on the market, while the government feels that these information services should be available to the public.

Recognition of market failure as a reason for information activities – Market failure is repeatedly considered a reason for government intervention in the provision of information services. For instance, Bruguière feels that the public bodies should intervene to respond to the inadequacy of the market,[66] while Kabel et al. state that public bodies performing public tasks are apparently fulfilling a public need that cannot be fulfilled (completely) by the market, entailing that public funds are needed to provide the service.[67] Sears calls market failure a 'valid policy rationale' for the provision of information services by the public bodies.[68]

Recognition of market failure as a limit to information activities – Failure of the market has been stated as a reason for the public bodies to provide information services, but sometimes it is also used in order to *limit* the activities of the public bodies. Bruguière feels that public bodies can offer information services *only* if the market fails to do so, indicating that if the private sector can sufficiently cover the needs for information services, the government does not need to step in.[69] This point of view is also reflected in French policy. The Circular Letter of 14 February 1994 on the dissemination of public sector data imposes market failure as one of the necessary conditions for the public bodies to create information services.[70]

Evaluation: market failure does not always require government intervention – The market failure criterion also gives rise to some remarks. First, while it may be easy to determine that the market is not providing the spatial information services of which the public bodies feel that they should be offered to the public, the response that the public bodies should give to this market failure is not so clear. It has several options to ensure that the spatial information services are offered, and

65. Huub Dijstelbloem & Paul De Beer, 'Een kwestie van selectie: de overheid en de informatievoorziening via internet', *Beleid en Maatschappij* 30, no. 4 (2003): 242–250, <www.pauldebeer.nl/documenten/artikelen/Een%20kwestie%20van%20selectie%201.7.doc>, 22 Feb. 2010 [in Dutch].

66. Jean-Michel Bruguière, *Les données publiques et le droit* (Paris: Editions Litec, 2002), 117 [in French]; Jean-Michel Bruguière, 'Droit des données publiques', s.d., 35, 37, <www.educnet.education.fr/chrgt/droitddp.pdf>, 22 Feb. 2010 [in French].

67. Jan J.C. Kabel et al., *Kennisinstellingen en informatiebeleid. Lusten en lasten van de publieke taak* (Amsterdam: Universiteit van Amsterdam, 2001), 16 [in Dutch].

68. Garry Sears, 'Geospatial Data Policy Study. Executive Summary', 2001, 26, <www.geoconnections.org/programsCommittees/proCom_policy/keyDocs/KPMG/KPMG_E.pdf>, 22 Feb. 2010.

69. Jean-Michel Bruguière, 'Droit des données publiques', s.d., 35, <www.educnet.education.fr/chrgt/droitddp.pdf>, 22 Feb. 2010 [in French].

70. Circulaire du 14 février 1994 relative à la diffusion des données publiques, *Journal Officiel* 19 Feb. 1994, 2864, <http://admi.net/jo/PRMG9400081C.html>, 22 Feb. 2010 [in French].

they do not necessarily mean that it should provide the information service as part of its public task. For instance, in the case of concerns that the market is not ensuring pluriformity, reasonable prices, or sufficient quality, the government might impose regulations on the private sector ensuring that these conditions are fulfilled. If the government is concerned that the private sector will not create a spatial information service due to insufficient demand, it may award subsidies to the private sector for creating a service.[71] The government may also decide that the public bodies can or should offer the information service themselves, but that still does not mean that this service is part of the public body's public task. It may offer the service on the market, with or without competition from the private sector, or it may offer the information service as part of its public task. The decision between both is not only based on market failure, but rather on the value the public bodies give to the spatial information service in question.

EVALUATION: MARKET FAILURE IS NOT A CRITERION IN ITSELF – From the previous argument follows the second issue with the criterion of market failure: it is only part of the reason why the public bodies would choose to provide a spatial information service. The fact that the market does not offer a particular information service does not automatically mean that a public body should create such a service. The public body will have an underlying concern that this information service is needed to ensure certain societal needs or interests. As Joffe puts it: 'Government performs those services not adequately fulfilled by the private market to provide for the general welfare'.[72] Hence, the market's inadequacy is essentially not the reason for the activities of the public bodies, but rather the circumstance that determines whether a public body has to take care of the societal needs or interests itself or not. These societal needs or interests have taken on many different characteristics in the literature, which may be summarized under the term *public interest*. This will be discussed in the next section.

---

71. Gerrit-Jan Zwenne, 'De informatieconsument en de markt voor overheidsinformatie', *Tijdschrift voor Consumentenrecht* 1997: 231 [in Dutch]; Katleen Janssen, Jo Steyaert & Roland Van Gompel, *Transparantie van overheidsinformatie in Vlaanderen. Beschouwingen en aanbevelingen* (Bruges: Die Keure, 2003), 214 [in Dutch]; Inigo Baten & Gijs Van Der Starre, *Elektronische toegankelijkheid van overheidsinformatie* (The Hague: Rathenau Instituut, 1996), 28 [in Dutch]; Joseph E. Stiglitz, Peter R. Orszag & Jonathan M. Orszag, 'The role of government in a digital age', 2000, 62, 75, <http://unpan1.un.org/intradoc/groups/public/documents/apcity/unpan002055.pdf>, 22 Feb. 2010; Wetenschappelijke Raad voor het Regeringsbeleid, *Het borgen van publiek belang* (The Hague: Sdu Uitgevers, 2000), 49, <www.wrr.nl/content.jsp?objectid=2545>, 22 Feb. 2010 [in Dutch].

72. Bruce Joffe, 'The GIS Data Sales Dilemma – Finding a Middle Ground', 1995, <www.opendataconsortium.org/article_gis_data_sales_dilemma.htm>, 22 Feb. 2010. See also Gerrit-Jan Zwenne, 'De informatieconsument en de markt voor overheidsinformatie', *Tijdschrift voor Consumentenrecht* 1997: 230–231 [in Dutch]; Wetenschappelijke Raad voor het Regeringsbeleid, *Het borgen van publiek belang* (The Hague: Sdu Uitgevers, 2000), 49, <www.wrr.nl/content.jsp?objectid=2545>, 22 Feb. 2010 [in Dutch].

4. Public Interest

a. General

Point of departure for the public task – The point of departure for the activities of the public bodies can be seen as the public interest.[73] According to Baten and Van Der Starre, public bodies should actively disseminate their information and provide information services when it is in the interest of good and democratic policy.[74] Democratic rights are also mentioned as a clear reason for a public task by Corbin[75] and by Dommering et al.[76]

Examples of public interests – Baten and Van Der Starre give a number of examples of public interest for which the public bodies should provide an information service, such as the functioning of the democratic system, ensuring that the citizens know their rights and obligations, guaranteeing unity of the state, or guaranteeing the reliability of information.[77] Lamouline also includes the guarantee of continuity and of the neutral character of information,[78] while Nouel refers to the stimulation of diversity.[79] Dijstelbloem and De Beer believe that there should be a clear social issue for the public bodies to act upon, concerning information that is not only important for individual citizens, but that is also of community interest.[80] Bovens sees a continuum: the more the nature of the information

73. Bertrand Nouel, 'Commercialisation des données publiques et concurrence', paper given at the European Commission Conference: Access to Public Information. A Key to Commercial Growth and Electronic Democracy (Stockholm, 1996): 8 [in French]. See also Jean-Michel Bruguière, 'Droit des données publiques', s.d., 35, <www.educnet.education.fr/chrgt/droitddp.pdf>, 22 Feb. 2010 [in French]; Andrew Gamble & Tony Wright, 'Introduction', in *Restating the State? Special issue of The Political Quarterly*, ed. Andrew Gamble & Tony Wright (Malden, Blackwell Publishing, 2004), 1–10.
74. Inigo Baten & Gijs Van Der Starre, *Elektronische toegankelijkheid van overheidsinformatie* (The Hague: Rathenau Instituut, 1996), 29 [in Dutch].
75. Chris Corbin, 'What is the Public Task with respect to ECPs and HIPs', 2007, <www.epsiplus.net/forum/forum/is_there_a_need_to_define_public_task>, 22 Feb. 2010.
76. Egbert J. Dommering, P. Bernt Hugenholtz & Jan C. Kabel, 'De overheid en het publiek domein van informatie voor wetenschappelijk onderzoek', in *De publieke dimensie van kennis*, ed. Huub Dijstelbloem & Cornelius J.M. Schuyt (The Hague: WRR, 2002), 282 [in Dutch].
77. Inigo Baten & Gijs Van Der Starre, *Elektronische toegankelijkheid van overheidsinformatie* (The Hague: Rathenau Instituut, 1996), 23–26 [in Dutch]. See also Huub Dijstelbloem & Paul De Beer, 'Een kwestie van selectie: de overheid en de informatievoorziening via internet', *Beleid en Maatschappij* 30, no. 4 (2003): 242–250, <www.pauldebeer.nl/documenten/artikelen/Een%20kwestie%20van%20selectie%201.7.doc>, 22 Feb. 2010 [in Dutch].
78. Christian Lamouline, 'Presentation of PUBLAW 3 findings. Legal assessment', paper given at the workshop on commercial and citizen's access to government information (Luxembourg, 1995), 13.
79. Bertrand Nouel, 'Commercialisation des données publiques et concurrence', paper given at the European Commission Conference: Access to Public Information. A Key to Commercial Growth and Electronic Democracy (Stockholm, 1996): 9 [in French].
80. Huub Dijstelbloem & Paul De Beer, 'Een kwestie van selectie: de overheid en de informatievoorziening via internet', *Beleid en Maatschappij* 30, no. 4 (2003): 242–250, <www.pauldebeer.nl/documenten/artikelen/Een%20kwestie%20van%20selectie%201.7.doc>, 22 Feb. 2010 [in Dutch].

in question is crucial to the social functioning of citizens, the more active a stance the government will have to assume.[81] Longworth indicates a number of social interests that may justify the provision of information services by the public bodies, such as environmental concerns, public health, consumer protection, social capital, e-governance, and so on.[82]

PUBLIC INTEREST V. COMMUNITY INTEREST – In its report on how to guarantee the public interest, the Dutch Wetenschappelijke Raad voor het Regeringsbeleid (WRR) makes a distinction between *individual* interests, *community* interests and the *public* interest.[83] Interests move from individual to the community level when their protection is wanted for society as a whole, for example, trains that are running, streets that are lit, the fight against pollution or poverty, the ensurance that there is enough bread for everyone, and so on. The entire society benefits from these things, not just individuals. However, for many of these interests, the government is not needed to ensure them. If there is no scarcity, there will be enough bread. Trains will run and poverty will be fought even if the government is not involved. Yet, the WRR wonders whether the quality of the bread would be sufficient without any government intervention, or whether trains would still run at unprofitable hours of the day. Apparently, not all community interests are guaranteed without intervention of the government, or at least not guaranteed as well as they should. This is the case for public goods, for instance, or if redistribution measures are needed. Therefore, the WRR makes an explicit distinction between community interests and the public interest. There is a public interest when the government concerns itself with a certain community interest because it feels that this interest will not be guaranteed properly otherwise.[84] In other words, an interest of the community becomes a public interest because government makes it a policy goal to look after this interest.

EVALUATION: CIRCULAR REASONING – Along with Sap, Vermeulen, and Zoethout, I believe that this may be a circular definition that is comparable to the reasoning with regard to the legal basis for the public task. The government determines what a public task is, and a public task is defined as what task the government considers so

81. Mark Bovens, 'Information Rights: Citizenship in the Information Society', *The Journal of Political Philosophy* 10, no. 3 (2002): 331; Jan J.C. Kabel et al., *Kennisinstellingen en informatiebeleid. Lusten en lasten van de publieke taak* (Amsterdam: Universiteit van Amsterdam, 2001), 68 [in Dutch].

82. Elizabeth Longworth, 'The role of public authorities in access to information: the broader and more efficient provision of public content'. Study prepared for the Third UNESCO International Congress on Ethical, Legal and Societal Challenges of Cyberspace INFOethics, 2000, 19–21, <http://unesdoc.unesco.org/images/0012/001210/121051e.pdf>, 22 Feb. 2010.

83. Wetenschappelijke Raad voor het Regeringsbeleid, *Het borgen van publiek belang* (The Hague: Sdu Uitgevers, 2000), 19, <www.wrr.nl/content.jsp?objectid=2545>, 22 Feb. 2010 [in Dutch]; KPMG, 'Ministerie van Waterstaat. Evaluatie van de wet op het KNMI', 2007, 48, <www.verkeerenwaterstaat.nl/Images/2008651%20bijlage_tcm195-231492.pdf>, 22 Feb. 2010 [in Dutch].

84. Wetenschappelijke Raad voor het Regeringsbeleid, *Het borgen van publiek belang* (The Hague: Sdu Uitgevers, 2000), 19, <www.wrr.nl/content.jsp?objectid=2545>, 22 Feb. 2010 [in Dutch].

important that it should deliver it itself. Merusk believes that '[p]ublic tasks may be treated as objectively existing tasks determined by public interests, which have been sensed as such by the legislator and stipulated as tasks of the state or given for fulfilment to different institutions of the society or voluntarily taken for fulfilment by the latter'.[85]

CRITICISM: POLITICAL QUESTION – In discussing the content of this public interest, the WRR immediately warns that this is essentially a political question which cannot be answered by scientific or theoretical reasoning alone.[86] It is in the process of political decision making that certain interests are indicated as public. To ensure legal certainty, the public interest should be formulated as specifically as possible, since vagueness will only lead to uncertainty. However, the WRR admits that it is impossible to provide full certainty, and in addition, there may be multiple public interests that may or may not contradict each other.[87] Hence, the criterion of public interest is very important to determine the public task of the public bodies to provide spatial information services, yet its general and vague character does not make this determination of the task much easier.

EXAMPLE: ACCESSIBILITY – One of the main expressions of the public interest with regard to the provision of information services is accessibility. In order for the citizens to fully benefit from information that should ensure their democratic rights or their social needs, they not only need the information to be available, but they also need it to be accessible. Such accessibility can often only be provided by processing public sector spatial data beyond the public bodies' own needs to perform their own tasks. Hence, accessibility is one of the main reasons of public interest – or at least a means to achieve the reasons of public interest – for a public task of providing spatial information services. The next section addresses the requirements for the public bodies to make their information accessible through the provision of information services.

b. Concretization of Public Interest: Accessibility

CONCRETIZATION OF PUBLIC INTEREST – Accessibility could be considered the concretization of the public interest need that should be strived for by public bodies with the provision of spatial information services. The public bodies should not only make their information available, but they should also ensure that this information is accessible. A certain level of accessibility to all citizens must be guaranteed by means of open and objective criteria.[88] Accessibility entails that the

85. Kalle Merusk, 'Assignment of public tasks to private legal persons. Legal bases and limits', *Juridica International* 5 (2000): 41, <www.juridica.ee/get_doc.php?id=327>, 22 Feb. 2010.
86. Wetenschappelijke Raad voor het Regeringsbeleid, *Het borgen van publiek belang* (The Hague: Sdu Uitgevers, 2000), 46, <www.wrr.nl/content.jsp?objectid=2545>, 22 Feb. 2010 [in Dutch].
87. *Ibid.*, 55–56.
88. Bertrand Nouel, 'Commercialisation des données publiques et concurrence', paper given at the European Commission Conference: Access to Public Information. A Key to Commercial Growth and Electronic Democracy (Stockholm, 1996): 9 [in French].

citizens can use information services with a relatively small effort[89] and with minimal barriers.[90] Yet, one should keep in mind that this accessibility is not a goal in itself, but should also improve the possibility of the citizens to pursue their other needs and interests.

OUTLINE – In this section I first describe the aspects of accessibility and then move on to the dilemma the public bodies face in ensuring accessibility.

i. Aspects of Accessibility

OVERVIEW OF ASPECTS – Accessibility is a concept that has several aspects. Beers names three: the legal availability of information, the practical accessibility, and the usability.[91] Bovens discerns comparable categories. First, the information has to be physically accessible. Second, the financial aspect requires that the cost of access should not bar large groups from such access. The third aspect relates to intellectual accessibility: the information must be well-organized and understandable, for instance by including hyperlinks, references, and search engines.[92] As for the intellectual accessibility, Bovens recognizes that this is 'on the one hand, a question of general literacy, basic education and computer training, on the other hand one of user friendliness and operating ease'.[93] While the former also depends on the addressee of the information service, the latter is in the hands of the public body. Baten and Van Der Starre make a more detailed distinction between five aspects of accessibility[94]:

- access to the medium on which the information is made available;
- the possibility or the means to find the information: the citizen should be able to find out that the information exists, where it is and how it can be found;

89. Harry Bouwman et al., 'Leuker kunnen we het niet maken, wel makkelijker. Over de toegankelijkheid van overheidsinformatie', in *Elektronische toegankelijkheid van overheidsinformatie*, ed. Inigo Baten & Gijs Van Der Starre (The Hague: Rathenau Instituut, 1996), 87 [in Dutch].
90. Katleen Janssen, Jo Steyaert & Roland Van Gompel, *Transparantie van overheidsinformatie in Vlaanderen. Beschouwingen en aanbevelingen* (Bruges: Die Keure, 2003), 35 [in Dutch].
91. Anton A.L. Beers, 'Openbaarheid van overheidsinformatie', in *Elektronische toegankelijkheid van overheidsinformatie*, ed. Inigo Baten & Gijs Van Der Starre (The Hague: Rathenau Instituut, 1996), 55 [in Dutch].
92. Mark Bovens, 'Information Rights: Citizenship in the Information Society', *The Journal of Political Philosophy* 10, no. 3 (2002): 330.
93. *Ibid.*, 334.
94. Inigo Baten & Gijs Van Der Starre, *Elektronische toegankelijkheid van overheidsinformatie* (The Hague: Rathenau Instituut, 1996), 21 [in Dutch]. Several other categorizations of the aspects of accessibility can be found, all more or less including the same facets. Bouwman et al. discern physical availability, financial, technical and cognitive accessibility (Harry Bouwman et al., 'Leuker kunnen we het niet maken, wel makkelijker. Over de toegankelijkheid van overheidsinformatie', in *Elektronische toegankelijkheid van overheidsinformatie*, ed. Inigo Baten & Gijs Van Der Starre (The Hague: Rathenau Instituut, 1996), 87 [in Dutch]). Dijstelbloem and De Beer refer to physical connection, skills to use computers and software and the ability to make meaningful use of the information, and unlocking the information (Huub Dijstelbloem & Paul De Beer, 'Een kwestie van selectie: de overheid en de informatievoorziening via internet', *Beleid en Maatschappij* 30, no. 4 (2003): 242–250, <www.pauldebeer.nl/documenten/artikelen/Een%20kwestie%20van%20selectie%201.7.doc>, 22 Feb. 2010

- the intelligible character of the information: the citizen has to be able to understand the information;
- the intelligible presentation of the content: this includes the form of access to the information, for example, the presentation, formatting, style, etc.;
- the price for consulting the information.

SUBJECTIVE CHARACTER OF ACCESSIBILITY – The accessibility demands that are put on an information service are determined by the target group the public body intends to reach. This shows the subjective character of accessibility.[95] It depends on the practical or technological means of access that are available to the target group, the level of education and skills, or the financial resources available to pay the possible charges for the information service. These differences can lead to a significant difference between the *information haves* and the *information have-nots*. These terms are mostly used in relation to the general digital divide debate, but they also apply to the specific discussion on the provision of accessible information services by the public bodies.[96] A lot of public sector data are only accessible for the highly

---

[in Dutch]). According to Bauler and Hecq, accessibility includes availability, comprehensiveness, diffusion, and potential for feedback (Tom Bauler & Walter Hecq, 'Some theoretical considerations in response to the claim after 'information for decision-making', paper given at the International Sustainable Development Research Conference, Nottingham, 2000, 2, <www.ulb.ac.be/igeat/igeat/ulb_igeat/upload/cedd/Bauler_leeds-paper00.pdf>, 22 Feb. 2010. The Dutch Ministry of the Interior lists six facets: the possibility to find information, the availability of the medium, manageability of the information, affordability, reliability and intelligibility (Ministry of the Interior, 'Naar toegankelijkheid van overheidsinformatie', 1997, 21, <www.minbzk.nl/aspx/download.aspx?file=/contents/pages/302/nota_ntvo_nl.pdf>, 22 Feb. 2010 [in Dutch]. Nouel indicates accessible information as information that is easily retrievable and obtainable for the public, without delays, and without extensive costs (Bertrand Nouel, 'Commercialisation des données publiques et concurrence', paper given at the European Commission Conference: Access to Public Information. A Key to Commercial Growth and Electronic Democracy (Stockholm, 1996): 8 [in French].

95. *Ibid.*
96. The United Nations Development Programme (UNDP) stated in 1999 that '[t]hose with income, education and – literally – connections have cheap and instantaneous access to information. The rest are left with uncertain, slow and costly access. When people in these two worlds live and compete side-by-side, the advantage of being connected will overpower the marginal and impoverished, cutting off their voices and concerns from the global conversation'. (UNDP, *Human Development Report 1999* (New York: Oxford University Press, 1999), 6. See Elizabeth Longworth, 'The role of public authorities in access to information: the broader and more efficient provision of public content'. Study prepared for the Third UNESCO International Congress on Ethical, Legal and Societal Challenges of Cyberspace INFOethics, 2000, 29–30, <http://unesdoc.unesco.org/images/0012/001210/121051e.pdf>, 22 Feb. 2010). The digital divide is a very broad term and represents a problem with many different facets, encompassing a wide spectrum of disparities based on race, gender, age, income, education, type of household, geographic location, physical abilities, and the level of economic development. The key goal of any digital divide policy should be to ensure that each person has meaningful access to information, via appropriate information and communication technologies (Peter K. Yu, 'Symposium – Bridging the Digital Divide: Equality in the Information Age. Introduction', *Cardozo Arts and Entertainment Law Journal* 20, no. 7–8 (2002): 32).

educated, leaving the majority of citizens unreached. While this problem has to be tackled by the public bodies themselves, an important role also lies with intermediary institutions such as the press, education, interest groups, and broadcasting corporations.[97]

ii. Measures to Ensure Accessibility

WAYS OF ENSURING ACCESSIBILITY – Information should be sufficiently accessible for everyone across the different facets. The government has a duty to ensure that information is as intelligible as possible for the ordinary citizen in order that the content of the message is understood.[98] There are several ways of doing this. All the necessary and relevant information can be collected and put in a logical order. The information can also be put in a broader context, leaving out unnecessary details; or be put in clear, simple, and understandable wording, etc.[99]

FORMAT AND MEDIA – The different aspects of accessibility that have been mentioned require different approaches from the public bodies. With respect to access to the media and the technological means to obtain the information services, initiatives are needed to include all citizens in the Information Society via the proliferation of information and communication technology, the Internet, and broadband access. Such initiatives have, for instance, been taken by the European Commission under the e-Accessibility initiative and the e-Inclusion campaign.[100] Next, making the presentation of the content more accessible can, for example, be done by following the World Wide Web Consortium's Web Content Accessibility Guidelines, which include suggestions on making websites easy to navigate, using symbols, or certain prints, etc.[101]

CONTENT – How to ensure accessibility of the content seems more difficult, as there are very few objective standards to measure this against. Creating a structured database with a search engine for citizens to find applicable legislation or find maps of their home area might easily be considered as a task for the public bodies. Is this also the case for creating graphs or figures representing measurements on air pollution, noise levels, or flood risks; or for summaries or explanations of these data or graphs; or for the combination of these data for a particular house or building? Do such services still fall under the denominator of increasing

97. Ministry of the Interior, 'Naar toegankelijkheid van overheidsinformatie', 1997, 21, retrieved from <www.minbzk.nl/aspx/download.aspx?file=/contents/pages/302/nota_ntvo_nl.pdf>, 22 Feb. 2010 [in Dutch]; Katleen Janssen, Jo Steyaert & Roland Van Gompel, *Transparantie van overheidsinformatie in Vlaanderen. Beschouwingen en aanbevelingen* (Bruges: Die Keure, 2003), 191 [in Dutch].

98. See Anton A.L. Beers, 'Openbaarheid van overheidsinformatie', in *Elektronische toegankelijkheid van overheidsinformatie*, ed. Inigo Baten & Gijs Van Der Starre (The Hague: Rathenau Instituut, 1996), 64 [in Dutch].

99. Katleen Janssen, Jo Steyaert & Roland Van Gompel, *Transparantie van overheidsinformatie in Vlaanderen. Beschouwingen en aanbevelingen* (Bruges: Die Keure, 2003), 35 [in Dutch].

100. See Commission of the European Communities, 'e-Inclusion', <http://ec.europa.eu/information_society/activities/einclusion/index_en.htm>, 22 Feb. 2010.

101. World Wide Web Consortium, 'Web Content Accessibility Guidelines (WCAG) Overview', <www.w3.org/WAI/intro/wcag.php>, 22 Feb. 2010.

accessibility? And is such a level of accessibility required from the public bodies? Is it sufficient that the average citizen can easily find and understand the public sector spatial data, or should the public bodies ensure that *all* citizens find the data comprehensible? While there may be some accessibility obligations that can clearly be considered to fall under the public tasks, others are much more disputable.

iii. The Dilemma in Ensuring Accessibility

OBJECTIVE INFORMATION V. SUBJECTIVE MANIPULATION – In making sure that the public information service is accessible to everyone, the public body will explain it, put it into a context, summarize it, simplify it, and so on. In doing so, the public body has to make sure that it does not harm the accuracy or the reliability of the information in any way, because it runs the risk of being reproved for using its information services to justify its policy choices.[102] Baten and Van Der Starre feel that the public sector faces a dilemma: if it does not act to make its information accessible, it will be criticized for the inaccessibility. If it does make its information more understandable, it will be accused of trying to influence the public.[103] As the limit between objective information and subjective manipulation is not always clear, the public should be able to verify the objectivity of the information by also having access to the public sector spatial data on which the information service is based.[104]

iv. Link with the Discussion on Raw Data and Added Value

ADDING VALUE FOR INCREASING ACCESSIBILITY – Chapter 4 already referred to the discussion on the distinction between raw data and added value information. This debate can be linked to the issue of accessibility, as making information accessible to the public will in most cases require adding value to the public sector spatial data beyond the public bodies' own needs to perform their tasks. As Newbery, Bently, and Pollock describe it, value-added information is 'information where value is added to raw data enhancing and facilitating its use and effectiveness for the user, for example through further manipulation, compilation and summarization into a more convenient form for the end-user, editing and/or further analysis and interpretation'.[105] Added value can take different forms. Mandelkern distinguishes between intellectual value through commenting and systematizing; documentary value consisting of the correlation of data or the compilation with other data; and purely technical or economic value when the information is presented on a

102. Katleen Janssen, Jo Steyaert & Roland Van Gompel, *Transparantie van overheidsinformatie in Vlaanderen. Beschouwingen en aanbevelingen* (Bruges: Die Keure, 2003), 191 [in Dutch].
103. Inigo Baten & Gijs Van Der Starre, *Elektronische toegankelijkheid van overheidsinformatie* (The Hague: Rathenau Instituut, 1996), 21 [in Dutch].
104. *Ibid.* See also Dag Wiese Schartum, 'Access to Government-Held Information: Challenges and Possibilities', *Journal of Information, Law and Technology* no. 1 (1998): 12, <www2.warwick.ac.uk/fac/soc/law/elj/jilt/1998_1/schartum/>, 22 Feb. 2010.
105. David Newbery, Lionel Bently & Rufus Pollock, 'Models of Public Sector Information Provision via Trading Funds', 2008, 10, <www.berr.gov.uk/files/file45136.pdf>, 22 Feb. 2010.

different carrier or adapted to a particular mode of communication or preservation method, without any intellectual or documentary ordering.[106] This is a comparable categorization to the aspects of accessibility.

VALUE CHAIN – The OECD sees different steps in the value chain of public sector spatial data. The first step is the creation of the data themselves as single data or information items. Next come aggregation and organization of numerous items of data in order to produce a complete or at least more comprehensive set of data. The third element of the value chain comprises functions such as data processing, editing, re-packaging, or remodelling. Finally, the fourth element includes marketing activities such as targeting the right customers, distribution, and delivery of the information services.[107]

RELATIVE CHARACTER OF ADDED-VALUE – Denying the public bodies the possibility to add value could prevent them from enhancing accessibility. Moreover, just like the concept of accessibility, added-value can mean different things to different target groups. What contains added-value for one person may still be raw data for another. Therefore, this distinction, sometimes referred to as dogmatic[108], should not be taken up in the debate on whether the public bodies should be allowed to provide spatial information services as part of their public task because it is not a sound, definite criterion. The United Kingdom's OPSI explains: 'It is reasonable to assume that, where documents most closely relate to [raw data], they will fall firmly within a public body's public task. Value-added information, however, is not automatically outside the public task, though it would be a reasonable working assumption that it was unless there are other persuasive factors arguing otherwise. At either end of the spectrum this is likely to be an easy decision; but there will, however, inevitably be grey areas'.[109]

SUMMARY – In summary, as long as the accessibility requirements are not made more concrete by detailed demands on the public bodies, it is hard to determine the scope of the public task in this issue. However, setting out such detailed demands is only possible when the government knows the needs of the citizen. As 'the citizen' does not exist, these needs will be different depending on so many elements that it is virtually impossible for the government to determine how far it should go in

106. Dieudonné Mandelkern & Bertrand Du Marais, 'Diffusion des données publiques et révolution numérique: rapport', 1999, 14–15, <www.ladocumentationfrancaise.fr/rapports-publics/994001620/index.shtml#>, 22 Feb. 2010 [in French]. See also Jean-Michel Bruguière, *Les données publiques et le droit* (Paris: Editions Litec, 2002), 40–42 [in French].

107. OECD Working Party on the Information Economy, 'Digital Broadband Content: Public Sector Information and Content', 2006, 10, <www.oecd.org/dataoecd/10/22/36481524.pdf>, 22 Feb. 2010.

108. See Johan Pas, 'The Commercialization of Government Information and the Proposal for a Directive COM (2002) 207 by the European Commission', *Murdoch University Electronic Journal of Law* 9, no. 4 (2002), <www.murdoch.edu.au/elaw/issues/v9n4/pas94_text.html>, 22 Feb. 2010.

109. Office of Public Sector Information, 'The Re-use of Public Sector Information: A Guide to the Regulations and Best Practice', 2005, 13, <www.opsi.gov.uk/advice/psi-regulations/advice-and-guidance/guide-to-psi-regulations-and-best-practice.doc>, 22 Feb. 2010.

providing accessible spatial information services. This is linked to the so-called dilemma of the public task: the public bodies should make their information accessible, but if they go too far they enter the market and compete with the private sector.

v. Dilemma of the Public Task

ORIGIN – The term *dilemma of the public task* was first used by the Dutch Ministry of the Interior in its policy paper Towards the Accessibility of Public Sector Information.[110] The policy paper described this dilemma as follows: when the public bodies provide access to the citizens to the public sector data they hold, these data have to be processed to make them suitable for the users. However, the public bodies should also stick to their public tasks and not process the data they hold beyond their own needs. This means that they cannot make the data available to the citizens in an accessible way. As a consequence, services that can be useful to society are not made because the public bodies are not allowed to make them and the private sector might not be willing or able to make them.[111]

BASIC PRINCIPLE – The basic assumption of the dilemma theory seems to be that public bodies that process their data to make them accessible for the public move outside of the scope of their public task. The public bodies should limit themselves to providing their data under the national freedom of information legislation. This is a very narrow view of the public task to provide information services.

BROADER CONTEXT – However, the dilemma of the public task can also be described in a wider sense. According to Burkert, the public bodies are caught between two poles. On the one hand, in the restrained model, the public sector develops and uses its information only in the framework of its public tasks and only to obtain its own goals. The information is only made available upon request, as is, and for charges that are determined by legislation or administrative law. In the luxuriant model, the public sector defines its role more extensively and it performs innovative activities and enters the market. In the former situation, the public sector is criticized for not delivering enough useful information against a cheap price. In the latter, the industry feels threatened and calls for subsidiarity of the state.[112] Dommering et al. speak about a deadlock: processing information is needed from a democratic viewpoint, but if the government wants to do it properly,

110. Ministry of the Interior, 'Naar toegankelijkheid van overheidsinformatie', 1997, 12, <www.minbzk.nl/aspx/download.aspx?file=/contents/pages/302/nota_ntvo_nl.pdf>, 22 Feb. 2010 [in Dutch]. See also Marijke Reinsma & H. Van Der Sluijs, 'Naar ruimere openbaarheid en een vrij gebruik van bestuurlijke informatie', 2002, 75–76, <www.minbzk.nl/aspx/download.aspx?file=/contents/pages/5006/RapportReinsmaenVanderSluijs.pdf>, 22 Feb. 2010 [in Dutch].

111. See Bastiaan Van Loenen et al., 'Open toegankelijkheidsbeleid voor geo-informatie vergeleken: het gras leek groener dan het was?', 2007, 7, <www.minbzk.nl/aspx/download.aspx?file=/contents/pages/87285/opentoegankelijkheidsbeleidvooroverheidvergeleken.pdf>, 22 Feb. 2010 [in Dutch].

112. Herbert Burkert, 'L'information du secteur public. Le secret, la transparence et le commerce', *Revue française d'administration publique* 72 (1994): 585 [in French].

it produces information services that can also be created by the private sector and enters into competition with the market.[113]

CONSEQUENCES – In a study for the European Commission on the sectors of geographic, meteorological, and legal information, Micus Management Consulting found that the offer of accessible information services by public bodies has 'in some cases made the business models of providers redundant in that information users turn to the services provided for free by publicly funded bodies rather than those of commercial service providers. However, new growth opportunities and innovation in other services compensate for this loss of economic activity. Society generally benefits from the wider dissemination and diffusion of legal and administrative information and these benefits also manifest themselves in terms of general growth in the economy'.[114]

EVALUATION – This demonstrates the difficulty of solving this dilemma of the public task. As it is difficult to apply sufficiently detailed criteria to determine whether the public bodies should make their information accessible through the provision of information services or that they are providing a service that should be done by the market, the only solution is to decide on a case-by-case basis. However, this creates considerable uncertainty for all stakeholders.

c. *Concretization of Public Interest: Types or Target Groups of Information Services*

ESSENTIAL INFORMATION? – The question can be asked of whether there are types of information for which an information service should be delivered by the public bodies. Again, this can be related to the public interest, which can include public participation, the protection of democratic or social rights, public safety, the protection of the environment, and so on. According to the Mandelkern report, such essential data should be presented in an easily and universally accessible form, that is,

113. Egbert J. Dommering, P. Bernt Hugenholtz & Jan C. Kabel, 'De overheid en het publiek domein van informatie voor wetenschappelijk onderzoek', in *De publieke dimensie van kennis*, ed. Huub Dijstelbloem & Cornelius J.M. Schuyt (The Hague: WRR, 2002), 288 [in Dutch]. See also Jan J.C. Kabel et al., *Kennisinstellingen en informatiebeleid. Lusten en lasten van de publieke taak* (Amsterdam: Universiteit van Amsterdam, 2001), 60 [in Dutch]; Corien Prins, 'Study on a Green Paper regulating Public Access to Government Information', paper given at the workshop on commercial and citizen's access to government information (Luxembourg, 1995), 37; Maeve McDonagh, 'Access to public sector information: the Australian experience', paper given at the European Commission Conference: Access to Public Information. A Key to Commercial Growth and Electronic Democracy (Stockholm, 1996): 2; Johan Pas, 'The Commercialization of Government Information and the Proposal for a Directive COM (2002) 207 by the European Commission', *Murdoch University Electronic Journal of Law* 9, no. 4 (2002), <www.murdoch.edu.au/elaw/issues/v9n4/pas94_text.html>, 22 Feb. 2010.

114. Micus Management Consulting Gmbh, 'Assessment of the Re-use of Public Sector Information (PSI) in the Geographical Information, Meteorological Information and Legal Information Sectors', 2008, 93–94, <http://ec.europa.eu/information_society/policy/psi/docs/pdfs/micus_report_december2008.pdf>, 22 Feb. 2010.

provided under the form of an information service as it was defined for the purpose of my analysis.[115] However, for some types of information, it may be easier to see a clear task for the public bodies, while for others, the public interest is not deemed so important.[116]

LEGAL AND ADMINISTRATIVE INFORMATION – The domain in which accessible information services are generally considered to be needed is the domain of legal and administrative information. Given the adagio that everyone is supposed to know the law, this law has to be easily accessible to everyone through databases with appropriate search possibilities, presentation and indexing. At the same time, one must admit that this view is fairly recent. Beers believes that considering the current state of technology the government can be expected to provide an accessible, up-to-date full text database of all legislation, while in the past this was not seen as a public task for the public bodies, but as a private sector activity.[117] Moreover, while it seems clear that legislation should be made accessible, is this also the case for jurisprudence, or literature explaining the legislation? This is already more difficult to judge. With regard to administrative information, services that are often mentioned are explanatory material on the activities and policies of the public bodies and on the rights citizens have and how they can exercise those rights.[118]

OTHER DOMAINS – For other categories of information, it is more difficult to state which information services should be made available. Within each category, some information will be more important for the citizens than other types, and this will depend on the needs of the particular circumstances that a citizen finds himself or herself in. In addition, some services may be important for some subgroups of citizens, while they are only nice to have for others. For instance, detailed forecasts on precipitation will be considered essential by farmers or by people living in areas prone to flooding, but less so by others. Information on health care, unemployment, education, and so on are all directed towards certain target groups. How big should these target groups be for a public body to consider making an information service in order to make information accessible for this target group? This is why legal and administrative information services are easily considered as part of the public task: every citizen needs them at a certain point. This may also be the case for general information on the weather or on traffic jams, but less so for long-term forecasts of sales of cookbooks or the history of the purchasing policy of army tents.

115. Dieudonné Mandelkern & Bertrand Du Marais, 'Diffusion des données publiques et révolution numérique: rapport', 1999, 96, <www.ladocumentationfrancaise.fr/rapports-publics/994001620/index.shtml#>, 22 Feb. 2010 [in French].

116. See George Papapavlou, 'Minutes', paper given at the workshop on commercial and citizen's access to government information (Luxembourg, 1995), 53.

117. Anton A.L. Beers, *Informatica Publica: publieke toegankelijkheid van elektronische overheidsinformatie* (The Hague: Rathenau Instituut, 1996), 10 [in Dutch]; Anton A.L. Beers, 'Openbaarheid van overheidsinformatie', in *Elektronische toegankelijkheid van overheidsinformatie*, ed. Inigo Baten & Gijs Van Der Starre (The Hague: Rathenau Instituut, 1996), 64–65 [in Dutch].

118. Cf. *supra*, Ch. 7, s. II.

The average citizen does not need to know this, but cookbook publishers or tent manufacturers might. Such specific information services are much harder to conceive as a public task of the public bodies.

SPECIALIZED SERVICES – Stiglitz et al. warn that caution should be exercised in providing information services for a targeted or specialized audience for a different reason, that is, the possible competition with the private sector. For instance, the government should provide search engines, but more specialized tasks, such as linking official information to related academic articles, should be left to the private sector.[119] This issue will be addressed later on in the chapter.

d. Concretization of Public Interest: Universal Service

OUTLINE – Some authors have suggested that the concept of universal service that was developed in the telecommunications sector can also be applied to information services.[120] This concept can also be seen as a concretization of public interest, as universal service is seen as a minimum service that has to be provided in order to ensure particular public interests. This section examines briefly how this relates to the issue of the public task of public bodies in the EU legal framework for the availability of public sector spatial data.

i. The Concept and Characteristics of Universal Service

GENERAL PRINCIPLE – The concept of universal service refers to services that are considered so essential by society that everyone should have permanent access to them under reasonable conditions, both from the perspective of quality and from a financial point of view.[121] In its legislative activities with regard to the

119. Joseph E. Stiglitz, Peter R. Orszag & Jonathan M. Orszag, 'The role of government in a digital age', 2000, 58–59, <http://unpan1.un.org/intradoc/groups/public/documents/apcity/unpan002055.pdf>, 22 Feb. 2010; Kirsti Nilsen, 'Economic Theory as It Applies to Statistics Canada: A Review of the Literature', 2007, 72, <www.chass.utoronto.ca/datalib/misc/Nilsen%20Economics%20Paper%202007%20final%20version.pdf>, 22 Feb. 2010.

120. Commission of the European Communities, *Publaw 3. Final report* (Luxembourg: CEC Legal Advisory Board, 1995), 82–92; Yves Poullet, 'Towards a universal public information service', paper given at the workshop on commercial and citizen's access to government information (Luxembourg, 1995), 15–28; Yves Poullet, 'Freedom and information highways or how to ensure electronic democracy', *Telematics and Informatics* 15 (1998): 163–180; Yves Poullet, 'Plaidoyer pour un ou des service(s) universel(s) d'informations publiques', paper given at the European Commission Conference: Access to Public Information. A Key to Commercial Growth and Electronic Democracy (Stockholm, 1996): 1–19 [in French]; Cécile De Terwangne, *Société de l'information et mission publique d'information*, Doctoral Thesis FUNDP (Namur: Faculté Universitaire de Notre Dame de Paix, 2001), 440–452 [in French]. For an overview of the concept of universal services in European telecommunications policy, see Alexandre De Streel & Robert Queck, 'Services d'intérêt économique général et communications électroniques', in *Les services d'intérêt économique général et l'Union européenne*, ed. Jean-Victor Louis & Stephane Rodrigues (Brussels: Bruylant, 2006), 339–384 [in French].

121. Peggy Valcke, *Digitale diversiteit. Convergentie van Media-, Telecommunicatie- en Mededingingsrecht* (Brussels: Larcier, 2004), 409 [in Dutch]; Iris Houben, 'Public service obligations: moral counterbalance of technical liberalization legislation', *European Review of*

telecommunications market, the European Commission defined it as 'a minimum set of services of specified quality which is available to all users independent of their geographic location and, in the light of specific national conditions, at an affordable price'.[122] Universal service can be seen as a safeguard against exclusion and as a means to correct the unrestrained free market: even when purely economic considerations would cause all market players to stop offering a particular service to a certain category of users, the government might compel one supplier to offer the services anyway, because it is deemed beneficial for society as a whole.[123]

CHARACTERISTICS – We can discern the following main characteristics of the universal service:

- Universality/affordability: all users must have access to the universal service at a reasonable price, even though this means that no profit will be

---

Private Law 16 (2008): 11–12; Mark Young, 'The future of universal service. Does it have one?', *International Journal of Law and Information Technology* 13, no. 2 (2005) (190–191;) Pascal Verhoest, 'The myth of universal service: hermeneutic considerations and political recommendations', *Media, Culture & Society* 22, no. 5 (2000): 599; Wouter Devroe, 'Universele dienstverlening als nieuwe manier van denken', *Sociaal-economische Wetgeving* 2000: 82–95 [in Dutch]; Alison Jones & Brenda Sufrin, *EC Competition Law: Text, Cases and Materials* (Oxford: Oxford University Press, 2008), 622; Colin Scott, 'Services of general interest in EC law: matching values to regulatory technique in the public and privatised sectors', *European Law Journal* 6, no. 4 (2000): 314; Alexandre De Streel & Robert Queck, 'Services d'intérêt économique général et communications électroniques', in *Les services d'intérêt économique général et l'Union européenne*, ed. Jean-Victor Louis & Stephane Rodrigues (Brussels: Bruylant, 2006), 346 [in French]; Wolf Sauter, 'Services of general economic interest and universal service in EU law', *European Law Review* 33, no. 2 (2008): 176–177.

122. Article 2.1.g of Directive 97/33/EC of the European Parliament and of the Council of 30 Jun. 1997 on interconnection in telecommunications with regard to ensuring universal service and interoperability through application of the principles of Open Network Provision (ONP), *OJ L* 199, 26 Jul. 1997, 32; Art. 2.1.f. of Directive 98/10/EC of the European Parliament and of the Council of 26 Feb. 1998 on the application of open network provision (ONP) to voice telephony and on universal service for telecommunications in a competitive environment, *OJ L* 101, 1 Apr. 1998, 24. See also Commission of the European Communities, *Communication. Services of General Interest in Europe*, COM (1996) 443 final, 11 Sep. 1996. At the time, the minimum set of services included access to the fixed telephone network at a fixed location; access to fixed public telephone services enabling users to make and receive national and international calls, supporting speech, facsimile and/or data communications; directory services; public pay phones; and certain measures for disabled users and users with special social needs. In the new 2002 Universal Service directive, the following list was included: access location to the public telephone network; access to publicly available telephone services at a fixed location to enable end-users to make and receive local, national, and international telephone calls, as well as facsimile and data communications; directory services; public pay telephones; and certain specific measures for disabled users such as those with low income or special social needs (Directive 2002/22/EC of the European Parliament and of the Council of 7 Mar. 2002 on universal service and users' rights relating to electronic communications networks and services (Universal Service Directive), *OJ L* 108, 24 Apr. 2002, 51).

123. Wouter Devroe, 'Universele dienstverlening als nieuwe manier van denken', *Sociaal-economische Wetgeving* 2000: 83 [in Dutch]; Peggy Valcke, *Digitale diversiteit. Convergentie van Media-, Telecommunicatie- en Mededingingsrecht* (Brussels: Larcier, 2004), 409 [in Dutch].

made on the delivery of the service to some users.[124] According to Poullet, the prices must be progressively geared towards costs, without this excluding a reasonable profit.[125] In any case, the prices have to be affordable, but not necessarily the same for everyone. Differentiation is possible within the limits of the affordable.[126]

- Equality: all users must have access regardless of their geographical location.
- Continuity: a continuous service of a defined quality must be guaranteed.[127]
- Proportionality: according to Poullet, a proportionality principle should be added. The information would have to be made available to the citizens using the same techniques as those employed internally with the public body's own agents.[128] However, this would only work if we assume that the public bodies will provide the universal services themselves. If this is not the case, it seems too demanding to require the private sector producers of the universal service to create a service that the citizens might not need or that might be very expensive to produce, just because the same service has to be available as the one that the public body uses internally.

Minimum obligations – On the basis of these principles, the government must analyze the social needs and the expectations of society to determine the scope of the universal service.[129] The European Commission sees the universal service as a minimum package. Hence, only the services can be selected that are essential for the citizens to participate in a democratic society. Devroe indicates the difficulty of this decision, given the plethora of forms, varieties, and quantities in which

124. Peggy Valcke, *Digitale diversiteit. Convergentie van Media-, Telecommunicatie- en Mededingingsrecht* (Brussels: Larcier, 2004), 409 [in Dutch].
125. Yves Poullet, 'Freedom and information highways or how to ensure electronic democracy', *Telematics and Informatics* 15 (1998): 178; Yves Poullet, 'Towards a universal public information service', paper given at the workshop on commercial and citizen's access to government information (Luxembourg, 1995), 81, 13.
126. Wouter Devroe, 'Universele dienstverlening als nieuwe manier van denken', *Sociaal-economische Wetgeving* 2000: 83 [in Dutch].
127. Yves Poullet, 'Towards a universal public information service', paper given at the workshop on commercial and citizen's access to government information (Luxembourg, 1995), 15–28.
128. Yves Poullet, 'Freedom and information highways or how to ensure electronic democracy', *Telematics and Informatics* 15 (1998): 176. See also Simone Van Der Hof, *Overheidsinformatie in de etalage. Belangen rondom de toegang tot overheidsinformatie* (Alphen a/d Rijn: Samsom Bedrijfsinformatie, 1997), 18–19 [in Dutch]; Alexandre De Streel & Robert Queck, 'Services d'intérêt économique général et communications électroniques', in *Les services d'intérêt économique général et l'Union européenne*, ed. Jean-Victor Louis & Stephane Rodrigues (Brussels: Bruylant, 2006), 357 [in French].
129. Cécile De Terwangne, *Société de l'information et mission publique d'information*, Doctoral Thesis FUNDP (Namur: Faculté Universitaire de Notre Dame de Paix, 2001), 442 [in French]. See also Yves Poullet, 'Freedom and information highways or how to ensure electronic democracy', *Telematics and Informatics* 15 (1998): 177.

services can be provided.[130] In any case, the focus is on the public interest and how this can be guaranteed by a universal service.

TRANSPARENCY – The decision should be made after an open debate.[131] When the service is determined, terms and conditions should be drawn up that enable the specification of the exact content of the universal service and the conditions that are put upon its delivery.[132] Whatever specifications and conditions have been set, they have to be transparent. For instance, this entails that calculating costs have to be known, which may require accountancy rules allowing the singling out of the universal service activity.[133] To safeguard the level playing field, an independent regulator should oversee the provision of the services.

ASSIGNMENT OF THE SERVICE PROVIDER – In addition to establishing the content of the universal service and the conditions under which it needs to be performed, the government also needs to determine who will be responsible for providing the service. It is not self-evident that this task should be assigned to a public body or a public undertaking, whether or not it has a monopoly. It may be well possible that a private market player will take up the task, or that multiple players are elected.[134] If more than one market party is selected to provide the service, this allows for competition in the provision of the universal service, increasing efficiency. No market player is *a priori* excluded from being designated as a provider of a universal service and all undertakings present on the market are eligible under an efficient, objective, transparent, and non-discriminatory designation mechanism.[135] A public procedure should be held so the public body assigning the service cannot just decide to offer the service to a particular market party or to provide the information service itself.

---

130. Wouter Devroe, 'Universele dienstverlening als nieuwe manier van denken', *Sociaal-economische Wetgeving* 2000: 83 [in Dutch].
131. Simone Van Der Hof, *Overheidsinformatie in de etalage. Belangen rondom de toegang tot overheidsinformatie* (Alphen a/d Rijn: Samsom Bedrijfsinformatie, 1997), 18–19 [in Dutch]; Yves Poullet, 'Plaidoyer pour un ou des service(s) universel(s) d'informations publiques', paper given at the European Commission Conference: Access to Public Information. A Key to Commercial Growth and Electronic Democracy (Stockholm, 1996): 12 [in French].
132. Yves Poullet, 'Freedom and information highways or how to ensure electronic democracy', *Telematics and Informatics* 15 (1998): 177; Yves Poullet, 'Towards a universal public information service, paper given at the workshop on commercial and citizen's access to government information (Luxembourg, 1995), 11.
133. Iris Houben, 'Public service obligations: moral counterbalance of technical liberalization legislation', *European Review of Private Law* 16 (2008): 12.
134. Wouter Devroe, 'Universele dienstverlening als nieuwe manier van denken', *Sociaal-economische Wetgeving* 2000: 83 [in Dutch]; Yves Poullet, 'Freedom and information highways or how to ensure electronic democracy', *Telematics and Informatics* 15 (1998): 177; Yves Poullet, 'Plaidoyer pour un ou des service(s) universel(s) d'informations publiques', paper given at the European Commission Conference: Access to Public Information. A Key to Commercial Growth and Electronic Democracy (Stockholm, 1996): 12 [in French].
135. Mira Burri Nenova, 'The new concept of universal service in a digital networked communications environment', *I/S: A Journal of Law and Policy for the Information Society* 3, no. 1 (2007): 128.

ii. Approaches to the Universal Service Concept

BACKGROUND – The concept of universal service originates from the United States. At the outset, in the 1930s, it was essentially about interconnection, while in the 1970s it was re-introduced under the basic idea of 'a telephone in every home'.[136] It was taken over by the European Commission in the liberalization process of the telecommunications market. The concept is generally associated with the dismantlement of government monopolies.[137] It has often been used as a means to convince opponents of liberalization of the market, who were worried that the interests of the public would not be taken into account properly on a fully competitive market.[138] Universal services are therefore sometimes referred to as 'consolation prize'[139] or a 'bargaining chip'.[140]

MINIMALISTIC VIEW – There are two views on universal service. One takes as a starting point that the service should be minimal and should restrict competition as little as possible. Some argue that it should only 'hold the fort until competition arrives'.[141] The proponents of this vision fear that extensive universal service obligations can be damaging to economic welfare, and that the universal service providers will be left open to 'cream-skimming' from their competitors in the market who are not subject to these obligations. On the other hand, new entrants to the market may also be worried about cross-subsidization of the activities of the assigned service provider.[142] This approach could be used to reduce the possibility to provide services on a non-market basis.[143]

BROADER VIEW – The other approach sees universal service more as a social right, which should not necessarily be restricted to the minimum, but should include whatever is necessary to ensure general welfare and the social rights of

136. Mark Young, 'The future of universal service. Does it have one?', *International Journal of Law and Information Technology* 13, no. 2 (2005): 190; Pascal Verhoest, 'The myth of universal service: hermeneutic considerations and political recommendations', *Media, Culture & Society* 22, no. 5 (2000): 595; Mira Burri Nenova, 'The new concept of universal service in a digital networked communications environment', *I/S: A Journal of Law and Policy for the Information Society* 3, no. 1 (2007): 120.
137. Wouter Devroe, 'Universele dienstverlening als nieuwe manier van denken', *Sociaal-economische Wetgeving* 2000: 84 [in Dutch].
138. See also Iris Houben, 'Public service obligations: moral counterbalance of technical liberalization legislation', *European Review of Private Law* 16 (2008): 11; Pascal Verhoest, 'The myth of universal service: hermeneutic considerations and political recommendations', *Media, Culture & Society* 22, no. 5 (2000): 604.
139. Wouter Devroe, 'Universele dienstverlening als nieuwe manier van denken', *Sociaal-economische Wetgeving* 2000: 92 [in Dutch].
140. Mira Burri Nenova, 'The new concept of universal service in a digital networked communications environment', *I/S: A Journal of Law and Policy for the Information Society* 3, no. 1 (2007): 124.
141. See Pascal Verhoest, 'The myth of universal service: hermeneutic considerations and political recommendations', *Media, Culture & Society* 22, no. 5 (2000): 602.
142. Mark Young, 'The future of universal service. Does it have one?', *International Journal of Law and Information Technology* 13, no. 2 (2005): 197–198.
143. Pascal Verhoest, 'The myth of universal service: hermeneutic considerations and political recommendations', *Media, Culture & Society* 22, no. 5 (2000): 596.

the citizen. With regard to telecommunications, for instance, communication was considered 'essential to democracy and an adjunct of freedom of speech, and access to telecommunications should not only be considered a social right but actively promoted as a positive liberty'.[144] This approach could be misused to promote monopolies or government intervention beyond where it is necessary.

iii. The Usefulness of the Concept for Spatial Information Services

SUITABLE CONCEPT FOR INFORMATION SERVICES – According to De Terwangne, in light of the information society, universal service needs to include infrastructure, but also information. Information services that can be thought of are local event calendars, job vacancies, opportunities for small and medium-sized enterprises, etc.[145] Earlier, Poullet found that the universal service concept was a good reaction to the Synergy Guidelines, which in his opinion were formulated too negatively, restricting the activity of the State so that market forces could operate optimally. A universal information service could affirm that information held by the public bodies plays a role in the functioning of the information society, and it could be a good means to ensure the synergy between the public and private sector on the information market.[146] A universal information service would make it possible for the public sector to maintain a role as regulator, while leaving the exploitation to others.

MINIMUM CONTENT IN ACCESS LEGISLATION – The minimum content of the universal information service has already been included in national access legislation, both with regard to providing access upon request and the dissemination of information of the public bodies' own motion. Hence, any universal service that would be set up would need to address additional information services that the government might deem necessary for the development of the democratic society. As Rodota stated, '[t]he positive duty of the public sector consists of giving the citizens not the minimum but the essential information, essential for performing their personal and civic tasks'.[147] This demonstrates the social approach to universal services that was mentioned above.

DEFINITION IS NOT POSSIBLE – Beyond the characteristics of universal service mentioned above, Poullet feels that it is impossible to give a unique definition of the universal information service, either in terms of content (raw data or value added data) or in terms of the technical parameters of access (paper support,

---

144. Mark Young, 'The future of universal service. Does it have one?', *International Journal of Law and Information Technology* 13, no. 2 (2005): 195.
145. Cécile De Terwangne, *Société de l'information et mission publique d'information*, Doctoral Thesis FUNDP (Namur: Faculté Universitaire de Notre Dame de Paix, 2001), 442 [in French].
146. Yves Poullet, 'Freedom and information highways or how to ensure electronic democracy', *Telematics and Informatics* 15 (1998): 176; Yves Poullet, 'Plaidoyer pour un ou des service(s) universel(s) d'informations publiques', paper given at the European Commission Conference: Access to Public Information. A Key to Commercial Growth and Electronic Democracy (Stockholm, 1996): 2, 8 [in French].
147. See George Papapavlou, 'Minutes', paper given at the workshop on commercial and citizen's access to government information (Luxembourg, 1995), 74.

electronic support, and so on). Only a definition in terms of the finalities and characteristics of such a universal service is possible.[148] This is a difficult starting point, if we consider that universal service obligations should be precisely defined and the conditions clearly set out under which the assigned provider needs to perform the service. Hence, the public body assigning the universal service will have to have a very clear idea of what it is asking. Some believe this will be even more difficult for information services than for other sectors. It requires a balancing act between the idea that the public bodies' role cannot be restricted to the margins of what is interesting for the private sector and the idea that this role can be extended to all kinds of information that is also offered by the private sector.[149] Yet, the universal information service will evolve with time, just like the scope of the existing universal services.[150] Young feels that 'in practice, universal service is principally about politics, and this will be true for any expansion in the future scope on universal service that might occur'.[151]

EVALUATION: TOO MUCH FOCUS ON INFRASTRUCTURE? – Lievrouw issues a general warning on applying the universal service idea to information: 'Universal service carries a good deal of telephone-related conceptual 'baggage' that needs unpacking if it is to be a useful principle for the Internet. We need to keep in mind that universal service was directed toward the diffusion of a single technological system and its services rather than to fulfil telephone customers' needs for information or communication: the emphasis was on the conduit rather than the content'.[152] She feels that in order to move universal service from telephone to new media and information, it needs to be looked at in an entirely new way. This would include taking into account individual capacity in addition to infrastructure issues: 'Universal service must have a training and education component to cultivate individual capacity'.[153] Blakemore and Craglia add that governments tend to look

148. Yves Poullet, 'Plaidoyer pour un ou des service(s) universel(s) d'informations publiques', paper given at the European Commission Conference: Access to Public Information. A Key to Commercial Growth and Electronic Democracy (Stockholm, 1996): 10 [in French]; Commission of the European Communities, *Publaw 3. Final report* (Luxembourg: CEC Legal Advisory Board, 1995), 86.
149. See George Papapavlou, 'Minutes', paper given at the workshop on commercial and citizen's access to government information (Luxembourg, 1995), 74.
150. *Ibid.*, 75.
151. Mark Young, 'The future of universal service. Does it have one?', *International Journal of Law and Information Technology* 13, no. 2 (2005): 203.
152. Leah A. Lievrouw, 'The Information environment and Universal Service', *The Information Society* 16, no. 2 (2000): 155. See also Pascal Verhoest, 'The myth of universal service: hermeneutic considerations and political recommendations', *Media, Culture & Society* 22, no. 5 (2000): 598, 606; Mira Burri Nenova, 'The new concept of universal service in a digital networked communications environment', *I/S: A Journal of Law and Policy for the Information Society* 3, no. 1 (2007):131–142.
153. Leah A. Lievrouw, 'The Information Environment and Universal Service', *The Information Society* 16, no. 2 (2000): 158. See also Pascal Verhoest, 'The myth of universal service: hermeneutic considerations and political recommendations', *Media, Culture & Society* 22, no. 5 (2000): 606.

at access to physical devices to increase access rather than to information itself: 'This is attractive in that it is a visible and tangible event, whereas the improvement of information-handling skills requires a much more sensitive and diverse approach'.[154] They worry that extending the concept from technology to public sector data may lead to even more complexities and policy failures, as any public sector data policy is constantly surrounded by funding debates.[155] In summary, the 'broadening of the concept of universal service does not improve political clarity. It renders its more apt for ideological misappropriation'.[156]

EVALUATION: NO SOLUTION TO THE QUESTION – While universal service can offer a number of requirements for the information service to fulfil, it still does not provide a definition that can be used to determine the scope of the public interest that is guaranteed and the public task that guarantees it. Hence, the use of the concept is limited.

5. Interim Summary

PUBLIC INTEREST AS A CENTRAL CRITERION – From what has been discussed above, it seems that neither market failure, legal basis, or the link to the core responsibilities of the public bodies are workable criteria for determining the scope of the public task. These criteria can all be brought back to the criterion of public interest. This can be seen as the point of departure for the activities of the public sector.

NOT USEFUL AS CRITERION – However, the general and evolving nature of the public interest concept does not immediately provide us with criteria that can be used in practice to determine the scope of the public task in the EU legal framework for the availability of public sector spatial data. Even though accessibility is already a more concrete application of the public interest criterion to the provision of spatial information services, it still is very dependent on evolutions in technology, socio-economic circumstances, and even personal factors such as an individual's level of understanding and intelligence. While the concept of universal service can give some boundary conditions to which information services provided under a public task should adhere (e.g., affordability, equality, and continuity), these boundaries still do not say anything about the content of the public task.

SUMMARY – Hence, it is difficult to determine the scope of the public task from the point of view of what services the public bodies *should* provide. The next section will take a brief look at the possibility to define it from the other end: what the public bodies *should not* do with regard to spatial information services.

154. Michael Blakemore & Max Craglia, 'Access to public-sector information in Europe: policy, rights, and obligations', *The Information Society* 22, no. 1 (2006): 19.
155. *Ibid.*, 20.
156. Pascal Verhoest, 'The myth of universal service: hermeneutic considerations and political recommendations', *Media, Culture & Society* 22, no. 5 (2000): 598, 607.

C. THE NEGATIVE APPROACH: WHAT THE PUBLIC BODIES SHOULD NOT DO

OUTLINE – While the previous section of this chapter examined the arguments that are often given for the public bodies to provide spatial information services to the public as part of their public task, this section looks at the arguments that are given for spatial information services *not* to be provided by the public bodies.

CAUTIOUS APPROACH – A review of national policy documents and literature shows that most institutions and authors are cautious about putting limits on the public tasks of the public bodies with regard to the provision of information services. Few voices can be heard truly forbidding information services delivered by public bodies. Some comments rather request the public bodies to consider the circumstances before creating an information service. The main concerns appear to be duplication of effort, inefficiency, and unfair competition.

REVIEW OF EXISTING SERVICES – As was mentioned before, the 1989 Synergy Guidelines had already indicated that before creating an information service, public bodies should consider whether a private sector service was already available that could be used or adapted to meet the purposes of the public body.[157] Exceptions to this principle could be made if the private sector was not willing or able to provide the service on reasonable terms; if the service was an inseparable part of the public task; or if a visibly neutral service was needed. If the public bodies did provide an information service, they should review regularly whether it was most appropriate for them to keep providing it, or whether the involvement of the private sector in their production or distribution was desirable, or possibly even the replacement of the information services by appropriate commercial services.[158] Hence, the role of the public bodies was only to react to market developments and look at the intentions of the private sector before acting.[159] As was discussed in the previous chapter, the Guidelines were criticized for their limited view on the role of the public bodies.

NO REPLICATION OF EXISTING SERVICES – However, this limited view can also be found in other documents. Earlier, the United Kingdom's Department of Trade and Industry had already stated that electronic information services should not be introduced in competition with existing private sector services, and if no service existed it should be considered first whether the private sector should introduce the service instead. In any case, information services should not be provided on a non-commercial basis where a similar commercial service exists.[160] This view was

157. Commission of the European Communities, *Synergy Guidelines*, 10.
158. *Ibid.*
159. Cécile De Terwangne, *Société de l'information et mission publique d'information*, Doctoral Thesis FUNDP (Namur: Faculté Universitaire de Notre Dame de Paix, 2001), 433 [in French].
160. See Zakaria A. Hadi & Neil McBride, 'The commercialisation of public sector information within UK government departments', *International Journal of Public Sector Management* 13 (2000): 553, 557; Brian Negin, 'The dissemination of government information. A comparison between England and the USA', 1996, <www.itpolicy.gov.il/topics_info/pubusuk-1.htm>, 22 Feb. 2010.

adapted to the Web 2.0 developments by Mayo and Steinberg, who felt that '[g]overnment should not attempt to replicate existing successful user-generated sites and crowd them out of the market'.[161] This could be considered as anti-competitive behaviour towards companies, or discourage non-profit organizations and their volunteers.

ONLY UNDER A LEGAL OBLIGATION – The French 1994 Circular regarding the dissemination of public sector data also had a limited view of the provision of information services by the public bodies. Public bodies should not get involved if they have no legal obligation to provide an information service, if the information has a legal value and if it is not secret. In those cases, the provision of information services should be left to the private sector. However, the Circular makes a distinction between public bodies whose mission is to disseminate information and all the other public bodies. The former can actually also provide the information services along with the private sector.[162]

NO UNNECESSARY MODIFICATIONS TO DATABASES – The Dutch approach at the end of the 1990s was similar: the public bodies should not make unnecessary modifications to their databases, which could be done by the private sector.[163] Such modifications are only permitted to the extent that they have a bearing on the exercise of the public task. Commercial use by the public sector itself should be avoided.

NO VALUE-ADDED ACTIVITY – Understandably, the private sector will share this view. According to Kennedy, in the Canadian information market, '[t]he general attitude of the information industry has been to favour limiting the role of government to the provision of basic infrastructure and standards-setting. The predominant view has been that governments should not engage in any type of value-added activity, both because this is perceived to represent unfair competition and because the private sector is much better equipped to fully develop the commercial potential of value-added products'.[164] However, Kennedy immediately

---

161. Ed Mayo & Tom Steinberg, 'The Power of Information: An independent Review', 2007, 24, <www.cabinetoffice.gov.uk/media/cabinetoffice/strategy/assets/power_information.pdf>, 22 Feb. 2010; Cabinet Office, 'The Government's Response to The Power of Information: An independent review by Ed Mayo & Tom Steinberg (2007)', 2007, 8, 11, <www.cabinetoffice.gov.uk/media/cabinetoffice/strategy/assets/power_information_response.pdf>, 22 Feb. 2010.

162. Circulaire du 14 février 1994 relative à la diffusion des données publiques, *Journal Officiel* 19 Feb. 1994, 2864, <http://admi.net/jo/PRMG9400081C.html>, 22 Feb. 2010 [in French]. See also Jean-Michel Bruguière, *Les données publiques et le droit* (Paris: Editions Litec, 2002), 116 [in French].

163. See Peter N. Weiss, 'Borders in Cyberspace: Conflicting Public Sector Information Policies and their Economic Impacts', in *Public Sector Information in the Digital Age, Between Markets, Public Management and Citizens' Rights*, ed. Georg Aichholzer & Herbert Burkert (Cheltenham: Edward Elgar Publishing Limited, 2004), 148, 155.

164. Ed Kennedy, 'A private sector perspective', in *Policy issues in modern cartography*, ed. D.R. Fraser Taylor (Oxford: Elsevier Science Ltd., 1998), 135–136. See also Harry Bouwman et al., 'Leuker kunnen we het niet maken, wel makkelijker. Over de toegankelijkheid van overheidsinformatie', in *Elektronische toegankelijkheid van overheidsinformatie*, ed. Inigo Baten & Gijs Van Der Starre (The Hague: Rathenau Instituut, 1996), 85–87 [in Dutch];

admits the problem of distinguishing between base and value-added products and also acknowledges that some private sector parties do not share this view.[165]

No exclusion of private sector activity – Yet, comparable arguments can also be found elsewhere. For instance, Prins feels that '[f]or competition purposes, it is of particular importance that the public sector does not act as to exclude or discourage private sector exploitation . . . The public sector should be reticent with respect to its dissemination activities, for it is also to be recognised, that all public sector dissemination activities eventually may threaten private sector exploitation'.[166] Therefore, the public bodies should only provide information services if this is considered an inseparable part of their public task.[167]

No specialized services – Stiglitz et al. feel that a public body should not necessarily be prohibited to provide an information service, but that '[t]he more specialized the benefit of a government information service (i.e. that adds value to the underlying data or information), the more cautious the government should be in providing it'.[168] The authors refer to the practice of the US National Weather Service, which provides general forecasts but leaves specialized forecasts and analyses for industrial clients to the private sector. They also believe that substantial caution should be exercised in entering markets in which private sector parties are already active. This is in line with, for example, the Synergy Guidelines and the Guidelines on Tradable Information. According to Stiglitz et al., '[t]he government should generally *not* enter markets to provide more competition to existing firms. If the government is concerned about the lack of competition in a market, it should use anti-trust and other tools to address the underlying barriers to such competition'.[169]

Practical arguments – While the previous opinions were more concerned with fair competition, Robinson et al. take a more practical point of view: 'The institutional workings of government make it systematically incapable of

---

University of Sheffield et al., 'MADAME – Methods for Access to Data and Metadata in Europe. Increasing Access to Public Sector Information: Guidelines of Best Practice' (Sheffield: University of Sheffield, 2000), 33; Economic Development and Infrastructure Committee, 'Inquiry into Improving Access to Victorian Public Sector Information and Data', 2009, 107–110, <www.parliament.vic.gov.au/edic/inquiries/access_to_PSI/final_report.html>, 22 Feb. 2010.

165. Ed Kennedy, 'A private sector perspective', in *Policy issues in modern cartography*, ed. D.R. Fraser Taylor (Oxford: Elsevier Science Ltd., 1998), 135–136.

166. Corien Prins, 'Study on a Green Paper regulating Public Access to Government Information', paper given at the workshop on commercial and citizen's access to government information (Luxembourg, 1995), 35.

167. Not everybody agrees to this. Replying to the presentation of Corien Prins on this topic, Stefan Rodota stated that the role of the public sector should not be reduced to filling the gaps left by the private sector (see George Papapavlou, 'Minutes', paper given at the workshop on commercial and citizen's access to government information (Luxembourg, 1995), 65.

168. Joseph E. Stiglitz, Peter R. Orszag & Jonathan M. Orszag, 'The role of government in a digital age', 2000, 58, <http://unpan1.un.org/intradoc/groups/public/documents/apcity/unpan002055.pdf>, 22 Feb. 2010.

169. *Ibid.*, 71.

adapting and improving web sites as fast as technology itself progresses. No one site can meet as many different needs as well as a range of privately provided options can. And the idea that government's single site for accessing data will be a well-designed one is optimistic at best'.[170] Schartum has another concern: the provision of information services by public bodies may lead to an excessive control by the government over its information, which may lead to manipulation of the public.[171]

SUMMARY – In summary, the public bodies are generally not prohibited from providing spatial information services. Only a few opinions appear to go in that direction. Most policy documents and literature only call for caution when entering the market, and for not duplicating existing services. Hence, the main focus lies on avoiding the distortion of competition on the market by public bodies providing spatial information services, even though these services are deemed to be in the public interest. However, this may lead to some confusion. Is there an absolute division between the public task and activities on the market, meaning that services in the public interest can never be considered as market activities, even though the private sector might offer comparable services? Or is there a grey zone between both, entailing that public interest may also be covered by a spatial information service on the market? This issue is addressed in the next section.

D. Public Task v. Market: An Absolute Demarcation Line?

RECAPITULATION – In the EU legal framework for the availability of public sector spatial data, a distinction is made between public tasks and other activities. If public sector spatial data are used for public task purposes, they have to be obtained under the rules for sharing. If they are used for other purposes, the PSI Directive is applicable to the requests for data. It is inherently assumed in this framework that public task activities are juxtaposed against commercial use of public sector spatial data for selling information services on the market. This can also be seen from the positive and negative criteria that have been used in the literature and policy documents to try to determine the scope of the public task with regard to the provision of spatial information services. However, such a distinction is anything but absolute. Public task and commercial or market activities are not exhaustive, nor are they mutually exclusive.

1. Not Exhaustive

WIDE DEFINITION OF RE-USE – While the PSI Directive attempts to address the activities of the information industry, its definition of re-use gives the directive a much

170. David Robinson et al., 'Government data and the invisible hand', *Yale Journal of Law and Technology* 11 (2009): 173.
171. Dag Wiese Schartum, 'Access to Government-Held Information: Challenges and Possibilities', *Journal of Information, Law and Technology* no. 1 (1998): 4, <www2.warwick.ac.uk/fac/soc/law/elj/jilt/1998_1/schartum/>, 22 Feb. 2010.

wider impact. Hence, re-use includes any use for commercial or non-commercial purposes outside of the public task. So, not all 'non-public task' activities are automatically commercial. For instance, non-governmental organizations can also perform activities that constitute re-use. These activities may be non-commercial, but they are outside of the public task.

EXAMPLE: RESEARCH AND EDUCATION – Activities that may also not be part of the public task, but might not be commercial either, include research and education activities. Research may be done by non-governmental organizations, by trade unions, or by other parties which cannot be categorized as a public body or as performers of a public task, but this does not automatically entail that this research is commercial. In addition, in some Member States private schools that are not state-funded might not be considered as performing a public task. But does this automatically mean that they are performing a commercial activity?

EXAMPLE: USER GENERATED CONTENT – Another example is the development of the Web 2.0 concept and the trend towards increasing user generated content on the Internet. This user generated content can often be based on public sector data. The United Kingdom's Power of Information Review recognized this phenomenon as an important element of re-use that should be considered just as much a priority as purely commercial re-use by the information industry. Examples are websites and fora regarding parenting, on public services, Wikipedia, and so on. Many such websites use data-mashing techniques, combining data from many different sources and using public sector spatial data 'like a glue to bind together disparate information'.[172] Steinberg and Mayo give the example of mtraffic,[173] a volunteer project based on BBC traffic information that makes traffic reports accessible via mobile phones, without a commercial purpose. While such use is taken into account under the 'economic purpose' of the PSI Directive as it was described previously, it is often disregarded in the debate on the scope and the limits of the public task. Following the Steinberg and Mayo report, the Ordnance Survey, with the support of OPSI, launched OpenSpace in February 2008, a platform allowing innovators to build Web 2.0 applications using Ordnance Survey data. The platform 'enables data mashers to access and use the best mapping data available free of charge, for non-commercial purposes' and has been widely used since its launch.[174]

EVALUATION – As the activities mentioned above are not performed on the market, the public bodies taking on the same activities as such data-mashers, non-profit organizations, researchers, or regular citizens would not distort the market. Yet, the public bodies would still be acting outside of their public task

172. Ed Mayo & Tom Steinberg, 'The Power of Information: An independent Review', 2007, 13, <www.cabinetoffice.gov.uk/media/cabinetoffice/strategy/assets/power_information.pdf>, 22 Feb. 2010.
173. See <www.mtraffic.org>, 22 Feb. 2010.
174. See Ordnance Survey, 'Welcome to the OS Openspace API', <http://openspace.ordnancesurvey.co.uk/openspace/>, 22 Feb. 2010. See also Office of Public Sector Information, 'Unlocking PSI potential. The United Kingdom Report on the Re-use of Public Sector Information', 2008, 34, <www.opsi.gov.uk/advice/psi-regulations/uk-report-re-use-psi-2008.pdf> 22 Feb. 2010.

and hindering innovation in the information society. However, as such non-commercial activities do not have such a tangible impact as public bodies competing with the private information industry, this issue is often disregarded – except by Mayo and Steinberg, who recommend that the public bodies should refrain from unnecessary duplicating user-generated initiatives.[175] Moreover, where competition law may be used to solve conflicts between public bodies and private companies operating on the market, there is no specific legal framework that can determine and sanction the behaviour of public bodies in the non-commercial area outside of the public task.

2. Not Mutually Exclusive

ASSUMPTION – In the EU legal framework for the availability of public sector spatial data, the activities performed by public bodies for the purpose of their public tasks are always separated from their commercial activities on the market. It is an underlying assumption of both the PSI Directive and the INSPIRE Directive that when the public bodies are performing commercial activities on the market, they are not performing a public task. For instance, Recital 9 of the PSI Directive states that: 'Activities falling outside the public task will typically include supply of documents that are produced and charged for exclusively on a commercial basis and in competition with others in the market'.

a. Selling Information Products as Part of the Public Task

i. General

PUBLIC TASK TO GENERATE REVENUE – However, the argument has been made that in some countries, selling spatial information services on the market can actually be a part of the public task. For instance, the United Kingdom's Trading Funds, such as the Ordnance Survey, the Land Registry, the Meteorological Office, and the Hydrographic Office, are required to develop profitable commercial outlets for their services in order to offset core and central overheads (and then they make a return to the Treasury).[176] The Review Board of the APPSI also 'does not take the view that any supply of products by a public sector body on a commercial basis must necessarily fall outside that body's public task'. However, the Board admits that: 'The boundary between what is within the public task and what lies outside it is difficult to discern, especially in the case of trading funds'.[177] The next section

175. Ed Mayo & Tom Steinberg, 'The Power of Information: An independent Review', 2007, 24, <www.cabinetoffice.gov.uk/media/cabinetoffice/strategy/assets/power_information.pdf>, 22 Feb. 2010.
176. Office of Public Sector Information, 'The Re-use of Public Sector Information: A Guide to the Regulations and Best Practice', 2005, 12, <www.opsi.gov.uk/advice/psi-regulations/advice-and-guidance/guide-to-psi-regulations-and-best-practice.doc>, 22 Feb. 2010.
177. Advisory Panel on Public Sector Information, 'The impact and success of legislation on the re-use of public sector information', 2008, 26, <www.appsi.gov.uk/reports/ec-submission-07-2008.doc>, 22 Feb. 2010.

presents a brief look at the debate on the scope of the public task of one of these Trading Funds, the Ordnance Survey – the United Kingdom's National Mapping Agency.

ii. The Example of the Ordnance Survey

COST RECOVERY OBLIGATIONS – As a Trading Fund, the Ordnance Survey is required to cover at least the majority of its own costs. A Trading Fund is established when the government judges that the revenue of an operation consists or could consist mainly of payments for goods or services provided in the course of the operations, rather than funding from the central government budget.[178] To gain the necessary revenue, the Ordnance Survey produces and sells large quantities of public sector spatial data, a variety of value-added products and services, and high resolution paper and digital maps. In 2008–2009, the Ordnance Survey had a turnover of United Kingdom Pounds (GBP) 116 million from the supply of information.[179]

COMMERCIAL ACTIVITIES AS A PUBLIC TASK? – Due to this requirement of generating a return on its assets, and the removal of the last element of public funding by the ending of the National Interest Mapping Services Agreement at the end of 2006, the Ordnance Survey is required to fund all its operations (both public task and non-public task) under its Trading Fund model, and consequently has to obtain a commercial return on all its activities. The distinction between its public service and commercial activity is therefore inherently blurred.[180] According to Intelligent Addressing, a private company that is both one of the main re-users of the Ordnance Survey spatial data and one of its main competitors, this 'even raises the possibility that the commercial activities of Ordnance Survey equate to their public duty'.[181]

SCOPE OF COMMERCIAL ACTIVITIES – The debate in the Parliament's Communities and Local Government Committee on the scope of the Ordnance Survey's public task also shows another issue with regard to determining the scope of the public task. The question is asked how to distinguish better between commercial activities and public tasks.[182] But what is more inherent to commercial activities than

178. David Newbery, Lionel Bently & Rufus Pollock, 'Models of Public Sector Information Provision via Trading Funds', 2008, 8, <www.berr.gov.uk/files/file45136.pdf>, 22 Feb. 2010.
179. Communities and Local Government Committee, 'Policy options for geographic information from Ordnance Survey. Consultation', 2009, 35, <www.communities.gov.uk/publications/corporate/ordnancesurveyconsultation>, 1, 22 Feb. 2010.
180. Ordnance Survey, 'Supplementary Memorandum to the Communities and Local Government Committee Fifth Report of Session 2007-08, HC 268', 2007, <www.publications.parliament.uk/pa/cm200708/cmselect/cmcomloc/268/268we20.htm>, 22 Feb. 2010.
181. Communities and Local Government Committee, 'Ordnance Survey, Fifth Report of Session 2007-08, HC 268', 2007, <www.publications.parliament.uk/pa/cm200708/cmselect/cmcomloc/268/26802.htm>, 22 Feb. 2010.
182. *Ibid.* See also Communities and Local Government Committee, 'Ordnance Survey: Government response to the Committee's Fifth Report, First Special Report of Session 2007-08, HC 516', 2008, <www.parliament.the-stationery-office.com/pa/cm200708/cmselect/cmcomloc/516/516.pdf>, 22 Feb. 2010; Stephen Saxby, 'Public sector information and re-use policy – where is the UK now?', paper given at the Second International Conference on Business, Law and Technology (New York, 2008), 17, <http://eprints.soton.ac.uk/52523/01/PSI_Re-use_Paper_28_April_2008_pdf.pdf>, 22 Feb. 2010.

obtaining a return on the performed activities? This raises the question on what exactly is meant by the term *commercial*. Does this indicate making a profit, or performing activities on the market? Does it have the same scope as *economic activities*? It is sufficient that the service is paid for by the user? The term *commercial* can mean many different things to many different users. In this way it complicates the discussion on the scope of the public task even further.

b. Market Services in the Public Interest

ASSUMPTION – As mentioned above, the division between sharing and re-use assumes that there is an absolute distinction between performing public tasks and re-using public sector spatial data to offer spatial information services on the market. This also seems to imply that if a particular information service is considered to be part of the task of a public body, it is not offered on the market. There is an absolute dichotomy between the public task and the market.

PUBLIC INTEREST ON THE MARKET? – However, sometimes references to a middle ground between the public task and the market can be found. It is felt that government can go on the market to ensure a particular public interest. For instance, Reinsma and Van Der Sluijs state that otherwise prohibited competition from public bodies could be justified to guarantee a continuous information service for an affordable price or to stimulate competition on a monopolistic market of products and services based on public sector data.[183] Bruguière has a comparable view that public bodies cannot participate on the market unless it is in the general interest.[184]

INTERMEDIATE CATEGORY – Hence, the public body is performing a public task, but it is doing so on the market. However, it does not have to obey the rules of the market. This seems difficult to reconcile with the distinction between access, re-use, and sharing in the EU legal framework for the availability of public sector spatial data. Yet, as such an intermediate category is also available in competition law, maybe the EU competition rules should be turned to for a solution? This will be discussed in the next section.

E. THE PUBLIC TASK DEBATE AND COMPETITION LAW

APPLICABILITY OF COMPETITION LAW TO SPATIAL INFORMATION SERVICES – When public sector spatial data is used by a public body for the creation and dissemination of

183. Marijke Reinsma & H. Van Der Sluijs, 'Naar ruimere openbaarheid en een vrij gebruik van bestuurlijke informatie', 2002, 90, <www.minbzk.nl/aspx/download.aspx?file=/contents/pages/5006/RapportReinsmaenVanderSluijs.pdf>, 22 Feb. 2010 [in Dutch]; Bertrand Nouel, 'Commercialisation des données publiques et concurrence', paper given at the European Commission Conference: Access to Public Information. A Key to Commercial Growth and Electronic Democracy (Stockholm, 1996): 8 [in French].
184. Jean-Michel Bruguière, *Les données publiques et le droit* (Paris: Editions Litec, 2002), 108 [in French].

spatial information services on the market, the public body has to compete fairly with the private companies on the market. It is often stated that in reality this is not the case. For instance, in the 2008 review of the PSI Directive, many stakeholders 'highlighted problems when competing commercially with public sector bodies, and notably when the latter is undertaking a statutory public task. Many respondents think they compete on unequal terms with the public sector body concerned, believing that limits and boundaries should be established on the role/limit of public bodies operating commercially in the market, and claim for greater transparency in these cases'.[185]

SAME CONDITIONS FOR PUBLIC AND PRIVATE BODIES – Public bodies on the market have to operate under the same conditions as their private sector competitors. This was already acknowledged in the 1989 Synergy Guidelines and the Green Paper on Public Sector Information. The PSI Directive made this explicit in its Article 10. As was discussed in Chapter 3, the first paragraph of Article 10 states that: 'Any applicable conditions for the re-use of documents shall be non-discriminatory for comparable categories of re-use'. Hence, this ensures that the public body acts fairly as a supplier of public sector spatial data. The second paragraph deals with fair competition by the public body as a user of its own public sector spatial data and as a market player. If documents are re-used by a public body as input for its commercial activities which fall outside the scope of its public tasks, the same charges and conditions apply to the supply of the documents for those activities as apply to other users.

COMPETITION LAW AS ADDITIONAL RULES FOR PUBLIC SECTOR SPATIAL DATA – These two provisions can be considered as applications of the rules of general competition law to the specific domain of public sector spatial data. Yet, the 2008 review of the PSI Directive shows that these rules are not sufficient to ensure fair competition. However, it has been suggested that the private sector should seek refuge in EU and national competition law in order to protect itself against anti-competitive behaviour and file a complaint with the appropriate competition authorities. Therefore, the next section of this chapter examines to what extent competition law could be used in disputes regarding the provision of spatial information services.

COMPETITION LAW AS INSPIRATION TO DEFINE THE PUBLIC TASK: ECONOMIC ACTIVITIES – Next, it should also be examined whether competition law can be used as an inspiration for defining the scope of the public task. The competition authorities have had to determine in many cases whether a public body offering a particular service could be considered as an undertaking subject to competition law. Can their findings be transposed to the sector of spatial information services? In the EU legal

185. Commission of the European Communities, Results of the online consultation of stakeholders 'Review of the PSI Directive', 2008, 2, <http://ec.europa.eu/information_society/policy/psi/docs/pdfs/online_consultation/report_psi_online_consultaion_stakeholders.pdf>, 22 Feb. 2010; Commission of the European Communities, *Staff working document. Accompanying document to the Communication from the Commission to the European Parliament, the Council, the European Economic and Social Committee and the Committee of the Regions on the re-use of Public Sector Information – Review of Directive 2003/98/EC*, SEC (2009) 597, 16, 7 May 2009.

framework regarding the availability of public sector spatial data, it is found that if a public body is performing activities outside of the public task, it is performing activities on the market and has to act according to the market rules. If it is performing a public task, it is not subject to these market rules. Can this distinction between the public task and other activities be aligned with the distinction of competition law based on the economic character of activities?

COMPETITION LAW AS INSPIRATION TO DEFINE THE PUBLIC TASK: SERVICES OF GENERAL ECONOMIC INTEREST – In the previous section, some nuances were brought to the distinction between the public task and market activities. It was stated that this distinction was not always considered absolute, as some found it possible that public bodies performed their public task on the market, entailing that the rules of the market would not be entirely applicable. While this seems to clash with the distinction between access, re-use, and sharing as it is set up in the appropriate European directives, this idea shows some similarity to the concept of services of general economic interest in competition law.[186] As Prosser describes it, there are two distinct issues: first, whether an activity is economic and thus falls within the field of application of EU competition law, and second, whether there are nevertheless good reasons for modifying the application of the competition rules.[187] Therefore, the following section on competition law will also try to examine whether this concept can help in the debate on the public task of public bodies to provide spatial information services.

IV. COMPETITION LAW

FOCUS ON EU COMPETITION LAW – Under the EU legal framework for the availability of public sector spatial data, public bodies offering their services on the market are subject to competition law and to the obligation not to distort the market. Competition law exists both on the national level and on the European level. As this book focuses on the legal framework at the level of the EU, it only addresses the competition rules as they can be found in Article 101 and following of the Treaty on the Functioning of the European Union (former Article 81 and following of the EC Treaty), and not national competition legislation. This entails that the rules will only be applicable if there is a distortion of competition that may affect trade between Member States. However, national competition rules are based on the EU rules and generally contain the same elements except for the cross-border effect of the distortion.

186. See also Commission of the European Communities, *Staff working document. Accompanying document to the Communication from the Commission to the European Parliament, the Council, the European Economic and Social Committee and the Committee of the Regions on the re-use of Public Sector Information – Review of Directive 2003/98/EC*, SEC (2009) 597, 27, 7 May 2009.

187. Tony Prosser, *The Limits of Competition Law, Markets and Public Services* (Oxford: Oxford University Press, 2005), 130.

LIMITED OVERVIEW OF COMPETITION RULES – There are a number of ways in which public bodies can distort the competition on the market.[188] Those ways will be addressed in this section, but it is not the intention to give a full overview of competition law. For this, the reader is referred to the extensive literature in this domain.[189] We will only discuss the issues that may come up when a public body enters the market to provide spatial information services in competition with the private sector.

OUTLINE – First, the question is asked when a public body falls under the field of application of competition law. In the second subsection, the possible distortions of the market by public bodies providing spatial information services are discussed. These distortions may consist of concluding agreements, making decisions, or conducting practices that restrict competition (Article 101 of the Treaty on the Functioning of the EU), or abusing a dominant position (Article 102 of the Treaty on the Functioning of the EU). The third subsection examines the situation where the public bodies could infringe competition rules by their actions as a public authority rather than an undertaking. Sometimes the funding of the provision of information services by a public body may be considered as state aid (Article 107 of the Treaty on the Functioning of the EU). Next, the Member States may infringe competition rules with measures they take towards their public undertakings or undertakings with special or exclusive rights. The fourth section looks at the specific case of services of general economic interest, because this case is related to the question on the non-exhaustive and non-exclusive relationship of the public task and the market that was discussed in the previous section. In each subsection, the rules of competition law are discussed briefly, followed by the application of these rules to the situation of spatial information services.

A. PUBLIC BODIES AS UNDERTAKINGS

1. Definition

ECONOMIC ACTIVITY – The competition rules in Article 101 and following of the Treaty on the Functioning of the EU are applicable to undertakings. However, the term *undertaking* is not defined. The European Court of Justice (ECJ) has

188. See Christian Lamouline, 'Presentation of Publaw 3 findings. Legal assessment', paper given at the workshop on commercial and citizen's access to government information (Luxembourg, 1995), 11–12; Dieudonné Mandelkern & Bertrand Du Marais, 'Diffusion des données publiques et révolution numérique: rapport', 1999, 67, <www.ladocumentationfrancaise.fr/rapports-publics/994001620/index.shtml#>, 22 Feb. 2010 [in French].

189. For example, Alison Jones & Brenda Sufrin, *EC Competition Law: Text, Cases and Materials* (Oxford: Oxford University Press, 2008), 1418 p.; Peter Roth & Vivien Rose (ed.), *Bellamy & Child European Community Law of Competition*. 6th edn (Oxford: Oxford University Press, 2008), 2252 p.; Richard Whish, *Competition law*, 6th edn (Oxford: Oxford University Press, 2009), 1006 p.; Jonathan Faull, *Faull and Nikpay: The EC Law of Competition* (Oxford: Oxford University Press, 2007), 2000 p.

described it as 'every entity engaged in an economic activity, regardless of the legal status of the entity or the way in which it is financed'.[190] Whether the undertaking makes a profit is irrelevant, as long as it carries out economic activities.[191] Hence, the fact that a service is provided free of charge does not automatically entail that it is not an economic activity. It is also not necessary for the entity to have legal personality under the law of the Member State where it operates.[192]

FUNCTIONAL APPROACH – The undertaking can supply either goods or services, but it has to qualify as an economic activity.[193] This is determined by considering the nature and the aim of the activity in question and how it is financed, that is, a functional approach. For instance, public placement offices are considered to be an undertaking, as employment procurement is deemed an economic activity that can also be carried out by private companies.[194] It is sufficient that the activities can be carried out by a private entity, even though this is not currently happening. As Jones and Sufrin describe it, the characteristic feature of an economic activity is '(1) the offering of goods or services on the market, (2) where that activity could, at least in principle, be carried on by a private undertaking in order to make profits'.[195] This will be the case for a very wide range of activities.

---

190. ECJ, Case C-41/90, *Höfner & Elser v. Macotron GmbH*, 23 Apr. 1991, para. 21, *European Court Reports* 1991, I-1979. See also ECJ, Case C-309/99, *Wouters and others*, 19 Feb. 2002, para. 46, *European Court Reports* 2002, I-1577.

191. ECJ, Case C-67/96, *Albany International BV v. Stichting Bedrijfspensioenfonds Textielindustrie*, 21 Sep. 1999, para. 85, *European Court Reports* 1999, I-5751; Commission Decision 1999/329/EC of 12 Apr. 1999 relating to a proceeding pursuant to Arts 85 and 86 of the EC Treaty and Arts 53 and 54 of the EEA Agreement (IV/D-1/30.373 – P & I Clubs, IGA and No. IV/D- 1/37.143 – P & I Clubs, Pooling Agreement), 12 Apr. 1999, *OJ L* 125, 19 May 1999, 12.

192. Commission decision 78/516/EEC of 26 May 1978 relating to a proceeding under Art. 85 of the EEC Treaty (IV/29.559 – RAI/UNITEL), *OJ L* 57, 15 Jun. 1978, 39; CFI, Cases T-217/03 and T-245/03, *Fédération nationale de la coopération bétail and viande v. Commission*, 13 Dec. 2006, *European Court reports* 2006, II-4987. See also Jurian Van Parys, 'Het begrip onderneming in mededingingsrecht: ziekenfonds als onderneming?', *Tijdschrift voor Belgische Mededinging* 2006: 60 [in Dutch]; Pierre De Bandt, 'Artikel 86, lid 2 EG-Verdrag en het begrip "diensten van algemeen economisch belang": meer dan een rechtsbasis voor een genuanceerde toepassing van de interne markt- en mededingingsregels op openbare diensten?', *Tijdschrift voor Belgisch Handelsrecht* 114, no. 3 (2008): 215 [in Dutch]; Wolf Sauter, 'Services of general economic interest and universal service obligations as an EU law framework for curative health care. TILEC Discussion Paper', 2007, 21–22, <http://papers.ssrn.com/sol3/papers.cfm?abstract_id=1013261>, 22 Feb. 2010; Markus Krajewski, 'Providing Legal Clarity and Securing Policy Space for Public Services through a Legal Framework for Services of General Economic Interest: Squaring the Circle?', *European Public Law* 14, no. 3 (2008): 387; Wolf Sauter, 'Services of general economic interest and universal service in EU law', *European Law Review* 33, no. 2 (2008): 181; Richard Whish, *Competition Law*, 6th edn (Oxford: Oxford University Press, 2009), 85.

193. Peter Roth & Vivien Rose (ed.), *Bellamy & Child European Community Law of Competition*. 6th edn (Oxford: Oxford University Press, 2008), 94.

194. ECJ, Case C-41/90, *Höfner & Elser v. Macotron GmbH*, 23 Apr. 1991, para. 23, *European Court Reports* 1991, I-1979.

195. Alison Jones & Brenda Sufrin, *EC Competition Law: Text, Cases and Materials* (Oxford: Oxford University Press, 2008), 128–129. See ECJ, Case C-475/99, *Ambulanz Glöckner v. Landkreis Südwestpflatz*, 25 Oct. 2001, para. 19, *European Court Reports* 2001, I-8089; ECJ,

2. The State or Public Bodies as Undertakings

Determining status of a public body is difficult – It is not an easy task to determine whether the State is performing an economic activity. The ECJ has stated that:

> In seeking to determine whether an activity carried on by the State or a State entity is of an economic nature, the Court is entering dangerous territory, since it must find a balance between the need to protect undistorted competition on the common market and respect for the powers of the Member States. The power of the State which is exercised in the political sphere is subject to democratic control. A different type of control is imposed on economic operators acting on a market: their conduct is governed by competition law. But there is no justification, when the State is acting as an economic operator, for relieving its actions of all control. On the contrary, it must observe the rule in such cases.[196]

Yet, while it may not be an easy task, the ECJ has been confronted with it on several occasions.

No exercise of official authority – Member States are not undertakings when they act in the performance of their sovereign or administrative functions.[197] The case law of the ECJ makes a sharp distinction between the economic activities and activities 'in the exercise of official authority'. The latter fall outside of the scope of competition law. An entity acts in the exercise of official authority where the activity in question is 'a task in the public interest which forms part of the

---

Case C-35/96 *Commission v. Italian Republic*, 18 Jun. 1998, para. 36, *European Court Reports* 1998, I-3851; ECJ, Case C-205/03 P, *FENIN v. Commission*, 11 Jul. 2006, para. 25, *European Court Reports* 2006, I-6295; ECJ, Case C-67/96, *Albany International BV v. Stichting Bedrijfspensioenfonds Textielindustrie*, 21 Sep. 1999, Jacobs AG, para. 311, *European Court Reports* 1999, I-5751; ECJ, Case C-180/98 to C-184/98, *Pavlov v. Stichting Pensioenfonds Medische Specialisten*, 12 Sep. 2000, para. 75, *European Court Reports* 2000, I-6451; CFI, Case T-155/04, *SELEX Sistemi Integrati v. Commission*, 12 Dec. 2006, para. 61, *European Court Reports* 2006, II-4797 (appeal C-113/07P, *SELEX Sistemi Integrati v. Commission*, 26 Mar. 2009). See also Jurian Van Parys, 'Het begrip onderneming in mededingingsrecht: ziekenfonds als onderneming?', *Tijdschrift voor Belgische Mededinging* 2006: 59–65, 60 [in Dutch]; Wolf Sauter, 'Services of general economic interest and universal service in EU law', *European Law Review* 33, no. 2 (2008): 181; Johan W. Van De Gronden, 'Purchasing care: economic activity or service of general (economic) interest', *European Competition Law Review* 25, no. 2 (2004): 87–88; Richard Whish, *Competition Law*, 6th edn (Oxford: Oxford University Press, 2009), 84–85.

196. ECJ, Case C-205/03 P, *FENIN v. Commission*, 11 Jul. 2006, Maduro AG, para. 26, *European Court Reports* 2006, I-6295.
197. See also Peter Roth & Vivien Rose (ed.), *Bellamy & Child European Community Law of Competition*. 6th edn (Oxford: Oxford University Press, 2008), 95; Pierre De Bandt, 'Artikel 86, lid 2 EG-Verdrag en het begrip "diensten van algemeen economisch belang": meer dan een rechtsbasis voor een genuanceerde toepassing van de interne markt- en mededingingsregels op openbare diensten?', *Tijdschrift voor Belgisch Handelsrecht* 114, no. 3 (2008): 215 [in Dutch]; Johan W. Van De Gronden, 'Purchasing care: economic activity or service of general (economic) interest', *European Competition Law Review* 25, no. 2 (2004): 88.

essential functions of the State' and where that activity 'is connected by its nature, its aim and the rules to which it is subject with the exercise of powers . . . which are typically those of a public authority'.[198] The European Commission refers to such activities as 'matters which are intrinsically prerogatives of the State'.[199] Examples of activities that have been considered of a public nature include the provision of air space management services[200]; pollution control services in a harbour[201]; the collection of taxes[202]; the giving of exclusive concessions for funeral services[203]; and compulsory social security schemes.[204]

NO SERVICES BASED ON SOCIAL SOLIDARITY – With regard to social services, such as social security, pension funds, and so on, another criterion is taken into account to determine whether an economic activity is being performed. If social solidarity prevails over the economic character of the activity, the entity will also not be qualified as an undertaking.[205]

OFFERING GOODS AND SERVICES ON THE MARKET – A public body is thought to be engaged in economic activities when it is offering goods or services that could also be offered on the market by a private firm to make a profit: 'If there were no possibility of a private undertaking carrying on a given activity, there would be no purpose in applying the competition rules to it'.[206] The following activities

198. ECJ, Case C-343/95, *Cali & Figli Srl v. Servizi ecologici porto di Genova SpA*, 18 Mar. 1997, para. 23, *European Court reports* 1997, I-1547. See also Richard Whish, *Competition law*, 6th edn (Oxford: Oxford University Press, 2009), 87–88.

199. Commission of the European Communities, *Green Paper on services of general interest*, COM (2003) 270 final, 14–15.

200. ECJ, Case C-364/92, *SAT Fluggesellschaft mbH v. Eurocontrol*, 19 Jan. 1994, *European Court Reports* 1994, I-43. See also Peter Roth & Vivien Rose (ed.), *Bellamy & Child European Community Law of Competition.* 6th edn (Oxford: Oxford University Press, 2008), 95.

201. ECJ, Case C-343/95, *Cali & Figli Srl v. Servizi ecologici porto di Genova SpA*, 18 Mar. 1997, *European Court reports* 1997, I-1547.

202. ECJ, Case C-207/01, *Altair Chimica SpA v. ENEL Distribuzione SpA*, 11 Sep. 2003, *European Court Reports* 2003, I-8875.

203. ECJ, Case 30/87, *Corinne Bodson v. Pompes Funèbres des Régions Libérées SA*, 4 May 1988, *European Court Reports* 1988, 2479.

204. ECJ, Case C-159/91 and C-160/91, *Poucet et Pistre v. Assurances Générales de France and Caisse Mutuelle Régionale du Languedoc-Roussillon*, 17 Feb. 1993, *European Court reports* 1993, I-637. See Richard Whish, *Competition law*, 6th edn (Oxford: Oxford University Press, 2009), 86–87.

205. ECJ, Case C-159/91 and C-160/91, *Poucet et Pistre v. Assurances Générales de France and Caisse Mutuelle Régionale du Languedoc-Roussillon*, 17 Feb. 1993, para. 18–19, *European Court reports* 1993, I-637; ECJ, Case C-355/00, *Freskot AE v. Elliniko Dimosio*, 22 May 2003, para. 77–79, *European Court reports* 2003, I-5263. See also Wolf Sauter, 'Services of general economic interest and universal service in EU law', *European Law Review* 33, no. 2 (2008): 183; Nina Boeger, 'Solidarity and EC Competition Law', *European Law Review* 32, no. 3 (2007): 319–340; Tony Prosser, *The Limits of Competition Law, Markets and Public Services* (Oxford: Oxford University Press, 2005), 128–130.

206. ECJ, Cases C-264/01, C-306/01, C-354/01 and C-355/01, *AOK Bundesverband and others v. Ichthyol-Gesellschaft Cordes, Hermani & Co and others*, 16 Mar. 2005, Jacobs AG, para. 27, *European Court Reports* 2004, I-2493.

have all been found to be economic: German state-run employment agencies;[207] independent customs agents in Italy;[208] medical aid organizations providing ambulance services in Germany;[209] and public television broadcasting organizations.[210]

SEPARATE LEGAL PERSONALITY NOT REQUIRED – As an undertaking does not need to have a separate legal personality, parts of a state's general administration that do not have their own legal personality may also be covered by the definition of *undertaking*.

BODIES PERFORMING MULTIPLE ACTIVITIES – It is possible that the state or a public body can be classified as an undertaking for some activities, but not for others. Each activity has to be analysed separately to determine whether it is an economic activity and whether it is subject to competition law.[211] Furthermore, the fact that an entity performing an economic activity has public service obligations imposed on it by the state may place it at a competitive disadvantage, but it does not deprive the activity of its economic character for other activities.[212]

3. Applicability to Spatial Information Services

OUTLINE – As was mentioned in the introduction, in each section, the discussion of the rules of EU competition law is followed by the application of these rules to the domain of spatial information services and the demonstration of this applicability by examples.

EXAMPLES – According to the Ravi Bedrijvenplatform, spatial information services may be produced as part of the public task, but some services may also be considered economic activities that are outside of the public task, such as the sale of geo-coding products by the commercial part of the Dutch Cadastre.[213] Other examples are the sale of DVDs with cycling routes or books with walking trails by the Belgian National Mapping Agency.[214]

207. ECJ, Case C-41/90, *Höfner & Elser v. Macotron GmbH*, 23 Apr. 1991, *European Court Reports* 1991, I-1979; ECJ, Case C-55/96, *Job Centre Coop. ARL*, 11 Dec. 1997, *European Court Reports* 1997, I-7119.
208. ECJ, Case C-35/96, *Commission v. Italian Republic*, 18 Jun. 1998, para. 36, *European Court Reports* 1998, I-3851.
209. ECJ, Case C-475/99, *Ambulanz Glöckner v. Landkreis Südwestpflatz*, 25 Oct. 2001, *European Court Reports* 2001, I-8089.
210. ECJ, Case 155/73, *Sacchi*, 30 Apr. 1974, *European Court Reports* 1974, 409.
211. ECJ, Case C-475/99, *Ambulanz Glöckner v. Landkreis Südwestpflatz*, 25 Oct. 2001, Jacobs AG, para. 72, *European Court Reports* 2001, I-8089; ECJ, Case 118/85, *Commission v. Italian Republic*, 16 Jun. 1987, para. 7, *European Court Reports* 1987, 2599; ECJ, Case C-82/01P, *Aéroports de Paris v. Commission*, 24 Oct. 2002, *European Court Reports* 2002, I-9297; Case T-155/05, *SELEX Sistemi Integrati v. Commission*, 12 Dec. 2006, para. 54, *European Court Reports* 2006, II-4797 (appeal C-113/07P, *SELEX Sistemi Integrati v. Commission*, 26 Mar. 2009).
212. Peter Roth & Vivien Rose (ed.), *Bellamy & Child European Community Law of Competition.* 6th edn (Oxford: Oxford University Press, 2008), 97.
213. Ravi Bedrijvenplatform, *Commercialisering van geo-informatie. Een drietal rapporten ter ondersteuning van de Ravi-discussie 1997* (Amersfoort: Ravi, 1997), 92 [in Dutch].
214. See <www.ngi.be/NL/NL0.shtm>, 22 Feb. 2010).

DIFFICULT TO DETERMINE THE NATURE OF ACTIVITIES – In some cases it may not be easy to determine whether the public body is performing an economic activity or whether it is acting as a public authority performing a sovereign or administrative function. However, considering the wide definition of economic activity as offering goods or services on the market which could also be offered by a private party, a considerable part of spatial information services could be considered as economic activities, and hence the public bodies performing them as undertakings. This could include services that are considered as public tasks. These services are not necessarily an act of official authority, but they may rather be economic activities that are considered to be in the interest of the public and therefore they should be part of the public task. This could have far-reaching consequences as to the application of competition law to these services. However, if an activity is connected to the exercise of powers by the public authority, as stated above, then it would not be considered an economic activity. Hence, the collection of public sector spatial data for the performance of the tasks of a public body will not be an economic activity, but the dissemination of the data and particularly the provision of spatial information services will be much less connected with the exercise of power as a public authority. A lot will depend on the interpretation of the competition authorities.

EXAMPLE: KONINKLIJK NEDERLANDS METEOROLOGISCH INSTITUUT – For instance, can the delivery of weather information by the national meteorological office be considered as an economic activity? This question was examined in a report on the role of KNMI, the Royal Dutch Meteorological Office. The task of KNMI was determined in the so-called Law on the Weather in 2002.[215] This law set out what the public tasks of the Meteorological Office were. The other activities were separated into a commercial entity and left to the market. In 2007, the validity of this division was examined.[216] The report showed that many activities of the meteorological office could be considered as an economic activity under EU competition law. This is the case because these activities concern the provision of information services that are either also being provided by the private sector or that could be provided by the private sector, but are not because the meteorological office has an exclusive right to perform them. Services that are also delivered by the private sector include the dissemination of general weather reports; weather expectations for road traffic, such as iciness of the roads and waterways; and historical meteorological information. Services that are currently exclusively assigned to the meteorological office, but that are contested by the private sector, include meteorological expectations for civilian air traffic, oil rigs, and harbours. The private sector claims that they can also provide such services. Hence, all the previous examples can be considered as economic activities under the EU competition rules.

215. Wet van 1 november 2001 houdende regeling van de taken voor de meteorologie en andere geofysische terreinen (Wet op het Koninklijk Nederlands Meteorologisch Instituut), <http://wetten.overheid.nl/BWBR0012952/>, 22 Feb. 2010 [in Dutch].

216. KPMG, 'Ministerie van Waterstaat. Evaluatie van de wet op het KNMI', 2007, 71, <www.verkeerenwaterstaat.nl/Images/2008651%20bijlage_tcm195-231492.pdf>, 22 Feb. 2010 [in Dutch].

Even for services that are seen as public tasks by both the public and the private sector, it could be questioned whether they can be considered as economic activities. For instance, the provision of weather information to the Ministry of Defence or for rescue missions by the coastguard or in case of disasters is generally seen as a public task of the meteorological office, but these services might also be provided by the private sector on the market. However, the connection with the exercise of official authority could also be argued, particularly in the defence and disaster cases.

EXAMPLE: *RIJKSWATERSSTAAT* – Another Dutch example that caused discussion is the national road database.[217] *Rijkswaterstaat*, the executive agency of the Dutch Ministry of Transport, Public Works and Water Management decided in 2006 to make this database freely available on the Internet for access and re-use. The discussion this caused among the different stakeholders shall be addressed later on in this chapter. However, here the question can already be addressed whether this should be considered as an economic activity and whether to this extent *Rijkswaterstaat* can be considered as an undertaking subject to EU competition law. On the one hand, the fact that private bodies already offered such road databases on the market before *Rijkswaterstaat* decided to make its own database publicly available, would indicate that this is an economic activity. On the other hand, an independent commission installed by the Ministry stated that there was a public interest considering the national programme for establishing authentic sources and the requirements from INSPIRE to make spatial data available.[218] In addition, the Minister stated that the data set was needed by policy makers and road management to perform their tasks properly, and by contractors working on the roads.[219] Hence, once again it seems logical that the creation of the database is connected to the exercise of official authority, and the same can be said for sharing the database between all the public bodies that need it. However, making the database publicly available for access and re-use by the public may be less connected to the official authority and leaning more towards the provision of a service that can also be done by the private sector.

EXAMPLE: ORDNANCE SURVEY – As was mentioned before, the activities of the Ordnance Survey have been on the political agenda in the United Kingdom for a number of years. The distinction between the public tasks and the commercial activities of the Ordnance Survey is considered blurry and confusing. Moreover, the Ordnance Survey is required to make a return on its activities. This makes one wonder whether even more activities of this mapping agency could be considered

217. See <www.rijkswaterstaat.nl/apps/geoservices/portaal/index-applicaties.html> 22 Feb. 2010.

218. Commissie Tweede Consultatieronde Vrijgeven NWB Bestand, 'Advies', 2006, 15, <www.verkeerenwaterstaat.nl/Images/Advies%20Cie%20tweede%20consultatie_tcm195-189308.pdf>, 22 Feb. 2010 [in Dutch].

219. Minister of Transport and Water Management, *Reply to questions from Members of Parliament Roemer and Roefs*, 16 Mar. 2007, question no. 2060707530, <http://parlando.sdu.nl/cgi/login/anonymous>, 22 Feb. 2010 [in Dutch]; Vaste Commissie voor Verkeer en Waterstaat, *Verslag van een algemeen overleg*, 30 Jan. 2008, 31 200 XII, no. 73, <http://parlando.sdu.nl/cgi/login/anonymous>, 22 Feb. 2010 [in Dutch].

as economic activities under EU competition law than would be the case in the previous examples. For example, under the 2004 Framework Document, the Ordnance Survey's objectives include the collection, portrayal, and distribution of the definitive record of the natural, built, and planned environment of Great Britain that meets customer needs and the national interest in the most effective manner; provide the underpinning framework for the government and the private sector to join up its spatial information; and advise the United Kingdom government on all aspects of survey, mapping and geographic information. While the activities that are related to the provision of public sector spatial data to the government for the performance of its public tasks can easily be seen as connected to the exercise of official authority, this is much more difficult for providing a framework for the private sector and 'meeting customer needs'. Can the activities in question be considered as economic activities? Other strategic objectives mentioned in the Framework document include: 'create, develop and maintain strategic and commercial partnerships that will add further value to the Ordnance Survey data and products; grow the geographic information market and champion the extended use and sharing of geographic information in the government, business and leisure communities; generate profitable revenue that will fund continuous improvement in database content, data structure, data delivery, up-to-dateness, fitness for purpose and accuracy'.[220] While it is true that economic purpose or making profit is not required in order for an activity to be considered as economic, it may be considered an extra indication that while these activities are part of the the Ordnance Survey's public task, they are also economic activities making the Ordnance Survey subject to competition law.

SUMMARY – The previous examples show that many spatial information services provided by public bodies may be designated as economic activities, even though they are considered as public tasks under national law and hence under the PSI Directive and the INSPIRE Directive. Thus, the performance of these public tasks may also be subject to competition law. The conflicts that this may cause can possibly be solved under the provision of Article 106(2) of the Treaty on the Functioning of the EU. This will be discussed later on in this chapter.

B. DISTORTIONS OF COMPETITION BY PUBLIC BODIES PROVIDING SPATIAL INFORMATION SERVICES AS UNDERTAKINGS

OUTLINE – When public bodies are found to be undertakings and fall under the application of competition law, they can distort competition on the market in a number of ways. These distortions are discussed in this section. First, only a brief description is provided of the concluding of agreements that distort competition,

220. Ordnance Survey, 'Framework document July 2004', 2004, 5, <www.ordnancesurvey.co.uk/oswebsite/aboutus/reports/frameworkdocument/docs/frameworkdocument2004.pdf>, 22 Feb. 2010.

because this situation is much less common than the following types of distortions. The second subsection addresses abuse of a dominant position on the market, which counts for the largest part of market distortions. For each of these distortions, the rules are explained, followed by the application of the rules to the situation of spatial information services.

1. Article 101 Treaty on the Functioning of the EU: Agreements between Undertakings

a. General

ARTICLE 101 OF THE TREATY ON THE FUNCTIONING OF THE EU – Article 101 of the Treaty on the Functioning of the EU (former Article 81 of the EC Treaty) prohibits agreements between undertakings, decisions by associations of undertakings, and concerted practices that may affect trade between Member States, which have as their object or effect the prevention, restriction, or distortion of competition within the common market. Particular examples that are given include:

- the direct or indirect fixing of purchasing or selling prices or any other trading conditions;
- the limiting or controlling of production, markets, technical development, or investment;
- the sharing of markets or sources of supply;
- the application of dissimilar conditions to equivalent transactions with other trading parties, thereby placing them at a competitive disadvantage;
- the subjection of the conclusion of contracts to acceptance by the other parties of supplementary obligations which, by their nature or according to commercial usage, have no connection with the subject of such contracts.

CONSEQUENCES – If an agreement or decision distorts competition in a way described in this list, it is automatically void. However, the European Commission can declare the prohibition of Article 101 inapplicable for agreements, decisions, or practices that contribute to improving the production or distribution of goods or to promoting technical or economic progress, while allowing customers a fair share of the resulting benefit. The agreements, decisions, or practices cannot impose on the undertakings concerned restrictions that are not indispensable to the attainment of these objectives or afford the undertakings the possibility of eliminating competition with respect to a substantial part of the products in question (Article 101(3) of the Treaty on the Functioning of the EU).

NEW PROCEDURE – Since 1 May 2004, it is no longer the exclusive prerogative of the Commission to apply Article 101(3) (former Article 81(3)). The national competition authorities or the national courts may now also apply Article 101(3) whenever an agreement's compatibility with the provision is raised. Moreover, the Commission can no longer give a prior exemption to an agreement to prevent it from being challenged.[221] Existing individual exemptions all had a time limit and will therefore expire in due course.[222]

Horizontal and vertical agreements – Article 101 precludes restrictive agreements of a horizontal nature (between parties operating at the same level of the economy, often competitors) and of a vertical nature (between parties operating at a different level, e.g., between a manufacturer and its distributor).[223]

b. Applicability to Spatial Information Services

Example: Eurogeographics – There are several groups or organizations of undertakings providing public sector spatial data that could be covered under Article 101 of the Treaty on the Functioning of the EU if they would distort competition. For instance, Eurogeographics is an association under French law that represents fifty-two National Mapping and Cadastral Agencies located in forty-three countries. Its objectives are to act as the official and united voice for its members, to promote the National Mapping Agencies' national and pan-European products and services, and to promote collaboration and sharing of best practice between its members.[224] As Eurogeographics and its members sell products or services that could most likely be delivered by private entities on the market, they can be classified as undertakings. Hence, if the members of Eurogeographics would enter into agreements about, for example, price fixing for their activities that could be qualified as economic, and those agreements would distort or limit competition, Article 101 of the Treaty on the Functioning of the EU might be applicable.

Example: Ecomet – Another example of a group or association of undertakings that might fall under Article 101 is Ecomet. This association was set up in 1995 as an Economic Interest Grouping under Belgian law and is located in Brussels. Its objective is 'to preserve the free and unrestricted exchange of meteorological information between the [National Meteorological Services] for their operational functions within the framework of WMO regulations and to ensure the widest availability of basic meteorological data and products for commercial applications. Further objectives of Ecomet are: to guarantee the access to meteorological data and products, be it for public or private sectors; to recover part of the infrastructural expenses of the European [National Meteorological Services] by a contribution from all commercial users, in order to maintain and improve the meteorological infrastructure'.[225] The members of Ecomet have agreed on a joint data licensing

221. Alison Jones & Brenda Sufrin, *EC Competition Law: Text, Cases and Materials* (Oxford: Oxford University Press, 2008), 124–125.
222. Richard Whish, *Competition law*, 6th edn (Oxford: Oxford University Press, 2009), 163.
223. Alison Jones & Brenda Sufrin, *EC Competition Law: Text, Cases and Materials* (Oxford: Oxford University Press, 2008), 121.
224. See Eurogeographics, 'Mission and Objectives', <www.eurogeographics.org/eng/01_mission.asp>, 22 Feb. 2010.
225. Ecomet, 'The Concept of ECOMET', <www.ecomet.eu/description.htm#concept>, 22 Feb. 2010.

policy, which includes standard licenses for different user groups, volume discounts, and special tariff schemes. However, the prices for the products are still determined by the National Meteorological Services.

EXAMPLE: ECOMET COMFORT LETTER – After four years of negotiations, the European Commission issued a comfort letter in 1999, exempting Ecomet from the application of Articles 81 and 82 of the EC Treaty (current Articles 101 and 102 of the Treaty on the Functioning of the EU). The Commission was satisfied that there was still adequate competition between the members of Ecomet, particularly on prices, since their commercial divisions competed with each other and the grouping operated in such a way as to permit fair competition with independent service providers.[226] The Commission also included a number of remarks and conditions that were more related to the application of Article 82 of the EC Treaty (current Article 102 of the Treaty on the Functioning of the EU) on the abuse of a dominant position. This shows that Article 101 and Article 102 sometimes are applicable together, that is, if the association of undertakings has a dominant position on the market. In that case it might restrict competition under Article 101 and at the same time abuse its dominant position under Article 102. Abuse of dominant position is addressed in the next section.

2. Article 102 Treaty on the Functioning of the EU: Abuse of a Dominant Position

a. General

ARTICLE 102 TREATY ON THE FUNCTIONING OF THE EU – Article 102 of Treaty on the Functioning of the EU (former Article 82 of the EC Treaty) deals with the abuse by an undertaking of its dominant position. It prohibits 'any abuse by one or more undertakings of a dominant position within the common market or in a substantial part of it . . . insofar as it may affect trade between Member States'. Article 102 gives a list of potential abuses:

- directly or indirectly imposing unfair purchase or selling prices or other unfair trading conditions;
- limiting production, markets or technical development to the prejudice of consumers;
- applying dissimilar conditions to equivalent transactions with other trading parties, thereby placing them at a competitive disadvantage;

226. Michael Kamps, 'Thunder and Lightning: Public Sector Information Policy Experiences of Private Meteorological Service Providers', in *Public Sector Information in the Digital Age, Between Markets, Public Management and Citizens' Rights*, ed. Georg Aichholzer & Herbert Burkert (Cheltenham: Edward Elgar Publishing Limited, 2004), 175–176. See Association of Environmental Data Users in Europe, 'Comfort Letter on the joint sale of meteorological products by Ecomet. Comments', 1999, <www.primet.org/index.php?Itemid=29&id=81&option=com_content&task=view>, 22 Feb. 2010.

- making the conclusion of contracts subject to acceptance by the other parties of supplementary obligations which, by their nature or according to commercial usage, have no connection with the subject of such contracts.

CONSTITUENT ELEMENTS – Hence, there are several constituent elements in order for a breach of article to be present: a dominant position must be held; this position must be held by one or more undertakings; there must be actual abuse of that dominant position; and there must be an actual or potential effect on trade between Member States.[227] As the definition of an *undertaking* has already been discussed, and describing the potential effect on trade between Member States is not of immediate importance for my analysis, I will only address the dominant position and its abuse.

i. Dominant Position

DEFINITION – A *dominant position* is defined by the ECJ as 'a position of economic strength enjoyed by an undertaking which enables it to hinder the maintenance of effective competition on the relevant market by allowing it to behave to an appreciable extent independently of its competitors and customers and ultimately of consumers'.[228] Dominance is not an objective concept, but should be seen in relation to the particular market the undertaking is operating in.[229]

CRITERIA – For determining whether an undertaking is dominant, the European Commission and the ECJ rely both on the market shares of the undertaking and on other factors indicating dominance, such as economies of scale and scope, privileged access to supply, the established position of the undertaking on the market, and other strategic barriers to entry.[230] When an undertaking has a monopoly, the dominance is clear. When this is not the case, the opinion of the ECJ is that the higher the market share is, the more likely dominance is found.[231] Other factors may include superior technology and efficiency, vertical integration, economies of scale and scope, access to key inputs, or access to financial resources.

---

227. Peter Roth & Vivien Rose (ed.), *Bellamy & Child European Community Law of Competition*. 6th edn (Oxford: Oxford University Press, 2008), 917.
228. CFI, Case T-203/01, *Manufacture française des pneumatiques Michelin v. Commission*, 30 Sep. 2003, para. 30, *European Court Reports* 2003, II-4071. See also ECJ, Case 85/76, *Hoffman-La Roche & Co. AG v. Commission*, 13 Feb. 1979, para. 38, *European Court reports* 1979, 461.
229. ECJ, Case 6/72, *Europemballage Corporation and Continental Can Company Inc. v. Commission*, 21 Feb. 1973, *European Court reports* 1973, 215. See Richard Whish, *Competition law*, 6th edn (Oxford: Oxford University Press, 2009), 174–183.
230. Alison Jones & Brenda Sufrin, *EC Competition Law: Text, Cases and Materials* (Oxford: Oxford University Press, 2008), 396; Richard Whish, *Competition law*, 6th edn (Oxford: Oxford University Press, 2009), 180–181.
231. Alison Jones & Brenda Sufrin, *EC Competition Law: Text, Cases and Materials* (Oxford: Oxford University Press, 2008), 397. See also A.A. Van Rossum, 'Oneerlijke concurrentie door de overheid', in *Overheid en onderneming. Tussen publiek en privaat ondernemen*, ed. Sebastianus C.J.J. Kortmann et al. (Deventer: W.E.J. Tjeenk Willink, 1998), 226 [in Dutch].

ii. Abuse of Dominant Position

Definition – Dominance itself is not prohibited; it is only the abuse of this dominance that can be challenged.[232] Article 102 Treaty on the Functioning of the EU does not provide a general definition of abuse, apart from the examples that are given. The definition of abuse used by the European Commission and the ECJ is that of the *Hoffman-La Roche* case:

> The concept of abuse is an objective concept relating to the behaviour of an undertaking in a dominant position which is such as to influence the structure of a market where, as a result of the very presence of the undertaking in question, the degree of competition is weakened and which, through recourse to methods different from those which condition normal competition in products or services on the basis of the transactions of commercial operators, has the effect of hindering the maintenance of the degree of competition still existing in the market or the growth of that competition.[233]

Special responsibility of dominant undertakings – According to the ECJ's case law, dominant undertakings have a 'special responsibility' towards the competitive process, due to the prejudice their activities may cause to competition in general and to the interests of competitors, suppliers, customers and consumers.[234] This entails that conduct on the market that may be tolerated in a normal competitive environment, may be considered to illegally distort competition if it is performed by a dominant undertaking.

Types of abuse – Generally, two types of abuse are discerned: on the one hand, '[a]n exclusionary abuse is conduct which has the economic effect of impeding effective competition on the relevant market (by forcing out or marginalising existing competitors and/or raising barriers to entry for potential new competitors). On the other hand, an exploitative abuse is conduct which is unfair or unreasonable towards those persons who depend on the dominant firm for the supply of goods or services on the relevant market'.[235] Another distinction can be made between abusive acts that affect competition on the market where dominance is enjoyed and acts that affect competition on a different market, for example, an undertaking is dominant on the market of raw material and its actions have an effect on the downstream market.[236]

---

232. Richard Whish, *Competition law*, 6th edn (Oxford: Oxford University Press, 2009), 198.
233. ECJ, Case 85/76, *Hoffman-La Roche & Co. AG v. Commission*, 13 Feb. 1979, para. 91, *European Court reports* 1979, 461.
234. CFI, Case T-203/01, *Manufacture française des pneumatiques Michelin v. Commission*, 30 Sep. 2003, *European Court Reports* 2003, II-4071; ECJ, Case 322/81, *NV Nederlandsche Banden Industrie Michelin v. Commission*, 9 Nov. 1983, para. 57, *European Court reports* 1983, 3461; CFI, Case T-111/96, *ITT Promedia NV v. Commission*, 17 Jul. 1998, para. 139, *European Court reports* 1998, II-2937.
235. Peter Roth & Vivien Rose (ed.), *Bellamy & Child European Community Law of Competition*. 6th edn (Oxford: Oxford University Press, 2008), 953. See also Alison Jones & Brenda Sufrin, *EC Competition Law: Text, Cases and Materials* (Oxford: Oxford University Press, 2008), 435; Richard Whish, *Competition law*, 6th edn (Oxford: Oxford University Press, 2009), 199.
236. Alison Jones & Brenda Sufrin, *EC Competition Law: Text, Cases and Materials* (Oxford: Oxford University Press, 2008), 436.

REFERRAL – Several examples of possible abusive behaviour can be given. For more information, the reader is referred to the extensive literature on the subject.[237] In the next subsection, only some examples of abusive acts will be given that are relevant in the framework of the provision of spatial information services by public bodies.

iii. Examples of Abusive Acts with Regard to Pricing

PRICE DISCRIMINATION – The most obvious types of abusive acts can be found with regard to pricing. For instance, price discrimination can constitute an abuse. This occurs where the same good or service is sold at different prices to different customers despite the same costs, or at the same price despite different costs. Not all price discrimination constitutes an abuse, but where the exact boundary lies, remains uncertain.[238] The discrimination may affect the competitors of the undertaking on the same market, but the impact may also be in the downstream market. Jones and Sufrin give an example: 'If a supplier sells a product to X cheaper than to Y, and X and Y are competing manufacturers who need the product as an input, then X's costs will be lower than Y's. The supplier will have distorted competition between them'.[239]

EXCESSIVE PRICING – Second, excessive pricing is an obvious way of exploiting a dominant position on the market. In the *United Brands* case, the ECJ accepted that excessive prices can constitute an abuse and that charging a price which has no relation to the product's economic value would be excessive.[240] But, Jones and Sufrin ask, '[w]hat is the economic value of a banana other than what a customer is prepared to pay for it?'[241]

PREDATORY PRICING – A third pricing issue that can be problematic is predatory pricing. This occurs when an undertaking prices its products so low that competitors cannot follow and are driven from the market. According to Roth and Rose,

237. For example, Alison Jones & Brenda Sufrin, *EC Competition Law: Text, Cases and Materials* (Oxford: Oxford University Press, 2008), 1418 p.; Peter Roth & Vivien Rose (ed.), *Bellamy & Child European Community Law of Competition.* 6th edn (Oxford: Oxford University Press, 2008), 2252 p.; Richard Whish, *Competition law*, 6th edn (Oxford: Oxford University Press, 2009), 1006 p.; Jonathan Faull, *Faull and Nikpay: The EC Law of Competition* (Oxford: Oxford University Press, 2007), 2000 p.

238. Peter Roth & Vivien Rose (ed.), *Bellamy & Child European Community Law of Competition.* 6th edn (Oxford: Oxford University Press, 2008), 963.

239. Alison Jones & Brenda Sufrin, *EC Competition Law: Text, Cases and Materials* (Oxford: Oxford University Press, 2008), 442; Peter Roth & Vivien Rose (ed.), *Bellamy & Child European Community Law of Competition.* 6th edn (Oxford: Oxford University Press, 2008), 963–965.

240. ECJ, Case 27/76, *United Brands Company and United Brands Continentaal BV v. Commission*, 14 Feb. 1978, para. 250, *European Court reports* 1978, 207. See also ECJ, Case 26/75, *General Motors Continental NV v. Commission*, 13 Nov. 1975, *European Court reports* 1975, 1367; ECJ, Case 30/87, *Corinne Bodson v. Pompes Funèbres des Régions Libérées SA*, 4 May 1988, *European Court Reports* 1988, 2479.

241. Alison Jones & Brenda Sufrin, *EC Competition Law: Text, Cases and Materials* (Oxford: Oxford University Press, 2008), 588.

in principle three main elements have to be present in order for predatory pricing to be determined: (i) a temporary, substantial reduction in price by the dominant firm in response to an anticipated or actual market entry or growth in sales by a smaller competitor; (ii) price levels that are in themselves unprofitable or barely profitable; and (iii) pricing 'aimed at' eliminating or disciplining a specific competitor.[242] Once the competitors are eliminated the undertaking can raise its prices again to monopoly levels and recoup the losses incurred. Jones and Sufrin warn that there is great controversy about predatory pricing.[243] The basic concept of predatory pricing is that an undertaking prices below cost.[244] However, it is usually very difficult to calculate the costs of a particular good or service, and the relationship between the costs and the prices. Moreover, in certain circumstances, pricing below cost can be a normal commercial practice. This situation is not always easy to distinguish from predatory pricing.

MARGIN SQUEEZING – A fourth dominant pricing practice is referred to as *price* or *margin squeezing*. If an undertaking operates in both upstream and downstream markets and has a dominant position in the upstream market, it can prevent its competitors from operating profitably downstream. The CFI described price squeezing as follows: 'Price squeezing may be said to take place when an undertaking which is in a dominant position on the market for an unprocessed product and itself uses part of its production for the manufacture of a more processed product, while at the same time selling off surplus unprocessed product on the market, sets the price at which it sells the unprocessed product at such a level that those who purchase it do not have a sufficient profit margin on the processing to remain competitive on the market for the processed product'.[245] Jones and Sufrin state that there are two basic ways to squeeze out the competitors: either the dominant undertaking prices its products in the downstream market very low, or it charges high prices to the competitors for the raw materials.[246] The former is linked to the previous example of predatory pricing. The essence of this type of abuse is not the conduct on the upstream market or the behaviour on the downstream market, but rather the relationship between the two.

242. Peter Roth & Vivien Rose (ed.), *Bellamy & Child European Community Law of Competition*. 6th edn (Oxford: Oxford University Press, 2008), 955.

243. Alison Jones & Brenda Sufrin, *EC Competition Law: Text, Cases and Materials* (Oxford: Oxford University Press, 2008), 445.

244. CFI, Case T-83/91, *Tetra Pak International SA v. Commission*, 6 Oct. 1994, *European Court reports* 1994 II-755; ECJ, Case C-333/94 P, *Tetra Pak International SA v. Commission*, 14 Nov. 1996, *European Court reports* 1996, I-5951; CFI, Case T-340/3, *France Télécom SA v. Commission*, 30 Jan. 2007, *European Court reports* 2007, II-107; ECJ, Case C-202/07P, *France Télécom SA v. Commission*, 2 Apr. 2009 (not yet published).

245. CFI, Case T-5/97, *Industrie des poudres sphériques v. Commission*, 30 Nov. 2000, para. 178, *European Court Reports* 2000, II-3755.

246. Alison Jones & Brenda Sufrin, *EC Competition Law: Text, Cases and Materials* (Oxford: Oxford University Press, 2008), 466.

iv. Example of Abusive Acts: Refusal to Supply

DEROGATION TO THE GENERAL PRINCIPLE OF FREEDOM OF CONTRACT – The refusal of a dominant undertaking to supply goods or services has been considered an abuse in some cases by the Commission and the ECJ.[247] The idea that an undertaking has a duty to supply and cannot refuse is contrary to basic principle of contract law that everyone should be free to enter into a contract with whomever he chooses. Yet there are situations in which the European Commission and the ECJ have limited that freedom.[248] While the ECJ has never said that all dominant undertakings have an absolute duty to supply all those who request them to do so[249], in some situations, not doing so has been considered an abuse. This also includes making supplies conditional on control over further processing or marketing, supplying only on discriminatory and unfair conditions, or making an offer that the supplier knows is unacceptable.[250]

EXAMPLES – The first case in which a refusal to supply was considered an abuse was *Commercial Solvents*. An abuse was found because of three elements: the dominant undertaking was using its dominant position on the raw material market to affect competition on the derivatives market; it refused to supply an existing customer because it wanted to compete with it downstream; and the refusal risked eliminating the customer from the downstream market.[251] Another good example is the *British Sugar* case, where the dominant supplier of industrial sugar refused to supply to a competitor in the market for retail sugar.[252]

ESSENTIAL FACILITIES – Through the case of abuse of dominance due to refusal to supply, the *essential facilities* doctrine was developed in the European Union. While the doctrine is contested, it deserves to be mentioned because it may be applicable to spatial information services. An essential facility was defined by the ECJ as 'a facility or infrastructure without access to which competitors cannot provide services to their customers'.[253] The concept has to be interpreted narrowly,

---

247. See Laurent Richer, 'Droit d'accès et service public', *Actualité juridique Droit Administratif*, 16 Jan. 2006, 74 [in French].
248. CFI, Case T-41/96, *Bayer AG v. Commission*, 26 Oct. 2000, para. 180, *European Court Reports* 2000, II-3383. See Alison Jones & Brenda Sufrin, *EC Competition Law: Text, Cases and Materials* (Oxford: Oxford University Press, 2008), 529; Peter Roth & Vivien Rose (ed.), *Bellamy & Child European Community Law of Competition.* 6th edn (Oxford: Oxford University Press, 2008), 1004.
249. Alison Jones & Brenda Sufrin, *EC Competition Law: Text, Cases and Materials* (Oxford: Oxford University Press, 2008), 529.
250. *Ibid.*; Peter Roth & Vivien Rose (ed.), *Bellamy & Child European Community Law of Competition.* 6th edn (Oxford: Oxford University Press, 2008), 1004.
251. ECJ, Case 6/73 and 7/73, *Istituto Chemioterapico Italiano S.p.A. and Commercial Solvents Corporation v. Commission*, 6 Mar. 1974, para. 25, *European Court reports* 1974, 223. See Alison Jones & Brenda Sufrin, *EC Competition Law: Text, Cases and Materials* (Oxford: Oxford University Press, 2008), 530.
252. Commission Decision 88/518/EEC of 18 Jul. 1988 relating to a proceeding under Art. 86 of the EEC Treaty (Case No IV/30.178 Napier Brown – British Sugar), *OJ L* 284, 19 Oct. 1988, 41.
253. Commission Decision 94/19/EC of 21 Dec. 1993 relating to a proceeding pursuant to Art. 86 of the EC Treaty (IV/34.689 – Sea Containers v. Stena Sealink – Interim measures), *OJ L* 15,

as it causes a severe interference with the rights of the dominant undertaking to choose whom it contracts with. The facility has to be essential, not merely desirable or more efficient.[254] According to the CFI, a facility is only essential if there are no substitutes. Mere advantage to the competitor is not sufficient.[255] The ECJ stressed that the refusal must be likely to eliminate all competition, not just make it harder. Access must also be indispensable, not desirable or convenient.[256] In addition, some practical difficulties remain. Some facilities may have limited capacity, such as harbours, so the question arises who has to get access.[257] Next, at what terms should access be given? If it is left to the dominant undertaking, it may charge an unreasonably high price. But can the Commission or the ECJ interfere and determine the price against which access should be given? In any case, an overextension of the essential facilities doctrine might lead to undertakings having to give access to assets they have developed at great expense over the years, discouraging them from further investments and innovation.

INTELLECTUAL PROPERTY RIGHTS – A possible abuse that has been linked to the essential facilities doctrine is the refusal to supply others with IPR.[258] Refusal to license IPR is not an abuse in itself but in exceptional circumstances it might lead to the infringement of Article 102 Treaty on the Functioning of the EU. For instance, the *Magill* case concerned copyright in television listings. The ECJ found an abuse of former Article 82 of the EC Treaty because of the following reasons: there was no substitute for the listings of the dominant undertakings (broadcasting corporations); the refusal to license the IPR prevented the appearance of a new product for which there was potential customer demand (a weekly overview of television listings for all channels); there was no justification for the refusal; and the dominant undertakings were excluding all competition on the secondary market of weekly listings and reserving the market to themselves.[259] In the *IMS Health*

18 Jan. 1994, 8; Commission Decision 94/119/EC of 21 Dec. 1993 concerning a refusal to grant access to the facilities of the port of Rødby (Denmark), *OJ L* 55, 26 Feb. 1994, 52.

254. See Alison Jones & Brenda Sufrin, *EC Competition Law: Text, Cases and Materials* (Oxford: Oxford University Press, 2008), 542; Jan J.C. Kabel et al., *Kennisinstellingen en informatiebeleid. Lusten en lasten van de publieke taak* (Amsterdam: Universiteit van Amsterdam, 2001), 49 [in Dutch].
255. CFI, Cases T-374/94, T-375/94, T-384/94 and T-388/94, *European Night Services and others v. Commission*, 15 Sep. 1998, para. 208, *European Court reports* 1998, II-3141.
256. ECJ, Case C-7/97, *Oscar Bronner GmbH & Co. KG v. Mediaprint Zeitungs- und Zeitschriftenverlag GmbH & Co. KG and others*, 26 Nov. 1998, para. 41, *European Court reports* 1998, I-7791. See Alison Jones & Brenda Sufrin, *EC Competition Law: Text, Cases and Materials* (Oxford: Oxford University Press, 2008), 551; Jan J.C. Kabel et al., *Kennisinstellingen en informatiebeleid. Lusten en lasten van de publieke taak* (Amsterdam: Universiteit van Amsterdam, 2001), 49 [in Dutch].
257. Alison Jones & Brenda Sufrin, *EC Competition Law: Text, Cases and Materials* (Oxford: Oxford University Press, 2008), 542; Peter Roth & Vivien Rose (ed.), *Bellamy & Child European Community Law of Competition.* 6th edn (Oxford: Oxford University Press, 2008), 1015.
258. Alison Jones & Brenda Sufrin, *EC Competition Law: Text, Cases and Materials* (Oxford: Oxford University Press, 2008), 553.
259. ECJ, Cases C-241/91P and C-242/91P, *Radio Telefis Eireann v. Commission*, 6 Apr. 1995, *European Court reports* 1995, I-743. See Alison Jones & Brenda Sufrin, *EC Competition Law:*

case, the ECJ reiterated the conditions: there must be a new product involved; access to the protected material must be indispensable so that the refusal will include *any* or *all* competition on a secondary market; and the refusal must be unjustified.[260]

b. Applicability to Spatial Information Services

INFORMATION SERVICES AS MONOPOLY ACTIVITIES – Several elements of what has been set out above on Article 102 of the Treaty on the Functioning of the EU may also be applicable to the provision of spatial information services if these can be considered as an economic activity. This is particularly the case because public bodies often have a natural or legal monopoly on the public sector spatial data they hold, which has an impact on their information activities on the market.[261]

PREDATORY PRICING – The provision of information services by public bodies can lead to abuse of a dominant position in a number of ways. A first possibility is predatory pricing. This may be caused by the public body wanting to provide cheaper information services than its competitors on the market, but also because the public body itself believes that it is providing information services under its public task, while a court decision may decide otherwise on a challenge. Such cross-subsidization to the disadvantage of private sector competitors may be an abuse.[262]

SOLUTION: MARKET PRICES? – Therefore, the United Kingdom's Treasury guidelines on charging stated that where similar information is produced by the private sector, charges should reflect market prices, to safeguard fair competition. 'The aim is a price which, taking account of all the circumstances, it is reasonable to consider fair to both the purchasing public and any competitors supplying

---

Text, Cases and Materials (Oxford: Oxford University Press, 2008), 557–559; Peter Roth & Vivien Rose (ed.), *Bellamy & Child European Community Law of Competition.* 6th edn (Oxford: Oxford University Press, 2008), 1010; Jan J.C. Kabel et al., *Kennisinstellingen en informatiebeleid. Lusten en lasten van de publieke taak* (Amsterdam: Universiteit van Amsterdam, 2001), 50–51 [in Dutch].

260. ECJ, Case C-418/01, *IMS Health GmbH & Co. OHG*, 29 Apr. 2004, para. 38, 52, *European Court Reports* 2004, I-5039.

261. Johan Pas & Bruno De Vuyst, 'Re-Establishing the Balance Between the Public and the Private Sector: Regulating Public Sector Information Commercialization in Europe', *Journal of Information, Law and Technology* no. 2 (2004): 6, <www2.warwick.ac.uk/fac/soc/law/elj/jilt/2004_2/pasanddevuyst>, 22 Feb. 2010; Oxera, 'Public information, private profit : how should government agencies compete ?', 2005, 2, <www.oxera.com/cmsDocuments/Agenda_Oct%2005/Public%20information% 20private %20profit.pdf>, 22 Feb. 2010.

262. François Salgé, 'Geographic Information Network in Europe (GINIE). Recommendations for action', in *Spatial data infrastructure and policy development in Europe and the United States*, ed. Bastiaan Van Loenen & Bas C. Kok (Delft: DUP Science, 2004), 40; Yvo Volman, 'The re-use of public sector information in the EU. Increasing the scope for cross-border exploitation', in *Spatial data infrastructure and policy development in Europe and the United States*, ed. Bastiaan Van Loenen & Bas C. Kok (Delft: DUP Science, 2004), 62.

similar products and services to the public'.[263] It admitted that cross-subsidies are allowable on a temporary basis and that a product might be charged less than break-even for a short period, but 'no service should ever be provided at less than its average variable costs' and 'prices should not be reduced in such a way to stifle competition'.[264]

Example: *Rijkswaterstaat* – A possible example of such abuse, even though it may not be intended as such, is the freely available Dutch national road database. As was mentioned above, this database is to be made available on the Internet by *Rijkswatersstaat*, free of charge for re-use by all interested parties. This initiative was subject to heavy protests from some private sector mapping companies that produce and sell roadmap data sets, who claimed that making the database available would ruin their business. They had invested large sums of money in the creation of their datasets and they would not be able to maintain their business model, as the companies they supplied to would most likely switch to the freely re-usable national database.[265] Van Eechoud and Van Der Wal called this situation an 'example of a potential market distortion problem caused by liberal dissemination policies' and stated that '[m]aking available data for free can be the equivalent to price-dumping'.[266] However, not all the conditions would be fulfilled to speak about dumping or predatory pricing. The charges are not temporarily low to be raised again when the competitors have left the market, nor can it be said that *Rijkswaterstaat* intends to drive the private competitors out of the market.

Example: marine information – Another type of abuse that may occur is price discrimination, which has an effect on a secondary market. For instance, with regard to marine information resources, the United Kingdom's Marine Information Council Working Group on Data Access report stated that: '[s]ome of the organisations that hold publicly funded marine information compete directly with commercial organizations in providing ''value added'' data products. These public

263. HM Treasury, 'Charges for information: when and how. Guidance for government departments and other Crown bodies', 2000, 4-5, 10, <www.hm-treasury.gov.uk/media/7/C/charging_for_info.pdf>, 22 Feb. 2010.
264. *Ibid.*, 11.
265. Algemeen Dagblad Press Article, 'Kaartenmakers: 'oneerlijk spel Eurlings', 28 Jan. 2008, <www.ad.nl/multi-media/article2005979.ece>, 22 Feb. 2010 [in Dutch]; PocketInfo.nl Press Article, 'Falkplan, NAVTEQ en TeleAtlas boos over vrijgave wegenbestand', 29 Jan. 2008, <www.pocketinfo.nl/nieuws/navigatie/9874/falkplan-navteq-en-teleatlas-boos-over-vrijgave-wegenbestand.html>, 22 Feb. 2010 [in Dutch]; Mireille Van Eechoud, 'Het hergebruik regime voor overheidsinformatie in de Wob, een tussenstand', *Mediaforum* 20, no. 1 (2008): 9 [in Dutch]; Bastiaan Van Loenen et al., 'Open toegankelijkheidsbeleid voor geo-informatie vergeleken: het gras leek groener dan het was?', 2007, 11, <www.minbzk.nl/aspx/download.aspx?file=/contents/pages/87285/opentoegankelijkheidsbeleidvooroverheidvergeleken.pdf>, 22 Feb. 2010 [in Dutch].
266. Mireille Van Eechoud & Brenda Van Der Wal, 'Creative commons licensing for public sector information. Opportunities and pitfalls', 2008, 62, <www.ivir.nl/publications/eechoud/CC_PublicSectorInformation_report_v3.pdf>, 22 Feb. 2010. See also Bastiaan Van Loenen et al., 'Beschikbaar stellen van geo-informatie bij Rijkswaterstaat. Analyse van de (on)mogelijkheden van het op korte termijn vrij beschikbaar stellen van vier geo-data sets', 2006, <www.bastiaanvanloenen.nl/pubs/2006_RWS_AGI_Eindrapport_TUD.pdf>, 22 Feb. 2010[in Dutch].

sector bodies have blanket access to publicly funded information within their own organization at no net cost. Public sector organizations holding the information effectively have free access whereas commercial organizations must pay for the source information'.[267] This type of activities could also be considered as a form of cross-subsidization: the preferential use of a database produced within the public task by the commercial arm of the public body, while competitors have to pay more for using the database.[268]

EXAMPLE: ENVIRONMENT AGENCY – Also in the United Kingdom, the Environment Agency was challenged in 2002 for possible margin squeeze abuse. The Environment Agency provided environmental data to third parties for use in compiling property-specific environmental risk reports, but it also launched its own service for property search reports in the downstream market. The private companies, which had become competitors, complained that the Environmental Agency was abusing its dominant position. OFT felt that there could be a possibility for margin squeezing if the margin between the price of the data and the price of the product was too small to enable a reasonable return on the downstream market, but that there was no sufficient evidence of this actually occurring to support the claim.[269] However, OFT found later that the Environment Agency was treating its customers and competitors fairly by imposing the same conditions on itself as on its competitors.[270] Yet, the Agency closed down the service in 2007, leaving its resellers to provide the service.

EXAMPLE: METEOSERVICES – Another case where a public body was challenged for its distortion of competition was the complaint of the Belgian company Meteoservices, which operates on the weather market. It filed a complaint against the Belgian Meteorological Institute with the National Competition Authority on the basis that the Institute charged excessive prices for its data, applied price discrimination, and charged dumping prices on the secondary market of information products, cross-subsidizing these with services delivered under the public task.[271] The complaint was filed in 2004 but dropped by the authority responsible for the investigation of the case. According to Weiss, comparable cases could be found in Switzerland, Germany, Finland, and Sweden.[272]

---

267. United Kingdom Marine Information Council Working Group on Data Access, 'Achieving optimal value from publicly funded marine information resources', 2003, 8, <www.infomarine.org/UKMIC-achieving_optimal_value.pdf>, 22 Feb. 2010.
268. See Yvo Volman, 'The re-use of public sector information in the EU. Increasing the scope for cross-border exploitation', in *Spatial data infrastructure and policy development in Europe and the United States*, ed. Bastiaan Van Loenen & Bas C. Kok (Delft: DUP Science, 2004), 62.
269. Office of Fair Trading, 'Competition Act 1998. Competition case closure summaries 2004', 2004, 3-4, <www.oft.gov.uk/shared_oft/ca98_case_closures/2004.pdf>, 22 Feb. 2010.
270. ePSIplus, 'PSIH Closes competitive service', 2007, <www.epsiplus.net/news/psih_closes_competitive_service>, 22 Feb. 2010.
271. Johan Jacques, 'Meteoservices, Belgium: Case against KMI. Progress report', <www.primet.org/documents/240206_johan_jaques_meteoservices_belgium.html>, 22 Feb. 2010.
272. Peter N. Weiss, 'Borders in Cyberspace: Conflicting Public Sector Information Policies and their Economic Impacts', in *Public Sector Information in the Digital Age, Between Markets, Public Management and Citizens' Rights*, ed. Georg Aichholzer & Herbert Burkert (Cheltenham: Edward Elgar Publishing Limited, 2004), 145–149.

Example: Ecomet – An example of a dominant position that has raised much discussion and much concern is the Ecomet association of National Meteorological Services (NMS). Previously it was mentioned that this association could possibly be covered under Article 101 of the Treaty on the Functioning of the EU, but this is also the case for Article 102, as the association has a dominant position on the market due to the quasi-monopolistic character of many of the data sets held by the NMS. In the 1999 comfort letter that was mentioned earlier, the European Commission also addressed some concerns with regard to the possible abuse of dominance by Ecomet. The Commission based its exemption partly on the finding that '[t]he Ecomet rules ensure equal treatment of independent service providers and the commercial divisions of the national meteorological institutes with regard to access to meteorological data and the invoicing of that data'.[273] In addition, '[i]n order to guarantee fair competition with independent service providers, the Ecomet rules require the national meteorological institutes to keep analytical accounts in which their commercial activities are entered separately and which thus ensure that there is no cross-subsidisation', and '[t]here is now an arbitration procedure for settling disputes between independent service providers and ECOMET members on any matter'.[274]

Ecomet: criticism – However, despite the Commission's exemption, Ecomet and its members have been severely criticized by the private sector. The Association of Environmental Data Users in Europe complained in 2000 that there was no fair competition, and that most commercial activities of the NMS did not pay the same prices for the raw material as their competitors did.[275] In 2004, Primet, the association of private sector suppliers of meteorological information services, questioned the European Commission on the new procedures for enforcement of EU competition law and asked for guidance on whether to launch a complaint against the National Meteorological Services and against the European Centre for Medium-Range Weather Forecasts.[276] Soderman, the chairman of Primet, pointed out in a meeting of the PSI Group that only some of the Ecomet members follow their own principle of equal treatment of independent service providers, dividing their commercial activities from the public task and keeping analytical accounts to ensure fair competition. Otten added that private sector companies in Europe are

---

273. Michael Kamps, 'Thunder and Lightning: Public Sector Information Policy Experiences of Private Meteorological Service Providers', in *Public Sector Information in the Digital Age, Between Markets, Public Management and Citizens' Rights*, ed. Georg Aichholzer & Herbert Burkert (Cheltenham: Edward Elgar Publishing Limited, 2004), 175–176. See Association of Environmental Data Users in Europe, 'Comfort Letter on the joint sale of meteorological products by Ecomet. Comments', 1999, <www.primet.org/index.php?Itemid=29&id=81&option=com_content&task=view>, 22 Feb. 2010.

274. *Ibid.*

275. See Association of Environmental Data Users in Europe, 'Comfort Letter on the joint sale of meteorological products by Ecomet. Comments', 1999, <www.primet.org/index.php?Itemid=29&id=81&option=com_content&task=view>, 22 Feb. 2010.

276. See letter from Anne-Margrete Wachtmeister of the European Commission Competition Directorate, <www.primet.org/index.php?option=com_content&task=view&id=125&Itemid=35>, 22 Feb. 2010.

sometimes not given access to meteorological data on conditions equal to the conditions available to competing public bodies, partly due to historical traditions. In his experience, the NMS protect their own interest in upholding a public sector monopoly, leading to the application of high charges and cross-subsidization.[277]

EXAMPLE: ORDNANCE SURVEY – The abuse of refusal to supply might also come into play in the case of the Ordnance Survey. According to the OFT report on commercial use of public information, 'Ordnance Survey provides limited access to its unrefined information and concentrates on developing refined information products itself, giving businesses fewer opportunities to add value and develop new refined information products'.[278] The report also found that the Ordnance Survey's license exception policy states that it may refuse applications 'to market a product whose intended use is the same as, or comparable to, that of any product marketed by Ordnance Survey itself or any product which Ordnance Survey intends to market'.[279] The OFT stated in the 2007 Communities and Local Government Committee review that since the publication of the report, this reference had been removed, but that it had not seen any specific change of policy in determining to whom Ordnance Survey grants a license.[280] OFT acknowledged that the Ordnance Survey bases its position on ECJ case law that is *IMS Health*, but it still considered it 'preferable for [Ordnance Survey] to adopt a more flexible licensing policy in this area; indeed, such a more flexible approach would be far more appropriate in relation to the government policy to increase the commercial use and re-use of public sector information'.[281] In the reply to the 2007 review, the government also stated that such a provision was never included in Ordnance Survey contracts or agreements, but that this only involved some comments on its website relating to

277. PSI Group, '5th meeting minutes', 2004, 4-5, <http://ec.europa.eu/information_ society/policy/psi/docs/pdfs/minutes_psi_group_meetings/5th_23_april_2004.pdf>, 22 Feb. 2010. See also OECD Working Party on the Information Economy, 'Digital Broadband Content: Public Sector Information and Content', 2006, 24–25, <www.oecd.org/dataoecd/10/22/36481524.pdf>, 22 Feb. 2010; OECD Working Party on the Information Economy, 'OECD Workshop on Public Sector Information held on 31 May 2006: Summary', 2006, 13–14, <www.oecd.org/dataoecd/54/35/36854745.pdf>, 22 Feb. 2010.

278. Office of Fair Trading, 'The Commercial Use of Public Information (CUPI)', 2006, 137, <www.oft.gov.uk/shared_oft/reports/consumer_protection/oft861.pdf>, 22 Feb. 2010. See also Communities and Local Government Committee, 'Ordnance Survey, Fifth Report of Session 2007-08, HC 268', <www.publications.parliament.uk/pa/cm200708/cmselect/cmcomloc/268/26802.htm>, 22 Feb. 2010.

279. Office of Fair Trading, 'The Commercial Use of Public Information (CUPI)', 2006, 137, <www.oft.gov.uk/shared_oft/reports/consumer_protection/oft861.pdf>, 22 Feb. 2010.

280. Communities and Local Government Committee, 'Ordnance Survey, Fifth Report of Session 2007-08, HC 268', <www.publications.parliament.uk/pa/cm200708/cmselect/cmcomloc/268/26802.htm>, 22 Feb. 2010. See also Stephen Saxby, 'Public sector information and re-use policy – where is the UK now?', paper given at the Second International Conference on Business, Law and Technology (New York, 2008), 17, <http://eprints.soton.ac.uk/52523/01/PSI_Re-use_Paper_28_April_2008_pdf.pdf>, 22 Feb. 2010.

281. Office of Fair Trading, 'The Commercial Use of Public Information (CUPI)', 2006, 137, <www.oft.gov.uk/shared_oft/reports/consumer_protection/oft861.pdf>, 22 Feb. 2010.

license exceptions.[282] In the spring of 2009, it was announced that the Ordnance Survey would work together with the Office of Public Sector Information to simplify its licensing policies. While in 2010, it was announced that the Ordnance Survey would make some of its data sets available free of charge, not much activity could be announced with regard to the licensing policies.

EXAMPLE: DIRECTION DE LA MÉTÉOROLOGIE NATIONALE – A refusal to supply that led to a court case was the non-delivery of aeronautic data by the French Meteorological Office, which had a monopoly on the data and a dominant position on the secondary market of information services. The reasons that were given for the refusal concerned security and the risks such dissemination would cause for aerial traffic. The French competition authority believed that there was no abuse of dominant position because the concern for security justified the refusal.[283] The Court of Appeal rejected this argument and found an abuse[284], but it was called back by the *Cour de Cassation*. The highest court did not consider the delivery of data by the meteorological service an economic activity and it also considered that the dissemination of information to the public was not a commercial activity, but a public service.[285] This decision was criticised by Bruguière, Nouel, and Teresi, who all stated that the meteorological office was performing an economic activity and competition law should have been applicable.[286]

EXAMPLE: SHOM – Another French example was the complaint of the *Groupement des editions de la presse nautique* against the Marine Hydrographic and Oceanographic Service (*Service Hydrographique et Océanographique de la Marine* [SHOM]) of the Ministry of Defence. SHOM was charged under an international treaty with the task of disseminating nautical information for security reasons. It decided to expand its activities to nautical information for leisure boating, and private publishers of the same information found this an abuse of SHOM's dominant position. They stated that it could charge very low prices due

282. Communities and Local Government Committee, 'Ordnance Survey: Government response to the Committee's Fifth Report, First Special Report of Session 2007-08, HC 516', 2008, <www.parliament.the-stationery-office.com/pa/cm200708/cmselect/cmcomloc/516/516.pdf>, 22 Feb. 2010.

283. Cons. Conc. Déc. No 92-D-35, 13 May 1992 relative à une saisine de la Société du journal téléphoné à l'encontre de la direction de la météorologie nationale, *BOCCRF no. 13*, 4 Jul. 1992, 224 [in French].

284. Cour D'Appel Paris, *Société du journal téléphoné v. Ministère de l'équipement, Direction de la météorologie nationale*, 18 Mar. 1993, *BOCCRF* no. 6, 26 Mar. 1993, 110 [in French].

285. Cour de Cassation, *Ministère de l'équipement, Direction de la météorologie nationale v. Société du journal téléphoné*, 12 Dec. 1995, Bull. Civ. IV, no. 301, 276 [in French].

286. Jean-Michel Bruguière, *Les données publiques et le droit* (Paris: Editions Litec, 2002), 127–135 [in French]; Bertrand Nouel, 'Commercialisation des données publiques et concurrence', paper given at the European Commission Conference: Access to Public Information. A Key to Commercial Growth and Electronic Democracy (Stockholm, 1996): 6-7 [in French]; Laurent Teresi, 'Données publiques et concurrence', *Concurrences* no. 4 (2005): 145 [in French]; Dieudonné Mandelkern & Bertrand Du Marais, 'Diffusion des données publiques et révolution numérique: rapport', 1999, 58–61, <www.ladocumentationfrancaise.fr/rapports-publics/994001620/index.shtml#>, 22 Feb. 2010 [in French].

to cross-subsidization, and that it used its status as a public body entrusted with the task of providing information to promote the new products. In addition, SHOM also used its status to hinder the access of its competitors to the underlying data.[287] Contrary to the previous case, an economic activity was recognized and competition law was deemed applicable to the service for leisure boating. Without explicitly referring to the essential facilities doctrine, the competition authority stated that SHOM could not use its IPR to refuse the provision of information to its competitors. In addition, it could not use its status as an official service to promote its own products. The competition authority did not see a problem with the cross-subsidization, because this was not in itself an abuse.[288]

C. Distortions of Competition by the State as a Sovereign Power

Outline – From what we have seen above, the State or public bodies can be qualified as undertakings if they perform an economic activity. In this way they can be subject to Articles 101 and 102 of the Treaty on the Functioning of the EU. Next to possible infringements of competition due to its own behaviour on the market, the State can also infringe upon competition law based on its actions as the State. These possible distortions are discussed in this section. First, competition can be distorted by the provision of state aid to particular undertakings. This is addressed by Article 107 of the Treaty on the Functioning of the EU (former Article 87 of the EC Treaty). In addition, Member States can take or maintain measures contrary to the Treaty on the Functioning of the EU with regard to public undertakings and undertakings with special or exclusive rights. This situation is subject to Article 106(1) of the Treaty on the Functioning of the EU (former Article 86(1) of the EC Treaty). For each type of distortion, the rules are explained and the applicability to spatial information services is examined.

1. Article 107 Treaty on the Functioning of the EU: State Aid

a. General

i. Principle: Prohibition of State Aid

Example – It may be that the public sector provides subsidies to a particular undertaking (whether it is public or private) to perform a service and in this way

287. Cons. Conc. Avis no. 97 A-10, 25 Feb. 1997 relatif à une demande d'avis présentée par le *Groupement des editions de la presse nautique* et portant sur des questions de concurrences soulevées par la politique éditoriale du service hydrographique de la marine, *BOCCRF* 18 Nov. 1997, 759 [in French].

288. *Ibid.* See also Laurent Teresi, 'Données publiques et concurrence', *Concurrences* no. 4 (2005): 147 [in French]; Jean-Michel Bruguière, *Les données publiques et le droit* (Paris: Editions Litec, 2002), 140–145 [in French].

distorts the market. This section briefly sets out the principles of state aid as they are envisaged by EU competition law. These principles are set out in Article 107 of the Treaty on the Functioning of the EU.

ARTICLE 107 OF THE TREATY ON THE FUNCTIONING OF THE EU – Under Article 107 of the Treaty on the Functioning of the EU (former Article 87 of the EC Treaty), any aid granted by a Member State or through State resources in any form whatsoever that distorts or threatens to distort competition by favouring certain undertakings or the production of certain goods is incompatible with the common market, insofar as it affects trade between Member States. Examples of state aid are investment grants[289], loans at reduced interest rates[290], preferential fiscal treatment[291], and cross-subsidies by a public undertaking[292]

EXCEPTIONS – While Article 107(1) provides the main rule that prohibits state aid, Article 107(2) and (3) allow for exceptions and for the European Commission to approve types of state aids that are generally beneficial.[293] The Commission acknowledged that '[s]tate aid measures can sometimes be effective tools for achieving objectives of common interest. They can correct market failures, thereby improving the functioning of markets and enhancing European competitiveness. They can also help promote e.g. social and regional cohesion, sustainable development and cultural diversity'.[294] Yet the Commission also warns that state aid should only be used when it is an appropriate instrument to reach a defined objective, and if it is proportionate and only distorts competition to the least possible extent.[295]

ii. Elements of the State Aid Concept

FIVE ELEMENTS – The term *aid* has to be interpreted broadly, including not only true subsidies, but also any form of intervention or assistance that has the same or similar effects as a subsidy. Hence, the concept of aid is determined based on the effects.[296] Roth and Rose sum up the five elements that make up a state aid as follows: '(i) an advantage; (ii) granted by a Member State or through State

289. ECJ, Case 730/79, *Philip Morris Holland BV v. Commission*, 17 Sep. 1980, *European Court reports* 1980, 2671.
290. ECJ, Case 323/82, *SA Intermills v. Commission*, 14 Nov. 1984, *European Court reports* 1984, 3809.
291. ECJ, Case C-387/92, *Banco de Crédito Industrial SA v. Ayuntamiento de Valencia*, 15 Mar. 1994, *European Court reports* 1994, I-877.
292. CFI, Case T-106/95, *Fédération française des sociétés d'assurances and others v. Commission*, 27 Feb. 1997, *European Court reports* 1997, II-229.
293. Peter Roth & Vivien Rose (ed.), *Bellamy & Child European Community Law of Competition*. 6th edn (Oxford: Oxford University Press, 2008), 1498.
294. Commission of the European Communities, *State Aid Action Plan. Less and better targeted state aid: a roadmap for state aid reform 2005-2009*, COM (2005) 107 final, 4, 7 Jun. 2005.
295. *Ibid.*
296. ECJ, Case 173/73, *Italian Republic v. Commission*, 2 Jul. 1974, para. 13, *European Court reports* 1974, 709; ECJ, Case 310/85, *Deufil GmbH & Co. KG v. Commission*, 24 Feb. 1987, para. 8, *European Court reports* 1987, 901; ECJ, Case C-480/98, *Kingdom of Spain v. Commission*, 12 Oct. 2000, para. 16, *European Court Reports* 2000, I-8717.

resources; (iii) favouring certain undertakings or the production of certain goods; (iv) distorting competition; and (v) affecting inter-State trade'.[297] The first three elements are addressed below. With regard to the fourth element, I can refer to the previous section on the distortion of competition by public bodies as undertakings. The fifth element will not be discussed here, because this examination would be too far-reaching.[298]

ADVANTAGE – With regard to the condition of an advantage, a distinction needs to be made between advantages that are intended to attain particular economic and social objectives of a Member State, and measures that are commercially justifiable, as a private undertaking or investor in a comparable position would adopt an equivalent measure.[299] Such justifiable measures are generally not considered to be state aids. The advantage can be of a permanent or temporary nature. A measure can still be an advantage, even though the beneficiary has contributed wholly or partially to its financing as a result of a state levy or if it has to do something in return, for example, rationalize production capacity.[300] Next, to determine whether the provision of funds to an undertaking may be an aid, it should be examined whether the terms under which the funds are provided go beyond what would be acceptable for a private investor operating under normal market conditions, having regard to the information available and foreseeable developments at that time. This *market economy investor* principle requires the examination of whether there will be an acceptable return on the provision of funds within a reasonable period of time.[301]

PROVIDED BY A MEMBER STATE OR THROUGH STATE RESOURCES – The second element of state aid is the provision of an advantage *by a Member State or through state resources.*[302] Advantages must be granted directly or indirectly through state

297. Peter Roth & Vivien Rose (ed.), *Bellamy & Child European Community Law of Competition.* 6th edn (Oxford: Oxford University Press, 2008), 1504.

298. For more information on the effect on trade between Member States, see e.g., Alison Jones & Brenda Sufrin, *EC Competition Law: Text, Cases and Materials* (Oxford: Oxford University Press, 2008), 1418 p.; Peter Roth & Vivien Rose (ed.), *Bellamy & Child European Community Law of Competition.* 6th edn (Oxford: Oxford University Press, 2008), 2252 p.; Richard Whish, *Competition law*, 6th edn (Oxford: Oxford University Press, 2009), 1006 p.; Jonathan Faull, *Faull and Nikpay: The EC Law of Competition* (Oxford: Oxford University Press, 2007), 2000 p.

299. ECJ, Case C-278/92, *Kingdom of Spain v. Commission*, 14 Sep. 1992, para. 21, *European Court reports* 1994, I-4103. See also Marc Hansen, Anne Van Ysendyck & Susanne Zuhlke, 'The Coming of Age of EC State Aid Law: a Review of the Principal Developments in 2002 and 2003', *European Competition Law Review* 25, no. 4 (2004): 203–211.

300. ECJ, Case 290/83, *Commission of the European Communities v. French Republic*, 30 Jan. 1985, para. 14, *European Court reports* 1985, 439; CFI, Cases T-197/97 and 198/97, *Weyl Beef Products BV and others v. Commission*, 31 Jan. 2001, para. 81, *European Court reports* 2001, II-303; ECJ, Case C-251/97, *French Republic v. Commission*, 5 Oct. 1999, para. 39-47, *European Court Reports* 1999, I-663.

301. See Peter Roth & Vivien Rose (ed.), *Bellamy & Child European Community Law of Competition.* 6th edn (Oxford: Oxford University Press, 2008), 1507–1508.

302. For an overview of the case law regarding the issue of the provision of an advantage through state resources, see Andrea Biondi, 'Some Reflections on the Notions of 'State Resources', in

resources and be imputable to the State. Hence, decentralized or privatized distribution of aid,[303] and aid by local or regional authorities, are also included.[304] It is sufficient that the State has control over the funds; they do not necessarily have to come from the state budget. The state resources do not necessarily have to be transferred to the undertaking, for example, tax exemptions or waivers of tax revenue.[305]

FAVOURING CERTAIN UNDERTAKINGS OR THE PRODUCTION OF CERTAIN GOODS – Thirdly, the measure has to *favour certain undertakings or the production of certain goods* in order for it to be classified as state aid. It cannot be a general measure for all undertakings or all sectors. The fact that some undertakings are treated differently to others is not necessarily an advantage that favours certain undertakings. Such different treatment is possible if it is objectively justified, for example by reference to the different regulatory conditions applicable to particular undertakings or activities.[306]

iii. Compatible State Aid

ARTICLE 107(2) OF THE TREATY ON THE FUNCTIONING OF THE EU – Not all state aid is necessarily illegal. Article 107(2) of the Treaty on the Functioning of the EU provides a list of types of aid that are considered compatible with the common market. This list includes:

- aid having a social character, granted to individual consumers, provided that such aid is granted without discrimination related to the origin of the products concerned;
- aid to make good the damage caused by natural disasters or exceptional occurrences;
- aid granted to the economy of certain areas of the Federal Republic of Germany affected by the division of Germany, in so far as such aid is required in order to compensate for the economic disadvantages caused by that division.

NO DISTORTION – If the advantages that are given to the undertakings or the producers of the goods fall under either of these three types of aid, they are considered not to distort the common market and they are allowed. Each of the categories has to be interpreted narrowly.[307]

European Community State Aid Law', *Fordham International Law Journal* 30 (2007): 1426–1448.

303. ECJ, Case C-482/99, *French Republic v. Commission*, 16 May 2002, opinion Jacobs AG, para. 54, *European Court Reports* 2002, I-4397.

304. ECJ, C-482/99, *French Republic v. Commission*, 16 May 2002, para. 34-38, *European Court Reports* 2002, I-4397.

305. ECJ, C-482/99, *French Republic v. Commission*, 16 May 2002, para. 36, opinion Jacobs AG, para. 34-38, *European Court Reports* 2002, I-4397.

306. Peter Roth & Vivien Rose (ed.), *Bellamy & Child European Community Law of Competition*, 6th edn (Oxford: Oxford University Press, 2008), 1519.

307. *Ibid.*, 1535.

ARTICLE 107(3) OF THE TREATY ON THE FUNCTIONING OF THE EU – Next, Article 107(3) of the Treaty on the Functioning of the EU lists a number of types of aid that may be considered compatible with the common market, if they are approved by the European Commission.[308] The following categories are included:

- aid to promote the economic development of areas where the standard of living is abnormally low or where there is serious underemployment;
- aid to promote the execution of an important project of common European interest or to remedy a serious disturbance in the economy of a Member State;
- aid to facilitate the development of certain economic activities or of certain economic areas, where such aid does not adversely affect trading conditions to an extent contrary to the common interest;
- aid to promote culture and heritage conservation where such aid does not affect trading conditions and competition in the Union to an extent that is contrary to the common interest;
- such other categories of aid as may be specified by decision of the Council acting by a qualified majority on a proposal from the Commission.

ROLE OF THE EUROPEAN COMMISSION – These categories have to be interpreted strictly. All planned aid has to be notified in advance to the Commission and approved by it, unless a block exemption has been accepted.[309] Block exemptions have, for instance, been accepted for regional aid; investment in SMEs and employment aid; aid for the creation of enterprises by female entrepreneurs; aid for environmental protection; and aid for research, development and innovation.[310] The procedure with regard to individual exemptions is set out in Article 108 of the Treaty on the Functioning of the EU (former Article 88 of the EC Treaty).

b. Applicability to Spatial Information Services

GREEN PAPER ON PUBLIC SECTOR INFORMATION – While no official complaints have been made about public bodies offering spatial information services enjoying incompatible state aid, the European Commission's Green Paper on Public Sector Information stated in 1999 that public financing for public bodies providing information services could constitute state aid.[311]

---

308. See Conor Quigley, 'The European Commission's State Aid Action Plan for 2005-2009', *International Company and Commercial Law Review* 16, no. 12 (2005): 485–489.
309. Peter Roth & Vivien Rose (ed.), *Bellamy & Child European Community Law of Competition.* 6th edn (Oxford: Oxford University Press, 2008), 1537. See also Pierre De Bandt, 'Artikel 86, lid 2 EG-Verdrag en het begrip 'diensten van algemeen economisch belang: meer dan een rechtsbasis voor een genuanceerde toepassing van de interne markt- en mededingingsregels op openbare diensten?', *Tijdschrift voor Belgisch Handelsrecht* 114, no. 3 (2008): 221 [in Dutch].
310. Commission Regulation 800/2008/EC of 6 Aug. 2008 declaring certain categories of aid compatible with the common market in application of Arts 87 and 88 of the Treaty (General block exemption Regulation), *OJ L* 214, 9 Aug. 2008, 3.
311. Commission of the European Communities, *Green Paper on Public Sector Information*, 15.

Example: Ordnance Survey – For instance, the provision of government funds to the British Ordnance Survey was questioned. The Locus association felt that, if it is accepted that many of the activities of the Ordnance Survey are qualified as economic (which they felt should be the case), government funding provided under the National Interest Mapping Services Agreement (NIMSA) might be construed as state aid, 'being government granted funding to OS without public procurement for the collection of data which can only be obtained by third parties through the acquisition of [Ordnance Survey] commercial products'.[312] NIMSA was a seven-year agreement between Communities and Local Government and the Ordnance Survey that came into effect on 1 April 1999. Under NIMSA, Communities and Local Government funded those activities that were not commercially viable but were in the national interest.[313] At the end of 2006 it was decided not to renew the agreement.[314] Pitt found that: 'Since the precise terms on which Ordnance Survey receive . . . Government funds . . . are not public, it is not possible for the public to be satisfied that Ordnance Survey has not received State aid from the UK Government to the detriment of third party suppliers'.[315] He had the same concern about the agreement between the Ordnance Survey and central government, but the new version of that agreement has been completed after a competitive tendering process[316], so these conditions can be verified by possible competitors and the general public.

c. *Transparency and Separation of Activities*

i. Transparency Directive

Transparency of financial relations – In the framework of the EU rules on state aid, the Commission has adopted measures relating to transparency of the financial relations between the Member States and their public undertakings and relating to the separation of economic and other activities. The main objective of Directive 80/723/EEC on the transparency of financial relations between Member States and public undertakings[317] was to verify that the public undertakings would not receive

312. Communities and Local Government Committee, 'Ordnance Survey, Fifth Report of Session 2007-08, HC 268', 2008, <www.publications.parliament.uk/pa/cm200708/cmselect/cmcomloc/268/26802.htm>, 22 Feb. 2010.
313. Communities and Local Government, 'National Interest Mapping Services Agreement (NIMSA)', <www.communities.gov.uk/corporate/researchandstatistics/research1/geographicinformationinitiatives/nimsa/>, 22 Feb. 2010.
314. See Communities and Local Government, 'The Future of the National Interest Mapping Services Agreement Beyond 2006', 2006, <www.communities.gov.uk/documents/planningandbuilding/pdf/154073.pdf>, 22 Feb. 2010.
315. Edward Pitt, 'Ordnance Survey – the ground rules', 2003, <www.epsiplus.net/cases/ordnance_survey>, 22 Feb. 2010.
316. Ordnance Survey, 'New Pan Government Agreement signed', <www.ordnancesurvey.co.uk/oswebsite/media/news/2009/February/pgasigned.html>, 22 Feb. 2010.
317. Commission Directive 80/723/EEC of 25 Jun. 1980 on the transparency of financial relations between Member States and public undertakings, *OJ L* 195, 29 Jul. 1980, 35.

hidden aids from public authorities. The directive was amended several times and replaced in 2006 with a consolidating directive.[318]

ARTICLE 1 TRANSPARENCY DIRECTIVE – Under Article 1 of the directive, the Member States have to ensure that the financial relations between public authorities and public undertakings are transparent so that the amount of funds made available to public undertakings and the use to which these funds are put emerge clearly. Next, undertakings enjoying a special or exclusive right under Article 86(1) (current Article 106(1) of the Treaty on the Functioning of the EU) or that are entrusted with the provision of a service of general economic interest under Article 86(2) of the EC Treaty (current Article 106(2) Treaty on the Functioning of the EU), that receive public compensation for these services and perform other activities at the same time, have to maintain separate accounts. This obligation aims to identify any cross-subsidization.[319] The following rules have to be applied:

- the internal accounts corresponding to different activities have to be separate;
- all costs and revenues have to be correctly assigned or allocated on the basis of consistently applied and objectively justifiable cost accounting principles;
- the cost accounting principles according to which separate accounts are maintained have to be clearly established.[320]

ii. Treaty on the Functioning of the EU

NO SPECIFIC RULES ON TRANSPARENCY – While there are no explicit rules in the Treaty on the Functioning of the EU on transparency or separation of activities, such obligations may still rest on the public bodies under the prohibition of Article 102 for undertakings with a dominant position to abuse that position. For instance, the cross-subsidization of market activities with funds that were made available for the performance of the public task can form an abuse.

SEPARATION OF ACTIVITIES REALLY NECESSARY? – Whether the activities should be kept separate within the public body or assigned to different bodies is subject to different opinions. According to the Dutch Cohen report, it is not possible to maintain sufficient safeguards for fair competition if public tasks and market activities are carried out within the same entity. The report therefore states that market activities have to be separated and moved to another legal entity.[321]

---

318. Commission Directive 2006/111/EC of 16 Nov. 2006 on the transparency of financial relations between Member States and public undertakings as well as on financial transparency within certain undertakings, *OJ L* 318, 17 Nov. 2006, 17.
319. Peter Roth & Vivien Rose (ed.), *Bellamy & Child European Community Law of Competition*. 6th edn (Oxford: Oxford University Press, 2008), 1511.
320. Article 4 of the Transparency directive.
321. KPMG, 'Ministerie van Waterstaat. Evaluatie van de wet op het KNMI', 2007, 15, <www.verkeerenwaterstaat.nl/Images/2008651%20bijlage_tcm195-231492.pdf>, 22 Feb. 2010 [in Dutch].

The Swedish Agency for Administrative Development shared this view.[322] However, the United Kingdom's OFT felt that separating the two functions while retaining the operation of the public body as one entity should be sufficient.[323]

iii. Applicability to Spatial Information Services

Separation of activities – The issue of separating the public task and the commercial activities of public bodies has come up regularly with regard to the provision of information services. Among others, Weiss has pleaded that '[t]he most sensible solution is to separate commercial activities into truly commercial entities separate from the government and adopt open access policies. Separation of commercial activities would be the basis not only for an open market in accordance with European competition law, but also guarantee market structures with maximum overall potential'.[324]

---

322. Statskontoret, 'Competition at the Public/Private Interface', 2005, <www.statskontoret.se/upload/Publikationer/2005/200519A.pdf>, 22 Feb. 2010. See also Peter N. Weiss, 'Borders in Cyberspace: Conflicting Public Sector Information Policies and their Economic Impacts', in Georg Aichholzer & Herbert Burkert (ed.), *Public Sector Information in the Digital Age, Between Markets, Public Management and Citizens' Rights*, (Cheltenham: Edward Elgar Publishing Limited, 2004), 157; Herbert Burkert & Peter N. Weiss, 'Towards a Blueprint for a Policy on Public Sector Information', in *Public Sector Information in the Digital Age, Between Markets, Public Management and Citizens' Rights*, ed. Georg Aichholzer & Herbert Burkert (Cheltenham: Edward Elgar Publishing Limited, 2004), 330; Estelle Derclaye, 'Does the Directive on the Re-use of Public Sector Information affect the State's database sui generis right?', in *Knowledge rights – Legal, societal and related technological aspect. Knowright Symposium, 18-19 September 2008*, ed. Jens Gaster, Erich Schweighofer & Peter Sint (Austrian Computer Society, 2008), 144; Micus Management Consulting Gmbh, 'Assessment of the Re-use of Public Sector Information (PSI) in the Geographical Information, Meteorological Information and Legal Information Sectors', 2008, 60, <http://ec.europa.eu/information_society/policy/psi/docs/pdfs/micus_report_december2008.pdf>, 22 Feb. 2010.
323. Chris Corbin, 'PSI Pricing 1 – ePSIplus Thematic Meeting Report', 2007, 13, <www.epsiplus.net/reports/epsiplus_thematic_meeting_reports/fin_mtg_1>, 22 Feb. 2010.
324. Peter N. Weiss, 'Borders in Cyberspace: Conflicting Public Sector Information Policies and their Economic Impacts', in *Public Sector Information in the Digital Age, Between Markets, Public Management and Citizens' Rights*, ed. Georg Aichholzer & Herbert Burkert (Cheltenham: Edward Elgar Publishing Limited, 2004), 157; Herbert Burkert & Peter N. Weiss, 'Towards a Blueprint for a Policy on Public Sector Information', in *Public Sector Information in the Digital Age, Between Markets, Public Management and Citizens' Rights*, ed. Georg Aichholzer & Herbert Burkert (Cheltenham: Edward Elgar Publishing Limited, 2004), 330; Micus Management Consulting Gmbh, 'Assessment of the Re-use of Public Sector Information (PSI) in the Geographical Information, Meteorological Information and Legal Information Sectors', 2008, 60, <http://ec.europa.eu/information_society/policy/psi/docs/pdfs/micus_report_december2008.pdf>, 22 Feb. 2010; Estelle Derclaye, 'Does the Directive on the Re-use of Public Sector Information affect the State's database sui generis right?', in *Knowledge rights – Legal, societal and related technological aspect. Knowright Symposium, 18-19 September 2008*, ed. Jens Gaster, Erich Schweighofer & Peter Sint (Austrian Computer Society, 2008), 144; Chris Corbin, 'PSI Pricing 1 – ePSIplus Thematic Meeting Report', 2007, 13, <www.epsiplus.net/reports/epsiplus_thematic_meeting_reports/fin_mtg_1>, 22 Feb. 2010; A.A. Van Rossum, 'Oneerlijke concurrentie door de overheid', in *Overheid en onderneming. Tussen publiek en privaat ondernemen*, ed. Sebastianus C.J.J. Kortmann et al. (Deventer: W.E.J. Tjeenk Willink, 1998), 236 [in Dutch].

EXAMPLE: KNMI – In 1999, the commercial activities of KNMI, the Dutch Meteorological Office, were separated into a private entity, and in 2002 the so-called Law on the Weather determined its remaining public tasks.[325] This was intended to clarify the position of the Meteorological Office with regard to its private competitors and stimulate the development of the market for meteorological products. However, a 2008 KPMG report evaluating this piece of legislation found that the separation of commercial activities and public task activities may have been a good idea, but that the demarcation line might have to be moved already. Due to technological and socio-economic developments, the Meteorological Office was found to be performing market activities that are not or should not be part of the public task as it is described in the law.[326]

EXAMPLE: *LANTMATERIET* – The Swedish National Land Survey *(Lantmateriet)* underwent the same process. The Swedish Agency for Administrative Development *(Statskontoret)* found that *Lantmateriet* had an unfair competitive advantage over emerging commercial firms and obscured the demarcation between government and private activities. The commercial arm had to be separated and privatized, subject to open public audit and oversight, while the data held by *Lantmateriet* should be available to the public and to competing private sector entities.[327] The same recommendation was made in a later report for the Swedish Meteorological and Hydrological Institute. Of course, recommending a separation also requires a definition of the public task of the remaining public body. To this end, *Lantmateriet* set out its mission and tasks in its strategic plan.[328]

EXAMPLE: TRADING FUNDS – In the United Kingdom, separation of the Trading Funds' commercial activities and its public tasks into separate entities has never been considered by the government. However, OFT recommended that public bodies should separate their unrefined and refined information, operate in a more transparent and fair manner, and improve their accounting practices and quality of service.[329]

---

325. Wet van 1 november 2001 houdende regeling van de taken voor de meteorologie en andere geofysische terreinen (Wet op het Koninklijk Nederlands Meteorologisch Instituut), <http://wetten.overheid.nl/BWBR0012952/>, 22 Feb. 2010 [in Dutch]; KPMG, 'Ministerie van Waterstaat. Evaluatie van de wet op het KNMI', 2007, 16, <www.verkeerenwaterstaat.nl/Images/2008651%20bijlage_tcm195-231492.pdf>, 22 Feb. 2010 [in Dutch].

326. KPMG, 'Ministerie van Waterstaat. Evaluatie van de wet op het KNMI', 2007, 31, <www.verkeerenwaterstaat.nl/Images/2008651%20bijlage_tcm195-231492.pdf>, 22 Feb. 2010 [in Dutch].

327. Statskontoret, 'Competition at the Public/Private Interface', 2005, 14, <www.statskontoret.se/upload/Publikationer/2005/200519A.pdf>, 22 Feb. 2010. See also Peter N. Weiss, 'Borders in Cyberspace: Conflicting Public Sector Information Policies and their Economic Impacts', in *Public Sector Information in the Digital Age, Between Markets, Public Management and Citizens' Rights*, ed. Georg Aichholzer & Herbert Burkert (Cheltenham: Edward Elgar Publishing Limited, 2004), 147.

328. See Lantmateriet, 'Landmateriet's Strategic Plan 2006-2010', 2006, 18 p., <www.lantmateriet.se/upload/filer/om_lantmateriet/strategiskaplanen_engelska.pdf>, 22 Feb. 2010.

329. Office of Public Sector Information, 'The United Kingdom Implementation of the European Directive on the Re-use of Public Sector Information – the first two years', 2007, 22–23, <www.opsi.gov.uk/advice/psi-regulations/uk-implementation-first-years.pdf>, 22 Feb. 2010.

EXAMPLE: ORDNANCE SURVEY – For the United Kingdom's mapping agency, the Ordnance Survey, the issue of separation had already come up in 2002 when the Transport, Local Government and the Regions Committee of the House of Commons started a first inquiry into the Ordnance Survey's operations.[330] The report found that: '[i]f Ordnance Survey wants to enter into commercial activities we can see no reason why it should not do so, but the two activities ought to be separately accounted for and its commercial arm should pay the same copyright fees as any other organisation/competitor'.[331] OFT and the Communities and Local Government Committee once again recommended the separation of OS's public services and commercial operations, in 2006 and 2007 respectively.[332] However, it was also admitted that the line between both was blurry, and the clear separation of the different activities was not an easy task. In April 2009, the Ordnance Survey, in cooperation with OPSI, issued its new business strategy, which included plans to establish a subsidiary company that would operate on the market and offer new and innovative services to government, business and individual customers. This subsidiary will operate on the same terms in relation to the Ordnance Survey as any commercial rival.[333] The business strategy is being fully developed by April 2010.

2. Article 106(1) of the Treaty on the Functioning of the EU: Measures Regarding Public Undertakings and Undertakings with Special or Exclusive Rights

a. *Introduction*

ARTICLE 106(1) OF THE TREATY ON THE FUNCTIONING OF THE EU – Article 106(1) of the Treaty on the Functioning of the EU (former Article 86(1) of the EC Treaty) prohibits Member States from enacting or maintaining in force any measure regarding public undertakings and undertakings granted with special or exclusive rights, contrary to the rules of the Treaty on the Functioning of the EU, particularly the sections on competition and discrimination on the grounds of nationality (Article 18 Treaty on the Functioning of the EU – former Article 12 of the EC Treaty). This provision is often considered as an application of Article 10 of the EC

330. See Communities and Local Government, 'The Future of the National Interest Mapping Services Agreement Beyond 2006', 2006, <www.communities.gov.uk/documents/planningandbuilding/pdf/154073.pdf>, 22 Feb. 2010.

331. *Ibid.*, See also Edward Pitt, 'Ordnance Survey – the ground rules', 2003, <www.epsiplus.net/cases/ordnance_survey>, 22 Feb. 2010.

332. Office of Fair Trading, 'The Commercial Use of Public Information (CUPI)', 2006, 12, 45, <www.oft.gov.uk/shared_oft/reports/consumer_protection/oft861.pdf>, 22 Feb. 2010; Communities and Local Government Committee, *Ordnance Survey*, Fifth Report of Session 2007-08, HC 268, <www.publications.parliament.uk/pa/cm200708/cmselect/cmcomloc/268/26802.htm>, 22 Feb. 2010.

333. Ordnance Survey Business Strategy, 'Goal 5: Enhance value through the creation of an innovative trading entity, <http://strategy.ordnancesurvey.co.uk/goal-5-enhance-value-through-trading-entity/>, 22 Feb. 2010.

Treaty (repealed by the Lisbon Treaty), which addressed the so-called Community loyalty and requires the Member States to abstain from any measure that could jeopardize the attainment of the objectives of the EC Treaty.[334]

NO SEPARATE INFRINGEMENT – Article 106(1) can only be infringed upon in conjunction with another provision of the Treaty on the Functioning of the EU, such as Articles 101 and 102, Article 34 (former Article 28 of the EC Treaty) on the free movement of goods,[335] and Article 56 (former Article 49 of the EC Treaty) on the free movement of services.[336] Hence, if the Member States infringe upon those articles in relation to public undertakings or undertakings with special or exclusive rights, they also infringe upon Article 106(1). Jones and Sufrin indicate that Article 106 is particularly relevant to Article 102, as public undertakings and undertakings with special or exclusive rights frequently hold a dominant position.[337]

b. Public Undertakings and Undertakings with Special or Exclusive Rights

PUBLIC UNDERTAKING – *Public undertaking* has its own meaning under EU law, independent of any national definition, 'because Article 86(1) [current Article 106(1)] would be deprived of its effect if Member States were free to choose their own conception of public undertaking'.[338] Article 106(1) of the Treaty on the Functioning of the EU does not have a definition of *public undertakings*, but a definition can be found in the 1980 Transparency Directive: 'any undertaking over which the public authorities may exercise directly or indirectly a dominant influence by virtue of their ownership of it, their financial participation therein, or the rules which govern it'.[339] A dominant influence from the public authorities is presumed when these authorities, directly or indirectly in relation to an undertaking, either

334. ECJ, Case 267/86, *Pascal Van Eycke v. ASPA NV.*, 21 Sep. 1988, para. 16, *European Court reports* 1988, 4769; ECJ, Cases C-140/94, C-141/94 and C-142/94, *IP SpA v. Comune di Bassano del Grappa v. Comune di Chiogga*, para. 15, 17 Oct. 1995, *European Court reports* 1995, I-3257; Peter Roth & Vivien Rose (ed.), *Bellamy & Child European Community Law of Competition.* 6th edn (Oxford: Oxford University Press, 2008), 1038, 1051; Alison Jones & Brenda Sufrin, *EC Competition Law: Text, Cases and Materials* (Oxford: Oxford University Press, 2008), 623.

335. ECJ, Case 18/88, *Régie des télégraphes et des téléphones v. GB-Inno-BM SA.*, 13 Dec. 1991, *European Court reports* 1991, I-5941; ECJ, Case 202/88, *French Republic v. Commission*, 19 Mar. 1991, *European Court Reports* 1991, I-1223.

336. ECJ, Case C-266/96, *Corsica Ferries France SA v. Gruppo Antichi Ormeggiatori del porto di Genova Coop. arl*, 18 Jun. 1998, *European Court reports* 1998, I-3949; ECJ, Case C-271/90, *Kingdom of Spain, Kingdom of Belgium and Italian Republic v. Commission*, 17 Nov. 1992, *European Court reports* 1992, I-5833.

337. Alison Jones & Brenda Sufrin, *EC Competition Law: Text, Cases and Materials* (Oxford: Oxford University Press, 2008), 624. See also Tony Prosser, *The Limits of Competition Law, Markets and Public Services* (Oxford: Oxford University Press, 2005), 126.

338. Alison Jones & Brenda Sufrin, *EC Competition Law: Text, Cases and Materials* (Oxford: Oxford University Press, 2008), 626.

339. Article 2 of the Transparency directive.

hold the major part of the undertaking's subscribed capital; control the majority of the votes attaching to shares issued by the undertakings; or can appoint more than half of the members of the undertaking's administrative, managerial, or supervisory body. It does not matter whether this public undertaking is a separate legal entity or whether it is part of the State's central administration.[340] This definition was challenged by some Member States before the ECJ, because they felt that the European Commission was not entitled to elaborate on the concept. The ECJ did not agree with the Member States and upheld the definition.[341]

SPECIAL OR EXCLUSIVE RIGHTS – Undertakings to which the Member States have granted special or exclusive rights may not have a dominant position on the market, but they are included in Article 106(1) because a Member State might be tempted to adopt measures that would favour these undertakings and in this way distort competition.[342] According to Advocate General Jacobs in the *Ambulanz Glöckner* case, three elements determine whether a special or exclusive right is present: a right is granted by the authorities of a Member State; the right is granted to one undertaking or to a limited number of undertakings; and it substantially affects the ability of other undertakings to exercise the economic activity in question in the same geographical area under substantially equivalent conditions.[343] The ECJ followed this proposal and stated that special and exclusive rights may be found to exist where 'protection is conferred by a legislative measure on a limited number of undertakings which may substantially affect the ability of other undertakings to exercise the economic activity in question in the same geographical area under substantially equivalent conditions'.[344]

SPECIAL V. EXCLUSIVE – *Special* and *exclusive* rights are not synonymous. The European Commission initially defined them jointly, but the ECJ decided that a distinction should be made between both categories.[345] Exclusive rights exist 'where a monopoly has been granted by the State to one entity to engage in a particular economic activity on an exclusive basis',[346] for example, broadcasting monopolies,[347] undertakings granted the exclusive right to operate employment

340. ECJ, Case 118/85, *Commission v. Italian Republic*, 16 Jun. 1987, para. 7, *European Court Reports* 1987, 2599; ECJ, Case C-69/91, *Criminal proceedings against Francine Gillon, née Decoster*, 27 Oct. 1993, para. 15, *European Court Reports* 1993, I-5335.
341. ECJ, Cases 188/80, 189/80 and 190/80, *French Republic, Italian Republic and United Kingdom of Great Britain and Northern Ireland v. Commission of the European Communities*, 6 Jul. 1982, *European Court reports* 1982, 2545.
342. Peter Roth & Vivien Rose (ed.), *Bellamy & Child European Community Law of Competition*. 6th edn (Oxford: Oxford University Press, 2008), 1039.
343. ECJ, Case C-475/99, *Ambulanz Glöckner v. Landkreis Südwestpflatz*, 25 Oct. 2001, Jacobs AG, para. 87, *European Court Reports* 2001, I-8089.
344. ECJ, Case C-475/99, *Ambulanz Glöckner v. Landkreis Südwestpflatz*, 25 Oct. 2001, para. 24, *European Court Reports* 2001, I-8089.
345. Jose Luis Buendia Sierra, *Exclusive rights and state monopolies under EC law* (Oxford: Oxford University Press, 1999), 64–65.
346. Alison Jones & Brenda Sufrin, *EC Competition Law: Text, Cases and Materials* (Oxford: Oxford University Press, 2008), 627.
347. ECJ, Case 155/73, *Sacchi*, 30 Apr. 1974, *European Court Reports* 1974, 409.

recruitment services,[348] undertakings with an exclusive right to operate on a certain air route,[349] or undertakings with a monopoly on the supply of unloading services at a port.[350]

DEFINITION OF SPECIAL RIGHTS – To define the concept of *special rights*, Buendia Sierra and Jones and Sufrin look to the field of telecommunications.[351] In the preamble to the 1994 Directive on satellite communications, it is stated that special rights (in the field of telecommunication services) are 'rights that are granted by a Member State to a limited number of undertakings, through any legislative, regulatory or administrative instrument which, within a given geographical area,

- limits to two or more the number of such undertakings, otherwise than according to objective, proportional and non-discriminatory criteria, or
- designates, otherwise than according to such criteria, several competing undertakings, or
- confers on any undertaking or undertakings, otherwise than according to such criteria, legal or regulatory advantages which substantially affect the ability of any other undertaking to engage in any of the abovementioned activities in the same geographical area under substantially equivalent conditions'.[352]

GENERAL APPLICABILITY – According to Jones and Sufrin, '[a]lthough this definition is set out in a directive relating to telecommunications there is no reason to suppose that it is not also applicable in other fields. What it does is to identify as "special rights" those which are given by the state to a limited number of undertakings chosen in a subjective and discretionary manner'.[353] Special rights can be found when there is a limitation of the number of undertakings which can benefit from rights granted by the state; where the exercise of state discretion plays a part in the designation of the operators; or where special advantages are conferred by the state on an enterprise or a number of enterprises.[354]

---

348. ECJ, Case C-41/90, *Höfner & Elser v. Macotron GmbH*, 23 Apr. 1991, *European Court Reports* 1991, I-1979.
349. ECJ, Case 66/86, *Ahmed Saeed Flugreisen and Silver Line Reisebüro GmbH v. Zentrale zur Bekämpfung unlauteren Wettbewerbs e.V.*, 11 Apr. 1989, *European Court Reports* 1989, 803.
350. ECJ, Case C-179/90, *Merci convenzionali porto di Genova SpA v. Siderurgica Gabrielli SpA.*, 10 Dec. 1991, *European Court Reports* 1991, 5889.
351. Jose Luis Buendia Sierra, *Exclusive rights and state monopolies under EC law* (Oxford: Oxford University Press, 1999), 65; Alison Jones & Brenda Sufrin, *EC Competition Law: Text, Cases and Materials* (Oxford: Oxford University Press, 2008), 627.
352. Commission Directive 94/46/EC of 13 Oct. 1994 amending Directive 88/301/EEC and Directive 90/388/EEC in particular with regard to satellite communications, *OJ L* 268, 19 Oct. 1994, 15.
353. Alison Jones & Brenda Sufrin, *EC Competition Law: Text, Cases and Materials* (Oxford: Oxford University Press, 2008), 628.
354. Jose Luis Buendia Sierra, *Exclusive rights and state monopolies under EC law* (Oxford: Oxford University Press, 1999), 66–69; Tony Prosser, *The Limits of Competition Law, Markets and Public Services* (Oxford: Oxford University Press, 2005), 131.

c. Measures Contrary to the Treaty on the Functioning of the EU

BROAD INTERPRETATION – The term *measures* has to be interpreted broadly. It includes legislative acts, administrative decisions, the exercise of shareholder's rights, and non-binding recommendations.[355] There are several ways in which such measures may distort competition. In some cases the measures *are* an infringement of Article 86(1) in themselves, while in other cases the measures *result* in an infringement.

THE GRANT OF THE RIGHT ITSELF IS NOT AN INFRINGEMENT – The ECJ has stated in *Höfner v. Macrotron* that the fact of creating a dominant position by granting exclusive rights is as such not incompatible with the EC Treaty. A Member State will only infringe upon the Treaty if the undertaking in question, merely by exercising the exclusive rights, cannot avoid abusing its dominant position.[356] In the *ERT* case, the ECJ took a more stringent approach: an infringement of Article 86(1) (current Article 106(1) Treaty on the Functioning of the EU) was found not because the violation of the Treaty by the undertaking was unavoidable due to the state measures, but because the undertaking *was led* to infringe upon Article 82 (current Article 102 of the Treaty on the Functioning of the EU).[357] The *Merci Convenzionali* judgment followed this approach: a Member State infringes upon Article 82 (and Article 86(1)) when it grants rights that are liable to *induce* the undertaking to commit abuses.[358] In the *Corbeau* case, the ECJ went even further. The Court suggested that the mere grant of special or exclusive rights could in itself be a violation of Article 86(1), because it would deprive the competition rules of their effectiveness.[359] However, in later cases, the ECJ returned to the view that Article 86(1) was only violated when the undertaking could not avoid abusing its dominant position or where it was led to commit an abuse merely by exercising its rights.[360]

355. ECJ, Case 202/88, *French Republic v. Commission*, 19 Mar. 1991, para. 22, *European Court reports* 1991, I-1223.
356. ECJ, Case C-41/90, *Höfner & Elser v. Macotron GmbH*, 23 Apr. 1991, para. 29, *European Court Reports* 1991, I-1979.
357. ECJ, Case C-260/89, *Elliniki Radiophonia Tiléorassi AE and others v. Dimotiki Etairia Pliroforissis and others*, 18 Jun. 1991, para. 37-38, *European Court reports* 1991, I-2925. See Alison Jones & Brenda Sufrin, *EC Competition Law: Text, Cases and Materials* (Oxford: Oxford University Press, 2008), 633–634.
358. ECJ, Case C-179/90, *Merci convenzionali porto di Genova SpA v. Siderurgica Gabrielli SpA.*, 10 Dec. 1991, paras 17 and 19, *European Court Reports* 1991, 5889.
359. ECJ, Case C-320/91, *Criminal proceedings against Paul Corbeau*, 19 May 1993, para. 11-14, *European Court reports* 1993, I-2533. See Alison Jones & Brenda Sufrin, *EC Competition Law: Text, Cases and Materials* (Oxford: Oxford University Press, 2008), 638–640; Tony Prosser, *The Limits of Competition Law, Markets and Public Services* (Oxford: Oxford University Press, 2005), 131–132.
360. ECJ, Case C-323/93, *Société Civile Agricole du Centre d'Insémination de la Crespelle v. Coopérative d'Elevage et d'Insémination Artificielle du Département de la Mayenne*, 5 Oct. 1994, para. 18, *European Court reports* 1994, I-5077; ECJ, Case C-55/96, *Job Centre Coop. ARL*, 11 Dec. 1997, para. 29, *European Court Reports* 1997, I-7119; ECJ, Case C-163/96,

EXAMPLES – Some state measures granting exclusive rights have been considered to distort competition because the undertakings holding the rights to provide a certain service were unable to satisfy the demand for this service. This formed a breach of Article 102 and Article 106(1) of the Treaty on the Functioning of the EU.[361] Another type of measures that has been found contrary to the Treaty by the ECJ are measures assisting a public undertaking or an undertaking that has exclusive rights to extend its market power into a secondary or neighbouring market without objective justification, with the aim of eliminating competition.[362] An infringement was also found if the national measures created a conflict of interest, because the undertaking that was holder of the special or exclusive rights was able to control aspects of its competitors' activities as a regulator, and in this way combined economic activities with regulatory functions.[363] Article 106(1) was also considered to be violated if a state measure sets the tariffs to be charged by the undertaking with special or exclusive rights and this is contrary to the Treaty, for example, price discrimination.[364] Finally, measures may also lead to a refusal of supply in violation of Article 102.[365]

Criminal proceedings against Silvano Raso and others, 12 Feb. 1998, para. 27, *European Court reports* 1998 I-533; ECJ, Case C-266/96, *Corsica Ferries France SA v. Gruppo Antichi Ormeggiatori del porto di Genova Coop. arl*, 18 Jun. 1998, *European Court reports* 1998, I-3949; ECJ, Case C-67/96, *Albany International BV v. Stichting Bedrijfspensioenfonds Textielindustrie*, 21 Sep. 1999, para. 93, *European Court Reports* 1999, I-5751; ECJ, Case C-475/99, *Ambulanz Glöckner v. Landkreis Südwestpflatz*, 25 Oct. 2001, para. 39, *European Court Reports* 2001, I-8089.

361. ECJ, Case C-41/90, *Höfner & Elser v. Macotron GmbH*, 23 Apr. 1991, para. 31, *European Court Reports* 1991, I-1979; ECJ, Case C-180/98 to C-184/98, *Pavlov v. Stichting Pensioenfonds Medische Specialisten*, 12 Sep. 2000, para. 190, *European Court Reports* 2000, I-6451; ECJ, Case C-55/96, *Job Centre Coop. ARL*, 11 Dec. 1997, para. 35, *European Court Reports* 1997, I-7119; ECJ, Case C-258/98, *Giovanni Carra and Others*, 8 Jun. 2000, para. 17, *European Court Reports* 2000, I-4217.
362. ECJ, Case 18/88, *Régie des télégraphes et des téléphones v. GB-Inno-BM SA.*, 13 Dec. 1991, *European Court reports* 1991, I-5941; ECJ, Case C-320/91, *Criminal proceedings against Paul Corbeau*, 19 May 1993, *European Court reports* 1993, I-2533; ECJ, C-475/99, *Ambulanz Glöckner v. Landkreis Südwestpflatz*, 25 Oct. 2001, *European Court Reports* 2001, I-8089. See Peter Roth & Vivien Rose (ed.), *Bellamy & Child European Community Law of Competition*. 6th edn (Oxford: Oxford University Press, 2008), 1044.
363. ECJ, Case C-163/96, *Criminal proceedings against Silvano Raso and others*, 12 Feb. 1998, *European Court reports* 1998 I-533; ECJ, Case C-260/89, *Elliniki Radiophonia Tiléorassi AE and others v. Dimotiki Etairia Pliroforissis and others*, 18 Jun. 1991, *European Court reports* 1991, I-2925; ECJ, Case 18/88, *Régie des télégraphes et des téléphones v. GB-Inno-BM SA.*, 13 Dec. 1991, *European Court reports* 1991, I-5941.
364. ECJ, Case C-242/95, *GT-Link A/S v. De Danske Statsbaner*, 17 Jul. 1997, *European Court reports* 1997, I-4449; ECJ, Case C-266/96, *Corsica Ferries France SA v. Gruppo Antichi Ormeggiatori del porto di Genova Coop. arl*, 18 Jun. 1998, *European Court reports* 1998, I-3949; ECJ, Case C-163/99, *Portuguese Republic v. Commission*, 29 Mar. 2001, *European Court reports* 2001, 2613.
365. Commission Decision 94/119/EC of 21 Dec. 1993 concerning a refusal to grant access to the facilities of the port of Rødby (Denmark), *OJ L* 55, 26 Feb. 1994, 52.

d. Enforcement

ARTICLE 106(3) OF THE TREATY ON THE FUNCTIONING OF THE EU – The European Commission may take action against the Member States under Article 106(3) of the Treaty on the Functioning of the EU (former Article 86(3) of the EC Treaty) by way of a decision or a directive. This is independent of the action the Commission might also start against the undertaking with exclusive or special rights under Article 101 or 102 of the Treaty on the Functioning of the EU, when applicable.

e. Applicability to Spatial Information Services

i. The Practice in the Member States

PUBLIC UNDERTAKINGS – Article 106(1) of the Treaty on the Functioning of the EU can be of considerable importance for the provision of spatial information services by public bodies. If these public bodies can be categorized as a public undertaking or an undertaking that has been granted a special or exclusive right, the Member States have to ensure that they do not take any measures that lead these undertakings to infringe upon competition law. From the descriptions given above, many public bodies providing spatial information services can be considered as public undertakings. If they abuse their position as described in the examples above and this abuse was caused by the rules that are imposed on them by the government, then the Member State can also be charged with infringement of Article 106(1). If a part of central government is considered to be an undertaking, which is possible as described above, this leads to the situation that the state itself is infringing upon both Article 102 and Article 106(1), the former as an undertaking and the latter in its capacity as the State.

SPECIAL OR EXCLUSIVE RIGHTS – In addition, the public bodies themselves could also give an exclusive or special right to a private undertaking to disseminate their information[366] and in this way limit competition by other private undertakings. According to Volman, '[e]xclusive deals may have short term advantages, since they guarantee a stable income for a certain period of time, but at the same time they prevent others from getting the information at the same conditions'.[367] Moreover, Perritt argues that the argument that such exclusive arrangements are efficient is rarely true, because the information market develops so quickly that it is not possible for one undertaking to meet the expectations of so many different market segments.[368] Therefore, De Terwangne states that the selection procedure of the

366. Cécile De Terwangne et al., *L'accès à l'information administrative et la commercialisation des données publiques: analyse théorique et pratique du problème d'accès, pour le citoyen en général, aux informations détenues par le service public* (Brussels: Story-Scientia, 1994), 112–113 [in French].

367. Yvo Volman, 'Exploitation of Public Sector Information in the Context of the *eEurope* Action Plan', in *Public Sector Information in the Digital Age, Between Markets, Public Management and Citizens' Rights*, ed. Georg Aichholzer & Herbert Burkert (Cheltenham: Edward Elgar Publishing Limited, 2004), 97.

368. Henry Perrit Jr., 'Should local governments sell local spatial databases through state monopolies?', *Jurimetrics Journal* 35 (1994–1995): 454–455.

private partners has to comply with the rules on public procurement and the assignment of concessions, and no privileges or discriminatory clauses can be included in the agreements.[369]

PUBLIC-PRIVATE PARTNERSHIPS – The question of exclusive rights on the provision of spatial information services has also come up a number of times with regard to the suggestion that public-private partnerships may be an appropriate way to organize the provision of spatial information services.[370] In setting up such partnerships, the private sector will of course be interested in becoming a privileged partner and obtaining the exclusive right to disseminate the information service. However, this includes the risk of unfair competition, because all other private partners will be refused access to the public sector spatial data. Such a partnership was, for instance, set up for the production of the digital large scale map of the Netherlands, and for the Dutch Cadastre, which had an exclusive agreement with its own commercial arm (this is no longer the case).[371]

ii. PSI Directive

EU POLICY DOCUMENTS – It has always been generally accepted that exclusive rights given to an undertaking to distribute public sector spatial data or provide spatial information services should not lead to infringements of the rules of EU competition law. The 1989 Synergy Guidelines already stated that: 'Contracts or other arrangements with private sector database providers or host services should not grant exclusive rights if they lead to distortion of competition'.[372] Yet, the 1996 PIRA report found a large number of exclusive agreements.[373] The report believed

369. Cécile De Terwangne et al., *L'accès à l'information administrative et la commercialisation des données publiques: analyse théorique et pratique du problème d'accès, pour le citoyen en général, aux informations détenues par le service public* (Brussels: Story-Scientia, 1994), 112–113 [in French].
370. HM Treasury, 'Cross Cutting Review of the Knowledge Economy. Review of Government Information', 2000, <www.hm-treasury.gov.uk/spend_sr00_ccr.htm>, 22 Feb. 2010; William S. Holland, 'Copyright, Licensing and Cost Recovery for Geographic and Land Information System Data: A Legal, Economic and Policy Analysis', 1997, <www.spatial.maine.edu/~onsrud/tempe/holland.htm>, 22 Feb. 2010; Harry Bouwman et al., 'Leuker kunnen we het niet maken, wel makkelijker. Over de toegankelijkheid van overheidsinformatie', in *Elektronische toegankelijkheid van overheidsinformatie*, ed. Inigo Baten and Gijs Van Der Starre (The Hague: Rathenau Instituut, 1996), 85–87 [in Dutch]. See also Hendrik J. De Ru, 'The Commercial Use of Government Controlled Information: A Guideline for Regulatory Policy Choices', in *Information Law towards the 21st Century*, ed. Willem F. Korthals Altes (Deventer: Kluwer, 1992), 218.
371. Mireille Van Eechoud, 'The Commercialization of Public Sector Information: Delineating the Issues', in *The Future of the Public Domain. Identifying the Commons in Information law*, ed. Lucie Guibault & P. Bernt Hugenholtz (Alphen a/d Rijn: Kluwer Law International, 2006), 287–289.
372. Commission of the European Communities, *Synergy Guidelines*, 5. See also Johan Pas, 'The Commercialization of Government Information and the Proposal for a Directive COM (2002) 207 by the European Commission', *Murdoch University Electronic Journal of Law* 9, no. 4 (2002), <www.murdoch.edu.au/elaw/issues/v9n4/pas94_text.html>, 22 Feb. 2010.
373. PIRA International, 'Commercial Exploitation of Europe's Public Sector Information. Final Report', 2000, 78, 15, <http://ec.europa.eu/information_society/policy/psi/docs/pdfs/pira_study/commercial_final_report.pdf>, 22 Feb. 2010.

that such exclusive deals for developing information services gave one player in the market an unfair advantage. However, it was admitted that there 'may be special situations in which exclusive arrangements may be the only way to ensure that a special service can be provided for an important but niche market'.[374] The European Commission's Green Paper on Public Sector Information and its 2001 Communication repeated the concern regarding exclusivity: such arrangements should be banned if they unreasonably restrict the re-use of PSI.[375]

GENERAL PROHIBITION OF EXCLUSIVE ARRANGEMENTS – This is still in line with the general provisions of competition law, under which the mere grant of an exclusive or special right is not an infringement, but only the inducement by such a right of infringing behaviour by the undertakings in question. However, the PSI Directive goes further and takes the opposite approach: all exclusive agreements should be prohibited under Article 11, not just the ones that infringe competition law. An exception to this prohibition can only be made where an exclusive right is necessary for the provision of a service in the public interest. In that case, the validity of the reason for granting such an exclusive right has to subject to regular review, and in any event, be reviewed every three years. The agreements have to be transparent and made public. Existing exclusive agreements that could not be justified under the exception had to be terminated by 31 December 2008 at the latest.

COMMISSION ACTIONS ON EXCLUSIVE ARRANGEMENTS – In June 2008, the European Commission stated that exclusive arrangements were still in place in the Member States under the form of digitization initiatives, public-private partnerships, outsourcing arrangements, aggregation of data sets in national registers, and agreements with national publishers. Member States still had until the end of 2008 to take action, but from 2009 onwards, '[t]he European Commission intends to take action – through the European Courts if required to ensure Member States have complied with the requirements of the Directive 2003/98/EC with respect to exclusive agreements'.[376] According to APPSI, the Commission may have a lot of work. The Panel found that it 'is hard to identify exclusive licenses, where they do exist, and to quantify the problem'. Therefore, 'APPSI recommends that wherever exclusive arrangements remain, a justification should be published, and that all such information about exclusive arrangements should appear in one place'.[377] In this way it will be easier to determine whether an exclusive arrangement could be allowed because it is in the public interest. The scope of this public

374. *Ibid.*, 104.

375. Commission of the European Communities, *Green Paper on Public Sector Information*, 15; Commission of the European Communities, *eEurope 2002 Communication*, 12.

376. See ePSIplus, 'Locating Exclusive agreements', 13 Aug. 2008, <www.epsiplus.net/news/locating_exclusive_agreements>, 22 Feb. 2010.

377. Advisory Panel on Public Sector Information, 'The impact and success of legislation on the re-use of public sector information', 2008, 32, <www.appsi.gov.uk/reports/ec-submission-07-2008.doc>, 22 Feb. 2010.

interest is not determined in the PSI Directive, but Recital 20 refers to the concept of services of general economic interest as a justification for an exclusive right. This concept will be addressed in the next section.

PUBLIC UNDERTAKINGS – While the PSI Directive is fully applicable to undertakings with special or exclusive rights, it states in Recital 10 of the Preamble that public undertakings do not fall under the definition of public sector body or body governed by public law, and hence do not fall under the PSI Directive. Hence, the rules on exclusive agreements of the PSI Directive do not apply to these public undertakings. However, they of course remain subject to Article 106(1) of the Treaty on the Functioning of the EU.

D. SPECIAL CASE: SERVICES OF GENERAL (ECONOMIC) INTEREST

RECENT OBJECT OF ATTENTION – Services of general (economic) interest have received a lot of attention from the European institutions in the last two decades.[378] However, this attention does not mean that the concept of services of general interest and the principles how these services should be treated have been made sufficiently clear.[379]

OUTLINE – As a starting point, it should be pointed out that services of general interest are not the same as services of general *economic* interest. The latter are mentioned in the Treaty on the Functioning of the EU, while the former term has been introduced by the European Commission in its 1996 Communication on Services of General Interest.[380] This section first addresses the provisions of the Treaty on the Functioning of the EU on services of general economic interest, and then moves on to the treatment of the concept of services of general (economic) interest by the legislator. Next, the link between services of general (economic) interest and spatial information services is examined.

---

378. Before the 1990s, not much importance was attached to the provisions of the EC Treaty with regard to services of general economic interest. The interest grew with the increasing liberalization tendencies of the European Commission. See Alison Jones & Brenda Sufrin, *EC Competition Law: Text, Cases and Materials* (Oxford: Oxford University Press, 2008), 616; Liyang Hou, 'Uncovering the veil of Article 86(2)EC', 2007, <http://papers.ssrn.com/sol3/papers.cfm?abstract_id=1025407>, 22 Feb. 2010; Laurent Richer, 'Droit d'accès et service public', *Actualité juridique Droit Administratif*, 16 Jan. 2006, 73 [in French]; Wolf Sauter, 'Services of general economic interest and universal service in EU law', *European Law Review* 33, no. 2 (2008): 170; Wouter Devroe, 'Universele dienstverlening als nieuwe manier van denken', *Sociaal-economische Wetgeving* 2000: 89 [in Dutch]; Tony Prosser, *The Limits of Competition Law, Markets and Public Services* (Oxford: Oxford University Press, 2005), 133.
379. Stephane Rodrigues, 'Les qualifications concurrentes des activités d'intérêt général en droit communautaire', *Actualité juridique Droit Administratif*, 16 Jan. 2006, 84 [in French].
380. Commission of the European Communities, *Communication. Services of General Interest in Europe*, COM (1996) 443 final, 11 Sep. 1996.

1. Article 106(2) Treaty on the Functioning of the EU

a. Introduction

ARTICLE 106(2) OF THE TREATY ON THE FUNCTIONING OF THE EU – Article 106(2) of the Treaty on the Functioning of the EU (former Article 86(2) of the EC Treaty) provides a derogation from the competition rules for undertakings entrusted with the operation of services of general economic interest or having the character of a revenue-producing monopoly. In principle, such undertakings remain subject to the rules of competition, insofar as the application of these rules does not obstruct the performance, in law or in fact, of the particular tasks that are assigned to them. If this is not possible, they are exempted from the application of competition law. In any case, however, the development of trade must not be affected to such an extent as would be contrary to the interests of the Union.

BALANCE BETWEEN COMPETITION AND PUBLIC INTEREST – Hence, Article 106(2) can in some cases be used to justify a breach of the rules of Article 101 or 102 or measures granting special or exclusive rights contrary to the Treaty on the Functioning of the EU under Article 106(1). As the ECJ stated, '[t]hat provision seeks to reconcile the Member States' interests in using certain undertakings, in particular in the public sector, as an instrument of economic or fiscal policy with the Union's interest in ensuring compliance with the rules on competition and the preservation of the unity of the Common market'.[381] Hence, Article 106(2) tries to find a balance between the safeguarding of the provision of services in the public interest with the proper functioning of the market.[382]

> Nuance of dichotomy between competition and public service – Sauter indicates the importance of Article 106(2) to avoid characterizing EU competition law as a binary system, classifying entities either as undertakings (i.e., carrying out economic activities bearing economic risk), or not, with important consequences in terms of the legal obligations that follow:
>
> – As soon as services are provided by undertakings the competition rules apply in full force;

381. ECJ, Case 202/88, *French Republic v. Commission*, 19 Mar. 1991, *European Court Reports* 1991, para. 12, I-1223; ECJ, Case C-159/94, *Commission v. French Republic*, 23 Oct. 1997, para. 12, *European Court reports* 1997, I-5815; ECJ, Cases C-115/97, C-116/97 and C-117/97, *Brentjens' Handelsonderneming BV v. Stichting Bedrijfspensioenfonds voor de Handel in Bouwmaterialen*, 21 Sep. 1999, para. 103, *European Court reports* 1999, I-6025; ECJ, Case C-219/97, *Maatschappij Drijvende Bokken BV v. Stichting Pensioenfonds voor de Vervoer- en Havenbedrijven*, 21 Sep. 1999, para. 93, *European Court reports* 1999, I-6121.

382. It should be noted that these two objectives are not necessarily contradictory. In many cases the provision of a service on the market and their submission to competition rules leads to improved services, lower prices and increasing choices for the consumers (see Commission of the European Communities, *Communication. Services of General Interest in Europe*, COM (1996) 443 final, 11 Sep. 1996; Wolf Sauter, 'Services of general economic interest and universal service obligations as an EU law framework for curative health care. TILEC Discussion Paper', 2007, 23, <http://papers.ssrn.com/sol3/papers.cfm?abstract_id=1013261>, 22 Feb. 2010.

– If on the other hand the entities concerned are not undertakings . . . they are excluded from the competition rules altogether.

This may occur while the actual services concerned are very similar or even identical and has been the basis for much dissatisfaction with the EU legal framework . . . Article 86(2) EC-Treaty [current Article 106(2) of the Treaty on the Functioning of the EU] offers a way out of this binary system because it allows proportionate restrictions on competition to be imposed to the benefit of undertakings charged with services of general economic interest. It allows a tailor-made solution for each service of general economic interest'.[383] Hence, services of general economic interest are performed on the market but are not fully subject to the market rules.

b. Definitions

OUTLINE – Article 106(2) of the Treaty on the Functioning of the EU provides derogation from the competition rules for undertakings that are entrusted with a service of general economic interest or that have the character of a revenue producing monopoly. These two types of services need to be clarified.

i. Revenue Producing Monopoly

LITTLE ATTENTION – While undertakings providing services of general economic interest have received a lot of attention in the case law and in literature, this is less so for undertakings that have the character of a revenue-producing monopoly. Such undertakings exploit their exclusive rights to raise revenue for the State and may constitute commercial monopolies falling under Article 37 of the Treaty on the Functioning of the EU (former Article 31 of the EC Treaty).[384]

ii. Service of General Economic Interest

CASUISTIC APPROACH – The term *service of general economic interest* was not defined in the Treaty or in secondary legislation. However, is has been the topic of a large body of case law from the ECJ. The Court has held that it is a Community concept and must be uniformly applied by the Member States.[385] However, the inconsistent and casuistic approach of the ECJ makes it difficult to deduct a definition or a general vision on the scope of services of general economic interest.

---

383. Wolf Sauter, 'Services of general economic interest and universal service obligations as an EU law framework for curative health care. TILEC Discussion Paper', 2007, 3, <http://papers.ssrn.com/sol3/papers.cfm?abstract_id=1013261>, 22 Feb. 2010; Wolf Sauter, 'Services of general economic interest and universal service in EU law', *European Law Review* 33, no. 2 (2008): 168. See also Leonor Moral Soriano, 'How Proportionate Should Anti-Competitive State Intervention Be?', *European Law Review* 28, no. 1 (2003): 114.
384. Alison Jones & Brenda Sufrin, *EC Competition Law: Text, Cases and Materials* (Oxford: Oxford University Press, 2008), 651.
385. ECJ, Case 10/71, *Ministère public luxembourgeois v. Madeleine Muller and others*, 14 Jul. 1971, para. 14-15, *European Court reports* 1971, 723.

Jones and Sufrin feel that this may be inevitable.[386] De Bandt even finds a definition undesirable, because of the considerable differences between the Member States with regard to the values and priorities of the public sector and the intervention of the state on the market.[387] The Member States have to maintain the discretion and the margin of appreciation to create a service of general economic interest when they feel it is needed. The downside of this approach is the uncertainty that is caused on the market.

MARGIN OF APPRECIATION FOR THE MEMBER STATES – The lack of a definition may indeed indicate the freedom that is left to the Member States to determine within their national legal order what they consider a service of general economic interest.[388] This freedom is greater in the sectors where the European Commission has not taken any harmonization initiatives. Such initiatives have for instance been taken in the sectors of electronic communications, gas, and electricity. This may lead to different results when the ECJ has to determine whether an undertaking providing a service of general economic interest can be exempted from the application of competition law. In general, the Member States have a considerable margin of appreciation, and the Commission can only intervene when there is a manifest error of judgment.[389]

386. Alison Jones & Brenda Sufrin, *EC Competition Law: Text, Cases and Materials* (Oxford: Oxford University Press, 2008), 671. See also Liyang Hou, 'Uncovering the veil of Article. 86(2)EC', 2007, <http://papers.ssrn.com/sol3/papers.cfm?abstract_id=1025407>, 22 Feb. 2010; Jochen Meulman, 'Na het Groenboek over Diensten van Algemeen Belang: naar een horizontale aanpak van overheidsinterventie?', *Sociaal-Economische Wetgeving* 52, no. 3 (2004): 98-107, 100 [in Dutch]; Wouter Devroe, 'Universele dienstverlening als nieuwe manier van denken', *Sociaal-economische Wetgeving* 48 (2000): 86, 90 [in Dutch]; A.A. Van Rossum, 'Oneerlijke concurrentie door de overheid', in *Overheid en onderneming. Tussen publiek en privaat ondernemen*, ed. Sebastianus C.J.J. Kortmann et al. (Deventer: W.E.J. Tjeenk Willink, 1998), 227 [in Dutch].

387. Pierre De Bandt, 'Artikel 86, lid 2 EG-Verdrag en het begrip 'diensten van algemeen economisch belang: meer dan een rechtsbasis voor een genuanceerde toepassing van de interne markt- en mededingingsregels op openbare diensten?', *Tijdschrift voor Belgisch Handelsrecht* 114, no. 3 (2008): 216 [in Dutch]. See also Liyang Hou, 'Uncovering the veil of Article 86(2)EC', 2007, <http://papers.ssrn.com/sol3/papers.cfm?abstract_id=1025407>, 22 Feb. 2010. For more information on the different approaches in the Member States, see Tony Prosser, 'Services of General Economic Interest in Community Law: From Single Market to Citizenship Rights', paper given at the University of Rome (Rome, 2003), 1-3, <http://ecoservr.eco.uniroma1.it/pdf/Rome_lecture_Dece03.pdf>, 22 Feb. 2010; Michel Mangenot et al. *Public Administrations and Services of General Interest: What Kind of Europeanisation?* (Maastricht: European Institute of Public Administration, 2005), 65-94.

388. See Tony Prosser, *The Limits of Competition Law, Markets and Public Services* (Oxford: Oxford University Press, 2005), 133.

389. CFI, Case T-106/95, *Fédération française des sociétés d'assurances and others v. Commission*, 27 Feb. 1997, para. 99, *European Court reports* 1997, II-229; CFI, Case T-17/02, *Fred Olsen v. Commission*, 15 Jun. 2005, para. 216, *European Court reports* II-2031; CFI, Case T-289/03, *BUPA v. Commission*, 12 Feb. 2008, para. 169, *European Court reports* 2008, II-281. See also Wolf Sauter, 'Services of general economic interest and universal service obligations as an EU law framework for curative health care. TILEC Discussion Paper', 2007, 13, <http://papers.ssrn.com/sol3/papers.cfm?abstract_id=1013261>, 22 Feb. 2010; Stephane Rodrigues, 'Les qualifications concurrentes des activités d'intérêt

Dynamic character – A second reason is the dynamic character of the concept of services of general economic interest. Technological, social, economic, and other developments may change the scope of the services that should be provided outside of the rules of the market over time and place. Therefore, providing a list of services of general economic interest is not feasible or even desirable.[390] However, this does not mean that it is impossible to determine some common characteristics. While these characteristics may not be easily deducted from the ECJ's judgments, the European Commission has made some attempts at listing them. I will address this later on.

Common elements – A few remarks should be made with regard to the elements that make up the term *service of general economic interest*. First, the provision of a *service* is intended to cover 'the widest spectrum of activities',[391] including both services and goods.[392] Next, the service has to be *economic*, hence an economic activity as described in the section above on the definition of *undertaking*.[393] Thirdly, *general interest* entails that the services have to be provided in the public interest. According to Advocate General Jacobs in the *Dusseldorp* case,

---

général en droit communautaire', *Actualité juridique Droit Administratif*, 16 Jan. 2006, 86 [in French]; Markus Krajewski, 'Providing Legal Clarity and Securing Policy Space for Public Services through a Legal Framework for Services of General Economic Interest: Squaring the Circle?', *European Public Law* 14, no. 3 (2008): 379; Karen E. Arnon & Jan J.M. Sluijs, 'Het arrest BUPA: publieke diensten en marktwerking', *Nederlands Tijdschrift voor Europees Recht* no. 7/8 (2008): 207, 208, 210 [in Dutch]; Leigh Hancher & S.J.H. Evans, 'Altmark als katalysator: het Commissiepakket met alle antwoorden rond staatssteun en diensten van algemeen economisch belang?', *Nederlands Tijdschrift voor Economisch Recht* 12 no. 7 (2006): 153 [in Dutch]; Johan W. Van De Gronden, 'Purchasing care: economic activity or service of general (economic) interest', *European Competition Law Review* 25, no. 2 (2004): 89; Marc Hansen, Anne Van Ysendyck & Susanne Zuhlke, 'The Coming of Age of EC State Aid Law: a Review of the Principal Developments in 2002 and 2003', *European Competition Law Review* 25, no. 4 (2004): 219; Wolf Sauter, 'Services of general economic interest and universal service in EU law', *European Law Review* 33, no. 2 (2008): 174–175; Tony Prosser, *The Limits of Competition Law, Markets and Public Services* (Oxford: Oxford University Press, 2005), 134.

390. Wolf Sauter, 'Services of general economic interest and universal service obligations as an EU law framework for curative health care. TILEC Discussion Paper', 2007, 12–13, <http://papers.ssrn.com/sol3/papers.cfm?abstract_id=1013261>, 22 Feb. 2010. See also Commission of the European Communities, *Green Paper on services of general interest*, COM (2003) 270 final, 14–15, 11 Sep. 1996; Liyang Hou, 'Uncovering the veil of Article 86(2)EC', 2007, <http://papers.ssrn.com/sol3/papers.cfm?abstract_id=1025407>, 22 Feb. 2010; Wolf Sauter, 'Services of general economic interest and universal service in EU law', *European Law Review* 33, no. 2 (2008): 175.

391. Alison Jones & Brenda Sufrin, *EC Competition Law: Text, Cases and Materials* (Oxford: Oxford University Press, 2008), 653.

392. Liyang Hou, 'Uncovering the veil of Article 86(2)EC', 2007, <http://papers.ssrn.com/sol3/papers.cfm?abstract_id=1025407>, 22 Feb. 2010; Markus Krajewski, 'Providing Legal Clarity and Securing Policy Space for Public Services through a Legal Framework for Services of General Economic Interest: Squaring the Circle?', *European Public Law* 14, no. 3 (2008): 386.

393. Tony Prosser, *The Limits of Competition Law, Markets and Public Services* (Oxford: Oxford University Press, 2005), 133.

'[t]he reason for the assignment of particular tasks to undertakings is often that the tasks need to be undertaken in the public interest but might not be undertaken, usually for economic reasons, if the service were to be left to the private sector'.[394] In line with this, according to Sauter, an important reason for the entrustment of a service of general economic interest, is market failure: 'After all, it is only in those cases that the services concerned are not already provided to the requisite standard by the market (and/or by public authorities) that market parties must be entrusted with the provision of particular services in the public interest'.[395]

c. *Conditions of Application*

Outline – Even if the ECJ recognizes there is a service of general economic interest, three conditions still have to be fulfilled for Article 106(2) to apply. First, the undertaking has to be entrusted with the service of general economic interest. Next, there is a proportionality condition. The derogation from EU competition rules is only allowed insofar as the application of these rules does not obstruct the performance of the service. Finally, the development of trade must not be affected to such an extent as would be contrary to the interests of the Union.

i. Entrustment

Act of public authority – For an undertaking to be entrusted with a service, there must have been an act of public authority.[396] According to Jones and Sufrin, '[i]t can be a public or private undertaking, but the important thing is that the state has assigned it certain tasks by a positive act conferring on it certain functions or by granting it a concession'.[397] Activities performed by undertakings at their own

394. ECJ, Case, C-203/96, *Chemische Afvalstoffen Dusseldorp BV and Others v. Minister van Volkshuisvesting, Ruimtelijke Ordening en Milieubeheer*, 25 Jun. 1998, Opinion Jacobs AG, para. 105, *European Court reports* 1998, I-4075.

395. Wolf Sauter, 'Services of general economic interest and universal service obligations as an EU law framework for curative health care. TILEC Discussion Paper', 2007, 18, <http://papers.ssrn.com/sol3/papers.cfm?abstract_id=1013261>, 22 Feb. 2010.

396. ECJ, Case 66/86, *Ahmed Saeed Flugreisen and Silver Line Reisebüro GmbH v. Zentrale zur Bekämpfung unlauteren Wettbewerbs e.V.*, 11 Apr. 1989, para. 55, *European Court Reports* 1989, 803; ECJ, Case, C-203/96, *Chemische Afvalstoffen Dusseldorp BV and Others v. Minister van Volkshuisvesting, Ruimtelijke Ordening en Milieubeheer*, 25 Jun. 1998, *European Court reports* 1998, I-4075; CFI, Cases T-204/97 and T-270/97, *EPAC v. Commission*, 13 Jun. 2000, para. 126, *European Court reports* 2000, II-2267.

397. Alison Jones & Brenda Sufrin, *EC Competition Law: Text, Cases and Materials* (Oxford: Oxford University Press, 2008), 653. See also Wolf Sauter, 'Services of general economic interest and universal service obligations as an EU law framework for curative health care. TILEC Discussion Paper', 2007, 13–14, <http://papers.ssrn.com/sol3/papers.cfm?abstract_id=1013261>, 22 Feb. 2010; Liyang Hou, 'Uncovering the veil of Article 86(2)EC', 2007, <http://papers.ssrn.com/sol3/papers.cfm?abstract_id=1025407>, 22 Feb. 2010; Wolf Sauter, 'Services of general economic interest and universal service in EU law', *European Law Review* 33, no. 2 (2008): 184; Michel Mangenot et al. *Public Administrations and Services of General Interest: What Kind of Europeanisation?* (Maastricht: European Institute of Public Administration, 2005, 138–139; Marc Hansen, Anne Van Ysendyck & Susanne Zuhlke,

initiative, or merely approved, tolerated or endorsed by the public authority are not *entrusted* activities. Hence, in principle a specific assignment of a certain task in a formal manner is needed, for example by law, decree, administrative act, or public contract. In addition, the assigned task has to be described as precisely as possible, and not just in general terms.[398] De Bandt finds this evident: the undertaking generally gets exclusive or special rights or public funding specifically for the provision of the service of general economic interest, so its competitors should know the scope of this service.[399] Of course, such a precise description is not always easy, as it should also leave enough room for flexibility so that the undertaking does not have to request a change to the entrustment act for any modification in the activity.

NO ABSOLUTE REQUIREMENT OF AN EXPLICIT ACT – However, the European Commission has held that there is no absolute need for an explicit act of entrustment. In the absence of a specific act, the existence of a service of general economic interest can also be derived from the broader legal context.[400]

ii. Proportionality

EVOLUTION IN INTERPRETATION – Article 106(2) is an exception to the rules of the Treaty, so as a starting point it should be interpreted restrictively.[401] However, the

---

'The Coming of Age of EC State Aid Law: a Review of the Principal Developments in 2002 and 2003', *European Competition Law Review* 25, no. 4 (2004): 219; Karen E. Arnon & Jan J.M. Sluijs, 'Het arrest BUPA: publieke diensten en marktwerking', *Nederlands Tijdschrift voor Europees Recht* no. 7/8 (2008): 210 [in Dutch]; Johan W. Van De Gronden, 'Deel III: Bestaat er een coherente visie op de verhouding tussen diensten van algemeen economisch belang, de Dienstenrichtlijn en andere communautaire regelingen?', *Sociaal-Economische Wetgeving* 55, no. 10 (2007): 409–410 [in Dutch]; Jochen Meulman, 'Na het Groenboek over Diensten van Algemeen Belang: naar een horizontale aanpak van overheidsinterventie?', *Sociaal-Economische Wetgeving* 52, no. 3 (2004): 105 [in Dutch]; Peter Roth & Vivien Rose (ed.), *Bellamy & Child European Community Law of Competition.* 6th edn (Oxford: Oxford University Press, 2008), 1062–1063; A.A. Van Rossum, 'Oneerlijke concurrentie door de overheid', in *Overheid en onderneming. Tussen publiek en privaat ondernemen*, ed. Sebastianus C.J.J. Kortmann et al. (Deventer: W.E.J. Tjeenk Willink, 1998), 232 [in Dutch].

398. Pierre De Bandt, 'Artikel 86, lid 2 EG-Verdrag en het begrip 'diensten van algemeen economisch belang: meer dan een rechtsbasis voor een genuanceerde toepassing van de interne markt- en mededingingsregels op openbare diensten?', *Tijdschrift voor Belgisch Handelsrecht* 114 no. 3 (2008): 217 [in Dutch].

399. *Ibid.*

400. For example, Commission of the European Communities, State Aid measure N46/2003, *Risk equalisation scheme in the Irish health insurance market – Ireland*, 13 May 2003, *OJ C* 186, 6 Aug. 2003, 16; Commission of the European Communities, State Aid measures N 541/2004 and N 542/2004, *Risk equalisation system and retention of reserves – The Netherlands*, 3 May 2005, *OJ C* 324, 21 Dec. 2005, 30.

401. Liyang Hou, 'Uncovering the veil of Article 86(2)EC', 2007, <http://papers.ssrn.com/sol3/papers.cfm?abstract_id=1025407>, 22 Feb. 2010; Malcolm Ross, 'A Healthy Approach to Services of General Economic Interest? The BUPA Judgment of the Court of First Instance', *European Law Review* 34, no. 1 (2009): 130; Pierre De Bandt, 'Artikel 86, lid 2 EG-Verdrag en het begrip 'diensten van algemeen economisch belang: meer dan een rechtsbasis voor een genuanceerde toepassing van de interne markt- en mededingingsregels op openbare

interpretation by the ECJ of the requirement that the performance of the service of general economic interest is obstructed by the full application of competition law has become less strict over the years. This interpretation has given rise to much controversy.[402]

FROM A NARROW INTERPRETATION ... – Initially, the ECJ applied Article 86(2) (current Article 106(2)) very narrowly, giving full preference to the competition rules.[403] For instance, in *Höfner v. Macrotron*, the court stated that the derogation of Article 86(2) could not be applied and the service of general economic interest remained subject to the competition rules 'unless and to the extent to which it is shown that their application is incompatible with the discharge of its duties'.[404] Hence, in this approach only restrictions that were indispensable in order to achieve an objective of general interest were allowed.[405]

TO A MORE FLEXIBLE ONE – In later judgments, the ECJ became more flexible and accepted that some services need to be provided in the general interest and therefore need to be protected from the full force of competition law. The turning point can be found in the *Corbeau* case in 1993. The ECJ no longer considered it necessary that the survival of the undertaking itself would be threatened in the absence of the derogation.[406] It was enough to show that under the normal

diensten?', *Tijdschrift voor Belgisch Handelsrecht* 114 no. 3 (2008): 219 [in Dutch]; Wolf Sauter, 'Services of general economic interest and universal service obligations as an EU law framework for curative health care., TILEC Discussion Paper', 2007, 24, <http://papers.ssrn.com/sol3/papers.cfm?abstract_id=1013261>, 22 Feb. 2010; Berend J. Drijber, 'Mededinging en diensten van algemeen economisch belang', in *Actualiteiten in het Europese Mededingingsrecht. Asser Instituut Colloquium Europees Recht*, ed. Marjan Olfers and Karolien Pieters (T.M.C. Asser Press: The Hague, 2006), 33–34.

402. Tony Prosser, *The Limits of Competition Law, Markets and Public Services* (Oxford: Oxford University Press, 2005), 134.

403. See also Alison Jones & Brenda Sufrin, *EC Competition Law: Text, Cases and Materials* (Oxford: Oxford University Press, 2008), 651; Leonor Moral Soriano, 'How Proportionate Should Anti-Competitive State Intervention Be?', *European Law Review* 28, no. 1 (2003): 115–116; Johan W. Van De Gronden, 'Deel III: Bestaat er een coherente visie op de verhouding tussen diensten van algemeen economisch belang, de Dienstenrichtlijn en andere communautaire regelingen?', *Sociaal-Economische Wetgeving* 55, no. 10 (2007): 410 [in Dutch]; Tony Prosser, *The Limits of Competition Law, Markets and Public Services* (Oxford: Oxford University Press, 2005), 135.

404. ECJ, Case C-41/90, *Höfner & Elser v. Macrotron GmbH*, 23 Apr. 1991, para. 24, *European Court Reports* 1991, I-1979. See also ECJ, Case C-260/89, *Elliniki Radiophonia Tiléorassi AE and others v. Dimotiki Etairia Pliroforissis and others*, 18 Jun. 1991, *European Court reports* 1991, I-2925; ECJ, Case C-179/90, *Merci convenzionali porto di Genova SpA v. Siderurgica Gabrielli SpA.*, 10 Dec. 1991, *European Court Reports* 1991, 5889.

405. Jose Luis Buendia Sierra, *Exclusive rights and state monopolies under EC law* (Oxford: Oxford University Press, 1999), 304; Tony Prosser, *The Limits of Competition Law, Markets and Public Services* (Oxford: Oxford University Press, 2005), 135.

406. ECJ, Case C-320/91, *Criminal proceedings against Paul Corbeau*, 19 May 1993, para. 16, *European Court reports* 1993, I-2533. See Peter Roth & Vivien Rose (ed.), *Bellamy & Child European Community Law of Competition.* 6th edn (Oxford: Oxford University Press, 2008), 1066; Michel Mangenot et al., *Public Administrations and Services of General Interest: What Kind of Europeanisation?* (Maastricht: European Institute of Public Administration, 2005),

conditions of competition law, the undertaking would not be able to provide the service of general economic interest 'under economically acceptable conditions'[407] or conditions of 'economic equilibrium'.[408] In a later case, *Dusseldorp*, the ECJ returned to its previous stance and found that the Dutch government should show that the objective could not be equally achieved by other means than a monopoly, and that Article 86(2) could apply only if it was shown that without the contested measure the undertaking could not carry out its entrusted task.[409] However, in *Deutsche Post*, it reinstated its lenient approach by going back to the 'economically balanced conditions'.[410]

CREAM-SKIMMING – In many cases, these 'economically acceptable conditions' are related to the issue of granting exclusive rights to counter cherry-picking or cream-skimming. If the undertaking has an obligation to provide a service of general economic interest which may not be profitable, it cannot compete against its competitors who do not have to provide such a service, but can direct their efforts to the profitable parts of the market. This entails that in some cases, cross-subsidization may take place. This is allowed insofar as it is necessary to maintain the economic equilibrium.[411] In *Corbeau*, the ECJ stated that: 'the obligation on

---

65–94, 139; Iris Houben, 'Public service obligations: moral counterbalance of technical liberalization legislation', *European Review of Private Law* 16 (2008): 13–14; Tony Prosser, *The Limits of Competition Law, Markets and Public Services* (Oxford: Oxford University Press, 2005), 136–137.

407. ECJ, Case C-320/91, *Criminal proceedings against Paul Corbeau*, 19 May 1993, para. 16, *European Court reports* 1993, I-2533. See also ECJ, Case C-393/92 *Municipality of Almelo and others v. NV Energiebedrijf Ijsselmij*, 27 Apr. 1994, para. 49, *European Court Reports* 1994, I-1477; ECJ, Case C-67/96, *Albany International BV v. Stichting Bedrijfspensioenfonds Textielindustrie*, 21 Sep. 1999, para. 107, *European Court Reports* 1999, I-5751; ECJ, Case C-340/99, *TNT Traco SpA v. Poste Italiane SpA*, 17 May 2001, para. 54, *European Court Reports* 2001, I-4109.

408. ECJ, Case C-320/91, *Criminal proceedings against Paul Corbeau*, 19 May 1993, para. 17, *European Court Reports* 1993, I-2533. It is interesting to note the particular approach of the ECJ to the Corbeau case. First, it applies Art. 86(1) of the EC-Treaty (current Art. 106(1)) very strictly by stating that even establishing an exclusive right can be contrary to competition law. Then it goes on with a very lenient view on Art. 86(2) (current Art. 106(2)), stating that the service in question is a service of general economic interest that needs to be able to function in economically acceptable conditions for the competition rules to apply.

409. ECJ, Case C-203/96, *Chemische Afvalstoffen Dusseldorp BV and Others v. Minister van Volkshuisvesting, Ruimtelijke Ordening en Milieubeheer*, 25 Jun. 1998, para. 67, *European Court reports* 1998, I-4075.

410. ECJ, Case C-147/97 and C-148/97, *Deutsche Post AG v. Gesellschaft für Zahlungssysteme mbH GZS and Citicorp Kartenservice GmbH*, 10 Feb. 2000, para. 50, *European Court reports* 2000, I-825.

411. Alison Jones & Brenda Sufrin, *EC Competition Law: Text, Cases and Materials* (Oxford: Oxford University Press, 2008), 657. See ECJ, Case C-475/99, *Ambulanz Glöckner v. Landkreis Südwestpflatz*, 25 Oct. 2001, *European Court Reports* 2001, I-8089, where the extension of a monopoly was found justified to subsidize the service of general economic interest. See also Pierre De Bandt, 'Artikel 86, lid 2 EG-Verdrag en het begrip 'diensten van algemeen economisch belang: meer dan een rechtsbasis voor een genuanceerde

the part of the undertaking entrusted with that task to perform its services in conditions of economic equilibrium presupposes that it will be possible to offset less profitable sectors against the profitable sectors and hence justifies a restriction of competition from individual undertakings where the economically profitable sectors are concerned'.[412] However, some authors criticize the lack of economic reasoning underlying the decisions of the ECJ.[413]

BURDEN OF PROOF FOR THE MEMBER STATES – The burden of proof lies on the Member States, but as they have a margin of discretion to decide on the provision of services of general economic interest, they do not have to 'prove positively that no other conceivable measure, which by definition would be hypothetical, could enable those tasks to be performed under the same conditions'.[414] The Member States can choose whether they provide the services themselves or whether they entrust the service to a third (public or private) party. If they choose the latter option, the public procurement rules will generally apply.[415]

MARGIN OF APPRECIATION – As was mentioned above, this margin of discretion entails that Member States only have to prove that the measures they took contrary to competition rules are not manifestly disproportionate. However, if EU legislation has harmonized the rules in the sector with regard to the provision of such services, the Member States may only intervene based on the least restrictive

---

toepassing van de interne markt- en mededingingsregels op openbare diensten?', *Tijdschrift voor Belgisch Handelsrecht* 114, no. 3 (2008): 220 [in Dutch]; Johan W. Van De Gronden, 'Purchasing care: economic activity or service of general (economic) interest', *European Competition Law Review* 25, no. 2 (2004): 93. Van De Gronden criticizes the ECJ for focusing on cherry-picking as the only form of market failure in its judgments, disregarding other forms: Johan W. Van De Gronden, 'Deel III: Bestaat er een coherente visie op de verhouding tussen diensten van algemeen economisch belang, de Dienstenrichtlijn en andere communautaire regelingen?', *Sociaal-Economische Wetgeving* 55, no. 10 (2007): 410–411 [in Dutch].

412. ECJ, Case C-320/91, *Criminal proceedings against Paul Corbeau*, 19 May 1993, para. 17, *European Court reports* 1993, I-2533, ECJ, Case C-393/92, *Municipality of Almelo and others v. NV Energiebedrijf Ijsselmij*, 27 Apr. 1994, para. 46, *European Court Reports* 1994, I-1477; ECJ, Case C-157/94, *Commission v. Kingdom of the Netherlands*, 23 Oct. 1997, para. 35, *European Court reports* 1997, I-5699; ECJ, Case C-475/99, *Ambulanz Glöckner v. Landkreis Südwestpflatz*, 25 Oct. 2001, para. 57, *European Court Reports* 2001, I-8089. See also Leonor Moral Soriano, 'How Proportionate Should Anti-Competitive State Intervention Be?', *European Law Review* 28, no. 1 (2003): 117.
413. Leonor Moral Soriano, 'How Proportionate Should Anti-Competitive State Intervention Be?', *European Law Review* 28, no. 1 (2003): 117–120.
414. ECJ, Case C-157/94, *Commission v. Kingdom of the Netherlands*, 23 Oct. 1997, para. 58, *European Court Reports* 1997, I-5699. See also Pierre De Bandt, 'Artikel 86, lid 2 EG-Verdrag en het begrip 'diensten van algemeen economisch belang: meer dan een rechtsbasis voor een genuanceerde toepassing van de interne markt- en mededingingsregels op openbare diensten?', *Tijdschrift voor Belgisch Handelsrecht* 114, no. 3 (2008): 219 [in Dutch].
415. Wolf Sauter, 'Services of general economic interest and universal service obligations as an EU law framework for curative health care. TILEC Discussion Paper', 2007, 27–28, <http://papers.ssrn.com/sol3/papers.cfm?abstract_id=1013261>, 22 Feb. 2010.

means.[416] This harmonization has taken place mostly with regard to the network sectors and basic utilities.[417]

EXAMPLES – The sectors of networks and basic utilities are also the sectors where originally most services of general economic interest could be found, such as mail services, electricity supply, the operation of airlines, etc. However, Article 106(2) has also been applied to other sectors, for example ambulance services,[418] pension funds,[419] mooring services in ports,[420] the treatment of waste,[421] and the broadcasting of television services.[422]

iii. Effect on the Development of Trade

WEIGHING OF INTERESTS – An undertaking can only benefit from the exemption of Article 106(2) Treaty on the Functioning of the EU if the development of trade is not affected to such an extent as would be contrary to the interests of the Union. Hence, the particular task entrusted to the undertaking and the Union interests have to be weighed.[423] So far, this third condition has been of very little importance in the evaluation of services of general economic interest under Article 106(2).[424] Therefore, this condition is not discussed any further.

---

416. Wolf Sauter, 'Services of general economic interest and universal service obligations as an EU law framework for curative health care. TILEC Discussion Paper', 2007, 25, <http://papers.ssrn.com/sol3/papers.cfm?abstract_id=1013261>, 22 Feb. 2010. See also Johan W. Van De Gronden, 'Deel III: Bestaat er een coherente visie op de verhouding tussen diensten van algemeen economisch belang, de Dienstenrichtlijn en andere communautaire regelingen?', *Sociaal-Economische Wetgeving* 55, no. 10 (2007): 411–412 [in Dutch].
417. For example, Directive 2003/54/EC of the European Parliament and of the Council of 26 Jun. 2003 concerning common rules for the internal market in electricity and repealing Directive 96/92/EC, *OJ L* 176, 15 Jul. 1998, 37; Directive 2003/55/EC of the European Parliament and of the Council of 26 Jun. 2003 concerning common rules for the internal market in natural gas and repealing Directive 98/30/EC, *OJ L* 176, 15 Jul. 1998, 57; Directive 2002/22/EC of the European Parliament and of the Council of 7 Mar. 2002 on universal service and users' rights relating to electronic communications networks and services (Universal Service Directive), *OJ L* 108, 24 Apr. 2002, 51.
418. ECJ, Case C-475/99, *Ambulanz Glöckner v. Landkreis Südwestpflatz*, 25 Oct. 2001, *European Court Reports* 2001, I-8089.
419. ECJ, Case C-67/96, *Albany International BV v. Stichting Bedrijfspensioenfonds Textielindustrie*, 21 Sep. 1999, *European Court Reports* 1999, I-5751; ECJ, Cases C-115/97, C-116/97 and C-117/97, *Brentjens' Handelsonderneming BV v. Stichting Bedrijfspensioenfonds voor de Handel in Bouwmaterialen*, 21 Sep. 1999, *European Court Reports* 1999, I-6025.
420. ECJ, Case C-266/96, *Corsica Ferries France SA v. Gruppo Antichi Ormeggiatori del porto di Genova Coop. arl*, 18 Jun. 1998, *European Court Reports* 1998, I-3949.
421. ECJ, Case C-209/98, *Sydhavnens Sten & Grus ApS v. Københavns Kommune*, 23 May 2000, *European Court Reports* 2000, I-3743.
422. ECJ, Case 155/73, *Sacchi*, 30 Apr. 1974, *European Court Reports* 1974, 409; ECJ, Case C-260/89, *Elliniki Radiophonia Tiléorassi AE and others v. Dimotiki Etairia Pliroforissis and others*, 18 Jun. 1991, *European Court Reports* 1991, I-2925.
423. Peter Roth & Vivien Rose (ed.), *Bellamy & Child European Community Law of Competition* 6th edn (Oxford: Oxford University Press, 2008), 1068.
424. Alison Jones & Brenda Sufrin, *EC Competition Law: Text, Cases and Materials* (Oxford: Oxford University Press, 2008), 667.

d. *Enforcement*

ROLE OF THE EUROPEAN COMMISSION – The European Commission may take action in the form of decisions or directives under Article 106(3) in order to specify the meaning and extent of the derogation of Article 106(2) and set out rules for monitoring the fulfilment of the criteria for such derogation.[425] Using these instruments, it can deal with existing infringements and with the prevention of possible future infringements. Based on Article 106(3), the Commission has adopted rules on financial transparency of public undertakings[426] and rules on state aid with regard to public service compensation.[427]

e. *Interim Summary*

EVOLUTION OF THE ECJ'S VIEW – From what has been discussed above, it can be stated that the view of the ECJ on services of general economic interest has evolved from a very restrictive approach that any derogation from competition law is hindering the goals of the common market to a more open view on services of general economic interest. Services of general economic interest are now seen as an integral part of the common market, sometimes needed in order to ensure the public interest. Market and public interest are no longer seen as diametrically opposite. This evolution is confirmed by the legislative activities of the European institutions. These are addressed in the next section.

UNCERTAINTY REMAINS – However, as the ECJ decides on services of general economic interest on a case-by-case basis and has not developed a general theory on these services, the uncertainty about the scope of Article 106(2) Treaty on the Functioning of the EU remains. This uncertainty has only been increased by the introduction by the European Commission of the term *services of general interest.*

2. Legislative Activities

INITIATIVES OF THE EUROPEAN COMMISSION – While the ECJ has used a case-by-case approach to the conditions for the application of Article 86(2) of the EC Treaty (current Article 106(2) of the Treaty on the Functioning of the EU), the European Commission also took a number of initiatives to clarify the concept of services of general economic interest. First, it took the initiative for the inclusion of Article 16 in the EC Treaty (current Article 14 of the Treaty on the Functioning of the EU). Next, it also issued a number of Communications and Papers holding its own vision on services of general interest and services of general economic interest. These initiatives are discussed in this section.

---

425. Peter Roth & Vivien Rose (ed.), *Bellamy & Child European Community Law of Competition*. 6th edn (Oxford: Oxford University Press, 2008), 1068.
426. Cf. *supra*, s. IV.C.1.c.
427. Cf. *infra*, s. IV.D.2.c.

a. Article 16 of the EC Treaty: From Derogation to Obligation

ARTICLE 16 OF THE EC TREATY – Upon request of the European Commission, the Intergovernmental Conference included in the Treaty of Amsterdam in 1997 a new article on services of general economic interest.[428] Article 16 of the EC Treaty (current Article 14 of the Treaty on the Functioning of the EU) states that: 'Without prejudice to Articles 73, 86 and 87 [current Articles 93, 106, and 107], and given the place occupied by services of general economic interest in the shared values of the Union as well as their role in promoting social and territorial cohesion, the Community and the Member States, each within their respective powers and within the scope of application of this Treaty, shall take care that such services operate on the basis of principles and conditions which enable them to fulfil their missions'. A Declaration was added to the Treaty ensuring that Article 16 of the EC Treaty would be implemented 'with full respect for the jurisprudence of the Court of Justice, inter alia, as regards the principles of equality of treatment, quality and continuity of such services'.[429]

IMPORTANCE OF ARTICLE 16 – The fact that this provision on services of general economic interest was placed in the beginning of the EC Treaty among the fundamental principles of the EU shows the growing importance that is attached to these services. It also shows a change in perception: rather than a derogation from competition law or an impediment to the task of market creation,[430] services of general economic interest are seen as part of the tasks of the Member States and the EU. They have to ensure that the undertakings entrusted with a service of general economic interest are able to provide this service. Services of general economic interest are seen as a recognition of citizenship rights and 'interest has moved away from trying to limit the scope of such services to avoid infringing competition law and instead towards improving their delivery through applying concepts of good governance'.[431]

---

428. Commission of the European Communities, *Communication. Services of General Interest in Europe*, COM (1996) 443 final, 11 Sep. 1996. See also Peter Roth & Vivien Rose (ed.), *Bellamy & Child European Community Law of Competition*. 6th edn (Oxford: Oxford University Press, 2008), 1032; Alison Jones & Brenda Sufrin, *EC Competition Law: Text, Cases and Materials* (Oxford: Oxford University Press, 2008), 672 and following.
429. Declaration 11997D/AFI/DCL/13 on Art. 7d of the Treaty establishing the European Community, *OJ C* 340, 10 Nov. 1997, 133.
430. Tony Prosser, *The Limits of Competition Law, Markets and Public Services* (Oxford: Oxford University Press, 2005), 121.
431. Tony Prosser, 'Services of General Economic Interest in Community Law: From Single Market to Citizenship Rights', paper given at the University of Rome (Rome, 2003), 6, <http://ecoservr.eco.uniroma1.it/pdf/Rome_lecture_Dece03.pdf>, 22 Feb. 2010. See also Alison Jones & Brenda Sufrin, *EC Competition Law: Text, Cases and Materials* (Oxford: Oxford University Press, 2008), 673 Markus Krajewski, 'Providing Legal Clarity and Securing Policy Space for Public Services through a Legal Framework for Services of General Economic Interest: Squaring the Circle?', *European Public Law* 14, no. 3 (2008): 379–380; Stephane Rodrigues, 'Towards a general EC framework instrument related to SGEI? Political considerations and legal constraints', paper given at the International Conference 'The changing

Criticism on Article 16 – However, Article 16 of the EC Treaty has also been criticised because of its ambiguous drafting, allowing both a defensive interpretation from the Member States to maintain their own decision power, and the concept of a true Community service of general economic interest being developed.[432] Which interpretation is the right one will have to be determined by the ECJ.

European Constitution and Lisbon Treaty – The European Constitution also contained an article on services of general economic interest. As the Constitution was never accepted by all the Member States, the European Council agreed in 2007 to replace Article 16 of the EC Treaty by Article 14 in the Lisbon Treaty. This article contains an added paragraph giving the authority to the European Parliament and the Council to establish the principles and conditions based on which the undertakings should perform their services of general economic interest. This authority was without prejudice, however, to the competence of Member States to provide, commission and fund the services, in compliance with the EU rules. In addition, a protocol on services of general interest is added to the Treaty.[433] This Protocol recognizes

- the essential role and the wide discretion of national, regional and local authorities in providing, commissioning and organising services of general economic interest as closely as possibly to the needs of the users;
- the diversity between various services of general economic interest and the differences in the needs and preferences of users that may result from different geographical, social or cultural situations;
- a high level of quality, safety and affordability, equal treatment and the promotion of universal access and of user rights.

Signal of common approach? – Hence, on the one hand, the Protocol emphasizes the autonomy of the Member States, while on the other hand, it also includes a number of requirements that the services of general economic interest have to fulfil. According to Van de Gronden, the agreement between the Member States on these requirements shows that there is a basis for a common approach to the

---

framework of public services in Europe' (Potsdam, 2008), <www.uni-potsdam.de/jpkrajewski/Presentations/Rodrigues.PDF>, 22 Feb. 2010; Tony Prosser, *The Limits of Competition Law, Markets and Public Services* (Oxford: Oxford University Press, 2005), 138–139.

432. Alison Jones & Brenda Sufrin, *EC Competition Law: Text, Cases and Materials* (Oxford: Oxford University Press, 2008), 673; Iris Houben, 'Public service obligations: moral counterbalance of technical liberalization legislation', *European Review of Private Law* 16 (2008): 16–20. See also Wolf Sauter, 'Services of general economic interest and universal service in EU law', *European Law Review* 33, no. 2 (2008): 172; CEEP and CIRIEC, *Services of General Economic Interest in Europe. Regulation, Financing, Evaluation, Good Practices* (Liège, CIREC, 2000), 44–45; Tony Prosser, *The Limits of Competition Law, Markets and Public Services* (Oxford: Oxford University Press, 2005), 139; Jose Luis Buendia Sierra, *Exclusive rights and state monopolies under EC law* (Oxford: Oxford University Press, 1999), 329.

433. Protocol on Services of General Interest to the Lisbon Treaty, *OJ C* 306, 17 Dec. 2007, 159.

issue of services of general economic interest, even though the competences of the Member States remain intact.[434]

NON-ECONOMIC SERVICES OF GENERAL INTEREST – For the first time in the history of the EU Treaties, the Protocol also addresses non-economic services of general interest. It explicitly states that the competence of the Member States to provide, commission and organize non-economic services of general interest remains unaffected. These services will be addressed later on in this section.

CHARTER OF FUNDAMENTAL RIGHTS OF THE EUROPEAN UNION – It should also be mentioned that Article 36 of the Charter of Fundamental Rights of the European Union also contains a provision on services of general economic interest. This affirms the importance of access to services of general economic interest as a principle of the EU.[435]

b. The European Commission: From Services of General Economic Interest to Services of General Interest

i. Importance of Services of General Interest

ESSENTIAL ELEMENT OF THE EUROPEAN SOCIETY – The Commission first addressed services of general interest in a Communication from 1996, which led to the introduction of Article 16 in the EC Treaty. However, since then, the Commission has continued stressing the importance of services of general interest in several Communications, a Green Paper, and a White Paper.[436] These services are 'a part of the values shared by all European societies and form an essential element of the European model of society. Their role is essential for increasing quality of life for all citizens and for overcoming social exclusion and isolation'. In addition, they increase competitiveness and greater cohesion and their efficient and non-discriminatory provision is a condition for the smooth functioning of the Single Market and economic integration. Finally, the services are also considered a 'pillar of European citizenship', ensuring that citizens can enjoy their rights and providing an opportunity for dialogue and participation.[437] Rather than addressing each of

434. Johan W. Van De Gronden, 'Deel III: Bestaat er een coherente visie op de verhouding tussen diensten van algemeen economisch belang, de Dienstenrichtlijn en andere communautaire regelingen?', *Sociaal-Economische Wetgeving* 55, no. 10 (2007): 416 [in Dutch]. See also Wolf Sauter, 'Services of general economic interest and universal service in EU law', *European Law Review* 33, no. 2 (2008): 173–174; Stephane Rodrigues, 'Towards a general EC framework instrument related to SGEI? Political considerations and legal constraints', paper given at the International Conference 'The changing framework of public services in Europe' (Potsdam, 2008), <www.uni-potsdam.de/jpkrajewski/Presentations/Rodrigues.PDF>, 22 Feb. 2010; Wolf Sauter, 'Services of general economic interest and universal service obligations as an EU law framework for curative health care. TILEC Discussion Paper', 2007, 6, <http://papers.ssrn.com/sol3/papers.cfm?abstract_id=1013261>, 22 Feb. 2010.

435. Iris Houben, 'Public service obligations: moral counterbalance of technical liberalization legislation', *European Review of Private Law* 16 (2008): 20–23.

436. Commission of the European Communities, *Green Paper on services of general interest*, COM (2003) 270 final, 21 May 2003.

437. *Ibid.*, 3.

these documents in depth, this section will give an overview of the main elements of the Commission's view on services of general interest.

ii. Concept of Services of General Interest

DEFINITION – A first point of attention is that the Commission does not just address services of general economic interest as they are included in Article 106(2) of the Treaty on the Functioning of the EU, but addresses the wider category *services of general interest*. In its 2003 Green Paper, the Commission defines *services of general interest* as 'market and non-market services which the public authorities class as being of general interest and subject to specific public service obligations'.[438] Hence, it is a broader term that encompasses the services of general economic interest. Three types of services are included:

- Services of general economic interest provided by large network industries, for example telecommunications, postal services, electricity, gas and transport. In these sectors, the European Commission has undertaken many harmonization activities.
- Other services of general economic interest, for example waste management, water supply or public service broadcasting.
- Non-economic services and services without effect on trade. These are not subject to the competition and state aid rules of the Treaty on the Functioning of the EU.[439]

DYNAMIC DISTINCTION – Hence, the distinction between non-economic and economic services of general interest is important for the application of competition law. It is up to the Member States to determine and establish services of general interest. When these services consist of providing services or goods on a given market they are economic. According to the Commission, '[e]conomic and non-economic services can co-exist within the same sector and sometimes even be provided by the same organization'. In addition, '[t]he range of services that can be provided on a given market is subject to technological, economic and societal change and has evolved over time. As a consequence, the distinction between economic and non-economic activities has been dynamic and evolving, and in recent decades more and more activities have become of economic relevance. For an increasing number of services, this distinction has become blurred'.[440] Therefore, the Commission does not believe it desirable nor possible to develop a European definition of services of general interest. A case-by-case analysis is needed, like the ECJ approach.[441]

---

438. *Ibid.*, 6–7.
439. *Ibid.*, 10–11.
440. *Ibid.*, 14.
441. Commission of the European Communities, *Communication to the European Parliament, the Council, the European Economic and Social Committee and the Committee of the Regions accompanying the Communication on 'A single market for 21st century Europe'. Services of general interest, including social services of general interest: a new European commitment*, COM (2007) 725 final, 5, 20 Nov. 2007.

NO INITIATIVES FOR A LEGAL FRAMEWORK – The Green Paper launched a consultation on services of general interest, asking the stakeholders whether any horizontal initiatives should be taken with regard to these services on the level of the European Community, rather than the current practice of sector-specific legislation. Responses were divided and in its 2004 White Paper, the Commission decided not to go forward with a horizontal legal framework.[442] It stressed once again the role of national, regional and local authorities in defining, organising, financing, and monitoring services of general interest.

iii. No Clear Definition

PROBLEMATIC CHARACTER OF 'SERVICES OF GENERAL INTEREST' – The introduction of the term *services of general interest* by the European Commission is problematic on a number of levels. First, the term is not defined in the Treaty on the Functioning of the EU or secondary legislation.[443] Next, the Commission itself states that the distinction can be blurred and that services in the same sector may sometimes be economic and sometimes non-economic. Hence, 'the fundamental dividing line between "economic" and "non-economic" is blurred'.[444] The new term adds to the already existing uncertainty and confusion about the scope of services of general economic interest, universal service, public service obligations, and so on.[445] Moreover, while the Commission introduces *services of general interest* as an umbrella term, it focuses on services of general economic interest, leaving out the non-economic services.[446] This is logical, as the EU has no general competence to legislate on non-economic services, but Krajewski argues that by adopting a distinction, the Commission seems to imply that it has competence to act.[447]

---

442. See Markus Krajewski, 'Providing Legal Clarity and Securing Policy Space for Public Services through a Legal Framework for Services of General Economic Interest: Squaring the Circle?', *European Public Law* 14, no. 3 (2008): 382; Alison Jones & Brenda Sufrin, *EC Competition Law: Text, Cases and Materials* (Oxford: Oxford University Press, 2008), 677.
443. While Directive 2006/123/EC of the European Parliament and of the Council of 12 Dec. 2006 on services in the internal market (*OJ L* 376, 27 Dec. 2006, 36) used the term as a legal category, it did not define it.
444. Markus Krajewski, 'Providing Legal Clarity and Securing Policy Space for Public Services through a Legal Framework for Services of General Economic Interest: Squaring the Circle?', *European Public Law* 14, no. 3 (2008): 386.
445. Stephane Rodrigues, 'Les qualifications concurrentes des activités d'intérêt général en droit communautaire', *Actualité juridique Droit Administratif*, 16 Jan. 2006, 84 [in French]. See also Michel Mangenot et al., *Public Administrations and Services of General Interest: What Kind of Europeanisation?* (Maastricht: European Institute of Public Administration, 2005), 59–64.
446. Markus Krajewski, 'Providing Legal Clarity and Securing Policy Space for Public Services through a Legal Framework for Services of General Economic Interest: Squaring the Circle?', *European Public Law* 14, no. 3 (2008): 386. See also Jochen Meulman, 'Na het Groenboek over Diensten van Algemeen Belang: naar een horizontale aanpak van overheidsinterventie?', *Sociaal-Economische Wetgeving* 52 no. 3 (2004): 99 [in Dutch].
447. Markus Krajewski, 'Providing Legal Clarity and Securing Policy Space for Public Services through a Legal Framework for Services of General Economic Interest: Squaring the Circle?', *European Public Law* 14, no. 3 (2008): 397.

SOCIAL SERVICES OF GENERAL INTEREST – To complicate things even further, the European Commission introduced yet another term in a 2007 Communication: *social services of general interest*. However, the Commission admits that '[u]nder Community law, social services do not constitute a legally distinct category of service within services of general interest'. While the importance of social services and health services is emphasized in the Communication, no specific initiatives are taken with regard to social services of general interest.

iv. Some Elements of a Service of General Economic Interest

COMMON ELEMENTS – Even though a Community definition is deemed impossible and undesirable to the European Commission, the 2003 Green Paper lists some common elements that can be drawn on to define a useful concept of services of general *economic* interest. These elements include in particular: universal service, continuity, quality of service, and affordability, as well as user and consumer protection.[448] According to Van De Gronden, these elements form a first basis for an EU approach to services of general economic interest.[449]

UNIVERSAL SERVICE – The concept of universal service has been discussed before. As it can be defined as a 'set of general interest requirements ensuring that certain services are made available at a specified quality to all consumers and users throughout the territory of a Member State independently of geographic location and, in the light of specific national conditions, at an affordable price',[450] the link with services of general economic interest seems clear.

COMPARISON OF SERVICE OF GENERAL INTEREST AND UNIVERSAL SERVICE – At first thought, the common elements of the services of general economic interest mentioned in the Green Paper seem to be the same as those that make up a universal service: continuity, affordability, quality, and user and consumer protection. However, the concept of service of general economic interest is broader than the universal service.[451] The former can also include services that are in the general interest but that are not linked to territorial coverage or universal availability or access.[452]

448. Commission of the European Communities, *Green Paper on services of general interest*, COM (2003) 270 final, 15, 21 May 2003.
449. Johan W. Van De Gronden, 'Deel III: Bestaat er een coherente visie op de verhouding tussen diensten van algemeen economisch belang, de Dienstenrichtlijn en andere communautaire regelingen?', *Sociaal-Economische Wetgeving* 55, no. 10 (2007): 416 [in Dutch].
450. Commission of the European Communities, *Green Paper on services of general interest*, COM (2003) 270 final, 16, 21 May 2003.
451. See Wolf Sauter, 'Services of general economic interest and universal service obligations as an EU law framework for curative health care. TILEC Discussion Paper', 2007, 15–16, <http://papers.ssrn.com/sol3/papers.cfm?abstract_id=1013261>, 22 Feb. 2010. See also Markus Krajewski, 'Providing Legal Clarity and Securing Policy Space for Public Services through a Legal Framework for Services of General Economic Interest: Squaring the Circle?', *European Public Law* 14, no. 3 (2008): 387; Jochen Meulman, 'Na het Groenboek over Diensten van Algemeen Belang: naar een horizontale aanpak van overheidsinterventie?', *Sociaal-Economische Wetgeving*, 52 no. 3 (2004): 103–105 [in Dutch].
452. Jochen Meulman, 'Na het Groenboek over Diensten van Algemeen Belang: naar een horizontale aanpak van overheidsinterventie?', *Sociaal-Economische Wetgeving* 52, no. 3 (2004): 102 [in Dutch]; Karen E. Arnon & Jan J.M. Sluijs, 'Het arrest BUPA: publieke

c. Services of General Economic Interest and State Aid

ARTICLE 107 OF THE TREATY ON THE FUNCTIONING OF THE EU – Previously, the rules on state aid were discussed. These rules often also come into play when an undertaking is entrusted by the State with a service of general economic interest and receives funding or compensation for the provision of that service. This funding will have to comply with a number of conditions in order not to be qualified as an advantage that is distorting competition and that is prohibited under the rules on state aid of Article 107 of the Treaty on the Functioning of the EU.

ALTMARK CRITERIA – The provision to an undertaking of compensation for the carrying out of public service obligations does not constitute an advantage and state aid, under four conditions, commonly referred to as the *Altmark* criteria[453]:

- the recipient undertaking must actually have public service obligations to discharge and those obligations must have been clearly defined;
- the calculation parameters for the compensation must have been established in advance and in an objective and transparent manner;
- the compensation cannot exceed what is necessary to cover all or part of the costs incurred in the discharge of the public service obligations, taking into account any relevant receipts and allowing for a reasonable profit; and
- if the selected undertaking was not chosen in a public procurement procedure allowing the selection of the undertaking capable of providing the public service as the lowest cost to the community, the level of compensation must be determined on the basis of an analysis of the costs which a typical efficient undertaking with the means to meet the public service requirements would have incurred, taking into account any relevant receipts and allowing for a reasonable profit.[454]

CONSEQUENCES – If the criteria are not fulfilled, the compensation paid will be an advantage given to the undertaking that may fall under Article 107. De Bandt indicates, however, that the conditions of *Altmark* are unclear and there are few

diensten en marktwerking', *Nederlands Tijdschrift voor Europees Recht* no. 7/8 (2008): 210 [in Dutch].

453. ECJ, Case C-280/00, *Altmark Trans and Regierungspräsidium Magdeburg*, 24 Jul. 2003, para. 83-87, *European Court Reports* 2003, I-7747; ECJ, Case C-34/01, *Enirisorse SpA v. Ministero delle Finanze*, 27 Nov. 2003, para. 31, *European Court Reports* 2003, I-14243; ECJ, Case C-451/03, *Servizi Ausiliari Dottori Commercialisti Srl v. Giuseppe Calafiori*, 30 Mar. 2006, para. 51-71, *European Court Reports* 2006, I-2941. See also Markus Krajewski, 'Providing Legal Clarity and Securing Policy Space for Public Services through a Legal Framework for Services of General Economic Interest: Squaring the Circle?', *European Public Law* 14, no. 3 (2008): 389.

454. See also Peter Roth & Vivien Rose (ed.), *Bellamy & Child European Community Law of Competition*. 6th edn (Oxford: Oxford University Press, 2008), 1506–1507; Pierre De Bandt, 'Artikel 86, lid 2 EG-Verdrag en het begrip 'diensten van algemeen economisch belang: meer dan een rechtsbasis voor een genuanceerde toepassing van de interne markt- en mededingingsregels op openbare diensten?', *Tijdschrift voor Belgisch Handelsrecht* 114, no. 3 (2008): 221 [in Dutch].

Member States that have a clear view of all the funds or benefits that are given to the undertaking.[455] However, even though the benefits may be qualified as state aid, they may still be allowed by the European Commission under Article 108 of the Treaty on the Functioning of the EU.

EC PACKAGE – Following the *Altmark* judgment, the European Commission adopted an 'EC Package' with regard to services of general economic interest, consisting of three parts:

- commission Decision on the application of Article 86(2) [current Article 106(2)] of the Treaty to State aid in the form of public service compensation granted to certain undertakings entrusted with the operation of services of general economic interest;[456]
- community Framework for State Aid in the form of public service compensation;[457] and
- commission Directive 2005/81/EC amending Directive 80/723/EEC on the transparency of financial relations between Member States and public undertakings.[458]

THREE-STEP EVALUATION – The Commission follows three steps in its evaluation of compensation paid for the performance of services of general economic interest. First, it checks whether the *Altmark* conditions are fulfilled. If this is not the case, the compensation is qualified as state aid. This does not mean that it has to be reimbursed; the state aid may still be considered compatible with the rules of the Treaty on the Functioning of the EU. To determine whether this is the case, the Commission decision mentioned above is applied. If the conditions of this decision are fulfilled, the state aid is considered compatible and the Member States may grant it immediately, without having to notify the Commission. If the aid does not fall under the decision, the Framework can still consider it compatible with the Treaty on the Functioning of the EU. However, this aid will have to be

---

455. Pierre De Bandt, 'Artikel 86, lid 2 EG-Verdrag en het begrip 'diensten van algemeen economisch belang: meer dan een rechtsbasis voor een genuanceerde toepassing van de interne markt- en mededingingsregels op openbare diensten?', *Tijdschrift voor Belgisch Handelsrecht* 114, no. 3 (2008): 221–222 [in Dutch]. See also Markus Krajewski, 'Providing Legal Clarity and Securing Policy Space for Public Services through a Legal Framework for Services of General Economic Interest: Squaring the Circle?', *European Public Law* 14, no. 3 (2008): 390; Tony Prosser, 'Services of General Economic Interest in Community Law: From Single Market to Citizenship Rights', paper given at the University of Rome (Rome, 2003), 6, <http://ecoservr.eco.uniroma1.it/pdf/Rome_lecture_Dece03.pdf>, 22 Feb. 2010.

456. Commission Decision 2005/842/EC of 28 Nov. 2005 on the application of Art. 86(2) of the EC Treaty to State aid in the form of public service compensation granted to certain undertakings entrusted with the operation of services of general economic interest (notified under document number C(2005) 2673), *OJ L* 312, 29 Nov. 2005, 67.

457. Commission of the European Communities, *Community framework for State aid in the form of public service compensation*, *OJ L* 297, 29 Nov. 2005, 4.

458. Commission Directive 2006/111/EC of 16 Nov. 2006 on the transparency of financial relations between Member States and public undertakings as well as on financial transparency within certain undertakings, *OJ L* 318, 17 Nov. 2006, 17. Cf. *supra*, s. IV.C.1.c.

notified, so that the Commission can check the aid and possibly declare it compatible. Hence, the difference between aid falling under the decision and the framework is that aid falling under the former is automatically considered compatible and does not have to notified, while aid under the latter has to be checked individually by the Commission.

FIELD OF APPLICATION OF THE DECISION – According to its Article 2, the decision applies to the following four types of state aid:

- public service compensation granted to undertakings with an average annual turnover before tax, all activities included, of less than EUR 100 million during the two financial years preceding that in which the service of general economic interest was assigned, which receive annual compensation for the service in question of less than EUR 30 million;
- public service compensation granted to hospitals and social housing undertakings carrying out activities qualified as services of general economic interest by the Member State concerned;
- public service compensation for air or maritime links to islands on which average annual traffic during the two financial years preceding that in which the service of general economic interest was assigned does not exceed 300,000 passengers;
- public service compensation for airports and ports for which average annual traffic during the two financial years preceding that in which the service of general economic interest was assigned does not exceed 1,000,000 passengers, in the case of airports, and 300,000 passengers in the case of ports.

COMPATIBILITY REQUIREMENTS – The requirements for compatibility of the state aid are essentially the same under the decision and the framework:

- responsibility for the operation of the service of general economic interest has to be entrusted to the undertaking by way of one or more official acts, the form of which may be determined by the Member State. The act has to specify the nature and duration of the public service obligations; the undertaking and territory concerned; the nature of any exclusive or special rights assigned to the undertaking; the parameters for calculating, controlling and reviewing the compensation; and the arrangements for avoiding and repaying any overcompensation.
- the compensation cannot exceed what is necessary to cover the costs incurred in discharging the public service obligations, taking into account the relevant receipts and a reasonable profit on any own capital necessary for discharging those obligations. The compensation has to actually be used for the provision of the service of general economic interest. Requirements for calculation of the costs and revenues are included.
- member States have to perform regular checks to ensure that undertakings are not receiving overcompensation. Any overcompensation must be repaid by the undertaking.

STRICT APPLICATION – If these requirements are fulfilled, the state aid is compatible with Article 106(2) and the distortion of competition is allowed. However, the European Commission is very strict in applying the *Altmark* criteria, and the scope of application of the Commission Decision is quite small. Hence, very few state aids will escape the notification duty.[459]

3. Applicability to Spatial Information Services

a. Spatial Information Services of General Economic Interest

SPATIAL INFORMATION SERVICES AS ECONOMIC ACTIVITIES – As was described above, several spatial information services can be defined as economic activities, as they are services that could also be provided by the private sector on the market. In addition, while the creation and dissemination of public sector spatial data can be connected to the exercise of official authority, for the provision of information services this will not always be the case.

SPATIAL INFORMATION SERVICES AS SERVICES OF GENERAL ECONOMIC INTEREST – While such information services are economic activities, they may also be provided to guarantee a certain public interest. In such cases, one could consider them as services of general economic interest under Article 106(2) of the Treaty on the Functioning of the EU. For instance, the provision of the national road database on the Internet by the Dutch *Rijkswaterstaat* agency is an economic activity, as the same type of database is also disseminated by private companies on the market. However, *Rijkswaterstaat* deems it in the public interest that the database is available for everyone to be used for whatever purpose. Hence, one could state that this is a service of general economic interest.

SPATIAL INFORMATION SERVICES EXEMPTED FROM COMPETITION LAW – However, this does not mean that such a service is no longer subject to competition law. This would only be the case if the spatial information service cannot be provided under economically acceptable conditions. This may, for instance, be the case if an information service has to be funded by a public body outside of the budget it receives from government to perform its activities as an authority. In such cases, the service might not be possible if the public body does not have an exclusive right to offer the service, or if it cannot provide its public sector spatial data for its own spatial information service at a lower price than it charges its competitors, for example, through margin squeeze abuses and excessive pricing. Examples of such possible abuses were discussed previously with regard to Ecomet and the Belgian Meteorological Office. It seems less likely that a public body can complain about not being able to offer a service under economically acceptable conditions in the situations where it is being challenged for predatory pricing or charging dumping prices (provided of course that this is being done to drive the competition from the

459. Leigh Hancher & S.J.H. Evans, 'Altmark als katalysator: het Commissiepakket met alle antwoorden rond staatssteun en diensten van algemeen economisch belang?', *Nederlands Tijdschrift voor Economisch Recht* 12, no. 7 (2006): 158–159 [in Dutch].

market, otherwise there is no abuse of dominance). Hence, going back to the example of the Dutch national road database, such a service of general economic interest could still be provided under economically acceptable conditions even when subject to competition law. However, as *Rijkswaterstaat* does not have any intention of driving out its competitors with its predatory pricing, no concerns for infringement should exist.

b. The Public Task and Services of General Economic Interest

RECAPITULATION – Previously in this chapter, mention was already made of the relationship between the concept of public task from the EU legal framework for the availability of public sector spatial data on the one hand and the division between economic and non-economic activities from competition law on the other hand. The concept of services of general economic interest demonstrates the difference between the two approaches, even though the European Commission emphasizes the close relationship between them.[460]

RE-USE V. SHARING – As was explained in Chapter 2, the distinction between sharing and re-use is based on the public task. Re-use includes all commercial and non-commercial activities that are performed outside of the public task. If a public body is performing a public task, it is not entering the market. The market and the realm of public tasks are seen as mutually exclusive, entailing that competition law is not applicable to public task activities, but only to the activities outside of the public task.

NUANCING OF DICHOTOMY – Earlier in this chapter it was already stated that this dichotomy is not fully true, and that the EU legal framework for the availability of public sector spatial data gives the wrong impression. Neither mutual exclusiveness nor exhaustiveness are fully present. On the one hand, there are also non-market services that remain outside of the public task, such as re-use by non-profit organizations or for educational or research purposes. On the other hand, there may also be market activities that are part of the public task. This latter statement applies to the services of general economic interest.

SERVICES OF GENERAL ECONOMIC INTEREST AS PART OF THE PUBLIC TASK – Services of general economic interest are provided on the market, but they are essentially provided in the public interest. From what was discussed earlier in this chapter, the public interest is one of the essential criteria to determine the presence of a public task (even though this public interest itself may not be easy to define). Thus, services of general economic interest are part of the public task as it is considered in the PSI Directive and the INSPIRE Directive. This would mean that such public task services would still be subject to competition law, unless the performance of

460. Commission of the European Communities, *Staff working document. Accompanying document to the Communication from the Commission to the European Parliament, the Council, the European Economic and Social Committee and the Committee of the Regions on the re-use of Public Sector Information – Review of Directive 2003/98/EC*, SEC (2009) 597, 27, 7 May 2009.

the service is not possible under economically acceptable conditions. On the contrary, public tasks that are not economic activities would not be subject to competition law.

E. Interim Summary: Competition Law and Spatial Information Services

Spatial information services as economic activities subject to competition law – At the beginning of this chapter, it was stated that a large part of the spatial information services offered by public bodies could be considered as economic activities, and therefore these public bodies would be considered as undertakings subject to competition law. Hence, competition law is of considerable importance in the discussion on the public task. It puts limits on the behaviour of public bodies operating as an undertaking, which contribute to the attainment of the objectives of the EU legal framework for the availability of public sector spatial data, that is, the availability of data for all purposes under transparent and fair conditions. This is particularly the case for Article 102 of the Treaty on the Functioning of the EU (former Article 82 of the EC Treaty) on abuse of dominance, as many public bodies will have a monopoly on the data underlying their spatial information service and will be tempted to use that monopoly power to strengthen their own position on the downstream market. If public bodies provide spatial information services as an economic activity, they have to charge fair prices; they have to be careful in refusing to supply or making exclusive agreements; and so on. In addition, the Member States will have to consider carefully what rules they impose on these public bodies or what funds they attribute to them, in order not to infringe upon EU competition law in their capacity as Member States.

Competition law and the PSI Directive – A number of the obligations and prohibitions stemming from competition law can also be found in the PSI Directive, for example, the prohibition on exclusive arrangements, the obligations with regard to transparency, non-discrimination, and equal conditions for re-use between public bodies and the private sector. However, one might wonder whether the European institutions drawing the PSI Directive found competition law not completely sufficient to deal with these issues, as they made these provisions obligatory for all re-use. Under competition law, exclusive rights are only forbidden when they distort competition.

No solution to question of the scope of the public task – Yet, competition law cannot solve the main question: what is the distinction between the public task and other activities? It only moves the problem towards other definitions. First, it has to be determined whether a particular spatial information service is an economic activity or not. While there are clear guidelines from the EU case law, the ECJ admits that it still finds it challenging to determine the economic character of the activities of public bodies. Hence, competition law may solve a number of problems once it is determined to be applicable, but it does not solve this preliminary question: when is it applicable?

MARGIN OF APPRECIATION FOR THE MEMBER STATES – Moreover, the European Commission and the ECJ have made it clear that they essentially leave it to the Member States to decide whether an information service from a public body is a service of general economic interest. Hence, it is still up to the Member States to make this first decision, and only then competition law can be of interest. The Member States have radically different perceptions of the public service or services of general (economic) interest, so it is a great challenge for the ECJ to develop a harmonized concept of such services for the entire EU.

SUMMARY – In summary, the concepts and rules of competition law could contribute to determining the scope of the public task, but this contribution would be limited. In addition, the fact that the PSI Directive and the INSPIRE Directive contain a different distinction between the activities of providing spatial information services than the distinction made in competition law could make one question what the combination of both types of criteria would lead to. This exercise is done in the next section.

F. CONFRONTATION OF COMPETITION LAW WITH THE DIVISION IN THE EU LEGAL FRAMEWORK FOR THE AVAILABILITY OF PUBLIC SECTOR SPATIAL DATA

DIFFERENT DISTINCTIONS – A combination of previous statements on the exhaustiveness and the mutual exclusivity of the public task and the other (market) activities and of the relationship between the public task concept and the services of general economic interest leads to a different take on the distinction between public tasks and market activities as it is portrayed in the EU legal framework for the availability of public sector spatial data, at least with regard to the provision of spatial information services and the application of competition law.

MARKET ACTIVITIES INCLUDE ECONOMIC AND NON-ECONOMIC ACTIVITIES – First, the domain of market activities can be divided into different activities, which are often named as *commercial* or *non-commercial.* However, as these terms do not have a clear definition and are subject to discussion, it is preferred to use the distinction between economic activities and non-economic activities. Obviously, the re-use of public sector spatial data by the information industry for selling information products and services will be an economic activity, while the use of information in schools or for academic research may not be. The non-economic activities are not subject to competition law. The economic activities are always subject to competition law, without any exceptions. As both categories fall outside of the public task as determined by the EU legal framework on the availability of public sector spatial data, the provision of spatial data for these activities falls under the application of the PSI Directive.

PUBLIC TASK ACTIVITIES INCLUDE ECONOMIC AND NON-ECONOMIC ACTIVITIES – Next, the domain of public task activities can also be divided into two categories. First, there are the public task activities that are also non-economic activities under the provisions of competition law, because they are in the public interest and directly connected to the exercise of official authority. The European Commission refers to these services as non-economic services of general interest. The spatial information services that fall under this category are never subject to competition law. The second category includes public task activities that are in the public interest but that are not directly connected to the exercise of official authority and that qualify as economic activities. These are the services of general economic interest. This part of the public task activities is also subject to competition law, unless this obstructs the performance of the service in so far that it cannot be performed under economically acceptable conditions. As both categories fall under the notion of public task, the INSPIRE Directive is applicable to both.

CONSEQUENCE: CROSS-SUBSIDIES – This would mean for instance, that for a spatial information service, cross-subsidies would be possible for non-economic public task activities and non-economic market activities, because they are not subject to competition law. For economic market activities, this would also be possible unless it infringes competition law. For economic public task activities, it would be possible even if it infringes upon competition law, but only if it is necessary for the service to be performed under economically acceptable conditions. The same cross-subsidization is always a problem under Article 10 of the PSI Directive for the economic market activities, even if it does not infringe upon competition law. It is never a problem under the INSPIRE Directive for the economic public task activities, even though there might be an unjustified infringement of competition law.

CONSEQUENCE: EXCLUSIVE AGREEMENTS – A type of provisions that remains confusing even in this new categorization is the possibility for granting exclusive rights or exclusive agreements. Under the PSI Directive, this is not possible for any market activity, whether it is economic or non-economic. The exception to this is the provision of a service in the public interest. However, such a service in the public interest is part of the public task, so it would not fall under the provisions of re-use of the PSI Directive. Under competition law, there would be no limitations on the grant of exclusive agreements for non-economic activities, but there would be for economic activities. For the latter, the exclusive rights should not infringe competition law. For the economic public task activities, exclusive agreements are also possible, and they can infringe upon competition law in exceptional cases, where this infringement is necessary for the performance of the service in economically acceptable conditions. An exclusive right to provide spatial information services as a non-economic public task activity is always possible under competition law.

EVALUATION – The examples in the previous two paragraphs show that the criterion of the public task does not always combine well with the principles from competition law. The question can be asked why the European Commission did not seek more inspiration in competition law in the drafting of the PSI Directive (and consequently the INSPIRE Directive) in order to avoid additional interpretation difficulties. The distinction between economic and non-economic activities and the concept of services of general (economic) interest already requires the ECJ and the European and national competition authorities to perform a difficult balancing exercise for each case between the importance of a well-functioning market with a level playing field on the one hand, and the public interest on the other hand. Now the ECJ and the Member States are required to also consider the scope of the public task and to develop a separate framework for determining this scope.

SUMMARY – Hence, confrontation between the concepts developed in competition law and the distinction between the public task and market activities in the EU legal framework for the availability of public sector spatial data still does not give an answer to the question on the appropriate criteria to determine the scope of the public task. Combined with the lack of sufficiently clear and refined positive or negative criteria for defining the scope of the public task, we have to state that such a definition is not possible. It depends on political considerations and remains evolving, so it cannot be formulated with sufficient certainty as is needed by the stakeholders.

V. THE DEFINITION OF *PUBLIC TASK* REMAINS CONTINGENT

RECAPITULATION – The previous sections have tried to determine criteria for determining the scope of the public task. This was done first by looking at the different possible criteria that are provided in literature and policy documents. Next, competition law was examined in order to find out whether the criteria used for defining economic activities and services of general (economic) interest could give inspiration for determining the scope of the public task.

THE SCOPE OF THE PUBLIC TASK CANNOT BE DEFINED – As a result we could state that, while these criteria might make it possible to define the public task for a specific activity in a specific situation, the need for case-by-case decisions would make it impossible to define the scope of the public task in a sufficient manner to take away the uncertainty of public bodies and private parties on which spatial information services could or should be offered on the market. This is due to the ever evolving and political character of the public task. Moreover, it depends on who decides on the scope of the public task.

A. THE DEFINITION OF *PUBLIC TASK* CONTINUES TO EVOLVE

EVOLUTION OVER TIME – Whatever definition would be fabricated based on the criteria discussed previously in this chapter, one is always faced with the fact that this definition will only be valid for a limited period of time, although the length of this period can of course differ. The definition of *the public task* will never stop evolving and therefore will always be a moving target. As Reichard states it, '[t]here is no unambiguous and forever valid perception of a public service or a public task'.[461] Naschold and Von Otter emphasize this relative character of the

461. Christoph Reichard, 'New Institutional Arrangements of Public Service Delivery', in *The Governance of Services of General Interest between State, Market and Society*, ed.

public task concept over time: 'Empirical comparison of public sector activities reveals not only the wide variation between countries in the scope of governmental responsibilities; there is scarcely a single social "task" which has not been performed at some time or other both privately and publicly. In almost all cases there are no plausible criteria according to which to discriminate between public and private, governmental and non-governmental activities. It is generally agreed that there is no plausible or generally applicable theory defining the scope of the public sector'.[462]

Dependence on different factors – Many different factors play a role in determining the scope of the public task, and changes in these factors influence the view on the public task. For instance, the technological developments that have occurred in the span of the last three decades have had a big influence on what the public bodies can do with the data they hold on the one hand, and on what the public expects from public data on the other hand. As Stiglitz et al. state, '[e]xisting norms for government provision of goods and services need to be updated for the digital age'.[463]

Technological developments – The public bodies may develop new, previously unforeseen activities based on technological improvements. For instance, the KNMI's public task was laid down in the Law on the Weather in 2002. However, since then, technological developments have led to an expansion of its activities, such as the availability of higher resolution and more satellite images; improved quality of radar images; better software to create graphs or images; and the growth of the Internet as a dissemination channel. This has made it easier for the Meteorological Office to make graphs and has enabled the provision of detailed weather expectations for much smaller areas. These activities took much more resources a few years ago and they were seen as services that should be left to the private sector. Now, the Meteorological Office includes them in its tasks.[464] As technology

---

Christoph Reichard, Riccardo Mussari & Soren Kupke (Berlin: Wissenschaftlicher Verlag Berlin, 2006), 35–36. See also Jan P.H. Donner, *Staat in Beweging* (Sdu: The Hague, 1998), 118 p.

462. Frieder Naschold & Carsten Von Otter, *Public Sector Transformation. Rethinking markets and hierarchies in government* (Amsterdam: John Benjamins Publishing Company, 1996), 32–33. See also Kris Snijkers & Koen Verhoest, 'Het kerntakendebat: lessen voor de toekomst', *Vlaams Tijdschrift voor Overheidsmanagement* 8, no. 3 (2003): 20–25 [in Dutch]; Vito Tanzi, 'IMF Working Paper. The role of the state and the quality of the public sector', 2000, 6, <www.imf.org/external/pubs/ft/wp/2000/wp0036.pdf>, 22 Feb. 2010; Mirjam Plantinga & Albertjan Tollenaar, 'Public governance in the welfare state. The Consequences of Privatisation for Securing Public Interests in the History of the Dutch Welfare State', paper presented at the 4th ECPR conference (Pisa, 2007), 5, <http://papers.ssrn.com/sol3/Delivery.cfm/SSRN_ID1277882_code950891.pdf?abstractid=1103468&mirid=1>, 22 Feb. 2010.

463. Joseph E. Stiglitz, Peter R. Orszag & Jonathan M. Orszag, 'The role of government in a digital age', 2000, 50, <http://unpan1.un.org/intradoc/groups/public/documents/apcity/unpan002055.pdf>, 22 Feb. 2010.

464. KPMG, 'Ministerie van Waterstaat. Evaluatie van de wet op het KNMI', 2007, 29, <www.verkeerenwaterstaat.nl/Images/2008651%20bijlage_tcm195-231492.pdf>, 22 Feb. 2010 [in Dutch]. See also Robert Gellman, 'The American Model of Access to and Dissemination of

continues to develop, comparable changes may encourage the public bodies to develop more services that are beneficial to the public and that can be created very easily, 'with one click of the mouse'. Hence, it is very understandable that the public bodies do not see any harm in providing these services, but that the private sector will feel threatened by the public bodies taking over activities that used to be considered as typically private.

CHANGE IN USER EXPECTATIONS – Not only may the delivery of the services change, but so may the expectations of the users. According to Aichholzer and Tang, '[w]ith the wider diffusion of IT, and especially the Internet, the public has come to expect more availability of a greater variety of information, public and commercial'.[465] Beers gives the example in the field of legal information. In the past the government never saw the publication of the full coordinated texts of all legislation in force or the case law as its task. Considering the current state of technology, the government could be and is now being demanded to provide an accessible, up to date, full text database.[466] However, such databases had already been developed on the market by several legal publishers, which then had to adapt and expand their services in order to satisfy their customers.

OTHER CHANGES IN SOCIETY – Next to technological developments, there are of course many other evolutions that should be taken into account when determining the public task. Examples are economic developments, cultural legacies, globalization, increasing professionalism in society, security issues and terrorism, the growing attention for environmental sustainability, the financial crisis, and the concerns about the living costs of the public.[467] Developments in all these

---

Public Information' paper given at the European Commission Conference: Access to Public Information. A Key to Commercial Growth and Electronic Democracy (Stockholm, 1996): 9-10; Johan Pas, 'The Commercialization of Government Information and the Proposal for a Directive COM (2002) 207 by the European Commission', *Murdoch University Electronic Journal of Law* 9, no. 4 (2002), <www.murdoch.edu.au/elaw/issues/v9n4/pas94_text.html>, 22 Feb. 2010.

465. Georg Aichholzer & Puay Tang, 'Harnessing Public Sector Information for Greater Accessibility: Austria and the UK', in *Public Sector Information in the Digital Age, Between Markets, Public Management and Citizens' Rights*, ed. Georg Aichholzer & Herbert Burkert (Cheltenham: Edward Elgar Publishing Limited, 2004), 289.

466. Anton A.L. Beers, *Informatica Publica: publieke toegankelijkheid van elektronische overheidsinformatie* (The Hague: Rathenau Instituut, 1996),, 10 [in Dutch]; Anton A.L. Beers, 'Openbaarheid van overheidsinformatie', in *Elektronische toegankelijkheid van overheidsinformatie*, ed. Inigo Baten & Gijs Van Der Starre (The Hague: Rathenau Instituut, 1996), 64–66 [in Dutch].

467. See Wetenschappelijke Raad voor het Regeringsbeleid, *Het borgen van publiek belang* (The Hague: Sdu Uitgevers, 2000), 45, <www.wrr.nl/content.jsp?objectid=2545>, 22 Feb. 2010 [in Dutch]; KPMG, 'Ministerie van Waterstaat. Evaluatie van de wet op het KNMI', 2007, 30, <www.verkeerenwaterstaat.nl/Images/2008651%20bijlage_tcm195-231492.pdf>, 22 Feb. 2010 [in Dutch]; Joseph E. Stiglitz, Peter R. Orszag & Jonathan M. Orszag, 'The role of government in a digital age', 2000, 118–119, <http://unpan1.un.org/intradoc/groups/public/documents/apcity/unpan002055.pdf>, 22 Feb. 2010; Ben P. Vermeulen, 'De publieke taak: een veel-zijdig begrip', in *De publieke taak. Staatsrechtconferentie 2002, Vrije Universiteit Amsterdam*, ed. Jan W. Sap, Ben P. Vermeulen & Carla M. Zoethout (Amsterdam: Kluwer, 2003), 20 [in Dutch]. See also Anne-Lieke Mol, 'Voortdurend onderweg naar

areas influence the expectations of the public and the demands society will make on the public bodies. Hence, the demarcation of the public task is dependent on social consensus, which is in turn dependent on the developments mentioned above.[468] The continuous character of such developments requires a periodic re-evaluation of the scope of the public task: 'Such an approach will help to ensure that an activity that is appropriate initially does not expand into one that is inappropriate'.[469]

Differences between national views – In addition, the views of the public task may differ between countries. Steyger states that in national law, typical public tasks are seen as tasks where the general interest is so high that in principle they need to be performed by the government itself. The type of general interests that justify this can be different for each country. As Prosser states, some EU Member States attach great importance to their public services and consider the public tasks of these services as a matter of national identity.[470] For instance, guarding the dikes is one of the oldest public tasks in the Netherlands, while in other countries with different locations, topography, and hydrography, this may not be of great

---

morgen', in *Een kwestie van toegang. Bijdragen aan het debat over het publieke domein van de informatievoorziening*, ed. Inigo Baten & Jolien Ubacht (The Hague: Rathenau Instituut, 1995), 27–44 [in Dutch]. For more information on the development of the state, see Bert A. Rockman, 'Honey, I Shrank the State', in *Modern Systems of Government. Exploring the Role of Bureaucrats and Politicians*, ed. Ali Farazmand (London: Sage Publications Inc., 1997), 275–295; Carl Niebling, *Overheidsbemoeienis. Opkomst en ondergang van de verzorgingsstaat* (Rotterdam: Ad.Donker, 1999), 206 p. [in Dutch]; Gerry M. Van Veldhoven e.a. (ed.), *Overheidsbemoeienis. Opstellen uitgegeven ter gelegenheid van het 11de lustrum van de Katholieke Hogeschool te Tilburg* (Deventer: Kluwer, 1982), 314 p. [in Dutch]; Jan W. Sap, Ben P. Vermeulen & Carla M. Zoethout (ed.), *De publieke taak. Staatsrechtconferentie 2002, Vrije Universiteit Amsterdam* (Amsterdam: Kluwer, 2003), 217 p. [in Dutch]; Herman Matthijs, Frank Naert & Jef Vuchelen, *Handboek Openbare Financiën* (Antwerp: Intersentia, 2004), 636 p. [in Dutch]; Wetenschappelijke Raad voor het Regeringsbeleid, *De verzorgingsstaat herwogen*, 2006, 21–27, <www.wrr.nl/content.jsp?objectid=2658>, 22 Feb. 2010 [in Dutch]; Daniel Yergin & Joseph Stanislaw, *The Commanding Heights. The battle for the world economy* (New York: Touchstone, 2002), 488 p.; Bastiaan Van Loenen & Bas C. Kok, 'Spatial Data Infrastructures: Legal and Economic Issues', in *Spatial Data Infrastructure and Policy Development in Europe and the United States*, ed. Bastiaan Van Loenen & Bas C. Kok (Delft: DUP Science, 2004), 6; Hilary Wainwright, 'Reclaiming 'The Public' through the People', 2004, <www.tni.org/detail_page.phtml?act_id=2368&password=9999&publish=Y&username=guest@tni.org>, 22 Feb. 2010; Vito Tanzi, 'IMF Working Paper. The role of the state and the quality of the public sector', 2000, 6, <www.imf.org/external/pubs/ft/wp/2000/wp0036.pdf>, 22 Feb. 2010.

468. Frieder Naschold & Carsten Von Otter, *Public Sector Transformation. Rethinking markets and hierarchies in government* (Amsterdam: John Benjamins Publishing Company, 1996), 35.

469. Joseph E. Stiglitz, Peter R. Orszag & Jonathan M. Orszag, 'The role of government in a digital age', 2000, 50, <http://unpan1.un.org/intradoc/groups/public/documents/apcity/unpan002055.pdf>, 22 Feb. 2010. See also KPMG, 'Ministerie van Waterstaat. Evaluatie van de wet op het KNMI', 2007, 71, <www.verkeerenwaterstaat.nl/Images/2008651%20bijlage_tcm195-231492.pdf>, 22 Feb. 2010, 22 Feb. 2010 [in Dutch].

470. Tony Prosser, *The Limits of Competition Law, Markets and Public Services* (Oxford: Oxford University Press, 2005), 97, 122.

importance.[471] This different weighing of interests complicates the definition of an activity as a public task. Consequently, there is no absolutely universal criterion for the typical public task.[472]

B. THE DEFINITION OF *PUBLIC TASK* REMAINS POLITICAL

POLITICAL VISION IS DECISIVE – As these expectations and demands are represented in a democratic society by way of elections, the scope of the public task is inherently a political question. Hence, in most cases the political vision on the role of government will determine whether a particular activity is seen as part of the public task or not. Political and social science, historical and international comparisons, law, and economics can all contribute to this mission, but in the end it remains up to the politicians to decide.[473] As was discussed before, even if the market fails, it still remains a political decision whether a particular information service is sufficiently valuable for the public to be provided by the public sector.[474]

EVOLUTION IN POLITICAL VIEWS ON THE PUBLIC TASK – These political views can change. For instance, the view on the core tasks of the public sector in the classical minimalist, liberal concept of government was different from the broadly based view of state activities at the end of the twentieth century.[475] The classical liberal

471. Elies Steyger, 'Deel II: De publieke taak: diensten van algemeen (economisch) belang en de gevolgen van de richtlijn', *Sociaal-Economische Wetgeving* 55, no. 10 (2007): 387 [in Dutch]. See also Joseph E. Stiglitz, Peter R. Orszag & Jonathan M. Orszag, 'The role of government in a digital age', 2000, 30–31, <http://unpan1.un.org/intradoc/groups/public/documents/apcity/unpan002055.pdf>, 22 Feb. 2010.
472. Elies Steyger, 'Deel II: De publieke taak: diensten van algemeen (economisch) belang en de gevolgen van de richtlijn', *Sociaal-Economische Wetgeving* 55, no. 10 (2007): 387 [in Dutch]. See also Mirjam Plantinga & Albertjan Tollenaar, 'Public governance in the welfare state. The Consequences of Privatisation for Securing Public Interests in the History of the Dutch Welfare State', paper given at the *4th ECPR conference* (Pisa, 2007), 5, <http://papers.ssrn.com/sol3/Delivery.cfm/SSRN_ID1277882_code950891.pdf?abstractid=1103468&mirid=1>, 22 Feb. 2010.
473. Wetenschappelijke Raad voor het Regeringsbeleid, *Het borgen van publiek belang* (The Hague: Sdu Uitgevers, 2000), 46–47, <www.wrr.nl/content.jsp?objectid=2545>, 22 Feb. 2010 [in Dutch].
474. *Ibid.*
475. Frieder Naschold & Carsten Von Otter, *Public Sector Transformation. Rethinking markets and hierarchies in Government* (Amsterdam: John Benjamins Publishing Company, 1996), 32. See also Peter Landry, 'The Theory of Government', <www.blupete.com/Literature/Essays/BluePete/Government.htm>, 22 Feb. 2010; M. Shamsul Haque, 'The Diminishing Publicness of Public Service under the Current Mode of Governance', *Public Administration Review* 61, no. 1 (2001): 69–70, <http://profile.nus.edu.sg/fass/polhaque/par-public.pdf>, 22 Feb. 2010; John Kay, 'The State and the Market', *Political Quarterly* 75, no. 1 (2004): 76; Charles Leadbeater, 'Creating the Public Good', *Political Quarterly* 75, no. 1 (2004): 89; Gordon Brown, 'State and Market: Towards a Public Interest Test', *Political Quarterly* 74, no. 3 (2003): 269–270; Herman Matthijs, Frank Naert & Jef Vuchelen, *Handboek Openbare Financiën* (Antwerp: Intersentia, 2004), 6–7 [in Dutch]; Raymond Plant, 'Neo-liberalism and the

position has always been based on a minimalist conception of state responsibilities. With the expansion of the welfare state in the second half of the twentieth century, this position steadily declined in influence. Subsequently, however, the position has enjoyed a revival born out of the mounting criticism of precisely this expansion of government, evidence of 'government failure', levels of taxation and public spending considered excessively high, and crowding-out effects.[476] Supporting these evolutions, a plethora of '[n]ew doctrines for managing the state, new conceptions of the state and new ideas for organising the delivery of public services have abounded: new public management, the enabling state, the hollowing out of the state, the competition state, the market state and public-private partnerships are just a few of these'.[477] According to Plantinga and Tollenaar, the welfare state is currently developing to a regulatory welfare state, in which private elements are introduced to a formerly public system, and the role of the government shifts from that of a public provider to that of a regulator of the private market, and the welfare state develops into a state-regulated welfare market.[478]

Theory of the State: From Wohlfahrtstaat to Rechtsstaat', *Political Quarterly* 75, no. 1 (2004): 31; Andrew Gamble & Tony Wright, 'Introduction', in *Restating the State? Special issue of The Political Quarterly*, ed. Andrew Gamble & Tony Wright (Malden: Blackwell Publishing, 2004), 2; Wetenschappelijke Raad voor het Regeringsbeleid, *De verzorgingsstaat herwogen*, 2006, 284, <www.wrr.nl/content.jsp?objectid=3759>, 22 Feb. 2010 [in Dutch]; Vic Veldheer, 'Waar ligt het evenwicht tussen overheidsbemoeienis en marktwerking?', in *Altijd een antwoord. SCP Nieuwjaarsuitgave 2006*, ed. sociaal en Cultureel Planbureau (The Hague: Sociaal en Cultureel Planbureau, 2006), 54–55, <www.scp.nl/publicaties/boeken/9037702635/Altijd_een_antwoord.pdf>, 22 Feb. 2010 [in Dutch]; Mark R. Bokkerink, 'De uitoefening van publiekrechtelijke taken door privaatrechtelijke rechtspersonen', in *Overheid en onderneming. Tussen publiek en privaat ondernemen*, ed. C.J.J. Kortmann Sebastianus et al. (Deventer: W.E.J. Tjeenk Willink, 1998), 293–296 [in Dutch]; Joseph E. Stiglitz, Peter R. Orszag & Jonathan M. Orszag, 'The Role of Government in a Digital Age', 2000, 30–31, <http://unpan1.un.org/intradoc/groups/public/documents/apcity/unpan002055.pdf>, 22 Feb. 2010.

476. Frieder Naschold & Carsten Von Otter, *Public Sector Transformation. Rethinking markets and hierarchies in government* (Amsterdam: John Benjamins Publishing Company, 1996). See also Max Haller, Franz Hollinger & Otto Raubal, 'Leviathan or Welfare State? Attitudes toward the Role of Government in Six Advanced Western Nations', in *Attitudes to Inequality and the Role of Government*, ed. Duane F. Alwin et al. (The Hague: Sociaal en Cultureel Planbureau, 1990), 33, 56; Bert A. Rockman, 'Honey, I Shrank the State', in *Modern Systems of Government. Exploring the Role of Bureaucrats and Politicians*, ed. Ali Farazmand (London: Sage Publications Inc., 1997), 276; Ernst M.H. Hirsch Ballin, 'De mens in de sociale rechtsstaat?', in *Overheidsbemoeienis. Opstellen uitgegeven ter gelegenheid van het 11de lustrum van de Katholieke Hogeschool te Tilburg*, ed. X (Deventer: Kluwer, 1982), 27–28 [in Dutch].
477. Andrew Gamble & Tony Wright, 'Introduction', in *Restating the State? Special issue of The Political Quarterly*, ed. Andrew Gamble & Tony Wright (Malden: Blackwell Publishing, 2004), 1.
478. Mirjam Plantinga & Albertjan Tollenaar, 'Public Governance in the Welfare State. The Consequences of Privatisation for Securing Public Interests in the History of the Dutch Welfare State', paper given at the 4th ECPR conference (Pisa, 2007), 10, <http://papers.ssrn.com/sol3/Delivery.cfm/SSRN_ID1277882_code950891.pdf?abstractid=1103468&mirid=1>,

SUBJECT TO CONSENSUS – The political character of the decision on the scope of the public task means that it is subject to compromise. There may be a broad consensus on the content of a particular task, but this content may also be decided by a very small political majority, which could change very quickly. This could affect the effort and the resources assigned to developing the public task: if there is a broad chance of the decision being overturned, it may be considered wiser to only take some first minor steps towards the implementation of a service rather than taking irreversible steps, or even postpone any action.[479]

POSSIBILITY TO ADAPT – Even if the government has made a decision regarding the scope of the public task, it has to be able to adapt its position according to the changes in the needs of the public. This principle, based on the French concept of the law of adaptability of the public service,[480] entails that the government always has the right to unilaterally change the working of an organization or even abolish the service if it no longer deems it necessary.

C. THE SCOPE OF THE PUBLIC TASK DEPENDS ON WHO DEFINES IT

IMPORTANCE OF KNOWING THE DECISION-MAKER – Someone has to decide whether the provision of a spatial information service constitutes part of the public task or whether it is an activity that should be performed on the market. Hence, the scope of the public task is determined by the person or organization that makes this decision. Thus, it is important to know who defines the public task, so this person or organization can be held accountable.

1. Democratic Process

LEGAL BASIS – It is often stated that the public task is defined by its legal basis. However, as was discussed before, such a legal basis inherently means that another reason for laying down a public task in such a law is present. In essence, this decision is taken by the government. However, even though the decision on the scope of the public task is political, this does not mean that the government gets carte blanche. It will have to justify itself for its choices and its actions under the democratic process, and prove that these actions are legitimate, based on the public interests they are guaranteeing.[481]

---

22 Feb. 2010. See also Colin Scott, 'Services of General Interest in EC Law: Matching Values to Regulatory Technique in the Public and Privatised Sectors', *European Law Journal* 6, no. 4 (2000): 313.

479. Wetenschappelijke Raad voor het Regeringsbeleid, *Het borgen van publiek belang* (The Hague: Sdu Uitgevers, 2000), 56–58, <www.wrr.nl/content.jsp?objectid=2545>, 22 Feb. 2010 [in Dutch].

480. Marc Boes, *Cursus administratief recht* (Acco: Leuven, 2005), 128 [in Dutch].

481. Mark Bovens, 'Information Rights: Citizenship in the Information Society', *The Journal of Political Philosophy* 10, no. 3 (2002): 329.

INVOLVEMENT OF PARLIAMENT – First, this means that the government or the public bodies cannot decide for themselves what their own public task should be. As was mentioned before, this would lead to the public bodies 'marking their own exam papers'.[482] Parliament has to be involved, but it was also determined that a legal basis is not sufficient to determine the scope of the public task, because this holds the risk of a circular reasoning: if a service is considered to be a public task, it has to be laid down in a law; and because it is laid down in the law, it becomes a public task. Hence, the involvement of Parliament is merely a procedural guarantee, an endorsement of a public task that has already been decided.

CRITICISM – The problem of the democratic process is that it may be too slow to react adequately to changing situations in society.[483] The procedure of adopting a legal basis to determine the scope of the public task takes too long to allow both the public bodies and the private sector to take an approach to the provision of spatial information services that is appropriate for the particular circumstances.

2. Stakeholders Involvement

REPRESENTATION OF PUBLIC INTEREST – Not only the involvement of Parliament is important. As it was described earlier in this chapter and in the previous chapter, the scope of the public task should be based on the public interest. While this public interest is represented by an elected Parliament and representative government, more direct participation of the public should be organized. According to the Wetenschappelijke Raad voor het Regeringsbeleid, a consensus on the level of society would be a very good basis to make a decision on the public task.[484] De Terwangne agrees, because if public bodies were able to decide entirely by themselves which information services they provided, there would be the risk of manipulation and lopsided decisions that do not take into account all the interests involved.[485]

DIALOGUE WITH APPROPRIATE SECTOR – Therefore, a dialogue with the stakeholders has to be set up, ensuring proper representation of all the interest groups involved. Burkert feels that this should be organized by sector: 'The best process would be for each information sector, such as geographical information, to organize a dialogue with all interested parties on what would be the role of the

482. Communities and Local Government Committee, 'Ordnance Survey, Fifth Report of Session 2007-08, HC 268', <www.publications.parliament.uk/pa/cm200708/cmselect/cmcomloc/268/26802.htm>, 22 Feb. 2010.

483. See Cécile De Terwangne, *Société de l'information et mission publique d'information*, Doctoral Thesis FUNDP (Namur: Faculté Universitaire de Notre Dame de Paix, 2001), 296 [in French].

484. Wetenschappelijke Raad voor het Regeringsbeleid, *Het borgen van publiek belang* (The Hague: Sdu Uitgevers, 2000), 57, <www.wrr.nl/content.jsp?objectid=2545>, 22 Feb. 2010 [in Dutch].

485. Cécile De Terwangne, *Société de l'information et mission publique d'information*, Doctoral Thesis FUNDP (Namur: Faculté Universitaire de Notre Dame de Paix, 2001), 371–372 [in French].

state'.[486] According to De Terwangne, ad hoc committees and organs should be involved, the composition of which has to reflect a diversity of opinions, with representations of public bodies, and users, and possibly members of parliament and independent experts of the domain. These bodies should provide technical and legal support, coordination, and have a say in the scope of the public task.[487] However, these ad hoc committees can only provide one-off decisions from their particular point of view, without a more general overview of the developments in society. Therefore, a more permanent body might be needed.

3. Regulatory or Advisory Body

CONTINUOUS REPRESENTATION – The government is already subject to justification for its decisions in the democratic process, but this may not be sufficient, as this process generally means that the government can only be held accountable by the public once in every couple of years in the electoral process. Therefore, a regulatory or advisory body that can act continuously should be in place.

ADVICE AND REVIEW – Such a regulatory or advisory body has to be neutral and independent. It has to include representatives from both the public and the private sector, and be ensured of a structural working through sufficient funding. On the one hand, this body should provide advice and assistance in preparing the public bodies' policies for providing spatial information services. This could stimulate the reflection of the public bodies on their activity and possibly also the harmonization of their information policies. The body could, for instance, organize regular sessions where possible future tasks of the public bodies to provide spatial information services are evaluated and existing tasks are re-examined. The general public should also be involved in this process. On the other hand, this body should be able to handle complaints or concerns that a public body might be acting contrary to the rules of the PSI Directive, the INSPIRE Directive, or the Access Directive.[488] It could mediate and arbitrate between the public bodies and the private sector or the general public when needed.[489]

EXAMPLE: OPSI – Such advisory and regulatory bodies are already available in some countries,[490] with the main example being the United Kingdom's OPSI and

486. George Papapavlou, 'Minutes', paper given at the workshop on Commercial and Citizen's Access to Government Information (Luxembourg, 1995), 52.

487. Cécile De Terwangne, *Société de l'information et mission publique d'information*, Doctoral Thesis FUNDP (Namur: Faculté Universitaire de Notre Dame de Paix, 2001), 371–372 [in French].

488. See also Edward Pitt, 'Ordnance Survey – The Ground Rules', 2003, <www.epsiplus.net/cases/ordnance_survey>, 22 Feb. 2010.

489. François Salgé, 'Geographic Information Network in Europe (GINIE): Recommendations for Action', in *Spatial data infrastructure and policy development in Europe and the United States*, ed. Bastiaan Van Loenen & Bas C. Kok (Delft: DUP Science, 2004), 42–43.

490. For example, the *Commissie voor de toegang tot bestuursdocumenten* in Belgium or the *Beroepsinstantie voor de openbaarheid van bestuur* in the Flemish Region, the *Information Commissioner* in Slovenia, the *Commission d'Accès aux Documents Administratifs* in France.

APPSI. OPSI acts as a regulator 'at the heart of information policy, setting standards, providing a practical framework of best practice for opening up and encouraging the use of public sector information'.[491] However, the concern has risen that OPSI is not sufficiently independent to act as a regulatory body. According to OFT, OPSI is 'unusual in being subject to ministerial control, which does not give it the independence of the economic sector regulators in the UK. There may be circumstances where such a position affects the credibility of OPSI's independence in decision making and thus may represent a conflict of interest'. Another possible conflict of interest could be the dual role of OPSI as a review body for complaints and an advisory body for government. The OFT does acknowledge, however, that there concerns might be purely theoretical.[492]

EXAMPLE: APPSI – In addition, this may be mitigated by the presence of APPSI, which consists of members from a wide variety of backgrounds including information providers, re-users and consumers of public sector data, experts from academia and industry, representatives of producer and consumer groups, and representatives of the administrations.[493] Its role is to advise ministers and OPSI on changes and opportunities in the information industry, and to act as reviewer of the complaints procedures launched before OPSI.

CRITICISM – However, the problem still remains that the decision of an advisory or regulatory body will in many situations have to be taken on a case-by-case basis, which could still cause uncertainty for the public bodies and the private sector. Yet, an accessible representative body that can react swiftly to possible problems in the definition of the scope of the public task would already be a big step forward.

4. Level of Decision Making

APPROPRIATE LEVEL IS NOT CLEAR – While it can be determined that the decision on the public task is a political decision that has to be made by government, with the involvement of the stakeholders and a regulatory body, the question remains at what level this should be determined. Currently, this is not entirely clear.

THE EUROPEAN INSTITUTIONS? – As the intention of the PSI Directive, the INSPIRE Directive, and the Access Directive is to harmonize the EU legal framework for the availability of public sector spatial data, one would assume that the substantial criteria for determining which directive is applicable would also be determined by the European institutions. Yet, this has not happened, at least not for the public task criterion. The fact that the directives contain no clear definition of

491. The National Archives, 'A New Vision for The National Archives 2006-2011', 2006, 13, <www.nationalarchives.gov.uk/documents/vision.pdf>, 22 Feb. 2010.

492. Communities and Local Government Committee, 'Ordnance Survey, Fifth Report of Session 2007-08, HC 268', <www.publications.parliament.uk/pa/cm200708/cmselect/cmcomloc/268/26802.htm>, 22 Feb. 2010.

493. Advisory Panel on Public Sector Information, 'Response to the Consultation Paper on the Use and Sharing of Personal Information in the Public and Private Sectors', 2008, 1, <www.justice.gov.uk/docs/Advisory-Panel-on-Public-Sector-Information-Data-Sharing-Response.pdf>, 22 Feb. 2010.

public task was found by APPSI to 'militate against the harmonising intent' of the directives.[494] Therefore, APPSI believes that an adjustment to the directives, or at least guidance on the principles by the European Commission, would be helpful. This would assist in promoting a harmonized approach, which at present is impossible if the critical concept of 'public task' is simply left to be interpreted in accordance with local law, binding rules, 'common administrative practice' (accepting that there should be some scope for variation between Member States on what is defined as 'public task') or even departmental preference. APPSI suggests that this concept would best be elucidated by reference to a set of common, EU-wide principles, focusing on the discharge of a public need. Any such elucidation should take account of the plethora of terms used in this area, notably the various binary oppositions (such as 'raw'/'value-added', 'refined'/'unrefined', 'competed'/'non-competed', or 'upstream'/'downstream'), which use adds confusion rather than clarity. Within these high-level principles, the Member States would then define a process for defining the public task. Other Member States also found that the European Commission should take some action in clarifying the scope of the public task.[495]

NOT THE EUROPEAN INSTITUTIONS – However, the Commission stated that it could not do this, as this was a matter for the Member States and it would go beyond the subsidiarity principle to determine this for all the Member States. As this subject is intrinsically linked to national competences, and the national traditions are very different across the Member States, the Commission feels that it is not competent to judge on this. It only encourages the Member States to define public tasks in such a way that they allow maximum PSI re-use and will promote good practices and monitor developments in this area.[496] This is in line with the approach of the European Commission and the ECJ with regard to the services of general economic interest, as was discussed earlier in the chapter. The Commission and the ECJ do not want to involve themselves in the national matter of deciding that a public service should be provided. They only check whether this public service remains within the limits of the competition rules and their exemptions.

CRITICISM – However, if it is not possible to determine the scope of the public task on the level of the EU, what is the point of basing an entire legal framework that is intended to harmonize the laws and practices of the Member States on this distinction? As Davies states, '[i]t may be held in Luxembourg that the problem of

494. Advisory Panel on Public Sector Information, 'The Impact and Success of Legislation on the Re-use of Public Sector Information', 2008, 26, <www.appsi.gov.uk/reports/ec-submission-07-2008.doc>, 22 Feb. 2010.

495. Commission of the European Communities, 'Results of the MS Consultation on the Review of the Application of Directive 2003/98/EEC on the Re-use of Public Sector Information', 2008, 5, <http://ec.europa.eu/information_society/policy/psi/docs/consultations/replies/members_states/report_ms_questionnaire.pdf>, 22 Feb. 2010.

496. Commission of the European Communities, *Communication to the European Parliament, the Council, the European Economic and Social Committee and the Committee of the Regions. Re-use of Public Sector Information – Review of Directive 2003/98/EC*, COM (2009) 212 final, 9, 7 May 2009.

defining "public task" across Member States is too difficult to tackle at European level because of the substantial variations between national understandings on this point. If it is to be seen therefore as an issue for Member States to tackle individually (leaving aside all the complications that this presents for cross-border products etc.) the question for me is where, then, could I expect to look to see a definition of the public task of any given government, department, agency, local authority, etc'.[497]

VI. INTERIM SUMMARY

IMPORTANCE OF A DEFINITION OF THE SCOPE OF THE PUBLIC TASK – This chapter addressed the possibility to define the scope of the public task in a sufficient manner for the EU legal framework for the availability of public sector spatial data to have an optimal effect. The delineation of the scope of the public task was found to be important not only for the definition of re-use and sharing, but also to ensure a level playing field between public bodies offering spatial information services on the market and the private sector.

EXAMINATION OF CRITERIA FOR DEFINING THE PUBLIC TASK – In order to examine the possibility for a definition of the scope of the public task, both positive and negative criteria for determining this scope were examined. With regard to the positive criteria, it was found that they could in essence all be brought back to the criterion of public interest. However, the evolving nature of this public interest (and also of its concretization in accessibility) is not sufficient to determine the scope of the public task that can be used in practice to determine the relationship between spatial information services under the public task and spatial information services on the market. A definition of possible negative criteria proved to be even more difficult. Only very few direct indications were found of which spatial information services the public bodies should not provide and leave to the private sector. Mostly, the public bodies were asked to consider whether existing services are already offered on the market and whether their intervention is still needed. This would then have to be determined on the basis of the public interest, which leads back to the previous point. Hence, it seemed that a sufficiently clear definition of the scope of the public task is not possible.

NUANCING OF DICHOTOMY BETWEEN PUBLIC TASK AND MARKET ACTIVITIES – The problem of defining the scope of the public task is even intensified by the fact that the distinction between public task and the market as it is assumed in the EU legal framework on the availability of public sector spatial data is not an absolute distinction. On the one hand, it was found that the public task and the market are not exhaustive, because there are many non-commercial activities that exist outside of the public task. On the other hand, the public task and market

497. Robert Davies, 'Public Task Definition – How? Reply to "Is there a need to define public task?"', 2007, <www.epsiplus.net/forum/forum/is_there_a_need_to_define_public_task>, 22 Feb. 2010.

activities are not mutually exclusive, as in some Member States selling information products on the market may be seen as part of the public task, because the public bodies in question have to gain revenues for Treasury. In addition, the relationship between public interest and the market is also not necessarily a dichotomy. Public bodies may also provide a service on the market in order to secure a public interest. In such cases, the public bodies are performing a public task, but they are doing so on the market.

EXAMINATION OF COMPETITION LAW DOMAIN – Public bodies providing spatial information services are often accused of distorting competition law. In addition, the issue of public bodies performing activities on the market shows many similarities with the discussion on public bodies acting as undertakings under competition law. Therefore, it was deemed valuable to look at the domain of EU competition law for inspiration on the issue of the demarcation between the public task activities and market activities of public bodies and on the issue of market services in the public interest.

ELEMENTS OF COMPETITION LAW THAT WERE EXAMINED – The section on competition law discussed the position of public bodies providing spatial information services as undertakings and the possible distortions of the market that could follow from this qualification as undertakings. Next, it addressed the specific situation of services in the general (economic) interest, which is comparable to the situation under the EU legal framework for the availability of public sector spatial data where public task activities are performed on the market and spatial information services are provided on the market, but in the public interest.

NO SOLUTION TO THE QUESTION ON THE PUBLIC TASK – While EU competition law could be of service in determining the scope of the public task, it still cannot solve the question of the distinction between the public task and other activities in the EU legal framework for the availability of public sector spatial data. It only moves the problem towards other definitions, that is, undertaking, economic activities and services in the general (economic) interest, which are also considered challenging by the competition authorities and the ECJ.

DIFFICULTY TO COMBINE CONCEPTS – The combination of the concepts of public task and market activities of the EU legal framework on the availability of public sector spatial data on the one hand and the concepts of competition law on the other hand, shows many differences. This means that the one set of concepts cannot be transposed to the other, making the combined application of the EU legal framework for the availability of public sector spatial and competition law very difficult. In addition, the definition of services of general (economic) interest is to a great extent left to the Member States and decided on a case-by-case basis, so this does not help the development of a definition of public task that can ensure harmonization of the activities based on public sector spatial data throughout the EU.

CONCEPT OF *PUBLIC TASK* CANNOT BE DEFINED IN A SUFFICIENT MANNER – All the previous findings lead us to believe that the concept of the public task is too evolving and political to serve as a criterion for the distinction between re-use and sharing, and that the EU legal framework for the availability of public sector spatial data cannot provide sufficient clarity on this basis. In addition, the definition of the scope of the public task is dependent on the competent authority. While a number of safeguards can be included to ensure that this decision is based on actual developments in society and is taken in a sufficiently representative manner, the question remains how it can be ensured that such national decisions lead to the harmonization that was intended by the EU legal framework for the availability of public sector spatial data.

Chapter 9
Conclusion

OUTLINE – In this final chapter, all the findings on the problematic character of the EU legal framework for the availability of public sector spatial data are tied together. This is followed by a general conclusion and some further thoughts for the development of the framework.

I. HYPOTHESIS: THE PROBLEMATIC CHARACTER OF THE EU LEGAL FRAMEWORK

QUESTION – As a basis for my analysis, I questioned whether the distinction between access, re-use and sharing that determines the applicability of the Access Directive, the PSI Directive and the INSPIRE Directive can be made sufficiently clear in order for the EU legal framework for the availability of public sector spatial data to properly ensure the availability of these data. The assumption was that this is not the case, and that the distinction between access, re-use and sharing is problematic.

HYPOTHESIS – The distinction was felt to be problematic because of the criteria that are used to make the distinction. On the one hand, these criteria could not be used in practice to make a clear difference between different situations in order to determine whether the Access Directive, the PSI Directive or the INSPIRE Directive should be applied. On the other hand, the distinction was also assumed to show a more conceptual problem, as it is partly based on a criterion that is impossible to define with enough precision and certainty on the EU level to ensure a distinction that is clear enough for both the public bodies providing the data and the users of the data.

OUTLINE – In order to examine whether the hypothesis that was made in the first chapter is true, a number of steps needed to be taken. This section brings together the main results of the analysis that was done in the previous chapters.

II. OVERVIEW OF THE RESULTS

A. Access, Re-use and Sharing in the Access Directive, the PSI Directive and the INSPIRE Directive

> The distinctions between the rules in the Access Directive, the PSI Directive, and the INSPIRE Directive are based on the purpose of the use that is made of the public sector spatial data. Three types of use are distinguished: access, re-use, and sharing.

Common denominator: public sector spatial data – The focus of this book is on the EU legal framework for the availability of public sector spatial data, consisting of the Access Directive, the PSI Directive, and the INSPIRE Directive. The common denominator between these three directives are the spatial data sets held by public authorities under the INSPIRE Directive that also fall under the definition of environmental information and that can also be qualified as public sector documents held by public sector bodies. As these data, referred to as public sector spatial data, form a very large subset of public sector data to which all three directives are applicable, it is important to know the field of application of the Access Directive, the PSI Directive, and the INSPIRE Directive.

Distinction: purpose of use – The respective fields of application of the Access Directive, the PSI Directive, and the INSPIRE Directive are distinguished on the basis of the purpose of the use that is made of the public sector spatial data that are made available. First, *access* is defined as obtaining public sector spatial data for the purpose of exercising one's democratic rights or obligations. Access is related to public participation and the control of the effectiveness and efficiency of government. *Re-use* as it is addressed in the PSI Directive can be described as any use of public sector spatial data for commercial or non-commercial purposes outside of the public task, requiring the processing of the data beyond access needs. This definition is very broad, but it can be stated that these commercial or non-commercial purposes all entail the objective of the user to obtain a benefit that goes beyond the mere democratic purpose of checking up on the government. Third, *sharing* under the INSPIRE Directive involves any use of public sector spatial data for the purpose of performing public task activities. As the distinction between the three directives is based on the distinction between the purposes of the use, the three categories of access, re-use, and sharing should be clearly demarcated, without showing gaps or overlaps.

B. The Need for a Demarcation between Access, Re-use, and Sharing

Outline – The distinction between access, re-use, and sharing is based on the purpose of the use. However, one should also question whether this distinction is of any relevance. For instance, if the rules would be the same for access, re-use, and sharing, there would be no point in determining the exact difference between the three types of use. This relevance can be determined in two ways. On the one hand, the content of the Access Directive, the PSI Directive, and the INSPIRE Directive may show that the rights and obligations attached to access, re-use, and sharing of public sector spatial data are different. On the other hand, the impact of the actual differences between the rights and obligations can be assessed.

1. The Content of Access, Re-use, and Sharing

> The rules for access, re-use, and sharing are different, and some rules only address one type of use, so it is important to know when they apply.

NO COMPLETE MATCH BETWEEN DIRECTIVES AND PURPOSES – Examining the three directives that promote the availability of public sector spatial data, one immediately finds that these three directives do not match perfectly with the three types of use. While the PSI Directive addresses re-use and the Access Directive applies to access, the INSPIRE Directive is not limited to sharing, but also contains provisions on access.

a. Access

LEGAL INSTRUMENTS – Access to public sector spatial data is regulated by the Access Directive and Chapter IV of the INSPIRE Directive. The Access Directive focuses on environmental information held by public authorities and the provisions of the INSPIRE Directive on network services for spatial data sets held by or on behalf of public authorities.

ACCESS DIRECTIVE – The basic principles of the Access Directive can be summarized as follows. The obligations of the public authorities are twofold. On the one hand, they must allow access on request to their environmental information, and on the other hand, they must actively disseminate environmental information to the public. With regard to access upon request, the public bodies have to make available environmental information without the applicant having to state an interest. Measures have to be taken in order to facilitate the applicant's search for and access to the information. Such measures may include registers or lists of data held by public authorities or the designation of information officers. Access may be refused for a number of reasons, which have to be interpreted restrictively and weighed against the interests in disclosure. Access to the registers or lists of information and consultation of the information on site must be free of charge, while the supply of information can be charged for. Such charges cannot exceed a reasonable amount. In principle, this implies that the cost of producing the material in question can be charged, but in the case where the public authority provides the information on a commercial basis, and needs to charge to continue such provision, the reasonable amount may include market-based charges.

INSPIRE DIRECTIVE – Next, the INSPIRE Directive obliges the Member States to create a network of services, which have to be made available for access to the public. Hence, the INSPIRE Directive does not contain a direct obligation for the public authorities to make their spatial data sets available for access to the public, but rather a requirement to provide access to the services for these datasets, so an indirect right of access to the spatial data can be recognized. Access to the network services can be limited for the same reasons as access to environmental information can be restricted under the Access Directive. An exception is made for the discovery services, which can only be restricted from public access for reasons related to security. These discovery services cannot be charged for, while for other services this may be the case. View services in principle also have to be available free of charge, but an exception to this principle is possible. Charges can be made if they secure the maintenance of the spatial data sets and services, especially in cases involving very large volumes of frequently updated data. The other services can be charged for without exception. In order to ensure that the principally free of charge view services are not abused to circumvent the charges for downloading, the view services may be delivered in a format that prevents commercial re-use.

b. Re-use

LEGAL INSTRUMENT – The 2003 PSI Directive focuses on the economic aspects of public sector information, based on the European Commission's concern for the competitive position of the European information industry. The directive only has a limited objective of a minimum harmonization of the policies and practices for re-use in the EU Member States.

FIELD OF APPLICATION – The Member States and the public bodies are not under any obligation to make their public sector documents available for re-use. Only if they choose to do so, do they have to follow the rules and conditions imposed by the PSI Directive. A large number of documents are exempted from the field of application, such as documents on which third parties hold IPR, documents that are excluded from access under national access legislation, or documents held by public broadcasters, cultural institutions or research/educational institutions. In addition, the directive does not apply to documents that are created by the public sector bodies outside of their public task, but that are, for example, sold as commercial products on the market. The limited field of application of the PSI Directive was compensated somewhat by the broad definition of *re-use*, which essentially includes any use of public sector documents for commercial or non-commercial purposes outside of the public task. If a public body allows one type of re-use, all other types must also be allowed. In addition, use outside of the public task by public sector bodies themselves is also categorized as re-use. However, the PSI Directive does not define the concept of the public task, which defines the limits of re-use.

CHARGES – The PSI Directive only contains an upper limit to the charges that can be made if public sector documents are made available for re-use, leaving a considerable margin of discretion to the Member States and the public bodies.

The total income from allowing re-use cannot exceed the costs of collection, production, reproduction and dissemination, together with a reasonable return on investment. The charges have to be cost-based and marginal cost charging is encouraged.

TRANSPARENCY – The PSI Directive also encourages transparency in the conditions and charges that are made, and on the means of redress. Practical arrangements should be made to facilitate the search for re-usable information. However, the scope of these arrangements is not determined and only the examples of portal sites and asset lists of the main documents available for re-use are suggested.

LEVEL PLAYING FIELD – Finally, the PSI Directive also takes into account the need for fair competition and a level playing field. First, it prohibits discrimination between comparable categories of re-use. However, to what extent categories of re-use are comparable remains uncertain. The principle of non-discrimination also entails that public sector bodies that re-use their own documents for their commercial activities have to allow these documents to be re-used by their private competitors under the same conditions and charges. Finally, the requirement of a level playing field also entails that exclusive arrangements are no longer allowed under the PSI Directive, and re-use has to be open for all potential actors. The only exception to this is when an exclusive arrangement is necessary for the provision of a service in the public interest.

c. Sharing

LEGAL INSTRUMENT – The sharing of public sector spatial data for the performance of public tasks is dealt with in the INSPIRE Directive. The aim of the INSPIRE Directive is to establish an SDI for the purposes of EU environmental policies and policies or activities which may have an impact on the environment. The obligations of the Member States and the public authorities to share spatial data sets and services for the performance of public tasks relating to the environment are included in Chapter V of the directive.

MEASURES FOR SHARING – The Member States have to adopt measures for the sharing of spatial data sets and services between the public authorities so that these public authorities can use them for the purposes of their public tasks that may have an impact on the environment. These measures have to preclude any obstacles likely to cause restrictions at the point of use. This should guarantee that the public servant requiring data for the performance of a particular task can get these data on his computer when he needs them without any significant delays or formalities. The sharing of the spatial data sets and services should in principle happen via the network services that must be established by the Member States. However, the charges and the limitations that can be made with regard to sharing are different from the rules set for public access to the network services.

LIMITATIONS AND CHARGES – The availability of the spatial data sets and services for sharing can only be limited if it would compromise the course of justice, public security, national defence or international relations. Charges can be made for the spatial data sets and services, and licenses can be imposed, but they must be fully compatible with the general aim of facilitating sharing. The charges have to be kept to the minimum required to ensure the necessary quality and supply of the spatial data sets and services together with a reasonable return on investment. At the same time, the self-financing requirements of the public authorities supplying the spatial data sets and services have to be respected where possible.

d. *Comparison*

OBLIGATORY CHARACTER – A short comparison of some of the main rules on access, re-use, and sharing shows that many of the topics occur in the three directives, but the obligations for the public bodies with regard to these topics differ. For instance, while the Member States and public bodies are under an obligation to make their public sector spatial data available for access and sharing, they are free to choose whether they make these data available for re-use purposes.

LIMITATIONS – Access can be limited for a large number of reasons, while the possibilities for limiting sharing are much smaller. Exceptions to re-use are not included in the PSI Directive, as it is not compulsory for the public bodies to allow re-use. However, many documents are exempted from the field of application, which has a comparable result. The effect of these limitations and exemptions can entail a different outcome for the availability of public sector spatial data for access, re-use, or sharing.

CHARGES – Another example of the different approach to access, re-use, and sharing can be found in the rules on charging. While all the rules contain some reference to the reasonable character of the charges, their scope is different. Even within the field of access, one could question how the free of charge examination on site and the reasonable charges for the supply of environmental information relate to the charges for viewing services that are necessary to maintain the quality and availability of the data and the unlimited charging possibilities for download services under the INSPIRE Directive. The large margin of appreciation for the Member States on the charges for re-use and the limits on the charges for sharing can also lead to a very different result.

SPECIFIC RULES FOR ONE TYPE OF USE – Next, some rules are only applicable to a particular type of use, without a comparable rule in the other directives. For instance, when a public body re-uses its own documents, it has to make these documents available to other users under the same conditions and charges as it has applied to itself. There is no such obligation for a public body that further uses its spatial data for other public tasks than the purpose these data were collected for, to apply the same conditions (or most likely, the lack thereof) to other public bodies that want to use the data for their public tasks. Hence, the fact that a public body does not charge itself for further using its own data does not mean that it cannot charge other public bodies under the rules of the INSPIRE Directive.

SUMMARY – The examination of the rules on access, re-use, and sharing shows considerable differences between these rules. Therefore, knowing the difference between the types of use is important. This can be demonstrated by the example of the discussion on possible charges for the availability of public sector spatial data.

2. The Importance of the Demarcation

> The importance of the distinction between access, re-use, and sharing is twofold. On the one hand, the public bodies, the users, and the bodies of appeal need to know which rules to apply. On the other hand, the lawmakers or the policy makers need clear concepts to make informed decisions. The example of the discussion on charging for public sector spatial data shows this.

PRACTICAL AND CONCEPTUAL IMPORTANCE: EXAMPLE OF CHARGING – The example of the discussion on charging policies for the availability of public sector spatial data can demonstrate the importance of a clear distinction between access, re-use, and sharing. First, the rules are different, so in practice the users may pay different fees for different types of use. Next, there is also a conceptual problem. The debate on charging has been on the table for a number of years, but it has been coloured with ideologies and personal opinions, without much empirical evidence to back up the arguments. The discussion circles around the question whether the availability of public sector spatial data would benefit most from open access or cost recovery. Open access entails that public sector spatial data are made available free of charge or at the cost of dissemination. This approach assumes that the collection of the data is fully financed by central government funding. Cost recovery involves the gaining of revenues from making data available in order to maintain the collection of these data. In this situation, the public body making the public sector spatial data available has to provide for its own financial means, or at least a part thereof, and it has to 'sell' its data to do this.

EVALUATION – The arguments that are used in favour of either open access or cost recovery do not take into account the differences between access, re-use, and sharing. For instance, the argument that open access increases democratic participation is very valuable for access, as it relates to the democratic value of public sector data. However, it is less important for re-use, as this is not intended to improve the democratic rights and obligations of the citizens, but involves gaining certain benefits beyond access. Next, the argument for cost recovery that the taxpayer should not pay for the specific needs of a small group of users intending to make money with these public sector spatial data is easier to defend with regard to re-use, as this is mostly intended to make a profit. This argument cannot be defended with regard to access, as this should benefit a much larger group and is intended as a democratic right which the society as a whole has the responsibility to safeguard. It also does not work for sharing: the policies prepared or the services delivered by a public body that obtained public sector spatial data can serve a small group, but also the entire society. These are just a few examples of arguments that show that a clear distinction between the three categories could lift this discussion to a higher level, leading to a better informed debate on charging for the availability of public sector spatial data.

C. The Problematic Character of the Distinction Between Access, Re-use, and Sharing

> The distinction between access, re-use, and sharing is deemed problematic. This problematic character occurs on two levels: the level of the public bodies making available public sector spatial data and the level of the public bodies providing spatial information services.

Two levels where problems occur – A further analysis of the problematic character of the distinction between access, re-use, and sharing shows that this problematic character plays on two levels in the development chain of public sector spatial data. For these two levels, the question needs to be asked to which extent the distinction between access, re-use, and sharing really is unclear and what causes this lack of clarity.

Public bodies as providers of public sector spatial data – The first level involves the public bodies as the providers of public sector spatial data to their users. On the one hand, the distinction between access and re-use is unclear due to the wide definition of re-use. Public bodies may find it difficult to determine whether an applicant is requesting access or re-use. On the other hand, the distinction between re-use and sharing is problematic because of the difficulty in defining the criterion of the public task. This issue mostly occurs when public bodies are using their own data or data from other public bodies to create spatial information services. They could do this as part of their public task or they could do this as re-users on the market.

Public bodies as providers of spatial information services – This difference between sharing and re-use is essential on the second level, where the problematic character occurs, that is, the level of the public bodies providing spatial information services, or information services that require processing of the public sector spatial data beyond the public body's own needs. If the public bodies provide such information services as part of their public task, they are sharing data and the data can be obtained under the INSPIRE Directive. If they do so outside of the public task, they are re-using data. Under the existing EU legal framework for the availability of public sector spatial data, the latter situation would entail that the obligations on fair competition and a level playing field of the PSI Directive come into play and that the rules of competition law have to be complied with.

D. Public Bodies as Providers of Public Sector Spatial Data

Outline – The first level where the problematic character plays a role is when public bodies make available public sector spatial data for access, re-use, or sharing. A possible answer to the problem might be found in two steps. First, a clear delineation of the obligations of the public bodies can be sought in international and European legislation and policy documents applicable to public sector spatial data. Second, the underlying criteria for the distinction between access, re-use, and sharing can be examined closely for an answer.

1. Obligations to Make Public Sector Spatial Data Available in International and European Legislation and Policy Documents

> The international and European legislation and policy documents cannot provide criteria for the distinction between access, re-use, and sharing, and hence they cannot decrease the problematic character of this distinction. Other solutions need to be considered.

OBJECTIVE – One might wonder whether existing international and European legislation or policy documents than the three directives might provide clear descriptions of the obligation of the public bodies to make public sector spatial data available, and whether they can show a clear distinction between access, re-use, and sharing. If this would be the case, these legislative and policy documents could be used as a background and inspiration for determining the difference between access, re-use, and sharing as they are designed in the Access Directive, the PSI Directive, and the INSPIRE Directive.

MOST ATTENTION FOR ACCESS – In general, the documents imposing obligations on the Member States and the public authorities to make their data available for access are widespread and reasonably well-developed, while this is much less the case for the obligations to allow re-use and sharing. This could be explained by the more recent attention for the possibilities of re-use and sharing due to the developments of information and communication technology.

NO CLEAR DISTINCTION IN MANY DOCUMENTS – However, many of the documents address more than one type of use, and not all of them manage to clearly distinguish between these different types. When applying the definitions of access, re-use, and sharing to these legislative and policy documents, mismatches and overlaps become obvious. For example, the GEOSS initiative is focused on data sharing, yet its wording implies that the wide availability of data it prescribes also includes access and possibly re-use purposes. At the level of the EU, access and re-use have also sometimes been confused before the introduction of the PSI Directive. For instance, the Publaw reports looked at re-use as a requirement for access, but they also attempted to use access legislation for re-use purposes. Currently, the Global Monitoring for Environment and Security initiative addresses sharing but also intends to develop the market. It promotes cooperation between the public and private sector for the development of services, but it does not clearly make a difference between the services that are intended for environment and security policies and the services that should be part of the market. Hence, it is not clear whether these partnerships will be subject to re-use legislation or sharing obligations.

NO CRITERIA FOR DEFINING THE DISTINCTION – In summary, it could be stated that the international and European legislation and policy documents do not contribute greatly to a clear distinction between access, re-use, and sharing. This is partly due to the fact that the obligations for access are much more developed than the obligations for sharing and re-use, but mainly because the concepts of access, re-use, and sharing as they are maintained by the Access Directive, the PSI Directive, and the INSPIRE Directive cannot always be transposed to the international and European documents. This is even the case for the documents and initiatives from the EU itself. Hence, the international and European legislation and policy documents do not provide any criteria for determining the differences between access, re-use, and sharing and cannot help in decreasing the problematic character of this distinction. Therefore, another option should be examined: the possibility to determine criteria that can be be used to make the distinction between access, re-use, and sharing.

2. Defining the Distinction Between Access, Re-use, and Sharing

a. Indicators for the Type of Use

> Neither the type of user nor the type of data based on the content, the level of processing, or the format and volume can be used as a foolproof indicator for the presence of access, re-use, or sharing.

OBJECTIVE – As the international and European legislation and policy documents do not provide any more indications on how to interpret the concepts of access, re-use, and sharing and on how to make a clear distinction between them, other ways to determine whether a clear distinction between access, re-use, and sharing could be made should be considered.

SEARCH FOR INDICATORS – First, it can be considered whether clear indicators can be found that show whether access, re-use, or sharing is the purpose of a request for the availability of public sector spatial data. Four possible indicators come to mind: the type of user, the type of data based on the content, the level of processing of the data, and the format and volume of the data.

TYPE OF USER – The first indicator is the type of user. Can one assume that the type of user is always related to a particular type of use of public sector spatial data, that is, access to individual citizens, re-use to undertakings and sharing to public bodies? This is not the case: citizens can access or re-use data, public bodies can share but also re-use, and undertakings may also use data for different purposes.

CONTENT OF THE DATA – A second possible indicator could be the type of data based on the content. The question can be raised of whether the fact that the Access Directive addresses environmental information and the INSPIRE Directive deals with spatial data means that the types of use are linked to the types of data. This would entail that requests for environmental information have an access purpose and requests for spatial data a sharing purpose, while any other type of public sector data is generally requested for re-use. However, this indicator is also not appropriate, at least not for spatial data and environmental data. Both types of data are valuable and needed for access, re-use, and sharing. While the analysis was not repeated in detail for other types of data, one can assume that the indicator will not work either for other types of data, for example, research data, cultural data, business data, etc.

LEVEL OF PROCESSING OF THE DATA – A third indicator that deserves consideration is the level of processing of the data. Can it be assumed that re-users always need raw data, while access and sharing benefit more from value-added data? Once again, such an indicator cannot be used, because the broad definition of *re-use* may also entail that value-added data are used, and access and sharing may also require entire data sets to process. Moreover, added value is a relative term that in itself cannot serve as an indicator.

FORMAT AND VOLUME OF THE DATA – The fourth and final indicator involves the format and the volume of the data requested. One could assume that electronic data would be required for re-use, because of the needs of the information industry, while this would be less important for access and sharing. Next, re-use would involve large volumes of data or entire data sets, while access would be limited to a small set of data. Once again, this indicator cannot be maintained. Considering the developments of the information society, electronic data have become the focus for all types of use. In addition, re-users might sometimes need one small subset of data, while sharing and access may also require entire data sets to be combined with other datasets.

b. The Purpose of the Use

> The democratic objectives representing access, the socio-economic objectives representing re-use, and the objective of better decision making representing sharing cannot be clearly distinguished. Overlaps exist between the different purposes of use.

OBJECTIVE – As the indicators that were assumed to show for which purpose public sector spatial data was requested cannot provide sufficient certainty, I analysed whether the purpose itself can be clearly defined, or whether this actually causes the problematic character.

NO CLEAR DISTINCTION BETWEEN THE PURPOSES – The objectives that are regularly mentioned for the availability of public sector spatial data were examined under the categories of democratic objectives, socio-economic objectives, and the objective of better decision making, which should correlate with the categories of access, re-use, and sharing. An analysis of these objectives shows that a clear distinction between these objectives is not possible, putting into question the demarcation of access, re-use, and sharing. For instance, the democratic objective of fulfilling one's rights and obligations is difficult to distinguish from the socio-economic objective of societal growth, as these rights and obligations also involve the welfare of the citizens, their health and environment, their employment, etc. Hence, while public sector spatial data on these issues are needed for democratic purposes, at the same time they also fulfil an economic purpose. A comparable issue can be found with the socio-economic purpose of the development of communities, for example on-line fora and data-mashing activities. These can be considered to be economic, as many of these communities or activities have the intention of gaining an economic or personal benefit, but they are also a very good means of increasing public participation, as they lower the barriers for the citizens to voice their opinion. Finally, it should also be stated that the objective of better decision making by the government and the public bodies is difficult to see as an objective in itself, as these decisions are always directed towards increasing economic growth or improving the life of the citizens or their participation in society and democracy. Hence, the objectives for which public sector spatial data are made available cannot be clearly distinguished. This means that the distinction between access, re-use, and sharing cannot be clearly made.

3. Confirmation of the Problematic Character of the Distinction Between Access, Re-use, and Sharing in the Access Directive, the PSI Directive, and the INSPIRE Directive

> The distinction between access, re-use, and sharing cannot be defined clearly. On the one hand, the distinction between access and re-use is not clear and unenforceable. This is mainly due to the wide definition of re-use. On the other hand, the distinction between re-use and sharing is assumed to be unclear because the criterion of the public task cannot be defined in a sufficiently clear manner. This requires further research.

Objective – Based on the previous findings, we can go back to the distinction between access, re-use, and sharing to determine whether the problematic character of the distinction can be confirmed. This examination is divided into two parts: the distinction between access and re-use and the distinction between re-use and sharing.

Access v. re-use: problematic delineation – First, the relationship between access and re-use was originally assumed as depending on the democratic or economic character of the purpose of the use. Access involved transparency, public participation, and the place of the citizen in society. Re-use was based on the economic value of public sector spatial data and their use for exploitation purposes. The development of the market was the central thought of re-use. However, this distinction cannot be maintained. Two examples can demonstrate this. First, journalists are traditionally seen as the watchdogs of the government, with an explicit contribution to the democratic control of this government by the citizens. However, they are also under pressure of commercial interests and required to sell newspapers, magazines, TV programmes, etc. Hence, the question can be asked how requests from journalists should be treated. Should they be judged under access rules, as has been the tradition? Or should they fall under the new definition of re-use in the PSI Directive, which includes 'any use', including the use by journalists? While a preference may be given either way, it cannot be deducted clearly from the law. A second example involves lawyers. On the one hand, a lawyer helps his or her clients in exercising their fundamental rights, that is, the right to defend oneself and the right to a fair trial, which are inherent to a democratic state under the rule of law. On the other hand, the lawyer performs his or her job as any other market player, subject to competition from other lawyers. Should a request from a lawyer to obtain public sector spatial data be treated as a request for access or a request for re-use?

ACCESS V. RE-USE: PROBLEMATIC ENFORCEMENT – The previous issue is linked to the practical difficulty that the distinction between access and re-use cannot be enforced. On the one hand, a citizen may in many cases not know whether he or she is requesting access to public sector spatial data or re-use. In addition, even if the citizen is aware of the distinction and abuses the access rules for obtaining public sector spatial data for re-use purposes, it will be almost impossible for the public bodies to know this. The citizens cannot be asked to state their interests when requesting data for access, so the public bodies cannot evaluate the purposes before replying to requests. Moreover, it was stated above that any indicators for either access or re-use, for example the type of user or the volume of the data are not foolproof. Hence, the public bodies would be required to search for the actual re-use taking place to determine whether all requests were made correctly. This would cause a large burden on the public bodies, and it would be a disproportionate interference in the lives of the citizens.

ACCESS V. RE-USE: CAUSE OF PROBLEMATIC CHARACTER – The problematic character of the distinction between access and re-use is caused by the wide definition of re-use. Due to this wide definition, almost any request for public sector spatial data could be considered as re-use, as virtually all citizens have a personal interest in obtaining the data that goes beyond knowing the content for purely democratic interests in the activities of the government. However, interpreting re-use so widely would negate the right of access.

RE-USE V. SHARING: THE PUBLIC TASK – In addition to the relationship between access and re-use, the distinction between re-use and sharing also requires further examination. The distinction between these two types of use is based on the criterion of the public task. The main concern lies with the activities of the public bodies based on public sector spatial data. A public body can use these data for performing its public tasks. These tasks include the preparation, implementation and evaluation of policy, but also the provision of services to the citizens. Next, the public body can also create information services based on these data that are outside of its public task and provide these services on the market, possibly in competition with the private sector. In the former case, the public body can obtain the public sector spatial data under the INSPIRE Directive, in the latter the PSI Directive applies. In addition, it is assumed that competition law is applicable to the services offered by the public bodies on the market. Hence, it is important to define the scope of the public task, as this determines the applicability of the rules for sharing or re-use.

RE-USE V. SHARING: SEARCH FOR A DELINEATION – However, it can be questioned whether defining this public task is possible in a sufficient manner for the distinction between re-use and sharing to be clear. This question mostly comes into play on the level of the provision of services by the public bodies rather than the provision of public sector spatial data. Therefore, it was examined in two separate chapters dealing with the public bodies as providers of spatial information services, a role that is much contested by the private sector.

E. PUBLIC BODIES AS PROVIDERS OF SPATIAL INFORMATION SERVICES

OBJECTIVE – Spatial information services involve the provision by a public body of public sector spatial data that have been processed beyond the public body's own needs, entailing that another entity is the target of the service. Such services can either be considered re-use or sharing, depending on their public task character. I questioned whether this public task character can be defined in a sufficient manner to be used as a criterion for the application of the rules on re-use or sharing in the EU legal framework for the availability of public sector spatial data.

OUTLINE – This analysis was done in two steps. First, the international and European legislation and policy documents that were examined earlier were re-examined to determine if they can give an indication on the scope of the public task to provide information services. Second, possible criteria to define the public task were examined and inspiration was sought in the domain of competition law, which deals with comparable issues of separating economic and non-economic activities of public bodies.

1. The Public Task to Provide Spatial Information Services in International and European Legislation and Policy Documents

> The international and European policy documents do not provide much insight on which spatial information services should be part of the public task and which services should remain outside of it. They are not a sufficient basis to define the scope of the public task to provide information services. Other solutions need to be considered.

OBJECTIVE – Can the scope of the public task to provide spatial information services be determined based on the existing obligations of the public bodies to provide such information services? This can be determined in two ways. On the one hand, international and European legislation and policy documents can be analyzed to determine the scope of the public task from a positive point of view, that is, what information services should be provided. On the other hand, an analysis can also be made from the negative point of view, attempting to assess if the public task can also be delineated by determining what information services the public bodies should not provide.

OBLIGATIONS TO PROVIDE PUBLIC SECTOR INFORMATION SERVICES – From the analysis of the international and European legislation and policy documents, the scope of the public task appears very limited with regard to information services based on public sector data in general. While many of the documents require the Member States and the public body to provide useful information, raise awareness, or meet the needs of citizens, very few clearly delineated public tasks can be found. The only information services that are regularly included are the provision of information on the functioning and activities of the public bodies, (e.g., in the UNESCO Policy Guidelines and the Council of Europe Convention on Access to Official Documents), and information on how to exercise the right of access (e.g., in EC Regulation 1049/2001, the opinions of the UN Special Rapporteur on Freedom of Opinion and Expression).

Obligations to provide environmental information services – The documents on environmental information contain some more concrete requirements for information services, but in general they still only recognize limited public tasks. Environmental information is generally required to be useful, adequate, effectively accessible, or comprehensible, or to promote awareness. Concrete requirements for information services mostly address reports on the state of the environment, and information on the activities of the public bodies. However, the introduction of the SEIS initiative at the level of the EU, of which the purpose is exactly to provide information services, indicates that the importance of information services is growing.

Obligations to provide spatial information services – This can also be found with regard to information services based on spatial data. While the UNSDI initiative and INSPIRE do not directly contain concrete requirements for the public bodies to provide spatial information services, GEOSS requires information products to be made available, and the GMES initiative is focused on providing services of public interest. Hence, in general, one can state that very few concrete obligations to provide spatial information services can be found in the international and European legislation and policy documents, entailing that the scope of the public task cannot be defined based on these obligations. However, the provision of services is still in full evolution, so this might change in the future.

Limits on providing spatial information services – The public task of the public bodies to provide information services can also be examined from the negative point of view. No general agreement can be found on where the public task ends and what the public bodies cannot do as part of the public task. Most legislation and policy documents do not go any further than recognising that a border should be drawn between public task activities of public bodies and their other activities. Many of them also recognize that these other activities of the public body are subject to competition law and have to respect the level playing field. Only a few indications are found on limits to the public task, for example, the Synergy Guidelines that required the public bodies to stay away from information services that were already provided by the private sector, unless there was a need to provide an additional service. Hence, the analysis of the international and European legislation and policy documents from the negative point of view provides even less usable results than the positive approach.

2. The Possibility of Defining *the Public Task*

OUTLINE – The examination of the international and European legislation and policy documents cannot help in defining the scope of the public task. However, more insight into the possible of defining the public task is important to determine if the problematic character of the distinction can also be confirmed on the level of the provision of spatial information services. Therefore, other ways of defining the public task need to be considered. I first addressed criteria found in the literature for determining the scope of the public task and then looked at the domain of competition law for inspiration.

a. Criteria for Defining the Scope of the Public Task to Provide Spatial Information Services

> The scope of the public task to provide spatial information services cannot be defined in a sufficiently clear manner based on the criteria found in the literature. The positive criterion of the public interest, which ties all the other positive criteria together, is too wide and evolving. Negative criteria only warn the public bodies to be careful in entering the market.
>
> The combination of the criteria and the previous findings shows that the assumed dichotomy between spatial information services on the market and spatial information services as part of the public task is not absolute. The public task and market activities are not exhaustive, as non-market activities may also be outside of the public task. They are not mutually exclusive either, as information services can be provided on the market as part of the public task.

OUTLINE – Two main types of criteria can be considered for determining the scope of the public task, in line with the approach to the analysis of the international and European legislation and policy documents. First, the positive criteria that need to be fulfilled to accept the presence of a public task, and second, the negative criteria that determine that a public task is not present and that the public body in question is operating on the market.

POSITIVE CRITERIA: LEGAL BASIS – Four main positive criteria for determining the scope of the public task appear in the literature and policy documents. The first criterion is a legal basis. In this view, the scope of the public task to provide spatial information services includes what is laid down in the law as a public task. The main problem with defining the scope of the public task as the obligations that have been laid down in the law is that this forms a circular reasoning. If an information service of a public body is considered a public task, it is included in a law, and because it is in the law, the information service becomes a public task. Hence, this criterion cannot be used to determine the scope of the public task in a sufficiently clear way.

POSITIVE CRITERIA: LINK TO CORE RESPONSIBILITY AND MARKET FAILURE – The second and the third criteria are the link of the information service to the core responsibility and market failure. Neither of these criteria are useful for determining the scope of the public task, as they only move the problem without solving it. If the spatial information service has to be linked to the core responsibility of the public body in order to be a public task, the question still remains what this core responsibility consists of. Next, market failure cannot be considered as a sufficient reason either for introducing spatial information services, as it still needs to be evaluated whether the public bodies should step in and provide the services which the market is not willing or able to offer. This will not always be the case, but only if it is considered important in the public interest that such services are provided. Hence, the actual criterion to be considered is the public interest.

POSITIVE CRITERIA: PUBLIC INTEREST – This public interest forms the fourth criterion for determining the scope of the public task. It entails that spatial information services are considered a part of the public task if they are needed to ensure the general welfare. The public interest can be formulated in many ways: the safeguarding of democratic rights, but also social interests such as public health, social capital, environmental concerns, etc.

PUBLIC INTEREST: ACCESSIBILITY – With regard to spatial information services, the concept of public interest can be made more concrete in the requirement of accessibility of information. From this perspective, citizens must be able to use information services with minimal barriers or effort. However, the subjective character of accessibility makes it difficult to determine how far the spatial information services should go to take down these barriers. As accessibility not only includes conditions of physical and financial availability of information, but also of intellectual accessibility, the target group plays a very important role. From this perspective, the public bodies are confronted with the dilemma of the public task. On the one hand, they may be required to provide extensive spatial information services in order to satisfy the needs for accessibility for some target groups. On the other hand, they may be accused of stepping into the domain of the private sector, whose core business it is to provide such extensive or specialized spatial information services.

PUBLIC INTEREST: UNIVERSAL SERVICE – Another concretization of the public interest criterion can be found in the concept of universal service, which has been suggested in the literature as a good concept to be applied to information services. Universal service can be considered as a service that is so essential that everyone should have permanent access to it at an affordable price. The requirements of affordability and continuation can provide more concrete indications of the public task, but as universal service is considered as a way to guarantee the public interest, this public interest in the provision of the spatial information service still has to be defined.

EVALUATION – Hence, the public interest is the main positive criterion for determining the scope of the public task, to which the other criteria can be brought back. However, it is submitted that this public interest cannot be defined in a sufficiently specific way in order to serve as a criterion for the distinction between re-use and sharing. The definition is too dependent on political opinions and on the evolution of the public interest to be useful.

Negative criteria – As no sufficient positive criteria are found to determine the scope of the public task, the question can be asked of whether negative criteria can be determined instead. The negative criteria encountered in policy documents and in the literature are very limited. They mostly require the public bodies to refrain from discouraging private sector activity; not to duplicate existing efforts; or at least consider whether existing information services on the market already foresee in the public interest before setting up a new service. Hence, the main concern is that public bodies think twice before entering the market in order to ensure a public interest. However, this concern seems to conflict with the approach of the PSI Directive: either a public body performs a public task, ensuring a public interest, or it performs an activity on the market. The negative approach in policy and in the literature suggests that market activities can also be performed in the public interest. Combined with previous findings, this requires a re-examination of the demarcation line between public task activities and the market, which was assumed to be absolute.

Re-examination of the distinction between market and public task – Such a re-examination of the distinction between the public task and market activities shows that these are neither exhaustive nor mutually exclusive. The public task and market activities are not exhaustive because the wide definition of re-use also includes non-commercial activities. Hence, these activities remain outside of the public task, yet they cannot be seen as commercial or market activities either. Next, the public task and market activities are not mutually exclusive either. On the one hand, in some Member States, selling information services on the market is considered a part of the public task because this public task requires the public bodies to generate revenue for the Treasury. On the other hand, as was mentioned above, public bodies may also ensure public interests by the provision of information services on the market. Hence, the absolute demarcation between public task activities and market activities that was assumed to underlie the PSI Directive and the INSPIRE Directive cannot be maintained.

b. *Competition Law as a Source of Inspiration*

The joint application of the EU legal framework for the availability of public sector spatial data and competition law shows that the categories of activities do not match. Public task activities under the INSPIRE Directive can contain economic and non-economic activities under competition law, and market activities under the PSI Directive can also contain economic and non-economic activities.

Competition law cannot provide much inspiration for the question on the distinction between re-use and sharing and the definition of the public task. The difficulty in the joint application of the EU legal framework for the availability of public sector spatial data and competition law enhances the problematic character of the distinction between re-use and sharing.

OBJECTIVE – In order to examine how far this lack of demarcation can enhance the problematic character of the distinction between re-use and sharing, inspiration can be sought in the area of competition law. As it is generally assumed in policy documents and in the literature that competition law is applicable when public bodies re-use public sector spatial data and offer spatial information services outside of the public task, this topic needs to be addressed for a good overview of the legal framework. But its relevance goes beyond completeness of the overview. The domain of competition law is mostly important to analyze because it has had extensive experience in defining when public bodies are performing market activities, and in dealing with the issue of services in the public interest that are performed on the market. In competition law, these terms are respectively referred to as economic activities and services of general economic interest.

SPATIAL INFORMATION SERVICES AS ECONOMIC ACTIVITIES – As a starting point, one should examine when spatial information services can be qualified as economic activities under EU competition law, and consequently whether the public bodies providing these services are undertakings subject to the competition rules. Considering the wide definition and interpretation of the ECJ, many spatial information services can be considered economic activities, even though they may be considered as part of the public task by the Member States or the public bodies involved. Hence, it can be stated that the concept of economic activities does not match with the distinction between public task activities and market activities.

APPLICATION OF COMPETITION LAW TO SPATIAL INFORMATION SERVICES – As many spatial information services can be categorized as economic activities, the public bodies/undertakings providing them are also in the position to distort competition on the market. Several possible infractions of competition law can be found in the domain of spatial information services. On the one hand, the public bodies can infringe upon competition law as providers of spatial information services. Examples show that spatial information services provided by public bodies are regularly in danger of infringing upon Article 101 of the Treaty on the Functioning of the EU (former Article 81 of the EC Treaty) on agreements between undertakings that distort competition, and mostly Article 102 of the Treaty on the Functioning of the EU (former Article 82 of the EC Treaty) on abuse of dominance. On the other hand, public bodies can also infringe upon competition law operating as state authorities by providing state aid (Article 107 Treaty on the Functioning of the EU – former Article 87 of the EC Treaty) and by taking measures regarding public undertakings and undertakings with special or exclusive rights, under Article 106(1) of the Treaty on the Functioning of the EU (former Article 86(1) of the EC Treaty). Once again, many spatial information services offered by public bodies can be contrary to the Treaty on the Functioning of the EU, particularly with regard to special or exclusive rights that are held by the public bodies.

SERVICES OF GENERAL (ECONOMIC) INTEREST – Competition law can also provide inspiration on the issue of spatial information services that ensure a public interest and in this way can be considered as a public task, but that are offered on the market. In competition law, a comparable category of services exists, referred to as services of general economic interest. Undertakings entrusted with providing such services are subject to competition law, insofar as the application of the competition rules does not obstruct the performance, in law or in fact, of the particular tasks that are assigned to them (Article 106 Treaty on the Functioning of the EU – former Article 86 of the EC Treaty). Hence, exemptions from the application of competition law are possible, but only insofar as this is necessary for the spatial information services to be provided under economically acceptable conditions. However, the inspiration that can be found is limited, as the ECJ has struggled with the concept, and it judges on a case-by-case basis whether a service of general economic interest is present and whether it can be exempted from competition law. The Court leaves a great margin of appreciation to the Member States to determine their own services of general economic interest, and only intervenes in manifestly clear situations. In addition, the activities of the European Commission in the domain of services of general interest make the concept even more unclear. However, even though their scope is unclear, it can be stated that as a principle, services of general economic interest are part of the public task in the framework of re-use and sharing, as they ensure the public interest. This means that part of the public task can be subject to competition law, unless this would obstruct the performance of this task.

EVALUATION: COMPETITION LAW CANNOT SOLVE THE PUBLIC TASK QUESTION – As a result, one could state that while competition law is also applicable to spatial information services and deals with the same problems as the EU legal framework for the availability of public sector spatial data, its use is limited. First, competition law can also not solve the public task question in a general manner for a number of reasons. The first reason is that the ECJ finds it challenging to determine whether public bodies are performing economic activities. Moreover, it has not developed straightforward criteria to determine whether a service of general economic interest is provided by the public bodies. It also accepts that the Member States essentially define the content of the services of general economic interest.

EVALUATION: COMPETITION LAW CAUSES MORE COMPLICATIONS – Second, competition law complicates the debate even further, as its categorization of activities does not match with the division between public task activities and market activities under the PSI Directive and the INSPIRE Directive. On the one hand, market activities as envisioned under the PSI Directive include both economic and non-economic activities. The economic activities are subject to competition law, while the non-economic activities are not. On the other hand, the public task activities under the PSI Directive and the INSPIRE Directive also contain economic and non-economic activities. They include the public tasks that are connected with the exercise of official authority, which are not seen as economic activities under competition law, and the public tasks that could be seen as economic activities, that is, the services of general economic interest. Once again, the economic activities are subject to competition law, while the non-economic activities are not. Hence, this entails that the combination of the EU legal framework for the availability of public sector spatial data and EU competition law leads to a distinction between more categories of activities, which complicates the definition of criteria for their demarcation even further.

c. *Confirmation of the Problematic Character of the Distinction between Re-use and Sharing*

> The distinction between re-use and sharing is problematic, because the scope of the public task cannot be defined in a sufficiently clear manner. This is due to its ever-evolving, political, and inherently national character.

EVALUATION – The examination of possible criteria for determining the scope of the public task shows that this public task cannot be defined in a sufficiently clear way to determine the distinction between re-use and sharing in the EU legal framework for the availability of public sector spatial data. This is due to the ever-evolving and political character of the public task. This also is apparent in the comparable situation of services of general economic interest under competition law. The public task is inherently linked to nationally determined political visions on the role of the public sector, which change continuously due to developments in society, technology, economy, user expectations, etc. This entails that the scope of the public task depends on who defines it. The development of the EU legal framework for the availability of public sector spatial data would require such a definition to be provided at the EU level in order to ensure optimal harmonization. However, the European Commission has made it clear that defining the public task remains in the hands of the Member States.

III. FINAL CONCLUSIONS AND FURTHER THOUGHTS

A. CONFIRMATION OF THE HYPOTHESIS

> As a general conclusion, it can be stated that the analysis of the distinction between access, re-use, and sharing confirms the assumption that this distinction is problematic. On the one hand, the distinction between access and re-use is not clear and not enforceable due to the wide definition of re-use provided in the PSI Directive. On the other hand, the distinction between re-use and sharing is not clear because the criterion on which the distinction is based, the public task, cannot be defined in a sufficiently clear manner.

B. COMMENTS AND REFLECTIONS

> The results of the analysis are not only valid for the limited domain of public sector spatial data, but could be transposed to any type of public sector data for which access, re-use, and sharing are regulated.

APPLICABILITY OF THE RESULTS TO OTHER DOMAINS – It should be noted that, while my analysis only addresses the limited scope of the overlap between the Access

Directive, the PSI Directive and the INSPIRE Directive, that is public sector spatial data and spatial information services, its results can be transposed to all domains of public sector data. The problematic character of the distinction between access, re-use, and sharing has come to light for public sector spatial data in particular, because this is the only domain for which the EU has regulated the three types of use. If such regulation would be extended to other types of public sector data, the same problems would occur. This entails that the Member States may have to deal with the discussion on access, re-use, and sharing on a much broader scale than is represented here. This is definitely the case for the relationship between access and re-use, as most of the Member States have adopted access legislation for all their public sector data, based on their national constitutions or Council of Europe recommendations (and in the future the Convention on Access to Official Documents). This may also be the case for sharing obligations, as existing policies for the availability of public sector data between public bodies for performing their public tasks are also widely organized in the Member States, whether informally or in an official form.

1. Wide Definition of *Re-use*

a. Risk for the Right of Access

> The full application of the wide definition of *re-use* may cause the public bodies to focus their information policies on re-use rather than access. They must ensure that the existing rights of access to public sector spatial data are not harmed.

DIFFICULTY IN DISTINGUISHING ACCESS AND RE-USE – The definition of *re-use* laid down by the PSI Directive essentially includes any use of public sector spatial data for commercial or non-commercial purposes outside of the public task. This broad character makes it difficult to distinguish re-use from access because it is not always possible to determine whether a request for public sector data is made for the democratic purpose of checking up on the activities of the government or for the economic purpose of gaining a (personal or business) benefit. In addition, sometimes both purposes may be present at the same time. As was already stated, while this problem was only examined for the subject of public sector spatial data, it has a much greater impact in the Member States, because almost all of them recognize a general right of access to all types of public sector data, not just to spatial or environmental data. In this way, the relationship between the Access Directive, the INSPIRE Directive, and the PSI Directive might give the distorted image that this issue could only be encountered for public sector spatial data.

DANGER TO THE RIGHT OF ACCESS – The risk of the wide definition of *re-use* is that it might endanger the right of access. As almost any request for public sector spatial data will involve a personal interest of the applicant, an extensive application of the definition of re-use by the public bodies could categorize all these requests as re-use. This could endanger the right of access of the citizen, while such right is recognized as essential for democratic participation by national constitutions, the Council of Europe Convention on Access to Official Documents, the Aarhus Convention, Article 15 of the Treaty on the Functioning of the EU (former Article 255 EC-Treaty), etc. The public bodies may be tempted to develop their policies for making public sector spatial data available with a focus on re-use rather than access and make the formal requirements, conditions of use, or charges more demanding. This temptation is made even stronger by the fact that the public bodies often will not know whether access or re-use is intended, so they want to ensure that possible re-use is covered. The rules for access in the international treaties and national legislation provide a number of safeguards to ensure the right of access, for example, an applicant requesting access to public sector data cannot be asked to state his interest in the data. However, within the limits of these safeguards, the public bodies have to ensure that they maintain the status quo on the citizens' right to access to public sector data.

b. Revision of the Definition

> In the reconsideration of the definition of *re-use*, existing rights should be safeguarded either by including them under access or by developing an 'access-like right'. If a new definition is made, its scope should be carefully considered and tested to ensure that no discussion on the distinction between access and re-use remains. If a new definition is not desired, the possibility should be considered to remove the distinction and apply the same rules to access and re-use.

RE-ORIENTATION OF THE DEFINITION OF *RE-USE?* – As the wide definition of *re-use* complicates the application of the Access Directive and the PSI Directive, and it threatens to harm the right of access, it should be considered whether it is possible and worthwhile to change the definition of *re-use*. As the establishment of the legal framework for re-use was originally intended to promote the development of the information industry and the creation of cross-border or even pan-European information products or services, the question should be asked of whether the definition of *re-use* can be re-oriented towards its original target group. If re-use is only applicable to the development of information products and services based on public sector data, the distinction may be much easier to make.

CONSIDERATION FOR CURRENT RE-USERS – However, several considerations should be taken into account when reconsidering the definition of *re-use*. First, it should be considered whether it is possible to reduce a wide definition to a narrower one, and on what the consequences would be for existing re-use. A large group of re-users involves individual citizens, who use public sector data for online communities, data-mashing, forums, blogs, etc. In some countries, mainly the United Kingdom, these re-users are considered as one of the main target groups. It should be made sure that the rights such re-users have gained are not withdrawn by a reduction of the definition of re-use, leaving these re-users in a legal vacuum. It may be possible to include this kind of use under the national rules of access, as actually might have been the case before the introduction of the PSI Directive. If this is not possible, another way should be found to safeguard these rights, through an 'access-like' right that includes further use.

DEFINITION OF THE TARGET GROUP – In the consideration of a new definition of *re-use*, it will be vital to clearly determine the purpose and the addressees of the definition in order to avoid comparable problems as the current discussion on the distinction between access and re-use. One must admit that this is a great challenge: as was already discussed before, using the type of user as a criterion for determining re-use is difficult. It seems artificial to only include European-wide multinationals and not local SMEs, for instance. In addition, this would hinder the objective of the directive to encourage innovation. Could the volume, format, or level of processing of data be more useful as a criterion for the new definition of *re-use* than they were for the old one? Or could it be stated that re-use only involves commercial use, for instance? This latter option would require a clear definition of *commercial*, which has also been much contested. In any case, any change to the definition of *re-use* would have to be carefully considered.

ABOLITION OF THE DISTINCTION? – The question can also be asked if a new definition for *re-use* is needed. It may be that the challenge of finding a good definition can be avoided by imposing the same conditions for access and re-use, so that it is no longer needed to distinguish between them. Such a solution could work in two directions: either the conditions for access are applied to re-use, or vice versa. As was mentioned above, the latter option is to be avoided if this would curtail the right of access as it is currently enjoyed by the citizens. More likely the rules for re-use should be adapted to the standards for access. This would lead to a comparable situation as in the United States that access includes the right to use the obtained public sector data for any further purpose. However, it should be examined what the effect of such a change would be on the public bodies that are required to provide for their own funding or to gain a revenue for the Treasury. The sustainability of the production of data must be guaranteed, so what these bodies would lose in revenue from re-use should be compensated in another way if needed, but without causing market distortion. In summary, removing the distinction between access and re-use may be a solution, but the consequences of this option for all the stakeholders should be examined.

2. The Concept of the Public Task

IMPOSSIBILITY OF DEFINING *THE PUBLIC TASK* – The second problem with the distinction between access, re-use, and sharing lies in the concept of the public task, which distinguishes sharing from re-use. Sharing involves the use of public sector spatial data for the performance of the public task, while re-use includes any use outside of that public task. This problematic character mainly came into play in the assessment of the provision of spatial information services by the public bodies as either part of their public task or an activity on the market competing with the private sector. The public task was found to be impossible to define in a manner that guarantees sufficient certainty for the public bodies providing public sector spatial data, the users making spatial information services with these data, and the courts or regulatory bodies having to decide about the character of these services.

a. The Uncertain Character of the Public Task

> The uncertain character of the public task ensures the possibility of the public bodies to adapt to the needs in society. It should be studied whether the unclear concept of the public task in the provision of information services really hinders the development of the information market. Such a study should be based on empirical evidence but not hinder the cooperation between the public and private sector.

EVOLUTIVE AND POLITICAL CONCEPT – The scope of the public task is impossible to capture in a concrete definition, because it depends on the views on the role of the state and its relationship with the private sector, and on the developments in society, economy, technology, user demands, etc. These views are reflected in the composition of the elected bodies, which in their turn decide which of these developments will be acknowledged by new legislation, policies or programmes. Hence, the public task is an evolutive and a political concept, subject to the current opinion on what is in the public interest. Depending on the country, this current opinion may be rooted more or less strongly in national traditions or visions on the public interest and the role of the public sector.

ALLEGED PROBLEMS DUE TO THE LACK OF A DEFINITION – It should be considered whether this evolutive and political character of the scope of the public task is not just a problem in theory, but also in practice. After all, society will never stop developing, and the demands of the public will always change with this development. The public bodies should be able to adapt to these changing needs and change their legislation, policies, and practices accordingly. In this perspective, it is even undesirable that the public task is defined in a rigorous manner, taking away the possibility for quick responses to emerging public interests and needs. Yet, the uncertain character of the scope of the public task is often mentioned as an important barrier for the development of information services by the private sector. It is a regular complaint of the private sector that public bodies are unfairly competing with them in the provision of information services on the market, while hiding behind their public task. Such complaints have been encountered regularly since the beginning of the interest of the European Commission in regulating the re-use of public sector data in the 1980s.

NEED FOR EMPIRICAL EVIDENCE – However, it should be determined with more certainty that this concern of the private sector is valid and that the problem goes deeper than a few isolated cases where discussion might occur on the activities of the public sector. Many studies have been made on the market for information products and services based on public sector data, but most of them were based on questionnaires and interviews with stakeholders, hence on perceptions rather than facts. More empirical evidence should be collected to determine whether the role of the public sector in the information market truly causes so many discussions that it needs to be addressed. In this gathering of evidence, it should be taken into account that many of the private sector service providers are dependent on the public bodies for the delivery of their data and that a good cooperation between both sectors should remain the focal point.

b. Reduction of the Uncertain Character of the Public Task

> The uncertain character of the public task could be reduced by the establishment of a regulatory body. This regulatory body should have a double role: provide guidance on the relationship between the public and the private sector and deal with complaints from private sector parties.
>
> The regulatory body should have a representative and continuous character and also have the power to enforce its decisions on complaints. It should carefully consider its relationship with existing regulatory bodies on access and with the competition authorities.

REGULATORY BODY – Defining the scope of the public task may be difficult, but this does not mean that the stakeholders have to remain in complete uncertainty about this scope. In situations where discussions would arise between the public and the private sector about the scope of the activities of the public bodies, a regulatory body could help to reduce the uncertainty. While such regulatory bodies already exist in several countries, for example the French *Commission d'accès aux documents administratifs*, the United Kingdom's OPSI and APPSI, the Slovenian Information Officer, etc., these bodies still need to be developed further in the direction of the considerations below.

DOUBLE ROLE OF THE REGULATORY BODY – Regardless of its practical organization, such a regulatory body should have a double role. On the one hand, it should be able to provide guidance to the public and the private sector on their relationship and on how to proceed on the market for information services. Such a proactive approach could already reduce the possible frictions between the public and the private sector. On the other hand, it should deal with complaints from private sector parties who feel that the activities of the public bodies are hindering them from operating their business.

REPRESENTATIVE AND CONTINUOUS CHARACTER – The members of this regulatory body should be representatives of all the stakeholders, including public bodies, the information industry, non-commercial re-users, interest groups, and the citizens. This should ensure a balanced view on the information society and the information market, taking into account the business models of the private sector and the need for the public bodies to ensure public interest, leaving room for innovation and adaptation to the changing needs of society. In order to develop such a balanced view, the regulatory body should be guaranteed continuity and independence. This includes sufficient funding to maintain this continuity.

Relationship with other bodies – In order for the regulatory body to positively influence the coherence of the public bodies' information policies, it should be able to address issues relating to access, re-use, and sharing. As in many Member States regulatory bodies already exist that help guarantee the correct application of national access legislation, the relationship with these existing bodies should be considered carefully. It should be examined whether it is possible to join the different bodies into one entity, ensuring that all capacities to deal with the different fields remain present; or if this not possible, how the relationship between the bodies can be organized in order for the stakeholders to obtain a quick and satisfactory solution to their problems. In addition, the relationship with the competition authority should be considered carefully, as the question of the public task of the provision of information services often will be joined with concerns of unfair competition. A quick and light procedure should be developed for the cooperation between the regulatory body and the competition authorities.

Enforcement and creation of goodwill – If the regulatory body cannot enforce its decisions, its impact on the question of the public task will remain limited. Therefore, it is important that it has the power to take action if its decisions on complaints are not followed. However, its advisory position may be better served by working together with the public bodies and the private sector rather than imposing rules that need to be followed by the public bodies. In this part of its activities, the emphasis should be on creating goodwill from all the stakeholders involved.

c. The Public Task on the Level of the European Union

> As the PSI Directive and the INSPIRE Directive intend to harmonize the availability of public sector spatial data on the level of the EU, it is unfortunate that they are based on the inherently national concept of the public task.

EU harmonization of the legal framework – While the establishment of a regulatory body that can give a ruling on the scope of the public task could already be a step forward in the relationship between the public bodies and the private sector on the market, it cannot completely meet the existing concern about the concept of the public task in the PSI Directive and the INSPIRE Directive. The objective of the PSI Directive was to harmonize the conditions for re-use of public sector data in the EU in order to stimulate the European information industry and encourage the development of cross-border and pan-European information products and services. The INSPIRE Directive is also based on the idea that environmental policy does not stop at national borders and that the sharing of cross-border spatial data is needed in order to respond properly to environmental needs and challenges.

Based on a national criterion – In this view, it can be questioned why the European institutions chose a criterion that is inherently linked to national views to base their harmonization on. Even when the public task may no longer be a concern on the national level, the differences in opinions and traditions regarding this public task in the Member States will continue to hamper the harmonization that was intended by the PSI Directive and the INSPIRE Directive. While the differences between the national traditions are intrinsic to the structure of the EU, and a balance between the different national interests will always have to be found, it remains unfortunate that the choice was made for such an inherently national concept as the public task.

d. Reduction of the Uncertainty of the Public Task at the Level of the European Union

> An advisory group on the EU level should provide more guidance on the public task and try to assess whether a *European Union public task* could be agreed upon.

NO GUIDANCE FROM THE EUROPEAN COMMISSION – The possibility of more guidance on the public task on the level of the EU can be questioned. While this has been asked for by some Member States and by members of the private sector, the European Commission has stated that it does not intend to provide any rules or guidelines on the interpretation of a *European Union public task*. Other options to provide more certainty on the scope of the public task from the point of view of the entire EU should be considered.

EU ADVISORY GROUP ON THE PUBLIC TASK – While a regulatory body with the power to address complaints on the European level may be unacceptable for the Member States, the possibility of setting up an advisory body should be considered. Existing structures may be used, such as the PSI Group, but it should be made sure that the stakeholders are brought together from the three domains of access, re-use, and sharing, and that all stakeholders are represented, including the national public bodies, the European and national information industry, interest groups representing non-commercial users, etc. In this way it can be examined whether a *European Union public task* – with all the inherent evolutionary and uncertain characteristics – could be agreed upon. The exercise of discussing the scope of this public task should be regularly repeated.

3. Coherence with Other Legislation and Policies

a. Competition Law

> It is unfortunate that the domain of EU competition law was not taken more into account in the development of the rules on re-use. This could have shown the difficulty in using national concepts for regulating the EU-level and could have avoided the introduction of the public task as a new term.

COMPARISON WITH COMPETITION LAW – We found that the domain of EU competition law had to deal with similar issues as the question on the scope of the public task in the distinction between re-use and sharing in the EU legal framework for the availability of public sector spatial data. Competition law has had to determine whether public bodies are performing economic activities and in this way can be qualified as undertakings subject to the rules of competition law. In addition, it has been confronted with the issue of finding a balance between safeguarding free competition and the common and internal market on the one hand and the protection of public interests in the other hand. Hence, comparable notions and concepts can be found in this domain: economic activities, market, exercise of official authorities, services in the general (economic) interest, etc.

INPUT FOR LEGAL FRAMEWORK FOR THE AVAILABILITY OF PUBLIC SECTOR SPATIAL DATA – As it is generally accepted that the domain of competition law is also applicable to the provision of information services of public bodies on the market (provided that the conditions of competition law are fulfilled, of course), it could be questioned why the European institutions did not seek inspiration from this domain in assessing the scope and impact of the PSI Directive. The value of this could have been two-fold.

EXISTING FRAMEWORK AS BASIS – On the one hand, competition law could have offered an existing framework within which the discussion on the re-use of public sector data could have been held, while now it was merely accepted that the competition rules would apply if information services were offered on the market. Using existing terms and criteria would have made it easier for the transposing Member States, the users and the courts to determine the impact and interpretation of the directive.

CAUTION FOR NEW INITIATIVES – On the other hand, the experiences in competition law could have also cautioned the European institutions. The European competition authorities have had extensive experience in dealing with concepts that are inherently national in nature, such as services of general interest and services of general economic interest. It has been accepted that these concepts are essentially filled in by the Member States, with only a marginal assessment by the European competition authorities. This had led to many case-by-case decisions of the ECJ and great uncertainty about the scope of these concepts. It could be questioned whether introducing another concept struggling with the same issues was a good idea. It is a missed chance for harmony between different areas of EU law, and it would have saved the courts and authorities the trouble of having to interpret another contested term.

FOLLOW-UP – As the discussion on services of general (economic) interest is still ongoing, it may be valuable for the discussions on re-use and sharing to follow-up this discussion and assess its value for the question of the public task.

b. Other Information Policies and Legislation

> A comprehensive policy for making public sector spatial data available is needed. New initiatives should build on the existing ones, but should avoid taking over the problems of these existing initiatives. Therefore it is important that the problematic character of the distinction between access, re-use, and sharing is addressed in the near future.

Need for a comprehensive approach – The domain of public sector spatial data is a good example to show that the legislators and policy makers need to consider a comprehensive approach to the policy for making public sector data available, because many different initiatives are being taken, both on the level of the EU and on the global level. This indicates that a clear overview on the direction of the information policy is needed in order for this policy to guarantee the optimal availability of public sector data. While the European institutions have tried to do this with the Access Directive, the PSI Directive, and the INSPIRE Directive, their attempts were not always well thought-out. To avoid the impact of such shortcomings to be increased, the relationship with other plans such as the Shared Environmental Information System, Global Monitoring for Environment and Security should be addressed with careful attention. This could be extended to any future initiatives on access, re-use or sharing of public sector data.

Need for short-term solution – However, while the increasing effort of the new policies to build on the existing rules in the Access Directive, the PSI Directive, and the INSPIRE Directive should be applauded, this also means that these new policies will be confronted with the same problems as the current framework. To prevent that these new policies suffer from the same shortcomings it is important that the problematic character of the distinction between access, re-use, and sharing of public sector data is addressed in the near future.

Table of Legislation and Policy Documents

UNITED NATIONS

A. Conventions and Declarations

a. General Assembly

United Nations. *Universal Declaration on Human Rights,* 10 December 1948, <www.un.org/Overview/rights.html>, 22 February 2010.

United Nations. *International Covenant on Civil and Political Rights*, 16 December 1966, <www2.ohchr.org/english/law/ccpr.htm>, 22 February 2010.

United Nations. *International Covenant on Economic, Social and Cultural Rights*, 16 December 1966, <www2.ohchr.org/english/law/cescr.htm>, 22 February 2010.

United Nations. *Convention on Biological Diversity*, 5 June 1992, <www.biodiv.org/doc/legal/cbd-un-en.pdf>, 22 February 2010.

United Nations. *United Nations Convention on Climate Change*, 1992, <http://unfccc.int/resource/docs/convkp/conveng.pdf>, 22 February 2010.

United Nations. *United Nations Millennium Declaration*, 18 September 2000, <www.un.org/millennium/declaration/ares552e.pdf>, 22 February 2010.

b. Conference on Environment and Development

United Nations Conference on Environment and Development. *Rio Declaration on Environment and Development*, 13 June 1992, <www.unep.org/Documents.

Multilingual/Default.asp?DocumentID=78&ArticleID=1163>, 22 February 2010.

c. Economic Commission for Europe

United Nations Economic Commission for Europe. *Convention on Access to Information, Public Participation in Decision-making and Access to Justice in Environmental Matters*, Aarhus, Denmark, 25 June 1998, <www.unece.org/env/pp/documents/cep43e.pdf>, 22 February 2010.

United Nations Economic Commission for Europe. *Resolution on Access to Information, Public Participation in Decision-making and Access to Justice in Environmental Matters*, 25 June 1998, <www.unece.org/env/documents/1998/cep/cep.43.add.1.rev.1.e.pdf>, 22 February 2010.

B. Other Documents

a. Special Rapporteur on Freedom of Opinion and Expression

United Nations Special Rapporteur on Freedom of Opinion and Expression. *Promotion and Protection of the Right to Freedom of Opinion and Expression*, UN Doc. E/CN.4/1998/40, 28 January 1998, <http://daccess-ods.un.org/access.nsf/Get?Open&DS=E/CN.4/1998/40&Lang=E>, 22 February 2010.

United Nations Special Rapporteur on Freedom of Opinion and Expression. *Promotion and Protection of the Right to Freedom of Opinion and Expression*, UN Doc. E/CN.4/2000/63, 18 January 2000, <http://daccess-ods.un.org/access.nsf/Get?Open&DS=E/CN.4/2000/63&Lang=E>, 22 February 2010.

United Nations Special Rapporteur on Freedom of Opinion and Expression. *Civil and Political Rights, Including the Question of Freedom of Expression: The Right to Freedom of Expression and Opinion*, UN Doc. E/CN.4/2005/64, 2004, <http://daccess-ods.un.org/access.nsf/Get?Open&DS=E/CN.4/2005/64&Lang=E>, 22 February 2010.

b. United Nations Educational, Scientific and Cultural Organization (UNESCO)

UNESCO. Records of the General Conference. Twenty-ninth session, Paris, 21 October to 12 November 1997. Volume 1 Resolutions, <http://ulis2.unesco.org/images/0011/001102/110220e.pdf>, 22 February 2010.

UNESCO. Policy Guidelines for the Development and Promotion of Governmental Public Domain Information, Paris, UNESCO, 2004, <http://unesdoc.unesco.org/images/0013/001373/137363eo.pdf>, 22 February 2010.

c. United Nations Economic Commission for Europe

United Nations Economic Commission for Europe. *Guidelines on Access to Environmental Information and Public Participation in Environmental Decision-making*, 1995, <www.eea.europa.eu/publications/92-9167-020-0/page017.html>, 22 February 2010.

United Nations Economic Commission for Europe. *The Aarhus Convention: An Implementation Guide. Part 2*, 1998, <www.unece.org/env/pp/implementation%20 guide/english/part2.pdf>, 22 February 2010.

d. Geographic Information Working Group

United Nations Geographic Information Working Group. *UNSDI Compendium: A UNSDI Vision, Implementation Strategy and Reference Architecture*, 2007, <www.ungiwg.org/docs/unsdi/UNSDI_Compendium_13_02_2007.pdf>, 22 February 2010.

United Nations Geographic Information Working Group. *Strategy for Developing and Implementing a United Nations Spatial Data Infrastructure in Support of Humanitarian Response, Economic Development, Environmental Protection, Peace and Safety*, 2007, <www.ungiwg.org/docs/unsdi/UNSDI_Strategy_Implementation_Paper.pdf>, 22 February 2010.

United Nations Geographic Information Working Group. *Frascati Statement*, 2007, <www.unsdi.nl/downloads/frascatistatement.pdf>, 22 February 2010.

e. Group on Earth Observations

Group on Earth Observations. *The Global Earth Observation System of Systems (GEOSS) 10-Year Implementation Plan*, 2005, <www.earthobservations.org/documents/10-Year%20Implementation%20Plan.pdf>, 22 February 2010.

Group on Earth Observations. *Global Earth Observation System of Systems (GEOSS) 10-Year Implementation Plan Reference Document*, Noordwijk, ESA Publications Division, Noordwijk, 2005, <www.earthobservations.org/documents.shtml>, 22 February 2010.

Group on Earth Observations. *GEO 2007–2009 Work Plan. Toward Convergence*, 2007, <www.earthobservations.org/documents/wp0709_v6.pdf>, 22 February 2010.

Group on Earth Observations. *Draft Implementation Guidelines for the GEOSS Data Sharing Principles*, 2008, <www.earthobservations.org/geoss_dsp.shtml>, 22 February 2010

Group on Earth Observations. *GEO 2009–2011 Work Plan*, 2009, 7, <www.earthobservations.org/documents/work%20plan/geo_wp0911_rev2_091210.pdf>, 22 February 2010.

f. Others

United Nations Commission on Sustainable Development. *Agenda 21*, 1992, <www.un.org/esa/sustdev/documents/agenda21/english/agenda21toc.htm>, 22 February 2010.
United Nations Development Programme. *Human Development Report 1999*, 1999, <http://hdr.undp.org/en/media/HDR_1999_EN.pdf>, 22 February 2010.

EUROPEAN UNION

A. Treaties, Protocols, and Declarations

Treaty establishing the European Community, *OJ C* 325, 24 December 2002, 1.
Treaty on European Union, *OJ C* 115, 9 May 2008, 13.
Treaty on the Functioning of the European Union, *OJ C* 115, 9 May 2008, 47.
Protocol to the Treaty of Amsterdam on the system of public broadcasting in the Member States, *OJ C* 340, 10 November 1997, 109.
Declaration 11997D/AFI/DCL/13 on Article 7d of the Treaty establishing the European Community, *OJ C* 340, 10 November 1997, 133.
Charter of Fundamental Rights of the European Union, *OJ C* 303, 14 December 2007, 1.
Protocol on Services of General Interest to the Lisbon Treaty, *OJ C* 306, 17 December 2007, 159.

B. Union Acts

a. Directives

Council

Council Directive 90/220/EEC of 23 April 1990 on the deliberate release into the environment of genetically modified organisms, *OJ L* 117, 8 May 1990, 15.
Council Directive 90/313/EEC of 7 June 1990 on the freedom of access to information on the environment, *OJ L* 158, 23 June 1990, 56.
Council Directive 92/50/EEC of 18 June 1992 relating to the coordination of procedures for the award of public service contracts, *OJ L* 209, 24 July 1992, 1.
Council Directive 93/36/EEC of 14 June 1993 coordinating procedures for the award of public supply contracts, *OJ L* 199, 9 August 1993, 1.
Council Directive 93/37/EEC of 14 June 1993 concerning the coordination of procedures for the award of public works contracts, *OJ L* 199, 9 August 1993, 54.

Council Directive 96/61/EC of 24 September 1996 concerning integrated pollution prevention and control, *OJ L* 257, 10 October 1996, 26.
Council Directive 96/62/EC of 27 September 1996 on ambient air quality assessment and management, *OJ L* 296, 21 November 1996, 55.

Council and European Parliament

Directive 95/46/EC of the European Parliament and of the Council of 24 October 1995 on the protection of individuals with regard to the processing of personal data and on the free movement of such data, *OJ L* 281, 23 November 1995, 31.
Directive 96/9/EC of the European Parliament and of the Council of 11 March 1996 on the legal protection of databases, *OJ L* 77, 27 March 1996, 20.
Directive 97/33/EC of the European Parliament and of the Council of 30 June 1997 on interconnection in telecommunications with regard to ensuring universal service and interoperability through application of the principles of Open Network Provision (ONP), *OJ L* 199, 26 July 1997, 32.
Directive 98/4/EC of the European Parliament and of the Council of 16 February 1998 amending Directive 93/38/EEC coordinating the procurement procedures of entities operating in the water, energy, transport and telecommunications sectors, *OJ L* 101, 1 April 1998, 1.
Directive 2001/29/EC of the European Parliament and of the Council of 22 May 2001 on the harmonization of certain aspects of copyright and related rights in the information society, *OJ L* 167, 22 June 2001, 10.
Directive 2002/22/EC of the European Parliament and of the Council of 7 March 2002 on universal service and users' rights relating to electronic communications networks and services (Universal Service Directive), *OJ L* 108, 24 April 2002, 51.
Directive 2003/4/EC of the European Parliament and of the Council of 28 January 2003 on public access to environmental information and repealing Council Directive 90/313/EEC, *OJ L* 41, 14 February 2003, 26.
Directive 2003/54/EC of the European Parliament and of the Council of 26 June 2003 concerning common rules for the internal market in electricity and repealing Directive 96/92/EC, *OJ L* 176, 15 July 1998, 37.
Directive 2003/55/EC of the European Parliament and of the Council of 26 June 2003 concerning common rules for the internal market in natural gas and repealing Directive 98/30/EC, *OJ L* 176, 15 July 1998, 57.
Directive 2003/87/EC of the European Parliament and of the Council of 13 October 2003 establishing a scheme for greenhouse gas emission allowance trading within the Community and amending Council Directive 96/61/EC, *OJ L* 275, 25 October 2003, 32.
Directive 2003/98/EC of the European Parliament and of the Council of 17 November 2003 on the re-use of public sector information, *OJ L* 245, 31 December 2003, 90.
Directive 2004/17/EC of the European Parliament and of the Council of 31 March 2004 coordinating the procurement procedures of entities operating in the

water, energy, transport and postal services sectors, *OJ L* 134, 30 April 2004, 1.

Directive 2004/18/EC of the European Parliament and of the Council of 31 March 2004 on the coordination of procedures for the award of public works contracts, public supply contracts and public service contracts, *OJ L* 134, 30 April 2004, 114.

Directive 2006/123/EC of the European Parliament and of the Council of 12 December 2006 on services in the internal market, *OJ L* 376, 27 December 2006, 36.

Directive 2007/2/EC of the European Parliament and of the Council of 14 March 2007 establishing an Infrastructure for Spatial Information in the European Community (INSPIRE), *OJ L* 108, 25 April 2007, 1.

Directive 2008/1/EC of the European Parliament and of the Council of 15 January 2008 concerning integrated pollution prevention and control, *OJ L* 24, 29 January 2008, 8.

Commission

Commission Directive 80/723/EEC of 25 June 1980 on the transparency of financial relations between Member States and public undertakings, *OJ L* 195, 29 July 1980, 35.

Commission Directive 94/46/EC of 13 October 1994 amending Directive 88/301/ EEC and Directive 90/388/EEC in particular with regard to satellite communications, *OJ L* 268, 19 October 1994, 15.

Commission Directive 2006/111/EC of 16 November 2006 on the transparency of financial relations between Member States and public undertakings as well as on financial transparency within certain undertakings, *OJ L* 318, 17 November 2006, 17.

b. Regulations

Parliament and Council

Regulation 1049/2001/EC of the European Parliament and of the Council of 30 May 2001 regarding public access to European Parliament. Council and Commission documents, *OJ L* 145, 31 May 2001, 43.

Regulation 166/2006/EC of the European Parliament and of the Council of 18 January 2006 concerning the establishment of a European Pollutant Release and Transfer Register and amending Council Directives 91/689/EEC and 96/ 61/EC, *OJ L* 33, 4 February 2006, 1.

Regulation 1367/2006/EC of the European Parliament and of the Council of 6 September 2006 on the application of the provisions of the Aarhus Convention on Access to Information. Public Participation in Decision-making and Access to Justice in Environmental Matters to Community institutions and bodies, *OJ L* 264, 25 September 2006, 13.

Commission

Commission Regulation 2008/1205/EC of 3 December 2008 implementing Directive 2007/2/EC of the European Parliament and of the Council as regards metadata, *OJ L* 326, 4 December 2008, 12.
Commission Regulation 800/2008/EC of 6 August 2008 declaring certain categories of aid compatible with the common market in application of Article 87 and 88 of the Treaty (General block exemption Regulation), *OJ L* 214, 9 August 2008, 3.

c. Decisions

Council

Council Decision 93/731/EC of 20 December 1993 on public access to Council documents, *OJ L* 340, 31 December 1993, 43.
Council Decision 2001/840/EC of 29 November 2001 amending the Council's Rules of Procedure, *OJ L* 313, 30 November 2001, 40.

Council and European Parliament

Decision 1600/2002/EC of the European Parliament and of the Council of 22 July 2002 laying down the Sixth Community Environment Action Programme, *OJ L* 242, 10 September 2002, 1.
Decision 280/2004/EC of the European Parliament and of the Council of 11 February 2004 concerning a mechanism for monitoring Community greenhouse gas emissions and for implementing the Kyoto Protocol, *OJ L* 49, 19 February 2004, 1.

European Parliament

European Parliament Decision 97/632/EC, ECSC, Euratom of 10 July 1997 on public access to European Parliament documents, *OJ L* 263, 25 September 1997, 27.
European Parliament Bureau Decision of 28 November 2001 on Public Access to European Parliament Documents, *OJ C* 374, 29 December 2001, 1.

Commission

Commission decision 78/516/EEC of 26 May 1978 relating to a proceeding under Article 85 of the EEC Treaty (IV/29.559 – RAI/UNITEL), *OJ L* 57, 15 June 1978, 39.
Commission Decision 88/518/EEC of 18 July 1988 relating to a proceeding under Article 86 of the EEC Treaty (Case No IV/30.178 Napier Brown – British Sugar), *OJ L* 284, 19 October 1988, 41.

Commission Decision 94/19/EC of 21 December 1993 relating to a proceeding pursuant to Article 86 of the EC Treaty (IV/34.689 – *Sea Containers v. Stena Sealink* – Interim measures), *OJ L* 15, 18 January 1994, 8.
Commission Decision 94/119/EC of 21 December 1993 concerning a refusal to grant access to the facilities of the port of Rødby (Denmark), *OJ L* 55, 26 February 1994, 52.
Commission Decision 94/90/ECSC, EC, Euratom of 8 February 1994 on public access to Commission documents, *OJ L* 46, 18 February 1994, 58.
Commission Decision 1999/329/EC of 12 April 1999 relating to a proceeding pursuant to Articles 85 and 86 of the EC Treaty and Articles 53 and 54 of the EEA Agreement (IV/D-1/30.373 – P & I Clubs, IGA and No IV/D- 1/37.143 – P & I Clubs, Pooling Agreement), 12 April 1999, *OJ L* 125, 19 May 1999, 12.
Commission Decision 2000/479/EC of 17 July 2000 on the implementation of a European pollutant emission register (EPER) according to Article 15 of Council Directive 96/61/EC concerning integrated pollution prevention and control (IPPC), *OJ L* 192, 28 July 2000, 36.
Commission Decision 2001/937/EC, ECSC, Euratom of 5 December 2001 amending its rules of procedure, *OJ L* 345, 29 December 2001, 94.
Commission Decision 2005/842/EC of 28 November 2005 on the application of Article 86(2) of the EC Treaty to State aid in the form of public service compensation granted to certain undertakings entrusted with the operation of services of general economic interest (notified under document number C(2005) 2673), *OJ L* 312, 29 November 2005, 67.
Commission Decision 2006/291/EC, Euratom of 7 April 2006 on the re-use of Commission information, *OJ L* 107, 20 April 2006, 38.
Commission Decision 2008/963/EC of 9 December 2008 amending the Annexes to Directives 2004/17/EC and 2004/18/EC of the European Parliament and of the Council on public procurement procedures, as regards their lists of contracting entities and contracting authorities, *OJ L* 349, 24 December 2008, 1.

Others

Committee of the Regions Decision 64/2003 of 11 February 2003 on public access to Committee of the Regions documents, *OJ L* 160, 28 June 2003, 96.
European Economic and Social Committee Decision 2003/603/EC of 1 July 2003 on public access to European Economic and Social Committee (EESC) documents, *OJ L* 205, 14 August 2003, 19.
European Central Bank Decision ECB/2004/3 of 4 March 2004 on public access to European Central Bank documents, *OJ L* 80, 18 March 2004, 82.

d. Miscellaneous

Council and Commission of the European Communities, Code of conduct 93/730/EC of 6 December 1993 concerning public access to Council and Commission documents, *OJ L* 340, 31 December 1993, 43.

European Parliament and Council, Joint declaration relating to Regulation 1049/2001/EC of the European Parliament and of the Council of 30 May 2001 regarding public access to European Parliament, Council and Commission documents, *OJ L* 173, 27 June 2001, 5.

C. Preparatory Documents

a. **Commission Documents**

Proposals for Directives

Commission of the European Economic Community, Proposal for a Council Directive on the Freedom of Access to Information on the environment, COM (88) 484 final, 28 November 1988.

Commission of the European Communities, Proposal for a Directive of the European Parliament and of the Council on public access to environmental information, COM (2000) 402 final, 29 June 2000.

Commission of the European Communities, Amended proposal for a directive of the European Parliament and of the Council on public access to environmental information, COM (2001) 303 final, 6 June 2001.

Commission of the European Communities, Proposal for a Directive of the European Parliament and of the Council on the re-use and commercial exploitation of public sector documents, COM (2002) 207 final, 5 June 2002.

Commission of the European Communities, Amended proposal for a directive of the European Parliament and of the Council on the re-use and commercial exploitation of public sector information, COM (2003) 119 final, 17 March 2003.

Commission of the European Communities, Proposal for a directive of the European Parliament and of the Council establishing an infrastructure for spatial information in the Community (INSPIRE), (COM) 2004, 516 final, 23 July 2004.

Proposals for Regulations

Commission of the European Communities, Proposal for a Regulation of the European Parliament and of the Council regarding public access to European Parliament, Council and Commission documents, COM (2000) 30 final, 21 February 2000.

Commission of the European Communities, Proposal for a Regulation of the European Parliament and of the Council regarding public access to European Parliament, Council and Commission documents, COM (2008) 229 final, 30 April 2008.

Commission of the European Communities, Proposal for a Regulation of the European Parliament and of the Council on the European Earth observation programme (GMES) and its initial operations (2011–2013), COM (2009) 223 final, 20 May 2009.

Communications

Commission of the European Communities, Communication to the European Parliament pursuant to the second subparagraph of Article 251(2) of the EC Treaty concerning the common position of the Council on the adoption of a Directive of the European Parliament and of the Council on public access to environmental information, SEC (2002) 103 final, 31 January 2002.

Commission of the European Communities, Communication to the European Parliament pursuant to the second subparagraph of Article 251 (2) of the EC Treaty concerning the common position of the Council on the adoption of a directive on the re-use and commercial exploitation of public sector information, SEC (2003) 627 final, 28 May 2003.

Commission of the European Communities, Communication to the European Parliament pursuant to the second subparagraph of Article 251(2) of the EC Treaty concerning the common position of the Council on the adoption of a Directive of the European Parliament and of the Council establishing an infrastructure for spatial information in the Community (INSPIRE), COM (2006) 51 final, 10 February 2006.

Opinions

Commission of the European Communities, Opinion pursuant to Article 251(2), third subparagraph, point (c) of the EC Treaty, on the European Parliament's amendments to the Council's common position regarding the proposal for a Directive of the European Parliament and of the Council on public access to environmental information, COM (2002) 498 final, 5 September 2002.

Commission of the European Communities, Opinion pursuant to Article 251(2), third subparagraph, point (c) of the EC Treaty, on the European Parliament's amendments to the Council's common position regarding the proposal for a directive of the European Parliament and of the Council on the re-use of public sector information, amending the proposal of the Commission pursuant to Article 250(2) of the EC Treaty, COM (2003) 620 final, 15 October 2003.

Commission of the European Communities, Opinion pursuant to Article 251(2) third subparagraph, point (c) of the EC Treaty, on the European Parliament's amendments to the Council Common Position regarding the proposal for a Directive of the European Parliament and of the Council establishing an infrastructure for spatial information in the Community (INSPIRE), COM (2006) 484 final, 13 September 2006.

b. Council

Council

Council of the European Union, Common Position no. 24/2002 adopted on 28 January 2002 with a view to adopting Directive 2002/.../EC of the European

Parliament and of the Council of ... on public access to environmental information and repealing Council Directive 90/313/EEC, OJ C 113E, 14 May 2002, 1.

Council of the European Union, Common position adopted with a view to the adoption of a Directive of the European Parliament and of the Council on public access to environmental information and repealing Council Directive 90/313/EEC. Statement of the Council's Reasons, doc. nr. 11878/01, 28 January 2002, <http://register.consilium.europa.eu/pdf/en/01/st11/st11878-re01ad01.en01.pdf> 22 February 2010.

Council of the European Union, Common position adopted with a view to the adoption of a Directive of the European Parliament and of the Council on the re-use and commercial exploitation of public sector documents. Statement of the Council's reasons, doc. nr. 7946/1/03, 26 May 2003, <http://register.consilium.europa.eu/pdf/en/03/st07/st07946-re01ad01.en03.pdf>, 22 February 2010.

Council of the European Union, Projet de Procès-verbal de la 2536ème session du Conseil de l'Union européenne 'Environnement', tenue à Luxembourg le 27 octobre 2003, doc. nr. 14140/03, 18 November 2003, <http://register.consilium.eu.int/pdf/fr/03/st14/st14140.fr03.pdf>, 22 February 2010.

Council of the European Union, Common Position adopted by the Council with a view to the adoption of a Directive of the European Parliament and of the Council establishing an Infrastructure for Spatial Information in the European Community (INSPIRE). Statement of the Council's reasons, doc. nr. 12064/2/05, 23 January 2006, http://register.consilium.europa.eu/pdf/en/05/st12/st12064-re02ad01.en05.pdf, 22 February 2010.

Council of the European Union, Common Position adopted by the Council on 23 January 2006 with a view to the adoption of a Directive of the European Parliament and of the Council establishing an Infrastructure for Spatial Information in the European Community (INSPIRE), OJ C 126E, 30 May 2006, 16.

Presidency

Presidency of the Council of the European Union, Note to the Working Party on Telecommunications on the Proposal for a Directive of the European Parliament and of the Council on the re-use and commercial exploitation of public sector documents, doc. nr. 12761/02, 7 October 2002, <http://register.consilium.europa.eu/pdf/en/02/st12/st12761.en02.pdf>, 22 February 2010.

Presidency of the Council of the European Union, Note to the Working Party on Telecommunications and Information Society Services on the proposal for a Directive of the European Parliament and of the Council on the re-use and commercial exploitation of public sector documents, doc. nr. 5862/03, 24 January 2003, <http://register.consilium.europa.eu/pdf/en/03/st05/st05862.en03.pdf>, 22 February 2010.

Presidency of the Council of the European Union, Note to the Research Working Party. Proposal for a Regulation of the European Parliament and of the

Council on the European Earth observation programme (GMES) and its initial operations (2011–2013 – key points, doc. nr. 16176/09, 19 November 2009, <http://register.consilium.europa.eu/pdf/en/09/st16/st16176.en09.pdf>, 22 February 2010.

Presidency of the Council of the European Union, Note to the Delegations. Proposal for a Regulation of the European Parliament and of the Council on the European Earth observation programme (GMES) and its initial operations (2011–2013) – Presidency Compromise Text, doc. nr. 5233/10, 15 January 2010, <http://register.consilium.europa.eu/pdf/en/10/st05/st05233.en10.pdf>, 22 February 2010.

Environment Working Party

Environment Working Party, Preparation of the Council Environment on 18–19 December 2000. Proposal for a directive on public access to environmental information, doc.nr 13709/00, 24 November 2000, <http://register.consilium.europa.eu/pdf/en/00/st13/st13709.en00.pdf>, 22 February 2010.

Environment Working Party, Proposal for a Directive on public access to environmental information. Outcome of proceedings, doc. nr. 5451/01, 18 January 2001, <http://register.consilium.europa.eu/pdf/en/01/st05/st05451.en01.pdf>, 22 February 2010.

Working Party on Telecommunications

Working Party on Telecommunications, Proposal for a Directive of the European Parliament and of the Council on the re-use and commercial exploitation of public sector documents. Outcome of proceedings, doc. nr. 11095/02, 31 July 2002, <http://register.consilium.europa.eu/pdf/en/02/st11/st11095.en02.pdf>, 22 February 2010.

Working Party on Telecommunications, Proposal for a Directive of the European Parliament and of the Council on the re-use and commercial exploitation of public sector documents. Outcome of proceedings, doc. nr. 12633/02, 27 September 2002, <http://register.consilium.europa.eu/pdf/en/02/st12/st12633.en02.pdf>, 22 February 2010.

Working Party on Telecommunications, Proposal for a Directive of the European Parliament and of the Council on the re-use and commercial exploitation of public sector documents – examination of Articles 1 to 8. Outcome of proceedings, doc.nr. 13225/02, 10 October 2002, <http://register.consilium.europa.eu/pdf/en/02/st13/st13225.en02.pdf>, 22 February 2010.

Working Party on Telecommunications, Proposal for a Directive of the European Parliament and of the Council on the re-use and commercial exploitation of public sector documents. Outcome of proceedings, doc. nr. 13740/02, 31 October 2002, <http://register.consilium.europa.eu/pdf/en/02/st13/st13740.en02.pdf>, 22 February 2010.

General Secretariat

General Secretariat of the Council of the European Union, Note to the Delegations. Proposal for a directive of the European Parliament and of the Council establishing an infrastructure for spatial information in the Community (INSPIRE), doc. nr. 14454/04, 13 December 2004, <http://register.consilium.europa.eu/pdf/en/04/st14/st14454.en04.pdf>, 22 February 2010.

General Secretariat of the Council of the European Union, Note to the Delegations. Proposal for a directive of the European Parliament and of the Council establishing an infrastructure for spatial information in the Community (INSPIRE), doc. nr. 5648/05, 1 February 2005, <http://register.consilium.europa.eu/pdf/en/05/st05/st05648.en05.pdf>, 22 February 2010.

General Secretariat of the Council of the European Union, Note to the Delegations. Proposal for a directive of the European Parliament and of the Council establishing an infrastructure for spatial information in the Community (INSPIRE), doc. nr. 7488/05, 22 March 2005, <http://register.consilium.europa.eu/pdf/en/05/st07/st07488.en05.pdf>, 22 February 2010.

General Secretariat of the Council of the European Union, Note to the Working Party on the Environment. Proposal for a directive of the European Parliament and of the Council establishing an infrastructure for spatial information in the Community (INSPIRE), doc. nr. 7985/05, 3 May 2005, <http://register.consilium.europa.eu/pdf/en/05/st07/st07985.en05.pdf>, 22 February 2010.

General Secretariat of the Council of the European Union, Note to the Working Party on the Environment. Proposal for a directive of the European Parliament and of the Council establishing an infrastructure for spatial information in the Community (INSPIRE), doc. nr. 8891/05, 13 May 2005, <http://register.consilium.europa.eu/pdf/en/05/st08/st08891-co01.en05.pdf>, 22 February 2010.

General Secretariat of the Council of the European Union, Note to the Permanent Representatives Committee. Proposal for a directive of the European Parliament and of the Council establishing an infrastructure for spatial information in the Community (INSPIRE), doc. nr. 9112/05, 20 May 2005, <http://register.consilium.europa.eu/pdf/en/05/st09/st09112.en05.pdf>, 22 February 2010.

General Secretariat of the Council of the European Union, Report to the Delegations. Proposal for a Regulation of the European Parliament and of the Council regarding public access to European Parliament, Council and Commission documents (recast)., doc. nr. 5671/09, 26 January 2009, <http://register.consilium.europa.eu/pdf/en/09/st05/st05671.en09.pdf>, 22 February 2010.

General Secretariat of the Council of the European Union, Note to the Permanent Representatives Committee/Council. Proposal for a Regulation of the European Parliament and of the Council regarding public access to European Parliament, Council and Commission documents. Outcome of the European

Parliament's First Reading, doc. nr. 7394/09, 17 March 2009, <http://register.consilium.europa.eu/pdf/en/09/st07/st07394.en09.pdf>, 22 February 2010.

General Secretariat of the Council of the European Union, Report the Delegations. Proposal for a Regulation of the European Parliament and of the Council regarding public access to European Parliament, Council and Commission documents (recast), doc. nr. 9234/09, 8 May 2009, <http://register.consilium.europa.eu/pdf/en/09/st09/st09234.en09.pdf>, 22 February 2010.

General Secretariat of the Council of the European Union, Note to the delegations. Proposal for a Regulation of the European Parliament and of the Council regarding public access to European Parliament, Council and Commission documents (recast), doc. nr. 5461/10, 19 January 2010, <http://register.consilium.europa.eu/pdf/en/10/st05/st05461.en10.pdf>, 22 February 2010.

c. European Parliament

Plenary session

European Parliament, Legislative resolution on the proposal for a Directive of the European Parliament and of the Council on public access to environmental information, *OJ C* 343, 5 December 2001, 177.

European Parliament, Legislative resolution on the proposal for a Regulation of the European Parliament and of the Council regarding public access to European Parliament, Council and Commission documents, *OJ C* 27E, 31 January 2002, 19.

European Parliament, Position adopted at second reading on 30 May 2002 with a view to the adoption of European Parliament and Council Directive 2002/.../EC on public access to environmental information and repealing Council Directive 90/313/EEC, *OJ C* 187E, 7 August 2003, 118.

European Parliament, Legislative resolution on the proposal for a European Parliament and Council directive on the re-use and commercial exploitation of public sector documents, *OJ C* 43E, 19 February 2004, 222.

European Parliament, Legislative resolution on the Council common position for adopting a European Parliament and Council directive on the re-use of public sector documents, *OJ C* 77E, 26 March 2004, 314.

European Parliament, Position adopted at first reading on 7 June 2005 with a view to the adoption of European Parliament and Council Directive 2005/.../EC establishing an infrastructure for spatial information in the Community (INSPIRE), *OJ C* 124E, 25 May 2006, 132.

European Parliament, Position adopted at second reading on 13 June 2006 with a view to the adoption of Directive 2006/ .../EC of the European Parliament and of the Council establishing an Infrastructure for Spatial Information in the European Community (INSPIRE), *OJ C* 124E, 25 June 2006, 116.

European Parliament, Resolution with recommendations to the Commission on access to the institutions' texts, *OJ C* 293E, 2 December 2006, 151.

Committee on Citizens' Freedoms and Rights, Justice and Home Affairs

Committee on citizens' freedoms and rights, justice and home affairs, Report on the proposal for a Regulation of the European Parliament and of the Council regarding public access to European Parliament, Council and Commission documents, doc. nr. A5-0318/2000, 27 October 2000, <www.europarl.europa.eu/sides/getDoc.do?pubRef=-//EP//NONSGML+REPORT+A5-2000-0318+0+DOC+PDF+V0//EN&language=EN>, 22 February 2010.

Committee on the Environment, Public Health and Consumer Policy

Committee on the environment, public health and consumer policy, Report on the proposal for a directive of the European Parliament and of the Council on public access to environmental information, doc. nr. A5-0074/2001, 28 February 2001, <www.europarl.europa.eu/sides/getDoc.do?pubRef=-//EP//NONSGML+REPORT+A5-2001-0074+0+DOC+PDF+V0//EN&language=EN>, 22 February 2010.

Committee on the environment, public health and consumer policy, Recommendation for second reading on the Council Common Position for adopting a European Parliament and Council directive on public access to environmental information and repealing Council Directive 90/313/EEC, doc. nr. A5-0136/2002, 24 April 2002, <www.europarl.europa.eu/sides/getDoc.do?pubRef=-//EP//TEXT+REPORT+A5-2002-0136+0+NOT+XML+V0//EN>, 22 February 2010.

Committee on the environment, public health and consumer policy, Report on the proposal for a directive of the European Parliament and of the Council establishing an infrastructure of spatial information in the Community (INSPIRE), doc.nr. A6-0108/2005, 26 April 2005, <www.europarl.europa.eu/sides/getDoc.do?pubRef=-//EP//NONSGML+REPORT+A6-2005-0108+0+DOC+PDF+V0//EN&language=EN>, 22 February 2010.

Committee on the environment, public health and consumer policy, Recommendation for second reading on the Council common position for adopting a directive of the European Parliament and of the Council establishing an Infrastructure for Spatial Information in the European Community (INSPIRE), doc. nr. A6-0081/2006, 2 May 2006, <www.europarl.europa.eu/RegWeb/application/registre/advancedSearch.faces>, 22 February 2010.

Committee on Industry, External Trade, Research and Energy

Committee on industry, external trade, research and energy, Report on the proposal for a European Parliament and Council directive on the re-use and commercial exploitation of public sector documents, doc. nr. A5-0025/2003, 29 January 2003, <www.europarl.europa.eu/sides/getDoc.do?pubRef=-//

EP//NONSGML+REPORT+A5-2003-0025+0+DOC+PDF+V0//EN>, 22 February 2010.

d. Council and European Parliament

Council of the European Union AND Commission of the European Communities, *eEurope 2002: An Information Society For All. Action plan prepared for the Feira European Council 19–20 June 2000*, <http://ec.europa.eu/information_society/eeurope/2002/action_plan/pdf/actionplan_en.pdf>, 22 February 2010.

e. Economic and Social Committee

European Economic and Social Committee, Opinion on Public Sector Information: A key resource for Europe – Green Paper on public sector information in the information society, *OJ C* 169, 16 June 1999, 30.

European Economic and Social Committee, Opinion on the Proposal for a Directive of the European Parliament and of the Council on public access to environmental information, *OJ C* 116, 20 April 2001, 43.

European Economic and Social Committee, Opinion on the Proposal for a Directive of the European Parliament and of the Council on the re-use and commercial exploitation of public sector documents, *OJ C* 85, 8 April 2003, 25.

f. Committee of the Regions

Committee of the Regions, Opinion on the Proposal for a Directive of the European Parliament and the Council on Public Access to Environmental Information, *OJ C* 148, 18 May 2001, 9.

Committee of the Regions, Opinion of 21 November 2002 on the Communication from the Commission to the Council, the European Parliament, the Economic and Social Committee and the Committee of the Regions eEurope 2002: Creating a EU framework for the exploitation of public sector information; and the Proposal for a Directive of the European Parliament and of the Council on the re-use and commercial exploitation of public sector documents, *OJ C* 73, 26 March 2003, 38.

D. Other documents

a. Commission

Communications

Commission of the European Communities, *Communication. Services of General Interest in Europe*, COM (1996) 443 final, 11 September 1996.

Commission of the European Communities, *eEurope: An information society for all. Communication on a Commission Initiative for the Special European Council of Lisbon, 23 and 24 March 2000*, COM (99) 687 final, 8 December 1999.

Commission of the European Communities, *Communication to the Council, the European Parliament, the Economic and Social Committee and the Committee of the Regions. eEurope 2002: creating a EU framework for the exploitation of public sector information*, COM (2001) 607 final, 23 October 2001.

Commission of the European Communities, *Communication to the Council and the European Parliament. Global Monitoring for Environment and Security (GMES) – Outline GMES EC Action Plan (InitialPeriod: 2001–2003*, COM (2001) 609 final, 23 October 2001.

Commission of the European Communities, *Communication to the Council, the European Parliament, the Economic and the Social Committee and the Committee of the Regions. eEurope 2005: An information society for all. An Action Plan to be presented in view of the Sevilla European Council, 21/22 June 2002*, COM (2002) 263 final, 28 May 2002.

Commission of the European Communities, *Communication to the European Parliament and the Council. Global Monitoring for Environment and Security (GMES): Establishing a GMES capacity by 2008 (Action Plan 2004–2008)*, COM (2004) 65 final, 3 February 2004.

Commission of the European Communities, *Communication to the Council and the European Parliament. Global Monitoring for Environment and Security (GMES): From Concept to Reality*, COM (2005) 565 final, 8, 10 November 2005.

Commission of the European Communities, *Communication to the Council, the European Parliament, the European Economic and Social Committee and the Committee of the Regions. Towards a Shared Environmental Information System (SEIS)*, COM (2008) 46 final, 1 February 2008.

Commission of the European Communities, *Communication to the European Parliament, the Council, the European Economic and Social Committee and the Committee of the Regions. Global Monitoring for Environment and Security (GMES): We care for a safer planet*, COM (2008) 748 final, 12 November 2008.

Commission of the European Communities, *Communication to the European Parliament, the Council, the European Economic and Social Committee and the Committee of the Regions. Re-use of Public Sector Information – Review of Directive 2003/98/EC*, COM (2009) 212 final, 7 May 2009.

Commission of the European Communities, *Communication to the European Parliament, the Council, the European Economic and Social Committee and the Committee of the Regions*, COM (2009) 589 final, 28 October 2009.

Green Papers

Commission of the European Communities, Public sector information: a key resource for Europe. Green Paper on public sector information in the information society, COM (1998) 585 final, 20 January 1999.

Commission of the European Communities, Green Paper on services of general interest, COM (2003) 270 final, 21 May 2003.

Commission of the European Communities, Green Paper on European Transparency Initiative, COM (2006) 194 final, 3 May 2006.

Commission of the European Communities, Green Paper on Public Access to Documents Held by Institutions of the European Community, COM (2007) 185 final, 18 April 2007.

Reports

Commission of the European Communities. Report from the Commission to the Council and the European Parliament on the experience gained in the application of Council Directive 90/313/EEC of 7 June 1990 on freedom of access to information on the environment. COM (2000) 400 final, 29 June 2000.

Commission of the European Communities. Publaw. Subject Report: General Access to Information Legislation. Luxemburg, CEC Legal Advisory Board, 1991.

Commission of the European Communities. Publaw 2. A Report to the Commission of the European Communities on an Evaluation of the Implementation of the Commission's Guidelines for Improving the Synergy between the Public and Private Sectors of the Information Market. Luxembourg, CEC Legal Advisory Board, 1993.

Commission of the European Communities. Publaw 3. Final report. Luxembourg, CEC Legal Advisory Board, 1995.

Working Documents

Commission of the European Communities. *Towards a European Union framework for the exploitation of public sector information. Working document*, 21 January 2002, <ftp://ftp.cordis.europa.eu/pub/econtent/docs/consultation_0102.pdf>, 22 February 2010.

Commission of the European Communities. *Commission staff working document. Proposal for a Directive of the European Parliament and of the Council establishing an infrastructure for spatial information in the Community (INSPIRE) – Extended Impact Assessment.* SEC (2004) 980, 23 July 2004.

Commission of the European Communities. *Commission Staff Working Document. Report on the Outcome of the Public Consultation on the Review of Regulation (EC) No. 1049/2001 regarding public access to European Parliament, Council and Commission documents.* SEC (2008) 29/2, 11 January 2008.

Commission of the European Communities. *Staff working document. Accompanying document to the Communication from the Commission to the Council and the European Parliament. Towards a Shared Environmental Information*

System (SEIS) – Executive summary of the impact assessment. SEC (2008) 111, 1 February 2008.

Commission of the European Communities. *Staff working document. Accompanying document to the Communication from the Commission to the European Parliament, the Council, the European Economic and Social Committee and the Committee of the Regions on the re-use of Public Sector Information – Review of Directive 2003/98/EC*. SEC (2009) 597, 7 May 2009.

Miscellaneous

Commission of the European Communities, *Guidelines for improving the synergy between the public and the private sectors in the information market*, 1989, <http://ec.europa.eu/information_society/policy/psi/docs/pdfs/brochure/1989_public_sector_guidelines_en.pdf, 22 February 2010.

Commission of the European Communities, *GI2000: Towards a European Policy Framework for Geographic Information. A Communication from the Commission to the Council, to the European Parliament, to the Economic and Social Committee and to the Committee of the Regions. Draft*, 1997, <www.ec-gis.org/docs/F20680/MICROSOFT%20WORD%20-%20GI2000V10_7_98.PDF>, 22 February 2010.

Commission of the European Communities, *Consultation Paper on a Forthcoming EU Legal Initiative on Spatial Information for Community Policy-making and Implementation*, 2003, <www.ec-gis.org/docs/F6017/INSPIRE-INTERNET-CONSULTATIONPHASEII.PDF>, 22 February 2010.

Commission of the European Communities, State Aid measure N46/2003, *Risk equalisation scheme in the Irish health insurance market – Ireland*, 13 May 2003, *OJ C* 186, 6 August 2003, 16.

Commission of the European Communities, *State Aid Action Plan. Less and better targeted state aid: a roadmap for state aid reform 2005–2009*, COM (2005) 107 final, 7 June 2005.

Commission of the European Communities, *Community framework for State aid in the form of public service compensation*, *OJ L* 297, 29 November 2005, 4.

Commission of the European Communities, State Aid measures N 541/2004 and N 542/2004, *Risk equalisation system and retention of reserves – The Netherlands*, 3 May 2005, *OJ C* 324, 21 December 2005, 30.

Commission of the European Communities, *Results of the MS Consultation on the review of the application of Directive 2003/98/EEC on the re-use of public sector information*, 2008, <http://ec.europa.eu/information_society/policy/psi/docs/consultations/replies/members_states/report_ms_questionnaire.pdf>, 22 February 2010.

Commission of the European Communities, *Results of the online consultation of stakeholders 'Review of the PSI Directive'*, 2008, <http://ec.europa.eu/information_society/policy/psi/docs/pdfs/online_consultation/report_psi_online_consultaion_stakeholders.pdf>, 22 February 2010.

Working Groups and Drafting Teams

INSPIRE Reference Data and Metadata Working Group, *Reference Data and Metadata Position Paper*, 2002, <http://inspire.jrc.ec.europa.eu/reports/position_papers/inspire_rdm_pp_v4_3_en.pdf>, 22 February 2010.

INSPIRE Environmental Thematic Coordination Group, *Environmental Thematic User Needs Position paper*, 2002, <http://inspire.jrc.ec.europa.eu/reports/position_papers/inspire_etc_pp_v2_3_en.pdf>, 22 February 2010.

INSPIRE Data Policy and Legal Issues Working Group, *Position Paper*, 2003, <http://inspire.jrc.ec.europa.eu/index.cfm/pageid/541/downloadid/812>, 22 February 2010.

INSPIRE Implementing Structures and Funding Working Group, *Position paper*, 2003, <http://inspire.jrc.ec.europa.eu/reports/position_papers/inspire_isf_pp_v1_1_en.pdf>, 22 February 2010.

INSPIRE Task Force Scoping, *INSPIRE Scoping Paper*, 2004, <http://inspire.jrc.ec.europa.eu/reports/inspire_scoping24mar04.pdf>, 22 February 2010.

INSPIRE Drafting Team on Data Specifications, *Deliverable D2.3: Definition of Annex Themes and Scope*, 2008, <http://inspire.jrc.ec.europa.eu/reports/ImplementingRules/DataSpecifications/D2.3_Definition_of_Annex_Themes_and_scope_v3.0.pdf>, 22 February 2010.

INSPIRE Drafting Team on Network Services, *INSPIRE Network Services Architecture v. 3.0*, 2008, <http://inspire.jrc.ec.europa.eu/reports/ImplementingRules/network/D3_5_INSPIRE_NS_Architecture_v3-0.pdf>, 22 February 2010.

INSPIRE Drafting Team on Network Services, *D3.9 Draft Implementing Rules for Download Services v. 3.0.*, 2009, <http://inspire.jrc.ec.europa.eu/documents/Network_Services/INSPIRE%20Draft%20Implementing%20Rules%20Download%20Services%20(version%203.0).pdf>, 22 February 2010.

c. Others

Article 29 Data Protection Working Party, *Opinion 7/2003 on the re-use of public sector information and the protection of personal data – Striking the Balance*, 2003, <http://ec.europa.eu/justice_home/fsj/privacy/docs/wpdocs/2003/wp83_en.pdf>, 22 February 2010.

GMES Support Team, *GMES Vision and targets 2008 and Work progress of the Initial Period of the GMES Action Plan (2002–2003)*, 2003, <www.gmes.info/pages-principales/library/reference-documents/?no_cache=1&download=GMES_Vision_and_progress.pdf&did=61>, 22 February 2010.

Andrej Osterman, *Minutes of the 11th Meeting of the Public Sector Information Group*, Luxembourg, 6 May 2008, <http://ec.europa.eu/information_society/policy/psi/docs/pdfs/minutes_psi_group_meetings/11th.pdf>, 22 February 2010.

The President of the European Commission, Margot Wallström, Siim Kallas, Danuta Hübner & Mariann Fischer Boel, *Communication to the Commission proposing the launch of a European Transparency Initiative*, <http://ec.europa.eu/commission_barroso/kallas/doc/etik-communication_en.pdf>, 19 May 2009.

Wachtmeister, Anne-Margrete, *Letter from European Commission Competition Directorate*, <www.primet.org/index.php?option=com_content&task=view&id=125&Itemid=35>, 22 February 2010.

Wallstrom, Margot, Pedro Solbes Mira & Philippe Busquin, *Memorandum of understanding. Infrastructure for Spatial Information in Europe (INSPIRE),* 18 April 2002, <http://inspire.jrc.ec.europa.eu/reports/MoU.pdf>, 22 February 2010.

COUNCIL OF EUROPE

A. Conventions

Council of Europe, *Convention for the Protection of Human Rights and Fundamental Freedom,* 4 November 1950, <http://conventions.coe.int/Treaty/en/Treaties/Html/005.htm>, 22 February 2010.

Council of Europe, *Convention on Civil Liability for Damage Resulting from Activities Dangerous to the Environment,* 21 June 1993, <http://conventions.coe.int/Treaty/EN/Treaties/Html/150.htm>, 22 February 2010.

Council of Europe, *Convention on Access to Official Documents*, 27 November 2008, <https://wcd.coe.int/ViewDoc.jsp?id=1377737&Site=CM>, 22 February 2010.

B. Committee of Ministers

Council of Europe Committee of Ministers, *Recommendation No. R (81) 19 to Member States on the Access to Information held by public authorities*, 25 November 1981, <www.coe.int/t/e/legal_affairs/legal_co-operation/administrative_law_and_justice/Texts_&_Documents/Conv_Rec_Res/Recommendation(81)19.asp>, 22 February 2010.

Council of Europe Committee of Ministers, *Declaration on the freedom of expression and information*, 29 April 1982, <https://wcd.coe.int/com.instranet.InstraServlet?command=com.instranet.CmdBlobGet&InstranetImage=601273&SecMode=1&DocId=675536&Usage=2>, 22 February 2010.

Council of Europe Committee of Ministers, *Declaration on a European policy for new information technologies*, 7 May 1999, <https://wcd.coe.int/ViewDoc.jsp?id=448133>, 22 February 2010.

Council of Europe Committee of Ministers, *Recommendation Rec(2002)2 on access to official documents*, 21 February 2002, <www.coe.int/T/E/Human_rights/rec(2002)2_eng.pdf>, 22 February 2010.

Council of Europe Committee of Ministers, *Recommendation CM/Rec(2007)11 on promoting freedom of expression and information in the new information and communications environment*, 26 September 2007, <https://wcd.coe.int/ViewDoc.jsp?id=1188541>, 22 February 2010.

Council of Europe Committee of Ministers, *Recommendation CM/Rec(2007)16 on measures to promote the public service value of the Internet*, 7 November 2007, <https://wcd.coe.int/ViewDoc.jsp?id=1207291>, 22 February 2010.

C. Parliamentary Assembly

Council of Europe Parliamentary Assembly, *Recommendation 582 (1970) on mass communication media and human rights*, 23 January 1970, <http://assembly.coe.int/main.asp?Link=/documents/adoptedtext/ta70/erec582.htm>, 22 February 2010.

Council of Europe Parliamentary Assembly, *Recommendation 854 (1979) on access by the public to government records and freedom of information*, 1 February 1979, <http://assembly.coe.int/main.asp?Link=/documents/adoptedtext/ta79/erec854.htm>, 22 February 2010.

D. Others

Steering Committee for Human Rights, *Draft Convention of the Council of Europe on Access to Official Documents and Draft Explanatory Report*, Doc. nr. CDDH(2008)008 Add. III Bil., 3 April 2008, 4, <http://gyla.ge/foi/doc_upload/capmaign/3.%20evropis%20sabchos%20konvenciis%20proeqti%20inglisur%20enaze.pdf>, 22 February 2010.

Committtee on Legal Affairs and Human Rights of the Parliamentary Assembly, *Draft Council of Europe Convention on Access to Official Documents. Report*, 12 September 2008, <http://assembly.coe.int/Main.asp?link=/Documents/WorkingDocs/Doc08/EDOC11698.pdf>, 22 February 2010.

ORGANIZATION FOR ECONOMIC COOPERATION AND DEVELOPMENT

A. Declarations

OECD, *The Seoul Declaration for the future of the internet economy*, 2008, <www.oecd.org/dataoecd/49/28/40839436.pdf>, 22 February 2010.

B. Recommendations

OECD, *Recommendation of the Council on Environmental Information*, 3 April 1998, <http://webdomino1.oecd.org/horizontal/oecdacts.nsf/linkto/C(98)67>, 22 February 2010.

OECD, *Recommendation C(2008)36 of the Council for Enhanced Access and More Effective Use of Public Sector Information*, 30 April 2008, <www.oecd.org/dataoecd/0/27/40826024.pdf>, 22 February 2010.

C. Others

OECD Working Party on the Information Economy, *Digital Broadband Content: Public Sector Information and Content*, 2006, 82 p., <www.oecd.org/dataoecd/10/22/36481524.pdf>, 22 February 2010.

OECD Working Party on the Information Economy, *OECD Workshop on Public Sector Information held on 31 May 2006: Summary*, <www.oecd.org/dataoecd/54/35/36854745.pdf>, 22 February 2010.

OTHER INTERNATIONAL BODIES

OSCE Office of the Representative on Freedom of the Media, Comments on the draft European Convention on access to official documents of the Council of Europe, as prepared by the Group of Specialists on Access to Official Documents, 11 September 2007, <www.access-info.org/data/File/OSCE%20 Comments%20to%20 CoE%20access%20convention%2011SEP07.doc>, 22 February 2010.

UN Special Rapporteur on Freedom of Opinion and Expression, OSCE Representative on Freedom of the Media and OAS Special Rapporteur on Freedom of Expression, *International Mechanisms for Promoting Freedom of Expression. Joint Declaration*, 26 November 1999, <http://portal.unesco.org/ci/en/files/26152/12047264643UNFEX1999.pdf/UNFEX1999.pdf>, 22 February 2010.

UN Special Rapporteur on Freedom of Opinion and Expression, OSCE Representative on Freedom of the Media and OAS Special Rapporteur on Freedom of Expression, *International Mechanisms for Promoting Freedom of Expression. Joint Declaration*, 6 December 2004, <www.osce.org/item/3957.html>, 22 February 2010.

NATIONAL DOCUMENTS

A. France

Circulaire du 14 février 1994 relative à la diffusion des données publiques, *Journal Officiel* 19 February 1994, 2864, <http://admi.net/jo/PRMG9400081C.html>,25 May 2009.

B. Germany

Bundesministerium fur Umwelt, Naturschutz und Reaktorsicherheit, Bericht der Bundesrepublik Deutschland über die bei der Anwendung der Richtlinie

2003/4/EG über den Zugang der Offentlichkeit zu Umweltinformationen gewonnenen Erfahrungen, s.d., <www.bmu.de/umweltinformation/downloads/doc/45109.php>, 22 February 2010.

C. Greece

Ministry of the Environment, Energy and Climate Change, *National Report of Greece on Directive 2003/4/EC on Public Access to environmental information*, 2009, <www.minenv.gr/4/41/00/National%20Report%20on%20Directive%202003.pdf>, 22 February 2010.

D. Ireland

Department of Environment, Heritage and Local Government, *Review of Implementation of EU Directive 2003/4/EC on Public Access to Information on the Environment. Report by Ireland*, 2009, <www.environ.ie/en/AboutUs/AccesstoInformationontheEnvironment/PublicationsDocuments/FileDownLoad,22174,en.doc>, 22 February 2010.

E. The Netherlands

a. Legislation and Preparatory Documents

Wet van 1 november 2001 houdende regeling van de taken voor de meteorologie en andere geofysische terreinen (Wet op het Koninklijk Nederlands Meteorologisch Instituut), <http://wetten.overheid.nl/BWBR0012952/>, 22 February 2010.

Minister of Economics and Minister of Justice, *Voorstel van Wet – Regels omtrent marktactiviteiten van overheidsorganisaties en omtrent ondernemingen die van overheidswege over een bijzondere positie beschikken (Wet markt en overheid)*, Kamerdocumenten 2001–2002, 28 050, nr. 1–2, <www.ez.nl/dsresource?objectid=148231&type=PDF>, 22 February 2010.

Laurens J. Brinkhorst, *Intrekking voorstel van wet Markt en Overheid*, 9 April 2004, <www.ez.nl/dsresource?objectid=47703&type=PDF>, 22 February 2010.

b. Others

Ministry of the Interior, *Naar toegankelijkheid van overheidsinformatie*, 1997, <www.minbzk.nl/aspx/download.aspx?file=/contents/pages/302/nota_ntvo_nl.pdf>, 22 February 2010.

Ministry of Transport and Water Management, *Reply to questions from Members of Parliament Roemer and Roefs*, 16 March 2007, question no. 2060707530, <http://parlando.sdu.nl/cgi/login/anonymous>, 22 February 2010.

Vaste Commissie voor Verkeer en Waterstaat, *Verslag van een algemeen overleg*, 30 January 2008, 31 200 XII, no. 73, <http://parlando.sdu.nl/cgi/login/anonymous>, 22 February 2010.

F. SWEDEN

Lantmateriet, *Landmateriet's Strategic Plan 2006–2010*, 2006, <www.lantmateriet.se/upload/filer/om_lantmateriet/strategiskaplanen_engelska.pdf>, 22 February 2010.

Statskontoret, *Competition at the Public/Private Interface*, 2005, <www.statskontoret.se/upload/Publikationer/2005/200519A.pdf>, 22 February 2010.

G. UNITED KINGDOM

a. Legislation

Government Trading Funds Act 1973, 25 October 1973, amended by Government Trading Funds Act 1990.

b. Others

Advisory Panel for Public Sector Information

Advisory Panel on Public Sector Information, *The need for a government strategy on the re-use of public sector information*, 2007, <www.appsi.gov.uk/ministers/gov-strategy-psi.pdf>, 22 February 2010.

Advisory Panel on Public Sector Information, *Meeting minutes. 11 June 2007*, 2007, <www.appsi.gov.uk/minutes/2007-06-11.pdf>, 22 February 2010.

Advisory Panel on Public Sector Information, *EU Member States Consultation: Review of the EC Directive on the re-use of public sector information. Answers to questions submitted to Member States [and more generally]*, 2008, <www.appsi.gov.uk/reports/review-psi-directive-appsi-response.pdf>, 22 February 2010.

Advisory Panel on Public Sector Information, *Response to the Consultation Paper on the use and sharing of personal information in the public and private sectors*, 2008, <www.justice.gov.uk/docs/Advisory-Panel-on-Public-Sector-Information-Data-Sharing-Response.pdf>, 22 February 2010.

Advisory Panel on Public Sector Information, *The impact and success of legislation on the re-use of public sector information*, 2008, <www.appsi.gov.uk/reports/ec-submission-07-2008.doc>, 22 February 2010.

Advisory Panel on Public Sector Information, *Response to the consultation on the Power of Information Report*, 2009, <www.appsi.gov.uk/2009/02/16/APPSIsResponseToTheDraftPowerOfInformationTaskForceReport>, 22 February 2010.

Communities and Local Government Committee

Communities and Local Government Committee, *DCLG Annual Report 2006*, Third Report of Session 2006–07, HC 106, <www.publications.parliament.uk/pa/cm200607/cmselect/cmcomloc/106/106.pdf>, 22 February 2010.

Communities and Local Government Committee, *Ordnance Survey*, Fifth Report of Session 2007–08, HC 268, <www.publications.parliament.uk/pa/cm200708/cmselect/cmcomloc/268/26802.htm>, 22 February 2010.

Communities and Local Government Committee, *Ordnance Survey: Government response to the Committee's Fifth Report*, First Special Report of Session 2007–08, HC 516, <www.parliament.the-stationery-office.com/pa/cm200708/cmselect/cmcomloc/516/516.pdf>, 22 February 2010.

Communities and Local Government Committee, *Policy options for geographic information from Ordnance Survey. Consultation*, 2009, 35, <www.communities.gov.uk/publications/corporate/ordnancesurveyconsultation>, 22 February 2010.

HM Treasury

HM Treasury, *Cross Cutting Review of the Knowledge Economy. Review of Government Information*, 2000, <www.hm-treasury.gov.uk/spend_sr00_ccr.htm>, 22 February 2010.

HM Treasury, *Charges for information: when and how. Guidance for government departments and other Crown bodies*, 2001, <www.hm-treasury.gov.uk/media/7/C/charging_for_info.pdf>, 22 February 2010.

HM Treasury, *Stability and opportunity: building a strong, sustainable future. Budget 2008*, 2008, <www.hm-treasury.gov.uk/d/bud08_completereport.pdf>, 22 February 2010.

Office of Fair Trading

Office of Fair Trading, *Competition Act 1998. Competition case closure summaries 2004*, 2004, <www.oft.gov.uk/shared_oft/ca98_case_closures/2004.pdf>, 22 February 2010.

Office of Fair Trading, *The commercial use of public information (CUPI)*, 2006, <www.oft.gov.uk/shared_oft/reports/consumer_protection/oft861.pdf>, 22 February 2010.

Office of Public Sector Information

Office of Public Sector Information, *The Re-use of Public Sector Information: A Guide to the Regulations and Best Practice*, 2005, <www.opsi.gov.uk/advice/psi-regulations/advice-and-guidance/guide-to-psi-regulations-and-best-practice.doc, 22 February 2010.

Office of Public Sector Information, *The United Kingdom Implementation of the European Directive on the Re-use of Public Sector Information – the first two years*, 2007, <www.opsi.gov.uk/advice/psi-regulations/uk-implementation-first-years.pdf, 22 February 2010.

Office of Public Sector Information, *PSI Guidance – Link between Access and Re-use*, 2008, <www.opsi.gov.uk/advice/psi-regulations/advice-and-guidance/psi-guidance-notes/links-between-access-and-re-use.htm, 22 February 2010.

Office of Public Sector Information, *Unlocking PSI potential. The United Kingdom Report on the Re-use of Public Sector Information 2008*, 2008, <www.opsi.gov.uk/advice/psi-regulations/uk-report-re-use-psi-2008.pdf>, 22 February 2010.

Ordnance Survey

Ordnance Survey, *Framework document July 2004*, 2004, <www.ordnancesurvey.co.uk/oswebsite/aboutus/reports/frameworkdocument/docs/frameworkdocument2004.pdf>, 22 February 2010.

Ordnance Survey, *Annual Report and Accounts 2006–07*, 2007, <www.ordnancesurvey.co.uk/oswebsite/aboutus/reports/annualreport/06–07/index.html>, 22 February 2010.

Ordnance Survey, *Supplementary Memorandum to the Communities and Local Government Committee Fifth Report of Session 2007–08, HC 268*, 2007, <www.publications.parliament.uk/pa/cm200708/cmselect/cmcomloc/268/268we20.htm>, 22 February 2010.

Miscellaneous

Advisory Panel on Crown Copyright, *Aligning the FOI and PSI initiatives in the UK*, 2004, <www.appsi.gov.uk/reports/aligning-FOI-and-PSI.pdf>, 22 February 2010.

Cabinet Office, *The Government's Response to The Power of Information: An independent review by Ed Mayo and Tom Steinberg (2007)*, 2007, <www.cabinetoffice.gov.uk/media/cabinetoffice/strategy/assets/power_information_response.pdf>, 22 February 2010.

Cabinet Office and Department of Trade and Industry, *Consultation document – Proposal for Regulations implementing a Directive of the European Parliament and Council on the re-use of public sector information*, 2005, <www.opsi.gov.uk/advice/psi-regulations/consultations/regulation/consultation-document-2004-12.pdf>, 22 February 2010.

Communities and Local Government, *The Future of the National Interest Mapping Services Agreement Beyond 2006*, 2006, <www.communities.gov.uk/documents/planningandbuilding/pdf/154073.pdf>, 22 February 2010.

Department for Environment, Food and Rural Affairs, *Public access to environmental information. Experience gained in the application of Directive 2003/4/*

EC, 2009, <www.defra.gov.uk/corporate/policy/opengov/eir/pdf/commission-report-uk.pdf>, 22 February 2010.

Her Majesty's Stationery Office, *Future Management of Crown Copyright*, 1999, <www.opsi.gov.uk/advice/crown-copyright/future-management-of-crown-copyright.pdf>, 22 February 2010.

Her Majesty's Stationery Office, *Consultation Document. Proposal for Regulations implementing a Directive of the European Parliament and Council on the re-use of public sector information*, 2004, <www.berr.gov.uk/files/file16304.doc>, 22 February 2010

Joint Information Systems Committee and Office of Public Sector Information, *The Power of Information Review and Unlocking UK Government Information – A Consultative Forum*, 2008, <www.opsi.gov.uk/advice/poi/poi-consultative-forum.pdf>, 22 February 2010.

Power of Information Taskforce, *Power of Information Taskforce Report. February 2009*, 14–16, <www.cabinetoffice.gov.uk/reports/power_of_information.aspx>, 22 February 2010.

Rhind, David, *Letter to Michael Wills MP, Minister of Justice – APPSI's recommendations to progress the re-use of public sector information*, 21 July 2008, <www.epsiplus.net/content/download/14491/181308/file/minister-letter-ec-submission-07-2008.pdf>, 22 February 2010.

Secretary of State for Communities and Local Government, *Communities in control: real people, real power*, 2008, <www.communities.gov.uk/documents/communities/pdf/886045.pdf>, 22 February 2010.

The National Archives, *A new Vision for The National Archives 2006–2011*, 2006, <www.nationalarchives.gov.uk/documents/vision.pdf>, 22 February 2010.

United Kingdom Marine Information Council Working Group on Data Access, *Achieving optimal value from publicly funded marine information resources*, 2003, <www.infomarine.org/UKMIC-achieving_optimal_value.pdf>, 22 February 2010.

H. United States

Clinton, William J., Executive Order 12906. Co-ordinating Geographic Data Acquisition and Access: The National Spatial Data Infrastructure, 1994, <http://govinfo.library.unt.edu/npr/library/direct/orders/20fa.html>, 22 February 2010.

Table of Cases

EUROPEAN COURT OF JUSTICE (ECJ)

All cases can be found at <http://curia.europa.eu/jcms/jcms/j_6/home>.

ECJ, Case 10/71, *Ministère public luxembourgeois v. Madeleine Muller and others*, 14 July 1971, *European Court Reports* 1971, 723.

ECJ, Case 6/72, *Europemballage Corporation and Continental Can Company Inc. v. Commission*, 21 February 1973, *European Court Reports* 1973, 215.

ECJ, Case 6/73 and 7/73, *Istituto Chemioterapico Italiano S.p.A. and Commercial Solvents Corporation v. Commission*, 6 March 1974, *European Court Reports* 1974, 223.

ECJ, Case 155/73, *Sacchi*, 30 April 1974, *European Court Reports* 1974, 409.

ECJ, Case 173/73, *Italian Republic v. Commission*, 2 July 1974, *European Court Reports* 1974, 709.

ECJ, Case 26/75, *General Motors Continental NV v. Commission*, 13 November 1975, *European Court Reports* 1975, 1367.

ECJ, Case 27/76, *United Brands Company and United Brands Continentaal BV v. Commission*, 14 February 1978, *European Court Reports* 1978, 207.

ECJ, Case 85/76, *Hoffman-La Roche & Co. AG v Commission*, 13 February 1979, *European Court Reports* 1979, 461.

ECJ, Case 730/79, *Philip Morris Holland BV v. Commission*, 17 September 1980, *European Court reports* 1980, 2671.

ECJ, Cases 188/80, 189/80 and 190/80, *French Republic, Italian Republic and United Kingdom of Great Britain and Northern Ireland v. Commission of the European Communities*, 6 July 1982, *European Court Reports* 1982, 2545.

ECJ, Case 322/81, *NV Nederlandsche Banden Industrie Michelin v. Commission*, 9 November 1983, *European Court Reports* 1983, 3461.

ECJ, Case 323/82, *SA Intermills v. Commission*, 14 November 1984, *European Court Reports* 1984, 3809.
ECJ, Case 290/83, *Commission of the European Communities v. French Republic*, 30 January 1985, *European Court Reports* 1985, 439.
ECJ, Case 118/85, *Commission v. Italian Republic*, 16 June 1987, *European Court Reports* 1987, 2599.
ECJ, Case 310/85, *Deufil GmbH & Co. KG v. Commission*, 24 February 1987, *European Court Reports* 1987, 901.
ECJ, Case 30/87, *Corinne Bodson v. Pompes Funèbres des Régions Libérées SA*, 4 May 1988, *European Court Reports* 1988, 2479.
ECJ, Case C-31/87, *Gebroeders Beentjes v. The Netherlands*, 20 September 1988, *European Court Reports* 1988, 4635.
ECJ, Case 66/86, *Ahmed Saeed Flugreisen and Silver Line Reisebüro GmbH v. Zentrale zur Bekämpfung unlauteren Wettbewerbs e.V.*, 11 April 1989, *European Court Reports* 1989, 803.
ECJ, Case 202/88, *French Republic v. Commission*, 19 March 1991, *European Court Reports* 1991, 1223.
ECJ, Case C-41/90, *Höfner & Elser v. Macotron GmbH*, 23 April 1991, *European Court Reports* 1991, I-1979.
ECJ, Case C-260/89, *Elliniki Radiophonia Tiléorassi AE and others v. Dimotiki Etairia Pliroforissis and others*, 18 June 1991, *European Court Reports* 1991, I-2925.
ECJ, Case C-179/90, *Merci convenzionali porto di Genova SpA v. Siderurgica Gabrielli SpA.*, 10 December 1991, *European Court Reports* 1991, I-5889.
ECJ, Case 18/88, *Régie des télégraphes et des téléphones v. GB-Inno-BM SA.*, 13 December 1991, *European Court Reports* 1991, I-5941.
ECJ, Case C-271/90, *Kingdom of Spain, Kingdom of Belgium and Italian Republic v. Commission*, 17 November 1992, *European Court Reports* 1992, I-5833.
ECJ, Case C-159/91 and C-160/91, *Poucet et Pistre v. Assurances Générales de France and and Caisse Mutuelle Régionale du Languedoc-Roussillon*, 17 February 1993, *European Court Reports* 1993, I-637.
ECJ, Case C-320/91, *Criminal proceedings against Paul Corbeau*, 19 May 1993, *European Court Reports* 1993, I-2533.
ECJ, Case C-69/91, *Criminal proceedings against Francine Gillon, née Decoster*, 27 October 1993, *European Court Reports* 1993, I-5335.
ECJ, Case C-278/92, *Kingdom of Spain v. Commission*, 14 September 1992, *European Court Reports* 1994, I-4103.
ECJ, Case C-364/92, *SAT Fluggesellschaft mbH v. Eurocontrol*, 19 January 1994, *European Court Reports* 1994, I-43.
ECJ, Case C-387/92, *Banco de Crédito Industrial SA v. Ayuntamiento de Valencia*, 15 March 1994, *European Court reports* 1994, I-877.
ECJ, Case C-393/92, *Municipality of Almelo and others v. NV Energiebedrijf Ijsselmij*, 27 April 1994, *European Court Reports* 1994, I-1477.

ECJ, Case C-323/93, *Société Civile Agricole du Centre d'Insémination de la Crespelle v. Coopérative d'Elevage et d'Insémination Artificielle du Département de la Mayenne*, 5 October 1994, *European Court Reports* 1994, I-5077.

ECJ, Cases C-241/91P and C-242/91P, *Radio Telefis Eireann v. Commission*, 6 April 1995, *European Court Reports* 1995, I-743.

ECJ, Cases C-140/94, C-141/94 and C-142/94, *IP SpA v. Comune di Bassano del Grappa v. Comune di Chiogga*, 17 October 1995, *European Court reports* 1995, I-3257.

ECJ, Case C-58/95, *Kingdom of the Netherlands v. Council of the European Union*, 20 April 1996, *European Court Reports* 1996, I-2169.

ECJ, Case C-333/94 P, *Tetra Pak International SA v. Commission*, 14 November 1996, *European Court Reports* 1996, I-5951.

ECJ, C-343/95, *Cali & Figli Srl v. Servizi ecologici porto di Genova SpA*, 18 March 1997, *European Court Reports* 1997, I-1547.

ECJ, Case C-242/95, *GT-Link A/S v. De Danske Statsbaner*, 17 July 1997, *European Court Reports* 1997, I-4449.

ECJ, Case C-157/94, *Commission v. Kingdom of the Netherlands*, 23 October 1997, *European Court Reports* 1997, I-5699.

ECJ, Case C-55/96, *Job Centre Coop. ARL*, 11 December 1997, *European Court Reports* 1997, I-7119.

ECJ, Case C-44/96, *Mannesmann Anlagenbau Austria AG and Others v. Strohal Rotationsdruck GesmbH*, 15 January 1998, *European Court Reports* 1998, I-73.

ECJ, Case C-163/96, *Criminal proceedings against Silvano Raso and others*, 12 February 1998, *European Court Reports* 1998, I-533.

ECJ, Case C-321/96, *Mecklenburg v. Kreis Pinneberg der Landrat*, 17 June 1998, *European Court Reports* 1998, I-3809.

ECJ, Case C-35/96, *Commission v. Italian Republic*, 18 June 1998, *European Court Reports* 1998, I-3851.

ECJ, C-266/96, *Corsica Ferries France SA v. Gruppo Antichi Ormeggiatori del porto di Genova Coop. arl*, 18 June 1998, *European Court Reports* 1998, I-3949.

ECJ, Case, C-203/96, *Chemische Afvalstoffen Dusseldorp BV and Others v. Minister van Volkshuisvesting, Ruimtelijke Ordening en Milieubeheer*, 25 June 1998, *European Court Reports* 1998, I-4075.

ECJ, Case C-323/96, *Commission v. Kingdom of Belgium*, 17 September 1998, *European Court Reports* 1998, I-5063.

ECJ, Case C-360/96, *Gemeente Arnhem and Gemeente Rheden v. BFI Holding BV*, 10 November 1998, *European Court Reports* 1998, I-6821.

ECJ, Case C-7/97, *Oscar Bronner GmbH & Co. KG v. Mediaprint Zeitungs- und Zeitschriftenverlag GmbH & Co. KG and others*, 26 November 1998, *European Court Reports* 1998, I-7791.

ECJ, Case C-306/97, *Connemara Machine Turf Co. Ltd v. Coillte Teoranta*, 17 December 1998, *European Court Reports* 1998, I-8761.

ECJ, Case C-353/96, *Commission v. Ireland*, 17 December 1998, *European Court Reports* 1998, I-8565.
ECJ, Case C-217/97, *Commission of the European Communities v. Federal Republic of Germany*, 9 September 1999, *European Court Reports* 1999, I-5087.
ECJ, Case C-251/97, *French Republic v. Commission*, 5 October 1999, *European Court Reports* 1999, I-663.
ECJ, Case C-147/97 and C-148/97, *Deutsche Post AG v. Gesellschaft für Zahlungssysteme mbH GZS and Citicorp Kartenservice GmbH*, 10 February 2000, *European Court Reports* 2000, I-825.
ECJ, Case C-209/98, *Sydhavnens Sten & Grus ApS v. Københavns Kommune*, 23 May 2000, *European Court Reports* 2000, I-3743.
ECJ, Case C-180/98 to C-184/98, *Pavlov v. Stichting Pensioenfonds Medische Specialisten*, 12 September 2000, *European Court Reports* 2000, I-6451.
ECJ, Case C-380/98, *The Queen v. HM Treasury Ex p. the University of Cambridge*, 3 October 2000, *European Court Reports* 2000, I-8035.
ECJ, Case C-480/98, *Kingdom of Spain v. Commission*, 12 October 2000, *European Court Reports* 2000, I-8717.
ECJ, Case C-237/99, *Commission v. French Republic*, 1 February 2001, *European Court Reports* 2001, I-939.
ECJ, Case C-163/99, *Portuguese Republic v. Commission*, 29 March 2001, *European Court Reports* 2001, I-2613.
ECJ, Case 223/99, *Agorà Srl and Excelsior Snc di Pedrotti Bruna & C. V. Ente Autonomo Fiera Internazionale di Milano and Ciftat Soc. coop. arl.*, 10 May 2001, *European Court Reports* 2001, I-3605.
ECJ, Case C-340/99, *TNT Traco SpA v. Poste Italiane SpA*, 17 May 2001, *European Court Reports* 2001, I-4109.
ECJ, Case C-475/99, *Ambulanz Glöckner v. Landkreis Südwestpflatz*, 25 October 2001, *European Court Reports* 2001, I-8089.
ECJ, Case C-353/99, *Council of the European Union v. Hautala*, 6 December 2001, *European Court Reports* 2001, I-9594.
ECJ, Case C-309/99, *Wouters and others*, 19 February 2002, *European Court Reports* 2002, I-1577.
ECJ, Case C-482/99, *French Republic v. Commission*, 16 May 2002, *European Court Reports* 2002, I-4397.
ECJ, C-82/01P, *Aéroports de Paris v. Commission*, 24 October 2002, *European Court Reports* 2002, I-9297.
ECJ, Case C-470/99, *Universale-Bau et al.*, 12 December 2002, *European Court Reports* 2002, I-11617.
ECJ, Case C-373/00, *Adolf Truley GmbH v. Bestattung Wien GmbH*, 27 February 2003, *European Court Reports* 2003, I-1931.
ECJ, Case C-214/00, *Commission v. the Kingdom of Spain*, 15 May 2003, *European Court Reports* 2003, I-4667.
ECJ, Case C-355/00, *Freskot AE v. Elliniko Dimosio*, 22 May 2003, *European Court Reports* 2003, I-5263.

ECJ, Case C-316/01, *Eva Glawischnig v. Bundesminister für soziale Sicherheit und Generationen*, 12 June 2003, *European Court Reports* 2003, I-5995.
ECJ, Case C-233/00, *Commission v. French Republic*, 26 June 2003, *European Court Reports* 2003, I-6625.
ECJ, Case C-280/00, *Altmark Trans and Regierungspräsidium Magdeburg*, 24 July 2003, *European Court Reports* 2003, I-7747.
ECJ, Case C-207/01, *Altair Chimica SpA v. ENEL Distribuzione SpA*, 11 September 2003, *European Court Reports* 2003, I-8875.
ECJ, Case C-283/00, *Commission v. Kingdom of Spain*, 16 October 2003, *European Court Reports* 2003, I-11697.
ECJ, Case C-34/01, *Enirisorse SpA v. Ministero delle Finanze*, 27 November 2003, *European Court Reports* 2003, I-14243.
ECJ, Case C-418/01, *IMS Health GmbH & Co. OHG*, 29 April 2004, *European Court Reports* 2004, I-5039.
ECJ, Cases C-264/01, C-306/01, C-354/01 and C-355/01, *AOK Bundesverband and others v. Ichthyol-Gesellschaft Cordes, Hermani & Co and others*, 16 March 2005, *European Court Reports* 2004, I-2493.
ECJ, Case C-451/03, *Servizi Ausiliari Dottori Commercialisti Srl v. Giuseppe Calafiori*, 30 March 2006, *European Court Reports* 2006, I-2941.
ECJ, Case C-205/03 P, *FENIN v. Commission*, 11 July 2006, *European Court Reports* 2006, I-6295.
ECJ, Case C-552/07, *Commune de Sausheim v. Pierre Azelvandre*, 17 February 2009 (not yet published in European Court Reports).
ECJ, C-202/07P, *France Télécom SA v. Commission*, 2 April 2009 (not yet published in European Court Reports).

COURT OF FIRST INSTANCE (CFI)

All cases can be found at <http://curia.europa.eu/jcms/jcms/j_6/home>.

CFI, Case T-83/91, *Tetra Pak International SA v. Commission*, 6 October 1994, *European Court Reports* 1994, II-755.
CFI, Case T-111/96, *ITT Promedia NV v. Commission*, 17 July 1998, *European Court Reports* 1998, II-2937.
CFI, Cases T-374/94, T-375/94, T-384/94 and T-388/94, *European Night Services and others v. Commission*, 15 September 1998, *European Court Reports* 1998, II-3141.
CFI, Case T-106/95, *Fédération française des sociétés d'assurances and others v. Commission*, 27 February 1997, *European Court Reports* 1997, II-229.
CFI, Cases T-204/97 and T-270/97, *EPAC v. Commission*, 13 June 2000, *European Court Reports* 2000, II-2267.
CFI, Case T-41/96, *Bayer AG v Commission*, 26 October 2000, *European Court Reports* 2000, II-3383.

CFI, Case T-5/97, *Industrie des poudres sphériques v. Commission*, 30 November 2000, *European Court Reports* 2000, II-3755.
CFI, Cases T-197/97 and 198/97, *Weyl Beef Products BV and others v. Commission*, 31 January 2001, *European Court Reports* 2001, II-303.
CFI, Case T-76/02, *Mara Messina v. Commission of the European Communities*, 17 September 2003, *European Court Reports* 2003, II-3203.
CFI, Case T-203/01, *Manufacture française des pneumatiques Michelin v. Commission*, 30 September 2003, *European Court Reports* 2003, II-4071.
CFI, Case T-168/02, *IFAW Internationaler Tierschutz-Fonds v. Commission of the European Communities*, 30 November 2004, *European Court Reports* 2004, II-4135.
CFI, T-187/03, *Isabella Scippacercola v. Commission of the European Communities*, 17 March 2005, *European Court Reports* 2005, II-1063.
CFI, Case T-17/02, *Fred Olsen v. Commission*, 15 June 2005, *European Court Reports* 2005, II-2031.
CFI, Case T-236/06, *Landtag Schleswig-Holstein v. Commission*, 1 September 2006, *European Court Reports* 2008, II-461.
CFI, Case T-155/05, *SELEX Sistemi Integrati v. Commission*, 12 December 2006, *European Court Reports* 2006, II-4797.
CFI, Cases T-217/03 and T-245/03, *Fédération nationale de la coopération bétail and viande v. Commission*, 13 December 2006, *European Court Reports* 2006, II-4987.
CFI, Case T-340/3, *France Télécom SA v. Commission*, 30 January 2007, *European Court Reports* 2007, II-107.
CFI, Case T-289/03, *BUPA v. Commission*, 12 February 2008, *European Court Reports* 2008, II-281.
CFI, Case T-380/08, *Kingdom of the Netherlands v. Commission*, 9 September 2008, *OJ C* 285, 8 November 2008, 51 (pending case).

EUROPEAN COURT OF HUMAN RIGHTS

All cases can be found at <http://cmiskp.echr.coe.int/tkp197/search.asp?skin=hudoc-en>.

Sunday Times v. United Kingdom, 24 April 1979.
Lingens v. Austria, 8 July 1986.
Leander v. Sweden, 26 March 1987.
Gaskin v. United Kingdom, 7 July 1989.
Thorgeir Thorgeirson v. Iceland, 25 June 1992.
Open Door and Dublin Well Women v. Ireland, 29 October 1992.
Casado Coca v. Spain, 24 February 1994.
Jersild v. Denmark, 23 September 1994.
Guerra and others v. Italy, 19 February 1998.
Hertel v. Switzerland, 25 August 1998.

Roche v. United Kingdom, 23 May 2002.
Colombani and others v. France, 25 June 2002.
Cetin and others v. Turkey, 13 February 2003.
Tepe v. Turkey, 9 May 2003.
Eccleston v. United Kingdom, 18 May 2004.
Sirbu and others v. Moldova, 15 June 2004.
Ukrainian Media Group v. Ukraine, 29 March 2005.
Leyla Sahin v. Turkey, 10 November 2005.
Sdruzeni Johoceske Matky v. Czech Republic, 10 July 2006.
Leempoel & SA Cine Revue v. Belgium, 9 November 2006.
Társaság a Szabadságjogokért v. Hungary, 14 April 2009.
Kenedi v. Hungary, 26 May 2009.

FRANCE

Cour de Cassation, 12 December 1995, *Ministère de l'équipement, Direction de la météorologie nationale v. Société du journal téléphoné*, Bull. Civ. IV, no. 301, 276.
Cour D'Appel Paris, 18 March 1993, *Société du journal téléphoné v. Ministère de l'équipement*, Direction de la météorologie nationale, BOCCRF no. 6, 26 March 1993, 110.
Cons. Conc. Déc. No 92-D-35, 13 May 1992 relative à une saisine de la Société du journal téléphoné à l'encontre de la direction de la météorologie nationale, BOCCRF no. 13, 4 July 1992, 224.
Cons. Conc. Avis no. 97 A-10, 25 February 1997 relatif à une demande d'avis présentée par le Groupement des editions de la presse nautique et portant sur des questions de concurrences soulevées par la politique éditoriale du service hydrographique de la marine, BOCCRF 18 November 1997, 759.

List of References

BOOKS

Arrowsmith, Sue. *The Law of Public and Utilities Procurement.* London: Sweet & Maxwell, 2005.

Baker, John C. et al. *Mapping the Risks. Assessing the Homeland Security Implications of Publicly Available Geospatial Information.* Santa Monica: RAND Corporation, 2004.

Baten, Inigo & Van Der Starre. Gijs, *Elektronische toegankelijkheid van overheidsinformatie.* The Hague: Rathenau Instituut, 1996.

Beers, Anton A.L. *Informatica Publica: publieke toegankelijkheid van elektronische overheidsinformatie.* The Hague: Rathenau Instituut, 1996.

Boes, Marc. *Cursus administratief recht.* Acco: Leuven, 2005.

Bovis, Christopher. *EC Public Procurement: Case Law and Regulation.* London: Oxford University Press, 2006.

Buendia Sierra & Jose Luis. *Exclusive Rights and State Monopolies under EC Law.* Oxford: Oxford University Press, 1999.

Burkert, Herbert. *The Law and Policies of Commercialization of Public Sector Information in the EC.* Stockholm: Institutet för rättsinformatik, 1993.

Bruguière, Jean-Michel. *Les données publiques et le droit.* Paris: Editions Litec, 2002.

Buttgenbach, André. *Manuel de droit administrative, première partie*, 3rd edn. Brussels: Larcier, 1966.

CEEP & CIRIEC, *Services of General Economic Interest in Europe: Regulation, Financing, Evaluation, Good Practices.* Liège: CIREC, 2000.

Cho, George. *Geographic Information Systems and the Law: Mapping the Legal Frontiers.* Chicester: John Wiley & Sons Ltd, 2001.

Cho, George. *Geographic Information Science: Mastering the Legal Issues.* Chicester: John Wiley & Sons Ltd, 2005.

Crompvoets, Joep et al. (eds). *A Multi-view Framework to Assess Spatial Data Infrastructures.* Melbourne: The Melbourne University Press, 2008.

Croswell, Peter L. & Alex Wernher. *GIS Program Revenue Generation and Legal Issues in Public Sector Organizations.* Park Ridge: URISA, 2004.

Daalder, Eric J. *Toegang tot overheidsinformatie. Het grensvlak tussen openbaarheid en vertrouwelijkheid.* The Hague: Boom Juridische Uitgevers, 2005.

Deketelaere, Kurt & Frankie Schram. *Milieurechten van het publiek in de participatiedemocratie.* Bruges: Vanden Broele, 2002.

Delfosse, François. *L'état, pour quoi, pour qui?.* Louvain-la-Neuve: Centre de Recherches Sociologiques – UCL, 1986.

De Terwangne, Cécile, Herbert Burkert & Yves Poullet (eds). *Towards a Legal Framework for a Diffusion Policy for Data Held by the Public Sector.* Deventer: Kluwer, 1995.

De Terwangne, Cécile et al. *L'accès à l'information administrative et la commercialisation des données publiques: analyse théorique et pratique du problème d'accès, pour le citoyen en général, aux informations détenues par le service public.* Brussels: Story-Scientia, 1994.

Donahue, John M. *The Privatization Decision: Public Ends, Private Means.* New York: Basic Books, 1989.

Durviaux, Ann-Lawrence. *Logique de marché et marché public en droit communautaire. Analyse critique d'un système.* Brussels: Larcier, 2006.

Faull, Jonathan, *Faull and Nikpay: The EC Law of Competition.* Oxford: Oxford University Press, 2007.

Flamme, Philippe, Maurice-André Flamme & Claude Dardenne. *Les marchés publics européens et belges*, Brussels: Larcier, 2005.

Groot, Richard & John D. McLaughlin (eds). *Geospatial Data Infrastructure: Concepts, Cases and Good Practice.* Oxford: Oxford University Press, 2000.

Hugenholtz, P. Bernt, *Auteursrecht op informatie: auteursrechtelijke bescherming van feitelijke gegevens en gegevensverzamelingen in Nederland, de Verenigde Staten en West-Duitsland: een rechtsvergelijkend onderzoek.* Deventer: Kluwer, 1989.

Janssen, Katleen, Jo Steyaert & Roland Van Gompel. *Transparantie van overheidsinformatie in Vlaanderen. Beschouwingen en aanbevelingen.* Bruges: Die Keure, 2003.

Johnson, M. Glen & Janusz Symonides. *The Universal Declaration of Human Rights: a History of Its Creation and Implementation 1948–1998.* Paris: UNESCO, 1998.

Jones, Alison & Brenda Sufrin. *EC Competition Law: Text, Cases and Materials.* 3rd edn. Oxford: Oxford University Press, 2008.

Kabel, Jan J.C. et al. *Kennisinstellingen en informatiebeleid. Lusten en lasten van de publieke taak.* Amsterdam: Universiteit van Amsterdam, 2001.

Kabel, Jan & Mireille Van Eechoud. *Prijsbepaling voor elektronische overheidsinformatie.* Deventer: Kluwer, 1998.

Kingma, Bruce R. *The Economics of Information: A Guide to Economic and Cost-Benefit Analysis for Information Professionals.* 2nd edn. Englewood: Libraries Unlimited, 2001.

Kranenborg, Herke and Wim Voermans. *Access to Information in the European Union. A Comparative Analysis of EC and Member State Legislation.* Groningen: Europa Law Publishing, 2005.

Lane, Jan-Erik. *The Public Sector: Concepts, Models and Approaches.* London: SAGE Publications, 1995.

Longhorn, Roger A. & Michael Blakemore. *Geographic Information: Value, Pricing, Production, and Consumption.* Boca Raton: CRC Press, 2008.

Longley, Paul A. et al. *Geographic Information Systems and Science.* Chicester: John Wiley and Sons Ltd, 2001.

Maes, Rudolf & Kathleen Jochmans (eds). *Inleiding tot de bestuurskunde.* Brussels: Studiecentrum Open Hoger Onderwijs, 1996.

Mangenot, Michel et al. *Public Administrations and Services of General Interest: What Kind of Europeanisation?.* Maastricht: European Institute of Public Administration, 2005.

Masser, Ian. *Building European Spatial Data Infrastructures.* Redlands: ESRI Press, 2007.

Matthijs, Herman. *Bestuurswetenschappen. De overheid, haar structuren en haar werking.* Brussels: Politeia, 2003.

Matthijs, Herman, Frank Naert & Jef Vuchelen. *Handboek Openbare Financiën.* Antwerp: Intersentia, 2004.

Naschold, Frieder & Carsten Von Otter. *Public Sector Transformation: Rethinking Markets and Hierarchies in Government.* Amsterdam: John Benjamins Publishing Company, 1996.

National Research Council. *Bits of Power. Issues in Global Access to Scientific Data.* Washington: National Academies Press, 1997.

National Research Council. *Licensing Geographic Data and Services.* Washington: National Academies Press, 2004.

Niebling, Carl. *Overheidsbemoeienis. Opkomst en ondergang van de verzorgingsstaat.* Rotterdam: Ad. Donker, 1999.

Onsrud, Harlan J. (ed.). *Research and Theory in Advancing Spatial Data Infrastructure Concepts.* Redlands: ESRI Press, 2007.

Paradissis, Jean-Jacques. *The Right to Access Environmental Information.* Saarbrucken: GDM Verlag, 2009.

Prosser, Tony. *The Limits of Competition Law: Markets and Public Services.* Oxford: Oxford University Press, 2005.

Rosenthal, Ute et al. *Openbaar bestuur. Beleid, organisatie en politiek.* Alphen a/d Rijn: Samson, 1996.

Roth, Peter & Vivien Rose (eds). *Bellamy & Child European Community Law of Competition.* 6th edn. Oxford: Oxford University Press, 2008.

Sands, Philippe. *Principles of International Environmental Law.* 2nd edn. Cambridge: Cambridge University Press, 2003.

Sap, Jan W., Ben P. Vermeulen & Carla M. Zoethout. *De publieke taak. Staatsrecht-conferentie 2002, Vrije Universiteit Amsterdam.* Amsterdam: Kluwer, 2003.

Schram, Frankie. *De publieke toegang tot milieu-informatie in internationaal en internrechtelijk perspectief.* Brussels: De Boeck & Larcier, 2008.

Sopers, Jan M.M. *De bakens verzet. Publieke taakorganisaties in verandering.* Assen: Uitgeverij Van Gorcum, 1999.

Steiner, Henry & Philip Alston. *International Human Rights in Context: Law, Politics, Morals.* 2nd edn. Oxford: Oxford University Press, 2000.

Sudre, Frédéric. *Droit Européen et International des Droits de l'Homme.* 2nd edn. Paris: Presses Universitaires de France, 2006.

Sui, Daniel Z. (ed.). *Geospatial Technologies and Homeland Security: Research Frontiers and Future Challenges.* Dordrecht: Springer Netherlands, 2008.

Uhlir, Paul F. et al. *The Socioeconomic Effects of Public Sector Information on Digital Networks.* Washington: National Academies Press, 2009.

Uyttendaele, Caroline. *Openbare informatie. Het juridisch statuut in een convergerende media-omgeving.* Antwerp – Apeldoorn: Maklu, 2002.

Valcke, Peggy. *Digitale diversiteit. Convergentie van Media-, Telecommunicatie- en Mededingingsrecht.* Brussels: Larcier, 2004.

Van der Heijden, Barend & Bahia Tahzib-Lie. *Reflections on the Universal Declaration of Human Rights: a Fiftieth Anniversary Anthology.* The Hague: Nijhoff, 1998.

Van Der Hof, Simone. *Overheidsinformatie in de etalage. Belangen rondom de toegang tot overheidsinformatie.* Alphen a/d Rijn: Samsom Bedrijfsinformatie, 1997.

Van Eechoud, Mireille & Jan Kabel. *Prijsbepaling voor elektronische overheidsinformatie.* Deventer: Kluwer, 1998.

Van Impe, Herman. *De Staat, wat is dat eigenlijk? Staat, democratie en recht.* Antwerp: Kluwer Rechtswetenschappen, 1990.

Walsh, Kieron. *Public Services and Market Mechanisms: Competition, Contracting and the New Public Management.* Basingstoke: Palgrave Macmillan, 1995.

Van Veldhoven, Gerry M. et al. (eds). *Overheidsbemoeienis. Opstellen uitgegeven ter gelegenheid van het 11de lustrum van de Katholieke Hogeschool te Tilburg.* Deventer: Kluwer, 1982.

Wells Branscomb, Anne. *Who Owns Information? from Privacy to Public Access.* New York: Basic Books, 1994.

Whish, Richard. *Competition law.* 6th edn. Oxford: Oxford University Press, 2009.

Yergin, Daniel & Joseph Stanislaw. *The Commanding Heights: The battle for the World Economy.* New York: Touchstone, 2002.

PHD THESES

De Terwangne, Cécile. *Société de l'information et mission publique d'information, Doctoral Thesis FUNDP.* Namur: Facultés Universitaires de Notre Dame de Paix, 2001.

Mason, Renate. *Developing Australian Spatial Data Policies – Existing Practices and Future Strategies, Doctoral Thesis University of New South Wales*, 2000. <http://handle.unsw.edu.au/1959.4/18646>, 22 February 2010.

Rajabifard, Abbas. *Diffusion of Regional Spatial Data Infrastructures: With particular reference to Asia and the Pacific, Doctoral Thesis University of Melbourne*, 2002. <www.sli.unimelb.edu.au/research/publications/Rajabifard_thesis.pdf>, 22 February 2010.

Schram, Frankie. *Begrenzingen aan de openbaarheid van bestuur: een analyse en vergelijking van de internrechtelijke openbaarheidsregelingen in het Belgische recht, Doctoral Thesis K.U. Leuven.* Leuven: K.U. Leuven, 2001.

Van Loenen, Bastiaan. *Developing Geographic Information Infrastructures: The Role of Information Policies.* Delft: DUP Science, 2006.

CONTRIBUTIONS IN COLLECTIVE WORKS

Ackermans-Wijn, Johanna C.E. 'De overheidsonderneming en Europese aanbesteding'. In *Overheid en onderneming. Tussen publiek en privaat ondernemen*, edited by Sebastianus C.J.J. Kortmann et al. Deventer: W.E.J. Tjeenk Willink, 1998.

Aichholzer, Georg & Puay Tang. 'Harnessing Public Sector Information for Greater Accessibility: Austria and the UK'. In *Public Sector Information in the Digital Age. Between Markets, Public Management and Citizens' Rights*, edited by Georg Aichholzer & Herbert Burkert. Cheltenham: Edward Elgar Publishing Limited, 2004.

Arrow, Kenneth J. 'The Economics of Information'. In *The computer Age: A Twenty Year Review*, edited by Michael Dertouzos & Joe Moses. Cambridge: MIT Press, 1979.

Asante, Kwabena O. et al. 'Spatial Data Infrastructures In Management of Natural Disasters'. In *Research and Theory in Advancing Spatial Data Infrastructure Concepts*, edited by Harlan J. Onsrud. ESRI Press: Redlands California, 2007.

Bargmann, Monika, Gerald Pfeifer & Boris Piwinger. 'A Citizen's Perspective on Public Sector Information'. In *Public Sector Information in the Digital Age: Between Markets, Public Management and Citizens' Rights*, edited by Georg Aichholzer & Herbert Burkert. Cheltenham: Edward Elgar Publishing Limited, 2004.

Beers, Anton A.L. 'Openbaarheid van overheidsinformatie'. In *Elektronische toegankelijkheid van overheidsinformatie*, edited by Inigo Baten & Gijs Van Der Starre. The Hague: Rathenau Instituut, 1996.

Beers, Ton A.L. 'Public Access to Government Information towards the 21st Century'. In *Information Law towards the 21st Century*, edited by Willem, F. Korthals Altes. Deventer: Kluwer, 1992.

Bokkerink, Marc. R. 'De uitoefening van publiekrechtelijke taken door privaatrechtelijke rechtspersonen'. In *Overheid en onderneming. Tussen publiek en privaat ondernemen*, edited by Sebastianus C.J.J. Kortmann et al. Deventer: W.E.J. Tjeenk Willink, 1998.

Bouwman, Harry et al. 'Leuker kunnen we het niet maken, wel makkelijker. Over de toegankelijkheid van overheidsinformatie'. In *Elektronische toegankelijkheid van overheidsinformatie*, edited by Inigo Baten & Gijs Van Der Starre. The Hague: Rathenau Instituut, 1996.

Burkert, Herbert. 'The Legitimacy of Commercialization of Public Sector Information as a Research Topic'. In *Towards a Legal Framework for a Diffusion Policy for Data Held by the Public Sector*, edited by Cécile De Terwangne, Herbert Burkert & Yves Poullet. Deventer: Kluwer Law and Taxation, 1995.

Burkert, Herbert. 'The Mechanics of Public Sector Information'. In *Public Sector Information in the Digital Age: Between Markets, Public Management and Citizens' Rights*, edited by Georg Aichholzer & Herbert Burkert. Cheltenham: Edward Elgar Publishing Limited, 2004.

Burkert, Herbert & Peter N. Weiss. 'Towards a Blueprint for a Policy on Public Sector Information'. In *Public Sector Information in the Digital Age: Between Markets, Public Management and Citizens' Rights*, edited by Georg Aichholzer & Herbert Burkert. Cheltenham: Edward Elgar Publishing Limited, 2004.

Corey, Mark. 'Digital Data, Copyright, Commercialization and Related Issues: A Canadian Perspective'. In *Policy Issues in Modern Cartography*, edited by D.R. Fraser Taylor. Oxford: Elsevier Science Ltd, 1998.

Craglia, Massimo & Michael Blakemore. 'Access Models for Public Sector Information: The Spatial Data Context'. In *Public Sector Information in the Digital Age: Between Markets, Public Management and Citizens' Rights*, edited by Georg Aichholzer & Herbert Burkert. Cheltenham: Edward Elgar Publishing Limited, 2004.

Craglia, Max & Alessandro Annoni. 'INSPIRE: An Innovative Approach to the Development of Spatial Data Infrastructures In Europe'. In *Research and Theory in Advancing Spatial Data Infrastructure Concepts*, edited by Harlan J. Onsrud. ESRI Press: Redlands California, 2007.

Déjeant-Pons, Maguelonne. 'Le Conseil de l'Europe et l'Accès à l'Information Environnementale: la Convention Européenne des Droits de l'Homme'. In *Ten Years of Access to Environmental Information in International, European and Belgian Law: Stock-Taking and Perspectives*, edited by Christine Larssen. Brussels: Bruylant, 2003.

Déjeant-Pons, Maguelonne. 'La Convention européenne des droits de l'homme et le droit à l'information en matière d'environnement'. In *La Convention européenne des droits de l'homme : développements récents et nouveaux défis*, edited by Jean-François Flauss & Michel De Salvia. Brussels: Bruylant, 1997.

Derclaye, Estelle. 'Does the Directive on the Re-use of Public Sector Information affect the State's database sui generis right?'. In *Knowledge Rights – Legal, Societal and Related Technological Aspect. Knowright Symposium, 18–19 September 2008*, edited by Jens Gaster, Erich Schweighofer & Peter Sint. Vienna: Austrian Computer Society, 2008.

De Ru, Hendrik J. 'The Commercial Use of Government Controlled Information: A Guideline for Regulatory Policy Choices'. In *Information Law towards the 21st Century*, edited by F. Korthals Altes Willem. Deventer: Kluwer, 1992.

De Streel, Alexandre & Robert Queck. 'Services d'intérêt économique général et communications électroniques'. In *Les services d'intérêt économique général et l'Union européenne*, edited by Jean-Victor Louis & Stéphane Rodrigues. Brussels: Bruylant, 2006.

De Wit, Onno. 'Langs lijnen van geleidelijkheid: de telefonie in Nederland als object van staatszorg, 1877–1989'. In *Toeval of noodzaak? Geschiedenis van de overheidsbemoeienis met de informatievoorziening*, edited by Rathenau Instituut. The Hague: Rathenau Instituut, 1995.

Dommering, Egbert J., P. Bernt Hugenholtz & Jan J.C. Kabel. 'De overheid en het publiek domein van informatie voor wetenschappelijk onderzoek'. In *De publieke dimensie van kennis*, edited by Huub Dijstelbloem & Cornelius J.M. Schuyt. The Hague: WRR, 2002.

Dommering, Egbert J. 'Inleiding'. In *Informatierecht. Fundamentele rechten voor de informatiesamenleving*, edited by Egbert J. Dommering et al. Amsterdam: Otto Cramwinckel, 2000.

Dommering, Egbert J. 'Informatie, communicatie en technologie'. In *Informatierecht. Fundamentele rechten voor de informatiesamenleving*, edited by Egbert J. Dommering et al. Amsterdam: Otto Cramwinckel, 2000.

Dommering, Egbert J. & Lodewijk, F. Asccher. 'Europese en Internationale Aspecten'. In *Informatierecht. Fundamentele rechten voor de informatiesamenleving*, edited by Egbert J. Dommering et al. Amsterdam: Otto Cramwinckel, 2000.

Dommering, Egbert J. 'Informatie tussen economie en cultuur'. In *Informatierecht. Fundamentele rechten voor de informatiesamenleving*, edited by Egbert J. Dommering et al. Amsterdam: Otto Cramwinckel, 2000.

Drijber, Berend J. 'Mededinging en diensten van algemeen economisch belang'. In *Actualiteiten in het Europese Mededingingsrecht. Asser Instituut Colloquium Europees Recht*, edited by MarJan Olfers & Karolien T.M.C. Pieters. Asser Press: The Hague 2005.

Dubuisson, François. 'Quand le "Droit" du Public à l'Information Rencontre les "Droits" sur l'Information: Vrai ou Faux Débat?'. In *Droit d'Auteur et Liberté d'Expression. Regards Francophones, d'Europe et d'Ailleurs*, edited by Alain Strowel & Françoise Tulkens. Brussels: Larcier, 2006.

Epstein, Earl F. 'Legal aspects of GIS'. In *Geographical Information Systems. Principles and applications*, edited by David Maguire, Michael Goodchild & David Rhind. Essex: Longman Scientific and Technical, 1993.

Gamble, Andrew & Tony Wright. 'Introduction'. In *Restating the State? Special issue of The Political Quarterly*, edited by Andrew Gamble & Tony Wright. Malden: Blackwell Publishing, 2004.

Gutwirth, Serge. 'Het juridisch statuut van informatie'. In *Technologie en recht*, edited by Paul De Vroede. Antwerp: Kluwer Rechtswetenschappen, 1987.

Haller, Max, Franz Hollinger & Otto Raubal. 'Leviathan or Welfare State? Attitudes toward the Role of Government in Six Advanced Western Nations'. In *Attitudes to Inequality and the Role of Government*, edited by Duane F. Alwin et al. The Hague: Sociaal en Cultureel Planbureau, 1990.

Hallo, Ralph E. 'Directive 90/313/EEC On the Freedom of Access to Information on the Environment: Its Implementation and Implications'. In *Access to Environmental Information in Europe: The Implementation and Implications of Directive 90/313/EEC*, edited by Ralph E. Hallo. London: Kluwer Law International, 1996.

Hallo, Ralph E. 'The EU Institutions'. In *Access to Environmental Information in Europe: The Implementation and Implications of Directive 90/313/EEC*, edited by Ralph E. Hallo. London: Kluwer Law International, 1996.

Hallo, Ralph E. 'The Right of Access to Environmental Information: Directive 90/313/EEC And Other Measures'. In *Développements récents du droit européen de l'environnement : droits des citoyens et des associations dans le droit européen de l'environnement*, edited by Michel Paques & Patricia Van Pelt. Antwerp: Story-Scientia, 1998.

Hanappi-Egger, Edeltraud. 'Cultural Heritage: The Conflict between Commercialisation and Public Ownership'. In *Public Sector Information in the Digital Age: Between Markets, Public Management and Citizens' Rights*, edited by Georg Aichholzer & Herbert Burkert. Cheltenham: Edward Elgar Publishing Limited, 2004.

Hengstschlager, Johannes. 'Privatisierung von Verwaltungsaufgaben'. In *Veröffentlichungen der Vereinigung Deutscher Staatsrechtlehrer 54*, edited by Vereinigung der Deutschen Staatsrechtlehrer. Berlin: De Gruyter, 1995.

Heyvaert, Veerle. 'Wie wordt er wijzer van? Enkele bedenkingen bij het recht op toegang tot milieu-informatie'. In *Ten Years of Access to Environmental Information in International, European and Belgian Law: Stock-Taking and Perspectives*, edited by Christine Larssen. Brussels: Bruylant, 2003.

Heijne, G. 'Perspectieven op informatiepolitiek. Een verslag gebaseerd op vier workshops'. In *Een kwestie van toegang. Bijdragen aan het debat over het publieke domein van de informatievoorziening*, edited by Inigo Baten & Jolien Ubacht. The Hague: Rathenau Instituut, 1995.

Hirsch Ballin, Ernst M. H. 'De mens in de sociale rechtsstaat?'. In *Overheidsbemoeienis. Opstellen uitgegeven ter gelegenheid van het 11de lustrum van de Katholieke Hogeschool te Tilburg*, edited by Gerry M. Van Veldhoven et al. Deventer: Kluwer, 1982.

Jadot, Benoit & Christine Larssen. 'L'accès à la justice en matière d'environnement au regard de la convention d'Aarhus'. In *L'accès à la justice en matière d'environnement*, edited by Christine Larssen & Marc Pallemaerts. Brussels: Bruylant, 2005.

Janssen, Katleen and Dumortier, Jos. 'Legal Framework for a European Union Spatial Data Infrastructure: Uncrossing the Wires'. In *Research and Theory in Advancing Spatial Data Infrastructure Concepts*, edited by Harlan J. Onsrud. ESRI Press: Redlands California, 2007.

Kamps, Michael. 'Thunder and Lightning: Public Sector Information Policy Experiences of Private Meteorological Service Providers'. In *Public Sector Information in the Digital Age. Between Markets, Public Management and Citizens' Rights*, edited by Georg Aichholzer & Herbert Burkert. Cheltenham: Edward Elgar Publishing Limited, 2004.

Kaul, Inge, Isabelle Grunberg & Marc A. Stern. 'Defining Global Public Goods'. In *Global Public Goods*, edited by Inge Kaul, Isabelle Grunberg & Marc A. Stern. Oxford: Oxford University Press, 1999.

Kennedy, Ed. 'A Private Sector Perspective'. In *Policy Issues in Modern Cartography*, edited by D.R. Fraser Taylor. Oxford: Elsevier Science Ltd, 1998.

Kortteinen, Uhani, Kristian Mynnti & Lauri Hannikainen. 'Article 19'. In *The Universal Declaration of Human Rights: A Common Standard of Achievement*, edited by Alfredsson Gudmundur & Eide Asbjorn. The Hague: Martinus Nijhoff Publishers, 1999.

Kramer, Ludwig. 'La directive 90/313/EEC sur l'accès à l'information en matière de l'environnement: genèse et perspectives d'application'. In *The right to environmental information*, edited by Marc Pallemaerts. Brussels: Story-Scientia, 1991.

Kubicek, Herbert. 'Third-Generation Freedom of Information in the Context of E-Government: the Case of Bremen, Germany'. In *Public Sector Information in the Digital Age. Between Markets, Public Management and Citizens' Rights*, edited by Georg Aichholzer & Herbert Burkert. Cheltenham: Edward Elgar Publishing Limited, 2004.

Larssen, Christine. 'L'Accès aux Informations sur L'Environnement en Droit International: la Convention d'Aarhus'. In *Ten Years of Access to Environmental Information in International, European and Belgian Law: Stock-Taking and Perspectives*, edited by Christine Larssen. Brussels: Bruylant, 2003.

Marquand, David. 'False Friend: the State and the Public Domain'. In *Restating the State? Special issue of The Political Quarterly*, edited by Andrew Gamble & Tony Wright. Malden: Blackwell Publishing, 2004.

Mason, Anthony. 'The Relationship between Freedom of Expression and Freedom of Information'. In *Freedom of Expression and freedom of information: Essays in honour of Sir David Williams*, edited by Jack Beatson & Yvonne Cripps. Oxford: Oxford University Press, 2000.

McNutt, Patrick. 'Public Goods and Club Goods'. In *Encyclopedia of Law and Economics*, edited by Boudewijn Bouckaert & Gerrit De Geest. Cheltenham: Edward Elgar Publishing Limited, 2000.

Mol, Anne-Lieke. 'Voortdurend onderweg naar morgen'. In *Een kwestie van toegang. Bijdragen aan het debat over het publieke domein van de informatievoorziening*, edited by Inigo Baten & Jolien Ubacht. The Hague: Rathenau Instituut, 1995.

Onsrud, Harlan J. 'Legal Access to Geographic Information: Measuring Losses or Developing Responses?'. In *Information, Place and Cyberspace: Issues in Accessibility*, edited by Donald G. Janelle & David Charles Hodge.

Berlin: Springer-Verlag, 2000. <www.spatial.maine.edu/%7Eonsrud/pubs/LegalAccessToGI.pdf>, 22 February 2010.

Onsrud, Harlan J. 'The Tragedy of the Information Commons'. In *Policy Issues in Modern Cartography*, edited by D.R. Fraser Taylor. Oxford: Elsevier Science Ltd, 1998.

Onsrud, Harlan et al. 'The Future of the Spatial Information Infrastructure'. In *A Research Agenda for Geographic Information Science*, edited by Robert B. McMaster & Usery E. Lynn. Boca Raton: CRC Press, 2004. <www.spatial.maine.edu/%7eonsrud/pubs/chapter8preprint.pdf>, 22 February 2010.

Onsrud, Harlan J. 'Liability in the Use of Geographic Information Systems and Geographic Data Sets'. In *Geographic Information Systems: Principles, Techniques, Management, and Applications*, edited by David Maguire et al. New York: Wiley, 1999. <www.spatial.maine.edu/~onsrud/pubs/liabilityabstract40.htm>, 22 February 2010.

Onsrud, Harlan J. 'Role of Law in Impeding and Facilitating the Sharing of Geographic Information'. In Gerard *Sharing Geographic Information*, edited by Onsrud, Harlan J. and Rushton. New Brunswick: Rutgers CUPR Press, 1995. <www.spatial.maine.edu/%7eonsrud/pubs/roleoflaw36.pdf>, 22 February 2010.

Prins, Corien. 'Directive 2003/98/EC of the European Parliament and of the Council'. In *Concise European IT law*, edited by Corien Prins. Alphen a/d Rijn: Kluwer Law International, 2006.

Prins, Corien. 'Access to Public Sector Information. In Need of Consitutional Recognition?'. In *Public Sector Information in the Digital Age: Between Markets, Public Management and Citizens' Rights*, edited by Georg Aichholzer & Herbert Burkert. Cheltenham: Edward Elgar Publishing Limited, 2004.

Raab, Charles D. 'Privacy Issues as Limits to Accesss'. In *Public Sector Information in the Digital Age. Between Markets, Public Management and Citizens' Rights*, edited by Georg Aichholzer & Herbert Burkert. Cheltenham: Edward Elgar Publishing Limited, 2004.

Reichard, Christoph. 'New Institutional Arrangements of Public Service Delivery'. In *The Governance of Services of General Interest between State, Market and Society*, edited by Christoph Reichard, Ricardo Mussari & Sören Kupke. Berlin: Wissenschaftlicher Verlag Berlin, 2006.

Rockman, Bert A. 'Honey, I Shrank the State'. In *Modern Systems of Government. Exploring the Role of Bureaucrats and Politicians*, edited by Ali Farazmand. London: Sage Publications Inc., 1997.

Salgé, François. 'Geographic Information Network in Europe (GINIE). Recommendations for action'. In *Spatial data infrastructure and policy development in Europe and the United States*, edited by Bastiaan Van Loenen & Bas C. Kok. Delft: DUP Science, 2004.

Stiglitz, Joseph E. 'Knowledge as a Global Public Good'. In *Global Public Goods*, edited by Inge Kaul, Isabelle Grunberg & Marc A. Stern. Oxford: Oxford University Press, 1999.

Danny Vandenbroucke, Katleen Janssen & Jos Van Orshoven. 'INSPIRE State of Play: Generic approach to assess the status of NSDIs'. In *A Multi-view*

Framework to Assess Spatial Data Infrastuctures, edited by Joep Crompvoets et al. Melbourne: The University of Melbourne, 2008.

Van Eechoud, Mireille. 'The Commercialization of Public Sector Information: Delineating the Issues'. In *The Future of the Public Domain. Identifying the Commons in Information law*, edited by Lucie Guibault & P. Bernt Hugenholtz. Alphen a/d Rijn: Kluwer Law International, 2006.

Van Loenen, Bastiaan & Bas C. Kok. 'Spatial Data Infrastructures: Legal and Economic Issues'. In *Spatial Data Infrastructure and Policy Development in Europe and the United States*, edited by Bastiaan Van Loenen & Bas C. Kok. Delft: DUP Science, 2004.

Van Loenen, Bastiaan and Jitske De Jong. 'SDIs and Privacy: Conflicting Interests of the Spatially Enabled Society'. In *Towards a Spatially Enabled Society*, edited by Abbas Rajabifard. Melbourne: University of Melbourne, 2007.

Van Loenen, Bastiaan et al. 'Privacy Versus National Security: the Impact of Privacy Law on the Use of Location Technology for National Security Purposes'. In *The European Information Society: Leading the way with Geo-Information Lecture Notes in Geoinformation and Cartography*, edited by Sara Irina Fabrikant & Monica Wachowicz. Berlin: Springer, 2007.

Van Rossum, A.A. 'Oneerlijke concurrentie door de overheid'. In *Overheid en onderneming. Tussen publiek en privaat ondernemen*, edited by Sebastianus C.J.J. Kortmann et al. Deventer: W.E.J. Tjeenk Willink, 1998.

Vermeulen, Ben P. 'De publieke taak: een veel-zijdig begrip'. In *De publieke taak. Staatsrechtconferentie 2002, Vrije Universiteit Amsterdam*, edited by Jan W. Sap, Ben P. Vermeulen & Carla M. Zoethout. Amsterdam: Kluwer, 2003.

Volman, Yvo. 'The Re-use of Public Sector Information in the EU: Increasing the Scope for Cross-Border Exploitation'. In *Spatial Data Infrastructure and policy development in Europe and the United States*, edited by Bastiaan Van Loenen and Bas C. Kok. Delft: DUP Science, 2004.

Volman, Yvo. 'Exploitation of Public Sector Information in the Context of the *eEurope* Action Plan'. In *Public Sector Information in the Digital Age: Between Markets, Public Management and Citizens' Rights*, edited by Georg Aichholzer & Herbert Burkert. Cheltenham: Edward Elgar Publishing Limited, 2004.

Voorhoof, Dirk. 'Artikel 10 – Vrijheid van meningsuiting'. In *Handboek EVRM. Deel 2. Artikelsgewijze commentaar*, edited by Johan Vande Lanotte & Yves Haeck. Antwerp: Intersentia, 2004.

Wallstrom, Margot. '10 Years of Access to Environmental Information in Eurpean, International and Internal Law: Evaluation and Perspectives'. In *Ten Years of Access to Environmental Information in International, European and Belgian Law: Stock-Taking and Perspectives*, edited by Christine Larssen. Brussels: Bruylant, 2003.

Wates, Jeremy. 'The UN ECE Guidelines and Draft Convention on Access to Environmental Information and Public Participation in Environmental Decisionmaking'. In *Access to Environmental Information in Europe: the*

Implementation and Implications of Directive 90/313/EEC, edited by Ralph E. Hallo. London: Kluwer Law International, 1996.

Weiss, Peter N. 'Borders in Cyberspace: Conflicting Public Sector Information Policies and their Economic Impacts'. In *Public Sector Information in the Digital Age: Between Markets, Public Management and Citizens' Rights*, edited by Georg Aichholzer & Herbert Burkert. Cheltenham: Edward Elgar Publishing Limited, 2004.

Williams, Rhiannon. 'Access to Documents Held by European Institutions with Particular Regard to Environmental Infringement Procedures, or the Emperor's New Clothes, and the Chimera of "Amicable Dispute Resolution"'. In *Ten Years of Access to Environmental Information in International, European and Belgian Law: Stock-Taking and Perspectives*, edited by Christine Larssen. Brussels: Bruylant, 2003.

Williamson, Ian, Abbas Rajabifard & Andrew Binns. 'The Role of Spatial Data Infrastructures in Establishing an Enabling Platform for Decision Making in Australia'. In *Research and Theory in Advancing Spatial Data Infrastructure Concepts*, edited by Harlan J. Onsrud. ESRI Press: Redlands California, 2007.

Zevenbergen, Jaap A. 'European Privacy Law and Its Effect on Location Information'. In *Location Privacy Workshop: Individual autonomy as a driver of designweb*, edited by Harlan Onsrud & Silvia Nittel. Maine: University of Maine, 2004.

ARTICLES IN JOURNALS AND REVIEWS

Ader, Thorsten & Max Schoental. 'Access to Information on Government Action, especially from the Media Point of View', *IRIS plus* no. 2 (2005): 2–8.

Arnon, Karen E. & Jan J.M. Sluijs. 'Het arrest BUPA: publieke diensten en marktwerking', *Nederlands Tijdschrift voor Europees Recht* no. 7/8 (2008): 206–213.

Batley, Richard. 'Public-Private Relationships and Performance in Service Provision', *Urban Studies* 33, no. 4–5 (1996): 723–751.

Biondi, Andrea. 'Some Reflections on the Notions of "State Resources" in European Community State Aid Law', *Fordham International Law Journal* 30 (2007): 1426–1448.

Birkinshaw, Patrick. 'Freedom of Information in the UK and Europe: Further Progress?', *Government Information Quarterly* 19, no. 1 (2002): 77–86.

Birkinshaw, Patrick. 'Freedom of Information and Openness: Fundamental Human Rights?', *Administrative Law Review* 58 (2006): 177–218.

Blakemore, Michael & Max Craglia. 'Access to Public-sector Information in Europe: Policy, Rights, and Obligations', *The Information Society* 22, no. 1 (2006): 13–24.

Boeger, Nina. 'Solidarity and EC Competition Law', *European Law Review* 32, no. 3 (2007): 319–340.

Bovens, Mark. 'Information Rights: Citizenship in the Information Society', *The Journal of Political Philosophy* 10, no. 3 (2002): 317–341.

Bovis, Christopher H. 'Developing Public Procurement Regulation: Jurisprudence and Its Influence on Law Making', *Common Market Law Review* 43, no. 2 (2006): 461–495.

Brown, Gordon. 'State and Market: Towards a Public Interest Test', *Political Quarterly* 74, no. 3 (2003): 266–284.

Brown, Kathryn. 'Mapping the Future', *Science* 298, no. 5600 (2002): 1874–1875.

Burdon, Mark. 'Commercializing Public Sector Information: Privacy and Security Concerns', *IEEE Technology and Society Magazine* 28, no. 1 (2009): 34–40.

Burkert, Herbert. 'Public Sector Information: Towards a More Comprehensive Approach in Information Law?', *Journal of Law and Information Science* 3 (1992): 47–62.

Burkert, Herbert. 'L'information du secteur public. Le secret, la transparence et le commerce', *Revue française d'administration publique* 72 (1994): 581–591.

Burri Nenova, Mira. 'The New Concept of Universal Service in a Digital Networked Communications Environment', *I/S: A Journal of Law and Policy for the Information Society* 3, no. 1 (2007): 117–146.

Cabral, Pedro. 'Access to Member State Documents in EC Law', *European Law Review* 31, no. 3 (2006): 378–389.

Cahir, John. 'Understanding Information Laws: A Sociological Approach', *Journal of Information, Law and Technology* no. 3 (2002), <www2.warwick.ac.uk/fac/soc/law/elj/jilt/2002_3/cahir/>, 22 February 2010.

Clark, Mike. 'Fee or Free? The hidden costs of free public sector information', *Business Information Review* 24, no. 1 (2007): 49–59.

Coppel, Philip. 'Environmental Information: The New Regime', *Journal of Planning & Environment Law* (2005): 12–33.

Crompvoets, Joep et al. 'Assessing the Worldwide Developments of National Spatial Data Clearinghouses', *International Journal of Geographical Information Science* 18 (2004): 665–689.

Croner, Charles M. 'Public Health, GIS, and the Internet', *Annual Review of Public Health* 24 (2003): 57–82.

Curtin Deirdre M. 'Citizens' Fundamental Right of Access to EU Information: An Evolving Digital Passepartout?', *Common Market Law Review* 37, no. 1 (2000): 7–41.

De Baere, Rik. 'Openbaarheid van milieu-informatie', *Tijdschrift voor Milieurecht* 4 no. 2 (1995): 90–98.

De Bandt, Pierre. 'Artikel 86, lid 2 EG-Verdrag en het begrip 'diensten van algemeen economisch belang: meer dan een rechtsbasis voor een genuanceerde toepassing van de interne markt- en mededingingsregels op openbare diensten?', *Tijdschrift voor Belgisch Handelsrecht* 114 no. 3 (2008): 211–225.

De Gendt, Tom. 'Het Verdrag van Aarhus in Vlaanderen: Actieve Openbaarheid van Milieu-Informatie, Participatie van de Burger aan het Milieubeleid en Toegang tot de Rechter in Milieuzaken', *Milieu- en Energierecht* no. 4 (2001): 232–302.

De La Presle, Anne. 'Vers un statut des données publiques?', *Revue Brises* no. 12 (1998): 40–48.

De Terwangne, Cécile. 'La Convention Européenne des Droits de l'Homme et le Droit de Recevoir des Informations de la Part des Autorités Publiques', *Aménagement – Environnement* (1998) : 265–270.

De Vries, Marc. 'Implementatie van de EU-richtlijn hergebruik overheidsinformatie in de Wob. Beter ten halve gekeerd dan ten hele gedwaald', *Nederlands Juristenblad* 79 no. 39 (2004): 2038–2043.

Devroe, Wouter. 'Universele dienstverlening als nieuwe manier van denken', *Sociaal-economische Wetgeving* 48 (2000): 82–95.

Di Fabio, Udo. 'Privatisierung und Staatsvorbehalt', *Juristen Zeitung* no. 2 (1999): 585–592.

Dijstelbloem, Huub & Paul De Beer. 'Een kwestie van selectie: de overheid en de informatievoorziening via internet', *Beleid en Maatschappij* 30, no. 4 (2003): 242–250, <www.pauldebeer.nl/documenten/artikelen/Een%20kwestie%20van%20selectie%201.7.doc>, 22 February 2010.

Epstein, Earl F. 'Liability Insurance and the Use of Geographical Information', *International Journal of Geographical Information Science* 12, no. 3 (1998): 203–214.

Fatome, Etienne & Laurent Richier. 'Un triomphe de l'interprétation fonctionnelle ou comment devenir un organisme de droit public', *Actualités de la. Commande et des Contrats Publics* (2003): 63–65.

Gavouneli, Maria. 'Access to Environmental Information: Delimitation of a Right', *Tulane Environmental Law Journal* 13 (2000): 303–327.

Georgiadou, Yola & Richard Groot. 'Policy Development and Capacity Building for Geoinformation Provision: a Global Public Goods Perspective', *GIS Development* 6, no. 7 (2002): 33–40, <www.gisdevelopment.net/policy/international/interna014.htm>, 22 February 2010.

Getis, Arthur et al. 'Geographic Information Science and Crime Analysis', *Journal of the Urban and Regional Information Systems Association* 12, no. 2 (2000): 7–14.

Goodchild, Michael F. & Robert P. Haining. 'GIS and Spatial Data Analysis: Converging Perspectives', *Papers in Regional Science* no. 83 (2004): 363–385.

Groot, Richard. 'Spatial Data Infrastructure (SDI) for Sustainable Land Management', *ITC Journal* no. 3–4 (1997): 287–294.

Hadi, Zakaria A. & Neil McBride. 'The Commercialisation of Public Sector Information within UK Government Departments', *International Journal of Public Sector Management* 13 (2000): 552–570.

Hancher, Leigh & S.J.H. Evans. 'Altmark als katalysator: het Commissiepakket met alle antwoorden rond staatssteun en diensten van algemeen economisch belang?', *Nederlands Tijdschrift voor Economisch Recht* 12 no. 7 (2006): 153–159.

Hansen, Marc, Anne Van Ysendyck & Susanne Zuhlke. 'The Coming of Age of EC State Aid Law: A Review of the Principal Developments in 2002 and 2003', *European Competition Law Review* 25, no. 4 (2004): 202–233.

Hins, Wouter & Dirk Voorhoof. 'Access to State-Held Information as a Fundamental Right under the European Convention on Human Rights', *European Constitutional Law Review* 3, no. 1 (2007): 114–126.

Houben, Iris. Public Service Obligations: Moral Counterbalance of Technical Liberalization Legislation, *European Review of Private Law* 16 (2008): 7–27.

Hovine, Axelle. 'La directive du 17 Novembre 2003 concernant la réutilisation des informations du secteur public: présentation et préfiguration des incidences de transposition en droit français', *Gazette du Palais* (2004): 754–761.

Huisman, Kees & Jan Van De Lei. 'Wet FIDO en publieke taak onder Wet Dualisering', *Tijdschrift B&G* (2003): 34–35.

Janssen, Katleen. 'De exploitatie van overheidsinformatie: eindelijk een ontwerprichtlijn', *Mediarecht* no. 19 (2002): 107–122.

Janssen, Katleen & Jos Dumortier. 'Towards a European Framework for the Re-use of Public Sector Information: A Long and Winding Road', *International Journal of Law and Information Technology* 11, no. 2 (2003): 184–201.

Janssen, Katleen & Jan J.C. Kabel. 'Commercialisering van overheidsinformatie door de overheid: rechtspraak en wetgeving in België en Nederland. De honden blaffen, maar de karavaan trekt voort', *Computerrecht* no. 3 (2005): 117–129

Janssen, Katleen. 'Hergebruik van overheidsinformatie – binnenkort ook bij u in de winkel?', *Privacy & Informatie* 9 no. 2 (2006): 59–63.

Janssen, Katleen & Jos Dumortier. 'The Protection of Maps and Spatial Databases in Europe and the United States by Copyright and the Sui Generis Right', *John Marshall Journal of Computer and Information Law* 24 (2006): 195–225.

Johnson, Jeffrey P. 'Case Studies of Dissemination Policy in Local Government GIS Agencies', *Computer Environment and Urban Systems* 19, no. 5–6 (1996): 373–389.

Karayigit, Mustafa T. 'Under the Triangle Rules of Competition, State Aid and Public Procurement: Public Undertakings Entrusted with the Operation of Services of General Economic Interest', *European Competition Law Review* 30, no. 11 (2009): 542–564.

Kay, John. 'The State and the Market', *Political Quarterly* 75, no. 1 (2004): 74–87.

Klinkenberg, Brian. 'The True Cost of Spatial Data in Canada', *The Canadian Geographer* 47, no. 1 (2003): 37–49.

Klosterman, Richard E. 'The Appropriateness of Geographic Information Systems for Regional Planning in the Developing World', *Computers, Environment and Urban Systems* 19, no. 1 (1995): 1–13.

Krajewski, Markus. 'Providing Legal Clarity and Securing Policy Space for Public Services through a Legal Framework for Services of General Economic Interest: Squaring the Circle?', *European Public Law* 14, no. 3 (2008): 377–398.

Larssen, Christine. 'Het Verdrag van Aarhus en de toepassing ervan in het Belgisch recht – deel 1', *Tijdschrift voor Milieurecht* 14 (2005): 244–270.

Leadbeater, Charles. 'Creating the Public Good', *Political Quarterly* 75, no. 1 (2004): 88–99.

Leith, Philip & Karen McCullagh. 'Developing European Legal Information Markets based on Government Information', *International Journal of Law and Information Technology* 12, no. 3 (2004): 247–281.

Lievrouw, Leah A. 'The Information Environment and Universal Service', *The Information Society* 16, no. 2 (2000): 155–159.

Longhorn, Roger & Michael Blakemore. 'Re-visiting the Valuing and Pricing of Digital Geographic Information', *Journal of Digital Information* 4, no. 2 (2004), <http://journals.tdl.org/jodi/article/view/103/102>, 22 February 2010.

Love, James, 'Pricing of government information', *Journal of Government Information* 22, no. 5 (1995): 363–387.

Manunza, Elisabetta R. 'Privatised Services and the Concept of "Bodies Governed by Public Law" in EC directives on public procurement', *European Law Review* 28 (2003): 273–282.

Marguenaud, Jean-Pierre. 'La Convention d'Aarhus et la Convention européenne des droits de l'homme', *Revue Juridique de l'Environnement* (1999): 77–87.

Maskus, Keith E. & Jerome H. Reichman. 'The Globalization of Private Knowledge Goods and the Privatization of Global Public Goods', *Journal of International Economic Law* 7 (2004): 279–320.

Masser, Ian. 'All shapes and sizes: the first generation of National Spatial Data Infrastructures', *International Journal of Geographical Information Science* 13, no. 1 (1999): 67–84.

McAllister, Sean T. 'The Convention on Access to Information, Public Participation in Decision-making and Access to Justice in Environmental Matters', *Colorado Journal of International Environmental Law and Policy* 187 (1998): 187–198.

Mendel, Toby. 'Freedom of Information: An Internationally protected Human Right', *Comparative Media Law Journal* no. 1 (2003): 39–70.

Merusk, Kalle. 'Assignment of public tasks to private legal persons. Legal bases and limits', *Juridica International* 5 (2000): 40–48, <www.juridica.ee/get_doc.php?id=327>, 22 February 2010.

Mock, William. 'The Centrality of Information Law: A Rational Choice Discussion of Information Law and Transparency', *John Marshall Journal of Computer and Information Law* 17 (1999): 1069–1100.

Moral Soriano, Leonor. 'How Proportionate Should Anti-Competitive State Intervention Be?', *European Law Review* 28, no. 1 (2003): 112–123.

Noteboom, Carrie. 'Addressing the External Effects of Internal Environmental decisions: Public Access to Environmental Information in the International Law Commission's Draft Articles on the Prevention of Transboundary Harm', *New York University Environmental Law Journal* 12 (2003): 245–285.

Onsrud, Harlan J. 'In Support of Open Access for Publicly Held Geographic Information', *GIS Law* 1, no. 1 (1992): 3–6, <www.spatial.maine.edu/~onsrud/pubs/In_Support_OA.htm>, 22 February 2010.

Onsrud, Harlan J. 'In Support of Cost Recovery for Publicly Held Geographic Information', *GIS Law* 1, no. 2 (1992): 1–7, <www.spatial.maine.edu/%7eonsrud/pubs/Cost_Recovery_for_GIS.html>, 22 February 2010.

O'Reilly, James T. ' "Access to Records" Versus "Access to Evil": Should Disclosure Laws Consider Motives as a Barrier to Records Release?', *Kansas Journal of Law and Public Policy* no. 12 (2003): 559–574.

Onsrud, Harlan J., Jeffrey Johnson & Judy Winnecki. 'GIS Dissemination Policy: Two Surveys and a Suggested Approach', *Journal of the Urban and Regional Information Systems Association* 8 (1996): 8–23.

Onsrud, Harlan J., Jeff P. Johnson & Xavier Lopez. 'Protecting Personal Privacy in Using Geographic Information Systems', *Photogrammetric Engineering and Remote Sensing* 70 (1994): 1083–1095, <www.spatial.maine.edu/%7eonsrud/pubs/privacy16.pdf>, 22 February 2010.

Perritt, Henry H. Jr. & Christopher J. Lhulier. 'Information Access Rights Based on International Human Rights Law', *Buffalo Law Review* 45 (1997): 899–929.

Perritt, Henry, Jr. 'Should Local Governments Sell Local Spatial Databases through State Monopolies?', *Jurimetrics Journal* 35 (1994–1995): 449–469.

Perritt, Henry H. Jr. 'Sources of Rights to Access Public Information', *William and Mary Bill of Rights Journal* 4 (1995): 179–221.

Pettifer, Richard E.W. 'Towards a Stronger European Market in Applied Meteorology', *Meteorological Applications* 15, no. 2 (2008): 305–312.

Phillips, Jennifer. 'Information Liability: the Possible Chilling Effect of Tort Claims against Producers of Geographic Information Systems Data', *Florida State University Law Review* 26 (1999): 743–781.

Plant, Raymond. 'Neo-liberalism and the Theory of the State: From Wohlfahrtstaat to Rechtsstaat', *Political Quarterly* 75, no. 1 (2004): 24–37.

Poullet, Yves. 'Freedom and Information Highways or How to Ensure Electronic Democracy', *Telematics and Informatics* 15 (1998): 163–180.

Poullet, Yves. 'The Commercialization of Data Held by the Public Sector', *The Computer Law and Security Report* 9 (1993): 227–233.

Poupaert, Nathalie & Katleen Janssen. 'La directive du 17 Novembre 2003 concernant la réutilisation des informations du secteur public', *Revue du droit des technologies de l'information* no. 19 (2004) : 29–50.

Prieur, Michel. 'La Convention d'Aarhus, Instrument Universel de la Démocratie Environnementale', *Revue Juridique de l'Environnement* (1999): 9–29.

Quigley, Conor. 'The European Commission's State Aid Action Plan for 2005–2009', *International Company and Commercial Law Review* 16, no. 12 (2005): 485–489.

Rase, Daniel. 'Initiatives Taken by the Commission to Establish a European Geographic Information Infrastructure', *Statistical Journal of the United Nations ECE*, 2000: 141–147, <www.unece.org/stats/documents/2000/04/gis/6.e.pdf>, 22 February 2010.

Richer, Laurent. 'Droit d'accès et service public', *Actualité juridique Droit Administratif* 16 (2006) : 73–79.

Robinson, David et al. 'Government Data and the Invisible Hand', *Yale Journal of Law and Technology* 11 (2009): 160–175.

Rodenhoff, Vera. 'The Aarhus Convention and its Implications for the "Institutions" of the European Community', *RECIEL* 11, no. 3 (2002): 343–357.

Rodrigues, Stéphane. Les qualifications concurrentes des activités d'intérêt général en droit communautaire', *Actualité juridique Droit Administratif* 16 (2006): 84–90.

Ronai, Maurice. 'L'Etat comme machine informationnelle', *Revue Française d'Administration Publique* 72 (1994): 571–580.

Ross, Malcolm. 'A Healthy Approach to Services of General Economic Interest? The BUPA Judgment of the Court of First Instance', *European Law Review* 34, no. 1 (2009): 127–140.

Rowlands, Ian. 'Toward Public-Private Synergy in the European Information Services Market', *Journal of Government Information* 22, no. 3 (1995): 227–235.

Salkin, Patricia. 'GIS In an Age of Homeland Security: Accessing Public Information to Ensure a Sustainable Environment', *William and Mary Environmental Law and Policy Review* 33 (2005): 55–94.

Samuelson, Paul A. 'The Pure Theory of Public Expenditure', *Annual Review of Information Science and Technology* 36, no. 4 (1954): 387–389.

Sauter, Wolf. 'Services of General Economic Interest and Universal Service in EU law', *European Law Review* 33, no. 2 (2008): 167–193.

Saxby, Stephen. 'The Development of UK Government Policy Towards the Commercialization of Official Information', *International Journal of Law and Information Technology* 4, no.3 (1996): 199–233.

Schram, Frankie. 'Openbaarheid van Europese bestuursdocumenten', *Nieuw Juridisch Weekblad* no. 32 (2003): 582–592.

Schram, Frankie. 'Openbaarheid van bestuur: een voorbeeld van interactie tussen het Europees en nationaal publiekrecht', *Tijdschrift voor Bestuurswetenschappen en Publiekrecht* no. 3 (1999): 155–179.

Scott, Colin. 'Services of General Interest in EC Law: Matching Values to Regulatory Technique in the Public and Privatised Sectors', *European Law Journal* 6, no. 4 (2000): 310–325.

Seipel, Peter. 'The Technology of Insight: Computers and Informed Citizens', *Chicago-Kent Law Review* 69, no. 2 (1993): 417–457.

Shamsul Haque M. 'The Diminishing Publicness of Public Service under the Current Mode of Governance', *Public Administration Review* 61, no. 1 (2001): 65–82, <http://profile.nus.edu.sg/fass/polhaque/par-public.pdf>, 22 February 2010.

Silverstein, Michelle R. 'The Copyrightability of Factual Compilations: An Interpretation of Feist Through Cases of Maps and Numbers', *Annual Survey of American Law* 53 (1996): 147–218.

Snijkers, Kris & Koen Verhoest. 'Het kerntakendebat: lessen voor de toekomst', *Vlaams Tijdschrift voor Overheidsmanagement* 8 no. 3 (2003): 20–25.

Steyger, Elies. 'Een overheidsbegrip in het Europese recht?', *Nederlands Tijdschrift voor Bestuursrecht* 55 no. 7 (2005): 221–228.

Steyger, Elies. 'Deel II: De publieke taak: diensten van algemeen (economisch) belang en de gevolgen van de richtlijn', *Sociaal-Economische Wetgeving* no. 10 (2007): 379–398.

Teresi, Laurent. 'Données publiques et concurrence', *Concurrences* no. 4 (2005): 143–156.

Van Calster, Geert. 'Openbaarheid van bestuur in de Europese Unie: recente ontwikkelingen', *Rechtskundig Weekblad* no. 36 (1994–1995): 1209–1219.

Van De Gronden, Johan W. 'Purchasing Care: Economic Activity or Service of General (Economic) Interest', *European Competition Law Review* 25, no. 2 (2004): 87–94.

Van De Gronden, Johan W. 'Deel III: Bestaat er een coherente visie op de verhouding tussen diensten van algemeen economisch belang, de Dienstenrichtlijn en andere communautaire regelingen?', *Sociaal-Economische Wetgeving* 55 no. 10 (2007): 399–421.

Van Eechoud, Mireille. 'Nader tot U – Brussel verordent openbaarheid EU-documenten', *Mediaforum* 12 no. 3 (2000): 73–79.

Van Eechoud, Mireille. 'Het hergebruik regime voor overheidsinformatie in de Wob, een tussenstand', *Mediaforum* 20 no. 1 (2008): 2–10.

Vanheusden, Bernard. 'Vrije toegang tot milieu-informatie: een stand van zaken en de implicaties van het verdrag van Aarhus voor Europa en voor België', *Milieu- en Energierecht* 1999: 266–296.

Van Damme, Eric. 'Concurrentievervalsing door de overheid?', *Openbare Uitgaven* 30, no. 4 (1999): 194–203.

Van Loenen, Bastiaan & Jaap A. Zevenbergen. 'The Impact of the European Privacy Regime on Location Technology Development', *Journal of Location based services* 1 (2007): 165–178.

Van Parys, Jurian. 'Het begrip onderneming in mededingingsrecht: ziekenfonds als onderneming?', *Tijdschrift voor Belgische Mededinging* (2006): 59–65.

Verhoest, Pascal. 'The Myth of Universal Service: Hermeneutic Considerations and Political Recommendations', *Media, Culture & Society* 22, no. 5 (2000): 595–610.

Voorhoof, Dirk. 'Europees Mensenrechtenhof maakt (eindelijk) toepassing van artikel 10 EVRM in het kader van openbaarheid van bestuur en toegang tot bestuursdocumenten', *Mediaforum* 18 no. 10 (2006): 290–294.

Walwei, Ulrich. 'Arbeitsvermittlung aks öffentliche Aufgabe under privatwirtschaftliche Dienstleistung', *Mitteillungen aus der Arbeitsmarkt- und Berufsforschung (MittAB)* no. 1 (1996): 54–72.

Wilcox, William A. Jr. 'Access to Environmental Information in the United States and the United Kingdom', *Loyola of Los Angeles International and Comparative Law Review* 23 (2001): 121–247.

Williams, Rhodri. 'European Commission Decision 2008/963 updating the lists of contracting authorities under the procurement directives', *Public Procurement Law Review* (2009): 93–94.

Young, Mark. 'The Future of Universal Service. Does It Have One', *International Journal of Law and Information Technology* 13, no. 2 (2005): 188–205.

Yu, Peter K. 'Symposium – Bridging the Digital Divide: Equality in the Information Age. Introduction,' *Cardozo Arts and Entertainment Law Journal* 20 (2002): 1–52.

Zwenne, Gerrit-Jan. 'De informatieconsument en de markt voor overheidsinformatie,' *Tijdschrift voor Consumentenrecht* (1997): 223–231.

ARTICLES AND PAPERS PUBLISHED ON THE INTERNET

Andrews, William J. 'Nurturing the Global Information Commons: Public Access, Public Infrastructure', 1996, <http://library2.usask.ca/gic/v2n3/andrews/andrews.html>, 22 February 2010.

Association of Environmental Data Users in Europe. 'Comfort Letter on the Joint Sale of Meteorological Products by Ecomet. Comments', 1999, <www.primet.org/index.php?Itemid=29&id=81&option=com_content&task=view>, 22 February 2010.

Bastiat, Frédéric. 'L'Etat', <http://bastiat.org/fr/l_Etat.html>, 22 February 2010.

Bing, Jon. 'Managing copyright in a digital environment', <www.portlandpress.com/pp/books/online/tiepac/session2/ch3.htm>, 22 February 2010.

Blakemore, Michael. 'Financing the NGDI', <www.gisdevelopment.net/policy/international/interna012pf.htm>, 22 February 2010.

Buchwald, Cheryl. 'Canada in Context: an Overview of Information Policies in Four Industrialized Countries', 1995, <www3.fis.utoronto.ca/research/iprp/publications/wp/wp2.html>, 22 February 2010.

Clarke, Roger. 'Freedom of Information? The Internet as Harbinger of the New Dark Ages', *First Monday* 4, no. 11 (1999), <http://firstmonday.org/htbin/cgiwrap/bin/ojs/index.php/fm/issue/view/112>, 22 February 2010.

Craglia, Max. 'Introduction to the International Journal of Spatial Data Infra structures Research', *International Journal of Spatial Data Infrastructures Research* 1 (2006): 1–13, <http://ijsdir.jrc.ec.europa.eu/index.php/ijsdir/article/view/14/8>, 22 February 2010.

ePSIplus. 'PSIH Closes Competitive Service', 2007, <www.epsiplus.net/news/psih_closes_competitive_service>, 22 February 2010.

ePSIplus. 'Locating Exclusive Agreements', 2008, <www.epsiplus.net/news/locating_exclusive_agreements>, 22 February 2010.

Geospatial Information & Technology Association. 'Free of Fee: the Governmental Data Ownership Debate. GITA White Paper', 2005, <www.gita.org/resources/whitepapers/free_or_fee.pdf>, 22 February 2010.

Grus, Lukasz, Joep Crompvoets & Arnold K. Bregt. 'Multi-view SDI Assessment Framework', *International Journal of Spatial Data Infrastructures Research* 2 (2007): 33–53, <http://ijsdir.jrc.ec.europa.eu/index.php/ijsdir/article/view/27/21>, 22 February 2010.

Holland, William S. 'Copyright, Licensing and Cost Recovery for Geographic and Land Information System Data: A Legal, Economic and Policy Analysis', 1997, <www.spatial.maine.edu/~onsrud/tempe/holland.html>, 22 February 2010.

Hou, Liyang. 'Uncovering the veil of Article 86(2)EC', 2007, <http://papers.ssrn.com/sol3/papers.cfm?abstract_id=1025407>, 22 February 2010.

Jacques, Johan. 'Meteoservices, Belgium: Case against KMI. Progress report', <www.primet.org/documents/240206_johan_jaques_meteoservices_belgium.html>, 22 February 2010.

Joffe, Bruce. '10 Ways to Support Your GIS Without Selling Data', <www.opendataconsortium.org/documents/10Ways2SupportGIS-3.pdf>, 22 February 2010.

Joffe, Bruce. 'The GIS Data Sales Dilemma – Finding a Middle Ground', 1995, <www.opendataconsortium.org/article_gis_data_sales_dilemma.htm>, 22 February 2010.

Johnson, Jeff. P. & Harlan J. Onsrud. 'Is Cost Recovery Worthwhile?', 1995, <www.spatial.maine.edu/%7eonsrud/pubs/Cost_Recovery_Worthwhile.html>, 22 February 2010.

Karjala, Dennis S. 'Copyright in Electronic Maps', *Jurimetrics Journal* 35 (1995): 395–415, <http://homepages.law.asu.edu/~dkarjala/Articles/Jurimetrics1995.html>, 22 February 2010.

Landry, Peter. 'The Theory of Government', <www.blupete.com/Literature/Essays/BluePete/Government.htm>, 22 February 2010.

Locus Association. 'Submission to the European Commission on the Re-use of Public Sector Information Directive', 2008, <http://ec.europa.eu/information_society/policy/psi/online_consultation/stakeholders/index_en.htm>, 22 February 2010.

Maven, Richard. 'Interview: Online Mapping Guru Ed Parsons', <www.e-consultancy.com/news-blog/362454/interview-online-mapping-guru-ed-parsons.html>, 22 February 2010.

McDonagh, Maeve. 'Access to Public Information: A key to Commercial Growth and Electronic Democracy – Stockholm. Conference report', *Journal of Information, Law and Technology* no. 3 (1996), <www2.warwick.ac.uk/fac/soc/law/elj/jilt/1996_3/mcdonagh/>, 22 February 2010.

Negin, Brian. 'The Dissemination of Government Information: A Comparison between England and the USA', 1996, <www.itpolicy.gov.il/topics_info/pubusuk-1.htm>, 22 February 2010.

Onsrud, Harlan J. & Xavier Lopez. 'Intellectual Property Rights in Disseminating Digital Geographic Data, Products, and Services: Conflicts and Commonalities among European Union and United States Approaches', 1998, <www.spatial.maine.edu/~onsrud/harlan/Course/CPOnsrudLopez.html>, 22 February 2010.

Onsrud, Harlan J. 'Geographic Information Legal Issues', 2004, <www.spatial.maine.edu/~onsrud/pubs/GILegalIssues.html>, 22 February 2010.

Oxera. 'Public Information, Private Profit: How Should Government Agencies Compete ?', 2005, 2, <www.oxera.com/cmsDocuments/Agenda_Oct%2005/Public%20information%20private%20profit.pdf>, 22 February 2010.

Papapavlou, George. 'Public Sector Information Initiatives in the European Union', 1999, <http://webworld.unesco.org/infoethics2000/documents/paper_papapavlou.rtf>, 22 February 2010.

Pas, Johan. 'The Commercialization of Government Information and the Proposal for a Directive COM (2002) 207 by the European Commission', *Murdoch University Electronic Journal of Law* 9, no. 4 (2002), <www.murdoch.edu.au/elaw/issues/v9n4/pas94_text.html>, 22 February 2010.

Pas, Johan & Bruno De Vuyst. 'Re-Establishing the Balance Between the Public and the Private Sector: Regulating Public Sector Information Commercialization in Europe', *Journal of Information, Law and Technology* no. 2 (2004), <www.warwick.ac.uk/fac/soc/law/elj/jilt/2004_2/pasanddevuyst>, 22 February 2010.

Peters, Johannes A.F. & Reinout D. Vriesendorp. 'Insolvency of Public Entities Other Than the State under Dutch Law', *Electronic Journal of Comparative Law* 6 (2002), <www.ejcl.org/64/art64-23.pdf>, 22 February 2010.

Pitt, Edward. 'Ordnance Survey – The Ground Rules', 2003, <www.epsiplus.net/cases/ordnance_survey>, 22 February 2010.

Sand, Peter H. 'The Right to Know: Environmental Information Disclosure by Government and Industry', 2002, <www.inece.org/forumspublicaccess_sand.pdf>, 22 February 2010.

Veldheer, Vic. 'Waar ligt het evenwicht tussen overheidsbemoeienis en marktwerking?'. In *Altijd een antwoord. SCP Nieuwjaarsuitgave 2006*, edited 2006, <www.scp.nl/publicaties/boeken/9037702635/Altijd_een_antwoord.pdf>, 22 February 2010.

Wainwright, Hilary. 'Reclaiming "The Public" through the People', 2004, <www.tni.org/detail_page.phtml?act_id=2368&password=9999&publish=Y&username=guest@tni.org>, 22 February 2010.

Wiese Schartum, Dag. 'Access to Government-Held Information: Challenges and Possibilities', *Journal of Information, Law and Technology* no. 1 (1998), <www2.warwick.ac.uk/fac/soc/law/elj/jilt/1998_1/schartum/>, 22 February 2010.

REPORTS

Acil Tasman. 'The value of spatial information. The impact of modern spatial information technologies on the Australian economy', 2008, <www.crcsi.com.au/uploads/publications/PUBLICATION_324.pdf>, 22 February 2010.

BDO Consultants Bestuur en Management. 'Elektronische bestanden van het bestuur'. Groningen: BDO, 1998.

Berenschot and NEI. 'Welvaartseffecten van verschillende financieringsmethoden van elektronische gegevensbestanden', 2001, <www.epsiplus.net/media/files/final_eindrapport_onderzoek_welvaartseffecten_feb2001>, 22 February 2010.

Bruguière, Jean-Michel. 'Droit des données publiques', <www.educnet.education.fr/chrgt/droitddp.pdf>, 22 February 2010.

Corbin, Chris. PSI Pricing 1 – ePSIplus Thematic Meeting Report 2007, <www.epsiplus.net/reports/epsiplus_thematic_meeting_reports/fin_mtg_1>, 22 February 2010.

Commissie Computercriminaliteit. *Informatietechniek en Strafrecht.* The Hague: Sdu Uitgevers, 1987.

Commissie Grondrechten in het Digitale Tijdperk. 'Rapport Grondrechten in het Digitale Tijdperk', 2000, <www.ivir.nl/dossier/grondrechten/bronnen/rapport_gdt_5-00.pdf>, 22 February 2010.

Commissie Tweede Consultatieronde Vrijgeven NWB Bestand. 'Advies', 2006, <www.verkeerenwaterstaat.nl/Images/Advies%20Cie%20tweede%20consultatie_tcm195-189308.pdf>, 22 February 2010.

Davies, Rob et al. 'Economic and Social Impact of the Public Domain: EU cultural institutions and the PSI Directive', 2009, <http://ec.europa.eu/information_society/policy/psi/docs/pdfs/report/comm_0022_psi_v6.0.pdf>, 22 February 2010.

Dekkers, Makx et al. 'Measuring European Public Sector Information Resources (MEPSIR). Final Report', 2006, <http://ec.europa.eu/information_society/policy/psi/docs/pdfs/mepsir/final_report.pdf>, 22 February 2010.

De Vries, Marc. *Redress of PSI Re-use Disputes: The Missing Teeth?– ePSIplus Thematic Meeting Report*, 2008, <www.epsiplus.net/reports/epsiplus_thematic_meeting_reports/law_and_regulation_meeting_3_report>, 22 February 2010.

Economic Development and Infrastructure Committee. 'Inquiry into improving access to Victorian public sector information and data. Discussion Paper', 2008, <www.parliament.vic.gov.au/edic/inquiries/access_to_PSI/call_for_submissions.html>, 22 February 2010.

Economic Development and Infrastructure Committee, 'Inquiry into Improving Access to Victorian Public Sector Information and Data', 2009, <www.parliament.vic.gov.au/edic/inquiries/access_to_PSI/final_report.html>, 22 February 2010.

European Cente on Sustainable Politics for Human and Environmental Rights. 'The Aarhus Regulation: New Opportunities for Citizens in the EU and Beyond? Report of the ECOSPHERE Forum held in Brussels on 27 October 2006', <www.ecosphere.be/pdfflyers/ReportForumOct2006.pdf>, 22 February 2010.

Fitzgerald, Anne. 'Open Access Policies, Practices and Licensing: a Review of the Literature in Australia and Selected Jurisdictions'. Brisbane: Queensland University of Technology, 2009.

Franck, Andrew U. 'The Surveying Activities at the Austrian Federal Office for Metrology and Surveying: an Economic Analysis', 2003, <ftp://ftp.geoinfo.tuwien.ac.at/frank/reportbevfinalv31prn.pdf>, 22 February 2010.

Geographic Information Network in Europe. 'New Issues for the European GI Strategy: Public Sector Information', 2003, <www.ec-gis.org/ginie/doc/d293_issues_pse_fv3.pdf>, 22 February 2010.

GSDI Association. 'The SDI Cookbook v2.0', 2004, <www.gsdi.org/docs2004/Cookbook/cookbookV2.0.pdf>, 22 February 2010.

Hallo, Ralph E. 'Public Access to Environmental Information: European Environment Agency's Experts Corner Number 1997/1'. Luxembourg: Office for Official Publications of the European Communities, 1997.

International Standards Organisation. 'Standard 2382-1. Information technology – Vocabulary – Part 1: Fundamental terms', 1993, <www.iso.org/iso/en/CatalogueDetailPage.CatalogueDetail?CSNUMBER=7229>, 22 February 2010.

KPMG. 'Ministerie van Waterstaat. Evaluatie van de wet op het KNMI', 2007, <www.verkeerenwaterstaat.nl/Images/2008651%20bijlage_tcm195-231492.pdf>, 22 February 2010.

Longworth, Elizabeth. 'The Role of Public Authorities in Access to Information: the Broader and More Efficient Provision of Public Content'. Study Prepared for the Third UNESCO International Congress on Ethical, Legal and Societal Challenges of Cyberspace INFOethics', 2000, <http://unesdoc.unesco.org/images/0012/001210/121051e.pdf>, 22 February 2010.

Mandelkern, Dieudonné & Bertrand Du Marais. 'Diffusion des données publiques et révolution numérique: rapport', 1999, <www.ladocumentationfrancaise.fr/rapports-publics/994001620/index.shtml#>, 22 February 2010.

Mayo, Ed & Tom Steinberg. 'The Power of Information: An Independent Review', 2007, <www.cabinetoffice.gov.uk/media/cabinetoffice/strategy/assets/power_information.pdf>, 22 February 2010.

Micus Management Consulting Gmbh. 'Assessment of the Re-use of Public Sector Information (PSI) in the Geographical Information, Meteorological Information and Legal Information Sectors', 2008, <http://ec.europa.eu/information_society/policy/psi/docs/pdfs/micus_report_december2008.pdf>, 22 February 2010.

Newbery, David, Lionel Bently & Rufus Pollock. 'Models of Public Sector Information Provision via Trading Funds', 2008, <www.berr.gov.uk/files/file45136.pdf>, 22 February 2010.

Nilsen, Kirsti. 'Economic Theory as It Applies to Statistics Canada: a Review of the Literature', 2007, <www.chass.utoronto.ca/datalib/misc/Nilsen%20Economics%20Paper%202007%20final%20version.pdf>, 22 February 2010.

Oxera. 'The Economic Contribution of Ordnance Survey GB', 1999, <www.ordnancesurvey.co.uk/oswebsite/aboutus/reports/oxera/oxera.pdf>, 22 February 2010.

PIRA International. Commercial Exploitation of Europe's Public Sector Information. Final Report, 2000, <http://ec.europa.eu/information_society/policy/psi/docs/pdfs/pira_study/commercial_final_report.pdf>, 22 February 2010.

Ravi Bedrijvenplatform. 'Auteursrecht op geo-informatie'. Amersfoort: Ravi, 1994.

Ravi Bedrijvenplatform. 'Bescherming en verstrekking van geo-informatie'. Amersfoort: Ravi, 1995.

Ravi Bedrijvenplatform. 'Economische effecten van laagdrempelige beschikbaarstelling van overheidsinformatie'. Amersfoort: Ravi, 2000.

Ravi Bedrijvenplatform. 'Commercialisering van geo-informatie. Een drietal rapporten ter ondersteuning van de Ravi-discussie 1997'. Amersfoort: Ravi, 1997.

Reinsma, Marijke & H. Van Der Sluijs. 'Naar ruimere openbaarheid en een vrij gebruik van bestuurlijke informatie', 2002, <www.minbzk.nl/aspx/download.aspx?file=/contents/pages/5006/RapportReinsmaenVanderSluijs.pdf>, 22 February 2010.

Sauter, Wolf. 'Services of General Economic Interest and Universal Service Obligations as an EU Law Framework for Curative Health Care. Tilec Discussion Paper', 2007, <http://papers.ssrn.com/sol3/papers.cfm?abstract_id=1013261>, 22 February 2010.

Schaefer, Brett D. 'How the Scope of Government Shapes the Wealth of Nations', 2005, <www.heritage.org/research/tradeandforeignaid/hl925.cfm>, 22 February 2010.

Sears, Garry. 'Geospatial Data Policy Study. Executive Summary', 2001, <www.geoconnections.org/programsCommittees/proCom_policy/keyDocs/KPMG/KPMG_E.pdf>, 22 February 2010.

Stiglitz, Joseph E., Peter R. Orszag & Jonathan M. Orszag. 'The Role of Government in a Digital Age', 2000, <http://unpan1.un.org/intradoc/groups/public/documents/apcity/unpan002055.pdf>, 22 February 2010.

Tanzi, Vito. 'The Role of the State and the Quality of the Public Sector', IMF Working Paper, 2000, 4, <www.imf.org/external/pubs/ft/wp/2000/wp0036.pdf>, 22 February 2010.

University of Sheffield et al. 'MADAME – Methods for Access to Data and Metadata in Europe. Increasing Access to Public Sector Information: Guidelines of Best Practice'. Sheffield: University of Sheffield, 2000.

Van Eechoud, Mireille & Brenda Van Der Wal. 'Creative Commons Licensing for Public Sector Information. Opportunities and Pitfalls', 2008, <www.ivir.nl/publications/eechoud/CC_PublicSectorInformation_report_v3.pdf>, 22 February 2010.

Van Loenen, Bastiaan et al. Beschikbaar stellen van geo-informatie bij Rijkswaterstaat. Analyse van de (on)mogelijkheden van het op korte termijn vrij beschikbaar stellen van vier geo-data sets, 2006, <www.bastiaanvanloenen.nl/pubs/2006_RWS_AGI_Eindrapport_TUD.pdf>, 22 February 2010.

Van Loenen, Bastiaan et al. 'Open toegankelijkheidsbeleid voor geo-informatie vergeleken: het gras leek groener dan het was?', 2007, <www.minbzk.nl/aspx/download.aspx?file=/contents/pages/87285/opentoegankelijkheidsbeleidvooroverheidvergeleken.pdf>, 22 February 2010.

Waelde, Charlotte & Mags McGinley. *Designing a Licensing Strategy for Sharing and Re-use of Geospatial Data in the Academic Sector*, 2007, <Http://edina.ac.uk/projects/grade/gradeDigitalRightsIssues.pdf>, 22 February 2010.

Werkgroep Markt en Overheid. *Eindrapport*, 1997, <www.ez.nl/dsresource?objectid=148229&type=PDF>, 22 February 2010.

Wetenschappelijke Raad voor het Regeringsbeleid. *Het borgen van publiek belang*. The Hague: Sdu Uitgevers, 2000, <www.wrr.nl/content.jsp?objectid=2545>, 22 February 2010.

Wetenschappelijke Raad voor het Regeringsbeleid. *De verzorgingsstaat herwogen*, 2006, <www.wrr.nl/content.jsp?objectid=2658>, 22 February 2010.

CONFERENCE PRESENTATIONS, PAPERS AND SPEECHES

Annoni, Alessandro & Massimo Craglia. 'Towards a Directive Establishing an Infrastructure for Spatial Information in Europe (INSPIRE)'. Paper given at the GSDI-8 Conference. Cairo, 2005, <http://gsdidocs.org/gsdiconf/GSDI-8/papers/ts_47/ts47_01_annoni_graglia.pdf>, 22 February 2010.

Bauler, Tom & Walter Hecq. 'Some Theoretical Considerations in Response to the Claim after Information for Decision-making'. Paper Given at the International Sustainable Development Research Conference. Nottingham, 2000, <www.ulb.ac.be/igeat/igeat/ulb_igeat/upload/cedd/Bauler_leeds-paper00.pdf>, 22 February 2010.

Bishop, Cheryl Ann. 'Access to Information as a Human Right: Analysis of the United Nations Human Rights Committee Documents'. Paper Given at the Annual Meeting of the International Communication Association. Dresden, 2006, <www.allacademic.com/meta/p91640_index.html>, 22 February 2010.

Burkert, Herbert. 'Conclusions'. In Commission of the European Communities. Paper given at the Workshops on Commercial and Citizen's Access to Government Information. Luxembourg, 1995.

Burkert, Herbert. 'Offentliche Informationen – Privater Markt – Thesen zur staatlichen Informationsverantwortung'. International Symposium Public Data on the Private Market – New Regulations on the Re-use of Public Sector Information. Potsdam, 4–5 June 2007, <www.lda.brandenburg.de/media/lbm1.a.2628.de/Burkert_070730.pdf>, 22 February 2010.

Burkhardt, Jacqueline. 'Conclusions from Session 4. Public Access to Environmental Information: Current Practices'. Paper given at the the OECD Seminar Public Access to Environmental Information. Athens, 2000, <www.olis.oecd.org/olis/2000doc.nsf/LinkTo/NT00000F96/$FILE/00080429.PDF>, 22 February 2010.

Cagdas Aydinoglu, Arif, Philippe De Maeyer & Tahsin Yomralioglu. 'Examining European SDI Initiatives in Anticipation of a SDI Framework for Turkey'. Paper given at the GSDI-8 Conference. Cairo, 2005, <http://gsdidocs.org/gsdiconf/GSDI-8/papers/ts_31/ts31_04_aydinoglu_etal.pdf>, 22 February 2010.

Craglia, Massimo & Andrew Johnston. 'Assessing the Impacts of Spatial Data Infrastructures: Methods and Gaps'. Paper given at the 7th AGILE Conference on Geographic Information Science. Heraklion, 2004, <http://plone.itc.nl/agile_old/Conference/greece2004/papers/1-1-1_Craglia.pdf>, 22 February 2010.

Gellman, Robert. 'The American Model of Access to and Dissemination of Public Information'. Paper given at the European Commission Conference: Access to Public Information. A Key to Commercial Growth and Electronic Democracy. Stockholm, 1996.

Janssen, Katleen. 'INSPIRE and the PSI Directive: Public Task versus Commercial Activities?'. Paper given at the 11th EC-GIS Workshop, 'ESDI – Setting the framework'. Alghero, 2005, <www.ec-gis.org/Workshops/11ec-gis/papers/303janssen.pdf>, 22 February 2010.

Kaschl, Arno. 'GMES Global Monitoring for Environment and Security: Recent Developments'. Paper given at the INSPIRE Conference 2008. Maribor, 2008, <www.ec-gis.org/Workshops/inspire_2008/presentations/05_03_kaschl.pdf>, 22 February 2010.

Lamouline, Christian. 'Presentation of Publaw 3 Findings. Legal Assessment'. Paper given at the European Commission Conference: Access to Public Information. A Key to Commercial Growth and Electronic Democracy. Stockholm, 1996.

Lim A. Po & R. Hans. 'Good Governance in the Shared Public Domain'. Paper given at the Symposium on Good Governance, a Basic Condition for Economic Development, Towards a Surinamese Agenda. Paramaribo, 2003, <www.fhrinstitute.com/good_governance.htm>, 22 February 2010.

Litman, Jessica. 'Rights in Government-generated Data'. Paper given at the Conference on Law and Information Policy for Spatial Databases. Tempe, 1994, <www.spatial.maine.edu/~onsrud/tempe/litman.html>, 22 February 2010.

Longhorn, Roger A. 'Regional Geographic Information Policy: Fact or Fiction – The Case in Europe and Lesssons for GSDI'. Paper given at the GSDI-4 Conference. Capetown, 2000, <http://gsdidocs.org/docs2000/capetown/longhorn.rtf>, 22 February 2010.

Martin, John. 'Historical Background of EC Initiatives Relating to the Public/Private Sector Relationship in the Information Market'. Paper given at the Workshop on Synergy between the Public and Private Sectors in the Information Market. Luxembourg, 1993.

Matsunaga, Keene & Jack Dangermond. 'Promoting a Free Access or Minimal Cost of Dissemination Arrangement for Government-held Geographic Information Systems Data'. Paper given at the Conference on Law and Information Policy for Spatial Databases. Tempe, 1994, <www.spatial.maine.edu/~onsrud/tempe/matsunaga.html>, 22 February 2010.

McDonagh, Maeve. 'Access to Public Sector Information: The Australian Experience'. Paper given at the European Commission Conference: Access to Public Information. A Key to Commercial Growth and Electronic Democracy. Stockholm, 1996.

Nouel, Bertrand. 'Commercialisation des données publiques et concurrence'. Paper given at the European Commission Conference: Access to Public Information. A Key to Commercial Growth and Electronic Democracy. Stockholm, 1996.

OECD Secretariat. 'OECD: Overview and Main Issues'. Paper given at the the OECD Seminar Public Access to Environmental Information. Athens, 2000, <www.olis.oecd.org/olis/2000doc.nsf/LinkTo/NT00000F96/$FILE/00080429.PDF>, 22 February 2010.

OECD Secretariat. 'The OECD Recommendation and Other International Acts on Environmental Information'. Paper given at the the OECD Seminar Public Access to Environmental Information. Athens, 2000, <www.olis.oecd.org/olis/2000doc.nsf/LinkTo/NT00000F96/$FILE/00080429.PDF>, 22 February 2010.

Papapavlou, George. 'Minutes'. Paper given at the European Commission Conference: Access to Public Information. A Key to Commercial Growth and Electronic Democracy. Stockholm, 1996.

Pas, Johan. 'The Use and Re-use of Public Sector Information (PSI): Some Legal and Policy Reflections'. Paper given at the Conference of the Canadian Association for Information Science. Winnipeg, 2004, <www.cais-acsi.ca/proceedings/2004/pas_2004.pdf>, 22 February 2010.

Peterson Dando, Lori. 'A Case for the Commercialization of Public Information'. Paper given at the Conference on Law and Information Policy for Spatial Databases. Tempe, 1994, <www.spatial.maine.edu/%7eonsrud/tempe/dando.html>, 22 February 2010.

Plantinga, Mirjam & Albertjan Tollenaar. 'Public Governance in the Welfare State. The Consequences of Privatisation for Securing Public Interests in the History of the Dutch Welfare State'. Paper given at the 4th ECPR conference. Pisa, 2007, <http://papers.ssrn.com/sol3/Delivery.cfm/SSRN_ID1277882_code950891.pdf?abstractid=1103468&mirid=1>, 22 February 2010.

Poullet, Yves. 'Towards a Universal Public Information Service in Commission of the European Communities'. Paper given at the Workshops on Commercial and Citizen's Access to Government Information. Luxembourg, 1995.

Poullet, Yves. 'Plaidoyer pour un ou des service(s) universel(s) d'informations publiques'. Paper given at the European Commission Conference: Access to Public Information. A Key to Commercial Growth and Electronic Democracy. Stockholm, 1996.

Prins, Corien. 'Study on a Green Paper Regulating Public Access to Government Information'. Paper given at the Workshops on Commercial and Citizen's Access to Government Information. Luxembourg, 1995.

Prosser, Tony. 'Services of General Economic Interest in Community Law: From Single Market to Citizenship Rights'. Paper given at the University of Rome. Rome, 2003, <http://ecoservr.eco.uniroma1.it/pdf/Rome_lecture_Dece03.pdf>, 22 February 2010.

Rodrigues, Stéphane. 'Towards a General EC Framework Instrument Related to SGEI? Political Considerations and Legal Constraints'. Paper given at the International Conference 'The changing framework of public services in Europe'. Potsdam, 2008.

Rybaczuk, Krysia & Michael Blakemore. 'Selling Government Information. A Comparative Perspective on UK and USA Developments'. Paper given at the Conference on Law and Information Policy for Spatial Databases. Tempe, 1994, <www.spatial.maine.edu/%7eonsrud/tempe/rybaczuk.html>, 22 February 2010.

Saxby, Stephen. 'Public Sector Information and Re-use Policy – Where is the UK Now?'. Paper given at the Second International Conference on Business, Law and Technology. New York, 2008, <http://eprints.soton.ac.uk/52523/01/PSI_Re-use_Paper_28_April_2008_pdf.pdf>, 22 February 2010.

Teller, Anne. 'Conclusions of the Chair'. Paper given at the the OECD Seminar Public Access to Environmental Information. Athens, 2000, <www.olis.oecd.org/olis/2000doc.nsf/LinkTo/NT00000F96/$FILE/00080429.PDF>, 22 February 2010.

Vandenbroucke, Danny, Katleen Janssen & Jos Van Orshoven. 'INSPIRE State of Play Development of the NSDI in 32 European Countries between 2002 and 2007'. Paper given at the GSDI-10 Conference. St Augustine, 2008, <www.gsdi.org/gsdiconf/gsdi10/papers/TS1.5paper.pdf>, 05 February 2010.

Van Loenen, Bastiaan & Jaap Zevenbergen. 'One Access Policy for High-quality Geographic Information: Results from a US-EU Comparative Study'. Paper given at the GSDI-9 conference. Santiago de Chili, 2006, <http://gsdidocs.org/gsdiconf/GSDI-9/papers/TS35.4paper.pdf>, 22 February 2010.

Worlock, David R. 'Opportunities and Constraints of Public Data Commercialisation: A private company viewpoint', paper given at the European Commission Conference: Access to Public Information. A Key to Commercial Growth and Electronic Democracy. Stockholm, 1996.

PRESS ARTICLES

Intelligent Addressing Press Release. 'Public Sector Information Debate', 20 August 2008, <www.iahub.net/docs/1221131589821.pdf>, 22 February 2010.

Micus Management Consulting GMBH Press Release. 'New Markets for Geo-Information in Times of Google Earth: MICUS-Study Names Market Information, Market Volume and Success Factors', 11 June 2008, <www.micus.de/pdf/Press_Release-GIW-study.pdf>, 22 February 2010.

PocketInfo.nl Press Article. 'Falkplan, NAVTEQ en TeleAtlas boos over vrijgave wegenbestand', 29 January 2008, <www.pocketinfo.nl/nieuws/navigatie/9874/falkplan-navteq-en-teleatlas-boos-over-vrijgave-wegenbestand.html>, 22 February 2010.

Rapid Press Release. 'Commission Takes 5 Countries to Court of Justice for Failure on EU Rules to Re-use Public Sector Information – A Resource Worth up to €48 Billion', 2006, <http://europa.eu/rapid/pressReleasesAction.do?reference=IP/06/1891&format=HTML&aged=0&language=EN&guiLanguage=en>, 22 February 2010.

Rapid Press Release. 'Commission Proposes to Improve and Streamline the European System for Collecting, Analyzing and Reporting Environmental Information', 2008, <http://europa.eu/rapid/pressReleasesAction.do?reference=IP/08/185>, 22 February 2010.

Rapid Press Release. 'Kopernikus: Observing Our Planet for a Safer World', 2008, <http://europa.eu/rapid/pressReleasesAction.do?reference=IP/08/1330&format=HTML&aged=0&language=EN&guiLanguage=fr> 22 February 2010.

Rapid Press Release. 'Re-use of Public Sector Information: Commission Launches Infringement Proceedings against Poland and Sweden', 2008, <http://europa.eu/rapid/pressReleasesAction.do?reference=IP/08/1524&type=HTML&aged=0&language=EN&guiLanguage=en>, 22 February 2010.

Rapid Press Release. 'Re-use of Public Sector Information: Commission Starts Infringement against Italy', 2009, <http://europa.eu/rapid/pressReleases Action.do?reference=IP/09/425&format=HTML&aged=0&language=EN&guiLanguage=en>, 22 February 2010.

Rapid Press Release. 'Commission Steps up Legal Action against Poland for Incorrect Transposition of EU Public Sector Information Rules', 2009, <http://europa.eu/rapid/pressReleasesAction.do?reference=IP/09/1011&format=HTML&aged=0&language=EN&guiLanguage=en> on 25 June 2009.

BLOGS AND FORUM POSTS

Arthur, Charles. 'Where Is Ordnance Survey's Public Task Set Out, Exactly? and Why Is It Paying an External PR Company?', 2008, <www.freeourdata.org.uk/blog/?p=195>, 22 February 2010.

Corbin, Chris. 'Is There a Need to Define "Public Task" ', 2007, <www.epsiplus.net/forum/forum/is_there_a_need_to_define_public_task>, 22 February 2010.

Corbin, Chris. 'What Is the Public Task with Respect to ECPs and HIPs', 2007, <www.epsiplus.net/forum/forum/is_there_a_need_to_define_public_task>, 22 February 2010.

Davies, Robert. 'Public Task Definition – How? Reply to "Is There a Need to Define Public Task?" ', 2007, <www.epsiplus.net/forum/forum/is_there_a_need_to_define_public_task>, 22 February 2010.

Macdonald, Dan. 'Reply to "Your Starter for Ten, Define Public Task . . ." ', 2008, <www.edparsons.com/2008/05/your-starter-for-ten-define-public-task/>, 22 February 2010.

Parsons, Ed. 'Your Starter for Ten, Define Public Task . . . ', 2008, <www.edparsons.com/2008/05/your-starter-for-ten-define-public-task/>, 22 February 2010.

WEBSITES

Communities and Local Government. 'National Interest Mapping Services Agreement', <www.communities.gov.uk/corporate/researchandstatistics/research1/geographicinformationinitiatives/nimsa/>, 22 February 2010.

Council of Europe. '29th Conference of Council of Europe Ministers of Justice', <www.coe.int/t/dghl/standardsetting/minjust/mju29/>, 22 February 2010.

Council of Europe Committee of Ministers, <www.coe.int/t/cm/home_en.asp>, 22 February 2010.

Council of Europe Convention for the Protection of Human Rights and Fundamental Freedoms. 'Ratifications', <http://conventions.coe.int/Treaty/Commun/ChercheSig.asp?NT=005&CM=7&DF=7/14/2007&CL=ENG>, 22 February 2010.

Council of Europe. 'Council of Europe Convention on Access to Official Documents. Signatures and ratifications', http://conventions.coe.int/Treaty/Commun/ChercheSig.asp?NT=205&CM=8&DF=10/02/2010&CL=ENG>, 22 February 2010.

Council of Europe Parliamentary Assembly, <http://assembly.coe.int/default.asp>, 22 February 2010.

ECOMET. 'The Concept of ECOMET', <www.ecomet.eu/description.htm#concept>, 22 February 2010.

Eurogeographics. 'Mission and Objectives', <www.eurogeographics.org/eng/01_mission.asp>, 22 February 2010.

European Commission. 'Alphabetical Index of Terms Referring to the Environment', <http://ec.europa.eu/environment/abc.htm#E>, 22 February 2010.

European Commission. 'e-Inclusion', <http://ec.europa.eu/information_society/activities/einclusion/index_en.htm>, 22 February 2010.

European Commission. 'Public Sector Information – Exclusive Agreements', <http://ec.europa.eu/information_society/policy/psi/facilitating_reuse/exlusive_agreements/index_en.htm>, 22 February 2010.

European Commission. 'Public Sector Information – What's New', <http://ec.europa.eu/information_society/policy/psi/news/index_en.htm>, 22 February 2010.

European Commission. 'Replies to the 2001 Working Document Consultation', <http://cordis.europa.eu/econtent/psi/psi_reuse_consultation_replies.htm>, 22 February 2010.

European Commission. 'Replies to the 2008 PSI Review Consultation – Stakeholders', <http://ec.europa.eu/information_society/policy/psi/online_consultation/stakeholders/index_en.htm>, 22 February 2010.

European Commission. 'Replies to the 2008 PSI review consultation – Member States', <http://ec.europa.eu/information_society/policy/psi/online_consultation/member_states/index_en.htm>, 22 February 2010.

European Commission. 'SEIS', <http://ec.europa.eu/environment/seis/>, 22 February 2010.

GEO. 'About GEO', <www.earthobservations.org/about_geo.shtml>, 22 February 2010.

Group on Earth Observations. 'GEO Data Sharing Principles Implementation', <www.earthobservations.org/geoss_dsp.shtml>, 22 February 2010.

Group on Earth Observations. 'What is GEOSS', <www.earthobservations.org/geoss.shtml>, 22 February 2010.

GMES. 'Overview', <http://ec.europa.eu/gmes/overview.htm>, 22 February 2010.

Merriam Webster's Online Dictionary, <www.m-w.com/dictionary/>, 22 February 2010.

Mtraffic, <www.mtraffic.org>, 22 February 2010.

Nationaal Geografisch Instituut, <www.ngi.be/NL/NL0.shtm>, 22 February 2010.

Office of Public Sector Information. 'Click-use Licence', <www.opsi.gov.uk/click-use/index>, 22 February 2010.

OpenStreetMap, <www.openstreetmap.org>, 22 February 2010.

Ordnance Survey. 'New Pan Government Agreement signed', <www.ordnancesurvey.co.uk/oswebsite/media/news/2009/February/pgasigned.html>, 22 February 2010.
Ordnance Survey. 'Re-use of Public Sector Information', <www.ordnancesurvey.co.uk/oswebsite/aboutus/yourinforights/ropsi.html>, 22 February 2010.
Ordnance Survey. 'Welcome to the OS Openspace API', <http://openspace.ordnancesurvey.co.uk/openspace/>, 22 February 2010.
Ordnance Survey Business Strategy. 'Goal 5: Enhance Value through the Creation of an Innovative Trading Entity, <http://strategy.ordnancesurvey.co.uk/goal-5-enhance-value-through-trading-entity/>, 22 February 2010.
Rijkswaterstaat. 'Geoservices Portaal', <www.rijkswaterstaat.nl/apps/geoservices/portaal/index-applicaties.html>, 22 February 2010.
UNFFC. 'An Introduction to the United Nations Framework Convention on Climate Change (UNFCCC) and its Kyoto Protocol', <http://unfccc.int/press/fact_sheets/items/4978.php>, 22 February 2010.
United Nations Economic Commission for Europe. 'Aarhus Convention Participants', <www.unece.org/env/pp/ratification.htm>, 22 February 2010.
World Wide Web Consortium. 'Web Content Accessibility Guidelines (WCAG) Overview', <www.w3.org/WAI/intro/wcag.php>, 22 February 2010.

Index

1. Stephen J. Turner, *A Substantive Environmental Right: An Examination of the Legal Obligations of Decision-makers towards the Environment*, 2009 (ISBN 978-90-411-2815-7).
2. Helle Tegner Anker, Birgitte Egelund Olsen & Anita Rønne (eds), *Legal Systems and Wind Energy: A Comparative Perspective*, 2009 (ISBN 978-90-411-2831-7).
3. David Langlet, *Prior Informed Consent and Hazardous Trade: Regulating Trade in Hazardous Goods at the Intersection of Sovereignty, Free Trade and Environmental Protection*, 2009 (ISBN 978-90-411-2821-8).
4. Louis J. Kotzé and Alexander R. Paterson (eds), *The Role of the Judiciary in Environmental Governance: Comparative Perspectives*, 2009 (ISBN 978-90-411-2708-2).
5. Tuula Honkonen, *The Common but Differentiated Responsibility Principle in Multilateral Environmental Agreement's: Regulatory and Policy Aspects*, 2009 (ISBN 978-90-411-3153-9).
6. Barbara Pozzo (ed.), *The Implementation of the Seveso Directives in an Enlarged Europe: A Look into the Past and a challenge for the Future*, 2009 (ISBN 978-90-411-2854-6).
7. Henrik M. Inadomi, *Independent Power Projects in Developing Countries: Legal Investment Protection and Consequences for Development*, 2010 (ISBN 978-90-411-3178-2).
8. Nahid Islam, *The Law of Non-Navigational Uses of International Watercourses: Options for Regional Regime-Building in Asia*, 2010 (ISBN 978-90-411-3196-6).
9. Yasuhiro Shigeta, *International Judicial Control of Environmental Protection: Standard Setting, Compliance Control and the Development of International Environmental Law by the International Judiciary*, 2010 (ISBN 978-90-411-3151-5).
10. Katleen Janssen, *The Availability of Spatial and Environmental Data in the European Union: At the Crossroads between Public and Economic Interests*, 2010 (ISBN 978-90-411-3287-1).